Geophysical Monograph Series

217 **Deep Earth: Physics and Chemistry of the Lower Mantle and Core** *Hidenori Terasaki and Rebecca A. Fischer (Eds.)*
218 **Integrated Imaging of the Earth: Theory and Applications** *Max Moorkamp, Peter G. Lelievre, Niklas Linde, and Amir Khan (Eds.)*
219 **Plate Boundaries and Natural Hazards** *Joao Duarte and Wouter Schellart (Eds.)*
220 **Ionospheric Space Weather: Longitude and Hemispheric Dependences and Lower Atmosphere Forcing Timothy Fuller-Rowell,** *Endawoke Yizengaw, Patricia H. Doherty, and Sunanda Basu (Eds.)*
221 **Terrestrial Water Cycle and Climate Change Natural and Human-Induced Impacts** *Qiuhong Tang and Taikan Oki (Eds.)*
222 **Magnetosphere-Ionosphere Coupling in the Solar System** *Charles R. Chappell, Robert W. Schunk, Peter M. Banks, James L. Burch, and Richard M. Thorne (Eds.)*
223 **Natural Hazard Uncertainty Assessment: Modeling and Decision Support** *Karin Riley, Peter Webley, and Matthew Thompson (Eds.)*
224 **Hydrodynamics of Time-Periodic Groundwater Flow: Diffusion Waves in Porous Media** *Joe S. Depner and Todd C. Rasmussen (Auth.)*
225 **Active Global Seismology** *Ibrahim Cemen and Yucel Yilmaz (Eds.)*
226 **Climate Extremes** *Simon Wang (Ed.)*
227 **Fault Zone Dynamic Processes** *Marion Thomas (Ed.)*
228 **Flood Damage Survey and Assessment: New Insights from Research and Practice** *Daniela Molinari, Scira Menoni, and Francesco Ballio (Eds.)*
229 **Water-Energy-Food Nexus – Principles and Practices** *P. Abdul Salam, Sangam Shrestha, Vishnu Prasad Pandey, and Anil K Anal (Eds.)*
230 **Dawn–Dusk Asymmetries in Planetary Plasma Environments** *Stein Haaland, Andrei Rounov, and Colin Forsyth (Eds.)*
231 **Bioenergy and Land Use Change** *Zhangcai Qin, Umakant Mishra, and Astley Hastings (Eds.)*
232 **Microstructural Geochronology: Planetary Records Down to Atom Scale** *Desmond Moser, Fernando Corfu, James Darling, Steven Reddy, and Kimberly Tait (Eds.)*
233 **Global Flood Hazard: Applications in Modeling, Mapping and Forecasting** *Guy Schumann, Paul D. Bates, Giuseppe T. Aronica, and Heiko Apel (Eds.)*
234 **Pre-Earthquake Processes: A Multidisciplinary Approach to Earthquake Prediction Studies** *Dimitar Ouzounov, Sergey Pulinets, Katsumi Hattori, and Patrick Taylor (Eds.)*
235 **Electric Currents in Geospace and Beyond** *Andreas Keiling, Octav Marghitu, and Michael Wheatland (Eds.)*
236 **Quantifying Uncertainty in Subsurface Systems** *Celine Scheidt, Lewis Li, and Jef Caers (Eds.)*
237 **Petroleum Engineering** *Moshood Sanni (Ed.)*
238 **Geological Carbon Storage: Subsurface Seals and Caprock Integrity** *Stephanie Vialle, Jonathan Ajo-Franklin, and J. William Carey (Eds.)*
239 **Lithospheric Discontinuities** *Huaiyu Yuan and Barbara Romanowicz (Eds.)*
240 **Chemostratigraphy Across Major Chronological Eras** *Alcides N.Sial, Claudio Gaucher, Muthuvairavasamy Ramkumar, and Valderez Pinto Ferreira (Eds.)*
241 **Mathematical Geoenergy:Discovery, Depletion, and Renewal** *Paul Pukite, Dennis Coyne, and Daniel Challou (Eds.)*
242 **Ore Deposits: Origin, Exploration, and Exploitation** *Sophie Decree and Laurence Robb (Eds.)*
243 **Kuroshio Current: Physical, Biogeochemical and Ecosystem Dynamics** *Takeyoshi Nagai, Hiroaki Saito, Koji Suzuki, and Motomitsu Takahashi (Eds.)*
244 **Geomagnetically Induced Currents from the Sun to the Power Grid** *Jennifer L. Gannon, Andrei Swidinsky, and Zhonghua Xu (Eds.)*
245 **Shale: Subsurface Science and Engineering** *Thomas Dewers, Jason Heath, and Marcelo Sánchez (Eds.)*
246 **Submarine Landslides: Subaqueous Mass Transport Deposits From Outcrops to Seismic Profiles** *Kei Ogata, Andrea Festa, and Gian Andrea Pini (Eds.)*
247 **Iceland: Tectonics, Volcanics, and Glacial Features** *Tamie J. Jovanelly*
248 **Dayside Magnetosphere Interactions** *Qiugang Zong, Philippe Escoubet, David Sibeck, Guan Le, and Hui Zhang (Eds.)*
249 **Carbon in Earth's Interior** *Craig E. Manning, Jung-Fu Lin, and Wendy L. Mao (Eds.)*
250 **Nitrogen Overload: Environmental Degradation, Ramifications, and Economic Costs** *Brian G. Katz*
251 **Biogeochemical Cycles: Ecological Drivers and Environmental Impact** *Katerina Dontsova, Zsuzsanna Balogh-Brunstad, and Gaël Le Roux (Eds.)*
252 **Seismoelectric Exploration: Theory, Experiments, and Applications** *Niels Grobbe, André Revil, Zhenya Zhu, and Evert Slob (Eds.)*
253 **El Niño Southern Oscillation in a Changing Climate** *Michael J. McPhaden, Agus Santoso, and Wenju Cai (Eds.)*
254 **Dynamic Magma Evolution** *Francesco Vetere (Ed.)*
255 **Large Igneous Provinces: A Driver of Global Environmental and Biotic Changes** *Richard. E. Ernst, Alexander J. Dickson, and Andrey Bekker (Eds.)*
256 **Coastal Ecosystems in Transition: A Comparative Analysis of the Northern Adriatic and Chesapeake Bay** *Thomas C. Malone, Alenka Malej, and Jadran Faganeli (Eds.)*
257 **Hydrogeology, Chemical Weathering, and Soil Formation** *Allen Hunt, Markus Egli, and Boris Faybishenko (Eds.)*
258 **Solar Physics and Solar Wind** *Nour E. Raouafi and Angelos Vourlidas (Eds.)*
259 **Magnetospheres in the Solar System** *Romain Maggiolo, Nicolas André, Hiroshi Hasegawa, and Daniel T. Welling (Eds.)*
260 **Ionosphere Dynamics and Applications** *Chaosong Huang and Gang Lu (Eds.)*
261 **Upper Atmosphere Dynamics and Energetics** *Wenbin Wang and Yongliang Zhang (Eds.)*
262 **Space Weather Effects and Applications** *Anthea J. Coster, Philip J. Erickson, and Louis J. Lanzerotti (Eds.)*
263 **Mantle Convection and Surface Expressions** *Hauke Marquardt, Maxim Ballmer, Sanne Cottaar, and Jasper Konter (Eds.)*
264 **Crustal Magmatic System Evolution: Anatomy, Architecture, and Physico-Chemical Processes** *Matteo Masotta, Christoph Beier, and Silvio Mollo (Eds.)*
265 **Global Drought and Flood Observation, Modeling, and Prediction** *Huan Wu, Dennis P. Lettenmaier, Qiuhong Tang, and Philip J. Ward (Eds.)*
266 **Magma Redox Geochemistry** *Roberto Moretti and Daniel R. Neuville (Eds.)*

Geophysical Monograph 266

Magma Redox Geochemistry

Roberto Moretti
Daniel R. Neuville
Editors

This Work is a co-publication of the American Geophysical Union and John Wiley and Sons, Inc.

AGU
ADVANCING EARTH
AND SPACE SCIENCE

WILEY

This edition first published 2022

Published under the aegis of the AGU Publications Committee

Registered Office
John Wiley & Sons, Inc., 111 River Street, Hoboken, NJ 07030, USA

Editorial Office
111 River Street, Hoboken, NJ 07030, USA

For details of our global editorial offices, customer services, and more information about Wiley products visit us at www.wiley.com.

Wiley also publishes its books in a variety of electronic formats and by print-on-demand. Some content that appears in standard print versions of this book may not be available in other formats.

Library of Congress Cataloging-in-Publication Data

Names: Moretti, Roberto, editor. | Neuville, Daniel R., editor.
Title: Magma redox geochemistry / Roberto Moretti, Daniel R. Neuville, editors.
Description: Hoboken, NJ : Wiley-American Geophysical Union, [2022] | Series: Geophysical monograph series | Includes index.
Identifiers: LCCN 2021027133 (print) | LCCN 2021027134 (ebook) | ISBN 9781119473251 (cloth) | ISBN 9781119473299 (adobe pdf) | ISBN 9781119473244 (epub)
Subjects: LCSH: Oxidation-reduction reaction. | Magmas.
Classification: LCC QD63.O9 M34 2021 (print) | LCC QD63.O9 (ebook) | DDC 541/.393–dc23
LC record available at https://lccn.loc.gov/2021027133
LC ebook record available at https://lccn.loc.gov/2021027134

Cover Design: Wiley
Cover Image: Lava fountaining and lava flows emitted from a fissure that opened in the wall of the Enclos Fouqué caldera at Piton de la Fournaise on December 27, 2005; © Aline Peltier (IPGP/OVPF)

Set in 10/12pt Times New Roman by Straive, Pondicherry, India

10 9 8 7 6 5 4 3 2 1

CONTENTS

LIST OF CONTRIBUTORS

Róbert Arató
Bayerisches Geoinstitut
University of Bayreuth
Bayreuth, Germany

Paul D. Asimow
Division of Geological and Planetary Sciences
California Institute of Technology
Pasadena, CA, USA

Andreas Audétat
Bayerisches Geoinstitut
University of Bayreuth
Bayreuth, Germany

Sonja Aulbach
Institute of Geosciences
Goethe-Universität
Frankfurt am Main, Germany

Aaron S. Bell
Department of Geological Sciences
University of Colorado Boulder
Boulder, CO, USA

Andrew J. Berry
Research School of Earth Sciences
Australian National University
Canberra, ACT, Australia

Suzanne K. Birner
Division of Natural Sciences, Nursing,
and Mathematics
Berea College
Berea, KY, USA

Marc Blanchard
Géosciences Environnement Toulouse (GET)
Université de Toulouse
CNRS, IRD
Toulouse, France

Maryjo Brounce
Department of Earth and Planetary Sciences
University of California Riverside
Riverside, CA, USA

Antony D. Burnham
Research School of Earth Sciences
Australian National University
Canberra, ACT, Australia

Maria Rita Cicconi
Université de Paris
Institut de Physique du Globe de Paris
Paris, France;
and
Department of Materials Science and Engineering
Friedrich-Alexander-Universität Erlangen-Nürnberg
Erlangen, Germany

Elizabeth Cottrell
Department of Mineral Sciences
National Museum of Natural History
Smithsonian Institution
Washington, DC, USA

Nicolas Dauphas
Origins Laboratory
Department of the Geophysical Sciences
and Enrico Fermi Institute
The University of Chicago
Chicago, IL, USA

Fred A. Davis
Department of Earth and Environmental Sciences
University of Minnesota Duluth
Duluth, MN, USA

Baptiste Debret
Université de Paris
Institut de Physique du Globe de Paris
Paris, France;
and
Laboratoire G-Time
DGES
Université Libre de Bruxelles
Brussels, Belgium

Katy A. Evans
School of Earth and Planetary Sciences
Curtin University
Perth, Western Australia, Australia

Stephen F. Foley
Department of Earth and Environmental Sciences, and
ARC Centre of Excellence for Core to Crust Fluid Systems
Macquarie University
Sydney, New South Wales, Australia

Raul O.C. Fonseca
Institute of Geology, Mineralogy and Geophysics
Ruhr-University Bochum
Bochum, Germany

Grant S. Henderson
Department of Earth Sciences
University of Toronto
Toronto, Canada

Edward C. Inglis
Université de Paris
Institut de Physique du Globe de Paris
Paris, France

Katherine A. Kelley
Graduate School of Oceanography
University of Rhode Island
Narragansett, RI, USA

Charles Le Losq
Research School of Earth Sciences
Australian National University
Canberra, ACT, Australia;
and
Université de Paris
Institut de Physique du Globe de Paris
Paris, France

Guilherme Mallmann
Research School of Earth Sciences
Australian National University
Canberra, ACT, Australia

Nicole Métrich
Université de Paris
Institut de Physique du Globe de Paris
Paris, France

Roberto Moretti
Université de Paris
Institut de Physique du Globe de Paris
Paris, France

Frédéric Moynier
Université de Paris
Institut de Physique du Globe de Paris
Paris, France

Daniel R. Neuville
Université de Paris
Institut de Physique du Globe de Paris
Paris, France

Hugh St.C. O'Neill
Research School of Earth Sciences
Australian National University
Canberra, ACT, Australia

Paolo A. Sossi
Institute of Geochemistry and Petrology
ETH Zürich
Zürich, Switzerland;
and
Université de Paris
Institut de Physique du Globe de Paris
Paris, France

Vincenzo Stagno
Department of Earth Sciences
Sapienza University of Rome
Rome, Italy

Andy G. Tomkins
School of Earth, Atmosphere and Environment
Monash University
Melbourne, Victoria, Australia

Laura E. Waters
Department of Earth and Environmental Science
New Mexico Institute of Mining and Technology
Socorro, NM, USA

PREFACE

In chemistry, redox issues occupy a central role for our understanding of how chemical exchanges take place between substances. Redox reactions include all chemical reactions in which atoms have their oxidation state changed because of the transfer of electrons between chemical species. Such a transfer involves both a reduction process and a complementary oxidation process and the consequent rearranging of chemical bonds.

The word oxidation originally implied reaction with oxygen to form an oxide, since dioxygen (O_2) was historically the first recognized oxidizing agent. Later, the term was expanded to encompass oxygen-like substances that accomplished parallel chemical reactions. Ultimately, the meaning was generalized to include all processes involving loss of electrons. The word reduction originally referred to the loss in weight upon heating a metallic ore, such as a metal oxide to extract the metal. In other words, ore was "reduced" to metal. Antoine Lavoisier (1743–1794) showed that this loss of weight was due to the loss of oxygen as a gas. This etymological premise also summarizes the significance that for a long time the world of geology has been attributed to the concept of "redox." In high-temperature geology, where aqueous solutions can no more be defined, redox has an upmost importance to both the formation of minerals and the mobilization of metals. However, it is often intended as a measure of oxygen fugacity (fO_2), anchored to an assemblage of minerals or compounds which constrains its variations as a simple function of temperature.

The redox state is one of the master variables driving Earth-forming processes and since the dawn of geochemistry, knowledge of redox state has been essential to understanding the compositional makeup of our planet and the fundamental processes occurring from the core up to the atmosphere. Most of these processes involve the major transport agent of matter on Earth, i.e., magma. The social and economic impact of redox geochemistry is enormous, because of the control played on metal mobility, solubility, and availability by redox state of metals and ligands that may complex them, and because of the widespread use of redox indicators for environmental hazards assessment, such as the volcanic one, which assessment greatly benefits from studies on volcanic gas speciation, which in turn is controlled by redox. Volatile components and their redox control on volcanic degassing and metal mobility offer good examples of redox exchanges in high-temperature environments close to human experience. Knowledge of redox mechanisms acting in volcanism and hydrothermalism have a great impact on the socioeconomic development of human societies because of their key role in volcanic hazard assessment, geothermal energy exploration, and ore deposits formation.

As shown in this monograph, redox state is also an image of the magma composition, and the understanding of magma (particularly melt) physicochemical nature is the basic prerequisite to understand how redox exchanges work in deep Earth systems and to understand "who controls what." The latter is a difficult task, given the almost infinite conditions of temperature, pressure, and chemical composition relevant to igneous petrology. However, the study of redox state and related properties cannot be reduced to simple rule of thumbs or assumptions such as the existence of stoichiometric mineral assemblages buffering oxygen fugacity via solid-gas equilibria.

Knowledge of the redox potential (or alternatively, oxygen fugacity) at which a rock forms and evolves is relevant for interpreting the rock's history. However, the approach inherited by mineral chemistry has avoided for too long to assess the role of the major phase making up the magma: the silicate melt. Silicate melts have been very often considered as a simple reservoir of elements almost chemically inert and fully controlled by other phases (mainly solid) able to impose their redox state, (i.e., fO_2), which in turn was treated like a Maxwell demon. However, magma is the most important transport agent throughout our planet, buffering entire planetary sectors both thermally and chemically.

This volume shows the multiple concepts and approaches useful to the study of the complex interactions occurring between melts, crystals, and fluids that are behind magma formation, ascent, and evolution. By joining the description of magma physical chemistry with geological issues, the chapters of this book disclose the multifaceted implications that redox variables and their gradients have on magma evolution in time and on the dynamics of planet Earth, or in other words, it brings to the reader's attention the power of redox geodynamics.

This volume provide a comprehensive overview and a state-of-the-art treatment of technological and scientific advances in our understanding of redox geochemistry. Given the almost infinite conditions of temperature, pressure, and chemical composition relevant to igneous petrology and volcanology, the chapters represent a selection of topics able to give a unique picture of the "redox" continuum of the Earth's interiors.

Part I is composed of chapters about oxygen fugacity in Earth's accretion and across geodynamic settings, with a focus on the redox boundaries associated with mantle melting. Part II deals with the role of redox in magma differentiation, from the magma source up to the surface throughout volcanic processes, particularly degassing. Part III gives an overview on the tools and experimental and theoretical techniques to measure the redox state in melts and glasses and estimate the role of redox state on element and isotope partitioning; the major recent advances in understanding redox mechanisms affecting multivalent elements other than iron and innovative oxybarometers are presented.

This volume is the result of the sessions "Linking the Redox State of Silicate Melts to Magmatic Processes" at the Goldschmidt Conference in Paris in 2017 and "Oxygen fugacity and redox mechanisms in high- to low-temperature geochemical processes" at the AGU Fall Meeting in San Francisco in 2019. In writing up their papers, the authors have taken into consideration the discussions had during the two sessions.

The editors wish to thank all authors for their contributions and also acknowledge the assistance of the reviewers, whose conscientious efforts helped the authors to improve the quality of the chapters in this volume. The editors also wish to thank the Institut de Physique du Globe (Université de Paris) for its support, Aline Peltier (Institut de Physique du Globe de Paris – Observatoire Volcanologique du Piton de la Fournaise) for the cover image, the Volcanology, Geochemistry and Petrology (VGP) division of the American Geophysical Union (AGU), and the Commission on Physics of Mineral (CPM) of the International Mineralogical Association (IMA).

Roberto Moretti
Daniel R. Neuville
Université de Paris, Institut de Physique du Globe de Paris, France

1

Redox Equilibria: From Basic Concepts to the Magmatic Realm

Roberto Moretti[1,2] and Daniel R. Neuville[1]

ABSTRACT

The basic aspects of redox geochemistry are reviewed to provide a useful compendium of the redox connection between the aqueous-hydrothermal and igneous realms of Earth. The redox description of a system is intimately coupled to the knowledge of acid-base properties of the solvent in which redox exchanges take place. For magmas, and then silicate melts, approaches reporting the redox state were so far cantered around the sole concept of oxygen fugacity, fO_2. Mastering the concept of fO_2 in experimental and observational petrology was the key to constrain the processes behind the very large range of relative oxygen fugacity observed on Earth. Although current descriptions of silicate melts and magma thermodynamic properties are mainly based on oxides or mineral-like molecular components, disregarding the actual melt reactivity poses many limits in our understanding of the true chemical exchanges involving oxygen, iron and the other redox-sensitive elements. Because silicate melts, unlike aqueous solutions, lack of a full acid-base description, compositional dependencies are solved by means of empirical treatments based on oxides and their combinations. However, these can bias the interpretation of redox exchanges recorded in analyzed samples and used to identify the several processes (e.g., batch or fractional crystallization, elemental recycling, degassing, deep fluid infiltration) which characterize magma evolution and its geodynamic environment. This short compendium aims at stimulating the quest for a comprehensive and unifying picture of the acid-base and redox properties of melts from which we could extrinsic its reactivity in way similar to aqueous solutions and molten salts.

1.1. GENERAL ASPECTS AND RATIONALE

1.1.1. Oxidation Number, Electron Transfer, and Half-Reactions

Oxidation-reduction (redox) geochemistry studies those natural reactions occurring on Earth in which the transfer of electrons determines a change in the oxidation number of participating chemical species. Oxidation state (or oxidation number) measures the degree of oxidation of an atom in a substance and it is the (hypothetical) charge of an atom if all bonds to atoms of different elements were 100% ionic, with no covalent bond fraction. The oxidation state of an atom is indicated with Roman numerals, whereas Arabic numbers are used for the charge on compound (e.g., in SiO_4^{4-}, silicon has oxidation number IV and oxygen has –II, whereas 4– is the formal charge of the whole silicate ion). Charge and oxidation number are the same for monoatomic ions. The oxidation number can be positive, negative, or zero, and it can have a

[1]*Université de Paris, Institut de Physique du Globe de Paris, Paris, France*

[2]*Observatoire Volcanologique et Sismologique de Guadeloupe, Institut de Physique du Globe de Paris, Gourbeyre, France*

Magma Redox Geochemistry, Geophysical Monograph 266, First Edition.
Edited by Roberto Moretti and Daniel R. Neuville.

DOI: 10.1002/9781119473206.ch1

fractional value as well, and although very similar, it does not correspond to valency, which is an atom property and represents the number of bonds the atom needs to become stable, i.e., to complete either duplet or octet rule.

Many redox reactions are familiar to us, such as fire and combustion, rusting, and dissolution of metals. Transition metals and main group elements (e.g., N, halogens, O, S, C) have multiple oxidation states and important redox chemistry, which affect element distribution within the geochemical shells on Earth but also through the boundaries between such shells (e.g., Moretti et al., 2020a). For instance, it is the redox state of metals and ligands that complex them, which then determines (i) their "unlocking" from pristine reservoirs (e.g. minerals in which they occur at trace level); (ii) their mobility on Earth through carriers such as magma, water, or vapor; and eventually (iii) their accumulation and precipitations in new phases making up ore deposits.

Redox reactions involve a coupled transfer of electrons, so for any oxidation (loss of electrons) a reciprocal reduction (gain of electrons) occurs. Moreover, redox reactions naturally occurring on Earth involve a net chemical change that can be described not only via the exchange of electrons between ions or their complexes, but also of oxygen and/or hydrogen atoms and compounds that these can form (e.g., Cicconi et al., 2020a and references therein). Here are some examples:

$$2Fe_{(s)} + O_{2(g)} \Leftrightarrow 2FeO_{(s)} \tag{1.1}$$

$$2FeO_{(m)} + H_2O_{(g)} \Leftrightarrow Fe_2O_{3(m)} + H_{2(g)} \tag{1.2}$$

$$Fe^{3+}{}_{(aq)} + \tfrac{1}{2}H_{2(g)} \Leftrightarrow Fe^{2+}{}_{(aq)} + H^{+}{}_{(aq)} \tag{1.3}$$

$$4Fe^{2+}{}_{(aq)} + O_{2(g)} + 4H^{+}{}_{(aq)} \Leftrightarrow 4Fe^{3+}{}_{(aq)} + 2H_2O_{(aq)} \tag{1.4}$$

$$SO_{2(g)} + 3H_{2(g)} \Leftrightarrow H_2S_{(g)} + 2H_2O_{(g)} \tag{1.5}$$

In which the subscripts *s, m, aq*, and *g* refer to solid, melt, aqueous, and gas (including supercritical fluids) phases, respectively. In the five examples above, $O_{2(g)}$, $H_2O_{(g)}$, $Fe^{3+}{}_{(aq)}$, $O_{2(g)}$ and $SO_{2(g)}$ are oxidizing agents, whereas $Fe_{(s)}$, $FeO_{(l)}$, $H_{2(g)}$, $Fe^{2+}{}_{(aq)}$ are the reducing agents.

The word oxidation was introduced by Antoine Lavoisier (1777) and indicates the chemical mechanisms in which oxygen is consumed and added to a compound. The parallel mechanism in which some other compounds loose oxygen was called reduction. The oxidation number of reaction components in Reactions 1.1 to 1.5 shows that some elements have a greater affinity for electrons than others and that oxygen tends to appear with oxidation number –II in its compounds (including the O^{2-} species in its ionic compounds). In contrast, hydrogen tends to appear with oxidation number I. Besides, hydrogen and oxygen atoms and their compounds in reaction are related to ligands making up the most important matrixes on Earth, such as aqueous solutions (including highly concentrated saline solution), supercritical fluids and gases, silicate melts or oxide, and silicate minerals making up rocks. Except for rocks, these matrixes are also important transport agents and thus redox carriers within Earth geochemical reservoirs and throughout their boundaries. Therefore, nearly all redox exchanges on Earth are related to chemical transfers involving hydrogen- and/or oxygen-based half-reactions such as (Cicconi et al., 2020a,b):

$$O_2 + 4e^- \Leftrightarrow 2O^{2-} \tag{1.6}$$

$$2H^+ + 2e^- \Leftrightarrow H_2 \tag{1.7}$$

that we will call henceforth oxygen and hydrogen electrodes (e.g., Cicconi et al., 2020a; Moretti, 2020) and in which the electron exchange is made explicit. For example, Reaction 1.1 can be written as the sum of Reaction 1.6 and:

$$2Fe^0 \Leftrightarrow 2Fe^{2+} + 4e^- \tag{1.8}$$

plus the formation of FeO oxides from its ions,

$$2Fe^{2+} + 2O^{2-} \Leftrightarrow 2FeO \tag{1.9}$$

in which oxidation numbers and then formal charges of involved atoms do not vary.

In ore geochemistry, but also in metallurgical practice, a special mention must be made to redox mechanisms involving chalcophile elements and sulfide. Most often, relevant equilibria are written without the involvement of the medium in which they actually occur. Pyrite formation can result from the hydrothermal alteration of igneous pyrrhotite, but their equilibrium can be simply written in the sole Fe–S system as (Barton, 1970):

$$2FeS + S_2 \Leftrightarrow 2FeS_2 \tag{1.10}$$

In the Fe–S system, pyrite is not at the liquidus (pyrite does not melt), but as a conceptual exercise we can still relate its formation to the occurrence of the following fictitious half-reactions in the solid phase involving sulfide and polysulfide anions:

$$S_{2(g)} + 2e^- \Leftrightarrow 2S^{2-} \tag{1.11}$$

$$S_2{}^{2-} + 2e^- \Leftrightarrow 2S^{2-} \tag{1.12}$$

and their combination with iron, which appears in its cationic form Fe^{2+} in both sides of Reaction 1.10. In the liquid state, melts in which sulfide is the main or the sole anionic ligand are very scarcely represented on Earth and segregate from reducing sulfur oversaturated magmatic silicate melts (e.g., Li and Ripley, 2013 and references therein). Instead, sulfide melts are of interest in extractive metallurgy (e.g., Sokhanwaran et al., 2016).

1.1.2. The Redox Potential in Solutions and the Ligand Role

In redox reactions a potential difference drives the transfer electrons from an anode (negative electrode) to a cathode (positive electrode): oxidation occurs at the anode and reduction occurs at the cathode. Reactions are spontaneous in the direction of $\Delta G < 0$, which is also the direction in which the potential (defined as $E_{cathode} - E_{anode}$) is positive. In a redox reaction the anode is then the half-reaction written with electrons on the right and the cathode is the half-reaction with electrons appearing on the left side.

The electric work done by a spontaneous redox reaction, like in a galvanic cell ($E > 0$), is the (measurable) electromotive force of the reacting systems and equals the Gibbs free energy change (e.g. Ottonello, 1997) via the Nernst equation:

$$-nFE = \Delta G = \Delta G^0 + RT\ln\prod a_i = -nFE^0$$

$$+ 2.303RT\log\prod a_i = -nFE^0 + 2.303RT\log Q_i \quad (1.13)$$

with a_i the activity of the ith component participating in the redox exchange, F as the Faraday constant (96,485 Coulomb per mole), n the number of transferred electrons, and Q the activity product. In writing redox reactions, complete electrolytes are often used because the activity coefficients are measured without extra thermodynamic assumptions, but Equation 1.13 is normally used for reactions based on individual ions. To establish a potential scale for half-reactions, we keep using the convention that electrons are reported on the left-hand side of the reaction, that is, in the sense of reduction. The potentials of half-reactions can be added and subtracted, like free energies, to give an overall value for the reaction. It is also worth noting that by convention, it was decided to use a hydrogen-electrode-scale electric potential, by setting $E^0 = 0.0$ V for reaction 1.7 with the constituents in their standard state (e.g., Casey, 2017). This arbitrary decision implies that (i) the Gibbs energy for $H^+_{(aq)}$, the electron (e^-), and $H_{2(g)}$ are all 0.0 kJ/mol, and (ii) potential difference of reactions involving the hydrogen electrode (Reaction 1.7) are given by the other half-reaction completing the redox exchange.

The electrode potential values (E^0) hold at standard conditions: by definition, standard conditions mean that any dissolved species have concentrations of 1 m, any gaseous species have partial pressures of 1 bar, and the system is 25°C. Standard potentials represent the case where no current flows and the electrode reaction is reversible. Measuring a voltage is an indication that the system is out of equilibrium. Nernstian processes are characterized by fast electron transfer and are rate-limited by the diffusion of the electron-active species into the electrolyte. The system then spontaneously approaches equilibrium because negative and positive charged species can flow in opposite directions. At equilibrium, the voltage drops to zero and the current stops, like in dead batteries. The magnitude of the cell potential, $E^0 = E^0_{cathode} - E^0_{anode}$, may be viewed as the driving force for current flow in the circuit.

The hydrogen-electrode scale electric potential so defined, E (also indicated as E_h in aqueous solutions), is a measure of the oxidation state of a system at equilibrium relative to a hydrogen electrode. E is not a constant (for given T and P) but depends on the system composition via activities of ions entering a half redox reaction. When coupled to a compositional parameter of the system related to the activity of the ligand making up the solvent of interest, such as a_{H^+} for aqueous solutions, E can be used to establish a kind of phase diagram that shows which species (dissolved ion species, gases, or solids) will predominate among a chosen set in the system of interest (a solution) for a given temperature.

To easily understand all this, we can look at the reaction leading to the formation of liquid water:

$$2H_{2(g)} + O_{2(g)} \Leftrightarrow 2H_2O_{(aq)} \quad (1.14)$$

which is given by the sum of Reaction 1.7 (H^+/H_2 redox couple: the anode) and the following half-reaction (the cathode):

$$O_{2(g)} + 4H^+_{(aq)} + 4e^- \Leftrightarrow 2H_2O_{(aq)} \quad (1.15)$$

which is governed by the O_2/H_2O redox couple. The presence of protons in both Reactions 1.7 and 1.15 shows that the overall Reaction 1.14 is defined for acidic conditions ($pH < 7$). For neutral or basic conditions ($pH \geq 7$), Reaction 1.14 can be obtained from the following two half-reactions for H_2O/H_2 and O_2/OH^- couples, respectively:

$$2H_2O_{(aq)} + 2e^- \Leftrightarrow H_{2(g)} + 2OH^- \quad (1.16)$$

$$O_{2(g)} + 2H_2O + 4e^- \Leftrightarrow 4OH^- \quad (1.17)$$

Let us now deal with Reactions 1.7 and 1.15 occurring in the acidic medium (see, for example, Ottonello, 1997). The standard potential of Reaction 1.15 is $E^0_{16} = 1.228$ V and refers to a standard state of water in equilibrium at

T = 25°C and P = 1 bar with an atmosphere of pure O_2. From Equation 1.13 we obtain:

$$\begin{aligned} E_{15} &= 1.228 + 0.0591 \log aH^{+} + 0.0148 \log aO_2 \\ &= 1.228 - 0.0591 pH + 0.0148 \log fO_2 \end{aligned} \quad (1.18)$$

where a and f denote activity and fugacity, respectively, $pH = -\log a_{H^+}$ and it is considered that $aO_2 = fO_2/fO_2^0$ with $fO_2^0 = 1$ bar.

Similarly, the redox potential related to Reaction 1.7 is then:

$$E_7 = -0.0591 pH - 0.0295 \log fH_2 \quad (1.19)$$

Equations 1.18 and 1.19, which can also be derived for Reactions 1.16 and 1.17, can be used to trace E-pH diagrams (also called Pourbaix or predominance diagrams; Casey, 2017) limiting the stability field of water (Figure 1.1) and of any other systems in which E-pH relationships can be established from the reactions of intervening species (Figure 1.2). E-pH diagrams are most used for understanding the geochemical formation, corrosion and passivation, leaching and metal recovery, water treatment precipitation, and adsorption.

For the set of species of interest, E-pH diagrams show boundaries that are given by:

1. lines of negative slope that limit the stability field of water (Equations 1.18 and 1.19) or related to solid–solid phase changes in which paired electron–proton exchanges occur because the ligand (water) participates in reaction, such as in the case of the hematite–magnetite boundary in Figure 1.2:

$$3Fe_2O_3 + 2H^{+} + 2e^{-} \Leftrightarrow 2Fe_3O_4 + H_2O \quad (1.20)$$

Note that boundary slope is negative because protons and electrons appear on the same side of reaction. The protons/electrons ratio determines the slope value.

2. (rare) pH-dependent lines of positive slopes, and associated with electron–proton exchanges involving, for example, reduction of dissolved cations to the oxide with a lower oxidation number, e.g.

$$2Cu^{2+} + H_2O_{(aq)} + 2e^{-} \Leftrightarrow Cu_2O_{(s)} + 2H^{+} \quad (1.21)$$

Boundary slope is positive because protons and electons appear on different sides of reaction.

3. horizontal lines (pure electron exchange), such as in case of half-reaction

$$Fe^{3+} + e^{-} \Leftrightarrow Fe^{2+} \quad (1.22)$$

which participates with half-reaction 1.7 in giving Reaction 1.3 and does not involve explicitly the water solvent.

4. vertical lines, representing no change of oxidation state but only acid–base reactions (a sole exchange of protons for aqueous solutions), such as

$$2Fe^{3+} + 3H_2O \Leftrightarrow Fe_2O_{3(s)} + 6H^{+} \quad (1.23)$$

with the boundary plotted at the pH value for which $Q_{23} = K_{23}$ with $aFe_2O_{3(s)} = (aFe^{3+})^2 = 1$ (and also $aH_2O = 1$).

We now see that Reactions 1.3 and 1.4 are both related to half-reaction 1.22, but they refer to different redox conditions and then have different meanings. In Reaction 1.3 water is the oxidizing agent in acidic conditions, whereas it is the reducing agent in Reaction 1.4 (Appelo and

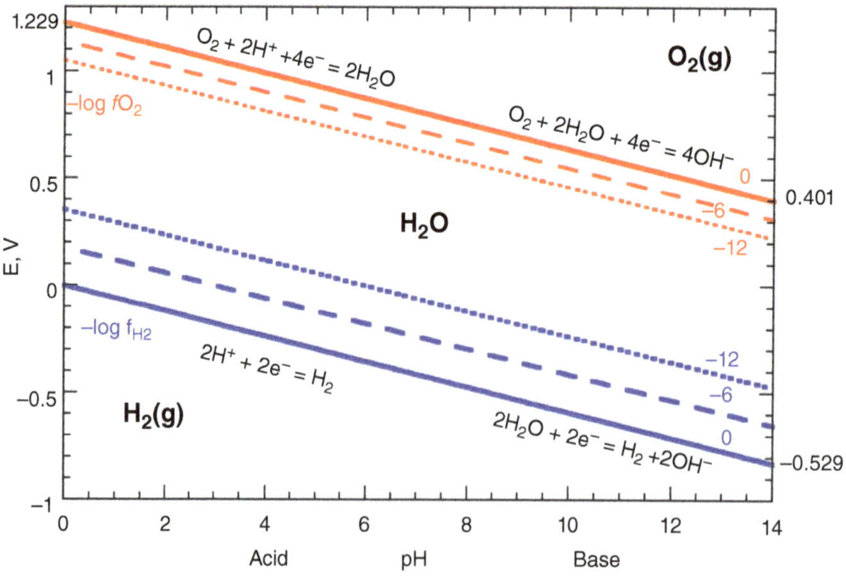

Figure 1.1 E-pH diagram reporting the stability of water at T = 25°C and P = 1 bar for different partial pressure of H_2 and O_2 (log-values).

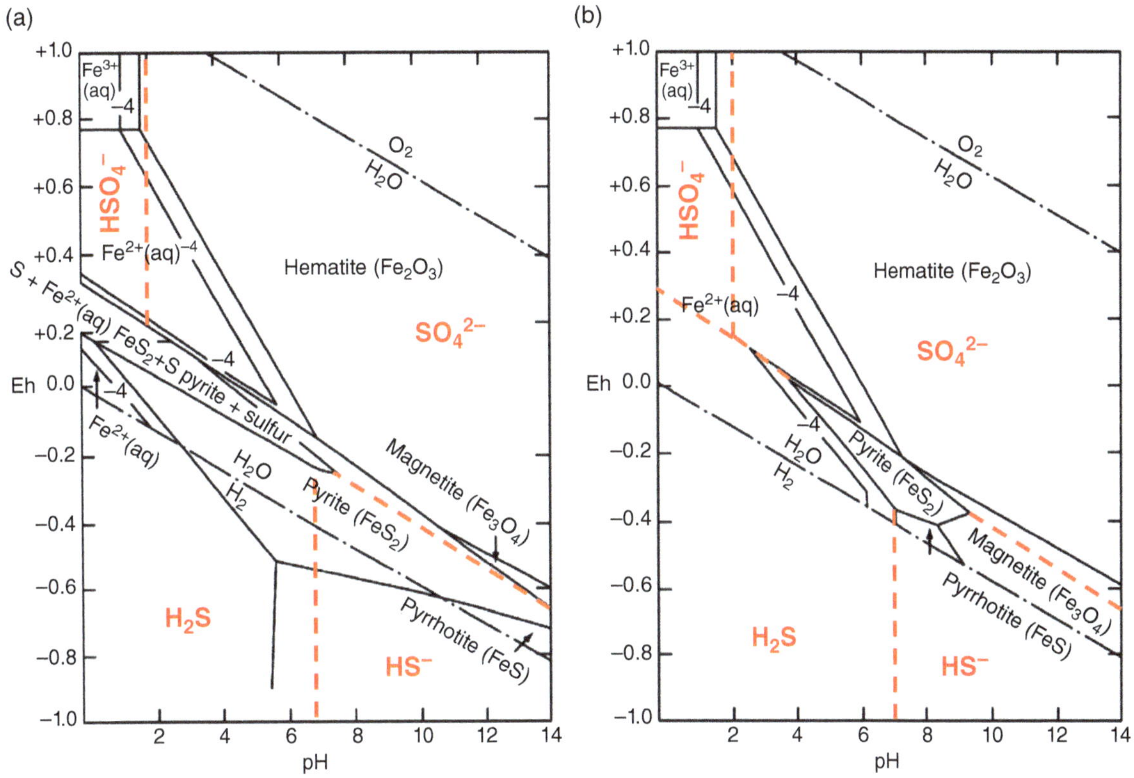

Figure 1.2 The E-pH (Pourbaix) diagram for the Fe-S-H_2O system at 25°C at 1 bar total pressure and for total dissolved sulfur activities of 0.1 (panel a) and 10^{-6} (panel b). Superimposed are the stability fields for H_2S, HS^-, HSO_4^{2-} and SO_4^{2-} dissolved species (red lines, traced from data in Biernat and Robins, 1969). Note how the field of stability of pyrite, FeS_2, shrinks and that of magnetite, Fe_3O_4, expands with decreasing total sulfur activity. Modifed from Vaughan (2005).

Postma, 1996), which represents a regulating mechanism of O_2, probably the oxidative alteration of rocks containing Fe^{2+}, in particular at oceanic ridges.

Basically, E-pH diagrams demonstrate that breaking of a redox reaction into half-reactions is one of the most powerful ideas in redox chemistry, which allows relating the electron transfer to the charge transfer associated with the speciation state and the acid–base behavior of the solvent. Superimposing E-pH diagrams allows a fast recognition of the existing chemical mechanism occurring in an electrolyte medium. For example, Figure 1.2 on the Fe-H-O-S system can be seen as the result of the superposition of stability diagrams for H-O-S and Fe-O-H system. The resulting diagram in Figure 1.2 shows that the pyrite–magnetite boundary has a negative slope due to half-reaction:

$$Fe_3O_{4(s)} + 6SO_4^{2-}{}_{(aq)} + 56H^+{}_{(aq)} + 44e^- \Leftrightarrow 3FeS_{2(s)} + 28H_2O_{(aq)} \quad (1.24)$$

but also a positive slope well visible in Figure 1.2b due to sulfur reduction and dissolution in water as HS^-:

$$3FeS_{2(s)} + 4H_2O_{(aq)} + 4e^- \Leftrightarrow Fe_3O_{4(s)} + 6HS^-{}_{(aq)} + 2H^+{}_{(aq)} \quad (1.25)$$

Reaction 1.25 implies of course a positive slope, because H^+ appears on the right side and electrons on the left. We can also appreciate the reduction of sulfur from pyrite to pyrrhotite at pH > 7:

$$FeS_{2(s)} + H^+ + 2e^- \Leftrightarrow FeS_{(s)} + HS^-{}_{(aq)} \quad (1.26)$$

which has a negative slope of –0.0295pH because the number of exchanged electrons is double than protons.

These concepts can then be transferred to other solvents in which ligand–metal exchanges lead to a different speciation state and are governed by a different notion of basicity, i.e. oxobasicity, such that (see Moretti, 2020 and references therein):

$$\text{Base} \Leftrightarrow \text{Acid} + O^{2-} \quad (1.27)$$

which can be also related to redox exchanges via the *normal oxygen electrode* (Equation 1.6), in the same way the normal hydrogen electrode (Reaction 1.7) can be put in relation with the Bronsted-Lowry definition of acid–base behaviour in aqueous solutions (see Moretti, 2020):

$$\text{Base} + H^+ \Leftrightarrow \text{Acid} \quad (1.28)$$

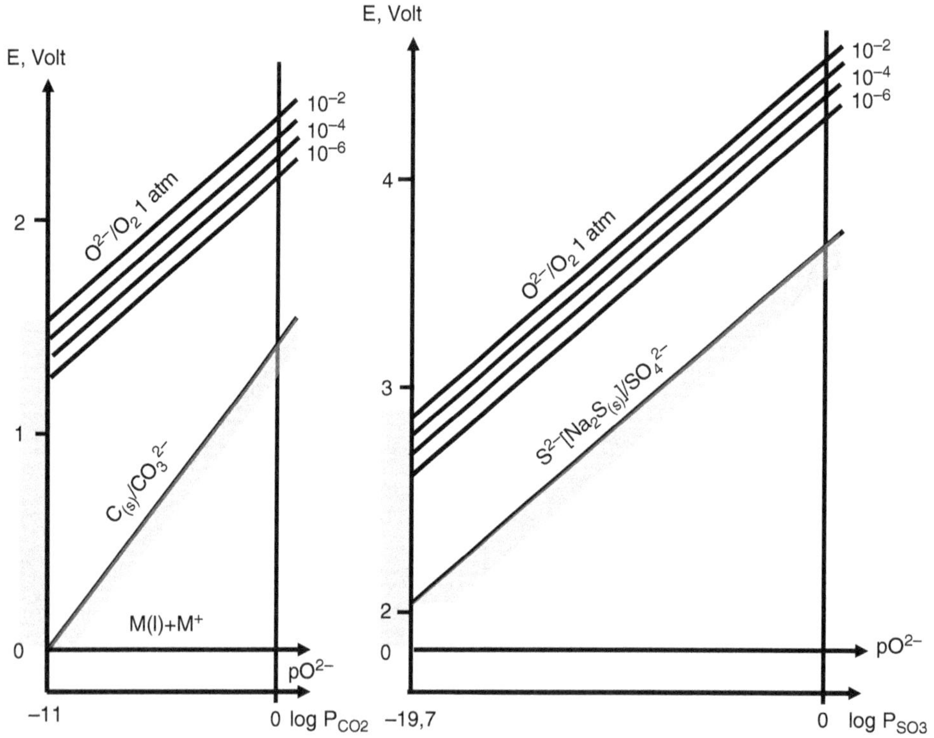

Figure 1.3 Limit of equilibrium potential-pO^{2-} graphs in molten alkali carbonates and sulfates, at 600°C (modified from Trémillon, 1974).

It is then possible to define pO^{2-} = -logaO^{2-} and introduce E-pO^{2-} diagrams, in which acid species will be located at high pO^{2-} values. These diagrams were first introduced by Littlewood (1962) to present the electrochemical behaviour of molten salt systems and provide an understanding of the stability fields of the different forms taken by metals in these systems. Reference potential for molten salt is chosen either from anion or from cation, but anion, making up the ligand, is normally selected because there may be several different cations in the system.

For molten solvent diagrams, such as carbonate and sulfate melts, the stability area of the bath depends on the salt itself and can be seen by using as examples oxyanion solvents (Figure 1.3). Limitations on the pO^{2-} scale of oxoacidity (Reaction 1.27) are given by the values of the Gibbs free energy of the formation reactions of alkali carbonates or sulfates at the liquid state, which depends on temperature as well as on pressure. On the basic side (low pO^{2-} side) the limit is imposed by the solubility threshold of the generic $M^{\nu+}O_{\nu/2}$ oxide in the electrolyte medium, i.e., pO^{2-}min ≈ $M^{\nu+}O_{\nu/2}$ solubility, whereas on the acidic side the limit is imposed by P_{CO2} or P_{SO3} = 1 bar. For example, it is 11 units in the case of the ternary eutectic Li_2CO_3+Na_2CO_3+K_2CO_3 at 600°C and 19.7 units in the case of the ternary eutectic Li_2SO_4 + Na_2SO_4 + K_2SO_4 at the same temperature (Trémillon 1974; Figure 1.3).

The upper stability limit is related to the $O^{-II}/O_{2(g)}$ redox system (Reaction 1.6), i.e., to the oxidation of CO_3^{2-} and SO_4^{2-} anions:

$$CO_3^{2-} \Leftrightarrow \tfrac{1}{2}O_{2(g)} + CO_{2(g)} + 2e^- \quad (1.29)$$

$$2SO_4^{2-} \Leftrightarrow \tfrac{1}{2}O_{2(g)} + S_2O_7^{2-} + 2e^- \quad (1.30)$$

which results from acid–base exchanges of the type:

$$CO_3^{2-} \Leftrightarrow CO_{2(g)} + O^{2-} \quad (1.31)$$

$$2SO_4^{2-} \Leftrightarrow S_2O_7^{2-} + O^{2-} \quad (1.32)$$

coupled to half-reaction 1.6.

Both Reactions 1.29 and 1.30 yield the E-pO^{2-} relationship:

$$E = E^0_{O_2} + \frac{2.3RT}{4F}\log P_{O_2} + \frac{2.3RT}{2F}pO^{2-} \quad (1.33)$$

The lower stability limit of the solvent can be given by either the reduction of alkaline cation in the corresponding metal:

$$M^{\nu+} + \nu e^- \Leftrightarrow M_{(l)} \quad (1.34)$$

whose potential is independent of pO^{2-}, or the reduction of CO_3^{2-} and SO_4^{2-} anions given by:

$$CO_3^{2-} + 4e^- \Leftrightarrow C_{(s)} + 3O^{2-} \quad (1.35)$$

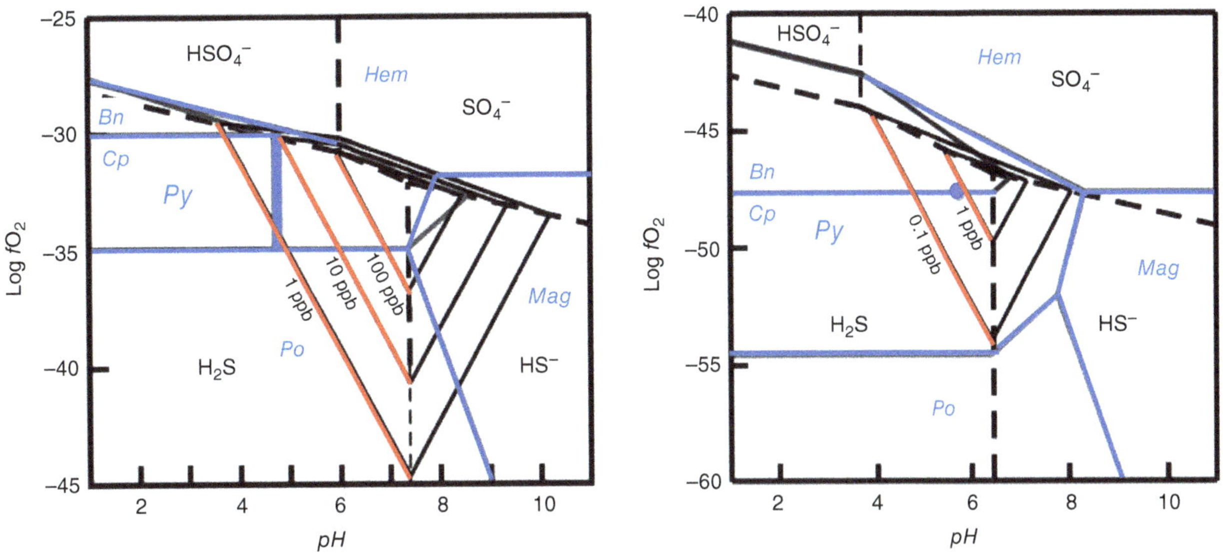

Figure 1.4 Log fO_2 - Ph diagrams for 290°C (left panel) and 145°C (right panel) at saturated vapor pressure, showing predominance fields for aqueous sulfur species (dashed lines), stability fields for Fe–O–S minerals and bornite–chalcopyrite (solid grey lines). The solubility contours (left panel: 1, 10, 100 ppm; right panel: 0.1, 1 ppm) are for gold in the form $Au(HS)_2^-$. Modified from Raymond et al. (2005).

$$SO_4^{2-} + 8e^- \Leftrightarrow S^{2-} + 4O^{2-} \quad (1.36)$$

and to which the following E-pO^{2-} relationships correspond (CO_3^{2-} and SO_4^{2-} anions having unitary activity):

$$E = E^0_{C/CO_3^{2-}} + \frac{3}{4}\cdot\frac{2.3RT}{F}pO^{2-} \quad (1.37)$$

$$E = E^0_{S^{2-}/SO_4^{2-}} - \frac{2.3RT}{8F}\log\left[S^{2-}\right] + \frac{2.3RT}{2F}pO^{2-} \quad (1.38)$$

Figure 1.3 shows the results on carbonate and sulfate melts (modified from Trémillon, 1974, and references therein). The utilizable regions appear as quadrilaterals on the E-pO^2 graph. If in sulfates, the region is a parallelogram similar to the E-pH region in aqueous solution, the theoretical range of potential in molten carbonates appears more restricted in an oxoacidic medium than in an oxobasic medium, because the lower limit varies versus pO^{2-} with a slope greater than that of the upper limit (Trémillon, 1974).

Silicate melts have been so far an underestimated electrolytic medium acting as a solvent for oxides. This is mainly because E-pO^{2-} diagrams cannot be based on predictive thermodynamics and physical chemistry assessments such as the dilute electrolyte concept and its developments in the case of previous solvents, aqueous solutions particularly (Allanore, 2015). In silicate melts, and more generally molten oxides, oxygen *tout-court* cannot be identified as the solvent, despite its abundance. Silicate melts are in fact a high-temperature highly interconnected (polymerized) matrix in which solvation units cannot be easily defined and both ionic and covalent bonds rule the reactive entities that make up the melt network. Because of this, some approaches have been formalized in terms of the Lewis acid–base definition (network formers and their oxides, such as SiO_2 and Al_2O_3 are acids; network modifiers and their oxides such as MgO, CaO, Na_2O are bases) by using electronegativity and optical basicity that allow distinguishing and calculating three types of oxygen (bridging, non-bridging, and so-called free oxygen) whose mixing determines the polymerization and the thermodynamic properties of the melt mixture as a function of composition (Toop and Samis, 1962a,b; Allanore, 2013, 2015 and references therein; Moretti, 2020 and references therein). Nevertheless, silicate melts still lack a fully developed acid–base framework formalizing the thermodynamic properties of reactive species formed during the solvolysis, as the solvent itself changes its polymerization properties upon introduction of other oxide components, which are highly soluble contrary to what observed for salts in aqueous solutions. The most general thermodynamic approaches postulate mineral-like molecular structures to interpolate existing data.

In molten silicates the electric charge is primarily transported by cations, whose contribution increases with concentration of network modifiers, hence basicity. The major element, oxygen in its three forms and particularly O^{2-} and O-based anionic complexes do not contribute substantially to the charge transport (Dickson and Dismukes, 1962; Dancy and Derge, 1966; Cook and Cooper, 1990, 2000; Cooper et al., 1996a,b; Magnien et al., 2006, 2008; Cochain et al., 2012, 2013; Le Losq et al. 2020). This is a striking difference when comparing silicate melts to previously described electrolytes, particularly with the aqueous electrolytes, in which the concentration of hydroxide ions or water is always large enough to sustain high current densities (Allanore, 2013). From a practical standpoint the anode reaction producing oxygen (the inverse of Reaction 1.6) is strongly impacted by the low transport of free oxide ions, that is free oxygen, which is at very low concentration. Besides, the anode design and technical performance would greatly benefit of the precise knowledge of oxygen physical chemistry in silicate melts, which for the moment is still too limited to narrow compositional ranges that have been investigated spectroscopically (Allanore, 2015).

Nevertheless, because of their nature, silicate melts can dissolve important amounts of metals. Besides, they exhibit a large range of thermal stability, with high temperature conditions that favor fast kinetics of redox exchanges. Upper temperature limits for electrochemical applications are given by the formation of gaseous silicon monoxide or by alkali oxide thermal decomposition in alkaline systems or by high vapor pressure for Mn-bearing systems (Allanore, 2015). In terms of transport properties, silicate melts are a solvent with high viscosity, a fact that is however compensated by the high diffusivity of the metal cation (Allanore, 2013), i.e., the cathode reactant.

Many measurements have however been carried out on melts of geological interest, also fostered by the interest in silicate electrolysis to produce on site metals and particularly molecular oxygen for terraformation of extraterrestrial planets (e.g., Haskin et al., 1992). Electrochemical series were then established in binary SiO_2-MO systems (e.g., Schreiber, 1987) but also in ternary joins such as the diopside one (Semkow and Haskin 1985; Colson et al., 1990) for redox exchanges of the type:

$$4M^{m+}_{(melt)} + 2nO^{2-} \Leftrightarrow 4M^{(m-n)+}_{(melt)} + nO_{2(gas)} \quad (1.39)$$

in which half-reactions of the type

$$M^{m+}_{(melt)} + ne^{-} \Leftrightarrow M^{(m-n)+}_{(melt)} \quad (1.40)$$

combine with Half-reaction 1.7. Nevertheless, such series do not consider the effect of the solvent, the melt and its structure, in determining the speciation state (e.g., anionic or cationic) following the definition at Reaction 1.27. The effect of the solvent also includes the amphoteric behaviour of some dissolved oxides such as Fe_2O_3 or Eu_2O_3, which can behave either as acids, yielding FeO_2^- (i.e., FeO_4^{5-} tetrahedral units) and EuO_2^- (i.e., EuO_4^{5-}) or bases, yielding Fe^{3+} and Eu^{3+} cations (Fraser, 1975; Ottonello et al., 2001; Moretti, 2005; Le Losq et al., 2020). The multiple speciation behaviours determined by pO^{2-} can be summarized by the following reaction mechanism (e.g., Moretti, 2005; Pinet et al., 2006):

$$MO_{x,melt}^{(2x-m)-} \Leftrightarrow MO_{y,melt}^{(2y-m+n)-} + \left(x - y - \frac{n}{2}\right)O^{2-} + \frac{n}{4}O_2 \quad (1.41)$$

Predominance and stability diagrams (e.g., E-pH, E-pO^{2-}, E-logfO_2, or logfO_2 vs. the log-fugacity of pH or other gaseous species in the system such as SO_2, CO_2) depend on the availability of good thermodynamic data and especially a well-established testament of acid-base properties of the investigated system and its solvent(s). For silicate melts and glasses, such a testament is represented by the oxobasicity scale from the Lux definition (Reaction 1.27). Electrochemical experiments should then be envisioned to complete and validate the database in order to ensure predictions about forming species and measure their activities.

1.2. OXYGEN FUGACITY: THE CENTRALITY OF AN ELUSIVE PARAMETER

Voltage E and oxygen fugacity (fO_2) are both measures of oxidation state. The relation between fO_2 and E for a given electrolytic medium can be established by the anode reaction where oxygen is produced. In the case of aqueous solutions, conversion is provided by half-reaction 1.15 and Equation 1.18. We can then replace E-pH diagrams with analogous logfO_2-pH diagrams. In this treatment, the actual speciation state of solutions is still the key to investigate the system, but half-reactions are not considered, and the equilibrium values of overall reactions are used, same as for activity plots. As for E-pH plots, boundaries will shift by varying the total amount of soluble elements in the electrolytic solution, hence the activity of dissolved ionic species or the corresponding gas fugacity (e.g. when carbonates or sulfides and sulfates are present).

Figure 1.4 shows that in logfO_2-pH diagrams phase, boundaries that in E-pH diagrams were dependent on pH only (and not on the concentration of dissolved ions) are horizontal. Moreover, a quick comparison between figures 2 (Eh-pH diagram) and 4 (logfO_2-pH diagram) also shows that on the logfO_2 vs. pH representation the pyrite–magnetite boundary, which appears only for pH > 7, maintains the positive slope. At the T conditions of Figure 1.4 this boundary is represented by the overall reaction:

$$3FeS_{2(S)} + 6H_2O_{(aq)} \Leftrightarrow Fe_3O_{4(s)} + 6HS^-_{(aq)} + 6H^+_{(aq)} + O_2 \quad (1.42)$$

Besides, at 145°C (water saturated conditions) the pyrite–pyrrhotine boundary is defined at pH > 7 and is positive because:

$$FeS_2 + 2H_2O \Leftrightarrow FeS + 2HS^- + 2H^+ + O_2 \quad (1.43)$$

whereas for pH < 7 the boundary is horizontal, given by the proton-free reaction:

$$FeS_2 + H_2O_{(aq)} \Leftrightarrow FeS + H_2S_{(aq)} + \tfrac{1}{2}O_2 \quad (1.44)$$

As we have already seen, when considering high-temperature non-aqueous (oxide) systems in the inner Earth geospheres, there is no acid–base framework and anchoring fO_2 or E to pH makes no sense in absence of the solvent liquid water.

To measure the chemical potential of redox exchanges in higher temperature geological systems, geoscientists turned their attention to molecular oxygen transfer among molecular components such as oxides or mineral-like macromolecular entities.

The practice between geoscientists becomes to assess criteria for fO_2 (or aO_2) estimations disconnected from the formal description of the acid–base character of magmas. In particular, techniques were established involving mineral phases coexisting in igneous rock to establish thermodynamic or empiric laws and trends from quenched glasses via indirect measurements, most often of spectroscopic nature (e.g., Neuville et al., 2020). This change of perspective reflects the obvious consideration that geoscientists deal with samples (solidified rocks) made accessible at Earth's surface and which represent the final snapshots at the end of a long thermal and chemical evolution, whose *a posteriori* reconstruction is the objective of the geochemical (*lato sensu*) investigation.

We may then say that for practical reasons geoscientists remained anchored to the original Lavoisier-like definition of oxidation occurring in combustion processes, related to the exchange of oxygen molecules. The fact that most of the chemical analyses were from techniques in which oxygen was not directly determined but allowed to give oxides has also further favored these approaches.

In this framework, a mutual exchange of knowledge has always characterized the field of geochemistry and petrology on one side and that of metal extraction in metallurgy in the other. Relations of the type

$$M^{\nu+}{}_{2/\nu}O_{(s/l)} \Leftrightarrow 2/\nu M_{(s/l)} + 1/2O_{2(g)} \quad (1.45)$$

with v the charge (positive) of the cation of the metal M in the corresponding oxide. Reaction 1.45 is the main target of extractive metallurgy (see also Reaction 1.34), but also sketches the ensemble of processes that occurred since early Earth's evolution to segregate the metallic core.

Ellingham diagrams (Ellingham 1944; Figure 1.5) are used in metallurgy to evaluate the ease of reduction of metal oxides, as well as chlorides, sulfides, and sulfates. The diagram shows the variation of the standard Gibbs free energy of formation, ΔG^0, with temperature for selected oxides and is used to predict the equilibrium temperature for reactions of the type of Reaction 1.45 and particularly the oxygen fugacity under which ore will be reduced to its metal. The standard Gibbs energy change of formation of a compound (the Gibbs energy change when one mole of a compound is formed from elements at P = 1 bar) is given by:

$$\Delta G_f^0 = \Delta H_f^0 - T\Delta S_f = \left[\Delta H_{298}^0 + \int_{298}^{T} Cp dT\right]$$

$$- T\left[\Delta S_{298} + \int_{298}^{T} \frac{\Delta Cp}{T} dT\right] = A + B \times T(K) \quad (1.46)$$

with R the universal gas constant and A and B constants.

To compare the relative stabilities of the various oxides, the Ellingham diagram is prepared for oxidation reactions involving one mole of oxygen. For the oxidation of a metal, ΔG^0 represents the chemical affinity of the metal for oxygen. When the magnitude of ΔG^0 is negative, the oxide phase is stable over the metal and oxygen gas. Furthermore, the more negative the value, the more stable the oxide is. The Ellingham diagram also indicates which element will reduce which metal oxide. The similarity between the electromotive force series (E^0) and the Ellingham diagram, which rates the tendency of metals to oxidize, should be easily recognized.

When both Me and $Me^{\nu+}{}_{2/\nu}O$ in Reaction 1.45 are in their standard states, the equilibrium constant, K_{45}, corresponding to this reaction can be expressed as:

$$K_{45} = aM^{2/\nu}{}_{(s/l)}/aM^{\nu+}{}_{2/\nu}O_{(s)} \cdot 1/aO_2 = (f^0O_2/fO_2)^{1/2} \quad (1.47)$$

where $fO_2{}^0$ is the pure gas component gas fugacity at standard state (in this case 1 bar and T of interest). If, at any temperature, the acting oxygen fugacity is greater than the calculated value from Equation 1.47, spontaneous oxidation of metal M occurs, while oxide $Me^{\nu+}{}_{2/\nu}O_{(s)}$ decomposes to metal Me and gaseous oxygen at the oxygen partial pressure less than the equilibrium value. In other words, an element is unstable, and its oxide is stable at higher oxygen potentials than its $\Delta G_f{}^0$–T line on the Ellingham diagram. Therefore, the larger negative value for $\Delta G_f{}^0$ an oxide has, the more stable it is. In the Ellingham diagram of Figure 1.4, it can be seen, for example, that the reduction of Cr_2O_3 by carbon is possible (from the thermodynamic standpoint) at temperature above 1250°C and at each reported temperature by aluminum. It is worth noting that the Ellingham line for the formation of carbon monoxide (CO) has a negative slope, while those of all other oxides have positive slopes. As a result, at sufficiently high

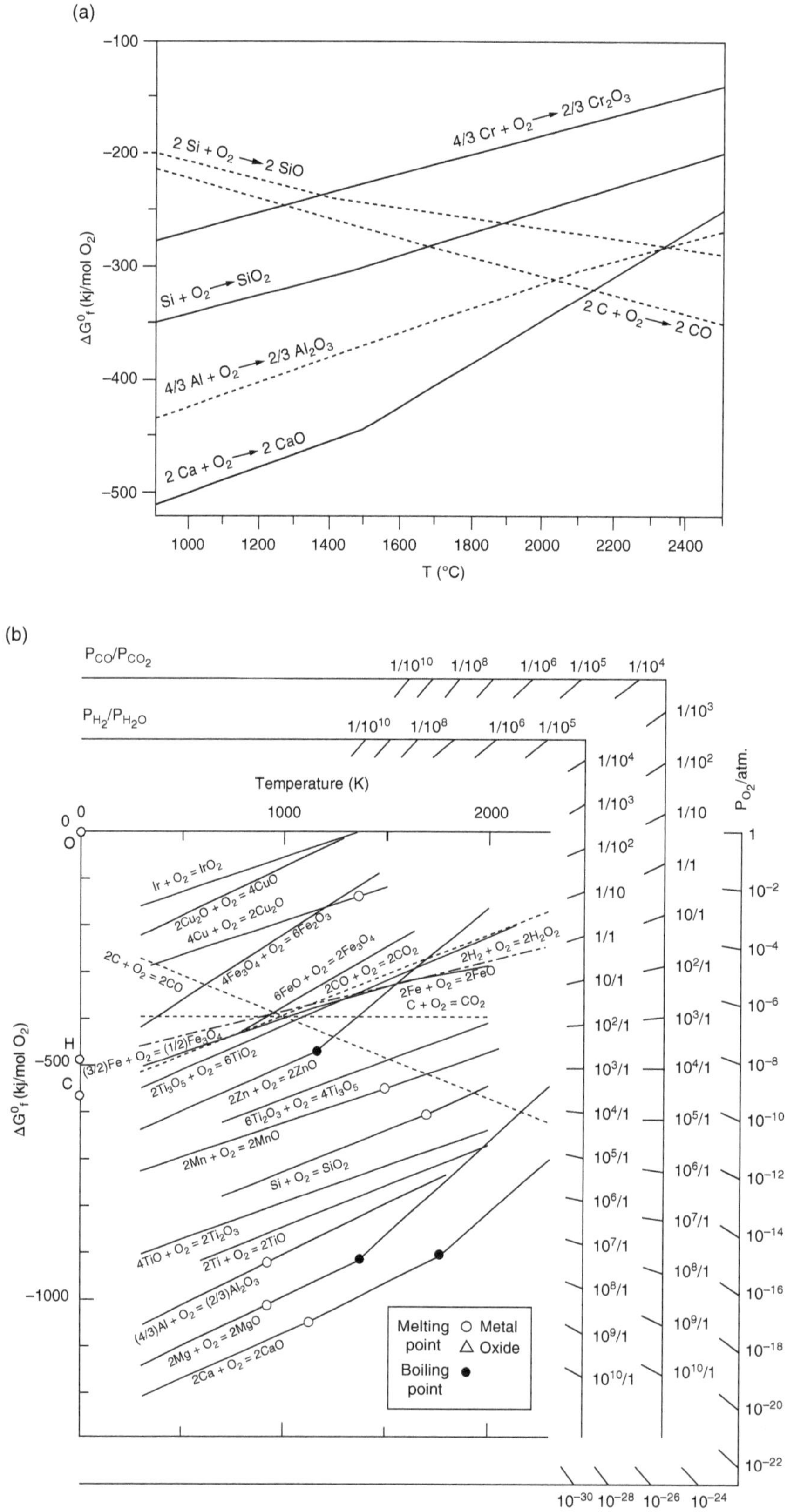

Figure 1.5 **(a)** Ellingham diagram for the main components of the melt/slag (solid lines) and possible reducing agents (dotted lines). The slopes of the lines representing ΔG_f^0–T relations change at the temperature in which the phase transformations of reactants or products occur. Modified from Zhang et al. (2014). **(b)** Full Ellingham diagram for some relevant oxides including CO_2 equilibria and normographic scales for oxygen fugacity and related quantities via CO/CO_2 and H_2/H_2O ratios at P_{tot} = 1 bar. The scale of P_{H_2}/P_{H_2O} ratio is used in the same manner as that of P_{CO}/P_{CO_2} ratio, except that the point H on the ordinate at T= 0 K is used instead of point C. Modified from Hasegawa (2014). All reactions' components are considered in their pure stable phase t 1 bar and T of interest.

temperatures, carbon will reduce even the most stable oxides.

Reaction 1.45 illustrates in fact how pairs of metals and their oxides, both having unitary activity, can be used as redox buffers, such that fO_2 values can be easily fixed at any temperature. Even in the presence of a third phase, such as silicate melts or any other liquid, gas–solid assemblages allow a straightforward application of Equation 1.47 to impose fO_2, unless solid phases are not refractory, and dissolve other components exchanges with the coexisting liquid. Reaction 1.1, involving metal iron and wustite, is a typical example (so-called IW buffer) of one of these gas–solid equilibria fixing fO_2.

In order to illustrate the effect of the fluid phase, refined versions of the Ellingham diagrams included normographic scales, which are designed so that the equilibrium oxygen partial pressure or the corresponding P_{H_2}/P_{H_2O} or P_{CO}/P_{CO_2} ratios between metal and its oxide can be read off directly at a given temperature by drawing the line connecting point O for P_{O_2}, C for P_{CO}/P_{CO_2}, or H for P_{H_2}/P_{H_2O} in Figure 1.5 and the condition of interest (Reaction 1.45 at T of interest) and then extending it to the corresponding normographic scale.

It is now quite obvious to see how both metallurgists and petrologists could then develop techniques to constrain fO_2 to investigate melting and sub-solidus conditions of oxides and silicates. After the seminal studies of Bowen and Schairer (1932, 1935) on $FeO–SiO_2$ and $MgO–FeO–SiO_2$ systems and in which the authors used iron crucibles in an inert (O_2-free) atmosphere to equilibrate the phases with metallic iron at very low but also unknown P_{O2}, early fO_2 control techniques were applied by Darken and Gurry (1945, 1946), who detailed the Fe–O system based on the use of CO and CO_2, or CO_2 and H_2, conveyed in a gas-mixer supplying a continuous mixture in definite constant volume proportions.

Later Eugster, in his early experiments to determine the phase relations of annite had to prevent its oxidation and formation of magnetite (Eugster 1957, 1959; Eugster and Wones 1962). He then developed the double capsule technique, with a Pt capsule containing the starting material, surrounded by a larger gold capsule. A metal–oxide or oxide–oxide pair plus H_2O was then placed between the two metal containers. In these experiments, reaction involving OH-bearing minerals and H_2O provides a fixed and known hydrogen fugacity (Eugster 1977), so through the dissociation Reaction 1.14 measurement of fH_2 allows calculating both fO_2 given the fH_2O at the experimental pressure (aH_2O = 1)of the internal capsule.

Since Eugster, many metal–oxide and oxide–oxide assemblages have been used in experimental petrology that have been called "redox buffers." These buffers have contributed to our understanding of the role of fO_2 on melt phase equilibria and mineral composition on Earth, also under volatile saturated conditions due to the possibility of evaluating fluid phase speciation at the experimental P and T conditions (e.g., Pichavant et al., 2007; Frost and McCammon, 2008. Mallmann and O'Neill, 2009; Feig et al., 2010). The results of experimental petrology made it possible to systematically collect glasses (quenched melts) to measure ratios of metals in their different oxidation states, particularly Fe^{II}/Fe^{III}, and relate such ratios to experimental P, T, and fO_2, glass composition, or other spectroscopic observations about glass/melt structure and the local coordination of targeted metals (Neuville, 2020 and references therein).

Obviously on Earth iron is the most abundant multivalent element and because of its speciation behaviour it gives rise to many reactions involving minerals and liquid (i.e., natural melts, particularly silicate ones). Reactions such as Reaction 1.1 are then useful for providing a scale with geological significance for fO_2 conditions that are recorded in rock and minerals formed in past and present Earth environments, from the core (in which Fe^0 dominates) up to shallow crust and through all igneous environments where it exists, and Fe^{II} and Fe^{III}. To provide systematics for the aO_2 conditions the following gas–solid reactions, other than Reaction 1.1, were then assessed (Figure 1.6):

$$4Fe_3O_4 + O_2 \Leftrightarrow 6Fe_2O_3 \text{ (magnetite – hematite; MH)} \quad (1.48)$$

$$3Fe_2SiO_4 + O_2 \Leftrightarrow 2Fe_3O_4 + 3SiO_2 \text{ (fayalite – magnetite – quartz; FMQ)} \quad (1.49)$$

$$3FeO + O_2 \Leftrightarrow Fe_3O_4 \text{ (wustite – magnetite; WM)} \quad (1.50)$$

$$SiO_2 + 2Fe + O_2 \Leftrightarrow Fe_2SiO_4 \text{ (iron – quartz – fayalite; IQF)} \quad (1.51)$$

All these equilibria have the interesting feature of displaying unitary activities for oxide component appearing as pure phases, such that their equilibrium constants simply describe the variations of O_2 activity (aO_2) with temperature:

$$\log fO_2 = A/T + B \quad (1.52)$$

or, by defining fugacity, with temperature and pressure:

$$\log fO_2 = A/T + B + C(P - 1)/T \quad (1.53)$$

with A, B, and C constants. The pressure term, C, allows computing directly fO_2 rather than aO_2 at the pressure of interest. Mineral assemblages making up Reactions 1.1 and 1.44 to 1.47 are not necessarily occurring in deep Earth and igneous environments, whereas the actual multi-component space of phases fix the aO_2 via multiple equilibria in which solid solutions and/or the presence of

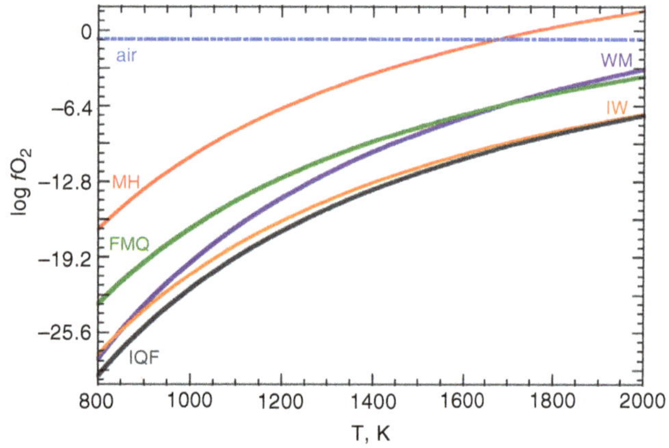

Figure 1.6 Common solid oxygen buffers used in petrology and geochemistry. The lines represent the fugacity-temperature conditions where the phases coexist stably. FMQ: fayalite, magnetite, quartz; IW: iron–wüstite; WM: wüstite–magnetite; HM magnetite–hematite. Oxygen fugacity values were computed for a total pressure of 1 bar. Also reported is the value of logfO_2 in air (PO_2 ~0.21 bar).

relatively mobile liquid (silicate melts) and/or supercritical fluids play a fundamental role.

However, when one of these mineral "buffers" is selected as a reference, the acting logfO_2 can be given as a relative value, without temperature:

$$\Delta buffer = \log fO_2 - \log fO_{2,\text{buffer}} \tag{1.54}$$

A relative fO_2 scale embodying temperature effects bears just a practical implication (tracking fO_2 variations with respect to a reference) but not a real meaning about logfO_2 evolution in igneous systems and related environments (e.g., Moretti and Steffansson, 2020). In particular, a common misconception was that in a system a given value of $\Delta buffer$ (e.g., ΔQFM = 0.5) could represent some kind of "magic number" characteristic of the whole "rock system" throughout its thermal and chemical evolution. In natural environments oxygen activity (hence fugacity) in fact varies to accommodate the compositional variations and the speciation state of the mineral/melt/fluid phases, and also when highly mobile volatile components are involved, such that fO_2 can thus be fixed by factors that are external to the system object of the thermodynamic description.

Similarly, the rock system evolution cannot be approximated by a unique Fe^{II}/Fe^{III} ratio, that the system had when completely molten. Indeed, two rocks that have ideally crystallized along the QFM buffer may have different proportions of fayalite ad magnetite because of the crystallization style but different Fe^{II}/Fe^{III} bulk ratios (Frost, 1991). However, in some systems, it is possible that the melt fixed the redox potential of the system via iron oxidation state, with the Fe^{II}/Fe^{III} ratio approaching unity (e.g. Moretti et al., 2013).

It is also important to recall that fO_2 is just a thermodynamic parameter used to conveniently report the oxidation state of a system, particularly when O_2 is not an existing gaseous species that could be detected if the system were accessible to measurements. This is clearly proved by the very low values reported in ordinates in Figure 1.6. Therefore, fO_2 turns out to be by-product of thermodynamic calculations applied to the analyses from natural samples, in which the true redox observables are the oxidation states of iron and other elements in minerals and liquids. The common practice is then to measure the concentration ratio of redox couples of multiple valence elements in melts (Fe^{II}/Fe^{III}, but also S^{-II}/S^{VI}, V^{III}/V^{V}, etc.) or in gases (e.g., H_2/H_2O, CO/CO_2, H_2S/SO_2) and relate them to fO_2 via thermodynamic calculations using appropriate standard state thermochemical data. As fO_2 is provided, its value is then anchored via Equation 1.54 to a given gas-solid buffer of the Reaction type 1.1 or 1.48 to 1.51. This is quite easy for gases, in which governing equilibria are directly solved if the gas analysis is provided (e.g., Giggenbach, 1980, 1987; Aiuppa et al., 2011 and references therein) and also for solid–solid equilibria, such as in case of coexisting iron–titanium oxide solid solutions titanomagnetite (Fe_3O_4-Fe_2TiO_4) and hemo-ilmenite (Fe_2O_3–$FeTiO_3$) (Buddington and Lindsley, 1964) or for peridotite assemblages in the mantle (e.g., Mattioli and Wood, 1988; Gudmundsson and Wood, 1995), in which the good thermodynamic characterization of solid solutions allows quite accurately treating component activities, based on mineral analyses.

On the contrary, it is much less straightforward for fO_2 estimates by oxidation states of Fe and/or S measured in glasses, as the oxybarometers derived by the study of synthetic quenched melts still suffer from too many empiric approaches. Because of their polymerized nature, silicate melts do not allow a precise distinction between solute and solvent like in aqueous solutions, where complexes and solvation shells can be easily defined in which covalence forces exhaust (see also Moretti et al., 2014). In fact, melt composition largely affects the ligand constitution and then the speciation state of redox-sensitive elements. In the case of the Fe^{II}/Fe^{III} ratio, the most common redox indicator for melts/glasses, the choice of components in reaction:

$$FeO_{(m)} + 1/4O_2 \Leftrightarrow FeO_{1.5(m)} \tag{1.55}$$

over a large compositional range (e.g., from mafic to silicic) does not offer the possibility to find accurate and internally consistent expressions for the activity coefficients of *oxide components* γFeO and $\gamma FeO_{1.5}$ (with $FeO_{1.5}$ conveniently replacing Fe_2O_3) that solve the reaction equilibrium constant:

$$\log fO_2 = 4\log\frac{XFe^{III}}{XFe^{II}} + 4\log\frac{\gamma FeO_{1.5}}{\gamma FeO} + \frac{\Delta G^0_{1,T} + \int_1^P \Delta V_{FeO_{1.5}-FeO}\,dP}{2.303RT} \quad (1.56)$$

in which the term within integral is the difference of partial molar volumes of iron components. Expansion of excess contribution to its Gibbs free energy of mixing is used to define γFeO and $\gamma FeO_{1.5}$ in melt mixtures. However, when these are adopted to solve Equation 1.56 for measured Fe^{II}/Fe^{III} values, they show success only over limited compositional datasets (Moretti, 2020 and references therein). Armstrong et al. (2019) calibrated volumes and interaction parameters of activity coefficients entering equation (1.56) for an andesitic melt (easily quenchable as a glass) with fO_2 buffered by the Ru-Ru_2O assemblage in the T-range 1673K to 2473K and for pressures up 23 GPa. Data fitting showed that volume term in Equation 1.56 turns from negative to positive for P > 10 GPa, which yielding iron oxidation with increasing pressure. The calibrated Equation 1.56 was then used by the authors to demonstrate how the mantle oxidized after the Earth's core started to form by a deep magma ocean with initial $Fe^{III}/Fe_{tot} = 0.04$ from which FeO disproportionated to Fe_2O_3 plus metallic iron at high temperature. The separation of Fe^0 to the core raised the oxidation state of the upper mantle and of exsolved gases that were forming the atmosphere (Armstrong et al., 2019).

The search for one general formulation for all melt compositions of interest in petrology and geochemistry led to empirical expressions, in which adjustable parameters are introduced without the formal rigor requested by Equation 1.56 (e.g. Kress and Carmichael, 1991). These formulations furnish quite accurate fO_2 values from measured Fe^{II}/Fe^{III} ratios within the compositional domain in which they have been calibrated. Besides, they often violate reaction stoichiometry and do not ensure internal consistency: if used to calculate activities they fail the application of the Gibbs-Duhem principle relating all component activities within the same phase (e.g., Lewis and Randall, 1961). Such expressions then treat fO_2 as a Maxwell's demon, doing what we need it to do to fit the calibration data and with the consequence that outside the calibration domain, all the unpredictable non-idealities are discharged on the fO_2 terms, resulting in biased calculations of fluid speciation, or other phase equilibria constraints.

As extensively treated in Moretti (2020), unpredictable non-idealities reflect counterintuitive behaviors that cannot be accounted for by activity coefficients used in the equilibrium constant of Reaction 1.55. It is well known that depending on melt composition alkali addition (i.e. decreasing pO^{2-}) can either oxidize or reduce iron in the melt. This occurs because of the change of speciation due to the amphoteric behavior of Fe^{III}, which depending on composition and then pO^{2-} can behave as either network former or modifier (see the conceptualization provided by Ottonello et al., 2001; Moretti, 2005, 2020; Le Losq et al., 2020; see also Reaction 1.41). Models that define the melt (oxo)acidity (Reaction 1.27) hence pO^{2-} (e.g; polymeric models based on the Toop and Samis mixing of bridging, non-bridging and free oxygens; see Moretti, 2020, and references therein) allows solving speciation and set activity-composition relations of ionic and molecular species, just as for aqueous solutions and molten salts.

The problem of unsolved compositional behaviors due to speciation, that are not accounted for by typical oxide-based approaches to mixtures, is exacerbated when dealing with the mutual exchanges involving iron and another redox-sensitive elements, such as sulfur. Sulfur-bearing melt species play a special role since the oxidation of sulfide to sulfate involves eight electrons: for any increment of the Fe^{III}/Fe^{II} redox ratio, there is an eight-fold increment for sulfur species (S^{-II}/S^{VI}; e.g., Moretti and Ottonello, 2003; Nash et al., 2019; Cicconi et al., 2020b; Moretti and Stefansson, 2020). Sulfur in magmas partitions between different phases (gas, solids such as pyrrhotite and anhydrite, and liquid as well, such as immiscible Fe–O–S liquids; Baker and Moretti, 2011 and references therein). The large electron transfer makes S^{-II}/S^{VI} a highly sensitive indicator to fO_2 changes in a narrow range (typically around QFM and NNO buffers in magmatic melts; Moretti, 2020 and references therein), whereas its effectiveness as a buffer of the redox potential is limited by the abundance of sulfur in magma, significantly lower than iron.

The modelling of joint Fe and S redox exchanges is still a major challenge which sees contrasting approaches (see Moretti, 2020). Formulations exist with various degrees of empiricism, but even those displaying better performances in exploring the $fO_2 - fS_2$ space of natural silicate melts (Moretti and Baker, 2008) should be carefully tested in reproducing phase diagrams involving multiple phases, including coexisting Fe–O–S melt, $FeS_{(s)}$, and $Fe_3O_{4(s)}$. Introduction of sulfur equilibria in petrogenetic grids would be a major step forward for modelling in igneous petrology. Besides, it would provide the liaison with processes occurring in late- to post-magmatic stages, prior to further cooling down to real hydrothermal conditions dominated by condensed water (see Figure 1.4). For late- to post-magmatic stages, such as in the case of porphyry-copper ore formation, $\log fO_2$-$\log fS_2$ diagrams. Figure 7 reliably summarizes phase relations in the Fe–S–O system, in a way similar to Figure 1.2 and 1.4. It is worth noting that the diagram in Figure 1.7 can also be seen as resulting from reactivity of a sub-solidus mixed iron

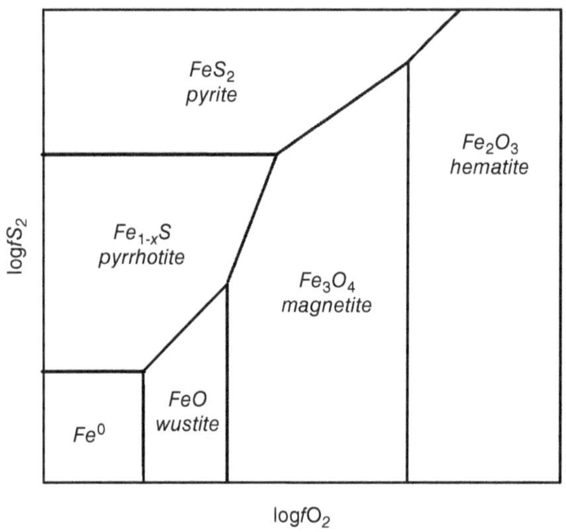

Figure 1.7 Two-redox potential fO_2-fS_2 diagram. The conformation of stability fields in the Fe–O–S space is essentially the same also for large fO_2 and fS_2 variations with temperature (redrawn from Nadoll et al., 2011).

molten oxide–sulfide, in which the two main ligands are O^{2-} and S^{2-} (half-reactions 1.6, 1.11, and 1.12).

A natural assemblage of pyrite + magnetite + pyrrhotite corresponds then to the triple point marked by a star in Figure 1.7, which at a given T is invariant for fO_2 and fS_2 values given by the simultaneous occurrence of Reaction 1.10 and:

$$Fe_3O_{4(s)} + 3FeS_{2(s)} \Leftrightarrow 6FeS_{(s)} + O_2 \quad (1.57)$$

that allow identifying the stable phase as a function of temperature and fugacities (or activities) of reference gas species. It is worth noting that in absence of water (no H in the system represented in Figure 1.7) the boundary between FeS_2 and FeS is a function of fS_2 only (see Reaction 1.10) but not of fO_2, as instead reported in Figure 1.4.

1.3. CONCLUDING REMARKS AND PERSPECTIVES

In this short compendium we show the redox features in aqueous-hydrothermal and igneous Earth. This allows the summarizing of the main redox features of a system, to show what we know of its equilibrium properties, but also what we do not know, especially for melts and magmas. We make a parallel between redox in magmas and redox in aqueous-hydrothermal solutions and show that what really changes is the way redox variables are reported. It is better described in aqueous solutions, via E-pH relations, because of the sound knowledge of acid–base properties, which allows good prediction of system reactivity with compositional changes and fosters applications in water-based geochemistry and industrial practice (e.g. corrosion and hydrometallurgy).

Just as the measurement of pH is the key to studies of acid–base reactions, those of reactions involving the O^{2-} ion in silicate melts naturally go through the measurement of pO^{2-}, a magnitude similar to pH and playing an identical role. Same as a pH indicator electrode, a pO^{2-} indicator electrode is the essential instrument to study acid–base properties of silicate melts. Technical challenges exist about this electrode as an instrument of analytical control operating directly in melts at high temperature and also as working assembly for anode reaction (inverse of Reaction 1.6). Besides, the low flux of oxide ion compared to cations is limited in melts and the lack of physical and chemical information about the molecular entities involved in the transport processes at each electrode and in the bulk of the electrolyte presents a conceptual difficulty for engineering the electrodes (Allanore et al., 2015).

Contrary to waters, in which E-pH pairs can be measured in the field by probe electrodes and then directly compared to theoretical assessments, magma-related samples do not offer the possibility to probe the conditions (temperature, pressure, gas composition, and also phase proportion) under which they equilibrated before becoming accessible to our observations. These conditions cannot be restored and must be calculated under strong assumptions, for example that glasses have preserved the same oxidation state of the melt from which they quenched. The impossibility to restore and measure the original system has surely contributed to overlooking the role played by acid–base properties and resulted in oxide-based redox descriptions of melts and magmas centred around the thermodynamic concept of fO_2, whose sole adoption was boosted by experimental petrology and the thermodynamic approach based on Ellingham's diagrams in extractive metallurgy. The acid–base concept was progressively relegated to qualitative assessments (e.g., silicic for acidic and mafic for basic).

With the exception of volcanic gases, where fO_2 (or related quantities such as fH_2) can be directly measured with solid-state O_2 electrodes (Aiuppa et al., 2011 and references therein), the common practice, recently boosted by advancements in microbeam spectroscopic techniques (e.g., XANES) is to measure masses of elements in their different oxidation state and then relate such ratios to fO_2 via approaches either based on thermodynamics or having a thermodynamic flavour to solve for the role of composition. Nevertheless, such approaches are valid only in limited compositional ranges and discharges on calculated fO_2 the non-ideal behaviours that are determined by speciation, hence by pO^{2-}, and which cannot be reproduced by interaction parameters of oxide components (see Moretti, 2020).

These unpredictable non-ideal behaviours are reflected by the shape of the excess Gibbs free energy of mixing, which is correctly reproduced only by ionic-polymeric approaches to silicate melts mixing properties (e.g., Mao et al., 2006; Hillert et al., 1985; Ottonello and Moretti, 2004), which formalize the role of composition in establishing the relationships between polymerization and redox state in the melt. These approaches could be in perspective used to generate pO^{2-} based phase diagrams, particularly $\log fO_2$-pO^{2-} diagrams analogous to $\log fO_2$-pH ones reported in Figure 1.4 for aqueous-hydrothermal solutions.

The joint description of acid–base properties and redox exchanges via predominance and stability diagrams would reduce ambiguities about the role of redox state and the "*who controls what*" dilemma: is fO_2 a rather elusive parameter, or is it the bulk composition in the system that fixes the observed redox state? Besides, they would help in providing a more complete and formal redox description of geodynamic settings, in particular of how the redox state of magmas is a reflection of the source signature or the result of subsequent modifications due to magma evolution via partial melting, fractional crystallization, assimilation, or degassing (Carmichael, 1991; Burgisser and Scaillet, 2007; Gaillard et al., 2015; Moretti and Stefansson, 2020 and references therein) whose understanding would be beneficial at the scale of a single volcanic system to improve the forecasting of eruptions at both open-conduit (generally basaltic) and closed-conduit (generally andesitic) volcanoes via gas monitoring. Besides, this would better constrain the role of redox state on Earth's evolution since its formation and accretion.

ACKNOWLEDGMENTS

We thank Alexander Pisch (SIMAP, CNRS, France) and Maria Rita Cicconi (FAU, Germany) for their valuable reviews. The precious support of AGU Books editorial staff is greatly acknowledged. This study contributes to the IdEx Université de Paris ANR-18-IDEX-0001.

REFERENCES

Aiuppa, A., Shinohara, H., Tamburello, G., Giudice, G., Liuzzo, M., & Moretti, R. (2011). Hydrogen in the gas plume of an open-vent volcano, Mount Etna, Italy. *Journal of Geophysical Research: Solid Earth*, *116*(B10). https://doi.org/10.1029/2011JB008461

Allanore A. (2013). Electrochemical engineering of anodic oxygen evolution in molten oxides. *Electrochimica Acta*, *110* (2013), 587–592.

Allanore A. (2015). Features and challenges of molten oxide electrolytes for metal extraction. *Journal of the Electrochemical Society*, *162*, E13–E22. https://doi.org/10.1149/2.0451501jes

Appelo C. A. J., & Postma, D. (1996). *Geochemistry, groundwater and pollution*. Rotterdam: Balkema. 536 pp.

Armstrong, K., Frost, D. J., McCammon, C. A., Rubie, D. C., & Ballaran, T. B. (2019). Deep magma ocean formation set the oxidation state of Earth's mantle. *Science*, *365* (6456), 903–906. doi: 10.1126/science.aax8376

Baker, D. R., & Moretti, R. (2011). Modeling the solubility of sulfur in magmas: a 50-year old geochemical challenge. *Reviews in Mineralogy and Geochemistry*, *73*(1), 167–213. https://doi.org/10.2138/rmg.2011.73.7

Barton, P. B., Jr. (1970). Sulfide petrology: Mineralogical Society of America Special Paper 3, 187–198.

Biernat R. J., & Robins, R. G. (1969). High temperature potential/pH diagrams for the sulfur-water system. *Electrochimica Acta* *14*, 809–820. https://doi.org/10.1016/0013-4686(69)87003-9

Bowen, N. L., & Schairer, J. F. (1932). The system, FeO-SiO 2. *American Journal of Science*, *141*, 177–213. https://doi.org/10.2475/ajs.s5-24.141.177

Bowen, N. L., & Schairer, J. F. (1935). The system MgO-FeO-SiO 2. *American Journal of Science*, *170*, 151–217. doi: 10.2475/ajs.s5-29.170.151

Buddington, A. F., & Lindsley, D. H. (1964). Iron-titanium oxide minerals and synthetic equivalents. *Journal of Petrology*, *5*, 310–357. https://doi.org/10.1093/petrology/5.2.310

Burgisser, A., & Scaillet, B. (2007). Redox evolution of a degassing magma rising to the surface. *Nature*, *445*(7124), 194–197. https://doi.org/10.1038/nature05509

Carmichael, I. S. (1991). The redox states of basic and silicic magmas: a reflection of their source regions? *Contributions to Mineralogy and Petrology*, *106*(2), 129–141. https://doi.org/10.1007/BF00306429

Casey H. W. (2017). Oxidation-Reduction Reactions and Eh-pH (Pourbaix) Diagrams. In: W.M. White (ed.), *Encyclopedia of Geochemistry*, doi:10.1007/978-3-319-39193-9_21-1

Cicconi, M. R., Moretti, R., & Neuville, D. R. (2020a). Earth's Electrodes. *Elements*, *16*, 3, 157–160. doi: 10.2138/gselements.16.3.157

Cicconi, M. R., Le Losq, C., Moretti, R., & Neuville, D. R. (2020b). Magmas are the largest repositories and carriers of Earth's redox processes. *Elements*, *16*, 3, 173–178. doi: 10.2138/gselements.16.3.173

Colson, R. O., Haskin, L. A., & Crane, D. (1990). Electrochemistry of cations in diopsidic melt: Determining diffusion rates and redox potentials from voltammetric curves. *Geochimica et Cosmochimica Acta*, *54*, 3353–3367. https://doi.org/10.1016/0016-7037(90)90290-2

Cochain B., Neuville D. R., Henderson G. S., McCammon C., Pinet O., & Richet, P. (2012). Iron content, redox state and structure of sodium borosilicate glasses: A Raman, Mössbauer and boron K-edge XANES spectroscopy study. *Journal of the American Ceramics Society*, *94*, 1–12. https://doi.org/10.1111/j.1551-2916.2011.05020.x

Cochain, B., Neuville, D. R., de Ligny, D., Malki, M., Testemale, D., Pinet O., & Richet P. (2013). Dynamics of iron-bearing borosilicate melts: Effects of melt structure and composition on viscosity, electrical conductivity and kinetics of redox reactions. *Journal of Non-Crystalline Solids*, *373–374*, 18–27. https://doi.org/10.1016/j.jnoncrysol.2013.04.006

Cook, G. B., & Cooper, R. F. (2000). Iron concentration and the physical processes of dynamic oxidation in alkaline earth aluminosilicate glass. *American Mineralogist*, *85*, 397–406. https://doi.org/10.2138/am-2000-0401

Cook, G. B., Cooper, R. F., & Wu, T. (1990). Chemical diffusion and crystalline nucleation during oxidation of ferrous ironbearing magnesium aluminosilicate glass. *Journal of Non-Crystalline Solids*, *120*, 207–222. https://doi.org/10.1016/0022-3093(90)90205-Z

Cooper, R. F., Fanselow, J. B., & Poker, D. B. (1996a). The mechanism of oxidation of a basaltic glass: chemical diffusion of network-modifying cations. *Geochimica et Cosmochimica Acta*, *60*(17), 3253–3265. https://doi.org/10.1016/0016-7037(96)00160-3

Cooper, R. F., Fanselow, J. B., Weber, J. K. R., Merkley, D. R., & Poker, D. B. (1996b). Dynamics of oxidation of a Fe2+-bearing aluminosilicate (basaltic) melt. *Science*, *274*, 1173–1176. doi: 10.1126/science.274.5290.1173

Darken, L., & Gurry, R. W. (1945). The system iron-oxygen. I. The wüstite field and related equilibria. *Journal of the American Chemical Society*, *67*(8), 1398–1412. https://doi.org/10.1021/ja01224a050

Darken, L., & Gurry, R. W. (1946). The system iron—oxygen. II. Equilibrium and thermodynamics of liquid oxide and other phases. *Journal of the American Chemical Society*, *68*(5), 798–816. https://doi.org/10.1021/ja01209a030

Dickson, W. R., & Dismukes, E. B. (1962). The electrolysis of FeO-CaO-SiO2 melts. *Transactions of the Metallurgical Society of AIME*, *224*, 505–511.

Dancy, E. A., & Derge, G. J. (1966). Electrical conductivity of FeOx-CaO slags. *Transactions of the Metallurgical Society of AIME*, *236*, 1642.

Ellingham, H. J. T. (1944). Reducibility of oxides and sulfides in metallurgical processes. *Journal of the Society of Chemical Industry*, *63*, 125–133.

Eugster, H. P. (1957). Heterogeneous reactions involving oxidation and reduction at high pressures and temperatures. *The Journal of Chemical Physics*, *26*(6), 1760–1761. https://doi.org/10.1063/1.1743626

Eugster, H. P. (1959). Reduction and oxidation in metamorphism. In: Abelson, P. H. (ed.) *Researches in Geochemistry*. Volume 1. New York: John Wiley & Sons, pp. 397–426.

Eugster, H. P. (1977). Compositions and thermodynamics of metamorphic solutions. In: Fraser, D. G. (ed.) *Thermodynamics in Geology*. Dordrecht: D. Reidel Publishing Company, pp. 183–202.

Eugster, H. P., & Wones, D. R. (1962). Stability relations of the ferruginous biotite, annite. *Journal of Petrology*, *3*, 82–125. https://doi.org/10.1093/petrology/3.1.82

Feig, S. T., Koepke, J., & Snow, J. E. (2010). Effect of oxygen fugacity and water on phase equilibria of a hydrous tholeiitic basalt. *Contributions to Mineralogy and Petrology*, *160*, 551–568. doi:10.1007/s00410-010-0493-3

Fraser, D. G. (1975). Activities of trace elements in silicate melts. *Geochimica et Cosmochimica Acta*, *39*(11), 1525–1530. https://doi.org/10.1016/0016-7037(75)90154-4

Frost, B. R. (1991). Introduction to oxygen fugacity and its petrologic importance. *Reviews in Mineralogy and Geochemistry*, *25*, 1–9. https://doi.org/10.1515/9781501508684-004

Frost, D. J., & McCammon, C. A. (2008). The redox state of Earth's mantle. *Annual Review of Earth and Planetary Science*, *36*, 389–420. https://doi.org/10.1146/annurev.earth.36.031207.124322

Gaillard, F., Scaillet, B., Pichavant, M., & Iacono-Marziano, G. (2015). The redox geodynamics linking basalts and their mantle sources through space and time. *Chemical Geology*, *418*, 217–233. https://doi.org/10.1016/j.chemgeo.2015.07.030

Giggenbach, W. F. (1980). Geothermal gas equilibria. *Geochimica et cosmochimica Acta*, *44*(12), 2021–2032. https://doi.org/10.1016/0016-7037(80)90200-8

Giggenbach, W. F. (1987). Redox processes governing the chemistry of fumarolic gas discharges from White Island, New Zealand. *Applied Geochemistry*, *2*(2), 143–161. https://doi.org/10.1016/0883-2927(87)90030-8

Gudmundsson, G., & Wood, B. J. (1995). Experimental tests of garnet peridotite oxygen barometry. *Contributions to Mineralogy and Petrology*, *119*(1), 56–67. https://doi.org/10.1007/BF00310717

Hasegawa, M. (2014) Ellingham Diagram. *Treatise on Process Metallurgy*, Volume *3*, 507–513. http://dx.doi.org/10.1016/B978-0-08-096986-2.00032-1

Haskin, L. A., Colson, R. O., Lindstrom, D. J., Lewis, R. H., & Semkow, K. W. (1992, September). Electrolytic smelting of lunar rock for oxygen, iron, and silicon. In: Nasa. Johnson Space Center. The Second Conference on Lunar Bases and Space Activities of the 21st Century, Volume 2, 411–422.

Hillert, M., Jansson, B. O., & Sundman, B. O. (1985). A two-sublattice model for molten solutions with different tendency for ionization. *Metallurgical Transactions A*, *16*(1), 261–266. https://doi.org/10.1007/BF02816052

Kress, V. C., & Carmichael, I. S. (1991). The compressibility of silicate liquids containing Fe_2O_3 and the effect of composition, temperature, oxygen fugacity and pressure on their redox states. *Contributions to Mineralogy and Petrology*, *108*(1–2), 82–92. https://doi.org/10.1007/BF00307328

Lavoisier A. (1777) Mémoire sur la combustion en général, Académie des sciences, Mémoires de l'Académie Royale, Paris, 592–600.

Le Losq, C., Moretti, R., Oppenheimer, C., & Neuville, D. R. (2020) In situ XANES study of the influence of varying temperature and oxygen fugacity on iron oxidation state and coordination in a phonolitic melt. *Contributions to Mineralogy and Petrology*, *175*, 64–77. doi: 10.1007/s00410-020-01701-4

Lewis, G. N., & Randall, M. (1961). *Thermodynamics*, 2nd Edition. Revised By Kenneth Pitzer and Leo Brewer. McGraw-Hill Book Company. 723pp.

Ripley, E. M., & Li, C. (2013). Sulfide saturation in mafic magmas: Is external sulfur required for magmatic Ni-Cu-(PGE) ore genesis? *Economic Geology*, *108*(1), 45–58. https://doi.org/10.2113/econgeo.108.1.45

Littlewood, R. (1962). Diagrammatic representation of the thermodynamics of metal-fused chloride systems. *Journal of the Electrochemical Society*, *109*, 525–534.

Magnien, V., Neuville, D. R., Cormier, L., Roux, J., Hazemann, J-L., Pinet, O., and Richet, P. (2006). Kinetics of iron redox reactions in silicate liquids: a high-temperature X-ray absorption and Raman spectroscopy study. *Journal of Nuclear Materials*, *352*, 190–195. https://doi.org/10.1016/j.jnucmat.2006.02.053

Magnien, V., Neuville, D. R., Cormier, L., Roux, J., Hazemann, J-L., de Ligny D., et al. (2008). Kinetics and mechanisms of iron redox reactions in silicate melts: The effects of temperature and alkali cations. *Geochimica et Cosmochimica Acta*, *72*, 2157–2168. https://doi.org/10.1016/j.gca.2008.02.007

Mallmann, G., & O'Neill, H. S. C. (2009). The crystal/melt partitioning of V during mantle melting as a function of oxygen fugacity compared with some other elements (Al, P, Ca, Sc, Ti, Cr, Fe, Ga, Y, Zr and Nb). *Journal of Petrology*, *50*(9), 1765–1794. doi:10.1093/petrology/egp053

Mao, H., Hillert, M., Selleby, M., & Sundman, B. (2006). Thermodynamic assessment of the $CaO–Al_2O_3–SiO_2$ system. *Journal of the American Ceramic Society*, *89*(1), 298–308. https://doi.org/10.1111/j.1551-2916.2005.00698.x

Mattioli, G. S., & Wood, B. J. (1988). Magnetite activities across the $MgAl_2O_4$-Fe_3O_4 spinel join, with application to thermobarometric estimates of upper mantle oxygen fugacity. *Contributions to Mineralogy and Petrology*, *98*(2), 148–162. https://doi.org/10.1007/BF00402108

Moretti, R. (2005). Polymerisation, basicity, oxidation state and their role in ionic modelling of silicate melts. *Annals of Geophysics*, *48*, 4/5, 583–608. https://doi.org/10.4401/ag-3221

Moretti, R. (2021). Ionic syntax and equilibrium approach to redox exchanges in melts: basic concepts and the case of iron and sulfur in degassing magmas. In: Moretti, R., and Neuville, D. R. (eds.) *Redox Magma Geochemistry. Geophysical Monograph Series 266*. American Geophysical Union.

Moretti, R., & Baker, D. R. (2008). Modeling of the interplay of fO2 and fS2 along the FeS-Silicate Melt equilibrium. *Chemical Geology*, *256*, 286–298. doi:10.1016/j.chemgeo.2008.06.055.

Moretti, R., & Ottonello, G. (2003). Polymerization and disproportionation of iron and sulfur in silicate melts: insights from an optical basicity-based approach. *Journal of Non-Crystalline Solids*, *323*, 111–119. https://doi.org/10.1016/S0022-3093(03)00297-7

Moretti, R., Arienzo, I., Civetta, L., Orsi, G., & Papale, P. (2013) Multiple magma degassing sources at an explosive volcano. *Earth and Planetary Sciences Letters*, *367*, 95–104. https://doi.org/10.1016/j.epsl.2013.02.013

Moretti, R., & Stefánsson, A. (2020). Volcanic and geothermal redox engines. *Elements: An International Magazine of Mineralogy, Geochemistry, and Petrology*, *16*(3), 179–184. https://doi.org/10.2138/gselements.16.3.179

Nadoll, P., Angerer, T., Mauk, J.L., French, D., & Walshe, J. (2014). The chemistry of hydrothermal magnetite: A review. *Ore Geology Reviews*, *61*, 1–32. https://doi.org/10.1016/j.oregeorev.2013.12.013

Nash, W. M., Smythe, D. J., & Wood, B. J. (2019). Compositional and temperature effects on sulfur speciation and solubility in silicate melts. *Earth and Planetary Science Letters*, *507*, 187–198. https://doi.org/10.1016/j.epsl.2018.12.006

Neuville, D. R., Cicconi, M. R., & Le Losq, C. (2021). How to measure the oxidation state of multivalent elements in minerals, glasses and melts? In: Moretti, R., and Neuville, D. R. *Magma Redox Geochemistry*. Geophysical Monograph Series 266. American Geophysical Union.

Ottonello, G. (1997). *Principles of Geochemistry*. Columbia University Press, 894 pp.

Ottonello, G., & Moretti, R. (2004). Lux-Flood basicity of binary silicate melts. *Journal of Physics and Chemistry of Solids*, *65* (8–9), 1609–1614. https://doi.org/10.1016/j.jpcs.2004.01.012

Ottonello, G., Moretti, R., Marini, L., & Zuccolini, M. V. (2001). Oxidation state of iron in silicate glasses and melts: a thermochemical model. *Chemical Geology*, *174*(1–3), 157–179. https://doi.org/10.1016/S0009-2541(00)00314-4

Pichavant, M., Costa, F., Burgisser, A., Scaillet, B., Martel, C., & Poussineau, S. (2007). Equilibration scales in silicic to intermediate magmas—implications for experimental studies. *Journal of Petrology*, *48*(10), 1955–1972. https://doi.org/10.1093/petrology/egm045

Pinet, O., Phalippou, J., & Di Nardo, C. (2006). Modeling the redox equilibrium of the Ce^{4+}/Ce^{3+} couple in silicate glass by voltammetry. *Journal of Non-Crystalline Solids*, *352*(50–51), 5382–5390. https://doi.org/10.1016/j.jnoncrysol.2006.08.034

Raymond, J., Williams-Jones, A. E., & Clark, J. R. (2005). Mineralization associated with scale and altered rock and pipe fragments from the Berlın geothermal field, El Salvador; implications for metal transport in natural systems. *Journal of Volcanology and Geothermal Research*, *145*, 81–96. doi: 10.1016/j.jvolgeores.2005.01.003

Schreiber, H. D. (1987). An electrochemical series of redox couples in silicate melts: a review and applications to geochemistry. *Journal of Geophysical Research: Solid Earth*, *92*(B9), 9225–9232. https://doi.org/10.1029/JB092iB09p09225

Sokhanvaran, S., Lee, S.-K., Lambotte, G., & Allanore, A. (2016). Electrochemistry of molten sulfides: Copper extraction from BaS-Cu_2S. *Journal of The Electrochemical Society*, *163*, D115–D120.

Semkow, K. W., & Haskin, L. A. (1985). Concentrations and behavior of oxygen and oxide ion in melts of composition $CaO{\cdot}MgO{\cdot}xSiO_2$. *Geochimica et Cosmochimica Acta*, *49*(9), 1897–1908. https://doi.org/10.1016/0016-7037(85)90084-5

Toop, G. W., & Samis, C. S. (1962a). Some new ionic concepts of silicate slags. *Canadian Metallurgical Quarterly*, *1*, 129–152. https://doi.org/10.1179/cmq.1962.1.2.129

Toop, G. W., & Samis, C. S. (1962b). Activities of ions in silicate melts. *Transactions of the Metallurgical Society of AIME*, *224*, 878–887.

Trémillon B. (1974). *Chemistry in non-aqueous solvents*. Dordrecht: D. Reidel Publishing Company. 285 pp.

Vaughan, D. J. (2005). Minerals/Sulphides. *Encyclopedia of Geology. Elsevier.* 574–586.

Zhang, J., Matsuura, H., & Tsukihashi, F. (2014). Processes for Recycling. *Treatise on Process Metallurgy*, Volume *3*, 1507–1561. http://dx.doi.org/10.1016/B978-0-08-096988-6.00036-5

Part I

Redox from the Earth's Accretion to Global Geodynamics

2

Redox Processes Before, During, and After Earth's Accretion Affecting the Deep Carbon Cycle

Vincenzo Stagno[1] and Sonja Aulbach[2]

ABSTRACT

The distribution of volatile elements between Earth's core, mantle, and atmosphere over the last 4.57 billion years has been controlled through redox reactions involving multivalent elements like Fe in minerals and melts, with important implications for the speciation of elements like carbon, oxygen, and hydrogen as fluids or through dissolution in melts. The redox state of Earth's interior was likely established during accretion of predominantly chondritic building blocks and late-stage meteoritic impact and concurrent formation of the metallic core. In contrast to thermodynamic predictions of decreasing mantle redox state with depth, some melt and mineral inclusions trapped in diamonds down to the mantle transition zone and exhumed with carbonate-rich magmas (kimberlites) reveal both the passage of oxidized CO_2-rich fluids and the occurrence of oxidized minerals. These are proof of the redox heterogeneities of the whole mantle that might have in part been inherited from early-stage processes but also testify to the recycling of surface-altered, oxidized materials, a process that became particularly efficient with Earth's transition to a stable plate tectonic regime. Here, we review the current knowledge of the (early-)mantle oxidation state, how this might have changed through space and time, its effect on the genesis of CO_2-rich magmas, and also highlight unresolved issues requiring further work.

2.1. THE REDOX STATE OF PLANETARY INTERIORS AND THE SPECIATION OF CARBON IN THE EARTH

The redox state of a system determines the oxidation state of multivalent elements contained therein (e.g., Fe^{2+} vs Fe^{3+}) and is gauged as the partial pressure (fugacity) of oxygen (fO_2). In the mantle, fO_2 is buffered by the coexisting Fe-bearing minerals mainly within peridotite rocks (Luth et al., 1990; Stagno et al., 2013). Oxygen fugacity also controls the speciation of multivalent volatiles like carbon and defines conditions under which reduced accessory minerals like carbide, graphite, and diamond can be oxidized to C^{4+}. The consequent bonding with an O_2 molecule causes lowering of the melting temperature of the locally surrounding rocks (i.e., *redox melting;* Taylor & Green, 1988; Stagno et al., 2013). Conversely, carbonate minerals and carbonated melts (carbonatites and kimberlites) can be reduced to elemental carbon as a consequence of the locally buffered fO_2 by Fe-bearing mineral assemblages (i.e., *redox freezing;* Rohrbach & Schmidt, 2011; Stagno et al., 2011; Jacob

[1]*Department of Earth Sciences, Sapienza University of Rome, Rome, Italy*

[2]*Institute of Geosciences, Goethe-Universität, Frankfurt am Main, Germany*

Magma Redox Geochemistry, Geophysical Monograph 266, First Edition.
Edited by Roberto Moretti and Daniel R. Neuville.

DOI: 10.1002/9781119473206.ch2

et al., 2016) that behave as an "oxygen pump" by reducing the surrounding environment. The redox state of the interior of Earth, as in other planets, is, therefore, expected to have modulated the speciation of volatile elements and set the conditions for the onset of melting of rocks at variable mantle depths to produce a large spectrum of magmas (Gudfinnsson & Presnall, 2005; Green, 2015). While some relatively cool and highly reactive small-fraction carbonated melts may experience thermal and chemical "death" through interaction with the lithosphere (McKenzie, 1989), those that ultimately reach the surface and degas to the atmosphere set, in turn, the conditions for planetary climates and, ultimately, habitability of our planet (Williams et al., 1992; Martin et al., 2007; Gaillard et al., 2011; Brune et al., 2017; Lammer et al., 2018).

To better understand the link between the redox state of planetary interiors and volatile cycles, it is worth recalling the astronomical observation of the super-Earth *55 Cancri e* (Madhusudhan et al., 2012), the interior of which was proposed to host mostly Fe, diamonds, and SiC with no evidence of free gaseous species, based on the measurements of astronomical parameters of mass and radius used to constrain the chemistry of the interior of exoplanets (Endl et al., 2012; Gillon et al., 2012). The abundance of reduced phases obviously inhibits the formation of oxidized gaseous phases, their circulation, and exsolution at any depth in the planetary interior, and also precludes the possibility of generation and migration of volatile-rich magmas to the surface of a planet, due to the low solubility of reduced oxygen-free volatile species in melts, such as elemental C. In contrast, the tectonic and volcanic activity on Earth through time appears to have been coupled to the speciation of oxidized volatiles like water and CO_2 both in terms of chemical composition and erupted rate of mantle-derived magmas (Tappe et al., 2018; Fuentes et al., 2019). So, while *55 Cancri e* can be considered a reduced end member among the so-far observed planetary bodies along the carbide-carbonate joint, Earth's mantle represents a relatively oxidized body that records a variable redox state in its interior, allowing the stability of volatiles in different forms.

The oxidation state of planetary interiors, the volatile cycle, and, therefore, the role of carbon (and water) in the origin of magmas are all strictly linked with (i) the redox state and thermal conditions of the condensates where the planetesimals formed within the protoplanetary disks; (ii) the intrinsic oxygen fugacity (i.e., the oxygen fugacity inferred by the coexisting condensated minerals) of the building blocks; (iii) the chemical composition of the impactors during early and late accretion; and (iv) early-planetary differentiation, in particular separation of a metallic core. These points are addressed in the following sections.

2.2. OXIDATION STATE OF EARTH'S BUILDING BLOCKS AND EARLY DIFFERENTIATION

The redox state and its evolution in Earth's interior are established as a result of (i) early accretion under the influence of variably oxidized building blocks and late-stage meteoritic impact; (ii) magma ocean(s) formation, core segregation, and mantle crystallization; (iii) establishment of mantle convection exhuming early-generated deep heterogeneities; and (iv) recycling of surface materials by (initially regional and episodic) subduction processes.

2.2.1. The Solar Nebula Redox State

The fO_2 of Chondrites. The redox condition of the early accretion was set by the Solar Nebula environment and the intrinsic fO_2 of the building blocks. The calcium, aluminum-rich inclusions (CAI) in meteorites represent the first crystalline materials that condensated in equilibrium with the gaseous Solar Nebula (Bouvier & Wadhwa, 2010). These inclusions are made of refractory minerals like spinel, fassaite-like cpx (clinopyroxene), melilite, perovskite, and anorthite, formed according to their condensation temperature (Grossman, 1972). They record, therefore, the redox state of the Solar Nebula, which is defined as the fO_2 at which the Solar Nebula vapor phase and the condensed mineral assemblage were equilibrated. Important fO_2 indicators are represented by those elements with multiple oxidation state (e.g., Fe and Ti). The relative abundance of the species with different oxidation state (e.g., the Ti^{3+}/Ti^{4+} ratio in pyroxene) correlates with the fO_2 largely buffered by the relative abundances of carbon, oxygen, and hydrogen in the cosmic gases. The application of the Ti^{3+}/Ti^{4+} ratio in pyroxenes is based on the experimental calibration of the following equilibria (Grossman et al., 2008)

$$\underset{\text{cpx}}{4CaTi^{4+}Al_2O_6} + \underset{\text{cpx}}{4CaMgSi_2O_6} = \underset{\text{cpx}}{4CaTi^{3+}AlSiO_6} + \underset{\text{Akermanite}}{2Ca_2MgSi_2O_7} + \underset{\text{spinel}}{2MgAl_2O_4} + O_2 \tag{2.1}$$

and

$$\underset{\text{cpx}}{4CaTi^{4+}Al_2O_6} + \underset{\text{cpx}}{2CaMgSi_2O_6} + \underset{\text{Ca-tschermakite}}{2CaAl_2SiO_6} = \underset{\text{cpx}}{4CaTi^{3+}AlSiO_6} + \underset{\text{gehlenite}}{2Ca_2Al_2SiO_7} + \underset{\text{spinel}}{2MgAl_2O_4} + O_2 \tag{2.2}$$

These two oxy-thermometers resemble the assemblage found in CV3 Allende meteorite and allow us to determine its fO_2, calculated at 1500 K and 10^{-3} atm, to be between

~–10 and –7 log units as normalized to the iron-wüstite (IW) oxygen buffer (Fig. 2.1) (Allende Prieto et al., 2002). At these low fO_2, metallic Fe stably alloys with Ni over a wide temperature range. However, the occurrence of fayalitic (i.e., Fe as Fe^{2+}) olivine inclusions in chondrules of ordinary chondrites (Fe# of 0.15) implies the presence of oxidized regions in the Solar Nebula. This heterogeneity in fO_2 of the Solar Nebula has been explained by the input of oxidized materials from beyond the snow line. This resulted in a local increase of the bulk oxygen content accounting for a fO_2 of IW-4.5, although an additional contribution of oxidized chondritic material cannot be discounted (Grossman et al., 2012; Rubie et al., 2015). A further fO_2 increase by 5–6 log units is recorded by rims of CAI inclusions from some CV3 meteorites (e.g., Allende & Leoville meteorites; Wark & Lovering, 1977) showing a refractory assemblage (i.e. spinel, perovskite, hibonite, melilite, Al-rich cpx, and forsterite) from which fO_2 can be estimated through the application of Mg isotopes (i.e. $^{25}Mg/^{24}Mg$) as a cosmo-barometer, and the Ti^{3+}/Ti^{4+} ratio in pyroxene. Such an fO_2 increase was estimated to occur over 100–300k years and marks the formation of disks consisting of chondritic dust from which proto-planets formed (Wark & Lovering 1977).

The fO_2 of Planetesimals and the Role of Volatiles. Planetesimals of km-size diameter formed by condensation and melting from the Solar Nebula. Thus, their chemical composition and oxidation state likely resemble those of the portion of Solar Nebula where they formed, taking into consideration the proposed correlation between heliocentric distance and H_2/H_2O and C/O ratios reflected in the oxidized nature of the planetesimals upon melting (Grossman et al., 2008). On the basis of N-body simulations of particle interaction and planetary accretion, the bulk composition of planetesimals and their redox state change with respect to the heliocentric distances in the Solar Nebula. The bodies formed at a distance less than 1.1–1.7 AU exposed to higher temperatures were highly reduced at IW-4 log units (high C/O ratio) at which all Fe would be stable as metal (plus some Si) whereas the bodies formed beyond 1.1–1.7 AU were more oxidized (low C/O and H_2/H_2O ratios) with a proportion of Fe as FeO dissolved in silicates (Rubie et al., 2015). Melting of the interior of planetesimals enhanced by C-O-H species might also contribute to variations of their redox state with consequent volatile depletion and oxidation (or reduction) of the surrounding mineral matrix before magma can reach the surface and degas

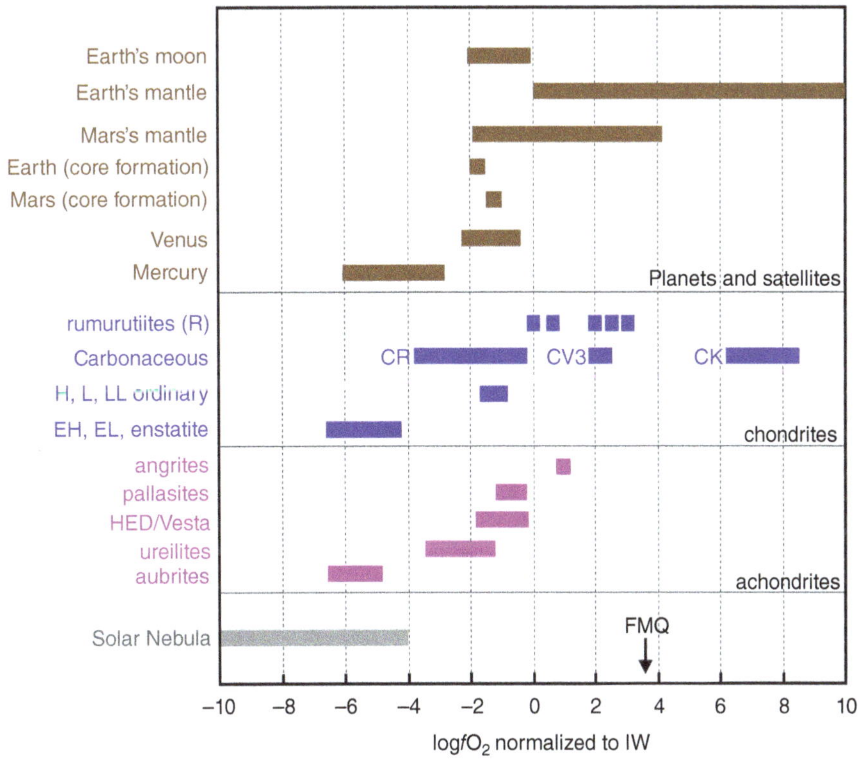

Figure 2.1 Range of oxygen fugacity estimates for Solar Nebula, achondrites, chondrites, and terrestrial planets (modified from Righter et al., 2016). The range of fO_2 of the Martian mantle refers to the intrinsic measurements of fO_2 in Martian basalts, while the fO_2 of Earth's mantle includes data for continental mantle xenoliths, abyssal peridotites, MORBs, arc lavas, and lamprophyres (Frost and McCammon, 2008).

(Fu et al., 2017). The variations in oxidation state of the planetesimals are inferred from the intrinsic oxygen fugacity estimated for various meteorite classes (Brett and Sato, 1984), as discussed in Section 2.2.1.3.

The Oxidation State of Earth's Interior, Other Planets, and Meteorites. Fig. 2.1 summarizes the estimated fO_2 of the Solar Nebula, achondrites, and chondrites along with estimates of the fO_2 for the interiors of planets including Mars, Mercury, Venus, Earth, and the Moon (Righter et al., 2016). Aubrites and achondrites are the most reduced materials, followed by ordinary chondrites and the oxidized carbonaceous chondrites. These partially overlap with the fO_2 determined for Mercury, Martian basalts, and Earth's mantle rocks; the latter, however, displays a uniquely broad range of values. Although there is geochemical evidence supporting bulk Earth's origin by accretion from enstatite chondrite material (Javoy, 1995), the fO_2 displayed by both appears quite different. Indeed, considerations of the Si isotopic variation and Mg/Si ratio of enstatite chondrites with respect to fertile mantle rocks (Fitoussi & Bourdon, 2012), as well as initial volatile content, are in favor of a multi-component mixing model involving more than one type of material, with a dominating enstatite chondrite (EH or EL) component, ~30% of carbonaceous chondrites, ~24% achondrites (angrites or eucrites), and a minor amount of ordinary L-type chondrites (Liebske & Khan, 2019). Importantly, it has been pointed out that Si isotopes are fractionated during core formation (e.g., Georg et al., 2007), nebular processes (Savage & Moynier, 2013), and evaporation during accretion (Pringle et al., 2014). Computational models suggest that the resulting fO_2 after accretion must have been low enough (IW-2) to allow core–mantle separation (Rubie et al., 2015; Schaefer & Fegley, 2017). Early Earth differentiation set a stratified redox interior, with a metallic core below and an FeO-bearing silicate mantle above the IW buffer. Core-mantle separation followed by magma ocean crystallization, meteorite impact processes (e.g., Late Heavy Bombardment) involving more oxidized impactors ("heterogeneous accretion"; e.g., O'Neill, 1991) and loss of primordial atmospheres are all likely to have played a major role in the gradula and stepwise oxidation of Earth's interior (Armstrong et al., 2019), although tight quantitative constraints remain elusive. Recent models suggest that the escape of H_2 from the early atmosphere must have been limited being the H_2/H_2O ratio < 0.3 in the gas released from the magma ocean, and this implies more oxidizing conditions (> IW+1) than those established during equilibrium with the core (Pahlevan et al. 2019).

2.3. MANTLE OXIDATION STATE OVER TIME AND ITS EFFECT ON THE C-O-H VOLATILE SPECIATION

The present-day large oxygen fugacity range in Earth's mantle (Fig. 2.1) reflects the evolution of the mantle through the pressure-driven crystallization of the magma ocean to form Fe-bearing minerals potentially able to set the fO_2 locally (e.g., bridgmanite, ferropericlase, majorite, garnet; McCammon, 2005), trigger volatile outgassing and ingassing by recycling. Given that there is controversy regarding the dominant oxygen-buffering species (e.g., Fe vs. C) even in the modern convecting mantle (e.g., Ballhaus & Frost, 1994; Cottrell & Kelley, 2013; Eguchi & Dasgupta, 2018), it is not surprising that to date it remains unclear how the redox state of the terrestrial mantle has evolved over time, and how this might have influenced volcanism, and ultimately, the atmospheric composition on Earth. Whether the chemical interaction between mantle rocks and volatile species like C played the most important role in determining conditions for the geological activity and habitability of our planet depends on the abundance of Fe (Fe^0, Fe^{2+}, and Fe^{3+}) species compared to the abundance of C in its different forms (C^0, CO, CO_2, CH_4) relevant in processes such as partial melting of rocks, melt migration, and diamond formation. However, three important variables influence the circulation of volatiles, among them carbon, in a planet: (i) the primordial volatile budget; (ii) the bulk oxygen fugacity of the early accreted terrestrial body; and (iii) the mineralogy of the planet's interior. The deep volatile cycle in any planet is inextricably linked to changes of these variables through time.

2.3.1. The Redox State of the Upper Mantle and the Mobilization of Deep C

Carbon can occur in the Earth's interior in different forms as elemental species (graphite and diamond), linked with iron as metal carbides (Fe_3C, Fe_7C_3, etc.), in association with hydrogen (e.g., methane CH_4 and a variety of aqueous organic and inorganic species) or in oxidized forms such as CO_2 carbonates (either liquid or solid). Its fluxing properties are strongly related to the availability of oxygen (redox state) within the surrounding mineral assemblages: elemental carbon passively follows the convection rates of the solid mantle while oxidized carbon present in fluids or melts has its own dynamics, moving through grain boundaries at rates ruled by the fluid mechanics and chemical interactions with minerals (Hammouda & Laporte, 2002; Stagno et al., 2019). The redox state of the convecting oceanic mantle and the continental lithosphere has been recently determined employing oxy-thermobarometry on natural rocks (mantle

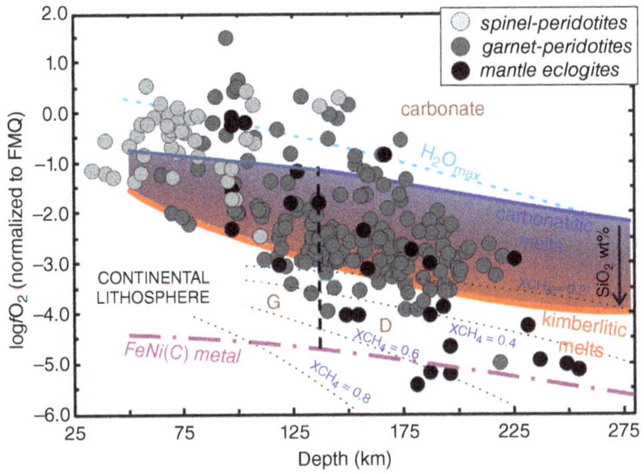

Figure 2.2 LogfO_2 (normalized to the FMQ buffer) calculated for mantle peridotite and eclogite xenoliths using oxy-thermobarometers by Stagno et al. (2013; 2015; 2019). The blue and red lines are the oxygen fugacity calculated along a cratonic geotherm (~44 mW·m^{-2}) where diamonds (graphite) coexist with carbonatitic and kimberlitic melts, respectively. The purple line is the FeNi precipitation curve (Frost and McCammon, 2008), while the blue dashed line crosses the fO_2 at which water is the main component of stable C–O–H fluids. At $fO_2 \leq$ FMQ-3 log units fluids become increasingly CH_4-rich (modified from Luth and Stachel, 2014).

xenoliths) from depths of 30–200 km, which is based on the knowledge of ferric/ferrous ratios in redox-sensitive silicates (Fig. 2.2; Frost & McCammon, 2008; Stagno et al., 2013; Stagno et al., 2015; Stagno, 2019). These studies point out that the upper mantle xenoliths, as sampled at the time of emplacement of the magmatic host rock (predominantly in the Phanerozoic), mostly fall in the diamond stability field (Fig. 2.2), and that fO_2 of the lithospheric mantle decreases with the depth to conditions where Fe(Ni) alloy is stable. A large fO_2 interval at a given depth is evident, reflecting small variations in the chemistry (or mineral mode) of the plotted rocks and/or the local effect of metasomatism. The experimental calibration of oxybarometers that relies on the Fe^{3+} of natural garnets can be used to estimate the fO_2 of ultrahigh-pressure metamorphic eclogites and cratonic mantle eclogite xenoliths (Stagno et al., 2015). These results are shown in Fig. 2.2 and demonstrate that peridotite and eclogite rocks record similar mantle redox states. A portion of the mantle likely coexisted with graphite/diamond along with C-O-H fluids (Luth & Stachel, 2014) and/or CO_2-bearing silicate melts such as kimberlite, whereas a small fraction coexisted with pure carbonatite (Stagno and Frost, 2010). On the other hand, geochemical evidence employing redox proxies involving multivalent elements, such as V, Ce, and Eu, has been used to claim the constancy of the mantle redox state over the last 3.8 billion years, to values where carbonate melts are expected to be stable. As a consequence, it has been hypothesized that a mantle oxidation event occurred during the crystallization of the proto-Earth from a magma ocean (Scaillet & Gaillard, 2011), which would imply that most of the deep carbon was mobilized to be released to the atmosphere. Basalts sampling the ambient mantle beneath spreading ridges (mid-ocean ridge basalts, MORB) are subducted after 200–250 Ma. Older equivalents may be sampled from rare obducted ophiolites, Archean examples of which are, however, extremely rare and have a controversial tectonic setting, as also applies to mafic rocks occurring in greenstone belts (Bickle et al., 1994). One way to derive the redox state of Archean ambient mantle is to determine the V/Sc ratios and $Fe^{3+}/\Sigma Fe$ of ancient basalts from the composition of well-characterized Archean eclogitic rocks with spreading ridge-derived mafic protoliths (Aulbach and Viljoen, 2015; Aulbach and Stagno, 2016; Aulbach et al., 2017; 2019). Although these studies find sympathetic behavior of both redox proxies and no evidence for a dominant control of metamorphism and partial melt extraction on V-Sc-$Fe^{3+}/\Sigma Fe$ systematics, more work is needed to investigate the redox-dependent partitioning of V and Fe^{3+} simulating the metamorphism of basaltic rocks coexisting with C-O-H fluids at high pressure and temperature. A complementary view of the deep convecting mantle sampled by thermo(chemical) plumes is provided by komatiites, which are hot, large-volume melts emplaced predominantly in the Archean and Paleoproterozoic, enabling investigation of the partitioning of V between olivine and melts (Nicklas et al., 2018; 2019) and Fe^{3+} /ΣFe of melt inclusions contained in komatiite olivine (Berry et al., 2018).

2.3.2. Is the fO_2 of the Transition Zone and Lower Mantle Recorded by Sublithospheric Diamonds?

Since no direct mantle rocks are available from depths greater than 200–250 km, the only tools to investigate the oxidation state at transition zone/lower mantle conditions and possible redox melting over time, in addition to the aforementioned mantle-derived magmas, are experimental data combined with analyses of inclusions in natural sublithospheric diamonds. These diamonds, brought to the surface by kimberlitic eruptions, are often characterized by mineral inclusions representative of the transition zone and lower mantle (cpx, ringwoodite; majorite, NAL, Ca-ferrite phase, bridgmanite, ferropericlase; Stachel et al., 2005; Pearson et al., 2014) that show geochemical evidence of the passage of oxidized C-O-H-rich fluids from which they might have originated (Kaminsky, 2012; Kiseeva et al., 2018; Smith et al., 2016; Thompson et al., 2016; Walter et al., 2008). Recent discoveries of hydrous ringwoodite (Pearson et al., 2014) and ice-VII

(Tschauner et al., 2018) in diamonds from the transition zone further highlight the role of volatiles (other than C) bearing fluids in sublithospheric diamond formation. The origin of these fluids is often linked to subducted slabs widely accepted to be responsible for the oxygenation of the mantle from its initially reduced state in the wake of accretion and core formation (Lécuyer & Ricard, 1999). However, recent experimental investigations show that minerals like cpx, garnet, and majorite, which contain redox-sensitive Fe and are the main constituents of metamorphosed oceanic crust, can incorporate large amounts of Fe^{3+} as result of increasing depth (i.e., pressure) and thus become significantly oxidized (Rohrbach et al., 2011; Kiseeva et al., 2018; Stagno et al., 2015, 2019). Such form of oxygen sequestration from minerals would result in reduced portions of the subducted slab, with diamond being the stable C form (i.e. redox freezing) and might explain the finding of large amounts of Fe^{3+} in majoritic inclusions trapped in diamonds (Kiseeva et al., 2018) as well as the occurrence of nanometer-sized Mg-ferrite inclusions in ferropericlase from superdeep diamonds (Harte et al., 1999; Wirth et al., 2014). While these natural observations open a scenario of a more oxidized rather than reduced Fe^0-saturated deep mantle as shown by thermodynamic and experimental predictions (Frost et al., 2004; Hu et al., 2019), some questions remain, such as the possibility that these inclusions do not reflect the whole-mantle redox state but only local redox reactions.

2.3.3. Temporal and Spatial Evolution of the Redox State of the Asthenospheric Mantle

The oxygen fugacity of the ambient mantle, source of MORBs, and the deep convecting mantle sporadically sampled by plumes and kimberlitic melts both control the speciation of volatiles and hence the locus of the peridotite solidus and the nature of the melts that are produced (Wyllie & Huang, 1975; Canil & Scarfe, 1990; Green, 2015). In addition, mantle fO_2 plays a role in mantle rheology and therefore dynamics, whereby more oxidizing conditions enhance concentrations and mobilities of lattice defects, resulting in a viscosity reduction (Cline et al., 2018). Given its importance, great efforts have been made to constrain mantle redox evolution, which is necessary to accurately capture mantle geodynamics and melting behaviour. For example, redox melting, whereby upwelling mantle will intersect the equilibrium C(graphite/diamond) + $2Fe_2O_3$(melt) = $4FeO + CO_2$ (both in the melt) due to increasing fO_2 with decreasing pressure, will occur in the depth interval of ~150 to 120 km at modern mantle redox conditions (Stagno et al., 2013), where MORB records fO_2 of ~0.4 ±0.4 log units below the fayalite–magnetite–quartz (FMQ) buffer (Frost & McCammon, 2008; Stagno, 2019). This process explains the production of kimberlites and some carbonatites, the rarity of which in the Archean has been ascribed to more reducing mantle conditions by 0.5 to 0.7 log units (Foley, 2011). Indeed, using the V/Sc redox proxy, a well-characterized mantle eclogite suite from Lace in the Kaapvaal craton, with demonstrated spreading ridge-derived origin, has yielded an fO_2 (FMQ) of −1.7±1.1 for the ambient mantle at ca. 3 Ga (Aulbach & Viljoen, 2015). Based on this value, redox melting occurs at substantially shallower depths (~100 to 80 km; Aulbach & Stagno, 2016) with the formation of carbonate–silicate magmas, precluding the formation of carbonatite below typically thicker continental lithosphere, and possibly restricting the formation of kimberlites to unusually oxidized and/or warm mantle regions. Using additional mantle and orogenic eclogite suites sampling Proterozoic to Archean spreading ridges, the ambient mantle was shown to have evolved across the Archean–Proterozoic boundary, from $\Delta \log fO_2$ of ~FMQ-1.19±0.33 to FMQ-0.26±0.44 (Aulbach & Stagno, 2016). This evolution is exactly mirrored by that recorded in komatiites, which show an increase in ~1.3 log units between 3.48 and 1.87 Ga (Nicklas et al., 2018, 2019; Fig. 2.3a). This increase and the absolute fO_2 values recorded across the Archean–Proterozoic are well within the range of, and therefore not in conflict with, proposed values after magma ocean solidification (~FMQ-3 to FMQ) obtained from thermodynamic and climate modelling (Pahlevan et al., 2019). On the other hand, the compositions of inclusions in sublithospheric diamonds testify to the strong chemical and redox gradients, ranging from carbonates to solidified iron–nickel–carbon–sulfur melts (Brenker et al., 2007; Kaminsky et al., 2015; Smith et al., 2016). Indeed, diamond formation is favoured by contrasting redox states, which are typically achieved when deeply subducted slabs containing oxidized components are juxtaposed with the predominantly reducing, metal-saturated mantle at depths > 250–300 km (Foley, 2011; Rohrbach & Schmidt, 2011). Thus, we take the relatively high fO_2 retrieved from the ferric iron content of majorite inclusions in sublithospheric diamonds (Kiseeva et al., 2018) to indicate local mantle conditions in the vicinity of recycled slabs that are not representative of the mantle as a whole.

2.4. THE MANTLE GREAT OXIDATION EVENT: FACT OR ARTEFACT?

The redox evolution of Earth's interior, as inferred from V/Sc (Fig. 2.3a), is central to the debate on the role of the mantle in the transition from an anoxic to an oxygenated atmosphere during the so-called Great Oxidation Event (GOE) around 2.4–2.2 Ga, which was preceded by "whiffs

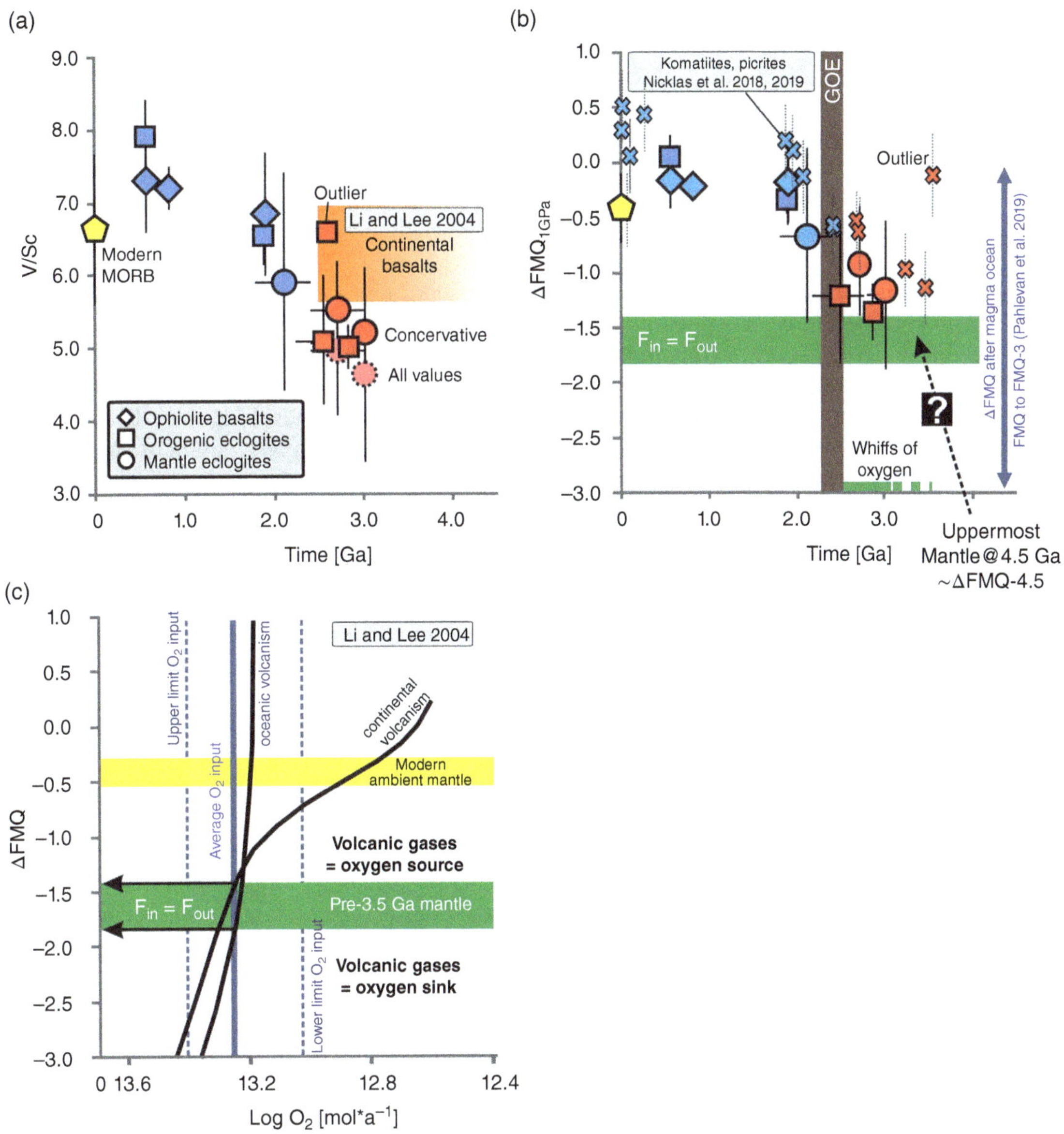

Figure 2.3 (a) V/Sc and (B) calculated LogfO_2 (normalized to the FMQ buffer) corrected to 1 GPa of mid-ocean ridge-derived (meta)basalts sampling the ambient convecting mantle (modified from Aulbach and Stagno, 2016), where error bars on V/Sc represent 1σ of the mean per sample suite, those on ΔFMQ are propagated 1σ errors of V/Sc, and those on the age represent age ranges or 1σ errors for isochron ages reported in the literature. In two suites, samples with the lowest V/Sc were excluded for a conservative estimate. Range of V/Sc for continental Archean basalts in (a) from Li and Lee (2004). Shown for comparison in (b) are fO_2 estimated by Nicklas et al. (2018; 2019) for picritic and komatiitic systems, based on the redox-dependent distribution of V between liquidus olivine and melt. Note that because of an unresolved inter-calibration issue, their MORB fO_2 estimate is offset to higher values (ΔFMQ + 0.60±0.15) from those derived in other works (~−0.4±0.4; Li and Lee, 2004; Frost and McCammon, 2008; +0.1; Berry et al., 2018), and all results were shifted down by one log unit for comparability to the V/Sc-derived MORB fO_2 of −0.4. Range of fO_2 in the magma ocean at ca. 4.5 Ga permitted based on chondritic D/H in the water ocean, which places an upper limit on H_2 escape from the atmosphere, from Pahlevan et al. (2019) is shown for comparison (vertical blue-grey arrow). Timing of the Great Oxidation Event (GOE) around 2.4 Ga and "whiffs of oxygen" from Lyons et al. (2014) (c) shows the speciation of continental and oceanic volcanic gases, with mixtures thereof lying between these two curves, as a function of fO_2 based on modern volcanic volatile fluxes (6.00 x 10^{12} for C, 1.15 x 10^{14} for H, and 2.13 x 10^{12} for S), illustrating the transition from such gases acting as sinks (reducing) and sources (oxidizing) of O_2 compared to the estimated average modern O_2 input (from burial of reduced components) and its upper and lower limit (modified from Li & Lee, 2004).

of oxygen" (Lyons et al., 2014) as shown in Fig. 2.3b. The mantle–asthenosphere connection was hypothesized early (Kasting et al., 1993; Holland, 2002), but subsequently dismissed by various workers finding only small or no difference in the redox state of Archean and post-Archean magmas (e.g., Canil, 1997; Delano, 2001; Li & Lee, 2004; Trail et al., 2011; Rollinson et al., 2017). In contrast, the data shown in Fig. 3.3c document a redox evolution based on samples derived from both ambient mantle and deeper "plume" mantle, which crosses the fO_2 region where volcanic gases become sources as opposed to sinks of O_2 when first whiffs of oxygen appeared (highlighted by the bar $F_{in} = F_{out}$ after Holland, 2002) at ca. 3.5–3.2 Ga, reaffirming a role for the mantle in atmospheric oxygenation. Although the relationship between the redox state of a magma and the gases released to the atmosphere at any point in time during magmatic activity is controlled by multiple parameters (e.g., Gaillard et al., 2011; Moussallam et al., 2019), it is inescapable that the oxidation state of the sum of the gases emitted over the course of volcanism must reflect that of the magma with which they were once in equilibrium. The cause of this apparent mantle redox evolution possibly goes back to Earth's origins. Once sufficient mass had been accreted, FeO disproportionated into Fe metal at high pressure, about 10% of which was extracted into the core, and Fe_2O_3, which was left behind in the magma ocean (Armstrong et al., 2019) and, thus, inherited by the rocky mantle where it partitioned into bridgmanite (Frost et al., 2004), corresponding to an increase in mantle O_2 (Wade & Wood 2005). This was followed by gradual exhumation of O_2-enriched lowermost material by mantle convection processes. This was put forward by both Aulbach and Stagno (2016) and Nicklas et al. (2019) as one plausible explanation for the fO_2 trends in their data. Early subduction of oxidized surficial material to raise mantle fO_2 was offered as an alternative process to explain the fO_2 trend (Nicklas et al., 2019). Interestingly, Archean placer diamonds formed between 3.5 and 3.1 Ga show geochemical markers of an origin by reduction of oxidized fluids accompanied by input of Archean sediments (Smart et al., 2016). However, mantle oxidation by subduction implies that O_2 would have to be taken from the ocean–atmosphere system and cause a corresponding reduction in order to produce the oxidizing material carried down. Similarly, subduction of reduced material (Duncan & Dasgupta, 2017) may cause a transient increase in atmospheric O_2, but, if not permanently sequestered into a "hidden" reservoir, would have a reducing effect on the mantle and ultimately cause it to release more reducing gases upon melting.

Regardless of the extent and timing of mantle oxidation, both the eclogite and the komatiite datasets contain outliers testifying to redox heterogeneity in the convecting mantle, possibly inherited from the aforementioned accretion and magma ocean processes and preserved in mantle regions that had not been remixed at the time of melt generation (Nicklas et al., 2019). This raises the question of the timescales needed to mix deep oxidized mantle regions into the upper convecting mantle. Gu et al. (2016) showed that an oxidized lower mantle assemblage left over after core formation would be less dense than its reduced equivalent, facilitating its ascent, which they modelled to be completed by 3.6 Ga ago, although the exact magnitude of the density contrast is debated (Liu et al., 2018). In contrast, the data shown in Fig. 3a–c suggest that this remixing occurred over longer timescales and into the Late Paleoproterozoic. Such long-lived heterogeneity is supported also from the preservation of anomalous compositions of the short-lived isotopes, which must date back to processes occurring during the first several 100 Ma, in magmas erupted throughout the Archean and even to this day (e.g., Rizo et al., 2016; Horan et al., 2018). Factors possibly contributing to retarded mantle homogenization include grain size (Foley & Rizo, 2017) and the strength of bridgmanite under lower mantle conditions (Girard et al., 2016). Vertical mixing of oxidized material left after core formation would eventually have been aided by the establishment of plate tectonics (Andrault et al., 2018), bearing in mind that the transition to sustained plate tectonics occurred late (3.2–2.5 Ga; Cawood et al., 2019) relative to the onset of upward mixing, sometime between the post-core formation establishment of fO_2 in the uppermost mantle of ~ΔFMQ-4.5 (Frost and McCammon, 2008) and the significantly more oxidizing source fO_2 recorded by 3.5 Ga komatiites (Fig. 2.3b; Nicklas et al., 2018).

Superposed on the upward mixing of post-core formation oxidized mantle regions would be heterogeneities due to late accretion and associated downward mixing of late-accreted material, which are difficult to quantify. For example, the abundance and partitioning of moderately siderophile elements apparently requires addition of increasingly oxidized material (Wood et al., 2006), whereas late meteoritic bombardment with reduced components has been inferred based on the Ce-in-zircon redox proxy (Yang et al., 2014). Wang and Becker (2013) suggest that volatile-rich objects were involved based on Se–Te systematics, whereas Fischer-Gödde and Kleine (2017) call for a reduced inner solar system source based on Ru isotopes. Thus, more work is required to quantitatively model the effects of downward mixing of late-accreted material on mantle fO_2 and to accurately capture the rise of post-core formation oxidized lower mantle regions in numerical models. Pinpointing this evolution is critical to constraining the composition of gases released from Hadaean and Archean magmas to the atmosphere, which has implications for a multitude of issues affecting habitability on Earth, from the production of prebiotic molecules to the Faint Young Sun paradox and finally, the transition to an oxygenated atmosphere.

ACKNOWLEDGMENTS

V.S. acknowledges financial support from Sapienza University of Rome through "Bandi di Ateneo 2016 and 2019." The Deutsche Forschungsgemeinschaft is acknowledged for funding under grant AU356/10 to S.A. The authors are grateful to two anonymous reviewers for their constructive comments.

REFERENCES

Allende Prieto, C., Lambert, D. L., & Asplund, M. (2002). A reappraisal of the solar photospheric C/O ratio. *Astrophysical Journal Letters*, *573*, L137–L140. https://doi.org/10.1086/342095

Andrault, D., Muñoz, M., Pesce, G., Cerantola, V., Chumakov, A., Kantor, I., et al. (2018). Large oxygen excess in the primitive mantle could be the source of the Great Oxygenation Event. *Geochemical Perspectives Letters*, *6*, 5–10. doi: 10.7185/geochemlet.1801

Armstrong, K., Frost, D. J., McCammon, C. A., Rubie, D. C, & Boffa Ballaran, T. (2019). Deep magma ocean formation set the oxidation state of Earth's mantle. *Science*, *365*, 6456, 903–906. doi: 10.1126/science.aax8376

Aulbach S., & Stagno V. (2016). Evidence for a reducing Archean ambient mantle and its effects on the carbon cycle. *Geology*, *44*, 9, 751–754. https://doi.org/10.1130/G38070.1

Aulbach, S., & Viljoen, K. S. (2015) Eclogite xenoliths from the Lace kimberlite, Kaapvaal craton: From convecting mantle source to palaeo-ocean floor and back. *Earth and Planetary Science Letters*, *431*, 274–286. https://doi.org/10.1016/j.epsl.2015.08.039

Aulbach, S., Woodland, A. B., Vasilyev, P., Galvez, M. E., & Viljoen, K. S. (2017). Effects of low-pressure igneous processes and subduction on $Fe^{3+}/\Sigma Fe$ and redox state of mantle eclogites from Lace (Kaapvaal craton). *Earth and Planetary Science Letters*, *474*, 283–295. https://doi.org/10.1016/j.epsl.2017.06.030

Aulbach, S., Woodland, A. B., Stern, R. A., Vasilyev, P., Heaman, L. M. & Viljoen, K. S. (2019). Evidence for a dominantly reducing Archaean ambient mantle from two redox proxies, and low oxygen fugacity of deeply subducted oceanic crust. Scientific Reports 9.

Ballhaus, C., Frost, B. R. (1994), The generation of oxidised CO2-bearing basaltic melts from reduced CH4-bearing upper mantle sources. *Geochimica et Cosmochimica Acta*, *58*, 4431–4440. https://doi.org/10.1016/0016-7037(94)90222-4

Berry, A. J., Stewart, G. A., O'Neill, H. S., Mallmann, G. F., & Mosselmans, J. F. (2018). A re-assessment of the oxidation state of iron in MORB glasses. *Earth and Planetary Science Letters*, *483*, 114–123. https://doi.org/10.1016/j.epsl.2017.11.032

Bickle, M. J., Nisbet, E. G., & Martin, A. (1994). Archaean greenstone belts are not oceanic crust? *Journal of Geology*, *102*, 121–138.

Bouvier, A., & Wadhwa, M. (2010) The age of the Solar System redefined by the oldest Pb–Pb age of a meteoritic inclusion. *Nature Geoscience*, *3*, 637–641. https://doi.org/10.1038/ngeo941

Brenker, F. E., Vollmer, C., Vincze, S., Vekemans, B., Szymanski, A., Janssens, K., et al. (2007). Carbonates from lower part of transition zone or even the lower mantle. *Earth and Planetary Science Letters*, *260*, 1–9. https://doi.org/10.1016/j.epsl.2007.02.038

Brett, R., & Sato, M. (1984). Intrinsic oxygen fugacity measurements on 7 chondrites, a pallasite and a tektite and the redox state of the meteorite parent bodies. *Geochimica et Cosmochimica Acta*, *48*, 111–120. https://doi.org/10.1016/0016-7037(84)90353-3

Brune, S., Williams, S. E., & Müller, R. D. (2017). Potential links between continental rifting, CO_2 degassing and climate change through time. *Nature Geoscience*, *10*, 941–946. doi: http://doi.org/10.1038/s41561-017-0003-6

Canil, D. (1997). Vanadium partitioning and the oxidation state of Archaean komatiite magmas. *Nature*, *389*, 842–845, doi:10.1038 /39860

Canil, D., & Scarfe, C. M. (1990). Phase relations in peridotite +CO_2 Systems to 12 GPa: Implications for the origin of kimberlite and carbonate stability in the Earth's upper mantle. *Journal of Geophysical Research*, *95*(B10), 15805–15816. doi:10.1029/JB095iB10p15805

Cawood, P. A., & Hawkesworth, C. J. (2019). Continental crustal volume, thickness and area, and their geodynamic implications. *Gondwana Research*, *66*, 116–125. https://doi.org/10.1016/j.gr.2018.11.001

Cline, C. J., Faul, U. H., David, E. C., Berry, A. J., & Jackson, I. S. (2018). Redox-influenced seismic properties of upper-mantle olivine. *Nature*, *555*, 355–358. https://doi.org/10.1038/nature25764

Cottrell E., & Kelley, K. A. (2013). Redox heterogeneity in mid-ocean ridge basalts as a function of mantle source. *Science*, *340*, 1314–1317. doi: 10.1126/science.1233299

Delano, J. W. (2001). Redox history of the Earth's interior since similar to 3900 Ma: Implications for prebiotic molecules. *Origins of Life and Evolution of the Biosphere*, *31*, 311–341. doi:10.1023/A:1011895600380.

Duncan, M. S., & Dasgupta, R. (2017). Rise of Earth's atmospheric oxygen controlled by efficient subduction of organic carbon. *Nature Geoscience*, *10*, 387–392. https://doi.org/10.1038/ngeo2939

Eguchi, J., & Dasgupta, R. (2018). Redox state of the convective mantle from CO_2-trace element systematics of oceanic basalts. *Geochemical Perspectives Letters*, *8*, 17–21. doi: 10.7185/geochemlet.1823

Endl, M., Robertson, P., Cochran, W. D., MacQueen, P. J., Brugamyer, E. J., Caldwell, C., et al. (2012). Revisiting ρ1 CANCRI e: A new mass determination of the transiting super-earth. *The Astrophysical Journal*, *759*, 1.

Fischer-Gödde, M., & Kleine, T. (2017) Ruthenium isotopic evidence for an inner Solar System origin of the late veneer. *Nature*, *541*, 525–527. https://doi.org/10.1038/nature21045

Fitoussi, C., & Bourdon, B. (2012). Silicon isotope evidence against an enstatite chondrite Earth. *Science*, *335*, 1477–1480. doi: 10.1126/science.1219509

Foley, S. F. (2011). A Reappraisal of Redox Melting in the Earth's Mantle as a Function of Tectonic Setting and Time. *Journal of Petrology*, *52*(7–8), 1363–1391. doi:https://doi.org/10.1093/petrology/egq061

Foley, B. J., & Rizo, H. (2017). Long-term preservation of early formed mantle heterogeneity by mobile lid convection: Importance of grainsize evolution. *Earth and Planetary Science Letters*, *475*, 94–105. https://doi.org/10.1016/j.epsl.2017.07.031

Frost, D. J., Liebske, C., Langenhorst, F., & McCammon, C. A. (2004). Experimental evidence for the existence of iron-rich metal in the Earth's lower mantle. *Nature*, *428*, 409–412. https://doi.org/10.1038/nature02413

Frost, D. J., & McCammon, C. A. (2008). The redox state of the Earth's mantle. *Annual Review of Earth and Planetary Science 36*, 389–420. https://doi.org/10.1146/annurev.earth.36.031207.124322

Fu, R. R., Young, E. D., Greenwood, R. C., Elkins-Tanton, L. T. (2017). Silicate melting and volatile loss during differentiation in planetesimals. In: Elkins-Tanton, L. T., Weiss, B. P. (Eds.), *Planetesimals: Early Differentiation and Consequences for Planets*. Cambridge: Cambridge University Press, pp. 115–135.

Fuentes, J. J., Crowley, J. W., Dasgupta, R., & Mitrovica, J. X. (2019). The influence of plate tectonic style on melt production and CO_2 outgassing flux at mid-ocean ridges. *Earth and Planetary Science Letters*, *511*, 154–163. doi: https://doi.org/10.1016/j.epsl.2019.01.020

Gaillard, F., Scaillet, B., & Arndt, N. T. (2011). Atmospheric oxygenation caused by a change in volcanic degassing pressure. *Nature*, *478*, 229–232. doi:10.1038/nature10460

Georg, R. B., Halliday, A. N., Schauble, E. A., Reynolds, B. C. (2007). Silicon in the Earth's core. *Nature*, *447*, 1102–1106. https://doi.org/10.1038/nature05927

Girard, J., Amulele, G., Farla, R., Mohiuddin, A., & Karato, S. (2016). Shear deformation of bridgmanite and magnesiowüstite aggregates at lower mantle conditions. *Science*, *351*, 144–147. doi: 10.1126/science.aad3113

Gillon, M., Demory, B.-O., Benneke, B., Valencia, D., Deming, D., Seager, S., et al. (2012). Improved precision on the radius of the nearby super-Earth 55 Cnc e. *Astronomy & Astrophysics*, *539*, A28. https://doi.org/10.1051/0004-6361/201118309

Green, D. H. (2015). Experimental petrology of peridotites, including effects of water and carbon on melting in the Earth's upper mantle. *Physics and Chemistry of Minerals*, *42*, 95–122. https://doi.org/10.1007/s00269-014-0729-2

Grossman, L. (1972). Condensation in the primitive solar nebula. *Geochimica et Cosmochimica Acta*, *36*, 597–619. https://doi.org/10.1016/0016-7037(72)90078-6

Grossman, L., Beckett, J. R., Fedkin, A. V., Simon, S. B., & Ciesla, F. J. (2008). Redox conditions in the Solar Nebula: Observational, experimental, and theoretical constraints. *Reviews in Mineralogy and Geochemistry*, *68* (1), 93–140. https://doi.org/10.2138/rmg.2008.68.7

Grossman, L., Fedkin, A. V., Simon, S. B. (2012). Formation of the first oxidized iron in the Solar System. *Meteoritics and Planetary Science*, *47*, 2160–2169. https://doi.org/10.1111/j.1945-5100.2012.01353.x

Gu, T., Li, M., McCammon, C., & Lee, K. K. (2016). Redox-induced lower mantle density contrast and effect on mantle structure and primitive oxygen. *Nature Geoscience*, *9*, 723–727. https://doi.org/10.1038/ngeo2772

Gudfinnsson, G. H., & Presnall, D. C. (2005). Continuous gradations among primary kimberlitic, carbonatitic, melilitic, basaltic, picritic, and komatiitic melts in equilibrium with garnet lherzolite at 3-8 GPa. *Journal of Petrology*, *46*, 1645–1659. https://doi.org/10.1093/petrology/egi029

Hammouda, T., & Laporte, D. (2000). Ultrafast mantle impregnation by carbonatite melts. *Geology*, *28*, 283–285. https://doi.org/10.1130/0091-7613(2000)28<283:UMIBCM>2.0.CO;2

Harte, B., Harris, J. W., Hutchison, M. T., Watt, G. R., & Wilding, M. C. (1999). Lower mantle mineral associations in diamonds from Sao Luiz, Brazil. *Mantle petrology: Field observations and high-pressure experimentation: A tribute to Francis R. (Joe) Boyd*, *6*, 125–153. 10.1180/minmag.1994.58A.1.201

Holland, H. D. (2002). Volcanic gases, black smokers, and the Great Oxidation Event. *Geochimica et Cosmochimica Acta*, *66*, 3811–3826. doi:10.1016/S0016-7037(02)00950-X

Horan, M. F., Carlson, R. W., Walker, R. J., Jackson, M., Garçon, M., & Norman, M. (2018). Tracking Hadean processes in modern basalts with 142-Neodymium. *Earth and Planetary Science Letters*, *484*, 184–191. doi:doi.org/10.1016/j.epsl.2017.12.017

Hu, Q., Kim, D. Y., Yang, W., Yang, L., Meng, Y., Zhang, L., & Mao, H. K. (2016). FeO_2 and FeOOH under deep lower-mantle conditions and Earth's oxygen–hydrogen cycles. *Nature*, *534*, 241–244. https://doi.org/10.1038/nature18018

Jacob, D. E., Piazolo, S., Schreiber, A., & Trimby, P. (2016). Redox-freezing and nucleation of diamond via magnetite formation in the Earth's mantle. *Nature Communications*, *7*, 11891. doi:10.1038/ncomms11891

Javoy, M. (1995). The integral enstatite chondrite model of the Earth. *Geophysical Research Letters*, *22*, 2219–2222. https://doi.org/10.1029/95GL02015

Kaminsky, F. (2012). Mineralogy of the Lower Mantle: A Review of "Super-Deep" Mineral Inclusions in Diamond. *Earth-Science Reviews*, *110*, 127–147. doi:10.1016/j.earscirev.2011.10.005

Kaminsky, F. V., Ryabchikov, I. D., McCammon, A. C., Longo, M., Abakumov, A. M., Turner, S., & Heidari, H. (2015). Oxidation potential in the Earth's lower mantle as recorded by ferropericlase inclusions in diamond. *Earth and Planetary Science Letters*, *417*, 49–56. doi:10.1016/j.epsl.2015.02.029

Kasting, J. F., Eggler, D. H., & Raeburn, S. P. (1993). Mantle redox evolution and the oxidation state of the Archean atmosphere. *Journal of Geology*, *101*, 245–257. doi:10.1086/648219

Kiseeva, E., Vasiukov, D. M., Wood, B. J., McCammon, C., Stachel, T., Bykov, M., et al. (2018). Oxidized iron in garnets from the mantle transition zone. *Nature Geoscience*, *11*, 144–147. https://doi.org/10.1038/s41561-017-0055-7

Lammer, H., Zerkle, A. L., Gebauer, S., Tosi, N., Noack, L., Scherf, M., et al. (2018). Origin and evolution of the atmospheres of early Venus, Earth and Mars. *The Astronomy and Astrophysics Review*, *26*, 2. https://doi.org/10.1007/s00159-018-0108-y

Lécuyer, C., & Ricard, Y. (1999). Long-term fluxes and budget of ferric iron: Implications for the redox state of Earth mantle and atmosphere. *Earth and Planetary Science Letters*, *165*, 197–211. https://doi.org/10.1016/S0012-821X(98)00267-2

Li, Z. X. A., & Lee, C. T. A. (2004). The constancy of upper mantle fO(2) through time inferred from V/Sc ratios in basalts. *Earth and Planetary Science Letters*, *228*, 483–493. doi:10.1016/j.epsl.2004.10.006.

Liebske, C., & Khan, A. (2019). On the principal building blocks of Mars and Earth. *Icarus*, *322*, 121–134. https://doi.org/10.1016/j.icarus.2019.01.014

Liu, J., Dorfman, S. M., Zhu, F. J., Li, J., Wang, Y., Zhang, D., et al. (2018). Valence and spin states of iron are invisible in Earth's lower mantle. *Nature Communications*, *9*, 1284. https://doi.org/10.1038/s41467-018-03671-5

Lyons, T. W., Reinhard, C. T., & Planavsky, N. J. (2014). The rise of oxygen in Earth's early ocean and atmosphere. *Nature*, *506*, 307–315. https://doi.org/10.1038/nature13068

Luth, R. W., & Stachel, T. (2014). The buffering capacity of lithospheric mantle: implications for diamond formation. *Contributions to Mineralogy and Petrology*, *168*, 1–12. https://doi.org/10.1007/s00410-014-1083-6

Madhusudhan, N. (2012). C/O ratio as a dimension for characterizing exoplanetary atmospheres. *Astrophysics Journal*, *758*, 36. doi:10.1088/0004-637X/758/1/36.

Martin, R. S., Mather, T. A., & Pyle D. M. (2007). Volcanic emissions and the early Earth atmosphere. *Geochimica et Cosmochimica Acta*, *71*, 3673–3685. https://doi.org/10.1016/j.gca.2007.04.035

McCammon, C. A. (2005). The paradox of mantle redox. *Science*, *308*, 807–808. 10.1126/science.1110532

McKenzie, D. (1989). Some remarks on the movement of small melt fractions in the mantle. *Earth and Planetary Science Letters*, *95*(1–2), 5372. https://doi.org/10.1016/0012-821X(89)90167-2

Moussallam, Y., Oppenheimer, C., & Scaillet, B. (2019). On the relationship between oxidation state and temperature of volcanic gas emissions. *Earth and Planetary Science Letters*, *520*, 260–267. https://doi.org/10.1016/j.epsl.2019.05.036

Nicklas R. W., Puchtel I. S., & Ash, R. D. (2018). Redox state of the Archean mantle: Evidence from V partitioning in 3.5–2.4 Ga komatiites. *Geochimica et Cosmochimica Acta*, *222*, 447–466. https://doi.org/10.1016/j.gca.2017.11.002

Nicklas, R. W., Puchtel, I. S., Ash, R. D., Piccoli, P. M., Hanski, E., Nisbet, E. G., et al. (2019). Secular mantle oxidation across the Archean-Proterozoic boundary: Evidence from V partitioning in komatiites and picrites. *Geochimica et Cosmochimica Acta*, *250*, 49–75. doi:https://doi.org/10.1016/j.gca.2019.01.037

O'Neill, H. St.C. (1991). The origin of the Moon and the early history of the Earth -a chemical model. Part 2: The Earth. *Geochimica et Cosmochimica Acta*, *55*, 1159–1172. https://doi.org/10.1016/0016-7037(91)90169-6

Pahlevan, K., Schaefer, L., & Hirschmann, M. M. (2019). Hydrogen isotopic evidence for early oxidation of silicate Earth. *Earth and Planetary Science Letters*, *115770*. https://doi.org/10.1016/j.epsl.2019.115770

Pearson, D. G., Brenker, F. E., Nestola, F., McNeill, J., Nasdala, L., Hutchison, M. T., et al. (2014). Hydrous mantle transition zone indicated by ringwoodite included within diamond. *Nature*, *507*, 221–224. doi:10.1038/nature13080.

Righter, K., Sutton, S. R., Danielson, L., Pando, K., & Newville, M. (2016). Redox variations in the inner solar system with new constraints from vanadium XANES in spinels. *American Mineralogist*, *101* (9), 1928–1942. doi: https://doi.org/10.2138/am-2016-5638

Rizo, H., Walker, R. J., Carlson, R. W., Horan, M. F., Mukhopadhyay, S., Manthos, V., et al. (2016). Preservation of Earth-forming events in the tungsten isotopic composition of modern flood basalts. *Science*, *352*, 809–812. 10.1126/science.aad8563

Rohrbach, A., Ballhaus, C., Ulmer, P., Golla-Schindler, U., & Schönbohm, D. (2011). Experimental evidence for a reduced metal-saturated upper mantle. *Journal of Petrology*, *52*, 717–731. https://doi.org/10.1093/petrology/egq101

Rohrbach, A., & Schmidt, M. W. (2011). Redox freezing and melting in the Earth's deep mantle resulting from carbon-iron redox coupling. *Nature*, *472*, 209–212. https://doi.org/10.1038/nature09899

Rollinson, H., Adetunji, J., Lenaz, D., & Szilas, K. (2017). Archaean chromitites show constant $Fe^{3+}/\Sigma Fe$ in Earth's asthenospheric mantle since 3.8 Ga. *Lithos*, *282–283*, 316—325. https://doi.org/10.1016/j.lithos.2017.03.020

Rubie, D. C., Jacobson, S. A., Morbidelli, A., O'Brien, D. P., Young, E. D., de Vries, J., et al. (2015). Accretion and differentiation of the terrestrial planets with implications for the compositions of early-formed Solar System bodies and accretion of water. *Icarus*, *248*, 89–108. https://doi.org/10.1016/j.icarus.2014.10.015

Scaillet, B., & Gaillard, F. (2011). Redox state of early magmas. *Nature*, *480*, 48–49. https://doi.org/10.1038/480048a

Schaefer, L., & Fegley, B. Jr. (2017). Redox states of initial atmospheres outgassed on rocky planets and planetesimals. *Astrophysical Journal Letters*, *843*, 120. https://doi.org/10.3847/1538-4357/aa784f

Smart, K. A., Tappe, S., Stern, R. A., Webb, S. J., & Ashwal, L. D. (2016). Early Archaean tectonics and mantle redox recorded in Witwatersrand diamonds. *Nature Geoscience*, *9*, 255–259. doi: 10.1038/NGEO2628

Smith, E. M., Shirey, S. B., Nestola, F., Bullock, E. S., Wang, J. H., Richardson, S. H., & Wang, W. Y. (2016). Large gem diamonds from metallic liquid in Earth's deep mantle. *Science*, *354*, 1403–1405. doi: 10.1126/science.aal1303

Stachel, T., Brey, G. P., & Harris, J. W. (2005). Inclusions in sublithospheric diamonds: glimpses of deep Earth. *Elements*, *1*, 73–78. https://doi.org/10.2113/gselements.1.2.73

Stagno, V. (2019). Carbon, carbonates and carbonatitic melts in the Earth's interior. *Journal of the Geological Society, London*, *176*, 375–387. doi:https://doi.org/10.1144/jgs2018-095

Stagno, V., & Frost, D. J. (2010). Carbon speciation in the asthenosphere: Experimental measurements of the redox conditions at which carbonate-bearing melts coexist with graphite or diamond in peridotite assemblages. *Earth and Planetary Science Letters*, *30*, 72–84. doi:10.1016/j.epsl.2010.09.038.

Stagno, V., Tange, Y., Miyajima, N., McCammon, C. A., Irifune, T., & Frost, D. J. (2011). The stability of magnesite in the transition zone and the lower mantle as function of oxygen fugacity. *Geophysical Research Letters*, *38*, L19309. https://doi.org/10.1029/2011GL049560

Stagno V., Ojwang, D. O., McCammon, C. A., & Frost, D. J. (2013). The oxidation state of the mantle and the extraction of carbon from Earth's interior. *Nature*, *493*, 84–88. https://doi.org/10.1038/nature11679

Stagno V., Frost, D. J., McCammon, C. A., Mohseni, H., & Fei, Y. (2015). The oxygen fugacity at which graphite or diamond forms from carbonate-bearing melts in eclogitic rocks. *Contributions to Mineralogy and Petrology*, *169*, 16. https://doi.org/10.1007/s00410-015-1111-1

Stagno, V., Cerantola, V., Aulbach, S., Lobanov, S., McCammon, C., & Merlini, M. (2019). Carbon-bearing phases throughout Earth's interior: Evolution through space and time. In: Orcutt, B., Daniel, I., & Dasgupta, R. (Eds.), *Deep Carbon: Past to Present*. Cambridge: Cambridge University Press. pp. 66–88.

Tappe, S., Smart, K. A, Torsvik, T. H., Massuyeau, M., & de Wit, M. C. J. (2018) Geodynamics of kimberlites on a cooling Earth: Clues to plate tectonic evolution and deep volatile cycles. *Earth and Planetary Science Letters*, *484*, 1–14. doi:10.1016/j.epsl.2017.12.013

Taylor, W. R., & Green, D. H. (1988). Measurement of reduced peridotite-C-O-H solidus and implications for redox melting of the mantle. *Nature*, *332*, 349–352. https://doi.org/10.1038/332349a0

Thomson, A. R., Walter, M. J., Kohn, S. C., & Brooker, R. A. (2016). Slab melting as a barrier to deep carbon subduction. *Nature*, *529*, 76–79. https://doi.org/10.1038/nature16174

Trail, D., Watson, E. B., & Tailby, N. D. (2011). The oxidation state of Hadean magmas and implications for early Earth's atmosphere. *Nature*, *480*, 79–82. https://doi.org/10.1038/nature10655

Tschauner, O., Huang, S., Greenberg, E., Prakapenka, V. B., Ma, C., Rossman, G. R., et al. (2018). Ice-VII inclusions in diamonds: Evidence for aqueous fluid in Earth's deep mantle. *Science*, *356*, 1136–1139. 10.1126/science.aao3030

Wade, J., & Wood, B. J. (2005). Core formation and the oxidation state of the Earth. *Earth and Planetary Science Letters*, *236*, 78–95. https://doi.org/10.1016/j.epsl.2005.05.017

Walter, M., Bulanova, G., Armstrong, L., Keshav, S., Blundy, J. D., Gudfinnsson, G., et al. (2008). Primary carbonatite melt from deeply subducted oceanic crust. *Nature*, *454*, 622–625. https://doi.org/10.1038/nature07132

Wang, Z., & Becker, H. A. (2013). Ratios of S, Se and Te in the silicate Earth require a volatile-rich late veneer. *Nature*, *499*, 328–331. https://doi.org/10.1038/nature12285

Wark, D. J., & Lovering, F. (1977). Marker events in the early evolution of the solar system: evidence from rims on Ca-Al-rich inclusion in carbonaceous chondrites. *Proceedings of the Lunar Planetary Science Conference*, *8th*, 95–112.

Williams, S. N, Schaefer, S. J., Marta Lucia Calvache, V., & Lopez, D. (1992). Global carbon dioxide emission to the atmosphere by volcanoes. *Geochimica et Cosmochimica Acta*, *56*(4), 1765–1770. doi:https://doi.org/10.1016/0016-7037(92)90243-C.

Wirth, R., Dobrzhineskaya, L., Harte, B., Schreiber, A., & Green, H. W. (2014). High-Fe (Mg, Fe)O inclusion in diamond apparently from the lowermost mantle. *Earth and Planetary Science Letters*, *404*, 365–375. https://doi.org/10.1016/j.epsl.2014.08.010

Wood, B. J., Walter, M. J., & Wade, J. H. (2006). Accretion of the Earth and segregation of its core. *Nature*, *441*, 825–833. https://doi.org/10.1038/nature04763

Wyllie, P. J., & Huang, W. L. (1975). Influence of mantle CO_2 in the generation of carbonatites and kimberlites. *Nature*, *257*, 297–299. https://doi.org/10.1038/257297a0

Yang, X., Gaillard, F., & Scaillet, B. (2014). A relatively reduced Hadean continental crust and implications for the early atmosphere and crustal rheology. *Earth and Planetary Science Letters*, *393*, 210–219. https://doi.org/10.1016/j.epsl.2014.02.056

3

Oxygen Fugacity Across Tectonic Settings

Elizabeth Cottrell[1], Suzanne K. Birner[2], Maryjo Brounce[3], Fred A. Davis[4], Laura E. Waters[5], and Katherine A. Kelley[6]

ABSTRACT

Experiment and observation have established the centrality of oxygen fugacity (fO_2) to determining the course of igneous differentiation, and so the development and application of oxybarometers have proliferated for more than half a century. The compositions of mineral, melt, and vapor phases determine the fO_2 that rocks record, and the activity models that underpin calculation of fO_2 from phase compositions have evolved over time. Likewise, analytical method development has made new sample categories available to oxybarometric interrogation. Here we compile published analytical data from lithologies that constrain fO_2 (n=860 volcanic rocks – lavas and tephras – and n=326 mantle lithologies – the majority peridotites) from ridges, back-arc basins, forearcs, arcs, and plumes. Because calculated fO_2 varies with choice of activity model, we recalculate fO_2 for each dataset from compositional data, applying the same set of activity models and methodologies for each data type. Additionally, we compile trace element concentrations (e.g., vanadium) which serve as an additional fO_2-proxy. The compiled data show that, on average, volcanic rocks and mantle rocks from the same tectonic setting yield similar fO_2s, but mantle lithologies span a much larger range in fO_2 than volcanics. Multiple Fe-based oxybarometric methods and vanadium partitioning vary with statistical significance as a function of tectonic setting, with fO_2 ridges < back arcs < arcs. Plume lithologies are more nuanced to interpret, but indicate fO_2s ≥ ridges. We discuss the processes that may shift fO_2 after melts and mantle lithologies physically separate from one another. We show that the effects of crystal fractionation and degassing on the fO_2 of volcanics are smaller than the differences in fO_2 between tectonic settings and that effects of subsolidus metamorphism on the fO_2 values recorded by mantle lithologies remain poorly understood. Finally, we lay out challenges and opportunities for future inquiry.

3.1. INTRODUCTION

The role that oxygen fugacity (fO_2) plays in producing the unique topography of Earth's surface, defined by high-standing continents and low-lying ocean basins, was recognized in the 1950s, when workers such as Kennedy (1955), Eugster (1957,1959), and Osborn (1959) established the link between fO_2 and the course of igneous differentiation. Crystallization under high fO_2 leads to the calc-alkaline magmatic series common on the continents, while crystallization under low fO_2 results in the tholeiitic magma series common in ocean basins. In order to

[1]Department of Mineral Sciences, National Museum of Natural History, Smithsonian Institution, Washington, DC, USA
[2]Division of Natural Sciences, Nursing, and Mathematics, Berea College, Berea, KY, USA
[3]Department of Earth and Planetary Sciences, University of California Riverside, Riverside, CA, USA
[4]Department of Earth and Environmental Sciences, University of Minnesota Duluth, Duluth, MN, USA
[5]Department of Earth and Environmental Science, New Mexico Institute of Mining and Technology, Socorro, NM, USA
[6]Graduate School of Oceanography, University of Rhode Island, Narragansett, RI, USA

Magma Redox Geochemistry, Geophysical Monograph 266, First Edition.
Edited by Roberto Moretti and Daniel R. Neuville.

DOI: 10.1002/9781119473206.ch3

connect these laboratory-based insights to natural rocks, petrologists began to develop proxies for fO_2, and to apply them in earnest. Through seminal contributions such as Eugster (1959), Haggerty (1976), Christie et al. (1986), Carmichael (1991), Wood et al. (1990), Frost and Lindsley (1992), Ballhaus (1993), Canil (1997), and innumerable others, petrologists began to quantify and map the link between tectonic environment and oxygen fugacity.

3.1.1. Theoretical Background

Oxygen fugacity, or fO_2, describes the potential for an element to occur in an oxidized or reduced state – that is, with a higher or lower charge. If oxygen were an ideal gas, its chemical potential (μ) would simply be related to its partial pressure (P) via

$$\mu_{O_2} = \mu^{\circ}_{O_2} + RTln\frac{P}{P_0} \tag{3.1}$$

Where $\mu^{\circ}_{O_2}$ is the standard state chemical potential of O_2, R is the gas constant, T is the temperature in Kelvin, and P_0 is the standard state pressure of pure O_2. Because no gas, and certainly no rock, behaves as an ideal gas, we substitute fugacity (f) for partial pressure, which corrects pressure for non-ideality, much as chemical activity corrects concentration for non-ideality. The chemical potential of oxygen is then

$$\mu_{O_2} = \mu^{\circ}_{O_2} + RTln\frac{f_{O_2}}{f^{\circ}_{O_2}} = \mu^{\circ}_{O_2} + RTlnf_{O_2} \tag{3.2}$$

when $f^{\circ}_{O_2} = 1$ (the fugacity of pure O_2 at 1 bar and T of interest). In this way, we can relate the free energy of any equilibria of pure phases involving O_2 to oxygen fugacity via the change in Gibbs free energy ($\Delta G^{\circ}_{reaction}$)

$$\Delta G^{\circ}_{reaction} = -RTlnK_{eq} \tag{3.3}$$

where K_{eq} is the equilibrium constant. When P and T are specified, such equilibria of pure phases fix the activity of O_2, or "buffer" the fO_2.

3.1.2. Fe-Based Oxybarometry

Equilibria involving iron are useful because iron is the most abundant multivalent element in the solid Earth and is present in the common rock-forming minerals. For example, the oxygen fugacity of the equilibrium reaction of the pure phases fayalite goes to ferrosilite plus magnetite:

$$\underset{(fayalite)}{6Fe_2SiO_4} + \underset{(oxygen)}{O_2} \leftrightarrow \underset{\text{(ferrosilite)}}{3Fe_2Si_2O_6} + \underset{(magnetite)}{2Fe_3O_4} \tag{3.4}$$

can be calculated from the free energy change of the reaction using Equation 3.3. The equilibrium constant can be written:

$$K_{eq} = \frac{\left(a^{opx}_{Fe_2Si_2O_6}\right)^3\left(a^{spl}_{Fe_3O_4}\right)^2}{\left(a^{olv}_{Fe_2SiO_4}\right)^6 f_{O_2}} \tag{3.5}$$

where $a^{mineral}_{component}$ represents the activity of the end-member component (e.g., ferrosilite) within the mineral phase (e.g., orthopyroxene). In the case of pure phases, the activity is equal to unity, and so the following relationship holds true:

$$logK_{eq} = -logf_{O_2} = \left(\frac{6\Delta G^{\circ}_{Fe_2SiO_4} - 2\Delta G^{\circ}_{Fe_3O_4} - 3\Delta G^{\circ}_{Fe_2Si_2O_6}}{ln(10)RT}\right) \tag{3.6}$$

In experimental systems, oxygen fugacity can be imposed by an invariant buffer reaction involving pure phases (activities equal to unity, for example nickel [Ni] and nickel oxide [NiO]). In natural systems, oxygen fugacity is determined by equilibria involving multi-component silicate minerals, melts, and gases; mineral phases are rarely found as pure end-member compositions. This requires us to relate mineral compositions to component activities. Accurate activity-composition models are therefore important when comparing the fO_2 recorded by a single equilibrium reaction across a wide range of compositions, and even more so when comparing the fO_2 recorded by different fO_2 equilibria – when the accuracy, and not just the precision, of each equilibrium reaction is paramount. We relate the measured compositions of natural phases to the activities of end-member components (e.g., the activity of pure magnetite in spinel, $a^{spinel}_{Fe_3O_4}$) via equations like

$$-logf_{O_2} = \left(\frac{\Delta G^{\circ}_{reaction(P,T)}}{ln\,(10)\,RT}\right) - 2loga^{spinel}_{Fe_3O_4} - 3loga^{opx}_{Fe_2Si_2O_6} + 6loga^{olivine}_{Fe_2SiO_4} \tag{3.7}$$

Equation 3.7 describes the spinel oxybarometer (Ballhaus et al., 1991; O'Neill & Wall, 1987; Wood & Virgo, 1989) and its form prompts a review of several important points.

Any calculation of fO_2 assumes that all phases that fix the activity of oxygen are present and in equilibrium. For example, fO_2 is not constrained by the spinel oxybarometer if orthopyroxene is absent from the assemblage (unless the activity of silica can be constrained by some other means). Likewise, the composition of magnetite records fO_2 only in the presence of ilmenite (Buddington & Lindsley, 1964). We can also see from the above equations that the activity of oxygen will rise with temperature such that it is convenient to refer to a reference buffer. In this contribution we reference the quartz-fayalite-magnetite (QFM) buffer (Frost, 1991) because its pressure-dependence parallels adiabatic melt ascent (Kress and Carmichael, 1991), facilitating comparison of magmas equilibrated at different pressures and temperatures.

It also follows from the thermodynamic treatment above that the bulk rock ratio of ferric to total iron ($Fe^{3+}/\Sigma Fe = Fe^{3+}/[Fe^{3+} + Fe^{2+}]$) in crystalline rocks cannot be equated to oxygen fugacity, because the former varies with mineral mode (extensive property) whereas the latter varies with component activity (intensive property). In melts (and glasses), steric effects are greatly reduced, such that melt (glass) $Fe^{3+}/\Sigma Fe$ ratios can be related to fO_2; however, this does not mean that the $Fe^{3+}/\Sigma Fe$ ratios of melts vary systematically with fO_2 independent of composition. With melts too, $Fe^{3+}/\Sigma Fe$ ratios must be related to fO_2 after considering composition because oxide components in the melt can stabilize or destabilize Fe^{3+} relative to Fe^{2+} at given fO_2 (e.g., Borisov et al., 2018; Kress & Carmichael, 1991; O'Neill et al., 2018).

Based on iron's role in both setting and monitoring fO_2 (Frost, 1991; D. J. Frost & McCammon, 2008), prior Fe-based compilations of fO_2 as a function of tectonic setting have relied on either spinel oxybarometry (e.g., Ballhaus, 1993; Wood et al., 1990), magnetite-ilmenite equilibria (e.g., Frost & Lindsley, 1992), or bulk rock and/or sediment $Fe^{3+}/\Sigma Fe$ ratios (e.g., Carmichael, 1991; Lecuyer & Ricard, 1999). In this chapter, we do not compile fO_2 calculated for bulk "glass" $Fe^{3+}/\Sigma Fe$ ratios because in most cases the metadata provided in publications are absent or insufficient to ensure that the samples have not suffered post-eruptive oxidation or are free of phenocrysts, both of which have been shown to compromise the fidelity of fO_2 proxies (Bezos & Humler, 2005; Brounce et al., 2017; Cottrell & Kelley, 2011; Grocke et al., 2016; Stolper & Bucholz, 2019; Bezos et al., 2021). Further, while very convenient for generating large datasets and informing box models (e.g., Brounce et al., 2019; Canil & Fellows, 2017; Evans, this volume; Evans & Tomkins, 2011; Evans et al., 2012; Lecuyer & Ricard, 1999; Stolper & Bucholz, 2019), applications of bulk "glass" or whole rock $Fe^{3+}/\Sigma Fe$ ratios are limited because the $Fe^{3+}/\Sigma Fe$ ratios of crystalline rocks and mixtures (like cumulates and sediments) cannot be directly equated to fO_2 – as detailed above.

3.1.3. Trace-Element Oxybarometry

In addition to iron, the periodic table offers us a wealth of elements that undergo valence state changes in response to changes in fO_2. The theory that underpins trace element oxybarometry is that fO_2-driven valence state changes will result in changes to mineral-melt partition coefficients. Thus, the concentration of a multivalent trace element in melt should reflect the residual mineralogy, the melt composition, and the pressure, temperature, and oxygen fugacity of the system during melting. In this way, the power of redox-sensitive minor and trace element partitioning can be harnessed to monitor the fO_2 of geologically relevant lithologies, and can offer advantages over Fe. Trace element concentrations offer advantages because they are (i) relatively straightforward to measure and (ii) should be less susceptible than formal valence state to modification after the melt separates from the residue. A wide range of redox-sensitive elements (e.g., Ti, V, Cr, Ce, Eu, U, Mo, W, Re) are discussed separately by Mallmann et al. (2021) within this volume, and we refer the interested reader to that chapter. Several trace elements, however, have been employed to infer the fO_2 of basalt source rocks (the mantle) as a function of tectonic setting (e.g., Bali et al., 2012; Bucholz & Kelemen, 2019; Canil, 1997; Laubier et al., 2014; Lee et al., 2005; Lee et al., 2010; Lee et al., 2012; Mallmann & O'Neill, 2009; Mallmann & O'Neill, 2013; Shervais, 1982). A review of all of these studies and proxies is beyond the scope of this chapter; however, we will discuss vanadium (V) – the most widely applied of these proxies. Vanadium valence can be V^{2+}, V^{3+}, V^{4+}, or V^{5+} in geologically relevant systems. The incompatibility of V^{3+} in mantle minerals is similar to many other trivalent elements, such as the rare earth elements, Sc, Y, and Ga, while V^{4+} and V^{5+} are more highly incompatible. Therefore, with all other parameters equal, melting peridotite at higher fO_2 should result in systematically higher melt V concentrations relative to other trivalent trace elements (Canil, 1997; Lee et al., 2003, 2005).

3.1.4. Other Oxybarometers

We emphasize that there are many additional, and very valuable, proxies for fO_2 that are not yet widely applied to volcanics from arc, ocean island, and mid-ocean ridge settings, but may be in the future, such as stable isotope proxies (Dauphas et al., 2009; Nebel et al., 2015; Williams et al., 2005). Similarly, microprobe peak energies (e.g., Carroll & Rutherford, 1988) and S X-ray Absorption Near Edge Structure spectra (e.g., Fleet et al., 2005) can determine sulfur oxidation state, but analyses from both techniques can be compromised by beam damage during analysis (e.g., Klimm et al., 2012; Rowe et al., 2007; Wilke et al., 2008), leaving only a handful of vetted analyses. For these, we point the interested reader to other chapters in this volume that describe some of these approaches in detail.

3.2. SAMPLE SELECTION, METHODOLOGY, AND DESIGN OF THIS STUDY

We can envision many valid approaches for compiling fO_2 determinations from the literature. In this contribution, we focus on the oxybarometric approach and apply oxybarometers that have all been cross-calibrated at 1 atmosphere under controlled atmosphere conditions at known fO_2. We compile published compositions from Earth's ridges, back arcs, forearcs, arcs, and plumes and recalculate the fO_2 values recorded by these volcanics (lavas and tephras) and mantle lithologies (mostly peridotites). We have opted to include samples wherein the original publications provide sufficient information to recalculate fO_2 (e.g., microprobe analyses) such that we may apply a single set of activity models and methodologies to all of the compositional data collected in disparate studies. This is essential, because the fO_2 calculated from a given composition may vary by up to a log unit depending on the activity model employed (e.g., Birner et al., 2017; Herd, 2008; Wood, 1990). Further, many publications do not report the formulation of the reference buffer (for example, QFM from O'Neill [1987] vs Frost [1991] vs Myers & Eugster [1983]), such that reported fO_2s may not be comparable.

We describe our methods, including chosen activity models, in detail in our methods appendix. In brief, we include widely applied oxybarometric techniques that have documented intra-technique consistency, and which require minimal metadata: magnetite-ilmenite pairs, spinel-oxybarometry, and glass spectroscopy. For example, Davis and Cottrell (2018) show that both the spinel oxybarometric method of Wood and Virgo (1989) and the Fe^{3+}/Fe^{2+} activity model of Kress and Carmichael (1991) return the experimentally imposed fO_2 of gas-mixing experiments containing spinel, olivine, and orthopyroxene in basaltic andesite liquid. Likewise, $Fe^{3+}/\Sigma Fe$ ratios of glasses converted to fO_2 via the Kress and Carmichael (1991) algorithm are consistent with magnetite-ilmenite oxybarometry using the activity model of Ghiorso and Evans (2008) (Crabtree & Lange, 2011; Waters & Lange, 2016). Table 3.1 summarizes the oxybarometric methods we apply and the studies we include. A more detailed discussion of alternative parameterizations for fO_2 can be found in the methods appendix. The compiled dataset can be downloaded as a data library (Cottrell et al., 2021). The dataset is unique in that we aggregate fO_2 data from different studies, different authors, and with different proxies, but we reprocess and quality check the data to facilitate global comparisons. This exercise also identifies which regions and tectonic settings lack direct fO_2 constraints (Figure 3.1).

We first present the Fe-proxy-based fO_2s recorded by rocks from each tectonic setting (ridge, back arc, arc, and plume). We discuss the processes that may perturb these proxies and the extent to which the fO_2 recorded by rock samples is representative of the mantle source. We then compare Fe-based proxies to the inferences that can be drawn from magmatic V concentrations across tectonic settings. Like all elements, the V concentrations of basaltic liquids change as fractional crystallization proceeds, necessitating the normalization of V concentration to another trace element. Although many studies normalize V concentrations to Scandium (Sc), Mallmann and O'Neill (2009) and Laubier et al. (2014) critically evaluate candidate trivalent cations and conclude that the V/Sc ratio does not remain constant as fractionation proceeds. V/Sc begins to rise as MgO falls below 8 wt.%, making this ratio unsuitable to evaluate primitive arc basalts (Laubier et al., 2014). Following Laubier et al. *(2014)* we choose Yb for normalization because it remains constant during crystallization of typical basalts down to < 6 wt.% MgO. We use the compilation of Gale et al. (2013a) for MORB and BABB chemistry, which ensures that samples derive from on-axis locations and that trace element analyses are of high quality. We use the compilation of Turner and Langmuir (2015) for arc basalt chemistry, which ensures that samples derive from arc-front volcanoes and, again, that trace element analyses are of high quality. Following Laubier et al. (2014), we filter these compilations for 6 wt.% < MgO < 12 wt.% to ensure that we are evaluating compositions that represent liquids (not cumulates) and that magnetite has not yet begun to crystallize (V is compatible in magnetite). We only evaluate compositions with Dy/Yb < 2 to avoid residual garnet in the source (Laubier et al., 2014). With these filters in place, we compile trace element data for 1294 MORB and 225 BABB, as well as 317 basalts from 37 arc-front volcanoes. We did not include plume-derived basalts in this comparison because the residual garnet in plume sources complicates interpretation of V/Yb. Yb (and Sc) are highly compatible in garnet, and the relationship between V/Yb (or V/Sc) and fO_2 is poorly defined (Canil, 2002; Davis et al., 2013; Lee et al., 2005).

3.3. RESULTS

Calculated fO_2 for each sample in this study is available in the data library associated with this publication (Cottrell et al., 2021).

3.3.1. Fe-Based Oxybarometry

Mid-Ocean Ridges. Mid-ocean ridges sample Earth's convecting mantle and represent the majority of volcanism on the planet. Both melts and mantle

Table 3.1 Results and Method Summary.

Tectonic setting	δQFM (s)	Lithology	n	Method	Activity-composition models and other method notes	Method error in fO_2 in log units	Notes	Data included
mid-ocean ridge	–0.17 (0.15)	basalt pillow glass	160	XANES	At 1atm. Kress & Carmichael, 1991; parameterization of fO_2 as a function of composition	0.27	Fe oxidation state determined using the Mössbauer calibration of Zhang et al., 2018	Cottrell & Kelley, 2011; Le Voyer et al., 2015 (except the 6 analyses of hydrothermally altered dredge 16D); Birner et al., 2018
mid-ocean ridge	0.19 (0.36)	basalt pillow glass	42	XANES	At 1atm. O'Neill et al., 2018; parameterization of fO_2 as a function of composition	0.52	Fe oxidation state determined using the Mössbauer calibration of Berry et al., 2018	O'Neill et al., 2018
mid-ocean ridge	0.16 (–)	lava	1	mag-ilm pairs	fO_2 at T recorded and 1 atm. Ghiorso & Evans 2008	0.25**	passes Bacon & Hirschmann, 1988, test for equilibrium	Mazzullo & Bence, 1976
mid-ocean ridge	0.31 (0.73)	peridotite	72	sp-oxybarometry	fO_2 at T and 0.6GPa. Mattioli & Wood, 1988; and Wood & Virgo, 1989; with aFe_3O_4 from Sack & Ghiorso, 1991a, 1991b; http://melts.ofm-research.org/CalcForms/index.html; Temperature from spinel-olivine Fe-Mg exchange thermometer of Li et al., 1995	Between about +0.6/–1.0 and ±0.2 depending on spinel $Fe^{3+}/\Sigma Fe$ ratio of spinel (see Methods Appendix and Davis et al., 2017)***	see Davis et al., 2017; and Birner et al. 2017; for discussions of a-X model choices	Bryndzia & Wood, 1990; Birner et al., 2018
back arc basin spreading center	0.22 (0.30)	submarine lava	37	XANES	at 1atm and 1200°C. Kress & Carmichael, 1991	0.27		Kelley & Cottrell, 2009; Brounce et al., 2014
arc front	0.96 (0.39)	basalt glass in olivine-hosted inclusions, basalt pillow glass	119	XANES	at 1atm and 1200°C. Kress & Carmichael, 1991	0.27		Bonnin-Mosbah, 2001; Kelley & Cottrell, 2009; Kelley & Cottrell, 2012; Brounce et al., 2014; Brounce et al., 2016 (data from Gaetani et al., 2012, excluded from average due to oxidative beam damage)
arc front	1.28 (0.64)	volcanics (lavas and tephra)	114	mag-ilm pairs	fO_2 at T recorded and 1 atm. Ghiroso & Evans, 2008	0.25**	passes Bacon & Hirschmann, 1988, test for equilibrium	Carmichael, 1967; Luhr & Carmichael, 1980; Wallace & Carmichael, 1994; Rutherford & Devine, 1996; Mandeville et al., 1996; Luhr, 2000; Coombs & Gardner, 2001; Devine et al., 2003; Costa et al., 2004; Grove et al., 2005; Larsen, 2006; Toothill et al., 2007; Izbekov et al., 2002; Browne et al., 2010; Baggerman & Debari, 2011; Crabtree & Lange, 2011; Stelten & Cooper, 2012; Arce et al., 2013; Waters & Lange, 2013; Howe et al., 2014; Frey & Lange, 2011; Muir et al., 2014; Waters et al., 2015; Grocke et al., 2016; Crabtree & Waters, 2017; Waters & Frey, 2018
arc front	0.96 (0.81)	peridotite xenoliths	47	sp-oxybarometry	fO_2 at T and 0.6GPa. Mattioli & Wood, 1988; and Wood & Virgo, 1989; with aFe_3O_4 from Sack & Ghiorso, 1991a, 1991b; http://melts.ofm-research.org/CalcForms/index.html; Temperature from spinel-olivine Fe-Mg exchange thermometer of Li et al., 1995	Between about +0.6/–1.0 and ±0.2 depending on spinel $Fe^{3+}/\Sigma Fe$ ratio of spinel (see Methods Appendix and Davis et al., 2017)***	see Davis et al., 2017; and Birner et al., 2017; for discussions of a-X model choices	Wood & Virgo, 1989; Canil, 1990; Brandon & Draper, 1996; Parkinson et al., 2003; Bénard et al., 2018

(continued overleaf)

Table 3.1 (*Continued*)

Tectonic setting	δQFM (s)	Lithology	n	Method	Activity-composition models and other method notes	Method error in fO_2 in log units	Notes	Data included
forearc	0.22 (0.75)	trench wall peridotite and peridotite xenoliths	64	sp-oxybarometry	fO_2 at T and 0.6GPa. Mattioli & Wood, 1988; and Wood & Virgo, 1989; with aFe_3O_4 from Sack & Ghiorso, 1991a, 1991b; http://melts.ofm-research.org/CalcForms/index.html; Temperature from spinel-olivine Fe-Mg exchange thermometer of Li et al., 1995	Between about +0.6/–1.0 and ±0.2 depending on spinel $Fe^{3+}/\Sigma Fe$ ratio of spinel (see Methods Appendix and Davis et al., 2017)***	see Davis et al., 2017; and Birner et al., 2017; for discussions of a-X model choices	Parkinson & Pearce, 1998; Pearce et al., 2000; Birner et al., 2017
plume	0.10* (0.71)	basalt pillow glass	334	XANES	at 1atm and 1200°C. Kress & Carmichael, 1991	0.59	Studies using the standard glasses of Cottrell et al., 2009, are recalculated using the Fe3+/Fe2+ ratios reported by Zhang et al., 2018	Brounce et al., 2017; Moussallam et al., 2014; Helz et al., 2017; Moussallam et al., 2016; Shorttle et al., 2015; Hartley et al., 2017; Moussallam et al., 2019
plume	–0.25 (0.55)	lavas	47	mag-ilm pairs	fO_2 at T recorded and 1 atm. Ghiorso & Evans, 2008	0.25**	passes Bacon & Hirschmann, 1988, test for equilibrium	Carmichael, 1967a,b; Anderson & Wright, 1972; Wolfe et al., 1997; Hasse et al., 1997; Gunnarsson et al., 1998; Beier et al., 2006; Genske et al., 2012; Portnyagin et al., 2012
plume	0.82 (1.40) at 0.6 GPa and 0.10 (1.42) at 2.5 GPa	peridotite and pyroxenite xenoliths	143	sp-oxybarometry	fO_2 at T and 0.6GPa. Mattioli & Wood, 1988; and Wood & Virgo, 1989; with aFe_3O_4 from Sack & Ghiorso, 1991a, 1991b; http://melts.ofm-research.org/CalcForms/index.html; Temperature from spinel-olivine Fe-Mg exchange thermometer of Li et al., 1995	Between about +1.2/–2.0 and ±0.4 depending on spinel $Fe^{3+}/\Sigma Fe$ ratio of spinel (see Methods Appendix and Davis et al., 2017)***	see Davis et al., 2017; and Birner et al. 2017; for discussions of a-X model choices	Abu El-Rus et al., 2006; Bonadiman et al., 2005; Davis et al., 2017; Grégoire et al., 2000; Hauri & Hart, 1994; Kyser et al., 1981; Neumann, 1991; Neumann et al., 1995; Neumann et al., 2002; Ryabchikov et al. 1995; Sen, 1987; Sen, 1988; Sen & Leeman, 1991; Sen & Presnall, 1986; Tracy, 1980; Wasilewski et al., 2017; Wulff-Pedersen et al. 1996

Note: *Authors of these studies infer higher fO_2 for primitive, near primary, melts based on these data: Mauna Kea >QFM 0.6 (Brounce et al., 2017); Kilauea QFM +0.4 to 0.7 (Helz et al., 2017, Moussallam et al., 2016); Iceland ~QFM + 0.4 (Shorttle et al., 2015; Hartley et al., 2017); Erebus ~QFM + 1.4 (Moussallam et al., 2014); Canary Islands ~QFM + 1.0 (Moussallam et al., 2019)

Note: **Magnetite-Ilmenite oxygen barometery errors reflect the average residual of model calcluations and the calibration dataset: (Ghiroso & Evans [2008] oxygen barometer-derived fO_2 – known fO_2 from calibration dataset), presented in supplemental material of Waters & Lange (2016)

Note: ***Uncertainty in fO_2 calculated from spinel oxybarometry is asymmetrical and decreases in magnitude as $Fe^{3+}/\Sigma Fe$ ratio of spinel increases. Spinels that have $Fe^{3+}/\Sigma Fe = 0.05$ have an uncertainty in log fO_2 at the high end listed and those with $Fe^{3+}/\Sigma Fe \geq 0.4$ at the low end. Hotspot residues, except four from Davis et al. (2017), are samples with spinel $Fe^{3+}/\Sigma Fe$ ratios determined without Mössbauer correction standards, which roughly doubles uncertainty compared to corrected analyses (Davis et al., 2017).

lithologies offer opportunities for oxybarometry. We begin with the volcanics.

Early estimates based on wet chemistry and magnetite–ilmenite pairs indicated that mid-ocean ridge basalts (MORBs) record fO_2s similar to QFM (Carmichael & Ghiorso, 1986; Haggerty, 1976). However, upon reexamining data from the literature compiled by Haggerty (1976), we found only one sample with multiple pairs of magnetite and ilmenite in equilibrium at magmatic temperatures according to Bacon and Hirschmann (1988), and that sample (15.6m cooling unit from DSDP Leg34: site 319A) records QFM+0.16 (±0.1) at 1232 (±37)°C (Mazzullo & Bence, 1976). Subsequent wet chemical work found that MORBs record fO_2s low enough to suggest graphite is a stable phase in the MORB source (i.e., ~QFM-1, Christie et al., 1986), but more recent wet-chemical work and Fe K-edge XANES analyses have revised average MORB fO_2 estimates back upwards to QFM (Bezos & Humler, 2005; Cottrell & Kelley, 2011; O'Neill et al., 2018; Zhang et al., 2018). Five recent studies determine $Fe^{3+}/\Sigma Fe$ ratios spectroscopically by XANES to determine the fO_2 of average MORB

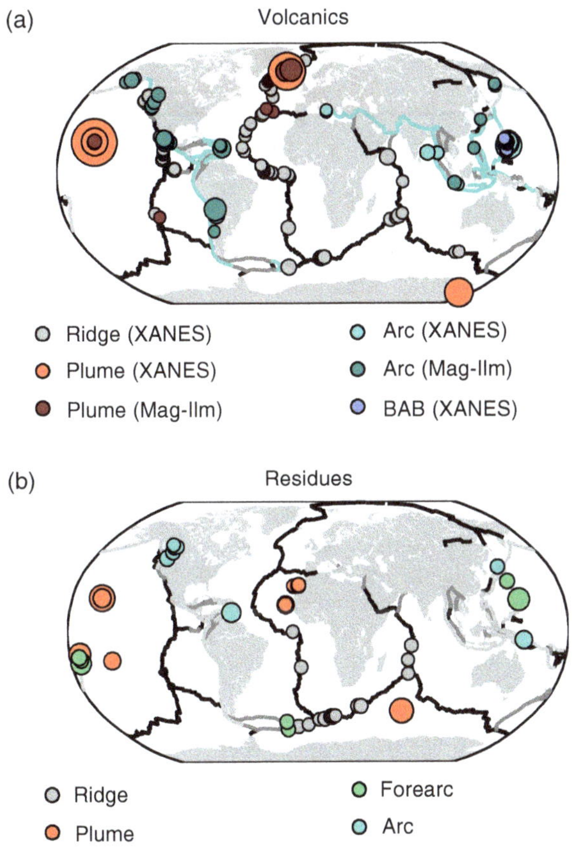

Figure 3.1 Locations of samples compiled in this study as a function of tectonic setting, lithology, and methodology. Symbol size scales linearly with the number of samples at a given locality.

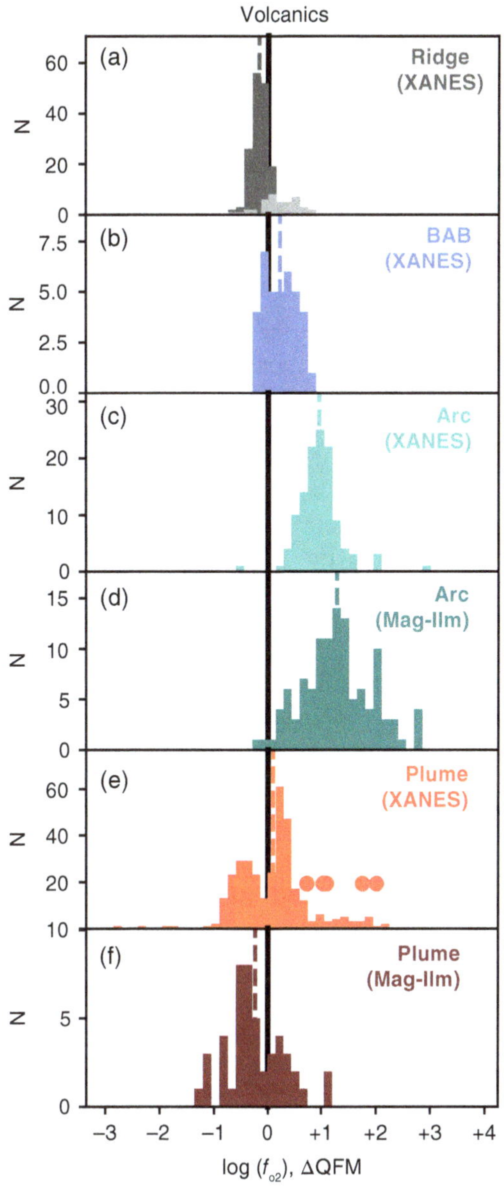

Figure 3.2 Distribution of fO_2 recorded by volcanics globally in different tectonic settings and by multiple methods of oxybarometry. We have recalculated the fO_2 recorded by each sample based on the reported chemical analyses except for the separate light gray dataset in panel (a), which are the observations as reported by O'Neill et al. (2018). The O'Neill et al. (2018) dataset was collected using a different set of primary standards, as described in our methods appendix. Vertical, dashed lines reflect calculated average values of fO_2. Note that volcanics in (e) include plume-affected ridge segments, which cause them to record bimodal fO_2; the fO_2s inferred for primitive plume magmas are higher than the average and we represent each plume's primitive magma fO_2 as a filled orange circle (as reported by those authors). See Table 3.1 and text for details.

(Fig. 3.1, Fig. 3.2a) (Birner et al., 2018; Cottrell & Kelley, 2011; Le Voyer et al., 2015; O'Neill et al., 2018; Zhang et al., 2018). Determinations for 166 MORB glasses that use the calibration of Zhang et al. (2018) find a narrow distribution around QFM –0.17±0.15 (all uncertainty is 1 standard deviation [σ] unless otherwise noted). Determinations for 42 MORB using the calibration of Berry et al. (2018) by O'Neill et al. (2018) return a mean of QFM +0.19 ±0.36. It is notable that O'Neill et al. (2018)'s corresponding $Fe^{3+}/\Sigma Fe$ ratios for average MORB are *lower* by ~0.04 than those from the global survey of Zhang et al. (2018), despite their equation to higher fO_2. The difference stems from O'Neill et al. (2018)'s application of a new compositional parameterization of fO_2, which we choose not to apply in this study (see Methods Appendix for a description and assessment of parameterizations). The important point for our purpose here is that, regardless of the value of the $Fe^{3+}/\Sigma Fe$ ratio of natural MORB, the Fe-XANES spectra of natural MORB glasses resemble the spectra of experimental MORB-composition glasses equilibrated at fO_2 similar to the QFM buffer (see Methods Appendix, Fig. S1), and there is general agreement among all recent spectroscopic studies that MORB glasses record QFM. The fO_2s recorded by average MORBs (7.58 wt.% MgO, Gale et al., 2013b) will be maxima with respect to the fO_2 of the mantle from which they derive, because Fe^{3+} is moderately incompatible during low-pressure fractional crystallization and average MORBs are not primary melts of the mantle ($Fe^{3+}/\Sigma Fe$ ratios increase by 0.03 as MgO falls from 10 to 5 wt.%; Cottrell & Kelley, 2011).

Oxybarometry of mid-ocean ridge peridotites has only been investigated in two studies (Birner et al., 2018 and Bryndzia & Wood, 1990) and in a handful of localities (Fig. 3.1). Global ridge peridotites record fO_2 = QFM +0.31 (±0.73), but we note that more than half of these data derive from a single ridge segment (Fig. 3.1, Fig. 3.3a). Birner et al. (2018) found that n=41 peridotites dredged from the Southwest Indian Ridge (SWIR) record QFM +0.61 (±0.63) at 0.6 GPa and the closure temperature of olivine-spinel exchange. This is significantly higher than the fO_2 recorded by basalts on the same segment (p-value < 0.01); however, the discrepancy disappears once the peridotites' conditions of last equilibration with basalt are considered (Birner et al., 2018). The method of projecting peridotite fO_2 to the PTX conditions of last equilibration with basalt considers three sub-solidus reactions that may alter the fO_2 recorded by the rock: Mg–Fe exchange between olivine and spinel, Al–Cr exchange between spinel and orthopyroxene, and a Tschermak reaction that produces spinel at the expense of olivine and Al-rich orthopyroxene during cooling. Although a dearth of knowledge as to ferric iron partitioning behavior between spinel and pyroxenes during cooling leads to

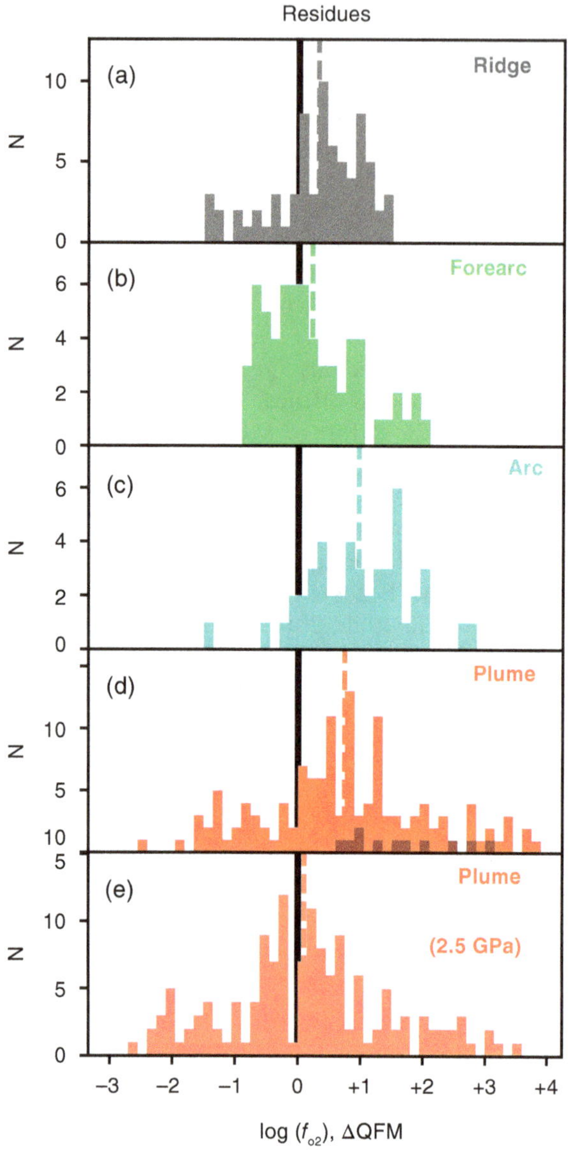

Figure 3.3 Distribution of fO_2 recorded by mantle lithologies (peridotites and olivine-orthopyroxene-spinel-bearing pyroxenites) globally in different tectonic settings. We have recalculated the fO_2 recorded by each sample at 0.6 GPa (2.5 GPa in [e]) and temperature recorded by spinel-olivine thermometry using the methodology of Birner et al. (2018) and Davis et al. (2017) based on the reported chemical analyses. All samples are peridotites except for (d) where the overlain histogram in red are pyroxenites. We caution against overinterpreting the wide range of xenolith fO_2s recovered at plumes due to (i) the near absence of samples from this setting characterized using Mössbauer-characterized spinel standards or Mössbauer spectroscopy; (ii) uncertainty in the barometry and metamorphic history of these samples (which will alter the fO_2 they record); and (iii) limited data by which we may judge the extent to which these lithospheric xenoliths record ridge versus plume fO_2.

significant uncertainty in the magnitude of projection (±0.5 log units, Birner et al., 2018), the direction of the model is to decrease recorded peridotite fO_2 values when projecting back to high temperature-pressure source conditions. This projection thus brings peridotite fO_2 values into closer agreement with basalt fO_2 values, suggesting that fO_2 values recorded by peridotites, without these corrections, may systematically overestimate the redox conditions of MORB-source mantle (Birner et al., 2018).

We highlight that peridotites along SWIR record five times greater range in fO_2 when compared to basalts dredged from the same segment. On the global scale, Bryndzia and Wood (1990) investigated the fO_2 of 35 ridge peridotites from 12 localities. When filtered to exclude four samples from two anomalous locations (the subaerial St. Paul's Rocks and the tectonically complex Mid-Cayman Rise), and recalculated according to the methods presented here, this sample set records fO_2 of QFM -0.08 ±0.68 and spans a range of nearly 2.5 orders of magnitude in fO_2 (Birner et al., 2018). When comparing Fig. 3.2a and 3.3a, we observe that global mid-ocean ridge volcanics display low variance relative to ridge peridotites. These limited data suggest that basalts may homogenize kilometer-scale redox heterogeneity in the upper mantle (Birner et al., 2018). Globally, ridge peridotites calculated at 0.6 GPa and the temperature of olivine-spinel closure record average fO_2s about half a log unit higher than basalts calculated at 1 bar and 1200 °C. Because we have not attempted here to account for subsolidus processes in the peridotites globally, comparisons between the two distributions should not be overinterpreted. A more comprehensive global peridotite dataset is required to evaluate the response of mantle residues to melt extraction and subsolidus re-equilibration.

Arcs and Back Arcs. Subduction influences the composition of mantle melts and arc volcanics, generating continental crust in the process (e.g., Elliott et al., 1997, Gill, 1981; Grove et al., 2012; Kelemen et al., 2003; Kelemen et al., 2007; Osborn, 1959; Plank & Langmuir, 1988; Stolper & Newman, 1994; Turner & Langmuir, 2015; Zimmer et al., 2010). Both melts and mantle lithologies offer opportunities for oxybarometry. We begin with the volcanics.

Seminal contributions by Carmichael (1991) and Frost and Lindsley (1992) surveyed the fO_2s recorded by arc rocks using wet-chemistry and magnetite-ilmenite pairs, respectively, and found that arc rocks record fO_2s up to several orders of magnitude higher than MORBs. Our compilation of 5 XANES spectroscopic studies (n=119 samples, Figure 3.2c) of olivine-hosted melt inclusions and submarine pillow glasses shows that arc basalts record, on average, QFM +0.96 (±0.39). One set of outliers from Cerro Negro record QFM +4.75 (±0.40) (Gaetani et al., 2012), but spectra from these hydrous samples have suffered from radiation-induced beam damage (Cottrell et al., 2018, Gaetani, pers. comm.) and are not included in our statistical analysis. Nearly 90% of samples with XANES measurements erupted through the thin crust (~25 km, Takahashi et al., 2007) of the active Mariana arc front (Fig. 3.1a), and thus there is significant location bias in this dataset.

Globally, magnetite–ilmenite pairs, from 114 arc lavas sampling 11 different arcs, record QFM +1.28 (±0.64) (Figure 3.2d; see methods appendix and the online data library associated with this contribution, Cottrell et al., 2021, for citations). These samples contain Fe–Ti oxides with compositions that record a range of temperatures (700–1085 °C), span a wide range of compositions (basaltic andesite to rhyolite) but are predominantly dacitic, and erupt through crust ranging from 25 to 66km thick.

The mean fO_2 recorded by olivine-hosted melt inclusions and submarine arc-front glasses (ΔQFM = 0.96 ±0.39, n=119) is slightly lower than that recorded by magnetite-ilmenite pairs (ΔQFM = 1.28±0.64 n=114) at the 95% confidence level ($t_{statistic}$ = 4.6, $t_{critical}$ = 2.0, degrees of freedom [DF] = 186, p-value < 0.001). (When we compare distribution means in this contribution, we will always apply a two-sample student's t-test with α= 0.05 [Krzywinski & Altman, 2013] for samples of unequal variance.) We caution that the datasets are not directly comparable because of the limited geographic distribution of the melt inclusion and submarine glass dataset; because there are no samples in common between the two distributions; and because the melt inclusions may reflect magma compositions that precede magnetite and ilmenite saturation. Thus, to first order, our global result is not inconsistent with the results of Waters and Lange (2016) and Crabtree and Lange (2011) who found congruence when they compared magnetite-ilmenite oxybarometry to wet-chemical titration on the same suite of very fresh aphyric lavas.

A more direct comparison can be made between the olivine-hosted melt inclusions and submarine glasses erupted along the active Mariana Arc, and back arc basin (BAB) glasses erupted at depth along the associated back arc spreading center: the Mariana Trough. Both datasets apply the same method (XANES) to arrive at fO_2 estimates, and both sample suites comprise basaltic to basaltic andesite glasses representing similar stages of differentiation in thin crust (similar MgO). The Mariana arc front samples record fO_2s that are on average 0.73 log units higher than the Mariana trough samples (Mariana arc: QFM+0.95±0.36 for n = 107 vs Mariana trough: QFM+0.22±0.30 for n=37, respectively) (Brounce et al., 2014; Kelley & Cottrell, 2009).

Another direct comparison can be made between submarine basalts in ridge (MORB) and BAB tectonic settings. Here we find that BAB from the Mariana trough (n=37) record significantly higher fO_2s than MORB globally (n=160) by 0.4 log units ($t_{statistic} = 7.8$, $t_{critical} = 2.0$, degrees of freedom [df] = 41, p-value << 0.001). This comparison is particularly germane for inferring the effect of subduction on mantle fO_2 because submarine back-arc ridges and mid-ocean ridges are tectonically similar and differences in their melt chemistry can be largely attributed to the influence of subduction (Stolper & Newman, 1994). As indices of subduction influence in Mariana trough lavas go from negligible to significant (e.g., as H_2O contents and the ratios of fluid mobile to fluid immobile incompatible trace elements increase), $Fe^{3+}/\Sigma Fe$ ratios (and fO_2s) also increase (Brounce et al., 2014; Kelley & Cottrell, 2009, 2012). In the Marianas, volcanics erupted over the course of the arc's maturation also record increasing fO_2 with increasing subduction influence. Modern arc tholeiites record similar fO_2s to the boninites that erupted during the early stages of slab influence on the mantle wedge; and both lithologies are more oxidized than the forearc basalts that tapped the mantle prior to slab influence (Brounce et al., 2021; Brounce et al., 2015). The volatile and trace element signals of subduction appear intimately tied to elevated fO_2s in space and in time.

Mantle lithologies recovered from arc settings comprise primarily forearc and arc peridotites. Forearc peridotites are exposed on trench walls and may sample ancient lithospheric mantle (Parkinson & Pearce, 1998), mantle wedge metamorphosed by the subducting slab (Fryer et al., 1985), or processes associated with subduction initiation (Birner et al., 2017). In comparison, arc peridotites rapidly ascend to the surface as xenoliths encased within their basaltic hosts at arc front volcanoes. The mean fO_2 recorded by forearc peridotites is statistically indistinguishable from the mean fO_2 recorded by ridge peridotites ($t_{statistic} = 0.73$, $t_{critical} = 2$, df = 131, p-value = 0.47) (Fig. 3.3). As discussed by Birner et al. (2017), this result contrasts with Parkinson and Pearce (1998)'s study of forearc peridotites from the Izu-Bonin subduction zone, primarily because we apply the spinel activity model of Sack and Ghiorso (1991) instead of Nell and Wood (1991). Yet, consistent with Parkinson and Pearce (1998), Birner et al. (2017) show that peridotites that have interacted with slab-influenced melts do yield elevated fO_2. This influence is additionally evident in the distribution of fO_2 recorded by arc xenoliths from five studies (Table 3.1), which lies significantly higher, by 0.65 log units, than ridge peridotites ($t_{stat} = 4.4$, $t_{crit} = 2.0$, df = 90, p-value << 0.001) or forearc peridotites. Another unique characteristic of sub-arc peridotites is the extended range of melt extraction they record. Spinel Cr#, commonly taken as a proxy for melt extraction, extends to much higher values (> 60) in sub-arc peridotites than in ridge peridotites. This extended range of melt extraction may provide an opportunity to investigate the relationship between extent of melting and fO_2. For example, Benard et al. (2018b) found a weak positive correlation (p-value > 0.06) between fO_2 and modal orthopyroxene, which they interpreted as evidence of fO_2 falling with melt extraction; however, the positive correlation between spinel Cr# and fO_2 in these same samples suggests the relationship between fO_2 and melt extraction may be more complicated. No correlation exists between fO_2 and orthopyroxene mode or spinel Cr# in the Tonga peridotites of Birner et al. (2017). More work is needed to better constrain the effects on fO_2 of extracting melt from the mantle.

Plumes. Mantle plumes are thermal upwellings that impinge on the lithosphere (French & Romanowicz, 2015; Montelli et al., 2006; Sleep, 1992). Capable of generating low degree mantle melts that more ably sample mantle heterogeneity, mantle plumes can produce ocean island basalts (OIB) (Dasgupta et al.; McKenzie & Onions, 1983; Stracke et al., 2005) and xenoliths wrenched from the lithospheric mantle (Frey & Roden, 1987). Both melts and xenoliths at ocean islands offer opportunities for oxybarometry; however, it remains challenging to interpret the relationship between lithospheric mantle peridotites and OIB. We begin with the volcanics.

Melt inclusions and submarine basalts erupting at and around the mantle plumes of Hawaii, Erebus, Iceland, and the Canary Islands record, on average, QFM +0.1 (7 XANES spectroscopic studies with n= 334 samples; Table 3.1, Fig. 3.2e). The Fe- XANES-based fO_2 in this case is higher than the record of magnetite–ilmenite pairs (n=47) by 0.35 log units ($t_{statistic} = 3.8$, $t_{critical} = 2$, df = 69, p-value = 0.003); however, we emphasize that these datasets have no samples in common. There is no meaningful difference between the fO_2s recorded by OIBs and MORBs. In the case of OIB volcanics, most XANES studies have either interrogated a geographic gradient in fO_2 (e.g., Shorttle et al., 2015) or the effect of differentiation (crystallization and degassing) on fO_2 (e.g. Brounce et al., 2017; Helz et al., 2017; Moussallam et al., 2016; Moussallam et al., 2014). We therefore observe a bimodal distribution of fO_2 recorded by plume-affected glasses (Fig. 3.2e). Glasses erupted along mid-ocean ridges that approach plumes, and glasses affected by degassing, record lower fO_2, while primitive and relatively undegassed melt inclusions record higher fO_2 (Fig. 3.2e, and see discussion). The authors of the detailed studies that have interrogated the fO_2 of plumes and plume-affected ridges have inferred plume mantle source fO_2s anywhere from 0.4 to 2 log units higher than the average fO_2 recorded by the volcanics (filled circles in Figure 3.2e). In all cases, the authors have suggested that the mantle sources of these OIBs are more oxidized than those of MORB, and that this may be due to incorporation of recycled components (e.g. Brounce et al., 2017; Moussallam et al., 2019; Helz et al., 2017; Moussallam et al., 2016; Moussallam et al., 2014; Shorttle et al., 2015). Investigators have drawn such inferences by projecting the glass $Fe^{3+}/\Sigma Fe$ ratios along compositional trends (e.g., to more primitive, less degassed, and more enriched compositions) or geographic trends (e.g., along a ridge toward a plume). We discuss some of these projections in greater detail in Section 3.4.1.1.

We have additionally compiled data from 13 studies to calculate the fO_2 of 143 ocean island xenoliths. Unlike all of the ridge peridotites and most of the arc peridotites, none of the published spinel compositions, save for Davis et al. (2017), were obtained using spinel standards with $Fe^{3+}/\Sigma Fe$ ratios independently characterized. Without using standards to correct spinel Fe3+/ΣFe ratios, uncertainties in log fO_2 are roughly double those of samples calculated from corrected spinel $Fe^{3+}/\Sigma Fe$ ratios (Davis et al., 2017, see Methods Appendix and Table 3.1). Despite, or because of, this limitation, we see that ocean island xenoliths record a wide range of fO_2, from QFM –2 to nearly QFM +4 with a mean equal to QFM +0.82 (±1.40). This is a much broader range, extending to lower fO_2, than compiled by Ballhaus (1993); the mean is within error of that compiled by Mallmann and O'Neill (2007), though again the variance is greater in the present compilation. Of the 143 xenoliths compiled here, 11 were identified by the original authors as pyroxenites (Sen, 1987; Tracy, 1980) which record a more oxidized mean fO_2 of QFM +1.44 (±0.63) than the whole set of OIB xenoliths (see red histogram overlain on Fig. 3.3d). The range of fO_2 recorded by OIB lavas falls within the range recorded by ocean island xenoliths; however, we cannot draw a genetic relationship between all lithospheric mantle xenoliths and the mantle melts that exhume them at plumes (compare Fig. 3.2e to 3.3d). Presumably, many of these xenoliths represent lithospheric mantle that has experienced limited chemical interaction with plume-derived melts. Others are likely products of melt-rock reaction between lithospheric peridotite and plume-generated melts. The pyroxenite xenoliths were especially likely to have been derived in this way (e.g., Sen & Leeman, 1991), but it is unclear how many of the peridotites were also influenced by plume-sourced melts. There is another caveat; we do not know the equilibration pressure at the closure temperature of these plume xenoliths. At ridges, we can reasonably infer pressure from peridotite thermometry because geothermal gradients are reasonably well-characterized. That is not the case in plume settings, where thermal gradients are radial as well as vertical (e.g., Farnetani & Hofmann, 2010). In Figure 3.3e we demonstrate how equilibration at 2.5 GPa, instead of 0.6 GPa, would shift fO_2 down by two thirds of a log unit. This illustrates our community's need for better peridotite mineral barometry.

We venture that a more appropriate comparison might be drawn between peridotitic ocean island xenoliths and ridge peridotites. Both may initiate as residues of melting at ridges, with the former transiting and cooling prior to interacting with a mantle plume. The range of fO_2s recorded by ocean island xenoliths encompasses the range of ridge peridotites but skews to higher fO_2s by 0.37 log units if we hold pressure constant at 0.6 GPa ($t_{stat} = 2.6$, $t_{crit} = 2.0$, df = 211, p-value = 0.01). Notably, spinel-olivine Fe-Mg exchange (Li et al., 1995) records higher temperatures in the ocean island xenoliths than in the ridge peridotites. This difference in temperature could be the result of heating of the oceanic lithosphere beneath oceanic islands by the plume (Ballhaus, 1993), or it could result from exhumation of these xenoliths from greater depths than the depth of last equilibration experienced by ridge peridotites. In the former case, temperature-dependent exchange reactions suggest that a mantle parcel preserving a record of hotter conditions should record lower fO_2 than a parcel at the same pressure that records cooler conditions (Birner et al., 2018). Consideration of these subsolidus reactions would thus predict ocean island xenoliths to be more reduced than ridge peridotites, in contrast to what we observe. If this interpretation is correct, then the difference in fO_2 between ridge peridotites and OIB xenolith source mantle prior to plume heating is even greater than the 0.6 GPa plots in Figure 3.3 suggest, perhaps driven by the interaction of some of these xenoliths with oxidized plume melts. In the latter case, changes in fO_2 due to changes in pressure would additionally have to be accounted for to make a direct comparison between ridge peridotites and OIB xenoliths. If no changes in mineral composition or mode are considered, the lower average fO_2 calculated assuming a higher pressure of equilibration (Fig. 3.3) suggests that if OIB xenoliths do generally sample deeper portions of the lithosphere than ridge peridotite, then average fO_2 of the two are comparable. While exchange reactions and modal changes during ascension and cooling may complicate this relationship, at this time, the data do not suggest that xenoliths recovered from plumes significantly differ in their fO_2 compared to peridotites recovered in the ridge setting. Constraints on pressure and effects of temperature on the fO_2 recorded by peridotites below their solidus remain poorly understood, and further work is needed to clarify the fO_2 signature of xenoliths entrained within plume lavas.

3.3.2. V/Yb Concentrations

V/Yb ratios range from 17–195 in MORB, 60–238 for BABB, and 65–422 in arcs (excluding one arc basalt with a ratio of > 800). We find average V/Yb concentrations in each tectonic setting that are all statistically distinct (p-values <<< 0.001) with V/Yb of MORB (93 ±17)

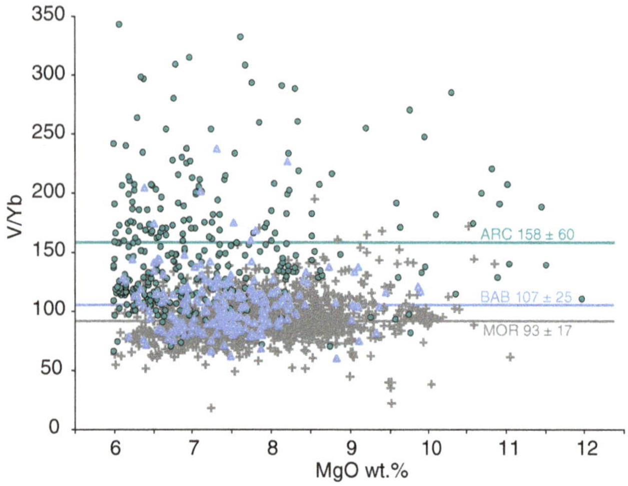

Figure 3.4 V/Yb ratios of ridge (gray "+" symbols, Gale et al., 2013), back-arc (blue triangles, Gale et al., 2013), and arc lavas (teal circles, Turner & Langmuir, 2015) as a function of weight percent MgO. We filtered each published data compilation for 6 wt.% < MgO < 12 wt.% and for Dy/Yb < 2 (Laubier et al., 2014). See text for details.

< BABB (107 ±25) < arcs (158 ±60). We illustrate this in Figure 3.4 with a plot of V/Yb ratios against MgO concentrations in ridge, back-arc, and arc settings. Translating trace element ratios measured in glass (magmatic liquid) to the oxygen fugacity of the source rock (residue) depends on having accurate, composition-dependent, mineral-melt partition coefficients, an accurate knowledge of the source composition, and an accurate melting model (e.g., Canil, 1997; Lee et al., 2003; Lee et al., 2005; Mallmann & O'Neill, 2009; Mallmann et al., 2019). Mallmann and O'Neill (2013), Nicklas et al. (2019), Mallmann et al. (2019), Bucholz and Kelemen (2019), and others have discussed the difficulty in translating V/Sc ratios or olivine-melt partition coefficients into source fO_2s. For example, the fO_2 of modern MORB, or Archean mantle, based on V partitioning is up to a log unit *higher* than that implied by modern MORB $Fe^{3+}/\Sigma Fe$ ratios (Mallmann and O'Neill, 2013; Nicklas et al., 2019). For these reasons, we do not calculate fO_2 for each tectonic setting based on the V/Yb ratio, but simply infer the relative oxygen fugacity of each tectonic setting based on the premise that basalt V/Yb will rise with the fO_2 of the mantle source that generated the basalt. Under these assumptions, it is clear that fO_2 ridges < fO_2 back arcs < fO_2 arcs.

3.4. DISCUSSION

Our compilation and reprocessing of analytical data from the literature yields a synoptic picture of the fO_2s recorded by volcanic and mantle rocks across tectonic settings. Some three decades have passed since the seminal compilations from last century (e.g., Ballhaus et al., 1991; Carmichael, 1991; Christie et al., 1986; Frost & Lindsley, 1992; Wood et al., 1990). The volume of data and geographic coverage have increased tremendously and the analytical techniques and activity models have evolved; however, a key finding from those studies remains robust today. Volcanic and mantle rocks from arc settings record significantly higher fO_2 than those from ridges. Recycling of oceanic crust and lithosphere back into the Earth at subduction zones generates arc volcanics and related mantle lithologies that record higher fO_2 relative to those recovered from ridges. Additional work, particularly in forearc and back-arc settings, support this observation by showing that fO_2 becomes elevated in proportion to the rock's subduction affinity (Benard et al., 2018; Birner et al., 2017; Brounce et al., 2014; Brounce et al., 2021; Brounce et al., 2015; Kelley & Cottrell, 2009; 2012; Parkinson & Arculus, 1999). This remains true for the back-arc basin basalts erupted at pressures > 200 bar and through a comparable crustal column to normal MORB. Moreover, in both the ridge and arc settings, melts (lavas and tephras) and mantle (peridotites and pyroxenites) both record an offset in oxygen fugacity of similar magnitude between the two settings. While these observations are robust, the mechanism by which subduction generates more oxidized lavas and associated mantle lithologies remains a matter of debate and is beyond the scope of this contribution to review (e.g., Andreani et al., 2013; Benard et al., 2018; Canil & Fellows, 2017; Carmichael, 1991; Chin et al., 2018; Debret et al., 2014; Evans, this volume; Farner & Lee, 2017; Foden et al., 2018; Gaillard et al., 2015; Kelley & Cottrell, 2009; Lecuyer & Ricard, 1999; Lee et al., 2005; Mungall, 2002; Nebel et al., 2015; Parkinson & Arculus, 1999; Tang et al., 2018; Tollan & Hermann, 2019; Williams et al., 2004; Wood et al., 1990).

3.4.1. Linking the fO_2 of Volcanics and Mantle Lithologies

Linking magmatic fO_2 to melt residue fO_2 and, ultimately, to the fO_2 of the subsolidus convecting mantle presents a grand challenge to the fields of petrology, experimental petrology, and modeling. When the mantle first begins to melt at infinitesimally small melt fractions, the composition of residual silicate phases will not change significantly, and the fO_2 of the unmelted, solid mantle source will impose fO_2 on low degree partial melts (whether silicate melts or carbonated silicate melts), and set the melt's $Fe^{3+}/\Sigma Fe$ ratio. However, at some point during adiabatic ascent, the melt fraction will grow to an extent that the composition of the solid phases are themselves modified, and the fO_2 of the assemblage (solid + melt) changes. Examples of anticipated changes during melting that could affect fO_2 include the exhaustion of reduced carbon or sulfur phases from the solid assemblage (Lee et al., 2012; Stagno et al., 2013), and an increase of the $Fe^{3+}/\Sigma Fe$ ratio of spinel resulting from Al/Fe^{3+} exchange with the melt that favors concentrating Fe^{3+} in the spinel (Ballhaus et al., 1991; Davis & Cottrell, 2018). At this point, the fO_2 of the mantle will change in ways that experiments and models have not yet elucidated. Indeed, experiments that investigate fO_2 as a variable in the context of partially melting the mantle are nascent (Ballhaus et al., 1991; Davis & Cottrell, 2018; Sorbadere et al., 2018), and stand as a challenge for the future (see Section 3.5). However, once melt and residue separate, melt fO_2 changes little relative to the QFM buffer during volatile-undersaturated adiabatic ascent (Kress & Carmichael, 1991). Thus, at ridges, near primary basaltic melts should serve as accurate proxies for average mantle source conditions (e.g., Carmichael, 1991); however, such accuracy may not extend to mantle rocks, and volcanic rocks from other settings.

Once melts become volatile saturated or begin to undergo assimilation-fractional crystallization processes within the crust, the relationship between mantle source and magmatic fO_2 may become more tenuous. Here we briefly review processes that may modify melt fO_2 signal from source to surface: degassing and crystal fractionation within thick crust.

Degassing. Observational studies that have investigated the relationship between degassing and magmatic fO_2 have thus far observed reduction of melt $Fe^{3+}/\Sigma Fe$ ratios in response to degassing. For example, numerous XANES studies on natural glasses have captured the potential for sulfur degassing to reduce melt $Fe^{3+}/\Sigma Fe$ ratios (Brounce et al., 2014; Brounce et al., 2017; Hartley et al., 2017; Helz et al., 2017; Kelley & Cottrell, 2012; Moussallam et al., 2016; Moussallam et al., 2014; Moussallam et al.; Shorttle et al., 2015). In order to gain further insight into the effect of C–O–H–S degassing on magmatic fO_2, we present results from the gas-melt equilibrium model D-Compress of Burgisser et al. (2015) (see Methods Appendix). Detailed explanations of how D-Compress treats fO_2 during degassing can be found in Burgisser et al. (2015), Moussallam et al. (2016), and Brounce et al. (2017). A general feature of the model output is a prediction of modest changes in fO_2 (< 0.2 log units), both positive and negative, as degassing proceeds until pressures fall below ~250 bar, at which point fO_2 falls sharply in all tectonic settings (Fig. 3.5). D-Compress therefore predicts that the fO_2 of MORB is negligibly affected by degassing (Fig. 3.5, gray line) owing to the fact that MORBs erupt at pressures greater than 200 bar and the dominant degassing volatile is, which has limited

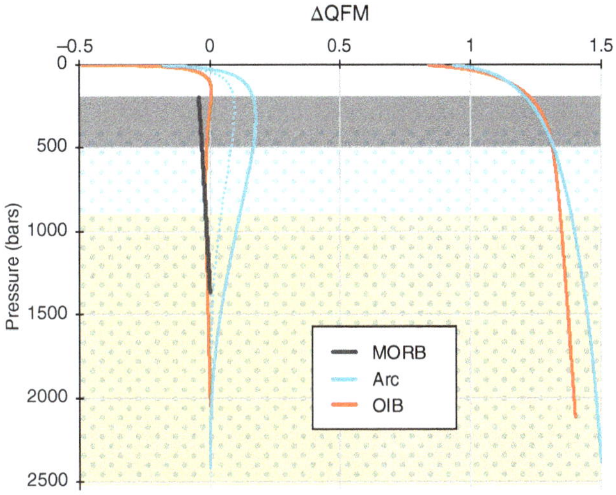

Figure 3.5 Progressive degassing of a C–O–H–S vapor in ridge, arc, and plume settings using the gas-melt equilibrium model D-Compress of Burgisser et al. (2015). For the model run shown by the dashed blue curve, we have set the solubility of H2 in silicate melt to zero; this assumption creates no resolvable difference for any of the other curves shown. For DCompress model details, see the Methods Appendix. Model runs marked OIB, Arc, and MORB are intended to mimic the degassing trajectories of typical magmas from each setting (see methods appendix for details). The OIB and arc scenarios were each run at QFM and near QFM +1.5 to demonstrate the impact of initial fO_2 on degassing trajectory. Orange shaded field shows estimated entrapment pressures for olivine-hosted melt inclusions in OIB based on water and CO_2 concentrations (Tucker et al., 2019). Teal stippled field shows minimum estimated entrapment pressures for olivine-hosted melt inclusions considering water concentrations alone (Plank et al., 2013), which will underestimate typical entrapment pressures. Gray shaded field shows eruption pressures for MORB, which are constrained by ridge depth.

power to shift magmatic fO_2 so long as the MORB source is more oxidized than graphite saturation (Cottrell and Kelley, 2011). Degassing-fO_2 pathways at arcs and subaerial plume volcanoes have greater potential to shift fO_2 because lavas and tephras can degas to 1 atmosphere total pressure. However, melt inclusions may be trapped at nearly any point along the degassing pathway, which may spare some arc and plume sample sets from recording strong modification. Typical water contents of arc melt inclusions yield a conservative estimate of entrapment pressures of 400–3000 bar (Plank et al., 2013). At these entrapment pressures, arc magmas with typical volatile concentrations may be slightly more or less oxidized, but within ~0.2 log units of their source, depending on their initial fO_2 (Fig. 3.5). The fact that arc melt inclusions become progressively reduced during degassing (e.g., Helz et al., 2017; Kelley & Cottrell, 2012; Moussallam et al., 2014; Moussallam et al., 2019) strongly indicates that the arc source is significantly more oxidized than MORB (compare teal arc degassing pathways initiated at QFM vs QFM +1.5 in Figure 3.5).

The much higher ratios of sulfur to water in undegassed OIB magmas lead to a larger magnitude of reduction as degassing proceeds in OIB settings compared to arc settings, because the reducing effects of degassing ~2000 ppm S are not as strongly offset by the slightly oxidizing effects of degassing H_2O, as is the case for H_2O-rich arc magmas (Fig. 3.5, orange lines). The model predictions in Fig. 3.5 that link degassing to reduction are consistent with observations of natural glass and melt inclusion suites, as reviewed above, and implies that the fO_2 recorded by plume-source glasses is typically a minimum (Brounce et al., 2017; Helz et al., 2017; Moussallam et al., 2019). Corroborating evidence is found when we compare the more primitive and less degassed plume glasses (orange circles, Fig. 3.2e) to magnetite–ilmenite oxybarometry in plume lavas (Fig. 3.2f). The latter record lower fO_2 on average, and we speculate that this may reflect magnetite and ilmenite crystallization further along the liquid line of descent, after significant degassing. Further, if fractional crystallization acts to oxidize iron in typical plume lavas, its effect is either counterbalanced by degassing or is not on display here as we compare these two proxies. The caveat is that these are global compilations; Fig. 3.2e and 3.2f do not have samples in common, and we need targeted studies to untangle these competing effects.

If we look to the mantle for further insights about plume fO_2, we find additional evidence that plumes may be oxidized, but the story remains nuanced. Pyroxenite xenoliths from OIB localities have been interpreted as the products of extensive refertilization of the lithosphere by plume-derived melts (Sen & Leeman, 1991). The relatively oxidized fO_2 recorded by these pyroxenite xenoliths provides further evidence that OIB source regions are more oxidized than the MORB source (Figure 3.3d). The opportunity to investigate the $Fe^{3+}/\Sigma Fe$ ratios of submarine and melt inclusion suites from arcs and plumes holds great future promise for reconstructing primitive melt compositions from partially degassed samples.

Crystal Fractionation. Cottrell and Kelley (2011) and numerous studies since (Birner et al., 2018; O'Neill et al., 2018; Shorttle et al., 2015), have shown how $Fe^{3+}/\Sigma Fe$ ratios increase during low-pressure crystal fractionation by a few percent, and that the fO_2 of MORB glass spans a smaller range of fO_2 (e.g., Fig. 3.2a) than indicated by wet-chemistry. However, extensive crystal fractionation and crustal assimilation are commonly observed in magmas that transit thick arc crust, and this has been invoked to shift magmatic fO_2 away from its mantle source (e.g., Chin et al., 2018; Grocke et al., 2016; Lee et al., 2005; Tang et al., 2018). The dearth of fO_2 studies on the rare basalts and olivine-hosted melt inclusions that transit the continental crust poses a challenge to the community. Insights can be gleaned, however, from extensive analytical work on magnetite–ilmenite pairs in more evolved arc rocks. In the arc crust, we observe that magnetite–ilmenite only precipitate once primary magmas have fractionated significant quantities of olivine and clinopyroxene (which remove Fe^{2+} from the melt), and magnetite–ilmenite pairs in arc lavas record slightly higher fO_2s than primitive arc glasses, in accordance with this expectation (see Results, Table 3.1, and Fig. 3.2). Within the BABB suite, silica and fO_2 do covary, although this relationship is demonstrably unrelated to crystal fractionation (Brounce et al., 2014). SiO_2 and fO_2 covary because of two independent phenomena: melting more hydrous mantle yields primary magmas with higher SiO_2 concentrations (Kushiro, 1972) and melting mantle with more subduction influence yields primary magmas with higher fO_2s and also more water (Kelley & Cottrell, 2009).

Figure 3.6 shows all of the volcanic data we have compiled as a function of both crustal thickness (from 7 to nearly 70 kilometers), SiO_2 concentration (a proxy for crystal fractionation, ranging from ~45 to > 75 wt.%), and SiO_2/Alkali ratios. We observe that magmas record fO_2s in excess of ~QFM +1 only when the crust is thicker than that found in the ridge setting. However, from volcanics erupted through ~25 km of oceanic crust to volcanics erupted through nearly 70 km of continental crust, we observe no global correlation between crustal thickness and fO_2. Neither do we observe a correlation between fO_2 and silica content or SiO_2/alkalis, within or among arcs. Figure 3.6 demonstrates that, to first order and on average, there is no simple relationship amongst the variables of crustal thickness, differentiation, and oxygen fugacity. Thus, while the relative influence of slab characteristics, the mantle wedge, and differentiation within the overlying crust on the geochemistry of arcs in the broadest sense remains an active area of research (Chin et al., 2018; Farner & Lee, 2017; Lee et al., 2013; Tang et al., 2018; Turner & Langmuir, 2015; Turner et al., 2016), thick crust and crystal fractionation are not necessary for the generation of oxidized magmas.

Inferences about Mantle fO_2 as a Function of Tectonic Setting. From the analysis above, we may conclude that neither degassing nor crystal fractionation can generate the increases we observe in $Fe^{3+}/\Sigma Fe$ ratios as we move from the mid-ocean ridge, to the back-arc, to the arc-front environment. The inability of these processes to greatly alter fO_2 is evidenced by the fact that magnetite–ilmenite oxybarometry on volcanics, and spinel-oxybarometry on the source mantle itself, also record increasing fO_2 as we

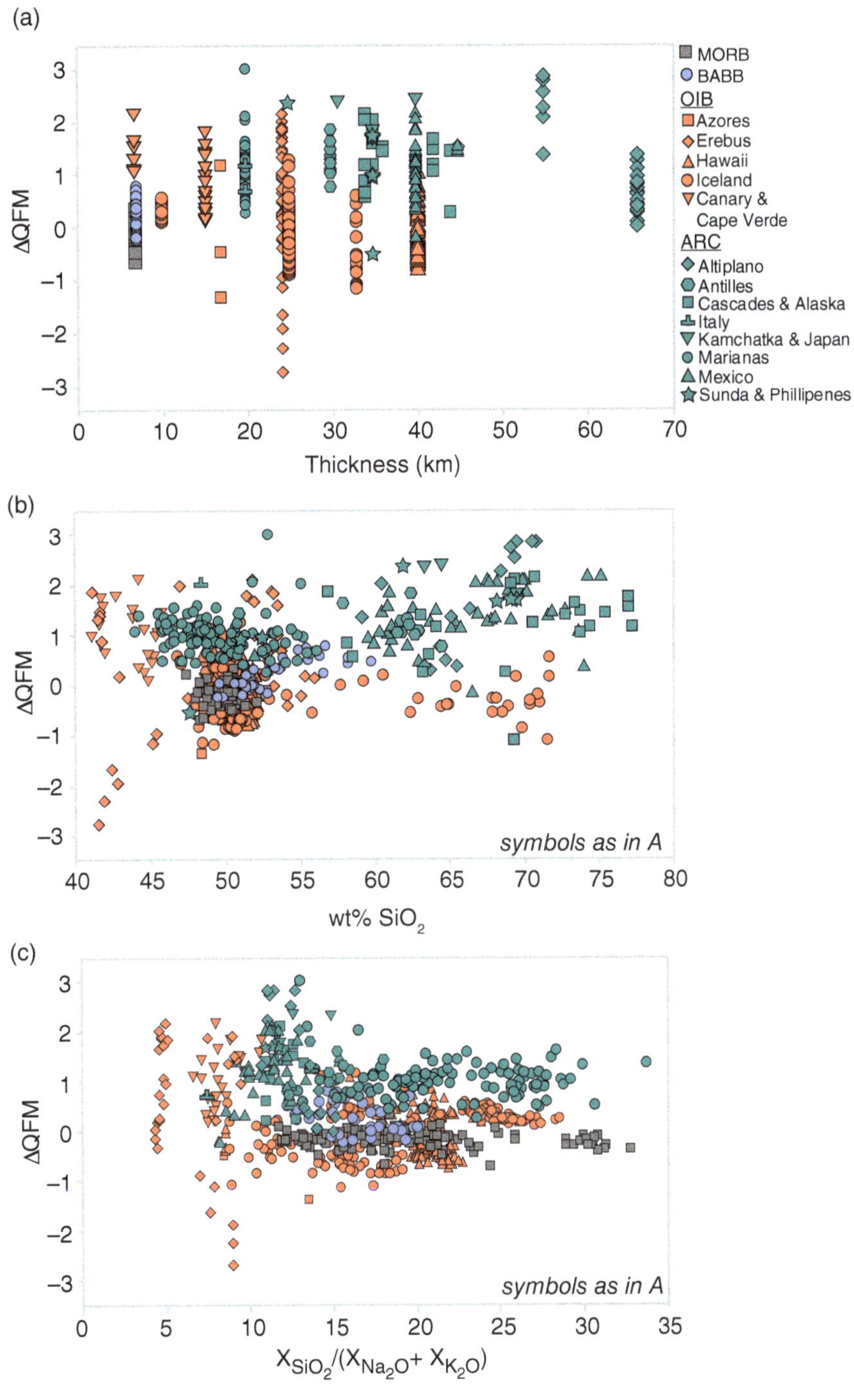

Figure 3.6 Magmatic oxygen fugacity as recorded by volcanics for samples in our compilation as a function of the crustal thickness (a) and SiO_2 (b), and the molar ratio of SiO_2/Na_2O+K_2O (c). We combine oxygen fugacities derived from melt inclusions with those from magnetite-ilmenite pairs for each tectonic setting in each panel; we do not distinguish results based on method. Crustal thickness taken from (Behn & Grove, 2015; Calvert et al., 2008; Chulick et al., 2013; Darbyshire et al., 2000; Das & Nolet, 1998; Ferrari et al., 2012; Finotello et al., 2011; Janiszewski et al., 2013; Levin et al., 2002; Manalo et al., 2015; McGlashan et al., 2008; Nicolich et al., 2000; Saiga et al., 2010; Sevilla et al., 2010; Spieker et al., 2018; Syuhada et al., 2016; Takahashi et al., 2007; Veenstra et al., 2006; Watts & ten Brink, 1989).

move from the ridge, to the back-arc, to the subduction-influenced forearc, to the arc-front setting. We conclude from this evidence that the mantle source itself must become oxidized as the influence of subduction increases. Indeed, *in situ* work by Kelley and Cottrell (2009) and numerous studies since (Brounce et al., 2014; Brounce et al., 2021; Brounce et al., 2015; Kelley & Cottrell, 2012), have linked the fO_2 recorded by submarine glasses and olivine-hosted melt inclusions to enrichment in slab-derived fluid-mobile incompatible trace elements. However, it is critical to consider other lines of evidence, such as trace element proxies, that have been argued to be more robust proxies for mantle fO_2 (e.g., Lee et al., 2005).

We compare our Fe-based oxybarometry results to those obtained from trace element partitioning. Several studies based on V/Sc, V/Y, V/Ti, V/Ga, Zn/Fe ratios, or Cu concentrations have concluded that the fO_2 recorded by arc volcanics are statistically indistinguishable from those recorded by MORB (Lee et al., 2005; Lee et al., 2010; Lee et al., 2012; Mallmann & O'Neill, 2013); this study, and others, have reached the opposite conclusion (Bucholz & Kelemen, 2019; Laubier et al., 2014; Shervais, 1982). It is beyond the scope of this contribution to translate the V/Yb ratios we have compiled into fO_2; however, the conclusion that V/Yb of MORB < BABB < arc basalts is robust (Figure 3.4). This implies that subduction-modified mantle is more oxidized than the MORB source mantle, consistent with Fe-based oxybarometry, for reasons that remain debated (e.g., Andreani et al., 2013; Benard et al., 2018; Canil & Fellows, 2017; Carmichael, 1991; Chin et al., 2018; Debret et al., 2014; Evans, this volume; Farner & Lee, 2017; Foden et al., 2018; Gaillard et al., 2015; Kelley & Cottrell, 2009; Lecuyer & Ricard, 1999; Lee et al., 2005; Mungall, 2002; Nebel et al., 2015; Parkinson & Arculus, 1999; Tang et al., 2018; Tollan & Hermann, 2019; Williams et al., 2004; Wood et al., 1990).

3.5. CONCLUSIONS AND FUTURE DIRECTIONS

Oxygen fugacity varies as a function of tectonic setting. We have shown that all estimators of magmatic fO_2 (XANES, magnetite-ilmentite pairs) and mantle source fO_2 (spinel oxybarometry, V/Yb ratio) show independently that the fO_2 of ridges < back-arcs < arcs. Inferences about plume fO_2 are strongly model dependent, and our study indicates that plume fO_2s range widely, but on average are similar to or higher than mid-ocean ridges. We also strongly conclude that mantle lithologies record a much greater range in fO_2 than volcanics.

The process of subduction generates elevated fO_2 in both melts and mantle lithologies, though the mechanism and locus of this shift remains debated. The extent to which arc rocks are oxidized relative to MORB does not correlate with crustal thickness or indices of crystal fractionation. Degassing may oxidize or reduce magmas to small extents (< 0.2 log units) so long as melts are captured at pressures > 500 bar, and the tendency for shallow degassing (< 200 bar) to reduce magmas appears universal across all tectonic settings. Plate recycling may also enable plumes to achieve their elevated fO_2 relative to mid-ocean ridges; however, when attempting to project back to near-primary compositions, the fO_2 of plume lithologies is uncertain. This is because plume volcanics have thus far only constrained fO_2 minima, and because plume xenoliths derive from lithospheric mantle that was generated at distal mid-ocean ridges that have subsequently metamorphosed and are variably overprinted by the passage of transiting plume-derived melts.

Several additional challenges confront a more complete understanding of oxygen fugacity as a function of tectonic setting. Geographic coverage of fO_2 estimates remains poor, with peridotites from ridges, volcanic rocks and xenoliths from plumes, and primitive volcanic rocks from arcs especially so. Further sampling of primitive melts and mantle lithologies from diverse tectonic environments is needed in order to illuminate the geodynamic and compositional origins of variable fO_2 across tectonic settings. Analytical challenges must still be overcome. Some of the most promising samples for the elucidation of redox processes – melt inclusions – are difficult to prepare and susceptible to radiation beam damage (Cottrell et al., 2018). Experiments and models are needed to gain insights into processes that may shift the fO_2s recorded by melts and residues during partial melting of the source and after melt and residue separate. Additional observations of natural samples and new experiments and models are required to ultimately connect the fO_2 recorded by partial melts, peridotites and pyroxenites, and the fO_2 of the solid convecting mantle.

ACKNOWLEDGMENTS

We thank Oliver Shorttle for providing his XANES data reprocessed using the updated calibration of Zhang et al. (2018). We are grateful to Paolo Sossi, Frank Spera, and an anonymous reviewer for thoughtful, detailed, and constructive comments that greatly improved this contribution. We thank Roberto Moretti and Daniel Neuville for organizing the monograph and for extending the invitation to submit. EC thanks the Lyda Hill Foundation for support during preparation of this manuscript.

METHODS APPENDIX

OXYGEN FUGACITY CALCULATIONS

Volcanics

a. Magnetite-Ilmenite Pairs

For arc and ocean island volcanics, we collected previously reported compositions of magnetite and ilmenite from 28 studies (Arce et al., 2013; Baggerman & DeBari, 2011; Beier et al., 2006; Browne et al., 2010; Carmichael, 1967, 1964; Coombs & Gardner, 2001; Costa et al., 2004; Crabtree & Lange, 2011; Crabtree & Waters, 2017; Devine et al., 2003; Ferrari et al., 2012; Frey and Lange, 2011; Genske et al., 2012; Grocke et al., 2017; Grove et al., 2005; Gunnarsson et al., 1998; Howe et al., 2014; Izbekov et al., 2002; Larsen, 2006; Muir et al., 2014; Portnyagin et al., 2012; Stelten & Cooper, 2012; Toothill et al., 2007; Wallace & Carmichael, 1994; Waters et al., 2015; Wolfe et al., 1997; Wright, 1972). We applied the equilibrium criteria of Bacon & Hirschman (1988) to all collected oxide compositions and only use those that pass for calculations of temperature and oxygen fugacity. (Equilibrium between ilmenite and magnetite is assessed based on comparison Mn and Mg partitioning between ilmenite and magnetite pairs with a dataset of magnetite-ilmenite pairs from natural volcanics inferred to be at equilibrium), Oxide compositions were input into the model of Ghiorso and Evans (2008) to obtain temperatures and oxygen fugacities. Ghiorso and Evans (2008) report their error for the parameterization of their model in terms of the residual energies (kJ) associated with the exchange and redox equilibria between magnetite and ilmenite. We evaluate the uncertainty of the model of Ghiorso & Evans (2008) by comparing modeled values of temperature and oxygen fugacity for experimentally grown iron oxide pairs with the reported experimental conditions (see Fig. 10 from Ghiorso & Evans, 2008). We find the uncertainty in oxygen fugacity and temperatures associated with model of Ghiorso and Evans (2008) is ±0.25 log units and ±45°C, respectively. All temperatures and oxygen fugacities obtained from magnetite-ilmenite pairs are shown in Figure A1.

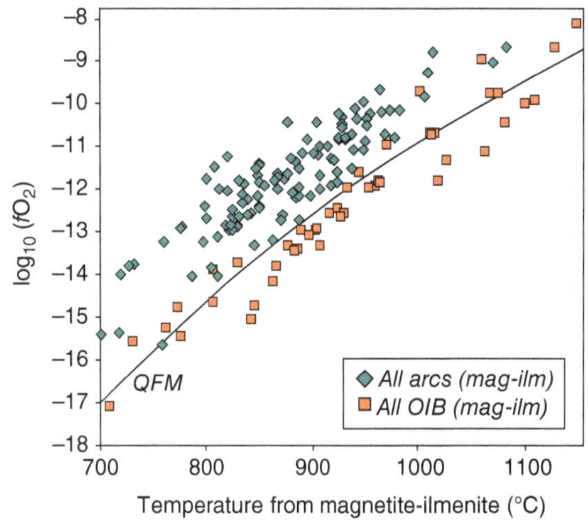

Figure A1 Temperatures and values of fO_2 from the model of Ghiorso and Evans (2008) and all magnetite–ilmenite pairs and used in this study.

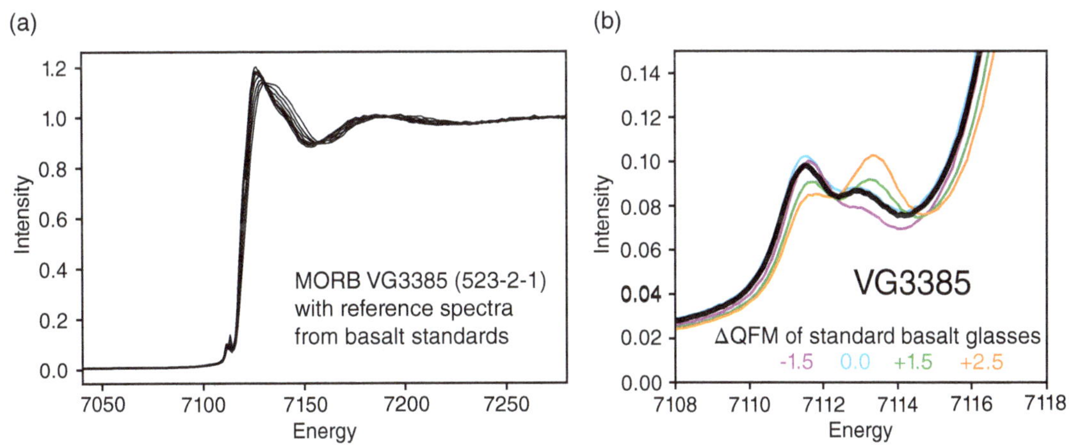

Figure A2 XANES spectra of MORB glass VG3385 with spectra of mid-ocean ridge basalt equilibrated over a range of fO_2, modified from Fig. 1 of Cottrell and Kelley (2011). The spectra of published MORB glasses fall between basalts equilibrated at furnace fO_2s of QFM-0.71 to QFM +0.87. XANES spectroscopy strongly indicates a narrow distribution of MORB fO_2 around QFM, regardless of the corresponding $Fe^{3+}/\Sigma Fe$ of the glasses.

b. $Fe^{3+}/\Sigma Fe$ ratios from XANES

We calculate magmatic fO_2s from measured $Fe^{3+}/\sum Fe$ ratios using Kress and Carmichael (1991) and referenced to the QFM oxygen buffer of Frost (1991) at one atmosphere and 1200 °C, using the major element compositions reported by each study. For studies that quantify $Fe^{3+}/\sum Fe$ ratios using the standard glasses of Cottrell et al. (2009; see Table 3.1), we have recalculated those $Fe^{3+}/\sum Fe$ ratios according to a revision of the Mössbauer-determined $Fe^{3+}/\sum Fe$ ratios of those standard glasses (Zhang et al., 2018). Oliver Shorttle (pers. comm.) provided us with his revised $Fe^{3+}/\sum Fe$ ratios based on the Zhang et al. (2018) update.

c. ***Application of Kress and Carmichael (1991)***

A subset of the data we compiled for this review reports glass $Fe^{3+}/\Sigma Fe$ ratios. Unlike mineral equilibria, we must relate glass $Fe^{3+}/\Sigma Fe$ ratios to fO_2 via an empirical model

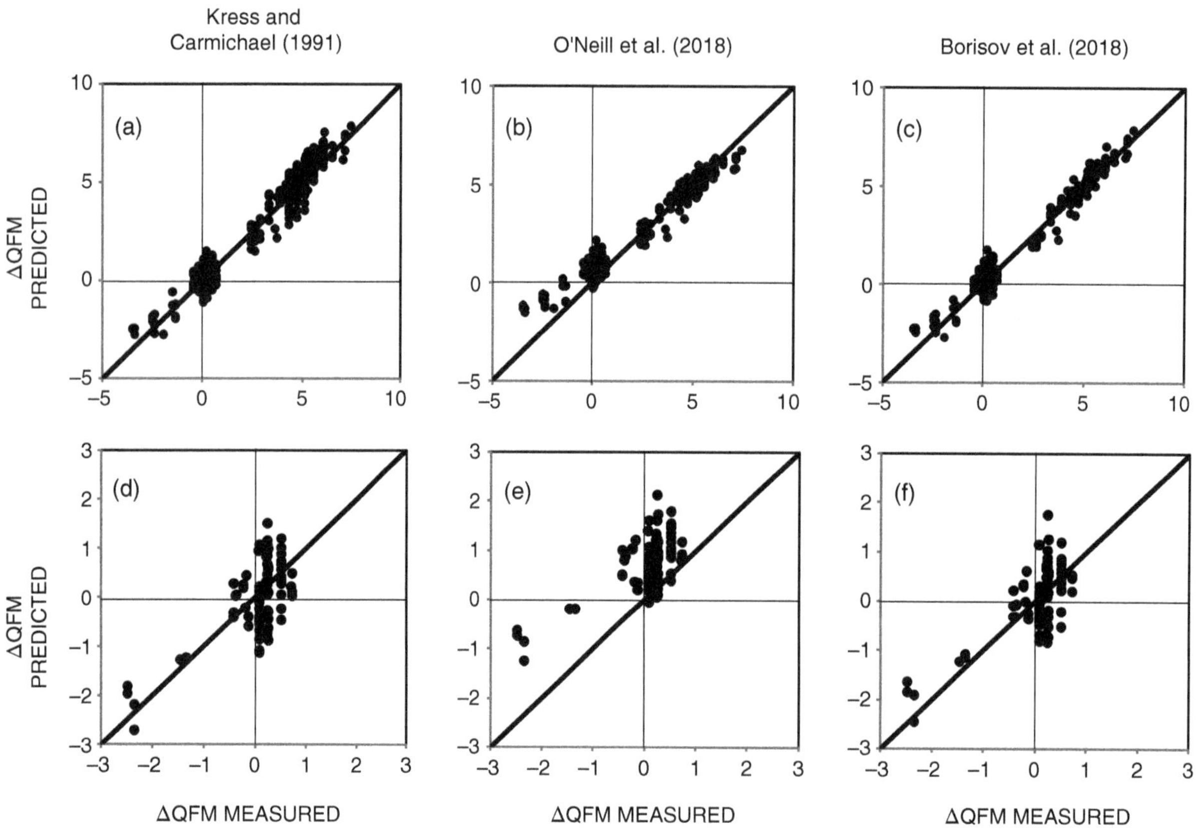

Figure A3 The measured experimental furnace fO_2 in log units relative to the QFM buffer for Borisov et al. (2018)'s recent compilation of 435 controlled-atmosphere experiments vs the fO_2 predicted by three fO_2 parameterizations (i.e. calculated fO_2 from the inputs of $Fe^{3+}/\Sigma Fe$ ratio, T, and major element composition). (a–c) Furnace fO_2 predicted by each parameterization for 435 compositions from QFM -3.3 to +7.3, and (d–f) for 98 "terrestrial" compositions (see Table A1) in the Earth-relevant fO_2 range of QFM -3 to +3. Panels (a) and (d) use the parameterization of Kress and Carmichael (1991); panels (b) and (e) use the parameterization of O'Neill et al. (2018); panels (c) and (f) use the parameterization of Borisov et al. (2018).

Table A1 Major element criteria for "terrestrial" lavas between QFM -3 and QFM +4.1 and "MORB-like" lavas between QFM -2 and QFM +2

	SiO_2	TiO_2	FeO	Na_2O	Al_2O_3	CaO	MgO	K_2O	MnO	P_2O_5
"terrestrial" N=98	42–78	0–4	0.1–18	1–6	11–23	0.1–15	1–14	0–6	0–0.5	0–2
MORB-like N=33	45–55	0.5–4	6–16	1.5–4	12–18	8–14	4–12	0–3	0–0.4	0–1

Major element ranges equal or exceed (Basaltic Volcanism Study Project, 1981) and (Ewart, 1979)

that accounts for composition. Several studies parameterize the relationship between $Fe^{3+}/\Sigma Fe$ ratio and fO_2 and a detailed comparison can be found in Borisov et al. (2018). For this compilation, we investigated those of Borisov et al. (2018), O'Neill et al. (2018), and Kress and Carmichael (1991). [During preparation of this manuscript, a typo in O'Neill et al. (2018) came to light; the coefficient for P_2O_5 in Eqn. 9b in the text of O'Neill et al. (2018) should be –0.018 not –0.18 as written. We use the correct equation here.]

The Borisov et al. (2018) and Kress and Carmichael (1991) models are both empirical parameterizations of hundreds of wet-chemical determinations of $Fe^{3+}/\Sigma Fe$ ratios of glasses of diverse compositions equilibrated in controlled-atmosphere experiments. O'Neill et al. (2018) heavily weights ("anchors") their calibration with the Mössbauer determinations of $Fe^{3+}/\Sigma Fe$ ratios of basalts (one basalt composition from Berry et al. (2018), two basalt compositions from Cottrell et al. (2009), but with the $Fe^{3+}/\Sigma Fe$ ratios "corrected" to be consistent with Berry et al. (2018), one low-Fe basalt composition from Jayasuriya et al. (2004), and one high-Fe sherggotite from Righter et al. (2013), but without that study's correction for recoilless fraction). They then derive the compositional terms from approximately the same database of wet-chemical results used in the Borisov et al. (2018) and Kress and Carmichael (1991) models, though O'Neill et al. (2018) uses only compositions with < 60 wt.% SiO2. Not included was the Mössbauer study of Zhang et al. (2018), which determined recoilless fraction using cryogenic Mössbauer. Correction for recoilless fraction reduces the $Fe^{3+}/\Sigma Fe$ ratios of Cottrell et al. (2009) by a few percent absolute, though this decrease is not equivalent to the "correction" applied by O'Neill et al. (2018). The Mössbauer studies of Zhang et al. (2018) and Berry et al. (2018) obtain fundamentally different results. We prefer the Mössbauer treatment of Zhang et al. (2018) because the methods applied in Berry et al. (2018) depend on assumptions we believe are flawed, including that highly reduced basalt is free of ferric iron (even under the most reducing conditions, Fe^0 coexists with substantial Fe^{3+} (Allen & Snow, 1955; Bowen & Schairer, 1932); that hyperfine parameters remain constant as $Fe^{3+}/\Sigma Fe$ ratio varies (there is ample evidence to the contrary, e.g., Mysen, 2006); and that center shifts > 0.6, at low quadrupole splitting, should be assigned to ferrous iron (this assertion is unsupported, see Zhang et al., 2018 for a discussion). Of course, when exploring the accuracy of a technique, it is advantageous to cross-calibrate. We note that the calibration of Zhang et al. (2018) yields an fO_2- $Fe^{3+}/\Sigma Fe$ ratio relationship that is the same within uncertainty as Kress and Carmichael (1991) model and Borisov et al. (2018) model, based on independent wet-chemical measurements (see also Partzsch et al., 2004), and spinel oxybarometry (Davis & Cottrell, 2018). Debate on these points must play out in the peer-reviewed literature and so for the purpose of this compilation, we take a different, agnostic, approach.

For our assessment, we take advantage of the fact that electrochemical sensors, the devices that monitor the fO_2 within gas-mixing furnaces, are accurate to better than ±0.1 log units in fO_2 and yield oxybarometric results consistent with independent calorimetric data, even accounting for potential interlaboratory biases due to poor calibration of the furnace hotspot (O'Neill & Pownceby, 1993). Taking advantage of this precision and accuracy, we use Borisov et al. (2018)'s recent compilation of 435 controlled-atmosphere experiments to assess the parameterizations; the same experimental database that provides the compositional terms in all three parameterizations. The 435 experiments have wet-chemical determinations of $Fe^{3+}/\Sigma Fe$ ratios, and so are independent of the aforementioned debate concerning Mössbauer spectroscopy. We calculated the furnace fO_2 predicted by each parameterization for 435 compositions from QFM -3.3 to +7.3, and for 98 "terrestrial" compositions (Table A1) in the Earth-relevant fO_2 range of QFM -3 to +4.1.

Because our inputs are the experimental temperatures, reported major elements, and reported wet-chemical determinations of $Fe^{3+}/\Sigma Fe$ ratios of the experiments, this analysis makes no assumptions about the accuracy of the data that underlie O'Neill et al. (2018), Borisov et al. (2018), or Kress and Carmichael (1991). This analysis only asks how well the three parameterizations predict the known furnace fO_2 of those 435 experiments given their independently-determined compositions. For the indicated terrestrial range, O'Neill et al. (2018)'s parameterization returns furnace fO_2s that are, on average, 0.56 (±0.55) log units higher than measured, Kress and Carmichael (1991)'s returns furnace fO_2s that are 0.09 (±0.58) lower than measured, and Borisov et al., (2018)'s returns 0.05 (±0.52) lower than measured. Standard errors on the estimates are reported in Table A2. We could therefore move forward confidently with either Kress and Carmichael (1991) or Borisov et al. (2018) but use the former simply because we had completed our analysis before Borisov et al. (2018) was published. Table A2 reports the standard error of each parameterization for the entire compilation and compositional subsets as defined in Table A1. Our analysis assumes that there is no systematic inaccuracy amongst the wet-chemical studies compiled by Borisov et al., (2018). O'Neill et al. (2018) raise the possibility that some wet-chemical determinations could be erroneous, and cite four suspect studies. Of these four, only two are included in the compilation of Borisov et al. (2018), and of these, 80% are from the study of Thornber et al. (1980). We therefore assessed whether

Table A2 Standard error (σ_{est}) of three fO_2 parameterizations.

	n=435	n=98	n=33
reference	(all expts compiled by [Borisov et al., 2018])	("terrestrial" lavas)	("MORB-like" lavas)
Kress & Carmichael, 1991	0.56	0.59	0.53
Hugh St C. O'Neill et al., 2018	0.58	0.79	0.8
Borisov et al., 2018	0.38	0.53	0.49

inclusion/exclusion of the Thornber et al. (1980) data would significantly impact our analysis. It does not. For example, excluding data from Thornber et al. (1980) from the terrestrial data set (n = 55 without Thornber) causes the standard error of O'Neill et al. (2018) parameterization to degrade to 0.84, while the standard error of Kress and Carmichael (1991) stays constant and that of Borisov et al. (2018) improves to 0.50.

*c. **Vanadium oxybarometry using V/Yb ratios.***
All method details provided in the main text.

Mantle Lithologies

We calculated the oxygen fugacity of mantle lithology (peridotites and olivine-orthopyroxene-spinel bearing pyroxenites) by spinel oxybarometry, following the procedures of Davis et al. (2017). This method uses phase equilibrium between olivine, orthopyroxene, and spinel to constrain the oxygen fugacity of the system.

Calculated oxygen fugacity values are highly dependent on mineral activity models. We have thus recalculated all literature data to use a single set of activity models. For olivine and orthopyroxene, we use the activity models cited in Wood and Virgo (1989). For spinel, we use the activity model developed by Sack and Ghiorso (1991a, b). This spinel activity model better reproduces the experimental data of Wood (1990) than do other commonly used spinel activity models such as those of Mattioli and Wood (1988) and Nell and Wood (1991) (see Davis et al., 2017, for further discussion). Additionally, the Sack and Ghiorso (1991a,b) model is more applicable to spinels at high Cr#, such as the arc and forearc peridotites reported in this work (see Birner et al., 2017, for further discussion).

The activity of magnetite in spinel is itself highly dependent on accurate determination of the ferric iron content within the spinel phase. The studies included in this compilation determine ferric iron content in spinel using either Mössbauer spectroscopy or electron probe microanalysis (EPMA). In the case of EPMA, ferric iron content cannot be determined directly and is instead calculated using stoichiometric constraints. The preferred method of determining ferric iron content in this manner involves correcting the values based on a set of calibration spinels, with ferric iron contents independently determined by Mössbauer, run at the beginning and end of each EPMA session (e.g., Wood & Virgo, 1989; Davis et al., 2017). For peridotites from ridges, arcs, and forearcs compiled in this study, we have only included data in which the $Fe^{3+}/\sum Fe$ ratio of spinel was determined via Mössbauer or corrected EPMA. In the case of xenoliths from OIB localities, we have chosen to additionally include a number of studies in which this correction was not applied, due to the paucity of measurements using spinel standards for correction. Uncertainty in fO_2 increases when uncorrected EPMA analyses of spinels are used to calculate fO_2, but the degree to which that uncertainty increases is dependent on the $Fe^{3+}/\sum Fe$ ratio of the spinel. Uncertainty in fO_2 is greater for spinels with lower $Fe^{3+}/\sum Fe$ ratios and lesser for spinels with higher $Fe^{3+}/\sum Fe$ ratios (Ballhaus et al. 1991; Davis et al. 2017). For example, fO_2 calculated from corrected EPMA analyses of spinels with $Fe^{3+}/\sum Fe$ = 0.10 has an fO_2 uncertainty of about +0.3/-0.4 log units, whereas the uncertainty roughly doubles for uncorrected spinel analyses. At $Fe^{3+}/\sum Fe$ > 0.35, fO_2 uncertainty is only about 0.1 log units for corrected EPMA analyses, and doubles to about 0.2 log units when the analyses are uncorrected. Therefore, the potential effects of including uncorrected analyses on the distribution of fO_2 recorded by peridotites from an oxidized setting is likely to be small.

The calculation of oxygen fugacity also depends highly on assumptions about the temperature and pressure of equilibration. In order to maintain consistency between datasets, we calculate the fO_2 values of all mantle lithologies at 0.6 GPa and the temperature recorded by spinel-olivine thermometry. Justification for this choice can be found in Birner et al. (2017) for forearc/arc peridotites and Birner et al. (2018) for mid-ocean ridge peridotites. Although we choose these values to maintain consistency, there is no rigorous method available to estimate pressure recorded by spinel peridotite xenoliths and no thermal model that can be easily applied to plume-influenced lithosphere that would allow recorded temperature to be related to a depth along a geotherm. OIB xenoliths could potentially have been exhumed from any depth within the spinel stability field. Assuming a maximum pressure of 2.5 GPa, the choice to calculate fO_2 at 0.6 GPa may lead to an overestimation of fO_2 relative to QFM by as much as 0.6 to 0.8 log units. This difference in fO_2 relative to QFM is owing to the differences in ΔV of the QFM reaction and the reaction underlying the spinel oxybarometer (fayalite-ferrosilite-magnetite).

Modeling in DCompress. We modeled the change in magmatic fO_2 with progressive degassing of a C-O-H-S vapor using the gas-melt equilibrium model of Burgisser et al. (2015). This thermodynamic model computes C, H, O, and S concentrations and speciation in coexisting gas and silicate melt as functions of pressure, temperature, melt composition, and fO_2, based on experimental calibrations of melt solubility and homogeneous equilibrium in the gas phase for H2, H2O, CO, CO_2, SO_2, H2S, and S2 species. The melt does not change in major element composition during degassing (i.e., there is no crystallization) and it is not permitted to precipitate separate sulfide or carbon phases.

We followed the methodology of Brounce et al. (2017) to compute the degassing trajectories, except that we used the DCompress default solubility models for C-O-H-S species. We used the default basalt composition and non-temperature dependent solubility relationships of H2, H2O, CO_2, H2S, and SO_2. We also executed model runs wherein we set the solubility of H2 in the silicate melt to zero in order to demonstrate how uncertainty in the speciation of H-species in silicate melts (e.g., finite solubility [Hirschmann et al., 2012; Mysen et al., 2011] vs no solubility ([Newcombe et al., 2017]) propagates into uncertainty in degassing trajectories, particularly those at relatively low fO_2. Among these simulations, only the scenario of an arc magma decompressing at QFM= 0 (i.e., H2O-rich magma in equilibrium with a gas phase containing non-negligible amounts of H2) was sensitive to this assumption (Fig. 3.5). All calculations are calculated as equilibrium (i.e., batch) isothermal decompression, at 1100 °C. The calculations intended to simulate MORB degassing were started at QFM and 1385 bar, with concentrations of volatiles similar to those calculated for globally representative primary MORB melts (Le Voyer et al., 2018) containing 0.2 wt.% H2O, 1100 ppm C_{O2}, and 1425 ppm S. Increasing CO_2 to several thousand ppm has no effect on the trajectories shown. The calculations intended to simulate OIB degassing were started at QFM +1.4 and 2115 bar, with concentrations of volatiles similar to those expected for undegassed Erebus melts (Mousallam et al., 2014) containing 1.5 wt% H_2O, 1710 ppm CO_2, and 2430 ppm S. The calculations intended to simulate arc degassing were started at QFM +1.5 and 2380 bar, with concentrations of volatiles similar to those observed in melt inclusions from Agrigan volcano, containing 4.5 wt.% H_2O, 800 ppm CO_2, and 2050 ppm S (e.g., Kelley & Cottrell, 2012). Melt chemistry (including fO_2) and gas phase compositions were calculated in 1 bar increments and stopped at 5 bars (total pressure).

REFERENCES

Allen, W., & Snow, R. (1955). The orthosilicate-iron oxide portion of the system CaO-"FeO"-SiO2. *Journal of the American Ceramic Society*, *38*(8), 264–272.

Anderson, A. T., & Wright, T. L. (1972). Phenocrysts and glass inclusions and their bearing on oxidation and mixing of basaltic magmas, Kilauea Volcano, Hawaii. *American Mineralogist*, *57*(1–2), 188–216.

Andreani, M., Munoz, M., Marcaillou, C., & Delacour, A. (2013). u-XANES study of iron redox state in serpentine during oceanic serpentinization. *Lithos*, *178*, 70–83. doi: 10.1016/j.lithos.2013.04.008

Arce, J. L., Gardner, J. E., & Macias, J. L. (2013). Pre-eruptive conditions of dacitic magma erupted during the 21.7 ka Plinian event at Nevado de Toluca volcano, Central Mexico. *Journal of Volcanology and Geothermic Research*, *249*, 49–65. doi: 10.1016/j.jvolgeores.2012.09.012

Bacon, C. R., & Hirschmann, M. M. (1988). Mg/Mn Partitioning as a Test for Equilibrium between Coexisting Fe-Ti Oxides. *American Mineralogist*, *73*(1–2), 57–61.

Baggerman, T. D., & DeBari, S. M. (2011). The generation of a diverse suite of Late Pleistocene and Holocene basalt through dacite lavas from the northern Cascade arc at Mount Baker, Washington. *Contributions to Mineralogy and Petrology*, *161*(1), 75–99. doi: 10.1007/s00410-010-0522-2

Bali, E., Keppler, H., & Audetat, A. (2012). The mobility of W and Mo in subduction zone fluids and the Mo–W–Th–U systematics of island arc magmas. *Earth and Planetary Science Letters*, *351–352*, 195–207. doi: 10.1016/j.epsl.2012.07.032

Ballhaus, C. (1993). Redox States of Lithospheric and Asthenospheric Upper-Mantle, *Contributions to Mineralogy and Petrology*, *114*(3), 331–348.

Ballhaus, C., Berry, R. F., & Green, D. H. (1991). High-pressure experimental calibration of the olivine-ortho-pyroxene-spinel oxygen geobarometer - implications for the oxidation-state of the upper mantle. *Contributions to Mineralogy and Petrology*, *107*(1), 27–40.

Basaltic Volcanism Study Project. (1981). *Basaltic volcanism of the terrestrial planets*, New York: Pergamon Press Inc. 1286 pp.

Behn, M. D., & Grove, T. L. (2015). Melting systematics in mid-ocean ridge basalts: Application of a plagioclase-spinel melting model to global variations in major element chemistry and crustal thickness. *Journal of Geophysical Research: Solid Earth*, *120*(7), 4863–4886.

Beier, C., Haase, K. M., & Hansteen, T. H. (2006). Magma evolution of the Sete Cidades volcano, Sao Miguel, Azores. *Journal of Petrology*, *47*(7), 1375–1411. doi: 10.1093/petrology/egl014

Benard, A., Klimm, K., Woodland, A. B., Arculus, R. J., Wilke, M., Botcharnikov, R. E., et al. (2018). Oxidising agents in sub-arc mantle melts link slab devolatilisation and arc magmas. *Nature Communications*, *9*. doi: 10.1038/s41467-018-05804-2.

Bénard, A., Woodland, A. B., Arculus, R. J., Nebel, O., & McAlpine, S. R. B. (2018). Variation in sub-arc mantle oxygen fugacity during partial melting recorded in refractory peridotite xenoliths from the West Bismarck Arc. *Chemical Geology*, *486*, 16–30. doi: 10.1016/j.chemgeo.2018.03.004

Berry, A. J., Stewart, G. A., O'Neill, H. S. C., Mallmann, G., & Mosselmans, J. F. W. (2018). A re-assessment of the oxidation state of iron in MORB glasses. *Earth and Planetary Science Letters*, *483*, 114–123. doi: https://doi.org/10.1016/j.epsl.2017.11.032

Bezos, A., Guivel, G., La, C., Fougeroux, T., & Humler, E. (2021). Unraveling the confusion over the iron oxidation state in MORB glasses. *Geochimica et Cosmochimica Acta*, *293*, 28–39. doi: https://doi.org/10.1016/j.gca.2020.10.004

Bezos, A., & Humler, E. (2005). The Fe3+/Sigma Fe ratios of MORB glasses and their implications for mantle melting. *Geochimica et Cosmochimica Acta*, *69*(3), 711–725.

Birner, S. K., Cottrell, E., Warren, J. M., Kelley, K. A., & Davis, F. A. (2018). Peridotites and basalts reveal broad congruence between two independent records of mantle fO_2 despite local redox heterogeneity. *Earth and Planetary Science Letters*, *494*, 172–189.

Birner, S. K., Warren, J. M., Cottrell, E., Davis, F. A., Kelley, K. A., & Falloon, T. J. (2017). Forearc peridotites from Tonga record heterogeneous oxidation of the mantle following subduction initiation. *Journal of Petrology*, *58*(9), 1755–1780.

Bonadiman, C., Beccaluva, L., Coltorti, M., & Siena, F. (2005). Kimberlite-like metasomatism and 'garnet signature'in spinel-peridotite xenoliths from Sal, Cape Verde Archipelago: relics of a subcontinental mantle domain within the Atlantic oceanic lithosphere? *Journal of Petrology*, *46*(12), 2465–2493.

Bonnin-Mosbah, M., Simionovici, A. S., Metrich, N., Duraud, J. P., Massare, D., & Dillmann, P. (2001). Iron oxidation states in silicate glass fragments and glass inclusions with a XANES micro-probe. *Journal of Non-Crystalline Solids*, *288*(1–3), 103–113.

Borisov, A., Behrens, H., & Holtz, F. (2018). Ferric/ferrous ratio in silicate melts: a new model for 1 atm data with special emphasis on the effects of melt composition. *Contributions to Mineralogy and Petrology*, *173*(12), doi: 10.1007/s00410-018-1524-8

Bowen, N. L., & Schairer, J. F. (1932). The System, FeO-SiO2, *American Journal of Science*, *24*(141), 177–213.

Brandon, A. D., & Draper, D. S. (1996). Constraints on the origin of the oxidation state of mantle overlying subduction zones: An example from Simcoe, Washington, USA. *Geochimica et Cosmochimica Acta*, *60*(10), 1739–1749.

Brounce, M., Kelley, K., & Cottrell, E. (2014). Variations in Fe3+/$\sum$ Fe of Mariana arc basalts and mantle wedge fO_2. *Journal of Petrology*, *55*(12), 2513–2536.

Brounce, M., Reagan, M., Kelley, K. A., Cottrell, E., Shimizu, K., & Almeev, R. (2021). Co-variation of slab tracers, volatiles, and oxidation during subduction initiation. *Geochemistry, Geophysics, Geosystems*. doi: https://doi.org/10.1029/2021GC009823

Brounce, M., Stolper, E., & Eiler, J. (2017). Redox variations in Mauna Kea lavas, the oxygen fugacity of the Hawaiian plume, and the role of volcanic gases in Earth's oxygenation. *Proceedings of the National Academy of Sciences of the United States of America*, *114*(34), 8997–9002. doi: https://10.1073/pnas.1619527114

Brounce, M., Cottrell, E., & Kelley, K. A. (2019). The redox budget of the Mariana subduction zone. *Earth and Planetary Science Letters*, *528*. doi: Doi.org/10.1016/j.epsl.2019.115859

Brounce, M., Kelley, K. A., Cottrell, E., & Reagan, M. K. (2015). Temporal evolution of mantle wedge oxygen fugacity during subduction initiation. *Geology*, *43*(9), 775–778. doi: 10.1130/G36742.1

Browne, B., Izbekov, P., Eichelberger, J., & Churikova, T. (2010). Pre-eruptive storage conditions of the Holocene dacite erupted from Kizimen Volcano, Kamchatka. *International Geology Review*, *52*(1), 95–110. doi: 10.1080/00206810903332413

Bryndzia, L. T., & Wood, B. J. (1990). Oxygen thermobarometry of abyssal spinel peridotites - The redox state and C-O-H volatile composition of the Earth's sub-oceanic upper mantle. *American Journal of Science*, *290*(10), 1093–1116.

Bucholz, C. E., & Kelemen, P. B. (2019). Oxygen fugacity at the base of the Talkeetna arc, Alaska. *Contributions to Mineralogy and Petrology*, *174*(10). doi: 10.1007/s00410-019-1609-z

Buddington, A., & Lindsley, D. (1964). Iron-titanium oxide minerals and synthetic equivalents. *Journal of Petrology*, *5* (2), 310–357.

Burgisser, A., Alletti, M., & Scaillet, B. (2015). Simulating the behavior of volatiles belonging to the C–O–H–S system in silicate melts under magmatic conditions with the software D-Compress. *Computers & Geosciences*, *79*, 1–14. doi: 10.1016/j.cageo.2015.03.002

Calvert, A. J., Klempere, S. L., Takahashi, N., & Kerr, B. C. (2008). Three-dimensional crustal structure of the Mariana island arc from seismic tomography. *Journal of Geophysical Research*, *113*, B01406.

Canil, D. (1990). Experimental study bearing on the absence of carbonate in mantle-derived xenoliths. *Geology*, *18*, 1011–1013.

Canil, D. (1997). Vanadium partitioning and the oxidation state of Archaean komatiite magmas. *Nature*, *389*(6653), 842–845.

Canil, D. (2002). Vanadium in peridotites, mantle redox and tectonic environments: Archean to present. *Earth and Planetary Science Letters*, *195*(1–2), 75–90.

Canil, D., & Fellows, S. A. (2017). Sulphide–sulphate stability and melting in subducted sediment and its role in arc mantle redox and chalcophile cycling in space and time. *Earth and Planetary Science Letters*, *470*, 73–86. doi: 10.1016/j.epsl.2017.04.028

Carmichael, I. S. E. (1967). The mineralogy of thingmuli, a tertiary volcano in Eastern Iceland. *American Mineralogist*, *52*, 1815–1841.

Carmichael, I. S. E. (1991). The redox states of basic and silicic magmas - a reflection of their source regions. *Contributions to Mineralogy and Petrology*, *106*(2), 129–141.

Carmichael, I. S. E., & Nicholls, J. (1967). Iron-titanium oxides and oxygen fugacities in volcanic rocks. *Journal of Geophysical Research*, *72*(18), 4665–4687.

Carmichael, I. S. E., & Ghiorso, M. S. (1986). Oxidation-reduction relations in basic magma - a case for homogeneous equilibria. *Earth and Planetary Science Letters*, *78*(2–3), 200–210.

Carroll, M. R., & Rutherford, M. J. (1988). Sulfur Speciation in Hydrous Experimental Glasses of Varying Oxidation-State - Results from Measured Wavelength Shifts of Sulfur X-Rays. *American Mineralogist*, *73*(7–8), 845–849.

Chin, E. J., Shimizu, K., Bybee, G. M., & Erdman, M. E. (2018). On the development of the calc-alkaline and tholeiitic magma series: A deep crustal cumulate perspective. *Earth and Planetary Science Letters*, *482*, 277–287. doi: 10.1016/j.epsl.2017.11.016

Christie, D. M., Carmichael, I. S. E., & Langmuir, C. H. (1986). Oxidation-states of Midocean Ridge basalt glasses. *Earth and Planetary Science Letters*, *79*(3–4), 397–411.

Chulick, G. S., Detweiler, S., & Mooney, W. D. (2013). Seismic structure of the crust and uppermost mantle of South America and surrounding oceanic basins. *Journal of South American Earth Sciences*, *42*, 260–276. doi: 10.1016/j.jsames.2012.06.002

Coombs, M. L., & Gardner, J. E. (2001). Shallow-storage conditions for the rhyolite of the 1912 eruption at Novarupta, Alaska. *Geology*, *29*(9), 775–778. doi: 10.1130/0091-7613(2001)029<0775:sscftr>2.0.co;2.

Cottrell, E., & Kelley, K. A. (2011). The oxidation state of Fe in MORB glasses and the oxygen fugacity of the upper mantle. *Earth and Planetary Science Letters*, *305*(3–4), 270–282. doi: 10.1016/j.epsl.2011.03.014

Cottrell, E., Kelley, K. A., Lanzirotti, A., & Fischer, R. A. (2009). High-precision determination of iron oxidation state in silicate glasses using XANES. *Chemical Geology*, *268*(3–4), 167–179. doi: 10.1016/j.chemgeo.2009.08.008

Cottrell, E., Birner, S. K., Brounce, M., Davis, F. A., Waters, L. E., & Kelley, K. A. (2021). Oxygen Fugacity Across Tectonic Settings, Version 1.0. Interdisciplinary Earth Data Alliance (IEDA). http://doi.org/10.26022/IEDA/111899

Cottrell, E., Lanzirotti, A., Mysen, B., Birner, S., Kelley, K. A., Botcharnikov, R., et al. (2018). A Mössbauer-based XANES calibration for hydrous basalt glasses reveals radiation-induced oxidation of Fe. *American Mineralogist: Journal of Earth and Planetary Materials*, *103*(4), 489–501.

Crabtree, S., & Lange, R. (2011). An evaluation of the effect of degassing on the oxidation state of hydrous andesite and dacite magmas: a comparison of pre- and post-eruptive Fe2+ concentrations. *Contributions to Mineralogy and Petrology*, *163*, 209–224. doi: 10.1007/s00410-011-0667-7

Crabtree, S. M., & Lange, R. A. (2011). Complex phenocryst textures and zoning patterns in andesites and dacites: Evidence of degassing-induced rapid crystallization? *Journal of Petrology*, *52*(1), 3–38. doi: 10.1093/petrology/egq067

Crabtree, S. M., & Waters, L. E. (2017). The petrologic history of the Sanganguey volcanic field, Nayarit, Mexico: Comparisons in a suite of crystal-rich and crystal-poor lavas. *Journal of Volcanology and Geothermic Research*, *336*, 51–67. doi: 10.1016/j.jvolgeores.2017.02.005.

Darbyshire, F. A., White, R. S., & Priestley, K. F. (2000). Structure of the crust and uppermost mantle of Iceland from a combined seismic and gravity study. *Earth and Planetary Science Letters*, *181*, 409–428.

Das, T., & Nolet, G. (1998). Crustal thickness map of the western United States by partitioned waveform inversion. *Journal of Geophysical Research: Solid Earth*, *103*(B12), 30021–30038.

Dasgupta, R., Jackson, M. G., & Lee, C.-T. A. (2010). Major element chemistry of ocean island basalts – Conditions of mantle melting and heterogeneity of mantle source. *Earth and Planetary Science Letters*, *289*(3–4), 377–392.

Dauphas, N., Craddock, P. R., Asimow, P. D., Bennett, V. C., Nutman, A. P., & Ohnenstetter, D. (2009). Iron isotopes may reveal the redox conditions of mantle melting from Archean to Present. *Earth and Planetary Science Letters*, *288*(1–2), 255–267.

Davis, F. A., & Cottrell, E. (2018). Experimental investigation of basalt and peridotite oxybarometers: implications for spinel thermodynamic models and Fe3+ compatibility during generation of upper mantle melts. *American Mineralogist*, *103*(7), 1056–1067. doi: http://doi.org/10.2138/am-2018-6280

Davis, F. A., Humayun, M., Hirschmann, M. M., & Cooper, R. S. (2013). Experimentally determined mineral/melt partitioning of first-row transition elements (FRTE) during partial melting of peridotite at 3GPa, *Geochimica et Cosmochimica Acta*, *104*, 232–260. doi: 10.1016/j.gca.2012.11.009

Davis, F. A., Cottrell, E., Birner, S. K., Warren, J. M., & Lopez, O. G. (2017). Revisiting the electron microprobe method of spinel-olivine-orthopyroxene oxybarometry applied to spinel peridotites. *American Mineralogist*, *102*(2), 421–435.

Debret, B., Andreani, M., Muñoz, M., Bolfan-Casanova, N., Carlut, J., Nicollet, C., et al. (2014). Evolution of Fe redox state in serpentine during subduction. *Earth and Planetary Science Letters*, *400*, 206–218. doi: 10.1016/j.epsl.2014.05.03

Devine, J. D., Rutherford, M. J., Norton, G. E., & Young, S. R. (2003). Magma storage region processes inferred from geochemistry of Fe-Ti oxides in andesitic magma, Soufriere Hills Volcano, Montserrat, WI. *Journal of Petrology*, *44*(8), 1375–1400. doi: 10.1093/petrology/44.8.1375

El-Rus, M. A. A., Neumann, E. R., & Peters, V. (2006). Serpentinization and dehydration in the upper mantle beneath Fuerteventura (eastern Canary Islands): Evidence from mantle xenoliths. *Lithos*, *89*(1), 24–46.

Elliott, T., Plank, T., Zindler, A., White, W., & Bourdon, B. (1997). Element transport from slab to volcanic front at the Mariana arc. *Journal of Geophysical Research: Solid Earth*, *102*(B7), 14991–15019.

Eugster, H. (1957). Heterogeneous reactions involving oxidation and reduction at high pressures and temperatures. *The Journal of Chemical Physics*, *26*(6), 1760–1761.

Eugster, H. P. (Ed.) (1959). *Oxidation and reduction in metamorphism*, New York: John Wiley & Sons. 397–426 pp.

Evans, K. A. (2021), Redox decoupling, redox budgets and magma recycling. In: D. R. Neuville and R. Moretti, (eds.) *AGU Geophysical Monograph Redox variables and mechanisms in magmatism and volcanism*. Wiley.

Evans, K. A., & Tomkins, A. G. (2011). The relationship between subduction zone redox budget and arc magma fertility. *Earth and Planetary Science Letters*, *308*, 401–409. doi: 10.1016/j.epsl.2011.06.009

Evans, K. A., Elburg, M. A., & Kamenetsky, V. S. (2012). Oxidation state of subarc mantle. *Geology*, *40*(9), 783–786. doi: 10.1130/g33037.1

Ewart, A. (1979). A review of the mineralogy and chemistry of Tertiary-recent dacitic, latitic, rhyolitic, and related salic volcanic rocks. *Developments in Petrology*, *6*, 13–121.

Farner, M. J., & Lee, C.-T. A. (2017). Effects of crustal thickness on magmatic differentiation in subduction zone volcanism: A global study. *Earth and Planetary Science Letters, 470*, 96–107. doi: 10.1016/j.epsl.2017.04.025

Farnetani, C. G., & Hofmann, A. W. (2010). Dynamics and internal structure of the Hawaiian plume. *Earth and Planetary Science Letters, 295*(1–2), 231–240. doi: 10.1016/j.epsl.2010.04.005

Ferrari, L., Orozco-Esquivel, T., Manea, V., & Manea, M. (2012). The dynamic history of the Trans-Mexican Volcanic Belt and the Mexico subduction zone. *Tectonophysics, 522*, 122–149. doi: 10.1016/j.tecto.2011.09.018

Finotello, M., Nyblade, A., Julià, J., Wiens, D. A., & Anandakrishnana, S. (2011). Crustal Vp-Vs ratios and thicknesses for Ross Island and the Transantarctic Mountain front, Antarctica. *Geophysical Journal International, 185*, 85–92.

Fleet, M. E., Liu, X., Harmer, S. L., & King, P. L. (2005). Sulfur K-edge XANES spectroscopy: Chemical state and content of sulfur in silicate glasses. *The Canadian Mineralogist, 43*(5), 1605–1618.

Foden, J., Sossi, P. A., & Nebel, O. (2018). Controls on the iron isotopic composition of global arc magmas. *Earth and Planetary Science Letters, 494*, 190–201. doi: 10.1016/j.epsl.2018.04.039

French, S. W., & Romanowicz, B. (2015). Broad plumes rooted at the base of the Earth's mantle beneath major hotspots. *Nature, 525*, 95–99. doi: 10.1038/nature14876

Frey, F., & Roden, M. F. (1987). The mantle source for Hawaiian Islands. Constraints from the lavas and ultramafic inclusions. In: Menzies, M. A., & Hawkesworth, C. J. (Eds.) *Mantle Metasomatism*. London: Academic Press. pp. 423–463.

Frey, H. M., & Lange, R. A. (2011). Phenocryst complexity in andesites and dacites from the Tequila volcanic field, Mexico: resolving the effects of degassing vs. magma mixing. *Contributions to Mineralogy and Petrology, 162*(2), 415–445. doi: 10.1007/s00410-010-0604-1

Frost, B. R. (Ed.) (1991). *Introduction to oxygen fugacity and its petrologic importance*, 1–9 pp. BookCrafters Inc., Chelsea, MI.

Frost, B. R., & Lindsley, D. H. (1992). Equilibria among Fe-Ti oxides, pyroxenes, olivine, and quartz 2. *Application American Mineralogist, 77*(9–10), 1004–1020.

Frost, D. J., & McCammon, C. A. (2008). The redox state of Earth's mantle. *Annual Review of Earth and Planetary Sciences, 36*(1), 389–420, doi: doi:10.1146/annurev.earth.36.031207.124322

Fryer, P., Ambos, E., & Hussong, D. (1985). Origin and emplacement of Mariana forearc seamounts. *Geology, 13* (11), 774–777.

Gaetani, G. A., O'Leary, J. A., Shimizu, N., Bucholz, C. E., & Newville, M. (2012). Rapid reequilibration of H2O and oxygen fugacity in olivine-hosted melt inclusions. *Geology, 40* (10), 915–918.

Gaillard, F., Scaillet, B., Pichavant, M., & Iacono-Marziano, G. (2015). The redox geodynamics linking basalts and their mantle sources through space and time. *Chemical Geology, 418*, 217–233. doi: 10.1016/j.chemgeo.2015.07.030

Gale, A., Laubier, M., Escrig, S., & Langmuir, C. H. (2013a). Constraints on melting processes and plume-ridge interaction from comprehensive study of the FAMOUS and North Famous segments, Mid-Atlantic Ridge. *Earth and Planetary Science Letters, 365*, 209–220. doi: 10.1016/j.epsl.2013.01.022

Gale, A., Dalton, C. A., Langmuir, C. H., Su, Y., & Schilling, J.-G. (2013b). The mean composition of ocean ridge basalts. *Geochemistry, Geophysics, Geosystems, 14*(3), 489–518. doi: 10.1029/2012gc004334

Genske, F. S., Turner, S. P., Beier, C., & Schaefer, B. F. (2012). The petrology and geochemistry of lavas from the Western Azores Islands of Flores and Corvo. *Journal of Petrology, 53*(8), 1673–1708. doi: 10.1093/petrology/egs029

Ghiorso, M. S., & Evans, B. W. (2008). Thermodynamics of rhombohedral oxide solid solutions and a revision of the FE-TI two-oxide geothermometer and oxygen-barometer. *American Journal of Science, 308*(9), 957–1039. doi: 10.2475/09.2008.01

Gill, J. B. (1981). *Orogenic Andesites and Plate Tectonics*, New York: Springer-Verlag. 390 pp.

Grégoire, M., Moine, B. N., O'Reilly, S. Y., Cottin, J. Y., & Giret, A. (2000). Trace element residence and partitioning in mantle xenoliths metasomatized by highly alkaline, silicate-and carbonate-rich melts (Kerguelen Islands, Indian Ocean). *Journal of Petrology, 41*(4), 477–509.

Grocke, S. B., Cottrell, E., de Silva, S., & Kelley, K. A. (2016). The role of crustal and eruptive processes versus source variations in controlling the oxidation state of iron in Central Andean magmas. *Earth and Planetary Science Letters, 440*, 92–104. doi: 10.1016/j.epsl.2016.01.026

Grove, T. L., Till, C. B., & Krawczynski, M. J. (2012). The role of H2O in subduction zone magmatism. *Annual Review of Earth and Planetary Sciences, 40*(1), 413–439. doi: 10.1146/annurev-earth-042711-105310

Grove, T. L., Baker, M. B., Price, R. C., Parman, S. W., Elkins-Tanton, L. T., Chatterjee, N., & Muntener, O. (2005). Magnesian andesite and dacite lavas from Mt. Shasta, northern California: products of fractional crystallization of H2O-rich mantle melts. *Contributions to Mineralogy and Petrology, 148* (5), 542–565. doi: 10.1007/s00410-004-0619-6

Gunnarsson, B., Marsh, B. D., & Taylor, H. P. (1998). Generation of Icelandic rhyolites:silicic lavas from Torfajökull central volcano, edited. *Journal of Volcanology and Geothermal Research, 83*(1–2), 1–45.

Haggerty, S. (1976). Opaque mineral oxides in terrestrial igneous rocks. *Oxide Minerals: Short Course Notes, 3*, 101–300.

Hartley, M. E., Shorttle, O., Maclennan, J., Moussallam, Y., & Edmonds, M. (2017). Olivine-hosted melt inclusions as an archive of redox heterogeneity in magmatic systems. *Earth and Planetary Science Letters, 479*, 192–205. doi: https://doi.org/10.1016/j.epsl.2017.09.029.

Hasse, K. M., Stoffers, P., & Dieter Garbe-Schönberg, C. (1997). The petrogenetic evolution of lavas from Easter Island and neighbouring seamounts, near-ridge hotspot volcanoes in the SE Pacific, edited. *Journal of Petrology, 38*(6), 785–813.

Hauri, E. H., & Hart, S. R. (1994). Constraints on melt migration from mantle plumes: a trace element study of peridotite xenoliths from Savai'i, Western Samoa. *Journal of Geophysical Research: Solid Earth*, *99*(B12), 24301–24321.

Helz, R., Cottrell, E., Brounce, M. N., & Kelley, K. A. (2017). Olivine-melt relationships and syneruptive redox variations in the 1959 eruption of K⊠lauea Volcano as revealed by XANES. *Journal of Volcanology and Geothermic Research*, *333*, 1–14.

Herd, C. D. K. (2008). Basalts as probes of planetary interior redox state. *Reviews in Mineralogy and Geochemistry*, *68*, 527–553.

Hirschmann, M., Withers, A., Ardia, P., & Foley, N. (2012). Solubility of molecular hydrogen in silicate melts and consequences for volatile evolution of terrestrial planets. *Earth and Planetary Science Letters*, *345*, 38–48.

Howe, T. M., Lindsay, J. M., Shane, P., Schmitt, A. K., & Stockli, D. F. (2014). Re-evaluation of the Roseau Tuff eruptive sequence and other Ignimbrites in Dominica, Lesser Antilles. *Journal of Quaternary Science*, *29*(6), 531–546. doi: 10.1002/jqs.2723

Izbekov, P. E., Eichelberger, J. C., Patino, L. C., Vogel, T. A., & Ivanov, B. V. (2002). Calcic cores of plagioclase phenocrysts in andesite from Karymsky volcano: Evidence for rapid introduction by basaltic replenishment. *Geology*, *30*(9), 799–802. doi: 10.1130/0091-7613(2002)030<0799:ccoppi>2.0.co;2

Janiszewski, H. A., Abers, G. A., Shillington, D. J., & Calkins, J. A. (2013). Crustal structure along the Aleutian island arc: New insights from receiver functions constrained by active-source data. *Geochemistry, Geophysics, Geosystems*, *14*(8), 2977–2992. doi: 10.1002/ggge.20211

Jayasuriya, K. D., O'Neill, H. S., Berry, A. J., & Campbell, S. J. (2004). A Mossbauer study of the oxidation state of Fe in silicate melts. *American Mineralogist*, *89*(11–12), 1597–1609.

Kelemen, P. B., Yogodzindki, G. M., & Scholl, D. W. (2003). Along-strike variation in the Aleutian Island Arc: Genesis of high Mg# Andesite and Implications for continental crust. *Geophysical Monograph 138*, 223–277. doi: 10.1029/138GM11

Kelemen, P. B., Hanghoj, K., & Greene, A. R. (2007). One view of the geochemistry of subduction-related magmatic arcs, with an emphasis on primitive andesite and lower crust. In Turekian, K., & Holland, H. (eds.) *Treatise on Geochemistry*, Elsevier Ltd. 1–70.

Kelley, K. A., & Cottrell, E. (2009). Water and the oxidation state of subduction zone magmas. *Science*, *325*(5940), 605–607. doi: 10.1126/science.1174156

Kelley, K. A., & Cottrell, E. (2012). The influence of magmatic differentiation on the oxidation state of Fe in a basaltic arc magma. *Earth and Planetary Science Letters*, *329*, 109–121. doi: 10.1016/j.epsl.2012.02.010

Kennedy, G. C. (1955). Some aspects of the role of water in rock melts. *Geological Society of America Special Paper*, *62*, 489–504.

Klimm, K., Kohn, S. C., O'Dell, L. A., Botcharnikov, R. E., & Smith, M. E. (2012). The dissolution mechanism of sulphur in hydrous silicate melts. I: Assessment of analytical techniques in determining the sulphur speciation in iron-free to iron-poor glasses. *Chemical Geology*, *322–323*, 237–249. doi: 10.1016/j.chemgeo.2012.04.027

Kress, V. C., & Carmichael, I. S. E. (1991). The compressibility of silicate liquids containing Fe2O3 and the effect of composition, temperature, oxygen fugacity and pressure on their redox states. *Contributions to Mineralogy and Petrology*, *108*, 82–92.

Krzywinski, M., & Altman, N. (2013). Significance, P values and t-tests. *Nature Methods*, *10*(11), 1041–1042. doi: 10.1038/nmeth.2698

Kushiro, I. (1972). Effect of water on composition of magmas formed at high pressures. *Journal of Petrology*, *13*(2), 311–334.

Kyser, T. K., O'Neil, J. R., & Carmichael, I. S. E. (1981). Oxygen isotope thermometry of basic lavas and mantle nodules. *Contributions to Mineralogy and Petrology*, *77*(1), 11–23.

Larsen, J. F. (2006). Rhyodacite magma storage conditions prior to the 3430 yBP caldera-forming eruption of Aniakchak volcano, Alaska. *Contributions to Mineralogy and Petrology*, *152*(4), 523–540. doi: 10.1007/s00410-006-0121-4

Laubier, M., Grove, T. L., & Langmuir, C. H. (2014). Trace element mineral/melt partitioning for basaltic and basaltic andesitic melts: An experimental and laser ICP-MS study with application to the oxidation state of mantle source regions. *Earth and Planetary Science Letters*, *392*, 265–278. doi: 10.1016/j.epsl.2014.01.053

Le Voyer, M., Cottrell, E., Kelley, K. A., Brounce, M., & Hauri, E. H. (2015). The effect of primary versus secondary processes on the volatile content of MORB glasses: An example from the equatorial Mid-Atlantic Ridge (5° N–3° S). *Journal of Geophysical Research: Solid Earth*, *120*(1), 125–144.

Le Voyer, M., Hauri, E. H., Cottrell, E., Kelley, K. A., Salters, V. J. M., Langmuir, C. H., et al. (2018). Carbon fluxes and primary magma CO2 contents along the global mid-ocean ridge system. *Geochemistry, Geophysics, Geosystems*, *20*(3), 1387–1424. doi: 10.1029/2018GC007630

Lecuyer, C., & Ricard, Y. (1999). Long-term fluxes and budget of ferric iron: implication for the redox states of the Earth's mantle and atmosphere. *Earth and Planetary Science Letters*, *165*(2), 197–211.

Lee, C.-T., Brandon, A. D., & Norman, M. (2003). Vanadium in peridotites as a proxy for paleo-fO(2) during partial melting: Prospects, limitations, and implications. *Geochimica Et Cosmochimica Acta*, *67*(16), 3045–3064.

Lee, C.-T., Leeman, W. P., Canil, D., & Li, Z.-X. A. (2005). Similar V/Sc Systematics in MORB and Arc Basalts: Implications for the Oxygen Fugacities of their Mantle Source Regions. *Journal of Petrology*, *46*(11), 2313–2336. doi: 10.1093/petrology/egi056

Lee, C.-T. A., Lee, T. C., & Wu, C.-T. (2013). Modeling the compositional evolution of recharging, evacuating, and fractionating (REFC) magma chambers: Implications for differentiation of arc magmas. *Geochimica et Cosmochimica Acta*, *143*, 8–22. doi: 10.1016/j.gca.2013.08.009.

Lee, C.-T. A., Luffi, P., Le Roux, V., Dasgupta, R., Albarede, F., & Leeman, W. P. (2010). The redox state of arc mantle using Zn/Fe systematics. *Nature*, *468*(7324), 681–685.

Lee, C. T. A., Luffi, P., Chin, E. J., Bouchet, R., Dasgupta, R., Morton, D. M., et al. (2012). Copper systematics in arc magmas and implications for crust-mantle differentiation. *Science*, *336*(6077), 64–68. doi: 10.1126/science.1217313

Levin, V., Park, J., Brandon, M., Lees, J., Peyton, V., Gordeev, E., & Ozerov, A. (2002). Crust and upper mantle of Kamchatka from teleseismic receiver functions. *Tectonophysics*, *358*(1–4), 233–265. doi: 10.1016/s0040-1951(02)00426-2

Li, J., Kornprobst, J., Vielzeuf, D., & Fabriès, J. (1995). An improved experimental calibration of the olivine-spinel geothermometer. *Chinese Journal of Geochemistry*, *14*(1), 68–77.

Luhr, J. F. (2000). The geology and petrology of Volcán San Juan (Nayarit, México) and the compositionally zoned Tepic Pumice. *Journal of Volcanology and Geothermal Research*, *95*, 109–156.

Luhr, J. F., & Carmichael, I. S. E. (1980). The Colima Volcanic complex, Mexico: I. Post-caldera andesite from Volcán Colima. *Contributions to Mineralogy and Petrology*, *71*, 343–372.

Mallmann, G., & O'Neill, H. S. C. (2007). The effect of oxygen fugacity on the partitioning of Re between crystals and silicate melt during mantle melting. *Geochimica et Cosmochimica Acta*, *71*(11), 2837–2857. doi: 10.1016/j.gca.2007.03.028

Mallmann, G., & O'Neill, H. S. C. (2009). The Crystal/Melt Partitioning of V during Mantle Melting as a Function of Oxygen Fugacity Compared with some other Elements (Al, P, Ca, Sc, Ti, Cr, Fe, Ga, Y, Zr and Nb). *Journal of Petrology*, *50*(9), 1765–1794. doi: 10.1093/petrology/egp053

Mallmann, G., & O'Neill, H. S. (2013). Calibration of an empirical thermometer and oxybarometer based on the partitioning of Sc, Y and V between olivine and silicate melt. *Journal of Petrology*, *54*(5), 933–949. doi: 10.1093/petrology/egt001

Mallmann, G., Burnham, A., & Fonseca, R. O. (2021). Mineral-melt partitioning of redox-sensitive elements. In: Neuville, D. R., Moretti, R. (eds.) *AGU Geophysical Monograph Redox variables and mechanisms in magmatism and volcanism*. Wiley.

Manalo, P. C., Dirnalanta, C.B., Faustino-Eslava, D. V., Ramos, N. T., Queano, K. L., & Yumul, G. P. (2015). Crustal thickness variation from a continental to an island arc terrane: Clues from the gravity signatures of the Central Philippines. *Journal of Asian Earth Sciences*, *104*, 205–214. doi: 10.1016/j.jseaes.2014.08.031

Mandeville, C. W., Carey, S., & Sigurdsson, H. (1996). Magma mixing, fractional crystallization and volatile degassing during the 1883 eruption of Krakatau volcano, Indonesia. *Journal of Volcanology and Geothermal Research*, *74*(3–4), 243–274.

Mazzullo, L. J., & Bence, A. (1976). Abyssal tholeiites from DSDP Leg 34: the Nazca plate. *Journal of Geophysical Research*, *81*(23), 4327–4351.

McGlashan, N., Brown, L., & Kay, S. M. (2008). Crustal thickness in the central Andes from teleseismically recorded depth phase precursors. *Geophysical Journal International*, *175*(3), 1013–1022. doi: 10.1111/j.1365-246X.2008.03897.x

McKenzie, D., & Onions, R. K. (1983). Mantle reservoirs and ocean island basalts. *Nature*, *301*(5897), 229–231.

Montelli, R., Nolet, G., Dahlen, F. A., & Masters, G. (2006). A catalogue of deep mantle plumes: New results from finite-frequency tomography. *Geochemistry, Geophysics, Geosystems*, *7*(11). doi: 10.1029/2006gc001248

Moussallam, Y., Edmonds, M., Scaillet, B., Peters, N., Gennaro, E., Sides, I., & Oppenheimer, C. (2016). The impact of degassing on the oxidation state of basaltic magmas: A case study of K⊠lauea volcano. *Earth and Planetary Science Letters*, *450*, 317–325.

Moussallam, Y., Oppenheimer, C., Scaillet, B., Gaillard, F., Kyle, P., Peters, N., et al. (2014). Tracking the changing oxidation state of Erebus magmas, from mantle to surface, driven by magma ascent and degassing. *Earth and Planetary Science Letters*, *393*, 200–209.

Moussallam, Y., Longpré, M.-A., McCammon, C., Gomez-Ulla, A., Rose-Koga, E. F., Scaillet, B., et al. (2019). Mantle plumes are oxidised. *Earth and Planetary Science Letters*, *527*. doi: 10.1016/j.epsl.2019.115798

Muir, D. D., Blundy, J. D., Rust, A. C., & Hickey, J. (2014). Experimental constraints on dacite pre-eruptive magma storage conditions beneath Uturuncu Volcano. *Journal of Petrology*, *55*(4), 749–767. doi: 10.1093/petrology/egu005

Mungall, J. E. (2002). Roasting the mantle: Slab melting and the genesis of major Au and Au-rich Cu deposits. *Geology*, *30*(10), 915–918.

Myers, C. E., & Eugster, H. P. (1983). The system Fe-Si-O: Oxygen buffer calibrations to 1500K. *Contributions to Mineralogy and Petrology*, *82*, 75–90.

Mysen, B. O. (2006). Redox equilibria of iron and silicate melt structure: Implications for olivine/melt element partitioning. *Geochimica et Cosmochimica Acta*, *70*(12), 3121–3138. doi: 10.1016/j.gca.2006.03.014

Mysen, B. O., Kumamoto, K., Cody, G. D., & Fogel, M. L. (2011). Solubility and solution mechanisms of C–O–H volatiles in silicate melt with variable redox conditions and melt composition at upper mantle temperatures and pressures. *Geochimica et Cosmochimica Acta*, *75*, 6183–6199. doi: 10.1016/j.gca.2011.07.035

Nebel, O., Sossi, P. A., Bénard, A., Wille, M., Vroon, P. Z., & Arculus, R. J. (2015). Redox-variability and controls in subduction zones from an iron-isotope perspective. *Earth and Planetary Science Letters*, *432*, 142–151. doi: 10.1016/j.epsl.2015.09.036

Nell, J., & Wood, B. J. (1991). High-temperature electrical measurements and thermodynamic properties of Fe_3O_4-$FeCr_2O_4$-$MgCr_2O_4$-$FeAl_2O_4$ spinels. *American Mineralogist*, *76*(3–4), 405–426.

Neumann, E.-R. (1991). Ultramafic and mafic xenoliths from Hierro, Canary Islands: evidence for melt infiltration in the upper mantle. *Contributions to Mineralogy and Petrology*, *106*(2), 236–252.

Neumann, E. R., Wulff-Pedersen, E., Pearson, N. J., & Spencer, E. A. (2002). Mantle xenoliths from Tenerife (Canary Islands): evidence for reactions between mantle peridotites and silicic carbonatite melts inducing Ca metasomatism. *Journal of Petrology*, *43*(5), 825–857.

Neumann, E. R., Wulff-Pedersen, E., Johnsen, K., Andersen, T., & Krogh, E. (1995). Petrogenesis of spinel harzburgite and dunite suite xenoliths from Lanzarote, eastern Canary

Islands: implications for the upper mantle. *Lithos*, *35*(1–2), 83–107.

Newcombe, M., Brett, A., Beckett, J., Baker, M., Newman, S., Guan, Y., et al. (2017). Solubility of water in lunar basalt at low PH2O. *Geochimica et Cosmochimica Acta*, *200*, 330–352.

Nicklas, R. W., Puchtel, I. S., Ash, R. D., Piccoli, P. M., Hanski, E., Nisbet, E. G., et al. (2019). Secular mantle oxidation across the Archean-Proterozoic boundary: Evidence from V partitioning in komatiites and picrites. *Geochimica et Cosmochimica Acta*, *250*, 49–75. doi: 10.1016/j.gca.2019.01.037

Nicolich, R., Laigle, M., Hirn, A., Cernobori, L., & Gallart, J. (2000). Crustal structure of the Ionian margin of Sicily: Etna Volcano in the Fram of regional evolution. *Tectonophysics*, *329*, 121–139.

O'Neill, H. S., & Pownceby, M. I. (1993). Thermodynamic data from redox reactions at high temperatures. 1. An experimental and theoretical assessment of the electrochemical method using stabilized zirconia electrolytes, with revised values for the Fe- FeO, Co-CoO, Ni-NiO and Cu-Cu_2O oxygen buffers, and new data for the W-WO_2 buffer. *Contributions to Mineralogy and Petrology*, *114*(3), 296–314.

O'Neill, H. S. C., & Wall, V. (1987). The olivine–orthopyroxene–spinel oxygen geobarometer, the nickel precipitation curve, and the oxygen fugacity of the Earth's upper mantle. *Journal of Petrology*, *28*(6), 1169–1191.

O'Neill, H. S. C., Berry, A. J., & Mallmann, G. (2018). The oxidation state of iron in Mid-Ocean Ridge Basaltic (MORB) glasses: Implications for their petrogenesis and oxygen fugacities. *Earth and Planetary Science Letters*, *504*, 152–162. doi: 10.1016/j.epsl.2018.10.002

O'Neill, H. S. C. (1987). Quartz-fayalite-rion and quartz-fayalite-magnetite equilibria and the free energy of formation of fayalite (Fe_2SiO_4) and magnetite (Fe_3O_4). *American Mineralogist*, *72*, 67–75.

Osborn, E. F. (1959). Role of oxygen pressure in the crystallization and differentiation of basaltic magma. *American Journal of Science*, *257*(9), 609–647.

Parkinson, I. J., & Pearce, J. A. (1998). Peridotites from the Izu–Bonin–Mariana forearc (ODP Leg 125): evidence for mantle melting and melt–mantle interaction in a supra-subduction zone setting. *Journal of Petrology*, *39*(9), 1577–1618.

Parkinson, I. J., & Arculus, R. J. (1999). The redox state of subduction zones: insights from arc-peridotites. *Chemical Geology*, *160*(4), 409–423.

Parkinson, I. J., Arculus, R. J., & Eggins, S. M. (2003). Peridotite xenoliths from Grenada, Lesser Antilles Island Arc. *Contributions to Mineralogy and Petrology*, *146*(2), 241–262. doi: 10.1007/s00410-003-0500-z

Partzsch, G. M., Lattard, D., & McCammon, C. (2004). Mössbauer spectroscopic determination of Fe3+/Fe2+ in synthetic basaltic glass: a test of empirical fO_2 equations under superliquidus and subliquidus conditions. *Contributions to Mineralogy and Petrology*, *147*(5), 565–580. doi: 10.1007/s00410-004-0571-5

Pearce, J. A., Barker, P. F., Edwards, S. J., Parkinson, I. J., & Leat, P. T. (2000). Geochemistry and tectonic significance of peridotites from the South Sandwich arc-basin system, South Atlantic. *Contributions to Mineralogy and Petrology*, *139*, 36–53.

Plank, T., & Langmuir, C. H. (1988). An evaluation of the global variations in the major element chemistry of arc basalts. *Earth and Planetary Science Letters*, *90*(4), 349–370.

Plank, T., Kelley, K. A., Zimmer, M. M., Hauri, E. H., & Wallace, P. J. (2013). Why do mafic arc magmas contain 4 wt% water on average? *Earth and Planetary Science Letters*, *364*, 168–179. doi: 10.1016/j.epsl.2012.11.044

Portnyagin, M., Hoernle, K., Storm, S., Mironov, N., van den Bogaard, C., & Botcharnikov, R. (2012). H2O-rich melt inclusions in fayalitic olivine from Hekla volcano: Implications for phase relationships in silicic systems and driving forces of explosive volcanism on Iceland. *Earth and Planetary Science Letters*, *357–358*(0), 337–346. doi: 10.1016/j.epsl.2012.09.047

Righter, K., Danielson, L. R., Pando, K., Morris, R. V., Graff, T. G., Agresti, D. G., et al. (2013). Redox systematics of martian magmas with implications for magnetite stability. *American Mineralogist*, *98*(4), 616–628.

Rowe, M. C., Kent, A. J. R., & Nielsen, R. L. (2007). Determination of sulfur speciation and oxidation state of olivine hosted melt inclusions. *Chemical Geology*, *236*(3–4), 303–322.

Rutherford, M. J., & Devine, J. D. (1996). Preeruption pressure–temperature conditions and volatiles in the 1991 dacitic magma of Mount Pinatubo. In: Punongbayan, R., & Newhall, C. G. (eds.) *Fire and Mud: Eruptions and Lahars of Mount Pinatubo, Phillipines*. University of Washington Press. 751–766.

Ryabchikov, J. D., Ntaflos, T., Kurat, G., & Kogarko, L. N. (1995). Glass-bearing xenoliths from Cape Verde: evidence for a hot rising mantle jet. *Mineralogy and Petrology*, *55*(4), 217–237.

Sack, R. O., & Ghiorso, M. S. (1991a). Chromian spinels as petrogenetic indicators: Thermo dynamics and petrological applications. *American Mineralogist*, *76*, 827–847.

Sack, R. O., & Ghiorso, M. S. (1991b), An internally consistent model for the thermodynamic properties of Fe−Mg-titanomagnetite-aluminate spinels. *Contributions to Mineralogy and Petrology*, *106*(4), 474–505.

Saiga, A., Matsumoto, S., Uehira, K., Matsushima, T., & Shimizu, H. (2010). Velocity structure in the crust beneath the Kyushu area. *Earth Planets and Space*, *62*(5), 449–462. doi: 10.5047/eps.2010.02.003

Sen, G. (1987). Xenoliths associated with the Hawaiian Hot Spot. *Mantle Xenoliths*, 359–375.

Sen, G. (1988). Petrogenesis of spinel lherzolite and pyroxenite suite xenoliths from the Koolau shield, Oahu, Hawaii: implications for petrology of the post-eruptive lithosphere beneath Oahu. *Contributions to Mineralogy and Petrology*, *100*(1), 61–91.

Sen, G., & Presnall, D. C. (1986). Petrogenesis of dunite xenoliths from Koolau volcano, Oahu, Hawaii: implications for Hawaiian volcanism. *Journal of Petrology*, *27*(1), 197–217.

Sen, G., & Leeman, W. P. (1991). Iron-rich lherzolitic xenoliths from Oahu: origin and implications for Hawaiian magma sources. *Earth and Planetary Science Letters*, *102*(1), 45–57.

Sevilla, W. I., Ammon, C. J., Voight, B., & De Angelis, S. (2010). Crustal structure beneath the Montserrat region of the Lesser Antilles island arc. *Geochemistry Geophysics Geosystems, 11*, 13. doi: 10.1029/2010gc003048

Shervais, J. W. (1982). Ti-V plots and the petrogenesis of modern and ophiolitic lavas. *Earth and Planetary Science Letters, 59*, 101–118.

Shorttle, O., Moussallam, Y., Hartley, M. E., Maclennan, J., Edmonds, M., & Murton, B. J. (2015). Fe-XANES analyses of Reykjanes Ridge basalts: Implications for oceanic crust's role in the solid Earth oxygen cycle. *Earth and Planetary Science Letters, 427*, 272–285. doi: 10.1016/j.epsl.2015.07.017

Sleep, N. H. (1992). Hotspots and mantle plumes. *Annual Review of Earth Planetary Science, 20*, 19–43.

Sorbadere, F., Laurenz, V., Frost, D. J., Wenz, M., Rosenthal, A., McCammon, C., & Rivard, C. (2018). The behaviour of ferric iron during partial melting of peridotite. *Geochimica et Cosmochimica Acta, 239*, 235–254. doi: 10.1016/j.gca.2018.07.019

Spieker, K., Rondenay, S., Ramalho, R., Thomas, C., & Helffrich, G. (2018). Constraints on the structure of the crust and lithosphere beneath the Azores Islands from tele seismic receiver functions. *Geophysical Journal International, 213*, 824–835.

Stagno, V., Ojwang, D. O., McCammon, C. A., & Frost, D. J. (2013). The oxidation state of the mantle and the extraction of carbon from Earth's interior. *Nature, 493*(7430), 84–88. doi: 10.1038/nature11679

Stelten, M. E., & Cooper, K. M. (2012). Constraints on the nature of the subvolcanic reservoir at South Sister volcano, Oregon from U-series dating combined with sub-crystal trace-element analysis of plagioclase and zircon. *Earth and Planetary Science Letters, 313*, 1–11. doi: 10.1016/j.epsl.2011.10.035

Stolper, D. A., & Bucholz, C. E. (2019). Neoproterozoic to early Phanerozoic rise in island arc redox state due to deep ocean oxygenation and increased marine sulfate levels. *Proceedings of the National Academy of Sciences of the United States of America, 116*(18), 8746–8755. doi: 10.1073/pnas.1821847116

Stolper, E., & Newman, S. (1994). The role of water in the petrogenesis of Mariana Trough magmas. *Earth and Planetary Science Letters, 121*(3–4), 293–325.

Stracke, A., Hofmann, A. W., & Hart, S. R. (2005). FOZO, HIMU, and the rest of the mantle zoo. *Geochemistry, Geophysics, Geosystems, 6*, Q05007.

Syuhada, S., Hananto, N. D., Abdullah, C. I., Puspito, N. T., Anggono, T., & Yudistira, T. (2016). Crustal structure along Sunda-Banda arc transition zone from teleseismic receiver functions. *Acta Geophysica, 64*, 2020–2049.

Takahashi, N., Kodaira, S., Klemperer, S. L., Tatsumi, Y., Kaneda, Y., & Suyehiro, K. (2007). Crustal structure and evolution of the Mariana intra-oceanic island arc. *Geology, 35*(3), 203–206.

Tang, M., Erdman, M., Eldridge, G., & Lee, C. A. (2018). The redox "filter" beneath magmatic orogens and the formation of continental crust. *Science Advances, 4*.

Thornber, C. R., Roeder, P. L., & Foster, J. R. (1980). The effect of composition on the ferric-ferrous ratio in basaltic liquids at atmospheric pressure. *Geochimica et Cosmochimica Acta, 44* (3), 525–532.

Tollan, P., & Hermann, J. (2019). Arc magmas oxidised by water dissociation and hydrogen incorporation in orthopyroxene. *Nature Geoscience, 12*(8), 667–671. doi: 10.1038/s41561-019-0411-x

Toothill, J., Williams, C. A., Macdonald, R., Turner, S. P., Rogers, N. W., Hawkesworth, C. J., et al. (2007). A complex petrogenesis for an arc magmatic suite, St Kitts, Lesser Antilles. *Journal of Petrology, 48*(1), 3–42. doi: 10.1093/petrology/egl052

Tracy, R. J. (1980). Petrology and genetic significance of an ultramafic xenolith suite from Tahiti. *Earth and Planetary Science Letters, 48*(1), 80–96.

Tucker, J. M., Hauri, E. H., Marske, J. P., Garcia, M. O., Trusdell, M. A., & Pietruszka, A. J. (2019). A high carbon content of the Hawaiian mantle from olivine-hosted melt inclusions. *Geochimica et Cosmochimica Acta, 254*, 156–172.

Turner, S. J., & Langmuir, C. H. (2015). The global chemical systematics of arc front stratovolcanoes: Evaluating the role of crustal processes. *Earth and Planetary Science Letters, 422*, 182–193. doi: 10.1016/j.epsl.2015.03.056

Turner, S. J., & Langmuir, C. (2015). What processes control the chemical compositions of arc front stratovolcanoes? *Geochemistry, Geophysics, Geosystems, 16*, 1865–1893. doi: 10.1002/2014GC005633

Turner, S. J., Langmuir, C., Katz, R. F., Dungan, M. A., & Escrig, S. (2016). Parental arc magma compositions dominantly controlled by mantle-wedge thermal structure. *Nature Geoscience, 9*, 772–776. doi: 10.1038/ngeo2788

Veenstra, E., Christensen, D. H., Abers, G. A., & Ferris, A. (2006). Crustal thickness variation in south-central Alaska. *Geology, 34*(9), 781–784. doi: 10.1130/g22615.1

Wallace, P. J., & Carmichael, I. S. E. (1994). S-Speciation in submarine basaltic glasses as determined by measurements of Sk-Alpha x-ray wavelength shifts. *American Mineralogist, 79* (1–2), 161–167.

Wasilewski, B., Doucet, L. S., Moine, B., Beunon, H., Delpech, G., Mattielli, N., et al. (2017). Ultra-refractory mantle within oceanic plateau: Petrology of the spinel harzburgites from Lac Michèle, Kerguelen Archipelago. *Lithos, 272*, 336–349.

Waters, L. E., & Lange, R. A. (2013). Crystal-poor, multiply saturated rhyolites (obsidians) from the Cascade and Mexican arcs: evidence of degassing-induced crystallization of phenocrysts. *Contributions to Mineralogy and Petrology, 166*(3), 731–754. doi: 10.1007/s00410-013-0919-9

Waters, L. E., & Lange, R. A. (2015). An updated calibration of the plagioclase-liquid hygrometer-thermometer applicable to basalts through rhyolites. *American Mineralogist, 100*(10), 2172–2184. doi: 10.2138/am-2015-5232

Waters, L. E., & Lange, R. A. (2016). No effect of H2O degassing on the oxidation state of magmatic liquids. *Earth and Planetary Science Letters, 447*, 48–59. doi: 10.1016/j.epsl.2016.04.030

Waters, L. E., & Frey, H. M. (2018). Crystal-poor rhyolites and rhyodacites from Volcan Tepetiltic, Mexico: Evidence for melt formation, crystallization and eruption over short timescales. *Journal of Volcanology and Geothermal Research, 361*, 36–50. doi: 10.1016/j.jvolgeores.2018.08.003

Watts, A. B., & ten Brink, U. S. (1989). Crustal structure, flexure, and subsidence history of the Hawaiian Islands. *Journal of Geophysical Research Solid Earth, 94*, 10473–10500.

Wilke, M., Jugo, P. J., Klimm, K., Susini, J., Botcharnikov, R., Kohn, S. C., & Janousch, M. (2008). The origin of S4+ detected in silicate glasses by XANES. *American Mineralogist*, *93*(1), 235–240. doi: 10.2138/am.2008.2765

Williams, H., Peslier, A., McCammon, C., Halliday, A., Levasseur, S., Teutsch, N., & Burg, J. (2005). Systematic iron isotope variations in mantle rocks and minerals: The effects of partial melting and oxygen fugacity. *Earth and Planetary Science Letters*, *235*(1–2), 435–452. doi: 10.1016/j.epsl.2005.04.020

Williams, H. M., McCammon, C. A., Peslier, A. H., Halliday, A. N., Teutsch, N., Levasseur, S., & Burg, J.-P. (2004). Iron isotope fractionation and the oxygen fugacity of the mantle. *Science*, *304*(5677), 1656–1659.

Wolfe, E. W., Wise, W. S., & Dalrymple, G. B. (1997). The geology and petrology of Mauna Kea Volcano, Hawaii – A study of postshield volcanism. USGS Professional Paper. pp. 1–129.

Wood, B. J. (1990). An experimental test of the spinel peridotite oxygen barometer. *Journal of Geophysical Research-Solid Earth and Planets*, *95*(B10), 15845–15851.

Wood, B. J., & Virgo, D. (1989). Upper mantle oxidation state – ferric iron contents of lherzoline spinels by Fe-57 Mossbauer spectroscopy and resultant oxygen fugacities. *Geochimica et Cosmochimica Acta*, *53*(6), 1277–1291.

Wood, B. J., Bryndzia, L. T., & Johnson, K. E. (1990). Mantle oxidation-state and its relationship to tectonic environment and fluid speciation. *Science*, *248*(4953), 337–345.

Wulff-Pedersen, E., Neumann, E.-R., & Jensen, B. á. (1996). The upper mantle under La Palma, Canary Islands: formation of Si- K- Na-rich melt and its importance as a metasomatic agent. *Contributions to Mineralogy and Petrology*, *125*(2–3), 113–139.

Zhang, H. L., Cottrell, E., Solheid, P. A., Kelley, K. A., & Hirschmann, M. M. (2018). Determination of Fe3+/SFe of XANES basaltic glass standards by Mössbauer spectroscopy and its application to the oxidation state of iron in MORB. *Chemical Geology*, *479*(2018), 166–175.

Zimmer, M. M., Plank, T., Hauri, E. H., Yogodzinski, G. M., Stelling, P., Larsen, J., et al. (2010). The role of water in generating the calc-alkaline trend: New volatile data for Aleutian magmas and a new tholeiitic index. *Journal of Petrology*, *51* (12), 2411–2444. doi: 10.1093/petrology/egq062

4

Redox Variables and Mechanisms in Subduction Magmatism and Volcanism

Katy A. Evans[1] and Andy G. Tomkins[2]

ABSTRACT

The redox state of magma produced at arcs affects the nature of continental crust, likelihood of ore deposit formation, geochemical cycling of key elements for life, and environmental impacts via magma degassing. Arc magmas are more oxidized than magmas produced at mid-ocean ridges. Proposed explanations include a source region oxidized by inputs from the subducting slab, effects of slab-derived components such as water, and/or processes that occur subsequent to magma production, such as differentiation and degassing. To distinguish between these possibilities, the capacity of the slab to oxidize or modify sub-arc mantle and the potential of differentiation and degassing to produce the observed oxidation are reviewed. Uncertainties remain, particularly relating to the application of experimental results to natural systems, but there is a growing consensus that the effects of differentiation and degassing cannot produce the observed degree of oxidation. It is harder to determine the relative contribution of the different slab-derived components to the production of oxidized arc magmas, but slab-derived sulfate produced by oxidation of sulfide-hosted sulfur by reduction of ferric iron remains a strong contender, as does the effect of water on melting relationships. Further work is necessary to test and refine these relationships.

4.1. INTRODUCTION

The oxidation state of subduction zone magmas and their source regions is a first order control on: (i) metal solubility and ore deposit formation (e.g., Jego et al., 2012; Mironov et al., 2018; Mungall, 2002; Sillitoe, 2008); (ii) continental crust production (e.g., Ishihara, 2004; Sato et al., 2004); (iii) cycling of redox-sensitive elements (e.g., K. A. Evans, 2012; Kelley et al., 2009), many of which are essential for life (e.g., Hayes et al., 2006); (iv) the extent of environmental impacts from arc magma degassing (e.g., Burgisser et al., 2007; Candela, 1986; Fiege et al., 2017; Mathez, 1984); and (v) the evolution of Earth's atmosphere (e.g., Duncan et al., 2017; Galvez et al., 2016; Lee et al., 2013; Oyarzun et al., 2008).

In spite of its importance, the oxidation state of sub-arc mantle (SAM) is poorly known – some studies suggest that SAM redox potential is similar to that of the mantle source of MORB (e.g., Lee et al., 2005, 2010, 2012), whereas a wealth of other studies, from Brandon et al. (1996) to Rielli et al. (2017) suggest otherwise. The complexity increases when temporal variation is considered. It has been hypothesized that the dramatic step-wise increase in oxygen abundance in the atmosphere over

[1]School of Earth and Planetary Sciences, Curtin University, Perth, Western Australia, Australia

[2]School of Earth, Atmosphere and Environment, Monash University, Melbourne, Victoria, Australia

Magma Redox Geochemistry, Geophysical Monograph 266, First Edition.
Edited by Roberto Moretti and Daniel R. Neuville.

DOI: 10.1002/9781119473206.ch4

the last three billion years, driven by the evolving biosphere (Campbell et al., 2010; Lyons et al., 2014), ultimately promoted a progressive increase in oxidation of the mantle at subduction zones, and that this controlled the oxidation state of arc magmas and thus created variation in ore deposit formation over time (K. A. Evans et al., 2014; Tomkins et al., 2015). Although recent work supports this suggestion (Stolper et al., 2018), the question of an increase in subduction-driven mantle oxidation over time is yet to be investigated thoroughly. Furthermore, although previous workers have used geochemical evidence to demonstrate that modern arc magmas show a correlation between oxidation state and the abundance of fluid-mobile species (e.g., Brounce et al., 2014; Kelley et al., 2009; Laubier et al., 2014), few studies have examined the geochemistry of oxidation by studying samples of fluid- and/or melt-altered ("metasomatized") mantle. This information gap fuels the long-lived debate as to whether oxidized arc magmas inherit their elevated oxidation state and metals from the mantle, or acquire them during migration through the crust (e.g., Lee et al., 2010; Parkinson et al., 1999).

In this chapter, we review the approaches used to investigate the redox state of arc magmas, their source zones, sources of oxidizing capacity, and controls on oxidation state variation throughout the system. The sequence of events that leads to the production of oxidized arc magmas is then described, starting with a discussion of sources of oxidizing capacity, then moving on to the means for transfer of that oxidizing capacity. The evidence for and against an oxidized source zone for arc magmas is then summarized and discussed, followed by a synthesis of the current knowledge regarding the role of redox in the formation of primary magmas in the sub-arc, and modification of those magmas by processes such as differentiation and degassing. Finally, we use insights gained from this overview to draw conclusions on the overall cycling of redox-sensitive elements associated with arc magmatism, and to propose opportunities for further research.

4.2. REDOX VARIABLES

4.2.1. Intensive and Extensive Redox Variables

The most common variable used to quantify the oxidation state of geological materials is oxygen fugacity (fO_2). Oxygen fugacity quantifies the availability of oxygen in a system and is equivalent to the concentration of oxygen in a hypothetical ideal gas in equilibrium with that system (e.g., Nordstrom et al.,1994). If fO_2 is high in a system then comparatively reduced material added to that system is likely to become oxidized, and vice versa. The concentration of free oxygen in most geological environments in the deep Earth is vanishingly small. For example, in the Earth's mantle, fO_2 is generally less than 10^{-10} bars (D. J. Frost et al., 2008). Because of these small values of fO_2, this parameter is usually presented in log units. The reality is that fO_2 is buffered by the mineral phases, fluid, and/or melt in most geological systems (Canil et al., 1994), and the fO_2 of these buffers varies significantly with temperature, so fO_2 is generally presented in relation to one of the well-known buffers (Figure 4.1). In this work and elsewhere, the fO_2 is presented in log units relative to an established buffer. For example, an fO_2 of NNO+2 refers to fO_2 that is two log units higher than the nickel-nickel oxide buffer, and FMQ−1 refers to fO_2 one log unit less than the fayalite-magnetite-quartz buffer. In this work, we discuss results reported relative to FMQ and NNO. The difference between the two buffers is a function of pressure and temperature, which are not always reported in sufficient detail to allow conversion between FMQ and NNO. Thus, results are reported relative to FMQ where possible, and approximately relative to FMQ when reported as NNO, using the assumption that NNO ~ FMQ+1.

Oxygen fugacity is an intensive variable, which means that it is a state variable that takes value in any phase within the system considered (Munster, 1970). While fO_2 is invaluable as a measure of the oxidation potential of a material, it is not sensitive to the

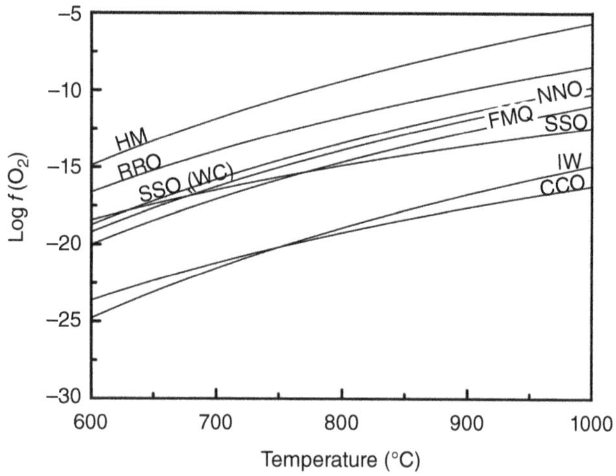

Figure 4.1 Positions of redox buffers at 1 bar. From Frost, (1991) unless otherwise noted. HM: hematite-magnetite; RRO: rhenium-rhenium oxide from Pownceby and O'Neill (1993); SSO: equal fractions of SO_4^{2-} and H_2S in melt according to Wallace and Carmichael (1994); NNO: nickel-nickel oxide; FMQ: fayalite-magnetite-quartz; SSO: equal fractions of H_2S and S_2 co-existing with water with an activity of 1, data from Evans & Tomkins (2011) and Holland and Powell (2012); IW: iron-wustite; CCO: graphite/CO/CO_2 equilibria from Eugster and French (1965).

magnitude of the capacity for oxidation of that material. For example, the fO_2 of a rock comprising 99% magnetite and 1% hematite is exactly the same as that of a rock comprising 1% magnetite and 99% hematite, but the buffering capacity of the two rocks is clearly very different. Oxygen fugacity is therefore of limited use when the magnitude, or capacity of that material for oxidation needs to be known, as it is when fluxes within open systems are considered. For such purposes, extensive variables, which are state variables that depend on the quantity of material, are required (e.g., Powell et al., 2005). An extensive variable devised to enable consideration of fluxes of a range of redox-sensitive elements is redox budget (*RB*) (K. A. Evans, 2006). The *RB* of a rock describes the numbers of moles of elements than can reduce or oxidize other elements, and is defined as the number of moles of electrons that need to be added to the rock to reach a given reference state:

$$RB = \sum_i n_i \xi_i \tag{4.1}$$

where n_i is the number of moles of redox state i present in the sample of interest, and ξ_i is the number of electrons required to take one mole of redox state i to the reference redox state. For example, the *RB* of two moles of FeO with respect to a reference state of Fe as Fe^{3+} and O as O^{2-} is –2 moles, because 2 moles of electrons would need to be removed to oxidize the Fe^{2+} to Fe^{3+}. It is also useful to be able to define *RB* for a unit mass or quantity, and for this, the term specific redox budget is used. Unless stated otherwise, the *RB*s referred to in this chapter are relative to a mantle-like reference state of Fe as Fe^{2+}, O as O^{-2}, S as S^{-2}, C as C^0, and H as H^+.

Redox budget is a useful variable for evaluation of open systems, where transfer of oxidizing capacity is of interest, because it allows quantification of fluxes, and the effects of those fluxes. An example of it's utility is provided by the work of Tumiati et al. (2015), who investigated the manganese ores of Praborna in the Italian Western Alps. These ores lie within a metasedimentary sequence incorporated into a subduction melange, and in contact, at their base, with sulfide- and magnetite-bearing basic rocks. The estimated fO_2 in the manganese ores is FMQ +12.7, much higher than that in the surrounding lithologies, was not associated with a particularly high *RB*, and is attributed to the unit following a different redox buffer during metamorphism to the surrounding lithologies. This case study provides an excellent example of a situation where localized bulk composition buffers fO_2 to very different values in neighboring lithologies with different bulk compositions, with minimal evidence of transfer of *RB* in response to, and across, the fO_2 gradient.

A further benefit of the use of *RB* is that it enables consideration of the combined effect of fluxes of redox-sensitive elements without the need to consider interchanges in valence between those elements explicitly. Some studies have used the redox budget concept and called the variable other names, such as $n(O_2)$ (e.g., Malaspina et al., 2017; Tumiati et al., 2015).

In closed systems, the *RB* is not affected by changes to the stoichiometry of the redox-budget bearing minerals. The stoichiometry of the mineral phases controls the distribution of the *RB* between phases, so, for example, changes to the thermodynamics of garnet at high pressures drive disproportionation of ferrous iron to form ferric-rich garnets and iron alloys (D. J. Frost & McCammon, 2008). However, the *RB* is unchanged so long as the system is closed. Consider three moles of FeO. The *RB* of these three moles, relative to a reference state of Fe as Fe^0 and O as O^{-2}, is 6 moles. If the three moles of FeO disproportionate to form a mole of Fe_2O_3 in garnet and a mole of Fe^0 in alloy, then the *RB* is still six moles. The tendency of *RB* to become mobile in open systems, on the other hand, is highly dependent on composition-dependent thermodynamic properties of the phases, and it is necessary to couple *RB* considerations with calculations using an intensive redox variable to quantify the consequences of open-system processes.

4.2.2. Uses of Redox-Sensitive Elements

Any redox-sensitive element can, in theory, transfer *RB* from one reservoir to another and thereby cause oxidation, or reduction, in the original and receiving reservoirs. There are a large number of redox-sensitive elements, and it would be time consuming to calculate *RB*s including all the elements, not least because the redox states of many of these elements, such as arsenic or antimony, are not well known or easy to measure. For this reason, *RB* calculations normally focus on the small number of elements that carry the bulk of the *RB* in any system (e.g., K. A. Evans, 2006, 2011, 2012; Gaillard et al., 2015; Malaspina et al., 2017; Tomkins et al., 2012). These are the elements that are present in significant concentrations, and that take different oxidation states readily in the geological processes considered. In the systems considered here, those elements are Fe, C, and S, although H is also important at sites of active serpentinization, where H_2 may carry significant *RB*. Manganese may also be significant for Mn-rich lithologies, such as metamorphosed Mn-rich sediments (Tumiati et al., 2015), although the global significance of such lithologies may be limited by their relatively small volume.

There are many other elements that can take multiple oxidation states, such as V, Mn, As, Cu, and Sb. Due to their typically low concentrations, these elements are not normally significant in consideration of the *RB*, but can be very useful in estimation of fO_2 as discussed in Section 4.2.3.

4.2.3. Measurement Techniques

Oxygen fugacity cannot be measured directly, because the concentration of free oxygen is minuscule compared to the overwhelmingly dominant O^{2-} abundance in silicate minerals. Also, even if the concentration of oxygen could be measured, changes in speciation during cooling or quenching of mantle-derived rocks and melts alter the free oxygen content. For this reason, there are a number of ways to estimate fO_2, some of which are described briefly here.

One of the most widely used techniques to determine fO_2 in mantle peridotites is through the equilibrium

$$\underset{\text{Olivine}}{6Fe_2SiO_4} + O_2 = \underset{\text{Opx}}{3Fe_2Si_2O_6} + \underset{\text{Spinel}}{2Fe_3O_4} \quad (4.2)$$

(e.g., Foley et al., 2006; Mattioli et al., 1988; Nell et al., 1991; Parkinson et al., 1999; Wood, 1990, 1991). Or, at greater depths:

$$\underset{\text{Garnet}}{2Fe_3Fe^{3+}{}_2Si_3O_{12}} = \underset{\text{Olivine}}{Fe_2SiO_4} + \underset{\text{Opx}}{4Fe_2Si_2O_6} + O_2 \quad (4.3)$$

(Gudmundsson et al., 1995). Application of this oxybarometer requires the ferric iron content of spinel or garnet to be known precisely. One way to measure ferric iron is via Mössbauer spectroscopy, but this technique is time consuming and the spatial resolution is large relative to the optimal micron scale spatial resolution. For this reason, other workers have used ferric iron contents calculated from EPMA (electron probe microanalysis) measurements, but this approach requires careful calibration because ferric iron is not determined directly (Ballhaus et al., 1991; Davis et al., 2017; Dyar et al., 1989; Hofer et al., 2006; Luhr et al., 1997; Wood et al., 1989). Without appropriate correction for other elements such as Cr in spinel, uncertainties in calculated fO_2 can be greater than ±1 log unit (Davis et al., 2017).

Estimation of fO_2 in glasses derived from the SAM provides a complementary approach to estimation of the fO_2 of a region from its peridotites. The Fe^{3+}/Fe_{tot} ratio of a glass can be measured directly using XANES (X-ray Absorption Near Edge Spectroscopy) if appropriate standards and calibrations are available (e.g., Berry et al., 2003, 2010, 2018; Kelley et al., 2009). The XANES technique provides an attractive method, in spite of the need for access to synchrotron radiation, because sample preparation is minimal, analysis time is relatively short, and the spatial resolution is much better than that of competing techniques such as whole rock dissolution and titration for ferrous iron, or Mössbauer spectroscopy, which are not discussed in detail here. However, fO_2 measurements in glasses can be problematic if there are quench effects, either as a result of the crystallization of phases with a different Fe^{3+}/Fe_{tot} to the melt, or to re-equilibration of redox pairs during cooling (Borisov, 2013).

The partitioning of iron isotopes is sensitive to redox conditions. The heavier Fe isotopes, ^{57}Fe and ^{56}Fe, partition preferentially into sites occupied by Fe^{3+} (Schoenberg et al., 2006). Iron isotopes therefore have the potential to provide a useful indication of source fO_2 (Dauphas et al., 2009). However, several processes can modify Fe^{3+}/Fe_{tot} in the source. These include stripping of Fe^{3+} by melting (Nebel et al., 2015), or sulfide saturation where ^{56}Fe partitions into the liquid or solid sulfide, producing an isotopically heavy silicate rock without any requirement for a change in bulk redox state (Williams et al., 2018). Such processes complicate the picture presented by iron isotope measurements to the extent that complementary techniques are necessary to utilize iron isotope data for this purpose.

Trace elements have been used extensively to estimate fO_2. For example, V/Sc ratios are thought to be sensitive to source redox conditions because V is sensitive to redox while Sc is not (Canil, 2002; Lee et al., 2005; Mallmann et al., 2009). More recent work has shown that Ce/Mo ratios are redox-sensitive in subducted slab lithologies. Molybdenum is hosted by sulfide and, to a lesser extent, rutile, whereas Ce is hosted by monazite in Ca-poor sediments and epidote in Ca-rich sediments, and carbonate when carbonate minerals are abundant (Skora et al., 2017). This incorporation of redox-sensitive elements means that sulfide and epidote stability are both dependent on redox state, so Ce/Mo ratios in arc lavas have the potential to record redox conditions and constrain the stable phases present in the subducting slab (Skora et al., 2017). Mineral partitioning approaches have also been applied; vanadium partitioning between melt and olivine has been measured in natural samples to show that lavas erupted from the Mutnovsky volcano equilibrated at FMQ+1.9–FMQ +2.3 (Shishkina et al., 2018).

Trace element approaches have recently been extended to ratios of platinum group elements (PGE) elements in arc magmas. Laser ablation–inductively coupled–mass spectroscopy (LA–ICP–MS) analysis of sulfides thought to have formed during oxidative mantle metasomatism has revealed unusual PGE patterns, interpreted as a record of transport of Os and Ru in oxidized fluids while Ir was immobile (Rielli & Tomkins, 2015). It is proposed that mantle metasomatism causes a distinctive fractionation of these elements that is not seen during normal partial melting of the mantle (cf. Lorand et al., 2008). Furthermore, this unusual signature is also found in porphyry Cu deposits, particularly those that are Au-rich. This result implies that PGE ratios can be used as an exploration tool for some types of ore deposits.

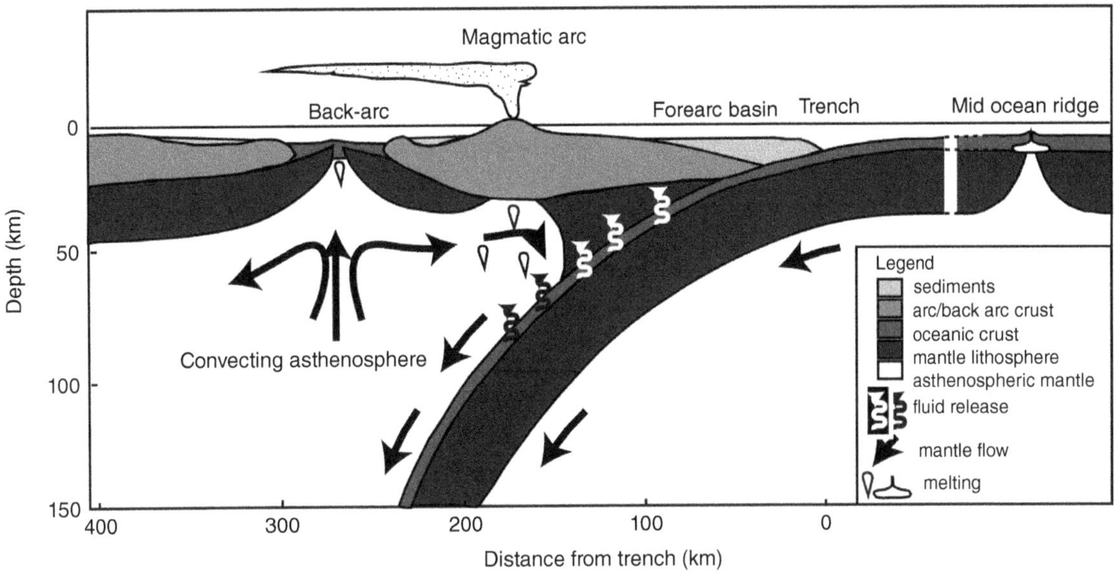

Figure 4.2 Conceptual model of subduction zone cycling. Redrawn and modified after Stern (2002).

4.3. MECHANISMS

The oxidation state of subduction-related magmas is affected by a suite of factors that include, but are not limited to, the oxidation state and hydration state of the subducting slab (Crossley et al., 2017; K. A. Evans et al., 2012; Tumiati et al., 2015), devolatilization of slab lithologies during subduction zone metamorphism (Debret, Andreani, et al., 2014; Debret, Koga, et al., 2014; Debret et al., 2017; Frezzotti et al., 2007, 2012; Malaspina et al., 2017; Pons et al., 2016), fluid/melt transfer within the subarc mantle (John et al., 2012; Spandler et al., 2011), conditions of primary magma formation, and modification of those magmas by differentiation and degassing prior to and during eruption (Figure 4.2). These processes are explored briefly here.

4.3.1. Processes Affecting the Slab Prior to Subduction

It has been suggested that redistribution of redox-sensitive elements within and from the subducting slab in response to subduction zone metamorphism affects the redox state and elemental composition of the overlying sub-arc mantle that is the source of arc magmas (e.g., K. A. Evans et al., 2011; Kelley et al., 2009). It is therefore of interest to consider the distribution of redox-sensitive elements in the oceanic lithosphere prior to subduction and the processes that control that distribution. Redox budget can be added to the subduction system by any redox-sensitive element in a redox state more oxidized that the average oxidation state of that element in typical asthenospheric mantle. For example, Fe^{3+} in hematite, C^{4+} in carbonate, S^{+6} in anhydrite, and S^{-1} in pyrite are all potential carriers of *RB* into subduction zones.

K. A. Evans (2012) describes the distribution of Fe, C, and S in the oceanic lithosphere in detail, including estimates of global *RB* fluxes into subduction zones. Redox budget fluxes, relative to a mantle reference state, are 3–9 × 10^{12} moles per year for sediments, 15–36 × 10^{12} moles per year for altered oceanic crust, and 2–46 × 10^{12} moles per year for serpentinized mantle lithosphere. The large ranges reflect the considerable uncertainties.

Fresh oceanic crust carries relatively little *RB*. Carbon and sulfur are present at very low concentrations in unaltered ocean crust, and the iron is mostly reduced. The proportion of ferric iron in MORB is ~0.1 +/- 0.02 (Berry et al., 2018; Le Losq et al., 2019). Hydrothermal circulation oxidizes iron and adds water, sulfate, and carbonate to the ocean crust and underlying lithospheric mantle (Alt, Garrido, et al., 2012; Alt et al., 2013; Schwarzenbach et al., 2012). The formation of magnetite in ultramafic rocks during serpentinization increases the number of moles of ferric iron in the rocks, and thus increases the *RB* (B. W. Evans, 2008; Honnorez, 2003; Lecuyer et al., 1999; Parkinson et al., 1999). Hydrothermal circulation primarily occurs at ridge-focused convection cells, and as the slab bends prior to subduction (Ranero et al., 2004; Stern, 2002).

The process of hydrothermal alteration of the oceanic crust and lithospheric mantle is complicated, with the sequence of reactions controlled by a range of interdependent parameters that include instantaneous and time-integrated water:rock ratio, the geometry of fluid flow, and porosity–permeability relationships (Bach et al., 2004; Frost et al., 2013; Klein & Bach, 2009; Klein et al., 2009; Schwarzenbach et al., 2009; Schwarzenbach et al., 2013; Schwarzenbach et al., 2012). For example, recent work by Klein et al. (2017) has provided a detailed record of the hydrothermal and subsequent low temperature alteration undergone by seawater-serpentinized peridotites dredged from the North Wall of the Puerto Rico Trench. The rocks there record localized serpentinization at antigorite-stable conditions, followed by pervasive, lower temperature serpentinization to produce lizardite. Subsequently, alteration to talc was accompanied by Fe-loss, and low temperature seawater alteration removed brucite and oxidized magnetite to hematite and goethite. These oxidized peridotites would have the potential to add *RB* to the subduction system in addition to what would be predicted from the mineralogy of the initial serpentinization alone.

In considering the alteration of oceanic crust, it is worth paying particular attention to sulfate, SO_4^{2-}, currently the second most abundant anion in seawater. Sulfate has potential to provide an important control on global redox cycling through subduction zones because it modifies the oxidation state of the oceanic crust prior to subduction in a way that no other element does. At mid-ocean ridges and at the slab bend at subduction zones, sulfate dissolved in seawater infiltrates oceanic crust (Alt et al., 2003; Alt et al., 2007). This sulfate addition drives oxidative metasomatism of the oceanic crust (Alt et al., 2003; Alt et al., 2007), which is controlled by the way that sulfur, carbon, and iron interact with one another and with H_2O. For example, when SO_4^{2-} in seawater reacts with the oceanic crust, a common redox reaction involves reduction of S^{6+} to S^{-1} in pyrite (FeS_2), and oxidation of Fe^{2+} from silicates to Fe^{3+} in magnetite and hematite (Fe_2O_3; cf. Barker et al., 2010), resulting in an increase in *RB*. Sulfate in seawater can also react with Ca from oceanic crust silicates to form $CaSO_4$ (anhydrite; Alt et al., 2003), which is a mineral that can be dissolved to liberate oxidized sulfur during later subduction, and so represents a way to store *RB*. Similarly, carbonate minerals are typically formed during alteration of the oceanic crust, and these can be dissolved during subduction to liberate CO_2 (Kelemen et al., 2015).

The deposition of carbonate-, pyrite-, and ferric-iron-bearing sediments also has the potential to add *RB* to the subducting ocean crust, but relatively few studies have quantified this potential, so this topic provides scope for further research.

A Note of Caution. Many studies have used evidence from ophiolitic peridotites to draw conclusions on subduction zone cycling of redox-sensitive elements under the assumption that these peridotites are representative of subducted mantle lithosphere (Debret, Andreani, et al., 2014; Debret et al., 2015; Pons et al., 2016). This approach is attractive, given the accessibility of ophiolite peridotites. However, geodynamic settings likely to produce ophiolites, such as slow and ultra-slow spreading ridges, transform faults, oceanic core complexes, and exhumed mantle at rifted continental margins are likely to be associated with extensive oceanic crust alteration and mantle serpentinization at high water:rock ratios (e.g., Beltrando et al., 2014; Lagabrielle et al., 2015). Crust and mantle lithosphere from these settings are therefore likely to carry a higher *RB* into subduction zones than rocks from more fluid-restricted settings. Ophiolites may also preserve back-arc basin crust, which is generated in a more oxidized suprasubduction zone setting than most subducted ocean crust (e.g., Kelley et al., 2009). In addition, ophiolite complexes only become abundant in the geological record after ~1000 Ma, implying that they do not necessarily provide a good representation of earlier plate tectonic processes (cf. Stern, 2005).

Indeed, a comparison of a suite of abyssal peridotites from the Mid Atlantic Ridge with ophiolite peridotites from New Caledonia thought to have been serpentinized primarily in the fore-arc or SAM has shown that abundances, mineral hosts, and oxidation states of the redox-sensitive elements, Fe, S, and C are different in the two sample suites, suggesting that caution is justified if ophiolitic samples are to be used as a proxy for subducted mantle lithosphere (Mothersole et al., 2017). Similarly, K. A. Evans et al. (2017) compared the suite of New Caledonia serpentinites studied by Mothersole et al. to serpentinites from the Zermatt Saas Zone ophiolite in the Western Alps, which is thought to include mantle from a subducted magma-poor rifted margin (Beltrando et al., 2014; Brovarone et al., 2014; Lagabrielle et al., 2015). Differences in the suites of stable alloy, oxide, and sulfide phases within the two suites of samples were attributed to differences in fluid:rock ratios and infiltrating fluid compositions. Trace elements also reflect differences in geodynamic setting (Deschamps et al., 2013; Spandler et al., 2013), and there is potential to use the redox-sensitive trace elements to determine the redox structure of the different settings. For example, Peters et al. (2017) demonstrated that alkali element/U ratios from mid-ocean ridge and fore-arc serpentinites are different, with much of this difference attributed to redox-driven differences in U mobility between the two environments.

4.3.2. Devolatilization of the Slab

The most commonly advanced explanation for the genesis of oxidized SAM is the addition of *RB* from the devolatilizing slab (e.g., K. A. Evans et al., 2011; Gaillard et al., 2015; Kelley et al., 2009; Lecuyer et al., 1999; Tomkins et al., 2015). It is proposed that the *RB* is carried by aqueous fluids, which can be quite silica-rich (Cruz et al., 2015), produced by hydrous minerals that break down during subduction. The oceanic crust is thought to be the most important fluid contributor to the sub-arc mantle during subduction (Kelemen et al., 2015), with the lithospheric mantle and sediment veneer playing contributory roles. In some settings and at deeper levels associated with migration of serpentinite-derived fluids through the hotter slab top, supercritical liquids may play an important role in transferring slab components into the mantle wedge (cf. Hermann et al., 2006; Spandler et al., 2013; Tomkins et al., 2015).

Calculations of the *RB* entering and leaving subduction zones indicate unequivocally that at least some of the subducted *RB* is added either to the sub-arc mantle or to the deep Earth (K. A. Evans, 2012). However, evidence to distinguish unambiguously between predominant oxidation of the sub-arc mantle or deep Earth oxidation, and the vector for *RB* transfer, is difficult to come by, given the inaccessible and ephemeral nature of slab-derived fluids. The extent to which supercritical liquids transport *RB* into the mantle wedge is not well known, but solute contents in these liquids can be higher than those of aqueous liquids, so they might be very important.

Water cannot, by itself, transfer *RB* effectively, because the concentration of dissolved oxygen is so low (Ballhaus et al., 1991). Solutes within slab-derived fluids are therefore required to transport *RB*, with sulfur, in its S^{6+} form as sulfate, deemed the most likely vector for *RB* transfer (Alt et al., 2013; Debret et al., 2013; Padron-Navarta et al., 2011; Pons et al., 2016). Oxidized sulfur is likely to be reduced to S^{-2} in the sub-arc mantle, thereby providing a recipient for 8 moles of electrons for each mole of sulfur reduced. Carbon, transported as C^{4+} in CO_2, also has the potential to transport *RB* that can oxidize the sub-arc mantle if it is reduced to elemental carbon C^0 or C^{-4} in methane when it interacts with the sub-arc mantle. Carbonate minerals are stable at sub-arc mantle conditions (Malaspina et al., 2012; Tumiati et al., 2013), but elemental carbon has been found in peridotites, either as diamond (Naemura et al., 2009; van Roermund, 2009) or graphite (Ferraris et al., 2004), and it has been suggested that redox-freezing of carbonate melts forms diamonds in the transition zone (Burnham et al., 2015). Indeed, the question of a role for carbon in redox-budget transfer, which was thought to be minimal (Connolly, 1995; Dasgupta et al., 2004; Malaspina et al., 2009; Poli et al., 2009; Yaxley et al., 2004) has been reopened (e.g., Blundy et al., 2010; Carter et al., 2018; Kelemen et al., 2015).

Mantle Lithosphere. The search for slab lithologies with the potential to release oxidizing fluid or melt components during subduction-induced devolatilization has focused largely on serpentinized peridotites, because serpentinization preceding subduction involves the addition of oxidized carbon, S^{-1} sulfur in pyrite, S^{6+} in serpentine (Debret et al., 2016), and the production of Fe^{3+} in serpentinite minerals and magnetite (B. W. Evans, 2008, 2010; B. R. Frost et al., 2007; Mothersole et al., 2017).

Debret and coworkers have documented reduction in bulk rock Fe^{3+}/Fe_{tot}, and a negative correlation between $\delta^{57}Fe$ and the bulk rock Fe^{3+}, and an associated decrease in magnetite mode with progression of high pressure–low temperature metamorphism for a suite of samples from four Alpine locations (Debret, Andreani, et al., 2014; Debret et al., 2015; Debret et al., 2016). These trends were also taken to indicate the release of sulfate-rich and/or hypersaline fluids, because the sulfate and chlorine in these fluids preferentially complex with the lighter iron isotope (Debret et al., 2016). A similar result was obtained from related work on zinc isotopes in the same sample suite (Pons et al., 2016). In these data, a progressive decrease in $\delta^{66}Zn$ with increasing metamorphic grade was shown to be correlated with a decrease in sulfur content. These observations, in conjunction with the tendency of zinc sulfate complexes to preferentially sequester ^{66}Zn, were used to suggest the release of oxidized, sulfate-rich serpentinite-derived fluids from the slab. These observations are consistent with a link between the observed reduction of ferric iron in serpentinites with progressive subduction metamorphism and transfer of electrons from originally sulfide-hosted sulfur, as suggested by Padron-Navarta et al. (2011) and Debret, Andreani, et al. (2014).

Given the widespread locations and differing geodynamic settings of serpentinization, it is difficult to be completely confident that the protoliths of the four sample localities were sufficiently similar for the above trends to be robust (cf. K. A. Evans et al., 2017). Notably, the results of a detailed study of a locality that spans the transition from antigorite-bearing serpentinite to chlorite-bearing harzburgite in a single outcrop located within the Cerro del Almirez Massif, Spain, are consistent with the other Debret et al. results (Debret et al., 2015). In the Cerro del Almirez example, bulk rock Fe^{3+}/Fe_{tot} decreases from 0.55 to 0.34–0.41 across the antigorite-out reaction, a change that can be accounted for largely by a decrease in the magnetite mode, although changes in the ferric iron content of antigorite are also likely to be significant. Hematite–ilmenite intergrowths in the higher grade rocks are also documented, and used to suggest that *RB* released from the rock could have been transported to oxidize the sub-arc mantle. However, more recent work by Bretscher et al. (2018) concludes that the same antigorite dehydration front at Cerro del Almirez represents a pre-existing oxidation front established upon serpentinization at the ocean floor, which emphasizes the necessity of careful characterization of the protolith. A study by Alt et al. (2013) documented an estimated decrease in the concentration of sulfur as sulfate of 260 ppm at the antigorite dehydration front; this decrease in sulfur concentration could be explained by the model of Bretscher et al. (2018), if the decrease in sulfur concentration was a pre-metamorphic feature related to seafloor serpentinization, or the Debret et al. model, if metamorphic iron oxidation was associated with sulfate loss.

It is interesting to compare the above results to those of Alt, Shanks, et al. (2012) who examined the sulfur and carbon contents of almost unmetamorphosed Apennine ophiolites and high pressure, low temperature antigorite serpentinites from the Voltri massif in the Ligurian Alps. The results show that the carbon or sulfur contents did not change significantly with progressive serpentinization. This observation supports the model of Bretscher et al. (2018) for Cerro del Almirez, although further work is required to constrain the systematic behavior of the redox-sensitive elements in this setting.

Ferric iron has not been much considered as a vector for transport of *RB* in subduction zones because it is sparingly soluble in most aqueous fluids (Simon et al., 2004; Wykes et al., 2008). However, subduction zone fluids may become quite silica-rich at high pressures, or might promote formation of supercritical liquids in deeper, hotter settings (Cruz et al., 2015; Manning, 1994), and ferric iron might be mobile to some extent in such fluids. Some evidence for the mobility of iron in the subducting slab is provided by the work of Crossley et al. (2017). They observed an increase in magnetite modes, and therefore in the ferric iron content, in high pressure ultramafic rocks from the Zermatt-Saas zone. The increase is associated texturally with the earliest phases of retrogression, and is consistent with the presence of an oxidizing fluid, potentially devolatilized from deeper lithologies.

Efforts have also been made to investigate the release of redox-sensitive elements in subduction zones using thermodynamic models, but these efforts are hindered by a lack of key data for species of interest. A study of the evolution of the *RB* of subducting serpentinized mantle lithosphere via thermodynamic modelling using THERMOCALC predicted the observed mineral assemblages but did not predict the decrease in the *RB* of subducting serpentinized mantle lithosphere that is necessary for *RB* to be released as oxidized species in aqueous fluids (K. A. Evans et al., 2015). This model replicated the observed mineral assemblages, but limitations are placed on the model results by a lack of data for some species. More recently, Debret et al. (2017) used theoretical chemical mass transfer calculations to calculate the redox state of fluids generated by the breakdown of antigorite to olivine, enstatite, and chlorite. The models suggested that reduction of ferric iron in antigorite to reduced iron in the product silicate phases generates a fluid with oxygen fugacities close to the hematite–magnetite buffer. These conditions are sufficiently oxidizing to stabilize sulfate (Figure 4.1). If such fluids infiltrated the SAM, then the *RB* transferred by sulfate in the fluids could produce an oxidized SAM.

The results of experimental work by Merkulova et al. (2017) are also consistent with the release of *RB* from subducted serpentinite during subduction zone metamorphism. The products of piston cylinder experiments on synthetic serpentinites record, with increasing pressure and temperature, conversion of ferric iron in magnetite and antigorite to ferrous iron in silicate phases, and conversion of S^{-1} in pyrite to S^{-2} in pyrrhotite. These authors suggested that these changes, which are associated with a decrease in the fO_2 in the serpentinite of several orders of magnitude, are complemented by the release of oxygen in fluids, and oxidation of neighboring lithologies. Merkulova et al. (2017) did not commit to the probability of sulfate release, because the aqueous sulfur species stable at their experimental conditions is sulfide or S_3^- rather than sulfate (Pokrovski et al., 2011; Pokrovski et al., 2009), and because water release was not simultaneous with the *RB* changes. The former point can be reconciled with evidence from rocks by the observation that the experiments were externally buffered by the experimental assembly, and that this may have produced conditions more reducing than those occurring in subduction zones. The second point, however, poses a significant obstacle to the transfer of *RB*, regardless of species, in aqueous fluids.

To summarize, there is a growing body of evidence that subducted mantle lithosphere loses *RB* during devolatilization and subduction zone metamorphism. However, the timing and magnitude of this loss, the vector for its transfer, and the ultimate destination of the fluids remain poorly understood.

Oceanic Crust. In early work on the cycling of redox-sensitive elements through subduction zones, ferric iron in magnetite hosted by hydrothermally altered oceanic crust was suggested to provide the prinicipal mechanism for addition of *RB* to subduction zones and ultimately to the deep Earth (Lecuyer et al., 1999). Since then, magnetite in serpentinized mantle lithosphere and the contribution of sulfate have been proposed to also play significant roles, but the potential for *RB* transfer from the oceanic crust remains both significant and poorly known.

K.A. Evans and Tomkins (2011) constructed a *RB* model of subducted oceanic crust, based on generalized estimates of the flux of redox-sensitive elements into subduction zones through geological time. They suggested that significant *RB* would have been released from dehydrating oceanic crust in the Phanerozoic, but not during the Archean or Proterozoic, based on a paucity of sulfate in the oceans that formed the older rocks. On this basis, they suggested that the concentration of porphyry Cu deposits (and related deposit types) in Phanerozoic rocks might be partly explained by enhanced oxidation of the sub-arc mantle, and thereby arc magmas.

More recently, 2 Ga evaporite deposits have been documented by Blättler et al. (2018) and it was proposed that the sulfate concentration of the oceans was of a similar magnitude to that of the present day. This contrasts with modeled sulfate concentrations several orders of magnitude less than those of the present day (e.g., Fakhraee et al., 2019). These lines of evidence could be reconciled if the evaporites described by Blättler et al. (2018) record higher sulfate concentrations in coastal settings while sulfate concentrations in the deeper ocean, and those relevant to hydrothermal alteration of the ocean crust, were sulfate poor.

Tomkins and Evans (2015) examined the stability of anhydrite in progressively metamorphosed oceanic crust by combining the results of experiments on the high pressure solubility of anhydrite from Newton and Manning (2005) with Perple_X modelling of fluid release across the blueschist–eclogite transition. They also examined pyrite stability by investigating sulfide assemblages in eclogites from the Zermat-Saas ophiolite belt, New Caledonia, and the Sulu Belt in China. They suggested that anhydrite dissolves at shallow depths in fluids released by dehydrating blueschists, and that these fluids can carry sulfate into the shallow sub-arc mantle. At greater depth, the dominant sulfur species is H_2S sourced from pyrite. They concluded that the release of progressively less oxidized fluid with increasing depth, followed by the release of supercritical silicate liquids from subducted slabs, might explain the observed relationship between ore deposits and the oxidation state of the associated magmas in the overlying arc. Ore deposits associated with oxidized magmas are commonly located close to the trench, whereas those associated with more reduced magmas are more distal to the trench.

A small number of studies of high pressure–low temperature metabasites have noted evidence of *RB* release from the subducting slab. Li et al. (2016) studied mafic and sedimentary rocks from the Akeyazi metamorphic terrane in the Southern Tianshan Orogen, China, which were subducted and metamorphosed at blueschist and eclogite conditions. These rocks contained complex assemblages of Cu-, Fe-, and Ti-bearing sulfides and oxides, and allowed conclusions to be drawn on the evolution of redox potential and *RB* during subduction and exhumation. The replacement of magnetite by pyrite during prograde to peak conditions was argued to record a decrease in fO_2 in conjunction with an increase in $f(S_2)$, and it was concluded that the fO_2 was close to FMQ. However, the *RB*, inferred from the author's use of an oxygen mass balance, did not change significantly from blueschist to eclogite facies conditions. This is consistent with internal buffering of fO_2 without any release of *RB*. In contrast, retrogression involved replacement of pyrite by magnetite, bornite by chalcocite and rutile by ilmenite, plus growth of epidote, which is rich in ferric iron. These assemblage changes are consistent with an increase in fO_2 to above the FMQ buffer (Ballhaus et al., 1991). The calculated *RB* is also higher in the retrogressed samples. The fluid responsible for this oxidation, and transfer of *RB*, was proposed to derive from dehydration of serpentinites in the subducting slab (Li et al., 2016).

Groppo et al. (2010) also reported a decrease in *RB* during subduction zone metamorphism of lawsonite eclogites from the Monviso meta-ophiolite in the Western Alps. The transformation from a lawsonite–eclogite to a talc-bearing eclogite was proposed to result in a decrease in the Fe^{3+}/Fe_{tot} of the bulk rock. However, the bulk rock Fe^{3+}/Fe_{tot} was not reported.

These results are mostly consistent with those of K. A. Evans et al. (2014), who measured sulfur isotopes in eclogite facies metabasalts from the Zermatt-Saas Zone in the Western Alps and the Pouebo terrane, New Caledonia. Values of $\delta^{34}S$ greater than zero record the involvement of seawater-derived sulfur, most likely incorporated during hydrothermal alteration of basalt on the sea floor prior to subduction. Prograde subduction metamorphism is thought to have involved some sulfur loss, but sulfide-growth occurred during the earliest stages of exhumation, recording the presence of sulfur-bearing fluids within the slab at depth. Heavy sulfur in these late sulfides was used to infer the presence of sulfate-bearing fluids within the slab at close to peak pressure, consistent with reports of the release of sulfate-bearing fluids from ultramafic lithologies (e.g., Pons et al., 2016)

Liu et al. (2016) presented a detailed study of an eclogitic sample from the Songduo eclogite, Tibet. Anhydrite and ferric-iron rich garnet and epidote, present in the earlier stages of the rock's metamorphic evolution, are thought to have transformed to an assemblage with pyrite, ferric-poor garnet, and epidote via release of an oxidizing fluid. However, the *RB* or proxies for the *RB* were not reported, so it is not clear if the ferric iron was simply transferred between phases. Further, the source of the sample was a tectonic slice of metabasalt surrounded by metasedimentary rocks, which calls the general applicability of the result into question, because the fluid might have been derived from the metasedimentary rocks.

The oxidation state of the slab during metamorphic fluid release is probably buffered by ferric–ferrous iron equilibria, because Fe is the most abundant redox modifier in this location. Therefore, Fe-bearing minerals are likely to control S speciation in devolatilized fluids. It has been suggested that the polysulfide ion, S_3^-, is likely to be stable in the fluid phase at moderate fO_2 conditions (between H_2S and HSO_4^- stability) in the sub-arc mantle (Pokrovski et al., 2011, 2015; Rielli et al., 2017). If so, investigations of metal speciation in slab-derived fluids should take the S_3^- complex into account. It has been suggested that the observed enrichment in soft metals, such as Au, Cu, and Mo, in arc magmas is caused by complexation with the polysulfide ion (Pokrovski et al., 2015). Involvement of the polysulfide ion is also consistent with Ba enrichment in arc magmas (Brounce et al., 2014; Brounce et al., 2015) and Ba addition to metasomatized mantle rocks, because Ba mobility is inhibited by the formation of barite in the presence of aqueous sulfate (Rielli et al., 2017). However, it should be noted that experiments on the stability of barite have only been performed at low pressures. Importantly, S_3^- might act as an oxidizing agent in the sub-arc mantle, where it might drive Fe oxidation through reduction of the S_3^- to H_2S and/or S^{2-} in sulfide minerals (Rielli et al., 2017).

Experimental work has contributed insights into sulfur and *RB* transfer from subducted oceanic crust. Jego et al. (2013) and Jego et al. (2014) describe the melting of a H_2O-saturated MORB with 1 wt.% sulfur at conditions equivalent to those at 60 to 100 km depth in subduction zones. Reduced partial melts of the experimental charges did not carry much sulfur, with sulfur contents of only a few hundreds of ppm. However, although the fluid phase in the experiments could not be analyzed directly, the authors used mass balance calculations to suggest that the aqueous phase could carry as much as 15 wt.% sulfur. If this is the case, and if aqueous fluids occur at these depths, then slab-derived fluids could transport sulfur from the slab to the SAM effectively at all fO_2, and an oxidized slab would not be a prerequisite for sulfur transfer. However, in natural systems, the subducting slab is likely to have dehydrated prior to the production of melt, so the existence of an aqueous fluid under these conditions requires another source, possibly colder lithologies elsewhere in the slab, or underlying serpentinites.

To summarize, there is some evidence that prograde subduction zone metamorphism of basaltic and gabbroic ocean crust releases *RB*, and that this may be carried in aqueous fluids by either the sulfate ion, S_3^-, or ferric iron in silica-rich aqueous fluids. Further work is necessary to constrain the magnitude of the effects of such fluxes, and the potential sources and sinks for aqueous fluids.

Sediments. Carbon as elemental carbon in organic material, or as C^{4+} in carbonate is common in subducted sediments (Kelemen et al., 2015; Plank et al., 1998; Rea et al., 1996). Oxidized and reduced iron and sulfur are also present (K. A. Evans, 2012).

Many workers have investigated the stability of carbonate in subduction zones (e.g., Connolly, 1995; Dasgupta et al., 2004; Malaspina et al., 2009; Poli et al., 2009; Yaxley et al., 2004) with the majority concluding, on the basis of observations of natural samples, thermodynamic modeling, and experiments, that the bulk of carbon is retained in the subducting slab as carbonate to depths beyond the sub-arc (Dasgupta et al., 2004, 2010; Galvez et al., 2015; Gorman et al., 2006; Kerrick et al., 1998; Molina et al., 2000; Poli et al., 2009; Thomsen et al., 2008; Tsuno et al., 2011; Yaxley et al., 1994, 2004). The experimental results are supported by carbonate inclusions in transition zone/lower mantle diamonds, which indicate that carbonate remains stable to beyond sub-arc depths (Bulanova et al., 2010; Burnham et al., 2016). More recently, it has been suggested that CO_2 is cycled through the sub-arc mantle at subduction zones (Bebout, 1995; Blundy et al., 2010; Carter et al., 2018; Kelemen et al., 2015), and this may be possible if carbonates are dissolved as carbonate ions rather than as CO_2 (Caciagli et al., 2003; Newton et al., 2010), if carbon is transported as reduced species such as CH_4 (Kelemen et al., 2015), or if carbonic melts are more important than previously thought (e.g., Poli, 2015, 2016).

There are reports of release of reduced species from the slab in sediment-derived fluids. For example, results from a study of diamonds from Dachine, French Guyana, are consistent with precipitation of diamonds from remobilized sedimentary carbon at depths of 200–250 km from fluids derived from slab serpentinites at the slab–mantle interface (Smith et al., 2016). The precipitation of diamond implies redox conditions more reducing than the CCO (graphite/CO_2) buffer at the site of precipitation, but the redox state of carbon in the transporting fluids is not discussed explicitly in their work. Diamonds are also noted by Frezzotti et al. (2011) in Mn-rich nodules in the UHP Lago di Cignana rocks in the Western Alps. Fluid inclusions in these diamonds contained bicarbonate, carbonate, and sulfate ions, as well as silica monomers. There is also evidence that reduction of C^{+4} in carbonate to C^0 in graphite or diamond occurs within subduction zones, albeit potentially only at lithological boundaries (Galvez, Beyssac, et al., 2013). Carbonate reduction provides another sink for electrons, so carbonate reduction could cause the release of *RB*, in the form of oxidized species, into a devolatilizing fluid. There is also evidence that oxidized carbon is dissolved in aqueous fluids and silica-rich aqueous fluids in subduction zones (Ferrando et al., 2017).

With respect to sulfur, Canil et al. (2017) used experiments and thermodynamic models based on the known composition of sediments in ocean drill cores to investigate the oxidation state of sulfur liberated from subducted sediment into the sub-arc mantle. The results suggest that calcic sediments subducted along colder geotherms tend to produce the most sulfate-rich melts. In addition, it is suggested that dissolution of carbonate in fluids released from devolatilizing sediments during subduction would promote addition of CO_2 to the overlying sub-arc mantle, and this also should vary as a function of sediment composition.

Sediment melts also contribute to the characteristics of arc rocks. Arc-related granitoids can be divided into the magnetite- and ilmenite-series, with the stable oxide indicating, approximately, redox condition during crystallization. Formation of ilmenite-series rocks is usually attributed to the involvement of graphitic sediments assimilated by granitic plutons at some location in the upper crust, with or without an extensional setting (Ishihara, 2004; Sato et al., 2004).

Overall, there is far less work on the consequences of subduction for the redox-sensitive element distribution in sediments than there is for the other lithologies, with most of the existing work focused on carbon. This knowledge gap presents an opportunity for further work.

4.3.3. Fluid/Melt Transfer and Mantle Metasomatism – Is the Sub-Arc Mantle More Oxidized than Mantle Elsewhere?

Studies of Fe^{3+}/Fe_{tot} in peridotite xenoliths and primitive arc glasses provide a comprehensive body of evidence suggesting that the sub-arc mantle is more oxidized than mantle elsewhere, and that the cause of elevated fO_2 is slab-derived fluids (Brounce et al., 2014; Brounce et al., 2015; K. A. Evans et al., 2012; Kelley et al., 2009; Parkinson et al., 2003). However, evidence from trace elements and iron isotopes suggest otherwise (Dauphas et al., 2009; Lee et al., 2003, 2005, 2010, 2012; Mallmann et al., 2009). The wide range of methods and massive effort expended emphasize both the significance and intractability of the question. There is still no consensus as to whether arc magmas inherit their high oxidization state from the mantle, or whether it is produced by differentiation and degassing during magma migration.

Any redox-sensitive element is capable of transferring *RB* to the sub-arc mantle, but the most likely candidates are iron, carbon, and sulfur (K. A. Evans, 2012). It has been suggested that O_2 produced by the dissociation of water released from the slab oxidizes the sub-arc mantle directly (Brandon et al., 1996). However, the solubility of oxygen is so low that this process could not produce the *RB* required to oxidize sub-arc mantle to the proposed extent. Ferric iron is sparingly soluble in aqueous fluids (Simon et al., 2004; Wykes et al., 2008), and is therefore unlikely to provide a direct agent for *RB* transfer, although iron reduction in the subducting slab, as documented by Debret and coworkers, among others, might produce *RB* that is transferred by sulfur or carbon.

Yes. The inaccessibility of the sub-arc mantle means that it is difficult to show conclusively that any measurement purported to relate to the sub-arc mantle redox state has not been overprinted by subsequent processes. However, a number of studies have presented evidence that the sub-arc mantle is variably oxidized relative to mantle elsewhere. Oxygen fugacity estimated from the ferric iron content of spinel in spinel peridotites from a variety of island arcs suggests that the sub-arc mantle is oxidized by 0.3 to 2 log fO_2 units relative to abyssal peridotites, which are thought to have equilibrated close to FMQ (e.g., Brandon et al., 1996; Parkinson et al., 1998, 1999, 2003; Wood et al., 1989, 1990). Similar results by a different means were obtained by Canil (2002), who combined whole rock V vs. Al_2O_3 measurements of peridotites from the Mexico subduction zone with a parameterized model of V partitioning. Canil's results suggest that the sub-arc source of the Mexican peridotites was 1–2 log units more oxidized than the source of abyssal peridotites. Other trace elements in peridotites have been used to differentiate between the different mantle settings. MORB harzburgites and dunites are distinguishable from suprasubduction zone harzburgites and dunites on the basis of Ga/Fe^{3+} vs. Ti/Fe^{3+}, with the processes that produce the difference attributed to the redox state of the source zone (Dare et al., 2009).

Primitive arc lavas have also been investigated to determine the redox state of their source. A number of these studies point to an oxidized source. For example, the presence of sulfur as sulfate in melt inclusions from Mount Pinatubo in the Philippines suggests that the sub-arc mantle is oxidized, and produces oxidized magmas that release SO_2 during eruption (de Hoog et al., 2004). K. A. Evans et al. (2012) used estimates of ferric iron in spinel in primitive arc lavas to infer that the fO_2 in the arc magma source zone is 1 to 4 log units higher than in the source zone of MORB or OIB. The elevated fO_2 of primary magmas from the Antilles arc, which have fO_2 up to FMQ+3, have also been attributed to an oxidized source region (Parkinson et al., 2003; Pichavant et al., 2007; Stamper et al., 2014). A detailed study of melt inclusions in samples from the Agrigan volcano, part of the arc that overlies the Marianas subduction zone, found evidence of elevated fO_2 in the source zone to those magmas (Kelley et al., 2012). Brounce et al. (2015) suggest that magmas produced during even the earliest stages of subduction on the Marianas arc record oxygen fugacities orders of magnitude higher than that of MORB. By 2–4 Ma after the first magmas are produced, fO_2 is thought to have increased by 1.3 log units. If this is the case, then the driver of oxidation for the arc magmas is, most likely, fluids/melts from the downgoing slab, and buffering phases are either not present in large quantities in the sub-arc mantle, or they are overwhelmed rapidly by slab components such as oxidized sulfur and CO_2. An alternative proxy for fO_2 in the mantle source region, V/Yb, was proposed by Laubier et al. (2014). They suggested that V/Yb is not modified by fractional crystallization of olivine, plagioclase, or clinopyroxene, on the basis of partition coefficient experiments at 0.1 MPa at 1150 °C to 1190 °C and fO_2 between the FMQ buffer and NNO+2 (~FMQ+3). Inspection of the global dataset reveals that V/Yb increases from MORB to back-arc basin basalts (BABB) to arc lavas, consistent with an overall, but spatially variable, increase in source fO_2 from FMQ to NNO+2 (~FMQ+3). These results are also consistent with sub-arc mantle fO_2 estimated by Lange et al. (1993) for lamprophyric lavas from Mexico, using the relationship of Kress et al. (1991) between fO_2 and the ferric:ferrous ratio of a silicic liquid. Redox considerations also provide an explanation for the commonly documented elevated U/Th ratios in primitive arc lavas (e.g., Hawkesworth et al., 1997; Pearce et al., 1995). U is effectively insoluble in aqueous fluids at the Fe–FeO buffer, but solubility increases to hundreds of ppm as salinity and fO_2 increase to the Re–ReO_2 buffer (Bali et al., 2011). The paucity of U and Th in unmetasomatized mantle, and the lack of an effect of partial melting on this ratio means that the U/Th ratio of either SAM peridotites or arc rocks has the potential to monitor the redox state of slab-derived fluids that may have contributed to that rock. The ratio of U/Th in arc rocks from Tonga and the Cascades were noted to be higher than those in samples from the East African rift, and calculated U/Th ratios match those observed for an average SAM oxidation state of around FMQ+1.48 (Bali et al., 2011). Bali et al. (2011) also proposed that a combination of U/Th with Mo/W ratios provides an even more powerful technique to interrogate arc magma source regions, with the potential to constrain redox conditions and fluid salinity and quantity, so long as the rutile content of the source is known (Bali et al., 2012).

There is also geophysical evidence of an oxidized sub-arc mantle. Experimental evidence suggests that mantle oxidation state is a significant control on the seismic properties of olivine, and that observed zones of high seismic attenuation above subduction zones may record an elevated fO_2 (Cline et al., 2018).

A number of studies point to the subducting slab as the direct or indirect cause of observed oxidation. Oxygen fugacity in the products of selected volcanoes in the Marianas arc correlate well with proxies for involvement of slab fluids such as Ba/La, but not with proxies for sediment contribution or previous melt extraction in the back arc (Brounce et al., 2014). Similarly, fO_2 in supra-subduction zone peridotites from the Ortaca area, Western Turkey, correlates with the Cr# of spinel – this relationship was inferred to record the formation of high Cr, low Ti chromitites from interactions between the mantle lithosphere and a percolating boninitic melt (Uysal et al., 2007). In a global study, Kelley et al. (2009) showed that the Fe^{3+}/Fe_{tot} of primitive undegassed basaltic glasses increased from MORB to BABB to arc rocks, and correlates with H_2O content and other proxies for slab-derived fluids. Elsewhere, analyses of dredge samples of forearc mantle from the Tonga Trench were used as evidence that the SAM is oxidized by infiltration of oxidized fluids released from the subducted slab (Birner et al., 2017). Here, measured trace element abundances and fO_2 define two groups of samples. The first group, which comprises the majority of dredged samples, is highly depleted, with fO_2 close to FMQ. These samples are inferred to have formed by high degrees of partial melting during subduction initiation. The second group records fO_2 of FMQ+1.7, coupled with enriched light rare earth elements (LREE). It is proposed that these samples underwent interaction with subduction-related fluids and melts (Birner et al., 2017). The Birner observations suggest that the SAM is heterogeneous with respect to redox state, and remains so during subduction, and that SAM oxidation is a direct consequence of subduction. An influx of *RB*, in the form of sulfate or other oxidized species, and water from the subducting slab, also explains the observations of Peslier et al. (2002), who documented a negative correlation between the water content of orthopyroxene and the fO_2 of xenoliths that host a suite of spinel peridotites from Mexico and Simcoe. These workers suggested that *RB* from the slab increases fO_2 in the sub-arc mantle, which decreases the solubility of water in orthopyroxene. Water from the slab, in addition to a small contribution of water from the dehydrating orthopyroxene, can then flux melting if temperatures are high enough (Peslier et al., 2002).

There is also some direct evidence of input of oxidized slab-derived components. Hattori et al. (2005) document As^{5+}, thought to be present in magnetite and antigorite, in the Tso Morari ecologites, which represent hydrated peridotites from the base of the SAM. The oxidized As is thought to have been released from the slab and overlying sediments, where it would have been adsorbed onto Fe-oxides (Hattori et al., 2005). Rielli et al. (2017) suggest that ferric-iron rich metasomatic garnet in samples from the Bardane peridotite, located in Norway's Western Gneiss Region, record oxidation of the sub-arc mantle by slab-derived sulfate- and carbon-bearing fluids, consistent with the results of Malaspina et al. (2010). The Bardane peridotite is thought to represent transition zone mantle accreted to the cratonic lithosphere and subsequently metasomatized by fluids from a subducting slab during the Caledonian Orogeny (Scambelluri et al., 2008, 2010). The high ferric iron content of the Bardane peridotite garnet compared to less metasomatized peridotites elsewhere in the Western Gneiss region, and the presence of sulfides, barite, graphite, and diamond as inclusions in the garnet were used as evidence that subduction oxidizes the Earth's mantle by cycling of sulfur and carbon (Rielli et al., 2017). Sulfides also occur outside the garnet (e.g., Scambelluri et al., 2010). Tantalizingly direct evidence of slab-derived components has also been documented, with analyses of precipitates from fluids included in high pressure minerals (e.g., Ferrando et al., 2005; Malaspina et al., 2017). Precipitated phases include ferric-iron rich spinel, amphibole, chlorite, hematite, and phlogopite. The oxidized nature of these minerals provides direct evidence of a high *RB* fluid at the slab–mantle interface.

Further evidence of the role played by slab fluids is provided by comparisons between arc rocks from different settings. For example, the northern Cascades arc overlies one of the hottest subduction zones, and it is thought that slab dehydration is mostly complete before the slab reaches sub-arc depths. The loss of water reduces the chances of slab melting, but if slab melts were produced here, or elsewhere, then they would be an effective vector for RB transfer, because such melts can carry iron in concentrations far higher the plausible ferric iron concentrations of aqueous fluids. However, without slab melt, there is no vector available for transfer of *RB* from the slab to the sub-arc mantle, and indeed, while there are elevated values for some proxies for slab involvement in the northern Cascades samples, such as the large-ion lithophile elements (LILE), the redox state of the rocks is between FMQ and FMQ+1.4, which is at the lower end of the range observed for arc lavas (Mullen et al., 2014). The Lesser Antilles, on the other hand, consume ocean crust produced at a slow-spreading centre (Figure 4.3), which is therefore expected to be highly hydrothermally altered (Stamper et al., 2014). Experimental determinations of fO_2 under liquidus conditions for these rocks suggest a source region fO_2 between NNO+3.2 (~FMQ+4.2) and NNO+7.8 (~FMQ+8.8) (Pichavant et al., 2007).

No. Other studies have applied trace element- and isotope-based approaches to the question of sub-arc redox state without finding any evidence that primary arc basalts are more oxidized than basalts produced at MOR (e.g., Dauphas et al., 2009; Lee et al., 2003; Lee et al., 2005; Lee et al., 2012; Lee et al., 2010; Mallmann et al., 2009). Approaches taken by these studies include experimental work and analyses of natural samples for V, V/Sc element ratios, iron isotopes, Zn/Fe ratios, and Cu concentrations.

It might be that global scatter, the limited detection limits and precision of the large available datasets, and the model dependency of results hinder recognition of statistically significant differences between SAM and mantle elsewhere (Gaillard et al., 2015). It has also been suggested that the evidence of oxidized arc basalts and not-oxidized primary arc basalts might be reconciled if oxidation of the basalts occurs at shallow depths after basalts have evolved past the point where they would be included in global data searches as "primitive" — generally taken as MgO > 8 wt.%, and if assimilation of slab-derived fluids carrying *RB* at these shallower depths could produce the observed oxidized arc magmas (Gaillard et al., 2015).

However, there are reasons for caution in interpretation of evidence apparently in favor of an oxidized SAM. For example, an experimental investigation of Fe^{3+}/Fe_{tot} ratios in olivine-hosted melt inclusions suggests that these inclusions are subject to migration of H^+ through the host olivine and re-equilibration of Fe^{3+}/Fe_{tot} with the external environment (Gaetani et al., 2012). Additionally, studies that correlate evidence of slab-derived fluid with oxidation do not point unequivocally at slab fluids as the direct source of the oxidation. The presence of water alters the position of the solidus so that the depth of mantle melting might be different in arcs to other mantle melting settings (Gaetani, 2016). The fO_2 of unaltered mantle decreases with depth in the mantle, where it is buffered by iron-bearing equilibria (e.g., D. J. Frost et al., 2008; Stagno et al., 2013), so the role of water and other slab-derived components is likely to be complex, with a combination of direct and indirect effects. Indeed, there are some records of reduction induced by infiltration of slab-derived fluids. An influx of H_2-rich serpentinization-related fluids from the slab-wedge interface has been proposed as the cause of native metal formation in otherwise oxidized sub-arc mantle peridotite xenoliths from the Avacha volcano in Kamchatka (Ishimaru et al., 2009), and $CH_4 + C + H_2$-bearing fluid inclusions were detected in olivine from the Qilian suture zone in northwest China (Song et al., 2009).

Further, if the fO_2 calculated from sub-arc-derived spinel harzburgite xenoliths is to be used as evidence for an oxidized source of arc magmas, then it is necessary to demonstrate that such xenoliths are the residue of melting that produced arc magmas. A study of refractory peridotite xenoliths from the West Bismarck arc suggests that the xenoliths interacted with oxidized arc magmas, but that the source of the arc magma was deeper than the xenoliths (Benard et al., 2018). However, it was proposed that the xenoliths underwent fluxed-melting in the presence of Si-rich slab-derived components, so the higher level xenoliths might be similar to the peridotite that produced the parent of the arc magmas at depth.

Resolution of the apparently contradictory evidence requires a careful approach that takes into account the complex array of factors that contributes to changes in

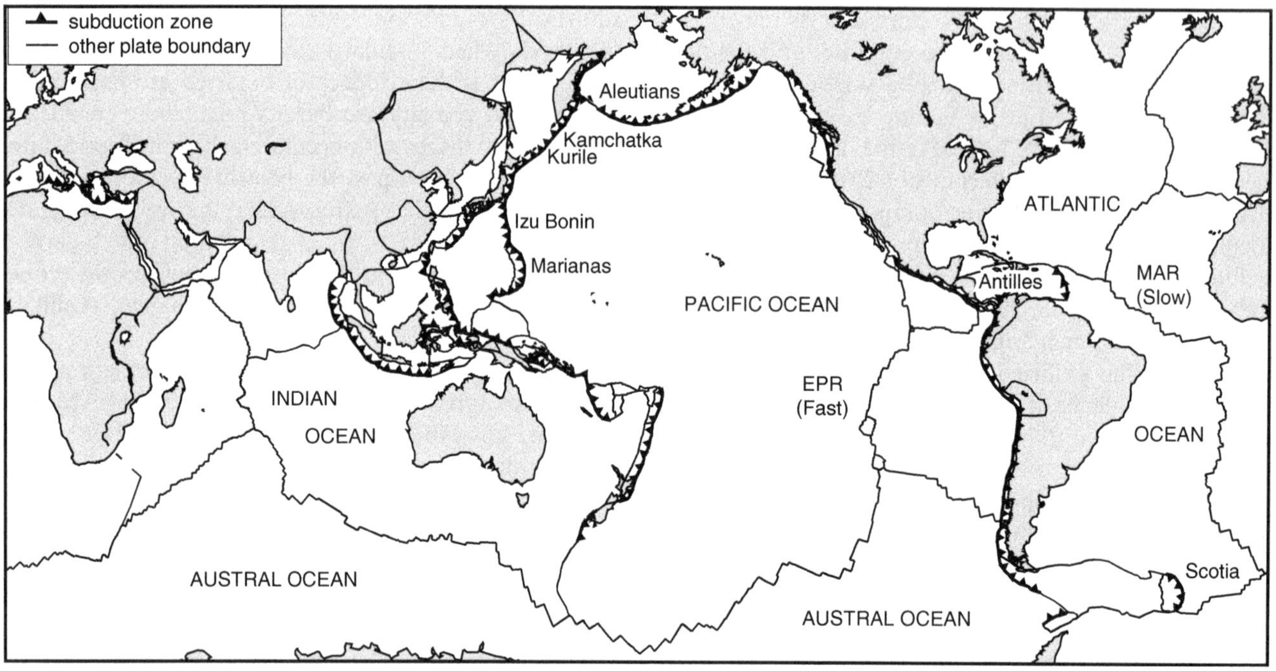

Figure 4.3 Location of volcanic arcs and spreading ridges mentioned in the text. Modified after Evans et al. (2012).

the *RB* of all the reservoirs involved in the formation of arc magmas.

4.3.4. Formation of Primary Magmas in the Sub-Arc Mantle

The formation of primary magmas in the SAM is complex, to say the least, and a comprehensive discussion of the topic is beyond the scope of this review – see Chapter 5 of this volume (Foley, 2021) for more detail. In a recent review, Gaillard et al. (2015) suggest that most melts produced in the mantle equilibrate within a relatively narrow window of redox state, between FMQ−1 and FMQ+1. Such constant redox state suggests that an effective redox buffer controls the redox state of most melts produced in the Earth's mantle and/or that the mantle is rapidly and effectively mixed. The latter has been excluded, based on mixing calculations (e.g., K. A. Evans, 2012). It has been proposed that the major buffer of oxidation state of the Earth's mantle is ferric/ferrous iron equilibria (Ballhaus et al., 1991). However, the buffering capacity of the garnet–orthopyroxene–olivine assemblage relies mostly on the capacity of garnet to receive ferric iron, which is limited (e.g., Stagno et al. [2013], and references therein). If mantle buffering is effective, then additional factors are required to produce the more oxidized melts proposed for arc magmas.

It is possible that volatiles added to the SAM from the slab overcome or modify existing buffers. Canil et al. (1994) suggest that the oxidation state of peridotites could be controlled by iron-, carbon-, hydrogen-, or sulfur-related buffers, depending on the composition and fO_2 of the system, and it has been proposed that reduction of CO_2 can increase mantle fO_2 as high as carbonate-bearing buffers, defined by the EMOD/G line (enstatite–magnesite–olivine–diamond/graphite) (Stagno et al., 2015). Rielli et al. (2018) found an association between depth of metasomatized mantle samples and oxidation state, with the deeper samples less oxidized. They suggested that many of the samples were sufficiently oxidized to require oxidation beyond the EMOD/G line, implying that oxidized sulfur was responsible. K. A. Evans et al. (2011) showed that sulfur, as sulfate, added in plausible quantities from the subducting slab, could cause oxidation of the observed magnitude within a few million years. In this case, redox would be buffered by SSO equilibria (Figure 4.1).

Here, to enable consideration of the redox state of subduction zone magmas, we discuss two first order controls on their redox state. The first is volatiles from the slab, which might contribute directly to the melt or contribute indirectly by altering the activities of redox-sensitive elements within the melt or metasomatize mantle that later melts, as described earlier. The second is refertilization of the SAM by basaltic magmas derived from sources other than the slab (Parkinson et al., 2003).

Role of Volatiles. Volatiles of interest in the consideration of arc magma production include CO_2, H_2O, H_2S, SO_2, and other species in the COHS system. Volatiles, like the alkali oxides Na_2O and K_2O, disrupt equilibria between bridging, non-bridging, and free oxygen within a melt (e.g., Fraser, 1977; Moretti, 2005), and therefore have the potential to alter the fO_2 of a melt (see Chapter 6, Moretti, 2021, this volume). The addition of *RB* from the slab also has the potential to induce redox melting, via the production of water or CO_2 (e.g., Foley, 2011; Song et al., 2009).

For this reason, the effect of volatiles on the redox state of silicate melts has been studied in detail. Moore et al. (1995) investigated the effect of H_2O on the redox state of iron in hydrous peralkaline rhyolites, and found no discernible relationship, consistent with an earlier study by Sisson et al. (1993). These results are also consistent with those of Gaillard et al. (2001) for melts produced at oxygen fugacities greater than NNO+1 (~FMQ+2). In other work by Botcharnikov et al. (2005), measurements of Fe^{3+}/Fe_{tot} in experimentally produced hydrous ferrobasalts were used to conclude that the water content of the melt affects Fe^{3+}/Fe_{tot}, but only within the range predicted by existing thermodynamic models for anhydrous basalts.

The results of other studies indicate that the magnitude of the effect of water on Fe^{3+}/Fe_{tot} should be measurable. The experimental results of Gaillard et al. (2001) required addition of a water-dependent term to the Kress et al. (1991) expression to allow prediction of Fe^{3+}/Fe_{tot} in rhyolitic melts produced at fO_2 less than NNO+1 (~FMQ+2). In another study, Gaillard, Pichavant, and Scaillet (2003) showed that the addition of water to reduced melts with 2–12 wt.% FeO_{tot} (~NNO−1.25) increased the activity coefficient of FeO from about 0.7 to 1, and a comparable but non-Henry's Law decrease in the activity coefficient of Fe_2O_3. The direction of these changes is consistent with that proposed by Gaillard et al. (2001). An increase in Fe^{3+}/Fe_{tot} with increasing H_2O content at constant fO_2 was observed by Schuessler et al. (2008) for hydrous phonotephrites. However, most of these experiments were performed on silica-rich compositions, so it is questionable whether the results should be extrapolated to the silica-poor compositions of primary mantle melts in the sub-arc mantle.

There are a number of possible reasons for the lack of consistency of experimental results. Redox control and water content are intimately linked by the constraints of the experimental approach. Any gain or loss of H_2 in response to buffering activity affects the water content of the charge, because H_2 is produced by the dissociation of water $H_2O = H_2 + \frac{1}{2} O_2$. If H_2 is gained or lost to the assembly, then the longer the experiment runs, the more H_2 is transferred to or from the experiment, with two important consequences: (i) that the water content of the initial charge is not necessarily that of the final experimental product; and (ii) that the water content of the charge changes constantly, and equilibrium might not be completely attained. Indeed, the kinetics of Fe^{2+}/Fe^{3+} re-equilibration are relatively slow (Gaillard et al., 2002), and might be slower than the rate of changes in water content in response to hydrogen loss or gain. Further, any formation of iron-bearing quench phases has the potential to alter Fe^{3+}/Fe_{tot} in the melt by preferentially fractionating one of the iron redox states. The formation of quench phases is common in experiments on mafic melt compositions (Parman et al., 2004), but the effects are seldom investigated in detail, with rare exceptions (Borisov, 2013).

Additionally, there are differences in the approaches taken to oxygen buffers for the experiments with potential implications for interpretation of the results. Some experiments are buffered using double capsule techniques (e.g., Jego et al., 2013, 2014; Poli et al., 2009); others are allowed to equilibrate with the assembly (e.g., Foley et al., 2009), or use gold capsules with short experimental time frames under the assumption that these prevent H_2 diffusion (Humphreys et al., 2015). Interpretation of the results is further hindered by a common lack of documentation of the extent of preservation or textural modification to the buffer assemblage in double capsule experiments.

Role of Refertilization. Oxidized sub-arc mantle might be produced by a range of fluids and over protracted timescales. The temperature of much of the SAM is thought to be too cool for fluid-saturated melting (Stern, 2002), which implies that slab-derived fluids drive solid-state metasomatism rather than melting of typical mantle compositions. At many localities that expose obducted mantle blocks, a range of styles of metasomatism are found (Kiseeva et al., 2017), and mantle xenoliths provide records of fluid-driven metasomatism. Alteration can be channelized or pervasive, and can cause extensive changes to the mineralogy, or subtle changes to mineral modes and compositions (O'Reilly et al., 2013). Typically, the addition of fluids produces refertilized mantle, which can be rich in hydrous minerals such as phlogopite and hornblende, or minerals that contribute strongly to the first stages of melting, such as clinopyroxene and garnet. Sieve textures in spinels might also record secondary metasomatic processes (Shaw et al., 2008). Under these circumstances, there is potential for refertilized mantle domains to be preserved for millions to billions of years before being melted during a later anomalous thermal event. If metasomatism involves oxidation through addition of oxidized sulfur or CO_2, the refertilized mantle domains might be able to generate the oxidized, metal-rich magmas that are ideal for ore deposit formation.

For example, ultramafic harzburgite, lherzolite, and dunite xenoliths from Grenada, which lies in the Lesser Antilles arc, show clear evidence for refertilization-related changes in redox state (Parkinson et al., 2003). It is suggested that depleted harzburgites were infiltrated by basalt, with a consequent increase in fO_2 from 1–2 log units above the FMQ buffer to 2–2.5 log units above the buffer. Bulk rock Fe^{3+}/Fe_{tot} is also higher in the Grenada xenoliths than in abyssal peridotites, suggesting that the elevated fO_2 is related to an increase in *RB* rather than, for example, a change in melting depth (Ballhaus et al., 1991).

Evidence for a redox-related role for refertilization is also provided by measurements of iron isotopes from primitive lavas from the Banda arc, Indonesia. The arc rock δ^{57}Fe values are less than typical δ^{57}Fe values from mid-ocean ridge settings (Nebel et al., 2015). Ferric iron is moderately incompatible during partial melting, so the observed isotope ratios are consistent with stripping of ferric iron from the SAM by repeated melting events to produce a SAM that is highly depleted in ^{57}Fe. Under these circumstances, any oxidized component in the source of arc lavas must be provided by refertilization of the SAM by slab-derived components that do not include iron.

While these studies provide excellent evidence that refertilization of the SAM plays a significant role in the production of oxidized arc magmas, only a relatively small proportion of studies in this research area address the redox aspect explicitly, providing a potentially productive avenue of research for further work.

4.3.5. Modification of Primary Magmas

A number of authors have proposed that the oxidized nature of arc magmas is due to processes during magma ascent and differentiation (Danyushevsky et al., 2000; Gaetani et al., 2012; Lee et al., 2005, 2010, 2012; Mallmann et al., 2009).

Differentiation. It has long been suggested that differentiation and fO_2 are intimately linked in arc magmatism, and these effects are discussed in detail in other chapters of this book (Chapter 6, Moretti, 2021; Chapter 8, Berry & O'Neill, 2021). One of the distinctive features in arc magmatism is that arc products often exhibit a calc-alkaline trend of silica-enrichment with relatively little iron enrichment, rather than the tholeiitic trend followed by most mid-ocean ridge basalts and ocean island basalts (Ishihara, 2004; Sato et al., 2004). A variety of causes for the different trends have been proposed, which include differences in the sequence of mineral fractionation (Grove et al., 1984), the presence of water (Zimmer et al., 2010), and differences in fO_2 (Miyashiro, 1974). The presence of water is thought to generate the calc-alkaline trend by suppressing plagioclase crystallization and enhancing the stability of Fe–Ti oxides (e.g., Grove et al., 1986; Osborn, 1959; Zimmer et al., 2010). Furthermore, Sisson et al. (1993) and Pichavant et al. (2007), note that high fO_2 and high Cr stabilize spinel at temperatures close to the liquidus of basaltic melts. A detailed study of Fe^{3+}/Fe_{tot} and water content in glasses from the arc rocks produced above the Marianas subduction zone confirm the link between calc-alkaline rocks and wetter, higher fO_2 magmas (Brounce et al., 2014). However, water and any associated *RB* transfer are not the only important factor. Calc-alkaline rocks are produced as the arc thickens, and as the SAM becomes depleted (e.g., Miyashiro, 1974), so factors related to these processes must also be considered.

During magma ascent, changes in iron speciation in the silicate melt driven by decompression result in small changes to fO_2 (Kress et al., 1991). Such changes are revealed by the results of Nebel et al. (2015) who interpreted an increase in δ^{57}Fe values of in samples from the Banda arc with magmatic differentiation as a record of progressive oxidation of the magma associated with fractionation of Fe^{2+}-bearing silicate minerals, though the magnitude of the oxidation effect on fO_2 was not quantified. The Fe^{3+}/Fe_{tot} ratios are also affected by fractionation of crystal phases, but these changes cannot explain observed changes of 2 to 3 log units in fO_2 observed in Marianas arc magmas by Brounce et al. (2014). Waters et al. (2016) suggest that arc magma's elevated oxidation state is imposed prior to degassing, and that crustal assimilation and crystal fractionation cannot explain why arc magmas are more oxidized than hydrous silicic melts from Iceland and Yellowstone. Similar conclusions were reached by Grocke et al. (2016), who investigated the effect of differentiation on magmatic fO_2 using a suite of 14 samples from the Central Volcanic Zone of the Andes. These authors found no evidence for systematic oxidation caused by either crustal assimilation or crystal fractionation; less evolved magmas produced closer to the trench, and with the greatest evidence of subducted components, recorded the highest fO_2. These conclusions are consistent with the work of Kelley et al. (2012) and Crabtree et al. (2012).

The consensus is that redox state and the presence of water provide a first order control on differentiation trends, and that differentiation can affect the *RB* of a magma, but not sufficiently to explain observed the magnitude of fO_2 elevation that is observed in arc products.

Degassing. Another way to produce an oxidized magma is to reduce the specific *RB* of the magma through loss of a phase with a specific *RB* less than that of the starting material (e.g., Carmichael, 1991; Carmichael et al., 1986; Mathez, 1984). This topic is discussed in detail elsewhere in this volume (Chapter 6 [Moretti, 2021]), but a brief mention of aspects relevant to subduction is made here. Reduced gases such as methane, H_2, and H_2S have low solubilities in magmas and in exsolved aqueous fluids and tend, therefore, to partition into the gaseous phase when pressures are sufficiently low to allow coexisting gaseous and aqueous phases rather than single phase supercritical fluids. The *RB* that can be carried by these species is limited by their solubility, but they can still carry the bulk of the *RB* in a fluid with few other redox-sensitive elements. Gases, due to their low density, are the most mobile phase in a magmatic system (Menand et al., 2007), and open system degassing from silicate melts, as proposed by Candela (1986), has the potential to increase the *RB* of the residual melt by several log units (Gaillard et al., 2015). However, for this process to be effective, the melt has to be sufficiently reduced to produce reduced species, which in practice means fO_2 should be less than FMQ+0.5 (Gaillard et al., 2015). Conversely, the loss of any species that hosts a redox-sensitive element which is oxidized relative to the same element in the source magma has the potential to decrease the *RB* of, and likely reduce, a degassing magma. Such reduction is proposed for the Agrigan volcano in the Marianas arc, where it is proposed that loss of SO_2 drove reduction of magmas (Kelley et al., 2012).

Whether a melt produces oxidized or reduced gases depends on composition, water content, and fO_2. Binder et al. (2018) showed that sulfur partitioned strongly into a fluid coexisting with a haplogranitic melt under reducing conditions, with $D_{fl/m}$ of 324 at the nickel–nickel oxide (NNO) buffer (~FMQ+1), and much less strongly, with $D_{fl/m}$ equal to 74 on the rhenium–rhenium oxide (RRO) buffer. The difference is attributed to changes in sulfur redox state in the melt; sulfur is much more soluble under more oxidizing conditions, and also to changes in the sulfur speciation in the fluid. Sulfur in the reduced fluids in the NNO (~FMQ+1) experiments is inferred to have been present as S^{-2} in HS^{-1} and H_2S, whereas sulfur in the fluids in the RRO experiments is present as S^{6+}. Similar results for basaltic compositions were calculated by Gaillard et al. (2015). Other parameters, such as the sodium content of the melt, also affect S partitioning. Sodium-rich melts in experiments were associated with lower $D_{fl/m}$ than sodium-poor experiments at otherwise identical conditions (Binder et al., 2018). This difference is attributed to the formation of S-complexes such as NaSH in the melt.

It has been proposed that the positive correlation between the proportion of ferric iron and water content in primitive arc lavas noted by Kelley et al. (2012), and others, is a consequence of degassing (e.g., Humphreys et al., 2015). Humphreys et al. (2015) used experiments to investigate the relationship between water content and oxidation state during the degassing and hydration of sulfur-free rhyolites. Loss of water from the melts increased fO_2, via the loss of H_2 into bubbles during decompression. However, it is not clear if this process occurs in nature to any great extent. In natural samples from the Mexican arc, Crabtree et al. (2012) found no evidence that loss of water during degassing magmas altered the fO_2, which remained around the NNO buffer (~FMQ+1) throughout the degassing process. Consistent with this result, a comparison of pre-eruptive magmatic ferrous iron concentrations, estimated from Fe–Ti oxides texturally linked to pre-eruptive phenocrysts with post eruptive ferrous iron suggest that H_2O degassing does not affect the ferric:ferrous iron ratio (Waters et al., 2016).

Redox Aspects of Ore Deposit Formation Associated with Arc Magmas. The search for arc segments associated with economic mineralization is an intensive and ongoing investigation involving multiple variables (e.g., J. R. Richards et al., 2007; J. P. Richards, 2011; Sillitoe, 2010). In spite of extensive study, the reason for the recurrent generation of major deposits in particular arc segments has not yet been identified with certainty, with parameters such as heterogeneously distributed metal pre-concentrations in source or host rocks, redox, and subduction geometry all under consideration (e.g., Sillitoe, 2010). This area of the literature is extensive, and is dealt with elsewhere in this book (Chapter 16 [Inglis & Moynier, 2021]), so only a limited number of aspects are discussed briefly here.

The fertility of an arc magma is directly related to its redox state and sulfur content via the stability of a sulfide phase, which may be a solid phase or a sulfide melt (e.g., Bataleva et al., 2016; Keays, 1995). Sulfide phases sequester metals of economic interest from the silicate melt effectively, and render the melt barren for the purposes of ore deposit formation. An understanding of the codependent evolution of sulfur contents and redox in melts and coexisting phases is therefore critical for prediction of the fertility of arc magmas (e.g., Mungall, 2002). See Chapter 10 for a more detailed discussion.

Magnetite also plays an important role. Jenner et al. (2010) showed that the concentrations of Au, Cu, and Ag in the parental melts to arc-like, oxidized magmas from the Manus basin are similar to those in MORB. However, the metals remain in the melts during differentiation of arc magmas because these melts, being more oxidized than MORB, are not saturated in a sulfide phase. Magnetite saturation, which occurs at around 60 wt.% SiO_2, is associated with a decrease in the metal content of the magmas, which is attributed to a transfer of electrons from ferrous iron to sulfate, producing ferric iron in magnetite and triggering sulfide saturation and metal sequestration (Sun et al., 2004).

In detail, the concentrations of metals in silicate melts vary as a function of the complex interplay between pressure, magmatic composition and evolution, sulfur content, and redox conditions. Taking Au as an example, it is known that the presence of sulfur enhances gold solubility in melts by one to two orders of magnitude (Jego et al., 2010, 2012, 2016). Redox also contributes, because sulfur is orders of magnitude more soluble in the S^{6+} state than as S^{-2} in sulfide (Jugo et al., 1999). Pressure provides another first order control, because the sulfate–sulfide boundary shifts to more oxidizing conditions as pressure increases (e.g., Jego et al., 2016; Matjuschkin et al., 2016). This effect is augmented by the decrease in mantle fO_2 with depth (e.g., D. J. Frost et al., 2008). Thus, under moderately oxidizing conditions, the stable sulfur-bearing species in melts generated at shallow depths would be sulfate, and the melt would have a relatively high capacity for gold. In contrast, melts generated at greater depths, even where the *RB* is relatively high, are likely to form in the sulfide stability field, and in this case, Au will sequester into sulfide-bearing phases rather than the silicate melt.

The need to track this complex evolution has led to the application of a range of techniques in attempts to constrain the conditions and consequences of arc magma evolution. For example, transition metal isotopes can track certain features of arc systems (Nebel et al., 2015; Sossi & Debret, 2021 [Chapter 15, this volume]). An example related to ore deposit formation is that of Williams et al. (2018), who performed a detailed study of the evolution of Fe isotopes associated with the differentiation of lavas erupted onto the island of Anatahan, which is part of the Marianas Arc in the West Pacific. The results record a multi-stage evolution of the Fe isotopes in response to sequestration of isotopically heavy iron into magnetite and subsequent initiation of the formation of sulfide melt or crystalline phases that sequestered isotopically light Fe. Under these circumstances, Cu and other chalcophile elements are removed from the melt into the sulfide phase relatively late in the magmatic evolution, an observation that has been explained previously by late stage exsolution of magmatic fluids and brines (Seo et al., 2012).

The growing body of knowledge regarding transfer of volatiles from the slab to the SAM has the potential to inform the study of arc-related ore deposits, by constraining the evolution of fO_2 and the sulfur content of the subarc mantle. Similarly, the wealth of ore-deposit studies has the potential to place constraints on conceptual models of SAM evolution. Currently, studies of the evolving SAM tend to focus on oceanic arcs to avoid the complication of continental crust involvement, whereas ore-deposit studies focus on arc segments with continental crust, because this is where accessible ore deposits form. Nevertheless, future integrated studies of transfer of material from slab to arc are likely to prove valuable.

4.4. DISCUSSION

4.4.1. Temporal Variation

Temporal variation in the composition of magmas produced at plate boundaries, and in the style of plate boundary ore deposits, has motivated speculation on the causes of these variations (e.g., de Oliveira et al., 2010; Goldfarb et al., 2010; Groves et al., 2007). One explanation is that the fO_2 of the Earth's mantle has evolved through geological time. Lee et al. (2003) compared V concentrations in ancient cratonic and young peridotites, but found no evidence that Archean cratonic mantle and modern oceanic mantle formed under different oxidation states. In other studies, it has been proposed that the onset of modern subduction is linked to the rise in oxygen (e.g., Duncan et al., 2017). However, as pointed out by Gaillard et al. (2015), it must also be considered that the overall effect of subduction is to reduce the *RB* of the atmosphere/ocean/crust system by transferring that *RB* to the deep Earth (K. A. Evans, 2012). However, even if oxidation state itself has varied little in the mantle over geological time, the cycling of redox-sensitive elements such as sulfur has changed in response to changing atmosphere and oceanic redox conditions (e.g., Canil et al., 2017), and these changes have the potential to drive temporal changes in the characteristics of arc magmas and associated ore deposits.

For example, K. A. Evans and Tomkins (2011) proposed that the change from sulfide-stable sediments to sulfate-stable during the Great Oxidation Event (GOE) initiated the oxidation of SAM, and that this change permitted the formation of modern style copper porphyry deposits. Along similar lines, Canil et al. (2017) noted that earlier in Earth's history, hot slabs associated with Fe-rich sediments favored sulfide stability over sulfate, and that this led to deep subduction and long-term mantle sequestration of the chalcophile elements. In the last 250 Ma, an increase in the abundance of carbonate-shelled plankton to produce Ca-rich sediments favored anhydrite stability and the cycling of sulfur as sulfate to sub-arc depths and shallower. Cycling of anhydrite-derived sulfur would (i) provide *RB* to the sub-arc mantle; (ii) provide sulfur ligands for transport of chalcophile elements; and (iii) reduce the budget of sulfide-hosted sulfur and hence the potential for sequestration of chalcophile elements in the deep mantle.

4.4.2. Opportunities

This review has identified a number of opportunities for further research, where the current knowledge and effort lags behind that in more frequently investigated areas. Attractive opportunities relate to novel trace element proxy approaches (e.g., Bali et al., 2011, 2012; Lafay et al., 2016; Laubier et al., 2014), the spatial coupling of fluid flow and element release (e.g., John et al., 2012; Spandler et al., 2011), the implications of new findings for carbon transport (e.g., Kelemen et al., 2015), the consequences of SAM refertilization for the redox state of arc magmas (e.g., Parkinson et al., 1999; Rielli et al., 2017), and changes to the *RB* of sediments during subduction zone metamorphism (e.g., Galvez, Beyssac, et al., 2013; Galvez, Martinez, et al., 2013).

New geochemical proxies for redox state and external contributors to arc magma sources are proposed regularly (e.g., Bali et al., 2012; Laubier et al., 2014). These proxies rely on assumptions such as the presence or absence of a particular phase in the source. Additionally, a number of redox-sensitive elements are potentially capable of providing information on subduction-related redox processes that are not currently much utilized for that purpose. These include As and Sb, with Sb observed to sequester into serpentine under laboratory conditions as Sb^{3+} or Sb^{5+}, depending on the redox conditions of serpentinization (e.g., Lafay et al., 2016). The applicability of this finding to natural systems, and its implications, have not been explored, and provide a significant opportunity for further work, because Sb, as one of the fluid mobile elements, can be used to trace a subduction fluid signature through the subduction to arc-magmatism cycle. The difficulty of redox state measurement has placed, and continues to place, an impediment to research on redox-sensitive element exchange and fluxes, but increasing access to synchrotrons and standards, plus ongoing method development for synchrotron radiation approaches are increasing the availability of such data (e.g., Berry et al., 2007, 2018; Danyushevsky et al., 2007; Yaxley et al., 2005).

Current knowledge suggests that the most abundant source of *RB* available for transfer to the SAM is the mantle lithosphere (Debret et al., 2016; Pons et al., 2016), and the volumetrically largest fluid source is the ocean crust. In this case, effective transfer of *RB* requires interaction of ocean crust-derived fluids with the sub-mantle lithosphere. Infiltration of external fluids into slab lithologies has been documented (e.g., K. A. Evans et al., 2014; John et al., 2012; Spandler et al., 2011), but the frequency and probability of such interactions have yet to be demonstrated. Further work could usefully be devoted to analysis of the proportion of *RB* released by mantle lithosphere that could plausibly become entrained in fluids derived from subducted oceanic crust.

Further opportunities are presented by new findings on potential vectors for carbon transport in arc environments. The results of decades of experiments and the observation of carbonate inclusions in transition zone/lower mantle diamonds suggest that carbonate remains stable to beyond sub-arc depths (Bulanova et al., 2010; Burnham et al., 2016; Dasgupta et al., 2004, 2010; Gorman et al., 2006; Kerrick et al., 1998; Molina et al., 2000; Poli et al., 2009; Thomsen et al., 2008; Tsuno et al., 2011; Yaxley et al., 1994, 2004), but more recent research suggests that carbonate may be mobilized by dissolution (Caciagli et al., 2003; Kelemen et al., 2015), or by incorporation into carbon-rich melts (Poli, 2015, 2016). The consequences of these findings, and examples from natural systems, have not yet been much explored.

Similarly, while refertilization is known to be common in SAM, the consequences of refertilization have been neglected relative to more commonly investigated aspects such as the consequences for trace elements, rare earths, and major element magma geochemistry (e.g., Becker et al., 2001; Kamber et al., 2002; Le Roux et al., 2014; Parkinson et al., 2003; Tollan et al., 2017; Ulrich et al., 2010; Uysal et al., 2015; van Acken et al., 2008; Yaxley et al., 1998). There is a wealth of information on the redox aspects of magma generation in the experimental literature (e.g., Gaetani et al., 2012; Gaillard, Pichavant, Mackwell, et al., 2003; Gaillard, Pichavant, & Scaillet, 2003; Gaillard et al., 2001, 2002; Klimm et al., 2012; Lesne et al., 2015; Merkulova et al., 2017; Poli et al., 2009; Tiraboschi et al., 2018) that could be coupled with the documented effects of refertilization in natural rocks to derive a more comprehensive understanding of the role played by SAM refertilization in the generation of oxidized magmas.

4.4.3. Synthesis

While there are many uncertainties, a consistent picture is beginning to emerge, that arc magmas are more oxidized, at source, than MORB, and that the reason for that oxidation is loss of *RB* from the subducting slab. The most invoked source of the bulk of this *RB* is the reduction of ferric iron in serpentinites, as documented by Debret and coworkers, and there is some evidence that sulfate is the transfer agent, consistent with the results of Pons et al. (2016) and Rielli et al. (2018). However, there are still contradictions in the evidence that require reconciliation, and knowledge gaps pertaining to a range of factors that need to be filled. We look forward to many decades of productive and fascinating research in this area.

ACKNOWLEDGMENTS

K. Evans acknowledges support from Australian Research Council Future Fellowship FT120100579. A. Tomkins acknowledges support from the Australian Research Council.

REFERENCES

Alt, J. C., Davidson, G. J., Teagle, D. A. H., & Karson, J. A. (2003). Isotopic composition of gypsum in the Macquarie Island ophiolite: Implications for the sulfur cycle and the subsurface biosphere in oceanic crust. *Geology*, *31*(6), 549–552. doi: 10.1130/0091-7613(2003)031<0549:icogit>2.0.co;2

Alt, J. C., Garrido, C. J., Shanks, W. C., Turchyn, A., Padron-Navarta, J. A., Sanchez-Vizcaino, V. L., et al. (2012). Recycling of water, carbon, and sulfur during subduction of serpentinites: A stable isotope study of Cerro del Almirez, Spain. *Earth and Planetary Science Letters*, *327*, 50–60. doi: 10.1016/j.epsl.2012.01.029

Alt, J. C., Schwarzenbach, E. M., Frueh-Green, G. L., Shanks, W. C., Bernasconi, S. M., Garrido, C. J., et al. (2013). The role of serpentinites in cycling of carbon and sulfur: Seafloor serpentinization and subduction metamorphism. *Lithos*, *178*, 40–54. doi: 10.1016/j.lithos.2012.12.006

Alt, J. C., Shanks, W. C., Bach, W., Paulick, H., Garrido, C. J., & Beaudoin, G. (2007). Hydrothermal alteration and microbial sulfate reduction in peridotite and gabbro exposed by detachment faulting at the Mid-Atlantic Ridge, 15 degrees 20 ' N (ODP Leg 209): A sulfur and oxygen isotope study. *Geochemistry Geophysics Geosystems*, *8*. doi: 10.1029/2007GC001617

Alt, J. C., Shanks, W. C., Crispini, L., Gaggero, L., Schwarzenbach, E. M., Fruh-Green, G. L., & Bernasconi, S. M. (2012). Uptake of carbon and sulfur during seafloor serpentinization and the effects of subduction metamorphism in Ligurian peridotites. *Chemical Geology*, *322*, 268–277. doi: 10.1016/j.chemgeo.2012.07.009

Bach, W., Garrido, C. J., Paulick, H., Harvey, J., & Rosner, M. (2004). Seawater-peridotite interactions: First insights from ODP Leg 209, MAR 15 degrees N. *Geochemistry Geophysics Geosystems*, *5*. doi: 10.1029/2004GC000744

Bali, E., Audetat, A., & Keppler, H. (2011). The mobility of U and Th in subduction zone fluids: an indicator of oxygen fugacity and fluid salinity. *Contributions to Mineralogy and Petrology*, *161*(4), 597–613. doi: 10.1007/s00410-010-0552-9

Bali, E., Keppler, H., & Audetat, A. (2012). The mobility of W and Mo in subduction zone fluids and the Mo-W-Th-U systematics of island arc magmas. *Earth and Planetary Science Letters*, *351*, 195–207. doi: 10.1016/j.epsl.2012.07.032

Ballhaus, C., Berry, R. F., & Green, D. H. (1991). High-pressure experimental calibration of the olivine-ortho-pyroxene-spinel oxygen geobarometer - implications for the oxidation-state of the upper mantle. *Contributions to Mineralogy and Petrology*, *107*(1), 27–40.

Barker, A. K., Coogan, L. A., & Gillis, K. M. (2010). Insights into the behaviour of sulphur in mid-ocean ridge axial hydrothermal systems from the composition of the sheeted dyke complex at Pito Deep. *Chemical Geology*, *275*(1–2), 105–115. doi: 10.1016/j.chemgeo.2010.05.003

Bataleva, Y. V., Palyanov, Y. N., Borzdov, Y. M., & Sobolev, N. V. (2016). Sulfidation of silicate mantle by reduced S-bearing metasomatic fluids and melts. *Geology*, *44*(4), 271–274. doi: 10.1130/g37477.1

Bebout, G. E. (1995). The impact of subduction-zone metamorphism on mantle-ocean chemical cycling. *Chemical Geology*, *126*(2), 191–218.

Becker, H., Shirey, S. B., & Carlson, R. W. (2001). Effects of melt percolation on the Re-Os systematics of peridotites from a Paleozoic convergent plate margin. *Earth and Planetary Science Letters*, *188*(1–2), 107–121. doi: 10.1016/s0012-821x(01)00308-9

Beltrando, M., Manatschal, G., Mohn, G., Dal Piaz, G. V., Brovarone, A. V., & Masini, E. (2014). Recognizing remnants of magma-poor rifted margins in high-pressure orogenic belts: The Alpine case study. *Earth-Science Reviews*, *131*, 88–115. doi: 10.1016/j.earscirev.2014.01.001

Benard, A., Woodland, A. B., Arculus, R. J., Nebel, O., & McAlpine, S. R. B. (2018). Variation in sub-arc mantle oxygen fugacity during partial melting recorded in refractory peridotite xenoliths from the West Bismarck Arc. *Chemical Geology*, *486*, 16–30. doi: 10.1016/j.chemgeo.2018.03.004

Berry, A. J., & O'Neill, H. St.C. (2021). Oxygen Content, Oxygen Fugacity, the Oxidation State of Iron, and Mid-Ocean Ridge Basalts. In: R. Moretti & D. Neuville (Eds.), *Magma Redox Geochemistry*. Geophysical Monograph Series. USA: American Geophysical Union.

Berry, A. J., Harris, A. C., Kamenetsky, V. S., Newville, M., & Sutton, S. R. (2007). The speciation of metals in natural fluid inclusions at temperatures up to 700 degrees C. *Geochimica Et Cosmochimica Acta*, *71*(15), A86–A86.

Berry, A. J., O'Neill, H. S., Jayasuriya, K. D., Campbell, S. J., & Foran, G. J. (2003). XANES calibrations for the oxidation state of iron in a silicate glass. *American Mineralogist*, *88*(7), 967–977.

Berry, A. J., Stewart, G. A., O'Neill, H. S. C., Mallmann, G., & Mosselmans, J. F. W. (2018). A re-assessment of the oxidation state of iron in MORB glasses. *Earth and Planetary Science Letters*, *483*, 114–123. doi: 10.1016/j.epsl.2017.11.032

Berry, A. J., Yaxley, Greg M., Woodland, Alan B., & Foran, Garry J. (2010). A XANES calibration for determining the oxidation state of iron in mantle garnet. *Chemical Geology*, *278*(1–2), 31–37.

Binder, B., Wenzel, T., & Keppler, H. (2018). The partitioning of sulfur between multicomponent aqueous fluids and felsic melts. *Contributions to Mineralogy and Petrology*, *173*(2). doi: 10.1007/s00410-018-1445-6

Birner, S. K., Warren, J. M., Cottrell, E., Davis, F. A., Kelley, K. A., & Falloon, T. J. (2017). Forearc Peridotites from Tonga Record Heterogeneous Oxidation of the Mantle following Subduction Initiation. *Journal of Petrology*, *58*(9), 1755–1779. doi: 10.1093/petrology/egx072

Blattler, C. L., Claire, M. W., Prave, A. R., Kirsimae, K., Higgins, J. A., Medvedev, P. V., et al. (2018). Two-billion-year-old evaporites capture Earth's great oxidation. *Science*, *360*(6386), 320–323. doi:10.1126/science.aar2687

Blundy, J., Cashman, K. V., Rust, A., & Witham, F. (2010). A case for CO2-rich arc magmas. *Earth and Planetary Science Letters*, *290*(3–4), 289–301. doi: 10.1016/j.epsl.2009.12.013

Borisov, A. A. (2013). Mutual interaction of redox pairs in silicate melts: V5+/V4+/V3+/V2+ tetrad and other equilibria. *Petrology*, *21*(4), 305–315. doi: 10.1134/s0869591113040036

Botcharnikov, R. E., Koepke, J., Holtz, F., McCammon, C., & Wilke, M. (2005). The effect of water activity on the oxidation and structural state of Fe in a ferro-basaltic melt. *Geochimica Et Cosmochimica Acta*, *69*(21), 5071–5085. doi: 10.1016/j.gca.2005.04.023

Brandon, A. D., & Draper, D. S. (1996). Constraints on the origin of the oxidation state of mantle overlying subduction zones: An example from Simcoe, Washington, USA. *Geochimica Et Cosmochimica Acta*, *60*(10), 1739–1749. doi: 10.1016/0016-7037(96)00056-7

Bretscher, A., Hermann, J., & Pettke, T. (2018). The influence of oceanic oxidation on serpentinite dehydration during subduction. *Earth and Planetary Science Letters*, *499*, 173–184. doi:10.1016/j.epsl.2018.07.017

Brounce, M. N., Kelley, K. A., & Cottrell, E. (2014). Variations in Fe3+/Sigma Fe of Mariana arc basalts and mantle wedge *f*O(2). *Journal of Petrology*, *55*(12), 2513–2536. doi: 10.1093/petrology/egu065

Brounce, M. N., Kelley, K. A., Cottrell, E., & Reagan, M. K. (2015). Temporal evolution of mantle wedge oxygen fugacity during subduction initiation. *Geology*, *43*(9), 775–778. doi: 10.1130/g36742.1

Brovarone, A. V., Picatto, M., Beyssac, O., Lagabrielle, Y., & Castelli, D. (2014). The blueschist-eclogite transition in the Alpine chain: P-T paths and the role of slow-spreading extensional structures in the evolution of HP-LT mountain belts. *Tectonophysics*, *615*, 96–121. doi: 10.1016/j.tecto.2014.01.001

Bulanova, G. P., Walter, M. J., Smith, C. B., Kohn, S. C., Armstrong, L. S., Blundy, J., & Gobbo, L. (2010). Mineral inclusions in sublithospheric diamonds from Collier 4 kimberlite pipe, Juina, Brazil: subducted protoliths, carbonated melts and primary kimberlite magmatism. *Contributions to Mineralogy and Petrology*, *160*(4), 489–510. doi:10.1007/s00410-010-0490-6

Burnham, A. D., Thomson, A. R., Bulanova, G. P., Kohn, S. C., Smith, C. B., & Walter, M. J. (2015). Stable isotope evidence for crustal recycling as recorded by superdeep diamonds. *Earth and Planetary Science Letters*, *432*, 374–380. doi:10.1016/j.epsl.2015.10.023

Burnham, A. D., Bulanova, G. P., Smith, C. B., Whitehead, S. C., Kohn, S. C., Gobbo, L., & Walter, M. J. (2016). Diamonds from the Machado River alluvial deposit, Rondonia, Brazil, derived from both lithospheric and sublithospheric mantle. *Lithos*, *265*, 199–213. doi:10.1016/j.lithos.2016.05.022

Burgisser, A., & Scaillet, B. (2007). Redox evolution of a degassing magma rising to the surface. *Nature*, *445*(7124), 194–197.

Caciagli, N.C., & Manning, C.E. (2003). The solubility of calcite in water at 6-16 kbar and 500-800 °C. *Contributions to Mineralogy and Petrology*, *146*, 275–285.

Campbell, I. H., & Squire, R. J. (2010). The mountains that triggered the Late Neoproterozoic increase in oxygen: The Second Great Oxidation Event. *Geochimica Et Cosmochimica Acta*, *74*(15), 4187–4206. doi: 10.1016/j.gca.2010.04.064

Candela, P. A. (1986). The evolution of aqueous vapor from silicate melts - effect on oxygen fugacity. *Geochimica Et Cosmochimica Acta*, *50*(6), 1205–1211. doi: 10.1016/0016-7037(86)90403-5

Canil, D. (2002). Vanadium in peridotites, mantle redox and tectonic environments: Archean to present. *Earth and Planetary Science Letters*, *195*(1–2), 75–90. doi: 10.1016/s0012-821x(01)00582-9

Canil, D., & Fellows, S. A. (2017). Sulphide-sulphate stability and melting in subducted sediment and its role in arc mantle redox and chalcophile cycling in space and time. *Earth and Planetary Science Letters*, *470*, 73–86. doi: 10.1016/j.epsl.2017.04.028

Canil, D., O'Neill, H. S., Pearson, D. G., Rudnick, R. L., McDonough, W. F., & Carswell, D. A. (1994). Ferric iron in peridotites and mantle oxidation-states. *Earth and Planetary Science Letters*, *123*(1–4), 205–220.

Carmichael, I. S. E. (1991). The redox states of basic and silicic magmas – a reflection of their source regions. *Contributions to Mineralogy and Petrology*, *106*(2), 129–141.

Carmichael, I. S. E., & Ghiorso, M. S. (1986). Oxidation-reduction relations in basic magma – a case for homogeneous equilibria. *Earth and Planetary Science Letters*, *78*(2–3), 200–210. doi: 10.1016/0012-821x(86)90061-0

Carter, L. B., & Dasgupta, R. (2018). Decarbonation in the Ca-Mg-Fe carbonate system at mid-crustal pressure as a function of temperature and assimilation with arc magmas – Implications for long-term climate. *Chemical Geology*, *492*, 30–48. doi: 10.1016/j.chemgeo.2018.05.024

Cline, C. J., Faul, U. H., David, E. C., Berry, A. J., & Jackson, I. (2018). Redox-influenced seismic properties of uppermantle olivine. *Nature*, *555*(7696), 355–358. doi: 10.1038/nature25764

Connolly, J. A. D. (1995). Phase-diagram methods for graphitic rocks and application to the system C-O-H-FeO-TiO2-SiO2. *Contributions to Mineralogy and Petrology*, *119*(1), 94–116.

Crabtree, S. M., & Lange, R. A. (2012). An evaluation of the effect of degassing on the oxidation state of hydrous andesite and dacite magmas: a comparison of pre- and post-eruptive Fe2+ concentrations. *Contributions to Mineralogy and Petrology*, *163*(2), 209–224. doi: 10.1007/s00410-011-0667-7

Crossley, R. J., Evans, K. A., Reddy, S. M., & Lester, G. M. (2017). Redistribution of iron and titanium in high-pressure ultramafic rocks. *Geochemistry, Geophysics, Geosystems*, *18*(11), 3869–3890.

Cruz, M. F., & Manning, C. E. (2015). Experimental determination of quartz solubility and melting in the system SiO2-H2O-NaCl at 15-20 kbar and 900-1100 degrees C: implications for silica polymerization and the formation of supercritical fluids. *Contributions to Mineralogy and Petrology*, *170*(4). doi: 10.1007/s00410-015-1187-7

Danyushevsky, L. V., Berry, A. J., O'Neill, H. S. C., Newville, M., & Sutton, S. R. (2007). Fe-3+/Fe-2+ of melt inclusions: Implications for melt H2O contents. *Geochimica Et Cosmochimica Acta*, *71*(15), A200–A200.

Danyushevsky, L. V., Della-Pasqua, F. N., & Sokolov, S. (2000). Re-equilibration of melt inclusions trapped by magnesian olivine phenocrysts from subduction-related magmas: petrological implications. *Contributions to Mineralogy and Petrology*, *138*(1), 68–83. doi: 10.1007/pl00007664

Dare, S. A. S., Pearce, J. A., McDonald, I., & Styles, M. T. (2009). Tectonic discrimination of peridotites using *f*O(2)-

Cr# and Ga-Ti-Fe-III systematics in chrome-spinel. *Chemical Geology*, *261*(3–4), 199–216. doi: 10.1016/j.chemgeo.2008.08.002

Dasgupta, R., & Hirschmann, M. M. (2010). The deep carbon cycle and melting in Earth's interior. *Earth and Planetary Science Letters*, *298*(1–2), 1–13. doi: 10.1016/j.epsl.2010.06.039

Dasgupta, R., Hirschmann, M. M., & Withers, A. C. (2004). Deep global cycling of carbon constrained by the solidus of anhydrous, carbonated eclogite under upper mantle conditions. *Earth and Planetary Science Letters*, *227*(1–2), 73–85.

Dauphas, N., Craddock, P. R., Asimow, P. D., Bennett, V. C., Nutman, A. P., & Ohnenstetter, D. (2009). Iron isotopes may reveal the redox conditions of mantle melting from Archean to Present. *Earth and Planetary Science Letters*, *288*(1–2), 255–267. doi: 10.1016/j.epsl.2009.09.029

Davis, F. A., Cottrell, E., Birner, S. K., Warren, J. M., & Lopez, O. G. (2017). Revisiting the electron microprobe method of spinel-olivine-orthopyroxene oxybarometry applied to spinel peridotites. *American Mineralogist*, *102*(1–2), 421–435. doi: 10.2138/am-2017-5823

de Hoog, J. C. M., Hattori, K. H., & Hoblitt, R. P. (2004). Oxidized sulfur-rich mafic magma at Mount Pinatubo, Philippines. *Contributions to Mineralogy and Petrology*, *146*(6), 750–761. doi: 10.1007/s00410-003-0532-4

de Oliveira, M. A., Dall'Agnol, R., & Scaillet, B. (2010). Petrological constraints on crystallization conditions of Mesoarchean sanukitoid rocks, Southeastern Amazonian craton, Brazil. *Journal of Petrology*, *51*(10), 2121–2148. doi: 10.1093/petrology/egq051

Debret, B., Andreani, M., Godard, M., Nicollet, C., Schwartz, S., & Lafay, R. (2013). Trace element behavior during serpentinization/de-serpentinization of an eclogitized oceanic lithosphere: A LA-ICPMS study of the Lanzo ultramafic massif (Western Alps). *Chemical Geology*, *357*, 117–133. doi: 10.1016/j.chemgeo.2013.08.025

Debret, B., Andreani, M., Munoz, M., Bolfan-Casanova, N., Carlut, J., Nicollet, C., et al. (2014). Evolution of Fe redox state in serpentine during subduction. *Earth and Planetary Science Letters*, *400*, 206–218. doi: 10.1016/j.epsl.2014.05.038

Debret, B., Bolfan-Casanova, N., Padron-Navarta, J. A., Martin-Hernandez, F., Andreani, M., Garrido, C. J., et al. (2015). Redox state of iron during high-pressure serpentinite dehydration. *Contributions to Mineralogy and Petrology*, *169*(4). doi: 10.1007/s00410-015-1130-y

Debret, B., Koga, K. T., Nicollet, C., Andreani, M., & Schwartz, S. (2014). F, Cl and S input via serpentinite in subduction zones: implications for the nature of the fluid released at depth. *Terra Nova*, *26*(2), 96–101. doi: 10.1111/ter.12074

Debret, B., Millet, M.-A, Pons, M.-L, Bouihol, P., Inglis, E., & Williams, H. (2016). Isotopic evidence for iron mobility during subduction. *Geology*, *44*, 215–218. https://doi.org/10.1130/G37565.1

Debret, B., & Sverjensky, D. A. (2017). Highly oxidising fluids generated during serpentinite breakdown in subduction zones. *Scientific Reports*, *7*. doi: 10.1038/s41598-017-09626-y

Deschamps, F., Godard, M., Guillot, S., & Hattori, K. (2013). Geochemistry of subduction zone serpentinites: A review. *Lithos*, *178*, 96–127. doi: 10.1016/j.lithos.2013.05.019

Duncan, M. S., & Dasgupta, R. (2017). Rise of Earth's atmospheric oxygen controlled by efficient subduction of organic carbon. *Nature Geoscience*, *10*(5), 387–392. doi: 10.1038/ngeo2939

Dyar, M. D., McGuire, A. V., & Ziegler, R. D. (1989). Redox equilibria and crystal-chemistry of coexisting minerals from spinel lherzolite mantle xenoliths. *American Mineralogist*, *74*(9–10), 969–980.

Evans, B.W. (2008). Control of the products of serpentinization by the Fe(2)Mg(1) exchange potential of olivine and orthopyroxene. *Journal of Petrology*, *49*(10), 1873–1887. doi: 10.1093/petrology/egn050

Evans, B.W. (2010). Lizardite versus antigorite serpentinite: Magnetite, hydrogen, and life(?). *Geology*, *38*(10), 879–882. doi: 10.1130/g31158.1

Evans, K. A. (2006). Redox decoupling and redox budgets: Conceptual tools for the study of earth systems. *Geology*, *34*(6), 489–492.

Evans, K. A. (2012). The redox budget of subduction zones. *Earth-Science Reviews*, *113*(1–2), 11–32. doi: 10.1016/j.earscirev.2012.03.003

Evans, K. A., Elburg, M.A., & Kamenetsky, V.S. (2012). Oxidation state of subarc mantle. *Geology*, *40*(9), 783–786. doi: 10.1130/g33037.1

Evans, K. A., Phillips, G. N., & Powell, R. (2006). Rock-buffering of auriferous fluids in altered rocks associated with the Golden Mile-style mineralization, Kalgoorlie gold field, Western Australia. *Economic Geology*, *101*(4), 805–817.

Evans, K. A., & Powell, R. (2015). The effect of subduction on the sulphur, carbon and redox budget of lithospheric mantle. *Journal of Metamorphic Geology*, *33*(6), 649–670. doi: 10.1111/jmg.12140

Evans, K. A., Reddy, S. M., Tomkins, A. G., Crossley, R. J., & Frost, B. R. (2017). Effects of geodynamic setting on the redox state of fluids released by subducted mantle lithosphere. *Lithos*, *278*, 26–42. doi: 10.1016/j.lithos.2016.12.023

Evans, K. A., Tomkins, A. G., Cliff, J., & Fiorentini, M. L. (2014). Insights into subduction zone sulfur recycling from isotopic analysis of eclogite-hosted sulfides. *Chemical Geology*, *365*, 1–19. doi: 10.1016/j.chemgeo.2013.11.026

Evans, K. A., & Tomkins, A.G. (2011). The relationship between subduction zone redox budget and arc magma fertility. *Earth and Planetary Science Letters*, *308*(3–4), 401–409. doi: 10.1016/j.epsl.2011.06.009

Fakhraee, M., Hancisse, O., Canfield, D. E., Crowe, S. A., & Katsev, S. (2019). Proterozoic seawater sulfate scarcity and the evolution of ocean-atmosphere chemistry. *Nature Geoscience*, *12*(5), 375–380. doi:10.1038/s41561-019-0351-5

Ferrando, S., Groppo, C., Frezzotti, M. L., Castelli, D., & Proyer, A. (2017). Dissolving dolomite in a stable UHP mineral assemblage: Evidence from Cal-Dol marbles of the Dora-Maira Massif (Italian Western Alps). *American Mineralogist*, *102*(1–2), 42–60. doi:10.2138/am-2017-5761

Ferrando, S., Frezzotti, M. L., Dallai, L., & Compagnoni, R. (2005). Multiphase solid inclusions in UHP rocks (Su-Lu, China): Remnants of supercritical silicate-rich aqueous fluids released during continental subduction. *Chemical Geology*, *223*(1–3), 68–81. doi: 10.1016/j.chemgeo.2005.01.029

Ferraris, C., Grobety, B., Fruh-Green, G. L., & Wessicken, R. (2004). Intergrowth of graphite within phlogopite from Finero ultramafic complex (Italian Western Alps): implications for mantle crystallization of primary-texture mica. *European Journal of Mineralogy*, *16*(6), 899–908. doi: 10.1127/0935-1221/2004/0016-0899

Fiege, A., Ruprecht, P., Simon, A. C., Bell, A. S., Gottlicher, J., Newville, M., et al. (2017). Calibration of Fe XANES for high-precision determination of Fe oxidation state in glasses: Comparison of new and existing results obtained at different synchrotron radiation sources. *American Mineralogist*, *102* (1–2), 369–380. doi: 10.2138/am-2017-5822

Foley, S. F. (2011). A reappraisal of redox melting in the earth's mantle as a function of tectonic setting and time. *Journal of Petrology*, *52*(7–8), 1363–1391. doi: 10.1093/petrology/egq061

Foley, S. F. (2021). Redox melting in the mantle. In: R. Moretti & D. Neuville (Eds.), *Magma Redox Geochemistry*. Geophysical Monograph Series. USA: American Geophysical Union.

Foley, S. F., Andronikov, A. V., Jacob, D. E., & Melzer, S. (2006). Evidence from Antarctic mantle peridotite xenoliths for changes in mineralogy, geochemistry and geothermal gradients beneath a developing rift. *Geochimica Et Cosmochimica Acta*, *70*(12), 3096–3120. doi: 10.1016/j.gca.2006.03.010

Foley, S. F., Yaxley, G. M., Rosenthal, A., Buhre, S., Kiseeva, E. S., Rapp, R. P., & Jacob, D. E. (2009). The composition of near-solidus melts of peridotite in the presence of CO2 and H2O between 40 and 60 kbar. *Lithos*, *112*, 274–283. doi: 10.1016/j.lithos.2009.03.020

Fraser, D. G. (1977). Thermodynamic properties of silicate melts. In Fraser, D. G. (Ed.), *NATO Advanced Study Institutes Series (Series C — Mathematical and Physical Sciences)*, vol *30*. Springer, Dordrecht. https://doi.org/10.1007/978-94-010-1252-2_15

Frezzotti, M. L., & Ferrando, S. (2007). Multiphase solid inclusions in ultrahigh-pressure metamorphic rocks: a petrographic approach. *Periodico Di Mineralogia*, *76*(2–3), 113–125.

Frezzotti, M. L., Ferrando, S., Tecce, F., & Castelli, D. (2012). Water content and nature of solutes in shallow-mantle fluids from fluid inclusions. *Earth and Planetary Science Letters*, *351*, 70–83. doi: 10.1016/j.epsl.2012.07.023

Frezzotti, M. L., Selverstone, J., Sharp, Z. D., & Compagnoni, R. (2011). Carbonate dissolution during subduction revealed by diamond-bearing rocks from the Alps. *Nature Geoscience*, *4*(10), 703–706.

Frost, B. R., & Beard, J. S. (2007). On silica activity and serpentinization. *Journal of Petrology*, *48*(7), 1351–1368. doi: 10.1093/petrology/egm021

Frost, B. R., Evans, K. A., Swapp, S. M., Beard, J. S., & Mothersole, F. E. (2013). The process of serpentinization in dunite from New Caledonia. *Lithos*, *178*, 24–39. doi:10.1016/j.lithos.2013.02.002

Frost, D. J., & McCammon, C. A. (2008). The redox state of Earth's mantle. *Annual Review of Earth and Planetary Sciences*, *36*, 389–420.

Gaetani, G. A. (2016). The behavior of Fe3+/Sigma Fe during partial melting of spinel lherzolite. *Geochimica Et Cosmochimica Acta*, *185*, 64–77. doi: 10.1016/j.gca.2016.03.019

Gaetani, G. A., O'Leary, J. A., Shimizu, N., Bucholz, C. E., & Newville, M. (2012). Rapid reequilibration of H_2O and oxygen fugacity in olivine-hosted melt inclusions. *Geology*, *40*(10), 915–918. doi: 10.1130/g32992.1

Gaillard, F., Pichavant, M., Mackwell, S., Champallier, R., Scaillet, B., & Mccammon, C. (2003). Chemical transfer during redox exchanges between H-2 and Fe-bearing silicate melts. *American Mineralogist*, *88*(2–3), 308–315.

Gaillard, F., Pichavant, M., & Scaillet, B. (2003). Experimental determination of activities of FeO and Fe_2O_3 components in hydrous silicic melts under oxidizing conditions. *Geochimica Et Cosmochimica Acta*, *67*(22), 4389–4409.

Gaillard, F., Scaillet, B., & Pichavant, M. (2002). Kinetics of iron oxidation-reduction in hydrous silicic melts. *American Mineralogist*, *87*(7), 829–837.

Gaillard, F., Scaillet, B., Pichavant, M., & Beny, J. L. (2001). The effect of water and *f*O(2) on the ferric-ferrous ratio of silicic melts. *Chemical Geology*, *174*(1–3), 255–273.

Gaillard, F., Scaillet, B., Pichavant, M., & Iacono-Marziano, G. (2015). The redox geodynamics linking basalts and their mantle sources through space and time. *Chemical Geology*, *418*, 217–233. doi: 10.1016/j.chemgeo.2015.07.030

Galvez, M. E., Beyssac, O., Martinez, I., Benzerara, K., Chaduteau, C., Malvoisin, B., & Malavieille, J. (2013). Graphite formation by carbonate reduction during subduction. *Nature Geoscience*, *6*(6), 473–477. doi: 10.1038/ngeo1827

Galvez, M. E., Connolly, J. A. D., & Manning, C. E. (2016). Implications for metal and volatile cycles from the pH of subduction zone fluids. *Nature*, *539*(7629), 420–424. doi: 10.1038/nature20103

Galvez, M. E., Manning, C. E., Connolly, J. A. D., & Rumble, D. (2015). The solubility of rocks in metamorphic fluids: A model for rock-dominated conditions to upper mantle pressure and temperature. *Earth and Planetary Science Letters*, *430*, 486–498. doi: 10.1016/j.epsl.2015.06.019

Galvez, M. E., Martinez, I., Beyssac, O., Benzerara, K., Agrinier, P., & Assayag, N. (2013). Metasomatism and graphite formation at a lithological interface in Malaspina (Alpine Corsica, France). *Contributions to Mineralogy and Petrology*, *166*(6), 1687–1708. doi: 10.1007/s00410-013-0949-3

Goldfarb, R. J., Bradley, D., & Leach, D. L. (2010). Secular variation in economic geology. *Economic Geology*, *105*(3), 459–465.

Gorman, P. J., Kerrick, D. M., & Connolly, J. A. D. (2006). Modeling open system metamorphic decarbonation of subducting slabs. *Geochemistry Geophysics Geosystems*, *7*. doi: 10.1029/2005GC001125

Grocke, S. B., Cottrell, E., de Silva, S., & Kelley, K. A. (2016). The role of crustal and eruptive processes versus source variations in controlling the oxidation state of iron in Central Andean magmas. *Earth and Planetary Science Letters*, *440*, 92–104. doi: 10.1016/j.epsl.2016.01.026

Groppo, C., & Castelli, D. (2010). Prograde P-T evolution of a lawsonite eclogite from the Monviso meta-ophiolite (Western Alps): Dehydration and redox reactions during subduction of oceanic FeTi-oxide gabbro. *Journal of Petrology*, *51*(12), 2489–2514. doi: 10.1093/petrology/egq065

Grove, T. L., & Baker, M. B. (1984). Phase-equilibrium controls on the tholeiitic versus calc-alkaline differentiation trends.

Journal of Geophysical Research, *89*(NB5), 3253–3274. doi: 10.1029/JB089iB05p03253

Grove, T. L., & Kinzler, R. J. (1986). Petrogenesis of andesites. *Annual Review of Earth and Planetary Sciences*, *14*, 417–454. doi: 10.1146/annurev.ea.14.050186.002221

Groves, David I., & Bierlein, Frank P. (2007). Geodynamic settings of mineral deposit systems. *Journal of the Geological Society*, *164*, 19–30.

Gudmundsson, G., & Wood, B. J. (1995). Experimental tests of garnet peridotite oxygen barometry. *Contributions to Mineralogy and Petrology*, *119*(1), 56–67. doi: 10.1007/bf00310717

Hattori, K., Takahashi, Y., Guillot, S., & Johanson, B. (2005). Occurrence of arsenic (V) in forearc mantle serpentinites based on X-ray absorption spectroscopy study. *Geochemica et Cosmochimica Acta*, *69*, 5585–5596.

Hawkesworth, C. J., Turner, S. P., McDermott, F., Peate, D. W., & vanCalsteren, P. (1997). U-Th isotopes in arc magmas: Implications for element transfer from the subducted crust. *Science*, *276*(5312), 551–555. doi: 10.1126/science.276.5312.551

Hayes, J. M., & Waldbauer, J. R. (2006). The carbon cycle and associated redox processes through time. *Philosophical Transactions of the Royal Society B-Biological Sciences*, *361*(1470), 931–950.

Hermann, J., Spandler, C., Hack, A., & Korsakov, A. V. (2006). Aqueous fluids and hydrous melts in high-pressure and ultra-high pressure rocks: Implications for element transfer in subduction zones. *Lithos*, *92*(3–4), 399–417.

Hofer, H. E., Brey, G. P., Yaxley, G. M., & Berry, A. J. (2006). Iron oxidation state determination in garnets by EPMA and XANES. *Geochimica Et Cosmochimica Acta*, *70*(18), A256–A256.

Honnorez, J. (2003). Hydrothermal alteration vs. ocean-floor metamorphism. A comparison between two case histories: the TAG hydrothermal mound (Mid-Atlantic Ridge) vs. DSDP/ODP Hole 504B (Equatorial East Pacific). *Comptes Rendus Geoscience*, *335*(10–11), 781–824.

Humphreys, M. C. S., Brooker, R. A., Fraser, D. G., Burgisser, A., Mangan, M. T., & McCammon, C. (2015). Coupled interactions between volatile activity and Fe oxidation state during arc crustal processes. *Journal of Petrology*, *56*(4), 795–814. doi: 10.1093/petrology/egv017

Inglis, E. C., & Moynier, F. (2019). Zinc and copper isotopes as tracers of redox processes. In R. Moretti & D. Neuville (Eds.), *Magma Redox Geochemistry*. USA: American Geophysical Union.

Ishihara, S. (2004). The redox state of granitoids relative to tectonic setting and earth history: The magnetite-ilmenite series 30 years later. In S. Ishihara, W. E. Stephens, S. L. Harley, M. Arima & T. Nakajima (Eds.), *Fifth Hutton Symposium on the Origin of Granites and Related Rocks* (Vol. 389, pp. 23–33).

Ishimaru, S., Arai, S., & Shukuno, H. (2009). Metal-saturated peridotite in the mantle wedge inferred from metal-bearing peridotite xenoliths from Avacha volcano, Kamchatka. *Earth and Planetary Science Letters*, *284*(3–4), 352–360. doi: 10.1016/j.epsl.2009.04.042

Jego, S., & Dasgupta, R. (2013). Fluid-present melting of sulfide-bearing ocean-crust: Experimental constraints on the transport of sulfur from subducting slab to mantle wedge. *Geochimica Et Cosmochimica Acta*, *110*, 106–134. doi: 10.1016/j.gca.2013.02.011

Jego, S., & Dasgupta, R. (2014). The fate of sulfur during fluid-present melting of subducting basaltic crust at variable oxygen fugacity. *Journal of Petrology*, *55*(6), 1019–1050. doi: 10.1093/petrology/egu016

Jego, S., Nakamura, M., Kimura, J. I., Iizuka, Y., Chang, Q., & Zellmer, G. F. (2016). Is gold solubility subject to pressure variations in ascending arc magmas? *Geochimica Et Cosmochimica Acta*, *188*, 224–243. doi: 10.1016/j.gca.2016.05.034

Jego, S., & Pichavant, M. (2012). Gold solubility in arc magmas: Experimental determination of the effect of sulfur at 1000 degrees C and 0.4 GPa. *Geochimica Et Cosmochimica Acta*, *84*, 560–592. doi: 10.1016/j.gca.2012.01.027

Jego, S., Pichavant, M., & Mavrogenes, J. A. (2010). Controls on gold solubility in arc magmas: An experimental study at 1000 degrees C and 4 kbar. *Geochimica Et Cosmochimica Acta*, *74*(7), 2165–2189.

Jenner, F. E., O'Neill, H. S. C., Arculus, R. J., & Mavrogenes, J. A. (2010). The magnetite crisis in the evolution of arc-related magmas and the initial concentration of Au, Ag and Cu. *Journal of Petrology*, *51*(12), 2445–2464.

John, T., Gussone, N., Podladchikov, Y. Y., Bebout, G. E., Dohmen, R., Halama, R., et al. (2012). Volcanic arcs fed by rapid pulsed fluid flow through subducting slabs. *Nature Geoscience*, *5*(7), 489–492. doi: 10.1038/ngeo1482

Jugo, P. J., Candela, P. A., & Piccoli, P. M. (1999). Magmatic sulfides and Au : Cu ratios in porphyry deposits: an experimental study of copper and gold partitioning at 850 degrees C, 100 MPa in a haplogranitic melt pyrrhotite intermediate solid solution gold metal assemblage, at gas saturation. *Lithos*, *46*(3), 573–589.

Kamber, B. S., Ewart, A., Collerson, K. D., Bruce, M. C., & McDonald, G. D. (2002). Fluid-mobile trace element constraints on the role of slab melting and implications for Archaean crustal growth models. *Contributions to Mineralogy and Petrology*, *144*(1), 38–56. doi: 10.1007/s00410-002-0374-5

Keays, R. R. (1995). The role of komatiitic and picritic magmatism and S-saturation in the formation of ore-deposits. *Lithos*, *34*(1–3), 1–18.

Kelemen, P. B., & Manning, C. E. (2015). Reevaluating carbon fluxes in subduction zones, what goes down, mostly comes up. *Proceedings of the National Academy of Sciences of the United States of America*, *112*(30), E3997–E4006. doi: 10.1073/pnas.1507889112

Kelley, K. A., & Cottrell, E. (2009). Water and the oxidation state of subduction zone magmas. *Science*, *325*(5940), 605–607.

Kelley, K. A., & Cottrell, E. (2012). The influence of magmatic differentiation on the oxidation state of Fe in a basaltic arc magma. *Earth and Planetary Science Letters*, *329*, 109–121. doi: 10.1016/j.epsl.2012.02.010

Kerrick, D. M., & Connolly, J. A. D. (1998). Subduction of ophicarbonates and recycling of CO_2 and H_2O. *Geology*, *26*(4), 375–378. doi: 10.1130/0091-7613(1998)026<0375:sooaro>2.3.co;2

Kiseeva, E. S., Kamenetsky, V. S., Yaxley, G. M., & Shee, S. R. (2017). Mantle melting versus mantle metasomatism – "The chicken or the egg" dilemma. *Chemical Geology*, *455*, 120–130. doi: 10.1016/j.chemgeo.2016.10.026

Klein, F., & Bach, W. (2009). Fe-Ni-Co-O-S phase relations in peridotite-seawater interactions. *Journal of Petrology*, *50*(1), 37–59. doi: 10.1093/petrology/egn071

Klein, F., Bach, Wolfgang, Joens, Niels, McCollom, Tom, Moskowitz, Bruce, & Berquo, Thelma. (2009). Iron partitioning and hydrogen generation during serpentinization of abyssal peridotites from 15 degrees N on the Mid-Atlantic Ridge. *Geochimica Et Cosmochimica Acta*, *73*(22), 6868–6893. doi: 10.1016/j.gca.2009.08.021

Klein, F., Marschall, H. R., Bowring, S. A., Humphris, S. E., & Horning, G. (2017). Mid-ocean ridge serpentinite in the Puerto Rico trench: From seafloor spreading to subduction. *Journal of Petrology*, *58*(9), 1729–1753. doi: 10.1093/petrology/egx071

Klimm, K., Kohn, S. C., & Botcharnikov, R. E. (2012). The dissolution mechanism of sulphur in hydrous silicate melts. II: Solubility and speciation of sulphur in hydrous silicate melts as a function of *f*O(2). *Chemical Geology*, *322*, 250–267. doi: 10.1016/j.chemgeo.2012.04.028

Kress, V. C., & Carmichael, I. S. E. (1991). The compressibility of silicate liquids containing Fe_2O_3 and the effect of composition, temperature, oxygen fugacity and pressure on their redox states. *Contributions to Mineralogy and Petrology*, *108*(1–2), 82–92. doi: 10.1007/bf00307328

Lafay, R., Montes-Hernandez, G., Janots, E., Munoz, M., Auzende, A. L., Gehin, A., et al. (2016). Experimental investigation of As, Sb and Cs behavior during olivine serpentinization in hydrothermal alkaline systems. *Geochimica Et Cosmochimica Acta*, *179*, 177–202. doi: 10.1016/j.gca.2016.02.014

Lagabrielle, Y., Brovarone, A. V., & Ildefonse, B. (2015). Fossil oceanic core complexes recognized in the blueschist metaophiolites of Western Alps and Corsica. *Earth-Science Reviews*, *141*, 1–26. doi: 10.1016/j.earscirev.2014.11.004

Lange, R. A., Carmichael, I. S. E., & Renne, P. R. (1993). Potassic volcanism near mono basin, california – evidence for high water and oxygen fugacities inherited from subduction. *Geology*, *21*(10), 949–952. doi: 10.1130/0091-7613(1993)021<0949:pvnmbc>2.3.co;2

Laubier, M., Grove, T. L., & Langmuir, C. H. (2014). Trace element mineral/melt partitioning for basaltic and basaltic andesitic melts: An experimental and laser ICP-MS study with application to the oxidation state of mantle source regions. *Earth and Planetary Science Letters*, *392*, 265–278. doi: 10.1016/j.epsl.2014.01.053

Le Losq, C., Berry, A. J., Kendrick, M. A., Neuville, D. R., & O'Neill, H. S. (2019). Determination of the oxidation state of iron in Mid-Ocean Ridge basalt glasses by Raman spectroscopy. *American Mineralogist*, *104*(7), 1032–1042. doi:10.2138/am-2019-6887

Le Roux, V., Dick, H. J. B., & Shimizu, N. (2014). Tracking flux melting and melt percolation in supra-subduction peridotites (Josephine ophiolite, USA). *Contributions to Mineralogy and Petrology*, *168*(4). doi: 10.1007/s00410-014-1064-9

Lecuyer, C., & Ricard, Y. (1999). Long-term fluxes and budget of ferric iron: implication for the redox states of the Earth's mantle and atmosphere. *Earth and Planetary Science Letters*, *165*(2), 197–211.

Lee, C. T. A., Brandon, A. D., & Norman, M. (2003). Vanadium in peridotites as a proxy for paleo-*f*O(2) during partial melting: Prospects, limitations, and implications. *Geochimica Et Cosmochimica Acta*, *67*(16), 3045–3064. doi: 10.1016/s0016-7037(03)00268-0

Lee, C. T. A., Leeman, W. P., Canil, D., & Li, Z. X. A. (2005). Similar V/Sc systematics in MORB and arc basalts: Implications for the oxygen fugacities of their mantle source regions. *Journal of Petrology*, *46*(11), 2313–2336. doi: 10.1093/petrology/egi056

Lee, C. T. A., Luffi, P., Chin, E. J., Bouchet, R., Dasgupta, R., Morton, D. M., et al. (2012). Copper systematics in arc magmas and implications for crust-mantle differentiation. *Science*, *336*(6077), 64–68. doi: 10.1126/science.1217313

Lee, C. T. A., Luffi, P., Le Roux, V., Dasgupta, R., Albarede, F., & Leeman, W. P. (2010). The redox state of arc mantle using Zn/Fe systematics. *Nature*, *468*(7324), 681–685.

Lee, C. T. A., Shen, B., Slotnick, B. S., Liao, K., Dickens, G. R., Yokoyama, Y., et al. (2013). Continental arc-island arc fluctuations, growth of crustal carbonates, and long-term climate change. *Geosphere*, *9*(1), 21–36. doi: 10.1130/ges00822.1

Lesne, P., Scaillet, B., & Pichavant, M. (2015). The solubility of sulfur in hydrous basaltic melts. *Chemical Geology*, *418*, 104–116. doi: 10.1016/j.chemgeo.2015.03.025

Li, J. L., Gao, J., Klemd, R., John, T., & Wang, X. S. (2016). Redox processes in subducting oceanic crust recorded by sulfide-bearing high-pressure rocks and veins (SW Tianshan, China). *Contributions to Mineralogy and Petrology*, *171*(8–9). doi: 10.1007/s00410-016-1284-2

Liu, Y., Santosh, M., Yuan, T. Y., Li, H. Q., & Li, T. F. (2016). Reduction of buried oxidized oceanic crust during subduction. *Gondwana Research*, *32*, 11–23. doi: 10.1016/j.gr.2015.02.014

Lorand, J. P., Luguet, A., & Alard, O. (2008). Platinum-group elements: A new set of key tracers for the earth's interior. *Elements*, *4*(4), 247–252. doi: 10.2113/gselements.4.4.247

Luhr, J. F., & ArandaGomez, J. J. (1997). Mexican peridotite xenoliths and tectonic terranes: Correlations among vent location, texture, temperature, pressure, and oxygen fugacity. *Journal of Petrology*, *38*(8), 1075–1112. doi: 10.1093/petrology/38.8.1075

Lyons, T. W., Reinhard, C. T., & Planavsky, N. J. (2014). The rise of oxygen in Earth's early ocean and atmosphere. *Nature*, *506*(7488), 307–315. doi: 10.1038/nature13068

Malaspina, N., Langenhorst, F., Tumiati, S., Campione, M., Frezzotti, M. L., & Poli, S. (2017). The redox budget of crust-derived fluid phases at the slab-mantle interface. *Geochimica Et Cosmochimica Acta*, *209*, 70–84. doi: 10.1016/j.gca.2017.04.004

Malaspina, N., Poli, S., & Fumagalli, P. (2009). The oxidation state of metasomatized mantle wedge: Insights from C-O-H-bearing garnet peridotite. *Journal of Petrology*, *50*(8), 1533–1552.

Malaspina, N., Scambelluri, M., Poli, S., Van Roermund, H. L. M., & Langenhorst, F. (2010). The oxidation state of mantle wedge majoritic garnet websterites metasomatised by C-bearing subduction fluids. *Earth and Planetary Science Letters*, *298*(3–4), 417–426. doi: 10.1016/j.epsl.2010.08.022

Malaspina, N., & Tumiati, S. (2012). The role of C-O-H and oxygen fugacity in subduction-zone garnet peridotites. *European Journal of Mineralogy*, *24*(4), 607–618. doi: 10.1127/0935-1221/2012/0024-2213

Mallmann, G., & O'Neill, H. S. C. (2009). The crystal/melt partitioning of V during mantle melting as a function of oxygen fugacity compared with some other elements (Al, P, Ca, Sc, Ti, Cr, Fe, Ga, Y, Zr and Nb). *Journal of Petrology*, *50*(9), 1765–1794.

Manning, C.E. (1994). The solubility of quartz in H_2O in the lower crust and upper mantle. *Geochemica et Cosmochimica Acta*, *58*(22), 4831–4839.

Mathez, E. A. (1984). Influence of degassing on oxidation-states of basaltic magmas. *Nature*, *310*(5976), 371–375. doi: 10.1038/310371a0

Matjuschkin, V., Blundy, J. D., & Brooker, R. A. (2016). The effect of pressure on sulphur speciation in mid-to deep-crustal arc magmas and implications for the formation of porphyry copper deposits. *Contributions to Mineralogy and Petrology*, *171*(7). doi: 10.1007/s00410-016-1274-4

Mattioli, G. S., & Wood, B. J. (1988). Magnetite activities across the $MgAl_2O_4$-Fe_3O_4 spinel join, with application to thermobarometric estimates of upper mantle oxygen fugacity. *Contributions to Mineralogy and Petrology*, *98*(2), 148–162. doi: 10.1007/bf00402108

Menand, T., & Phillips, J. C. (2007). Gas segregation in dykes and sills. *Journal of Volcanology and Geothermal Research*, *159*(4), 393–408. doi: 10.1016/j.jvolgeores.2006.08.003

Merkulova, M. V., Munoz, M., Brunet, F., Vidal, O., Hattori, K., Vantelon, D., et al. (2017). Experimental insight into redox transfer by iron- and sulfur-bearing serpentinite dehydration in subduction zones. *Earth and Planetary Science Letters*, *479*, 133–143. doi: 10.1016/j.epsl.2017.09.009

Mironov, N. L., & Portnyagin, M. V. (2018). Coupling of redox conditions of mantle melting and copper and sulfur contents in primary magmas of the Tolbachinsky Dol (Kamchatka) and Juan de Fuca Ridge (Pacific Ocean). *Petrology*, *26*(2), 145–166. doi: 10.1134/s0869591118020030

Miyashiro, A. (1974). Volcanic rock series in island arcs and active continental margins. *American Journal of Science*, *274*(4), 321–355. doi: 10.2475/ajs.274.4.321

Molina, J. F., & Poli, S. (2000). Carbonate stability and fluid composition in subducted oceanic crust: an experimental study on H2O-CO2-bearing basalts. *Earth and Planetary Science Letters*, *176*(3–4), 295–310. doi: 10.1016/s0012-821x(00)00021-2

Moore, G., Righter, K., & Carmichael, I. S. E. (1995). The effect of dissolved water on the oxidation-state of iron in natural silicate liquids. *Contributions to Mineralogy and Petrology*, *120*(2), 170–179.

Moretti, R. (2005). Polymerisation, basicity, oxidation state and their role in ionic modelling of silicate melts. *Annals of Geophysics*, *48*(4–5), 583–608.

Moretti, R. (2021). Ionic syntax and equilibrium approach to redox exchanges in melts: Basic concepts and the case of iron and sulfur in degassing magmas. In R. Moretti & D. Neuville (Eds.), *Magma Redox Geochemistry*. USA: American Geophysical Union.

Mothersole, F. E., Evans, K. A., & Frost, B. R. (2017). Abyssal and hydrated mantle wedge serpentinised peridotites: a comparison of the 15 ∘ 20 ′ N fracture zone and New Caledonia serpentinites. *Contributions to Mineralogy and Petrology*, *172*(8). doi: 10.1007/s00410-017-1381-x

Mullen, E. K., & McCallum, I. S. (2014). Origin of basalts in a hot subduction setting: Petrological and geochemical insights from Mt. Baker, Northern Cascade Arc. *Journal of Petrology*, *55*(2), 241–281. doi: 10.1093/petrology/egt064

Mungall, J. E. (2002). Roasting the mantle: Slab melting and the genesis of major Au and Au-rich Cu deposits. *Geology*, *30*(10), 915–918.

Munster, A. (1970). *Classical thermodynamics*. London: John Wiley and Sons.

Naemura, K., Hirajima, T., & Svojtka, M. (2009). The pressure-temperature path and the origin of phlogopite in spinel-garnet peridotites from the Blansky Les Massif of the Moldanubian Zone, Czech Republic. *Journal of Petrology*, *50*(10), 1795–1827. doi: 10.1093/petrology/egp052

Nebel, O., Sossi, P. A., Benard, A., Wille, M., Vroon, P. Z., & Arculus, R. J. (2015). Redox-variability and controls in subduction zones from an iron-isotope perspective. *Earth and Planetary Science Letters*, *432*, 142–151. doi: 10.1016/j.epsl.2015.09.036

Nell, J., & Wood, B. J. (1991). High-temperature electrical measurements and thermodynamic properties of Fe_3O_4--$FeCr_2O_4$-$MgCr_2O_4$-$FeAl_2O_4$ spinels. *American Mineralogist*, *76*(3–4), 405–426.

Newton, R. C., & Manning, C. E. (2010). Role of saline fluids in deep-crustal and upper-mantle metasomatism: insights from experimental studies. *Geofluids*, *10*(1–2), 58–72. doi: 10.1111/j.1468-8123.2009.00275.x

Nordstrom, D.K., & Munoz, J.L. (1994). *Geochemical thermodynamics*. London: Blackwell Scientific Publications.

O'Reilly, S. Y., & Griffin, W. L. (2013). Mantle Metasomatism. In D. E. Harlov & H. Austrheim (Eds.), *Metasomatism and the chemical transformation of rock* (pp. 471–533). Berlin, Heidelberg: Springer-Verlag.

Osborn, E.F. (1959). Role of oxygen pressure in the crystallization and differentiation of basaltic magma. *American Journal of Science*, *257*, 609–647.

Oyarzun, R., Lillo, J., & Oyarzun, J. (2008). No water, no cyanobacteria – No calc-alkaline magmas: Progressive oxidation of the early oceans may have contributed to modernize island-arc magmatism. *International Geology Review*, *50*(10), 885–894. doi: 10.2747/0020-6814.50.10.885

Padron-Navarta, J. A., Sanchez-Vizcaino, V. L., Garrido, C. J., & Gomez-Pugnaire, M. T. (2011). Metamorphic record of high-pressure dehydration of antigorite serpentinite to chlorite harzburgite in a subduction setting (Cerro del Almirez, Nevado-Filabride complex, Southern Spain). *Journal of Petrology*, *52*(10), 2047–2078. doi: 10.1093/petrology/egr039

Parkinson, I.J., & Arculus, R.J. (1999). The redox state of subduction zones: insights from arc-peridotites. *Chemical Geology, 160*(4), 409–423. doi: 10.1016/s0009-2541(99)00110-2

Parkinson, I.J., Arculus, R.J., & Eggins, S.M. (2003). Peridotite xenoliths from Grenada, Lesser Antilles Island Arc. *Contributions to Mineralogy and Petrology, 146*(2), 241–262. doi: 10.1007/s00410-003-0500-z

Parkinson, I.J., & Pearce, J. A. (1998). Peridotites from the Izu-Bonin-Mariana forearc (ODP leg 125): Evidence for mantle melting and melt-mantle interaction in a supra-subduction zone setting. *Journal of Petrology, 39*(9), 1577–1618.

Parman, S. W., & Grove, T. L. (2004). Harzburgite melting with and without H_2O: Experimental data and predictive modeling. *Journal of Geophysical Research-Solid Earth, 109*(B2). doi: 10.1029/2003jb002566

Pearce, J. A., & Peate, D. W. (1995). Tectonic implications of the composition of volcanic arc magmas. *Annual Review of Earth and Planetary Sciences, 23*, 251–285.

Peslier, A. H., Luhr, J. F., & Post, J. (2002). Low water contents in pyroxenes from spinel-peridotites of the oxidized, sub-arc mantle wedge. *Earth and Planetary Science Letters, 201*(1), 69–86.

Peters, D., Bretscher, A., John, T., Scambelluri, M., & Pettke, T. (2017). Fluid-mobile elements in serpentinites: Constraints on serpentinisation environments and element cycling in subduction zones. *Chemical Geology, 466*, 654–666. doi: 10.1016/j.chemgeo.2017.07.017

Pichavant, M., & Macdonald, R. (2007). Crystallization of primitive basaltic magmas at crustal pressures and genesis of the calc-alkaline igneous suite: experimental evidence from St Vincent, Lesser Antilles arc. *Contributions to Mineralogy and Petrology, 154*(5), 535–558. doi: 10.1007/s00410-007-0208-6

Plank, T., & Langmuir, C. H. (1998). The chemical composition of subducting sediment and its consequences for the crust and mantle. *Chemical Geology, 145*(3–4), 325–394.

Pokrovski, G. S., & Dubessy, J. (2015). Stability and abundance of the trisulfur radical ion S-3(-) in hydrothermal fluids. *Earth and Planetary Science Letters, 411*, 298–309. doi: 10.1016/j.epsl.2014.11.035

Pokrovski, G. S., & Dubrovinsky, Leonid S. (2011). The S-3(-) ion Is stable in geological fluids at elevated temperatures and pressures. *Science, 331*(6020), 1052–1054.

Pokrovski, G. S., Tagirov, B. R., Schott, J., Hazemann, J.-L., & Proux, O. (2009). A new view on gold speciation in sulfur-bearing hydrothermal fluids from in situ X-ray absorption spectroscopy and quantum-chemical modeling. *Geochimica Et Cosmochimica Acta, 73*(18), 5406–5427. doi: 10.1016/j.gca.2009.06.007

Poli, S. (2015). Carbon mobilized at shallow depths in subduction zones by carbonatitic liquids. *Nature Geoscience, 8*(8), 633–636. doi: 10.1038/ngeo2464

Poli, S. (2016). Melting carbonated epidote eclogites: carbonatites from subducting slabs. *Progress in Earth and Planetary Science, 3*. doi: 10.1186/s40645-016-0105-6

Poli, S., Franzolin, E., Fumagalli, P., & Crottini, A. (2009). The transport of carbon and hydrogen in subducted oceanic crust: An experimental study to 5 GPa. *Earth and Planetary Science Letters, 278*(3–4), 350–360. doi: 10.1016/j.epsl.2008.12.022

Pons, M. L., Debret, B., Bouilhol, P., Delacour, A., & Williams, H. (2016). Zinc isotope evidence for sulfate-rich fluid transfer across subduction zones. *Nature Communications, 7*. doi: 10.1038/ncomms13794

Powell, R., Guiraud, M., & White, R. W. (2005). Truth and beauty in metamorphic phase equilibria: Conjugate variables and phase diagrams. *Canadian Mineralogist, 43*, 21–33.

Ranero, C. R., & Sallares, V. (2004). Geophysical evidence for hydration of the crust and mantle of the Nazca plate during bending at the north Chile trench. *Geology, 32*(7), 549–552.

Rea, D.K., & Ruff, L.J. (1996). Composition and mass flux of sediment entering the world's subduction zones: implications for global sediment budgets, great earthqueakes, and volcanism. *Earth and Planetary Science Letters, 140*, 1–12.

Richards, J. R., & Kerrich, R. (2007). Special paper: Adakite-like rocks: Their diverse origins and questionable role in metallogenesis. *Economic Geology, 102*(4), 537–576.

Richards, Jeremy P. (2011). Magmatic to hydrothermal metal fluxes in convergent and collided margins. *Ore Geology Reviews, 40*(1), 1–26.

Rielli, A., Tomkins, A. G., Nebel, O., Brugger, J., Etschmann, B., & Paterson, D. (2018). Garnet peridotites reveal spatial and temporal changes in the oxidation potential of subduction. *Scientific Reports, 8*. doi:10.1038/s41598-018-34669-0

Rielli, A., Tomkins, A. G., Nebel, O., Brugger, J., Etschmann, B., Zhong, R., et al. (2017). Evidence of sub-arc mantle oxidation by sulphur and carbon. *Geochemical Perspectives Letters, 3*(2), 124–132. doi: 10.7185/geochemlet.1713

Rielli, A., & Tomkins, A. G. (2015). Understanding Cu-Au Metallogeny at Convergent Margins: Oxidation of the Sub-Arc Mantle and Genesis of Arc-Related Ore Deposits. Paper presented at the World class ore deposits: discovery to recovery, Hobart.

Sato, K., Kovalenko, S. V., Romanovsky, N. P., Nedachi, M., Berdnikov, N. V., & Ishihara, T. (2004). Crustal control on the redox state of granitoid magmas: tectonic implications from the granitoid and metallogenic provinces in the circum-Japan Sea Region. In S. Ishihara, W. E. Stephens, S. L. Harley, M. Arima & T. Nakajima (Eds.), *Fifth Hutton Symposium on the Origin of Granites and Related Rocks* (Vol. 389, pp. 319–337).

Scambelluri, M., Pettke, T., & van Roermund, H. L. M. (2008). Majoritic garnets monitor deep subduction fluid flow and mantle dynamics. *Geology, 36*(1), 59–62. doi: 10.1130/g24056a.1

Scambelluri, M., van Roermund, H. L. M., & Pettke, T. (2010). Mantle wedge peridotites Fossil reservoirs of deep subduction zone processes Inferences from high and ultrahigh-pressure rocks from Bardane (Western Norway) and Ulten (Italian Alps). *Lithos, 120*(1–2), 186–201. doi: 10.1016/j.lithos.2010.03.001

Schoenberg, R., & von Blanckenburg, F. (2006). Modes of planetary-scale Fe isotope fractionation. *Earth and Planetary Science Letters, 252*(3–4), 342–359. doi: 10.1016/j.epsl.2006.09.045

Schuessler, J. A., Botcharnikov, R. E., Behrens, H., Misiti, V., & Freda, C. (2008). Oxidation state of iron in hydrous phono-tephritic melts. *American Mineralogist*, *93*(10), 1493–1504. doi: 10.2138/am.2008.2795

Schwarzenbach, E. M., Frueh-Green, G. L., & Bernasconi, S. M. (2009). Sulfur and carbon geochemistry of carbonate-veined serpentinites from Liguria: A comparison to the Iberian Margin and Lost City. *Geochimica Et Cosmochimica Acta*, *73*(13), A1187–A1187.

Schwarzenbach, E. M., Fruh-Green, G. L., Bernasconi, S. M., Alt, J. C., & Plas, A. (2013). Serpentinization and carbon sequestration: A study of two ancient peridotite-hosted hydrothermal systems. *Chemical Geology*, *351*, 115–133. doi: 10.1016/j.chemgeo.2013.05.016

Schwarzenbach, E. M., Fruh-Green, G. L., Bernasconi, S. M., Alt, J. C., Shanks, W. C., Gaggero, L., & Crispini, L. (2012). Sulfur geochemistry of peridotite-hosted hydrothermal systems: Comparing the Ligurian ophiolites with oceanic serpentinites. *Geochimica Et Cosmochimica Acta*, *91*, 283–305. doi: 10.1016/j.gca.2012.05.021

Seo, J. H., Guillong, M., & Heinrich, C. A. (2012). Separation of molybdenum and copper in porphyry deposits: The roles of sulfur, redox, and pH in ore mineral deposition at Bingham Canyon. *Economic Geology*, *107*(2), 333–356. doi: 10.2113/econgeo.107.2.333

Shaw, C. S. J., & Dingwell, D. B. (2008). Experimental peridotite-melt reaction at one atmosphere: a textural and chemical study. *Contributions to Mineralogy and Petrology*, *155*(2), 199–214. doi: 10.1007/s00410-007-0237-1

Shishkina, T. A., Portnyagin, M. V., Botcharnikov, R. E., Almeev, R. R., Simonyan, A. V., Garbebe-Schonberg, D., et al. (2018). Experimental calibration and implications of olivine-melt vanadium oxybarometry for hydrous basaltic arc magmas. *American Mineralogist*, *103*(3), 369–383.

Sillitoe, R. H. (2008). Major gold deposits and belts of the North and South American Cordillera: Distribution, tectonomagmatic settings, and metallogenic considerations. *Economic Geology*, *103*(4), 663–687. doi: 10.2113/gsecongeo.103.4.663

Sillitoe, R. H. (2010). Porphyry copper systems. *Economic Geology*, *105*(1), 3–41.

Simon, A.C., Pettke, T., Candela, P. A., Piccoli, P. M., & Heinrich, C.A. (2004). Magnetite solubility and iron transport in magmatic-hydrothermal environments. *Geochemica et Cosmochimica Acta*, *68*, 4905–4914.

Sisson, T. W., & Grove, T. L. (1993). Experimental investigations of the role of H_2O in calc-alkaline differentiation and subduction zone magmatism. *Contributions to Mineralogy and Petrology*, *113*(2), 143–166. doi: 10.1007/bf00283225

Skora, S., Freymuth, H., Blundy, J., Elliott, T., & Guillong, M. (2017). An experimental study of the behaviour of cerium/molybdenum ratios during subduction: Implications for tracing the slab component in the Lesser Antilles and Mariana Arc. *Geochimica Et Cosmochimica Acta*, *212*, 133–155. doi: 10.1016/j.gca.2017.05.025

Smith, C. B., Walter, M. J., Bulanova, G. P., Mikhail, S., Burnham, A. D., Gobbo, L., & Kohn, S. C. (2016). Diamonds from Dachine, French Guiana: A unique record of early Proterozoic subduction. *Lithos*, *265*, 82–95. doi: 10.1016/j.lithos.2016.09.026

Song, S. G., Su, L., Niu, Y. L., Lai, Y., & Zhang, L. F. (2009). CH_4 inclusions in orogenic harzburgite: Evidence for reduced slab fluids and implication for redox melting in mantle wedge. *Geochimica Et Cosmochimica Acta*, *73*(6), 1737–1754. doi: 10.1016/j.gca.2008.12.008

Sossi, P. A., & Debret, B. (2019). The role of redox processes in determining the iron isotope compositions of minerals, melts, and fluids. In R. Moretti & D. Neuville (Eds.), *Magma Redox Geochemistry*. USA: American Geophysical Union.

Spandler, C., & Pirard, C. (2013). Element recycling from subducting slabs to arc crust: A review. *Lithos*, *170*, 208–223. doi: 10.1016/j.lithos.2013.02.016

Spandler, C., Pettke, T., & Rubatto, D. (2011). Internal and external fluid sources for eclogite-facies veins in the Monviso meta-ophiolite, Western Alps: implications for fluid flow in subduction zones. *Journal of Petrology*, *52*(6), 1207–1236.

Stagno, V., Ojwang, D. O., McCammon, C. A., & Frost, D. J. (2013). The oxidation state of the mantle and the extraction of carbon from Earth's interior. *Nature*, *493*(7430), 84–88. doi: 10.1038/nature11679

Stamper, C. C., Melekhova, E., Blundy, J. D., Arculus, R. J., Humphreys, M. C. S., & Brooker, R. A. (2014). Oxidised phase relations of a primitive basalt from Grenada, Lesser Antilles. *Contributions to Mineralogy and Petrology*, *167*(1). doi: 10.1007/s00410-013-0954-6

Stern, R. J. (2002). Subduction zones. *Reviews of Geophysics*, *40* (4). doi: 10.1029/2001rg000108

Stolper, D. A., & Keller, C. B. (2018). A record of deep-ocean dissolved O-2 from the oxidation state of iron in submarine basalts. *Nature*, *553*(7688), 323–327. doi: 10.1038/nature25009

Sun, W., Arculus, R.J., Kamenetsky, V.S., & Binns, R.A. (2004). Release of gold-bearing fluids in convergent margin magmas prompted by magnetite crystallization. *Nature*, *431*, 975–979.

Thomsen, T. B., & Schmidt, M. W. (2008). Melting of carbonated pelites at 2.5–5.0 GPa, silicate-carbonatite liquid immiscibility, and potassium-carbon metasomatism of the mantle. *Earth and Planetary Science Letters*, *267*(1–2), 17–31. doi: 10.1016/j.epsl.2007.11.027

Tiraboschi, C., Tumiati, S., Sverjensky, D., Pettke, T., Ulmer, P., & Poli, S. (2018). Experimental determination of magnesia and silica solubilities in graphite-saturated and redox-buffered high-pressure COH fluids in equilibrium with forsterite plus enstatite and magnesite plus enstatite. *Contributions to Mineralogy and Petrology*, *173*(1). doi: 10.1007/s00410-017-1427-0

Tollan, P. M. E., Dale, C. W., Hermann, J., Davidson, J. P., & Arculus, R. J. (2017). Generation and modification of the mantle wedge and lithosphere beneath the West Bismarck island arc: melting, metasomatism and thermal history of peridotite xenoliths from Ritter Island. *Journal of Petrology*, *58*(8), 1475–1510. doi: 10.1093/petrology/egx062

Tomkins, A. G., & Evans, K. A. (2015). Separate zones of sulfate and sulfide release from subducted mafic ocean crust. *Earth and Planetary Science Letters*, *428*, 73–83.

Tomkins, A. G., Rebryna, K. C., Weinberg, R. F., & Schaefer, B. F. (2012). Magmatic Sulfide Formation by Reduction of Oxidized Arc Basalt. *Journal of Petrology*, *53*(8), 1537–1567. doi: 10.1093/petrology/egs025

Tsuno, K., & Dasgupta, R. (2011). Melting phase relation of nominally anhydrous, carbonated pelitic-eclogite at 2.5–3.0 GPa and deep cycling of sedimentary carbon. *Contributions to Mineralogy and Petrology*, *161*(5), 743–763. doi: 10.1007/s00410-010-0560-9

Tumiati, S., Fumagalli, P., Tiraboschi, C., & Poli, S. (2013). An Experimental Study on COH-bearing Peridotite up to 3 center dot 2 GPa and Implications for Crust-Mantle Recycling. *Journal of Petrology*, *54*(3), 453–479. doi: 10.1093/petrology/egs074

Tumiati, S., Godard, G., Martin, S., Malaspina, N., & Poli, S. (2015). Ultra-oxidized rocks in subduction melanges? Decoupling between oxygen fugacity and oxygen availability in a Mn-rich metasomatic environment. *Lithos*, *226*, 116–130. doi: 10.1016/j.lithos.2014.12.008

Ulrich, M., Picard, C., Guillot, S., Chauvel, C., Cluzel, D., & Meffre, S. (2010). Multiple melting stages and refertilization as indicators for ridge to subduction formation: The New Caledonia ophiolite. *Lithos*, *115*(1–4), 223–236. doi: 10.1016/j.lithos.2009.12.011

Uysal, I., Ersoy, E. Y., Dilek, Y., Escayola, M., Sarifakioglu, E., Saka, S., & Hirata, T. (2015). Depletion and refertilization of the Tethyan oceanic upper mantle as revealed by the early Jurassic Refahiye ophiolite, NE Anatolia-Turkey. *Gondwana Research*, *27*(2), 594–611. doi: 10.1016/j.gr.2013.09.008

Uysal, I., Kaliwoda, M., Karsli, O., Tarkian, M., Sadiklar, M. B., & Ottley, C. J. (2007). Compositional variations as a result of partial melting and melt-peridotite interaction in an upper mantle section from the Ortaca Area, southwestern Turkey. *Canadian Mineralogist*, *45*, 1471–1493. doi: 10.3749/canmin.45.6.1471

van Acken, D., Becker, H., & Walker, R. J. (2008). Refertilization of Jurassic oceanic peridotites from the Tethys Ocean – Implications for the Re-Os systematics of the upper mantle. *Earth and Planetary Science Letters*, *268*(1–2), 171–181. doi: 10.1016/j.epsl.2008.01.002

van Roermund, H. (2009). Mantle-wedge garnet peridotites from the northernmost ultra-high pressure domain of the Western Gneiss Region, SW Norway. *European Journal of Mineralogy*, *21*(6), 1085–1096. doi: 10.1127/0935-1221/2009/0021-1976

Waters, L. E., & Lange, R. A. (2016). No effect of H_2O degassing on the oxidation state of magmatic liquids. *Earth and Planetary Science Letters*, *447*, 48–59. doi: 10.1016/j.epsl.2016.04.030

Williams, H. M., Prytulak, J., Woodhead, J. D., Kelley, K. A., Brounce, M., & Plank, T. (2018). Interplay of crystal fractionation, sulfide saturation and oxygen fugacity on the iron isotope composition of arc lavas: An example from the Marianas. *Geochimica Et Cosmochimica Acta*, *226*, 224–243. doi: 10.1016/j.gca.2018.02.008

Wood, B. J. (1990). An experimental test of the spinal peridotite oxygen barometer. *Journal of Geophysical Research-Solid Earth and Planets*, *95*(B10), 15845–15851. doi: 10.1029/JB095iB10p15845

Wood, B. J. (1991). Oxygen barometry of spinal peridotites. *Reviews in Mineralogy*, *25*, 417–431.

Wood, B. J., Bryndzia, L. T., & Johnson, K. E. (1990). Mantle oxidation-state and its relationship to tectonic environment and fluid speciation. *Science*, *248*(4953), 337–345.

Wood, B. J., & Virgo, D. (1989). Upper mantle oxidation-state - ferric iron contents of lherzolite spinels by Fe-57 Mössbauer-spectroscopy and resultant oxygen fugacities. *Geochimica Et Cosmochimica Acta*, *53*(6), 1277–1291.

Wykes, J. L., Newton, R. C., & Manning, C. E. (2008). Solubility of andradite, Ca3Fe2Si3O12, in a 10 mol% NaCl solution at 800 degrees C and 10 kbar: Implications for the metasomatic origin of grandite garnet in calc-silicate granulites. *American Mineralogist*, *93*(5–6), 886–892.

Yaxley, G. M., Berry, A. J., O'Neill, H. S. C., & Woodland, A. B. (2005). Calibration of XANES for determination of Fe3+/Sigma Fe in garnet. *Geochimica Et Cosmochimica Acta*, *69*(10), A828–A828.

Yaxley, G. M., & Brey, G. P. (2004). Phase relations of carbonate-bearing eclogite assemblages from 2.5 to 5.5 GPa: implications for petrogenesis of carbonatites. *Contributions to Mineralogy and Petrology*, *146*(5), 606–619.

Yaxley, G. M., & Green, D. H. (1994). Experimental demonstration of refractory carbonate-bearing eclogite and siliceous melt in the subduction regime. *Earth and Planetary Science Letters*, *128*(3–4), 313–325. doi: 10.1016/0012-821x(94)90153-8

Yaxley, G. M., & Green, D. H. (1998). Reactions between eclogite and peridotite: Mantle refertilisation by subduction of oceanic crust. *Schweizerische Mineralogische Und Petrographische Mitteilungen*, *78*(2), 243–255.

Zimmer, M. M., Plank, T., Hauri, E. H., Yogodzinski, G. M., Stelling, P., Larsen, J., et al. (2010). The role of water in generating the calc-alkaline trend: new volatile data for Aleutian magmas and a new Tholeiitic Index. *Journal of Petrology*, *51*(12), 2411–2444. doi: 10.1093/petrology/egq062

5

Redox Melting in the Mantle

Stephen F. Foley

ABSTRACT

Redox melting is caused by a change in oxidation state that is not necessarily coupled to a change in temperature or pressure, but hinges on the effects of COH volatiles on depressing the melting point. Hydrous redox melting (HRM) operates when fluids containing CH_4 and H_2O react to deposit solid carbon, and the increasing activity of water (aH_2O) that results causes a severe drop in the melting temperature. In more oxidized conditions, carbonated redox melting (CRM) results from the solidus with both H_2O and CO_2 being lower than that with H_2O alone. In many cases, carbon deposited during HRM provides the carbon reservoir for later carbonate-rich magmatism in the same or adjacent areas. HRM especially causes fractionation of redox conditions, with oxidized, H_2O-rich melts leaving behind reduced, dry residues, explaining cratonic lithosphere keels. This was the main driver of the oxidation of surface conditions in the late Archean. Incipient melting was restricted in the Archean, so that initial melting was not extensive despite higher mantle temperatures. The subduction environment became gradually more oxidized between 3.3 and 2.3 Ga. The CRM mechanism was inoperable until the late Archean, and redox melting of recycled lithospheric blocks in the deep mantle had to await oxidized conditions in subduction zones.

5.1. INTRODUCTION

Small amounts of volatile components consisting mainly of carbon, hydrogen, and oxygen (H_2O, CO_2, CH_4, and H_2) are the principal determining factors controlling partial melting of the Earth's mantle. Water, CO_2, and mixtures of them depress the melting point of peridotite in the upper mantle through several hundred degrees (Green, 1973; Olafsson & Eggler, 1983; Wallace & Green, 1988; Dasgupta & Hirschmann, 2006; Foley et al., 2009). This results in a considerable temperature range in which the percentage of melting is always low, named the "incipient melting" regime by Green and Falloon (1998), before major melting commences at much higher temperatures. The effect of this incipient melting regime is that volatile-rich melts, which have low viscosities and thus are very mobile, are commonly present in the mantle for extensive periods of time in low-temperature conditions where larger-volume basaltic melts cannot be generated. Volatile-rich, low-degree melts have therefore played a more significant role in the evolution of the mantle than has commonly been recognized.

The term "redox melting" was introduced by Taylor (1985) for partial melting that can occur without a change in temperature or pressure, resulting instead from a change in oxidation conditions in the presence of a migrating COH fluid. The oxidation of reduced H_2O + CH_4 fluids in contact with more oxidizing rocks causes an increase in water activity accompanied by the precipitation of solid carbon ($CH_4 + O_2 \rightarrow 2H_2O + C$), and the

Department of Earth and Environmental Sciences, and ARC Centre of Excellence for Core to Crust Fluid Systems, Macquarie University, Sydney, New South Wales, Australia

Magma Redox Geochemistry, Geophysical Monograph 266, First Edition.
Edited by Roberto Moretti and Daniel R. Neuville.

DOI: 10.1002/9781119473206.ch5

H_2O so produced depresses the melting point of the rock. Later, improved knowledge of the melting curves for mantle peridotite in the presence of mixed volatiles led to the recognition of a second redox melting mechanism by which solid carbon is oxidized to carbonate, leading to the formation of carbonate-rich melts (Foley, 2011). This occurs at still lower temperatures than the first type of redox melting, and may follow it chronologically in the same region.

The occurrence of redox melting by either mechanism is intimately associated with the incipient melting regime, because the redox melting process produces volatile-rich, low-degree melts. The wide variety of melts produced under these conditions, ranging from mildly alkaline silicate melts to carbonatites with less than 5 wt.% SiO_2, are known otherwise as the agents of mantle metasomatism. An improved understanding of low-degree melting and redox melting promises, therefore, to make inroads into the understanding of mantle metasomatism.

This contribution reviews the development of the redox melting concept and re-examines the geodynamic settings in which redox melting may be important. It concentrates on continental rifting and subduction, and also traces the importance of redox melting through geological time, relating it to the evolution of the mantle. Two points will be especially highlighted: (i) Redox melting will *usually* fractionate the oxidation states between melts and their residues, governed by the differential solubilities of the volatile species. Residues will thus be polarized in oxidation state, and a corollary is that the redox states of melts will *not* represent those of their source rocks, nor melting conditions. (ii) Redox states vary enormously from place to place, so that regional variations in oxidation state need to be examined. To speak of and expect to find a singular "oxidation state of the mantle" in any particular place is as unlikely as finding a garnet composition corresponding to average global garnet. Global averages such as the much discussed "oxidation state of the mantle" (e.g., O'Neill & Wall, 1987; Ionov & Wood, 1992; Frost & McCammon, 2008; Rohrbach et al., 2011) are only generalizations and may deflect attention from the importance and ubiquity of redox melting and freezing.

It is increasingly accepted that models of the mantle appealing to homogeneous peridotite are outdated and that other rocks types such as eclogite and pyroxenite, many of them containing hydrous minerals (O'Reilly & Griffin, 2013; Sobolev et al., 2005) may be common in some parts of the mantle (Foley, 1992a). This is also important for redox melting and freezing processes because of the differing buffering capacities of these rocks (Stachel and Luth, 2015; Beyer and Frost, 2017) and their lower melting points relative to their peridotite hosts (Foley, 1992a; Hirschmann & Stolper, 1996; Foley et al., 1999; Kogiso et al, 2004). This review will consider the effects of these other ultramafic lithologies on redox melting, as well as new evidence for extreme oxidation conditions, for more reduced conditions in the early Earth, and for the temporal evolution of redox states in the subduction process either side of the Great Oxidation Event at 2.3 Ga.

5.2. MANTLE MELTING WITH VOLATILE COMPONENTS

The fundamental basis for redox melting lies in the differential effects of COH volatiles on the melting point of mantle peridotite and other ultramafic mantle rocks. The melting curves for peridotite in the presence of various mixtures of COH volatiles are summarized in

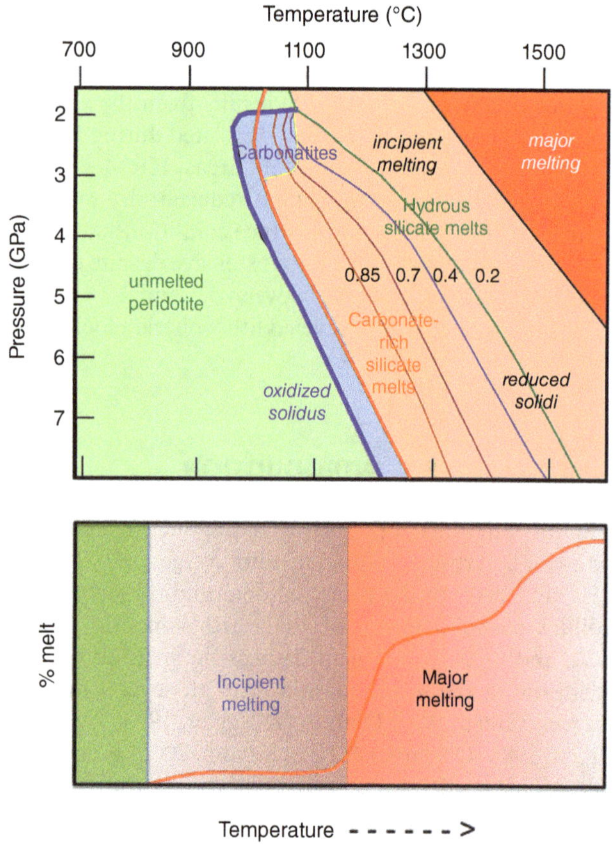

Figure 5.1 (a) Solidus curves for peridotite in dry conditions (black line) and with mixed volatiles; with H_2O and CO_2 (thick blue line), H_2O alone (thick red line), H_2O-undersaturated conditions (yellow line) and with CH_4+H_2O mixtures noted as water activities (thin coloured lines). (b) Illustration of low degrees of melting in the incipient melting regime which corresponds to the blue to buff-coloured regions in (a), in which melt compositions change from carbonate-rich, silicate-poor (blue) to carbonate-poor and SiO_2-richer.

Figure 5.1. With no volatiles present, peridotite melts at temperatures higher than 1300°C at mantle pressures, and the melting temperature increases rapidly with increasing pressure (black line in Figure 5.1; Green & Falloon, 1998; Walter, 1998). The relevance of the volatile-free melting curve for redox melting is limited to comparison of its high-temperature position relative to melting in the presence of volatiles.

Experimental studies of the effect of H_2O on the melting curve showed that the melting temperature is depressed through hundreds of degrees, resulting in the thick red curve in Figure 5.1 (Green, 1973). In water-undersaturated conditions with very small amounts of H_2O, the melting curve lies at slightly higher temperatures, particularly at pressures below 3 GPa because of the stability of amphibole. The position of this curve (thin yellow line in Figure 5.1) is approximate, because the stability of amphibole varies as a function of Mg# and alkali contents, being more stable where values for both of these factors are higher (Millhollen et al., 1974; Wallace & Green, 1991). The backbend in the curve at 3 GPa is governed by the maximum stability of amphibole; at higher pressures, the melting temperature drops sharply, and may be determined by melting of trace amounts of phlogopite wherever sufficient K_2O and H2O is present (Wyllie, 1977).

The melting curve for peridotite in the presence of CO_2 alone is not shown in Figure 5.1 as it is unlikely to apply in natural conditions. The blue curve shows the melting point in the presence of both CO_2 and H_2O. Carbon dioxide also depresses the solidus of peridotite by several hundred degrees (although less than H_2O), particularly at pressures above ~2 GPa because of the stability of solid carbonates (Olafsson & Eggler, 1983; Falloon & Green, 1989; Dalton & Presnall, 1998). In the presence of both H_2O and CO_2, the melting point is depressed even more than with H_2O alone (blue line in Figure 5.1), and a field for sodic carbonatite melt coexisting with amphibole occurs between 2 and 3 GPa (Wallace & Green, 1988). This carbonatite melt field thins considerably towards higher pressures, but a thin zone with carbonate-rich melt persists to at least 7 GPa (Foley et al., 2009; Pintér, 2019; Pintér et al., 2021) and is extrapolated to higher pressures assuming that it remains parallel to the better-determined position of peridotite + CO_2 (Dasgupta and Hirschmann, 2006; Ghosh et al., 2014).

Under reducing conditions close to and slightly above oxygen fugacities (fO_2) of the iron-wüstite buffer (IW), C–O–H fluids do not contain CO_2 but consist of CH_4+H_2O+H_2 (±C_2H_6) in proportions that vary widely over a relatively small range in oxygen fugacity (Figure 5.2). For example, at 5 GPa fluids consist of > 99% H_2O at the water maximum but about 75% CH_4, < 20% H_2O, and 2–3% each of H_2 and C_2H_6 at IW just 2 log fO_2 units lower. Because of this rapid variation in CH_4 and H_2O contents, a family of melting curves that vary over several hundred degrees at pressures of 3 GPa and greater is plotted in Figure 5.1 (Taylor & Green,

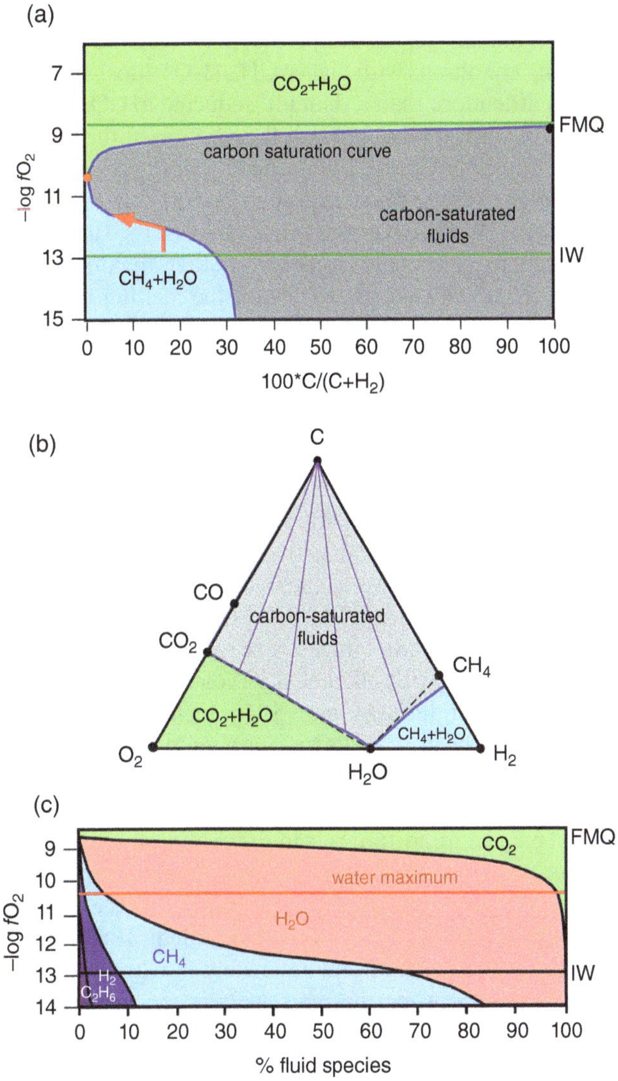

Figure 5.2 Two depictions of the reduced and oxidized fluid realms for COH fluid mixtures (a,b) and an illustration of species variation as function of oxygen fugacity (c; for 3 GPa and 1200°C). (a) The carbon saturation surface (blue line) moves to almost pure H_2O at intermediate fO_2 (red dot). The HRM mechanism is best seen in this diagram (red arrow); reduced fluids move along the saturation surface, depositing solid carbon to cause an increase in H_2O/CH_4 in the fluid, which may result in melting. (b) C–O–H system illustrating the same reduced (blue) and oxidized (green) fluid realms and carbon saturation surface (blue line). Tie lines illustrate coexisting solid carbon and fluid compositions. (c) The water maximum at intermediate fO_2 is replaced by H_2O+CO_2 mixtures to higher fO_2 and mixtures of H_2O and CH_4 with minor H_2 and C_2H_6 at lower fO_2.

1988). These are marked with different activity of H_2O (aH_2O): higher melting temperatures correspond to more reducing conditions with higher CH_4/H_2O ratios. Methane acts as little more than a dilutant, reducing aH_2O: it is the active role of H_2O that governs the position of the melting curve (Taylor & Green, 1988). Increasing oxygen fugacity conditions over the restricted range from IW to 2 log units above IW (IW+2) cover the entire range of mixtures from CH_4-rich to extremely H_2O-rich ("water maximum" in Figure 5.2c), with drastic effects on the melting temperature. Note that the preferred mantle melting curve of Green and Falloon (1998) is a combination of the curves for oxidizing and reducing conditions, constructed to correspond to falling oxygen fugacity with increasing pressure.

The combined effect of mixtures of H_2O, CO_2, and CH_4 is that they define a very variable extent of the incipient melting region (Green & Falloon, 1998; Foley & Pintér, 2018). These melts exist over the entire pressure-temperature range between the volatile-free melting curve and that for peridotite+H_2O+CO_2 (Figure 5.1a). Despite this extensive pressure-temperature range, only small proportions of melt with alkaline and volatile-rich compositions will be produced (Figure 5.1b), increasing from less than 5% melting at < 3 GPa to a maximum of around 25% at the base of cratonic lithosphere (Foley et al., 2009; Foley & Pintér, 2018). Melts from this incipient melting region play an important role in the evolution of the mantle, because they are mobile and have low viscosities due to the reduction of viscosity by dissolved volatile components (Mysen et al., 1982; Treiman & Schedl, 1983). Thus, although never voluminous, and rarely seen at the Earth's surface, they may be the most widespread melts in the mantle, where they are best known as the agents of mantle metasomatism (O'Reilly & Griffin, 2013) and as the major transport agents for volatile components (Foley & Fischer, 2017). Furthermore, incipient melting may progress to major melting, so that incipient melts are widespread, universal precursors to more voluminous melt production. Most voluminous major melts will therefore unavoidably contain chemical components from the incipient melting process.

5.3. REDOX MELTING

5.3.1. Development of the Concept

Redox melting refers to melting that occurs because of a contrast in oxidation state without any change in pressure or temperature. This will most commonly occur where a melt or fluid migrates into a rock with which its volatile speciation is incompatible, requiring an adjustment of the volatile speciation, which in turn affects the melting point of the rock. The term *redox melting* was originally applied to melting caused by the interaction of reduced CH_4- and H_2-bearing fluids with more oxidized rocks, resulting in an increase in aH_2O by the reaction of CH_4 with O_2 (Taylor, 1985). This process results in the precipitation of solid carbon and a rapid drop in the solidus temperature: it was thought to apply particularly to the interaction of reduced, mantle-derived H_2+CH_4 fluids with large oxidized blocks originating from the recycling of oceanic lithosphere in subduction zones (Taylor & Green, 1987; Green et al., 1987). The melting process corresponds to a change in the melting curve from the high-temperature part of the incipient melting zone in Figure 5.1, through the curves with increasing aH_2O towards the water-saturated curve (red line in Figure 5.1).

A second mechanism of redox melting in more oxidized conditions (Foley, 2011) was identified following high-pressure experiments that documented a lower melting point for peridotite with CO_2 and H_2O than with H_2O alone (Foley et al., 2009). Further oxidation beyond the water maximum (Figure 5.2) results in the conversion of solid carbon (graphite or diamond) to carbonate, which causes a further depression of the melting point and dissolves in the melt. This second mechanism was labelled CRM (carbonated redox melting) by Foley (2011), and the first HRM (hydrous redox melting). CRM relies on the oxidation of graphite or diamond that has accumulated in the lithosphere over time; much of this may be associated with earlier HRM activity.

Rohrbach and Schmidt (2011) coined the term *redox freezing* for the reverse of the CRM process: during the reaction of carbonatitic melts containing both CO_2 and H_2O with more reduced rocks, the melt freezes and carbonate in the melt solidifies as diamond or graphite. Redox freezing that is the converse of the HRM process may also occur where H_2O-rich melts react with reduced, carbon-bearing rocks: the melt would solidify and carbon would be consumed, generating CH_4-rich fluids. These freezing mechanisms should logically be referred to as CRF and HRF. We should note here that fluids also freeze, so that diamond formation may result from redox freezing without melts being involved at all. Figure 5.3 summarizes the variety of conditions over which the two redox melting mechanisms and redox freezing may occur. This also illustrates that redox fractionation may result: melting in an initially reduced mantle leaves a more reduced residue because of the incompatibility of H_2O and its transport away in the melt (Figure 5.3). A second melting episode (CRM) may occur because of *further* oxidation (Figure 5.3): this should result in a smaller redox fractionation between melt and residue, but also because of the high solubility of volatiles in the melt.

The operation of redox melting may be more common than has been recognized in the past, because the oxidation state of the upper mantle is not uniform, and greater

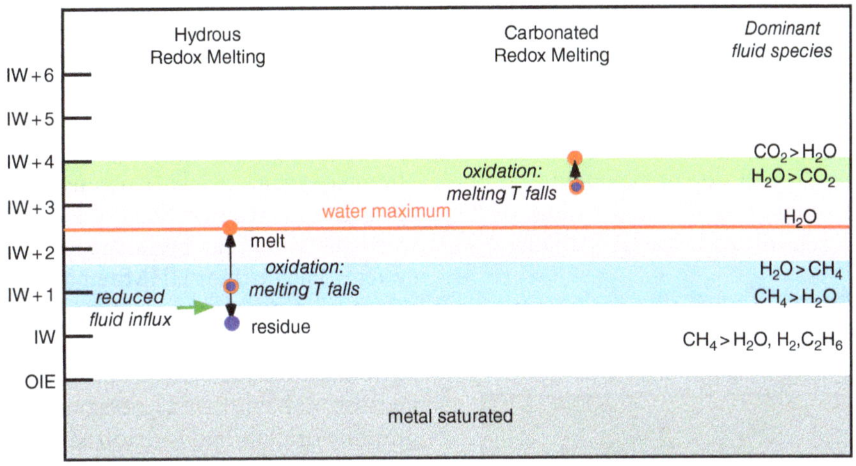

Figure 5.3 Illustration of the fO_2 levels at which HRM and CRM occur, related to COH fluid speciation (right). The two redox melting mechanisms occur either side of the water maximum (pink zone) in which all fluids are extremely H_2O-rich and coexist with solid carbon. The fractionation between melts (red dots) and residues (blue dots) are larger for HRM, which helps to cause the dry, depleted state of cratonic peridotites. Fractionation of redox states by CRM is much smaller.

variations in oxygen fugacity (fO_2) have been recognized in recent years (Section 5.5). These variations are considerable on the modern Earth as a function of depth and geodynamic setting, and the evolution of mantle oxidation states over time imply that melting and subduction conditions may have differed greatly in early Earth history. Redox melting and freezing of all types corresponds to many of the changes in mantle geochemistry and mineralogy generally referred to under the banner of "mantle metasomatism." Here, I consider melting due to redox changes alone, but just as incipient melts are incorporated in major melts as melting continues, redox melting will often be a forerunner of major decompression melting.

5.3.2. Redox Melting Mechanisms

The speciation of COH fluids at high pressures is shown in Figure 5.2, in which the first two panels show two commonly used and complementary illustrations of the existence of distinct regions for reduced CH_4+H_2O (blue) and oxidized H_2O+CO_2 fluids (green). Figure 5.2b shows fluid compositions in the system C-O-H: compositions within the grey areas consist of solid carbon and carbon-saturated fluids lying along the blue line, as illustrated by the family of tie-lines (Holloway, 1987; Ferry & Baumgartner, 1987; Holloway & Blank, 1994). Figure 5.2a plots the same information in a different way; it has the advantage of plotting fO_2 directly but cannot represent positions for individual fluid species or solid carbon effectively and so does not show tie-lines (Frost, 1979; Taylor & Green, 1987). Figure 5.2c illustrates the relative abundance of individual species as a function of fO_2, showing that fluids consist almost entirely of CO_2+H_2O in oxidized conditions just below the Fayalite-Magnetite-Quartz buffer (FMQ; ca. IW +4.5 log units), and become dominated by CH_4+H_2O with additional H_2 and C_2H_6 as fO_2 decreases. The relative positions of the oxygen buffers and the size of the areas for individual fluid species varies greatly with temperature and pressure.

HRM operates in relatively reduced conditions slightly above the iron–wüstite buffer at depths corresponding to the lower cratonic lithosphere. It can be expected to be particularly important for melt production in deeper levels of the "magmatic zone" of the upper mantle (150–250km) because of the decrease in fO_2 with increasing pressure, which should cause the prevalence of fO_2 conditions close to IW at these depths (Frost & McCammon, 2008; Rohrbach et al., 2011; Foley, 2011). On encountering more oxidized rocks, reduced fluids consisting principally of CH_4 > H_2O + H_2 will release water whilst precipitating carbon (red path in Figure 5.2a). The increase in water activity will cause melting if the solidus is depressed sufficiently to meet the geothermal gradient, which will be most common in low, cratonic geothermal gradients. Melting is thus an indirect product of the oxidation of methane to form water and solid carbon.

The effectiveness of the HRM mechanism hinges on the rapid change in the H_2O/CH_4 ratio of fluids in the fO_2 region 0.5 to 1.5 log units above IW (Figure 5.2a, c), and this restricts the operation of HRM to a narrow range of fO_2. The melting process is likely to begin only when relatively high aH_2O values are reached (> 0.70), because the melting curves for lower aH_2O (Figure 5.1) are unlikely to meet the geothermal gradient. Note, however, that CH_4+H_2O fluids are strongly non-ideal so that a small molar CH_4 content rapidly leads to lowering of aH_2O, emphasizing the role of changing CH_4/H_2O in the fluid (Woermann & Rosenhauer, 1985; Matveev et al., 1997).

At higher pressures than those corresponding to Figure 5.2, the water maximum widens and the range of fO_2 at which HRM can occur drops below IW (Figure 5.4).

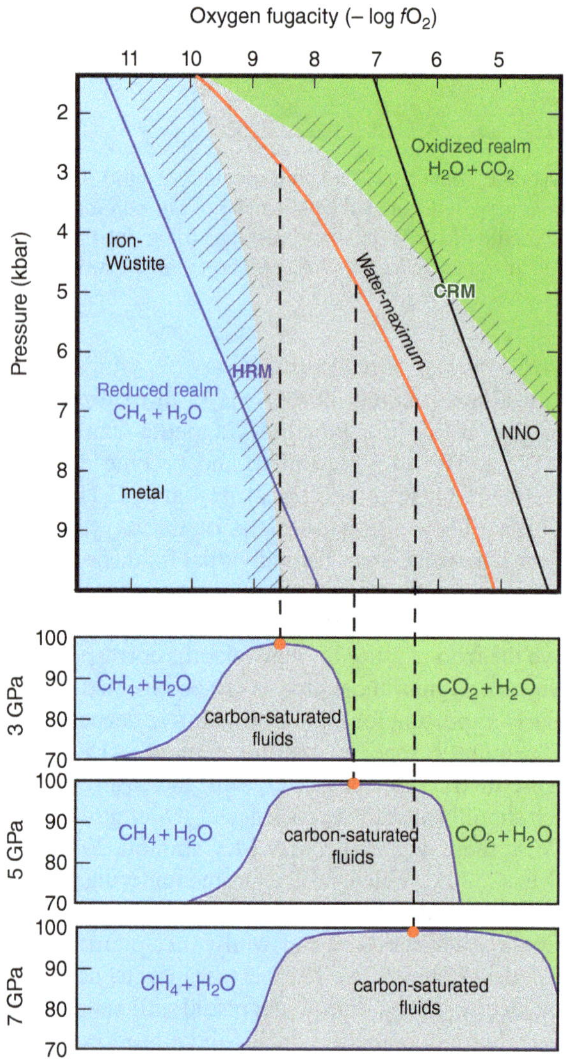

Figure 5.4 Illustration of pressure effects on the extent of redox fractionation during melting and the P-fO_2 regions for HRM and CRM. The grey area indicates fluids extremely rich in H_2O corresponding to lower panels for 3, 5, and 7 GPa. Hydrous redox melting (HRM) occurs at the reduced edge of this region and carbonated redox melting (CRM) at the oxidized end. The two RM mechanisms thus diverge in their fO_2 regions, meaning that redox melting fractionates redox conditions more strongly as pressure increases. CRM will only operate at pressures higher than 7 GPa where subducted blocks are recycled to deeper levels.

At pressures above ≈ 5 GPa, the exact position of the water maximum loses its meaning, as fluids consist of almost pure H_2O over several orders of magnitude in fO_2 (Woermann & Rosenhauer, 1985; lower panels in Figure 5.4). The water maximum is not a buffer, as only 50ppm oxygen is needed to change fO_2 through four orders of magnitude (Luth & Stachel, 2014), but movement away from this water maximum depends on the H/C ratio in the system. The field for HRM remains restricted to the lower fO_2 edge of this water maximum. For redox melting, the area of intermediate fO_2 (grey in Figure 5.4) is essentially barren. The shaded CRM field at much higher fO_2 occurs because of the further depression of the melting curve once graphite or diamond is oxidized to carbonate.

The occurrence of melting by HRM will depend on the availability of the volatile components carbon and hydrogen, which control the amount of water released by the redox mechanism. These will commonly be introduced by migrating fluids, but the simple juxtaposition of reduced and oxidized blocks may also lead to melting if they contain enough C and H. Recent summaries of the compositions of inclusions in diamonds indicate that the crystallisation of most diamonds occurs in reduced, CH_4-bearing conditions (Stachel et al., 2017), and may be related to the presence of iron carbide or metal in reduced environments (Palyanov et al., 2013; Smith et al., 2016). It is normally the fluids that control the fO_2 (Luth & Stachel, 2014), but their propensity to react with rocks and deposit diamonds depends on the identity and the Fe and C contents of the rocks. This is thought to promote the deposition of diamonds in eclogites relative to peridotites (Stachel & Luth, 2015; Beyer & Frost, 2017). A corollary of diamond deposition is the release of H_2O which may cause HRM in more oxidized, iron-rich rocks in the mantle such as eclogites and pyroxenites, and indeed any rocks that originate by the resolidification of incipient mantle melts.

The fractionation of oxidation states in the mantle will be accentuated by redox melting. Considering the path for HRM illustrated in Figure 5.2a, melts that form will contain abundant H_2O but not CH_4. This is because of the high solubility of H_2O in melts, whereas the solubility of CH_4 is limited to < 0.5wt.% (Taylor & Green, 1987; Mysen et al., 2009; Ardia et al., 2013). Thus, melts leaving the HRM melting region take with them a more oxidized volatile mixture and leave behind a more reduced residue. This casts doubt on the applicability of methods of estimation for mantle oxidation states that assume the oxidation state of volcanic melts corresponds to that of the source.

CRM results from a further increase in fO_2 from conditions corresponding to the water maximum, which causes a rapid increase in the ratio of CO_2 to H_2O in the fluid as long as the fluid remains saturated in carbon (Figure 5.2a, c). The possibility of this second redox melting mechanism was recognized after high-pressure experiments showed that the melting curve for peridotite with both H_2O and CO_2 is lower than for peridotite with either CO_2 or H_2O alone (Figure 5.1). This was first discovered with the carbonatite melt field at 2–3 GPa by Wallace and Green (1988), and shown by Foley et al. (2009) to extend to 6 GPa. It was recently confirmed to apply to a range of fertile and depleted peridotites (Pintér et al., 2021). CRM cannot occur at depths less than 65km because of the lack of carbonates in the peridotite (Brey et al., 1983), but it is now known that it could operate to depths of at least 200km. The rapid change from H_2O to CO_2-rich fluid speciation occurs $\approx$ 1 log unit below the Ni-NiO buffer (NNO) at 2–3 GPa, as confirmed experimentally by Tumiati et al. (2013), but at above NNO at higher pressures (Figure 5.4).

If little or no carbon is present in the rock or in the infiltrating fluid, then the carbon will soon be exhausted and the system cannot proceed to CO_2-rich conditions, giving the appearance of buffered conditions. The previous operation of HRM in the same rock would provide the ideal pre-treatment of the mantle for this second mechanism: the carbon deposited during HRM is available for oxidation to CO_2, which is then soluble in the melt formed by CRM. The abundance of diamonds in the lower cratonic lithosphere, documented by peridotite and eclogite xenoliths as well as mineral inclusions in diamonds (Boyd & Gurney, 1986; Stachel & Harris, 2008), means that the lower cratonic lithosphere may be especially susceptible to CRM.

The HRM and CRM mechanisms, particularly in combination, are the most likely cause of melting in the lower cratonic lithosphere. This is because changes in the speciation of COH volatiles are far more effective at changing the melting curve than are geodynamic effects on changing temperature on rapid timescales. Redox melting may indeed be the only way to initiate melting in the lithosphere at > 140km depth, as redox changes generate the volatile species needed to cause melting.

Redox freezing is the opposite of the CRM mechanism: the carbonate dissolved in carbonate-rich melts that migrate into a more reduced environment will be deposited as diamond (or graphite), offering an alternative formation mechanism for diamonds (Rohrbach & Schmidt, 2011). Stable isotope compositions of diamonds indicate the deposition of diamonds from both reduced and oxidized fluids or melts, with a predominance of reduced fluids (Palot et al., 2014; Stachel et al., 2017). Abundances of diamond in the lower cratonic lithosphere between 1 and 2 wt.% have recently been shown to be consistent with geochemical mass-balancing (Foley & Fischer, 2017) and fits of shear wave velocity profiles to mineral assemblages (Garber et al., 2021). These high contents of solid carbon attest to the operation of redox melting and freezing in this geodynamic environment.

5.4. COMPOSITIONS OF MELTS FORMED BY REDOX MELTING

The experimental determination of melt compositions in the mantle has concentrated on peridotite as the main rock of the mantle, and on the pressure range up to 4 GPa, because of the abundance of laboratories capable of reaching this pressure. Higher pressures require multi-anvil apparatuses that are less common and the conditions they achieve less well constrained. Our knowledge of peridotite melting without volatiles is summarized by Walter (1998) and Green and Falloon (1998). However, redox melting always involves mixtures of COH volatiles, which cause low-degree melting, and the information we have on compositions of volatile-rich melts is patchy and controversial. Furthermore, redox melting may occur preferentially where mineral assemblages other than peridotite are present, as these may be sites of contrasts in oxidation states and Fe, C, and S contents (Stachel & Luth, 2015). Currently, compositions of incipient melts are known better in oxidized conditions (H_2O+CO_2), but remain controversial due to quenching problems, whereas very little information is available for reduced conditions (CH_4+H_2O).

Many more experimental melt compositions have been determined for melting of peridotite with either H_2O or CO_2 alone (summarized by Foley & Pintér, 2018) than with H_2O and CO_2 together, despite these having limited direct applicability to redox melting. Some studies found a progression from carbonatitic melts with 5–6 wt.% SiO_2 to carbonated silicate melts with 30–35 wt.% SiO_2 (Dalton & Presnall, 1998; Dasgupta & Hirschmann, 2007; Brey et al., 2008; Ghosh et al., 2014), whereas others favoured an abrupt change (Moore & Wood, 1998; Dasgupta et al., 2007). The few experiments with both H_2O and CO_2 show similar features, with melt compositions progressing continuously from carbonatitic at < 10% melting to carbonate-bearing silicate melts similar to natural aillikites (carbonate-bearing ultramafic lamprophyres; Tappe et al., 2005) at 20–30% melting (Foley et al., 2009).

The carbonatite melt field is relatively wide between 2 and 3 GPa where it coexists with amphibole in peridotite with small amounts of both CO_2 and H_2O (Figure 5.1). These carbonatite melts are strongly enriched in Na_2O (5 wt.%; Wallace & Green, 1988) and many highly incompatible elements (Sweeney et al., 1992; Dasgupta & Hirschmann, 2006). At > 3 GPa, a thin zone of carbonatitic initial melts persists to at least 6 GPa, where Na_2O contents are lower (< 1.7 wt.%; Foley et al., 2009; Pintér et al., 2021), and decrease further to < 0.8 wt.% Na_2O at 10 GPa in experiments with CO_2 but no H_2O

(Brey et al., 2008). Current results show an abrupt change between carbonatite and SiO_2-understurated silicate melts only at 2–3 GPa; there is a transition to a continuum between carbonate and silicate melts as pressure increases above 4 GPa (Foley et al., 2009; Pintér et al., 2021).

In terms of resemblance of experimental melts to possible primary mantle-derived igneous rock types, most melts from the incipient melting regime with differing ratios of CO_2/H_2O probably never reach the surface, including all sodic carbonatitic melts from 2–3 GPa. Experiments with H_2O and CO_2, which are relevant to CRM, resemble aillikites at pressures of 4–6 GPa. These have SiO_2 contents intermediate between carbonatites and melilitites (15–30 wt.%; Tappe et al., 2006), but cannot form at lower pressures because their compositions fall in a miscibility gap that opens between carbonatitic melts with < 10 wt.% SiO_2 and melilitites with about 35 wt.% SiO_2 (Hamilton et al., 1979; Kjarsgaard & Peterson, 1991). The status of kimberlite is uncertain, largely because the composition of the primary magma and volatile contents are controversial (Mitchell, 1986; Kjarsgaard et al., 2009); their exceptionally high MgO contents may indicate sub-lithospheric pressures of origin.

For melting with CH_4+H_2O relevant to the HRM mechanism, most studies have concentrated on determining the solidus temperature or the solubility of volatiles in melts (Taylor & Green, 1988; Jakobsson & Holloway, 1986; Ardia et al., 2014), so that the composition of melts is very poorly constrained. The only melt compositions analyzed are from Jakobsson and Holloway (2008), who found SiO_2-saturated and exceptionally MgO-rich (25–33 wt.%) melts at 5–12.5 GPa. These melts are much richer in MgO than from dry peridotite at equivalent pressures (Walter, 1998). Litasov et al. (2014) calculated melt compositions by mass balance because of severe quenching, finding similarly high MgO contents (25–31 wt.%); the only notably nepheline-normative melt (7%) was at the highest pressure (12 GPa). With mixed CH_4+H_2O volatiles, melting is controlled by aH_2O, leading to the expectation that melts would increase in SiO_2-saturation as H_2O/CH_4 and fO_2 increase (Taylor & Green, 1987). This is because of the depolymerizing effect of water on silicate melt structure, expanding the olivine phase volume and so pushing melt compositions to compositions richer in SiO_2. At lower pressures, melts contain considerable normative feldspar due to higher Al_2O_3 contents, but remain SiO_2–saturated (Liu et al., 2018). Redox melting in reduced conditions becomes more likely once the H_2O/CH_4 (and thus aH_2O) climbs with concomitant deposition of carbon, so that melts are expected to be H_2O-rich. The rapid change in H_2O/CH_4 with changing fO_2 also makes interpretation of experimental melt compositions difficult, because they will vary rapidly over a small range of fO_2.

Amongst possible primary igneous rocks from H_2O-rich but CO_2-absent conditions, nephelinites and basanites result at < 3 GPa (Green, 1973; Millhollen et al., 1974) due to amphibole, which has a broadly similar composition to these melts, dominating the melting reaction (Foley et al., 1999). Many potassium-rich hydrous melts such as lamproites and minettes cannot be produced in CO_2-rich conditions, and probably originate in reduced, HRM-like melting conditions (Foley, 1989a, b). However, they will not be produced from peridotite alone, but require mixed source rocks including phlogopite-bearing assemblages (Foley, 1992a).

The melt products of HRM are less likely to be represented among igneous rocks at the surface because of the decrease in oxidation state with increasing pressure. For example, reduced, lamproitic melts are likely to be common at the base of thick cratonic lithosphere: their solidified products can be identified by their isotopic signals as components of alkaline magmas that originate by later CRM (Tappe et al., 2006; Rosenthal et al., 2009). These later melts are produced when the cratonic lithosphere is broken apart, facilitating their appearance at the surface (Foley, 2008; Foley and Fischer, 2017). Due to their origin by incipient melting, most products of mantle redox melting belong to the agents responsible for mantle metasomatism (in addition to fluids; Weiss et al., 2015), which range from potassium-rich, hydrous melts (HRM) to carbonate-rich low-SiO_2 melts (CRM).

5.5. THE OXIDATION STATE IN THE MANTLE LITHOSPHERE, ASTHENOSPHERE, AND SUBDUCTION ZONES

The operation of either mechanism of mantle redox melting depends on local to regional variations in the oxidation state of the mantle. The required contrast in oxidation states will frequently be realized by the movement of melts or fluids into rocks with a different oxidation state. It is now generally recognised that fO_2 in the mantle decreases with increasing depth (Frost & McCammon, 2008; Foley, 2011), although a new calibration suggests that the change may be less than previously thought (Stagno et al., 2013). Nevertheless, the convecting mantle is likely to be metal-saturated at depths greater than 200–300km (Figure 5.3; Rohrbach et al., 2011; Foley, 2011). In the deep lithosphere, redox conditions may vary more substantially on a smaller spatial scale due to the effects of redox fractionation during the melt depletion events that lead to cratonic stabilization (Figure 5.3), and the overprinting of these effects by episodic metasomatic events over the lifetime of the craton. Here, I concentrate on these more local differences as they will be the sites of redox melting, and on newer work published since the last review (Foley, 2011).

The oxidation state of the upper mantle is often generalized to a single value: average values tend to be emphasized whereas exceptions and regional variations – those that are important for the operation of redox melting – are not (Foley, 2011). This may be justified for oceanic regions, where melts are drawn and mixed from large volumes of asthenospheric mantle, which is itself homogenized by convection (McKenzie & Bickle, 1988). The resulting value for mid-ocean ridge basalts (MORB; FMQ–0.41 for MORB glasses; Bézos & Humler, 2005, which is IW+4-5) is uniform, but represents a well-mixed, near-surface value for the oceanic mantle. Furthermore, the differential solubilities of volatile components in melts means that oxidation states will be fractionated between melts and their residues (Figure 5.3), so that a value for MORB does not reflect that of its source. This is indicated in the lower and more variable value for abyssal peridotites (Bryndzia et al., 1989). Recent data confirm a much wider range for abyssal peridotites (3 log units) than for MORB from the same area (Birner et al., 2018), indicating substantial local variations.

Nevertheless, the near-surface oceanic mantle appears to be more uniform than most continental and subduction environments (Figure 5.5). Large variations are known from cratonic xenoliths (Kaapvaal 6 log units with an average of FMQ–2.8; Slave 4 log units, average FMQ-2.6; Baltic 2 log units, average FMQ-4.1; Foley, 2011). The range is caused partly by redox fractionation during the depletion events that helped to stabilize the cratons, resulting in reduced regions that are probably characteristic of the pre-metasomatic cratonic lithosphere. Recent studies of the redox effects of metasomatism have confirmed that it causes a range of redox changes up to 4 log units (Yaxley et al., 2017), and that diamond formation in the lower part of this range is an oxidative process (Weiss et al., 2018).

In the non-cratonic continental lithosphere (NCCL), evidence for more redox variation, particularly on a small scale, is accumulating; for example, fO_2 of FMQ–3 (ca. IW+2) has been identified for the Vitim xenoliths (Siberia) including disparate values for spinel- and garnet lherzolites (Goncharev & Ionov, 2012) in a region previously thought to be close to the average NCCL of FMQ–0.68 (Glaser et al., 1999; Foley, 2011). Metasomatic effects in the NCCL generally also cause oxidation (Uenver-Thiele et al., 2014, 2017), as they do in the cratonic lithosphere, although contrasting effects are seen in some cases, with some metasomatic events causing oxidation whilst others in the same area do not (Perinelli et al., 2012; Wang et al., 2013). The very large range in fO_2 for subduction regions (Figure 5.5) includes the generally oxidized environment indicated by xenoliths and volcanic oxidation states (McInnes et al., 2001; Parkinson et al., 2003), as well as reduced values of fO_2 (IW to IW+4; Song et al., 2009; Wang et al., 2009), and metal-bearing peridotites in Kamchatka (Ishimaru et al., 2009). Also

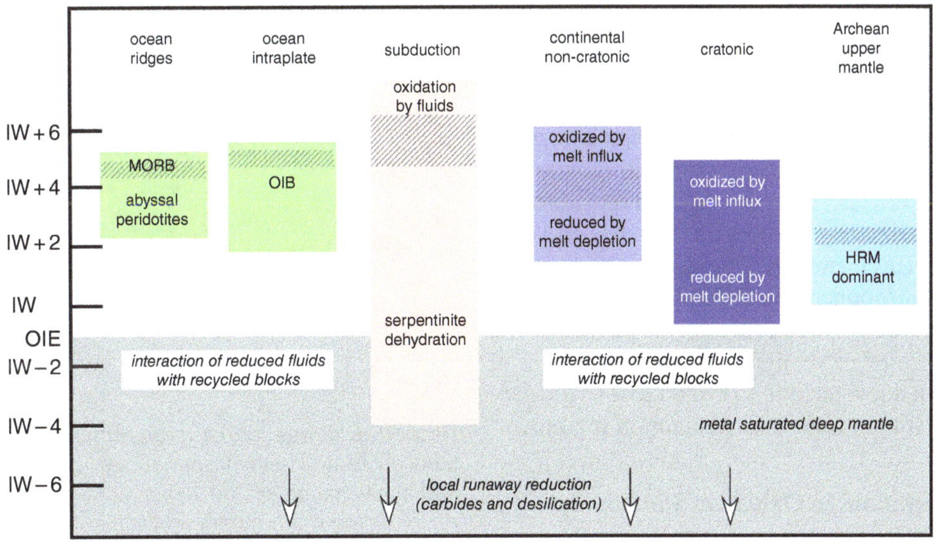

Figure 5.5 Ranges of redox states found in various geodynamic settings and in the upper Archean mantle (right). Coloured blocks show the approximate range and striped zones the commonest redox states within these settings. Subduction zones have the largest fO_2 ranges, and Archean upper mantle lay 1–2 log units below modern sub-ridge mantle. Metal-saturated conditions are shaded grey; The interaction of reduced fluids with oxidized recycled blocks can occur in a variety of settings, as can local extreme, runaway reduction that leads to desilication in conditions several log units below those shown in this figure (Griffin et al., 2018).

in subduction zones, recent work indicates local variations in oxidation state related to the distribution of passageways for fluids or melts from slab to mantle wedge (Birner et al., 2017; Benard et al., 2018).

The applicability of redox melting will depend on these local redox variations and on differences between juxtaposed rock types, which cause redox contrasts and variations in the redox buffering ability (Luth & Stachel, 2014). Eclogites in the cratonic lithosphere have been most thoroughly studied, showing a consistent range over at least 3 log units fO_2 for both Kaapvaal and Slave cratons, but which may contrast with the peridotite oxidation state at the same depth (Aulbach et al., 2017; Smart et al., 2017). The oxidation states of other non-peridotitic assemblages in the lithosphere are poorly known, but probably also highly variable: most contain hydrous minerals and crystallized from hydrous melts that were probably quite oxidized due to their H_2O content, but some are the products of crystallization in reduced environments, with "hydrous minerals" with low H_2O contents stabilized by high fluorine contents (Foley, 1989a; 1991).

5.6. DISCUSSION

Redox melting in the mantle is more important than commonly accepted, and in some geodynamic situations it may be a more important cause of melting than temperature or pressure changes. This is particularly true for the lower continental lithosphere and subducted materials, including both active subduction zones and large lithospheric blocks that have become stranded or partially recycled into the convecting mantle. Here, I update discussion of these in view of additional information from the last few years, and then consider temporal changes during Earth evolution. In these regions, the juxtaposition of rocks, melts, and fluids with contrasting lithologies and oxidation states may be common. The importance of redox melting is intimately linked with existence and evolution of the incipient melting regime: the changing speciation of volatile components with fO_2 controls both the extent in pressure and temperature of incipient melting and the occurrence of redox melting. Only the action of these volatiles can allow melting to occur close to the geothermal gradients in cratonic and subduction regions.

5.6.1. Spatial Variations in Oxidation State on the Modern Earth

Lower Cratonic Lithosphere. The general decrease in fO_2 with increasing depth in the mantle (Ballhaus & Frost, 1994; Frost & McCammon, 2008; Stagno et al., 2013) is only a large-scale cause of redox melting. Local differences in redox state are more likely due to specific events of melt loss and re-enrichment, reiterating the essential role of melt migration in the incipient melting regime in both redox melting and the preconditioning of mantle rocks for later redox melting. The depleted composition of most cratonic mantle peridotites results from substantial melting in the late Archean, when the mantle was more reduced than today. In these conditions, melting fractionates the redox state, concentrating H_2O in the melt, and leaving a more reduced residue behind (Figure 5.3). Low-degree melts introduced from below in different and most likely multiple later episodes are rich in H_2O and/or CO_2 and so re-oxidize the lower lithosphere to fO_2 values higher than the original undepleted starting point. The result is a complex redox picture containing several types of redox front, which is much more variable than a simple contrast across the lithosphere–asthenosphere boundary (Figure 5.6): (i) the depleted peridotite making up most of the cratonic mantle is more reduced than asthenospheric mantle it may come into contact with (Foley, 2011); (ii) the metasomatic front between influxing melts and this depleted peridotite will present a

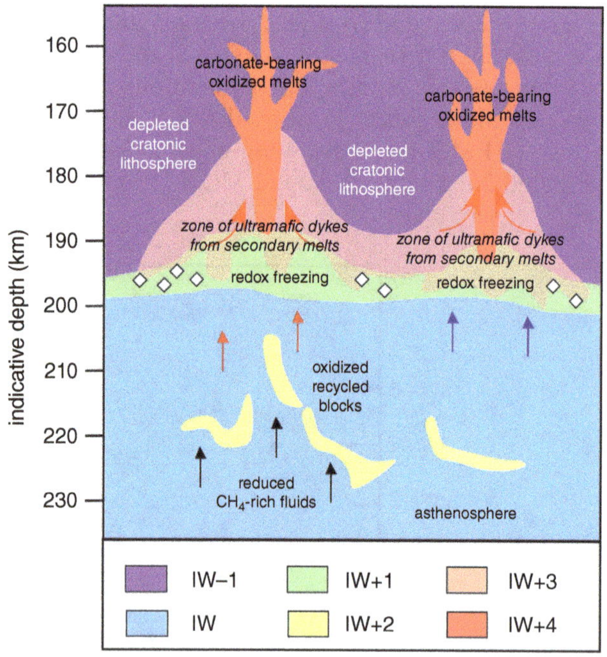

Figure 5.6 Illustration of redox fronts that develop between lithosphere and asthenosphere and during magmatic activity and rejuvenation of the lower lithosphere. Depths and fO_2 regions shown are merely indicative to illustrate processes at the cratonic lithosphere base. Most of the cratonic mantle remains dry and severely depleted; first melt influx causes abundant diamond formation by redox freezing, later melts progressively oxidize restricted parts of the lower lithosphere. Late redox melting is concentrated in these modified parts, culminating in carbonate-rich melts that derived their carbon from earlier redox freezing processes.

stronger redox contrast, frequently of several log units fO_2. These melts are concentrated in areas close to later magmatic activity (Foley, 2008), probably leaving much of the lithospheric volume unaffected (Griffin et al., 2013); (iii) many melts will solidify to produce dykes and patches of ultramafic rocks whose mineralogy, Fe content, and intrinsic redox state and buffering capacity contrasts with surrounding peridotite (Foley, 1992a, 2008; Luth & Stachel, 2014).

Re-enrichment of depleted cratonic peridotites may result in the growth of garnet and clinopyroxene, recognizable as introduced later by their spatial association in rocks that were depleted dunite and harzburgite in their pre-metasomatic state (Simon et al., 2007; Rehfeldt et al., 2008; Griffin et al., 2009), a process that has been termed stealth metasomatism by O'Reilly and Griffin (2013). Studies of mantle metasomatism used to focus on chemical and mineralogical details, but recently the characterization of fO_2 has improved greatly. Redox contrasts up to four log units fO_2 may be introduced within four-phase peridotites (Creighton et al., 2010; Yaxley et al., 2012, 2017) and even stronger variations may result where other mineral assemblages form. Some of the infiltrating melts are carbonatitic in nature (Shu and Brey, 2015). The resulting metasomatic assemblages are enriched in water and carbon, which in turn cause them to have lower melting points than surrounding drier rocks and to be the preferential locus of later melting events. Progressively more oxidized melts are thus funnelled towards the surface, eventually erupting as alkaline rock occurrences that delineate depleted block margins at depth. The redox freezing of oxidized melts that encounter reduced cratonic mantle may result in carbon concentrations up to several per cent (Sleep, 2009; Foley & Fischer, 2017; Garber et al., 2018), the reactivation of which explains the strong association of carbonate-rich igneous rocks with continental rifts (Foley & Fischer, 2017).

Inclusions in diamonds are especially useful in documenting the variable fO_2 in the lower cratonic lithosphere. Isotopes show that diamond-forming fluids range greatly in oxidation state (Palot et al., 2014), and a majority may be reduced (Stachel et al., 2017). This is consistent with mineral inclusions of cohenite, moissanite, and carbonates that reflect several log units difference in fO_2 on very small scales (Eggler & Lorand, 1994; Jacob et al., 2004, 2016; Palyanov et al., 2005), and with conclusions that some diamonds form in exceptionally reducing environments containing metals or carbides (Palyanov et al., 2013; Smith et al., 2016). Recently, it has been shown that extreme, runaway reduction caused by H_2-CH_4-rich fluids could result in the desilication of rocks (Griffin et al., 2018). The large variety of redox states increases the likelihood of redox melting as time progresses.

The Subduction Environment and Recycled Blocks. The fO_2 of peridotites from the mantle wedge above subducting slabs is more oxidized than in stable continental areas (McInnes et al., 2001; Parkinson et al., 2003). The average of available data is FMQ+0.5, including many at > FMQ+1 (IW > +6; Figure 5.5), particularly from oceanic island arcs. The pressure-corrected average is 2.3 log units above the asthenosphere value (Foley, 2011). This oxidized condition is usually attributed to the infiltration of water-rich fluids or melts originating in the subducting slab, and is more marked where the subducted slab is deeper and has released more water through dehydration (Birner et al., 2017). However, processes in the uppermost 200km of subduction zones are more important for modifying redox states above the slab than as the direct cause of redox melting, because only sedimentary material is likely to melt along subduction geotherms, even in volatile-rich conditions (Hermann & Spandler, 2008; Plank et al., 2009; Förster et al., 2019).

More reduced fluids (H_2 and H_2O) caused by the breakdown of serpentine in ultramafic rocks in deeper parts of the slab (Sleep et al., 2004) and of subducted organic matter will add to the cocktail of fluids with disparate redox states in subduction zones. Redox states in the residues will be increasingly polarized at higher pressure, because of widening of the H_2O-dominated fluid region (Figure 5.4), which leads to stronger reduction of reduced rocks and oxidation of oxidized rocks than occurs at lower pressures. This may be the cause of metals and methane-bearing fluid inclusions in xenoliths (Ishimaru et al., 2009; Song et al., 2009), contrasting with the more oxidized average for suprasubduction mantle. Extremely reduced conditions in Tibetan ophiolitic peridotites (carbides, nitrides, native elements; Griffin et al., 2016) are thought to originate in the transition zone prior to exhumation and thus to reflect deep subduction conditions rather than their original oceanic crustal characteristics.

The extremely reducing ambient mantle in the transition zone at 400–600km depth indicated by these results supports earlier suppositions that interaction between CH_4-rich fluids and oxidized parts of recycled blocks would result in melting by the HRM mechanism (Green et al., 1987). This mechanism may be important in generating melts beneath ocean islands (Figure 5.5), so that "hotspots" may more correctly be termed "wetspots" (Green et al., 2001). These recycled blocks will exhibit a large variation owing to near-surface hydrothermal alteration and to redox fractionation processes during subduction. Most redox melting events related to these blocks will be by HRM. Any initial melts with H_2O and CO_2 may freeze to deposit diamond due to the higher melting point on reduction to H_2O-rich conditions (Figure 5.1). Recycled blocks will be a significant source of reduced carbon and hydrogen (Poli et al., 2009), resulting in the release of CH_4 and H_2 that lead to H_2O release and thus potential melt production by HRM.

5.6.2. The Evolution of Mantle Oxidation States through Earth History

The pattern of redox melting depends on the oxidation state in various parts of the mantle from the Archean onwards, and especially on recycling processes that return crustal material to the mantle by subduction or delamination. This pattern must have changed during the Archean, associated with a major increase in continental crustal preservation that probably followed the onset of plate tectonics, and with the gradual oxidation of near-surface rock reservoirs that preceded the Great Oxidation Event (GOE) at around 2.3 Ga.

Redox State of the Mantle in the Archean. To arrive at the original fO_2 of the cratonic mantle lithosphere in the late Archean, we need to strip it of the later effects of metasomatic events, most of which have an oxidizing effect and may have increased the fO_2 through several log units (Creighton et al., 2010; Yaxley et al., 2012; 2017). The resulting picture is that the majority of the cratonic lithosphere volume was at the lower limit of the range shown in Figure 5.5 in the late Archean, and that conditions were most reducing at the deepest levels due to a combination of the pressure effect and of the greater fractionation of fO_2 by HRM at higher pressures (Figure 5.4).

This contrasts with claims for constant fO_2 through time based on V/Sc ratios (Li & Lee, 2004) and V and Cr contents (Canil, 1997; Delano, 2001) in volcanic rocks, which have been interpreted to show near-constancy of the redox state of the mantle since the early Archean. Recent results have reiterated these findings for olivines and chromites in komatiites as old as 3.8–3.5 Ga (Nicklas et al., 2016; 2018; Rollinson et al., 2017). However, this conclusion does not account for redox fractionation between melts and their sources, nor for the bias of the Archaean basalt and komatiite record towards continental regions.

It is important to acknowledge that the oxidation state of volcanic rocks and their minerals does not reflect that of the source because of the differential solubilities of volatile species. Redox states are fractionated during melting, especially when melting takes place in reducing conditions around or just above IW (Figure 5.3), which causes coupled reduction and drying of the residue because H_2O is transported away in the melt. The oxidation state displayed by such a melt is close to the water maximum, whereas the residue will be located at the underside of the carbon saturation curve (Figure 5.2), 2–4 log units below that of the melt, with increasing fO_2 fractionation towards higher pressures (Figure 5.4). Repeated partial melting is an important mechanism that leads to diversification of redox states in mantle rocks (Kadik, 1988; 1997), whereby several melting episodes will drive the oxidation state of the melt upwards; this may be misconstrued as indicating an oxidized source if only evidence from the melts is considered. Later melting events in oxidized conditions will lead to less fractionation, restricted to the range between the water maximum and CO_2-rich conditions, (Figs. 5.2,3). A more reduced Archean mantle is consistent with oxybarometry results for cratonic peridotites, which indicate a great range of redox states (Woodland & Peltonen, 1999; Kadik, 2003; Frost & McCammon, 2008).

The alleged constancy of redox state through time compares Archaean basalts with modern MORB in an attempt to access the fO_2 the uppermost asthenosphere, but the Archean rocks most likely originate in a continental margin or island arc setting, which is more oxidised than MORB on the modern Earth (Figure 5.5). A solution to this impasse is to study eclogite xenoliths that represent Archean *ocean crust* that was trapped in the cratonic lithosphere as it grew (Jacob et al., 1994; Jacob, 2004; Aulbach & Jacob, 2016). Early scanty data (Jacob & Foley, 1999; Barth et al., 2001; Jacob et al., 2005; Smart et al., 2009) indicated conditions slightly more reduced than MORB, and more recent data on V/Sc and $Fe^{3+}/\Sigma Fe$ in garnets from eclogites confirm a more reduced Archean upper mantle (Aulbach & Viljoen, 2015; Aulbach et al., 2017), which is now supported by measurements from Archean picrites and basalts (Aulbach & Stagno, 2016).

Although Archean melts may be more reduced partly because they are sourced deeper (Gaillard et al., 2015), it appears likely that the Archaean asthenospheric mantle was more reduced than the current sub-ridge mantle by 1–2 log units fO_2 at equivalent depth (Figure 5.5). Regional redox contrasts in the Archaean mantle may have been large, as they are on the modern Earth. It is important to recognize this, as assuming the Archean continental record to be representative of the asthenosphere can affect conclusions about the role of degassing in the oxygenation of the atmosphere (Brounce et al., 2017). Kasting et al. (1993) note that a constant mantle oxidation state since 3.8 Ga implies an oxygen-bearing Archaean atmosphere, which contrasts with evidence for the GOE – a major increase in atmospheric oxygen levels in the early Proterozoic (Pavlov & Kasting, 2002; Bekker et al., 2004). The ages of events recorded in cratonic peridotites and eclogites which provide this evidence for more reduced conditions in the Archean mantle are almost entirely from the Neoarchaean (< 2.8 Ga), raising the possibility that the mantle had progressively oxidized from still more reduced conditions during the Hadean to Mesoarchaean.

Evolution of Alkaline Melt Types as a Sensor of the Evolving Incipient Melting Regime. Many alkaline rocks cannot be partial melting products of peridotite, even with mixtures of volatile components, but require the melting of additional rocks in their sources (Foley, 1992a,b). The first appearance of several classes of alkaline rocks is consistent with the need for time to precondition the source with the necessary ingredients to allow generation of these alkaline melt types. This is intimately linked with the evolution of lithospheric reservoirs, which require sufficient time since the stabilization of the cratons in the late Archean to accumulate enriched reservoirs by magmatic events at the base of the lithosphere (Figure 5.6). Most carbonate-rich melts require pre-enrichment in CO_2, as depleted lithosphere would initially be devoid of H_2O and CO_2. Very little potassium can be accommodated in peridotite minerals, which rapidly leads to a strong association between K and H_2O in the form of phlogopitic mica. In the reduced conditions typical of early cratonic lithosphere, widespread HRM caused the production of mica-bearing lithologies, whereas carbon was deposited in reduced form as diamond. This carbon had to be oxidized during later magmatic events in order for carbonate-rich igneous rocks to reach the surface, which, despite a first appearance at 3.0 Ga (Bizarro et al., 2002), explains why carbonatites are rare until the Phanerozoic. The first appearances of kimberlites (2.0 Ga), ultramafic lamprophyres (1.9 Ga), and lamproites (1.4 Ga) are all notably younger than the cratonic lithosphere they are generated from, with hundreds of millions of years between craton stabilization and occurrence of the alkaline rocks. The only notable exception is shoshonites (2.7–2.8 Ga), but these are associated with subduction rather than the deep cratons.

The generation of various melt types from reduced to oxidized conditions can be traced in time in the Labrador Sea, which cuts through the North Atlantic craton, and in space in the western branch of the East African Rift: comparison of these two regions explains the association of melt types with depth and redox conditions of melting at the craton base (Foley, 2008). The first alkaline rocks in the Labrador Sea region are 1.4 Ga lamproites, which were generated in reduced conditions at the base of the craton before it was thinned (Tappe et al., 2007). Ultramafic lamprophyres and carbonatites followed in a later episode (590–560 Ma), in which accumulated carbon was oxidized and remobilized as carbonate in an episode of failed rifting that thinned the cratonic lithosphere appreciably. Ultramafic lamprophyres associated with this continental breakup are widespread on the North Atlantic Craton in eastern Canada, Greenland, and Scandinavia, and have recently been recognised in northern Scotland (Hutchinson et al., 2018). Uganda exhibits a similar range of rocks, but occurring contemporaneously with regional variation, with carbonatites and ultramafic-lamprophyre-like kamafugites occurring in the north where the base of the sub-rift lithosphere is still thickest (Rosenthal et al., 2009). Melting in oxidizing conditions to produce ultramafic lamprophyres at 140–200 km depth is consistent between these areas (Foley & Fischer, 2017).

Subduction Before the Great Oxidation Event (GOE). On the modern Earth, a large volume of material is returned to the mantle from the oxidized surface by subduction, which plays a major role in producing and mixing the variety of redox states in the upper mantle (Figure 5.5). The date of onset of subduction and plate tectonics is debated (Condie & Pease, 2008), but the large increase in production and preservation of continental crust in the Neoarchean and the associated cratonization of mantle lithosphere are consistent with a change in process that involved accumulation of crust at subduction zones from around 3.3 Ga onwards (Lee, 2006). This is around a billion years earlier than the GOE ($\approx$2.3 Ga; Pavlov & Kasting, 2002; Bekker et al., 2004), leaving a substantial proportion of Earth history with subduction of materials that were not as oxidized as they are today. Furthermore, accretionary and island-arc tectonic settings would have been much more common due to the much smaller volume of continental crust, consistent with a different composition of sediments at subduction zones, namely a much higher ratio of immature clastic and volcaniclastic sediments relative to shelf sediments (Eriksson et al., 1994). The persistence of this new environment may have led to the emergence of the continents and thus an increase in subaerial volcanism, which in turn assisted atmospheric oxygenation (Gaillard et al., 2011).

The change in oxidation state of subducting material is probably the major controlling factor in the gradual oxidation of internal processes that pre-empted the GOE at the Earth's surface in the early Proterozoic. The gradual oxidation of near-surface layers during these billion years, especially in the late Archaean and early Proterozoic, is documented in the appearance of abundant oxidized sediments and decrease in banded iron formations and uraninite deposits (Schidlowski 1981; Canfield, 2005). Given that the early mantle was more reduced (Section 5.6.2.1), the subduction process, if it happened in a similar manner to today, was not an oxidized process until the late Archaean, meaning that oxidized melting processes would have been volumetrically insignificant. Carbonate sediments (indeed shelf sediments of any kind) were rare, and carbon was stored mostly in reduced form in the mantle, meaning that the CRM mechanism first became important well into the Proterozoic. Recycled blocks would not have brought more oxidized material to great depths because of the lack of oxidized surface conditions, so that deep HRM would also have been less important in early subduction zones.

Subducting sediments in the Archean were less carbon-rich than on the modern Earth because of the rarity of both carbonate sediments and organic material. Nitrogen, however, may have been more abundant, because it is more soluble in melts at low fO_2, and so would be a more significant constituent of basalts (Libourel et al., 2003; Mysen et al., 2008). Sulfur would also behave differently, probably being retained to deeper levels in the subducted slab because it is released only at significantly higher temperatures in reduced form (Tomkins & Evans, 2015).

Some effects of the subduction process remain the same, however, including the fractionation of redox states due to differential loss of volatile components as either fluids or melts. Oceanic hydrothermal alteration causes serpentinization of basalts and ultramafic rocks, which polarizes oxidation states by the formation of magnetite that coexists with serpentine which releases hydrogen (Sleep et al., 2004). Degassing at high pressures in the deeper parts of the serpentinized slab leads to greater polarization of redox because the width of the H_2O-dominated fluid region widens to higher pressures (Figure 5.4).

Redox Melting Before the Cratons. The stabilization and growth of cratons in the late Archean provided the first opportunity for the large-scale persistence of crust and lithospheric mantle and, therefore, of the storage of fractionated redox states close to the Earth's surface. Before this, the generally more reduced state of the upper mantle (Section 6.2.1) will have prevented the CRM mechanism and returned almost all lithospheric components to the convecting mantle at a very young age. These will have been dominated by olivine- and pyroxene-rich cumulates with reduced oxidation states (Foley et al., 2003) rather than the oxidized and silica-richer eclogites characteristic of the late Archean onwards.

The more reduced upper mantle probably also had higher volatile contents in its less degassed state, so that HRM was widespread in the convecting mantle. Partial melting in reduced conditions concentrates H_2O in the melt, thus fractionating fO_2 between more oxidized melts and more reduced residues (Figure 5.3). This was the principal driving mechanism for the oxidation of the upper mantle through the Archean, which then became increasingly stored as disparate redox states once extensive continents grew in the late Archean.

On the modern Earth, the *magmatic zone* – the depth at which melting commonly occurs in the mantle – ranges from about 30km beneath mid-ocean ridges to around 250km in the lower reaches of cratonic lithosphere. Major melting is related to decompression of upwelling hot material and is therefore restricted to near-surface conditions because of the interplay of melting curves and geothermal gradients (Wyllie, 1988; McKenzie & Bickle, 1988). Thus, even for hot plumes, major melting occurs in the uppermost 120km and deeper melting requires H_2O and CO_2 to depress melting points sufficiently (Figure 5.1). In the early Archean, the mantle was hotter, which will have enabled deeper major melting (by perhaps 20–30km), which may have slightly accentuated reduced melting because of the pressure effect (Gaillard et al., 2015). Because the widening of the water maximum means that redox fractionation is greater at higher pressures (Figure 5.4), deeper melting on a hotter Earth may have helped to fractionate redox states. However, the more reduced state and the lack of stable cratonic lithosphere will have prevented melting at > 150km, because mixtures of H_2O and CO_2 will not have occurred. This explains the first crescendo of carbonatite occurrences (2.7–2.6 Ga; Cavell et al., 1992; Villeneuve & Relf, 1998) as an effect of melting conditions rather than preservation.

The change in redox state of the mantle and the diverse effects of different volatile components on the melting point of peridotite (Figure 5.1) have the counterintuitive effect that higher temperatures in the Archean mantle do not necessarily lead to much more extensive melting than later in Earth history. This is because of the severely limited extent of the incipient melting regime: Melting in the presence of H_2O+CO_2 mixtures did not apply and melting with H_2O+CH_4 was common only with high aH_2O and thus at relatively high temperatures (Figure 5.7). However, once the melting point was attained, there was a more limited temperature interval of incipient melting before major melting occurred. This carries over into a higher generally applicable solidus temperature for the mantle until the late Archean, which may contribute to the geochemical discontinuity noted for the end of the Archean (Keller & Schoene, 2012). This restriction of the extent of incipient melting must have implications for magmatic geochemistry that remain to be ascertained.

5.7. CLOSING COMMENTS

Redox melting in the mantle is more important than has been widely accepted and has changed through time because of the widely differing effects of different volatile components on melting temperatures and melt compositions. Regional differences on the modern Earth have been shown in the last few years to be more extreme than previously thought; for example, we have moved from a decades-long skepticism about the possible stability of

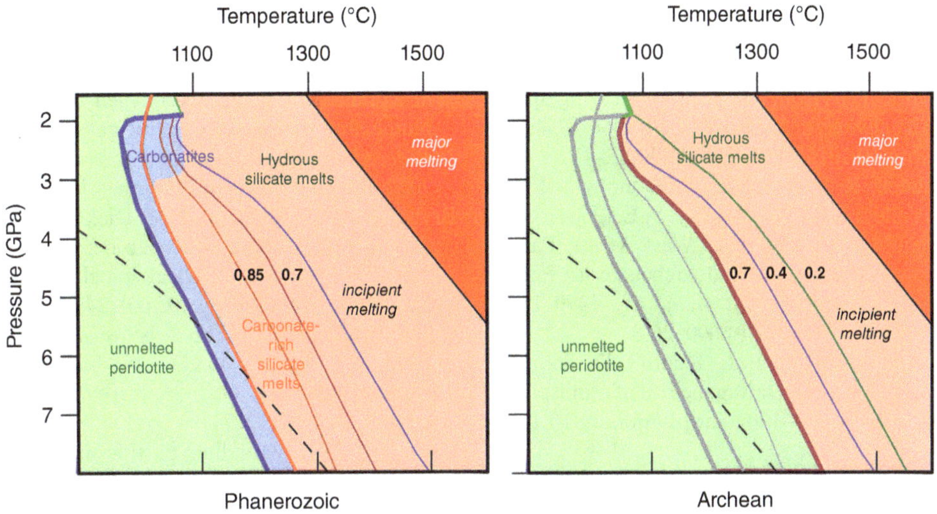

Figure 5.7 Comparison of likely widely applicable peridotite solidi in Phanerozoic and Archean times. Oxidized conditions are common in the Phanerozoic, so that solidi for CO_2+H_2O (blue line) and H_2O-saturated conditions may be commonly active. These were inactive in the Archean (right) so that the incipient melting zone was restricted to higher temperatures in a reduced mantle with CH_4+H_2O mixtures. Cratonic geotherm (dashed line) crosses the CO_2+H_2O solidus at modern lower lithosphere levels (left), but not the reduced solidus of the Archean (right). Thus, initial melts were not more common despite hotter conditions in the Archean.

moissanite in the mantle to hypotheses in which much more reduced conditions with widespread native metals are locally important (Griffin et al., 2018).

CRM was essentially inoperative until the late Archean when widespread continental crust and mantle lithosphere were formed. The CRM mechanism involves the oxidation of solid carbon accumulated in the lithosphere, largely by redox freezing, over time, and is therefore commonest late in Earth history during the reactivation of cratonic roots. This explains the strong association of carbonatites with continental rifts, and implies that CRM is a logical consequence of earlier HRM in the same area, which caused the enrichment in carbon as graphite or diamond (Foley & Fischer, 2017).

The subduction environment probably experienced greater changes in redox states than other tectonic environments, transitioning from reduced before the GOE to the most oxidizing geological conditions in the modern mantle. This carries considerable ramifications for the formation and location of metal resources: if the first billion years of subduction took place in reducing conditions, then the mobility of copper and gold would be much more restricted than today (Evans & Tomkins, 2011; Mironov & Portnyagin, 2018), whereas Pt and other siderophile elements would be much more mobile (Howarth et al., 2017; Bennett et al., 2014). The differences between subduction at 3.2 and 2.2 Ga may have been greater than between all tectonic environments existing on the modern Earth.

ACKNOWLEDGMENTS

This is contribution 1375 from the ARC Centre of Excellence for Core to Crust Fluid Systems. SF is funded by ARC grant FL180100134.

REFERENCES

Ardia, P., Hirschmann, M. M., Withers, A. C., & Stanley, B. D. (2013). Solubility of CH_4 in a synthetic basaltic melt, with applications to atmosphere-magma ocean-core partitioning of volatiles and to the evolution of the Martian atmosphere. *Geochimica et Cosmochimica Acta*, *114*, 52–71. https://doi.org/10.1016/j.gca.2013.03.028

Aulbach, S., & Viljoen, K. S. (2015). Eclogite xenoliths from the Lace kimberlite, Kaapvaal craton: from convecting mantle source to palaeo-ocean floor and back. *Earth and Planetary Science Letters*, *431*, 274–286. https://doi.org/10.1016/j.epsl.2015.08.039

Aulbach, S., & Jacob, D. E. (2016). Major- and trace-elements in cratonic mantle eclogites and pyroxenites reveal heterogeneous sources and metamorphic processing of low-pressure protoliths. *Lithos*, *262*, 586–605. https://doi.org/10.1016/j.lithos.2016.07.026

Aulbach, S., & Stagno, V. (2016). Evidence for a reducing Archean ambient mantle and its effects on the carbon cycle. *Geology*, *44*, 751–754. https://doi.org/10.1130/G38070.1

Aulbach, S., Woodland, A. B., Vasilyev, P., Galvez, M. E., & Viljoen, K. S. (2017). Effects of low-pressure igneous processes and subduction on Fe3+/ΣFe and redox state of mantle eclogites from Lace (Kaapvaal craton). *Earth and Planetary*

Science Letters, *474*, 283–295. https://doi.org/10.1016/j.epsl.2017.06.030

Ballhaus, C., & Frost, B. R. (1994). The generation of oxidized CO_2-bearing basaltic melts from reduced CH_4-bearing upper-mantle sources. *Geochimica et Cosmochimica Acta*, *58*, 4931–4940. https://doi.org/10.1016/0016-7037(94)90222-4

Barth, M. G., Rudnick, R. L., Horn, I., McDonough, W. F., Spicuzza, M. J., Valley, J. W., & Haggerty, S. E. (2001). Geochemistry of xenolithic eclogites from West Africa, Part I: A link between low MgO eclogites and Archean crust formation. *Geochimica et Cosmochimica Acta*, *65*, 1499–1527. https://doi.org/10.1016/S0016-7037(00)00626-8

Bekker, A., Holland, H. D., Wang, P. L., Rumble, D., Stein, H. J., Hannah, J. L., et al. (2004). Dating the rise of atmospheric oxygen. *Nature*, *427*, 117–120. https://doi.org/10.1038/nature02260

Bénard, A., Woodland, A. B., Arculus, R. J., Nebel, O., & McAlpine, S. R. B. (2018). Variation in sub-arc mantle oxygen fugacity during partial melting recorded in refractory peridotite xenoliths from the West Bismarck Arc. *Chemical Geology*, *486*, 16–30. https://doi.org/10.1016/j.chemgeo.2018.03.004

Bennett, N. R., Brenan, J. M., & and Koga, K. T. (2014). The solubility of platinum in silicate melt under reducing conditions: results from experiments without metal inclusions. *Geochimica et Cosmochimica Acta*, *133*, 422–442. https://doi.org/10.1016/j.gca.2014.02.046

Beyer, C., & Frost, D. J. (2017). The depth of sub-lithospheric diamond formation and the redistribution of carbon in the deep mantle. *Earth and Planetary Science Letters*, *461*, 30–39. https://doi.org/10.1016/j.epsl.2016.12.017

Bézos, A., & Humler, E. (2005). The Fe3+/Sigma Fe ratios of MORB glasses and their implications for mantle melting. *Geochimica et Cosmochimica Acta*, *69*, 711–725. https://doi.org/10.1016/j.gca.2004.07.026

Birner, S. K., Warren, J. M., Cottrell, E., Davis, F. A., Kelley, K. A., & Falloon, T. J. (2017). Forearc peridotites from Tonga record heterogeneous oxidation of the mantle following subduction. *Journal of Petrology*, *58*, 1755–1780. https://doi.org/10.1093/petrology/egx072

Birner, S. K., Cottrell, E., Warren, J. M., Kelley, K. A., & Davis, F. A. (2018). Peridotites and basalts reveal broad congruence between two independent records of mantle f_{O_2} despite local redox heterogeneity. *Earth and Planetary Science Letters*, *494*, 172–189. https://doi.org/10.1016/j.epsl.2018.04.035

Bizzarro, M., Simonetti, A., Stevenson, R. K., & David, J. (2002). Hf isotope evidence for a hidden mantle reservoir. *Geology*, *30*, 771–774. https://doi.org/10.1130/0091-7613(2002)030<0771:HIEFAH>2.0.CO;2

Boyd, F. R., & Gurney, J. J. (1986). Diamonds and the African lithosphere. *Science*, *272*, 472–477. https://doi.org/10.1126/science.232.4749.472

Brey G. P., Brice, W. R., Ellis, D. J., Green, D. H., Harris, K. L., & Ryabchikov, I. D. (1983). Pyroxene-carbonate reactions in the upper mantle. *Earth and Planetary Science Letters*, *62*, 63–84. https://doi.org/10.1016/0012-821X(83)90071-7

Brey, G. P, Bulatov, V. K., Girnis, A. V., & Lahaye, Y. (2008). Experimental melting of carbonated peridotite at 6-10 GPa. *Journal of Petrology*, *49*, 797–821. https://doi.org/10.1093/petrology/egn002

Brounce, M., Stolper, E. M., & Eiler, J. (2017). Redox variations in Mauna Kea lavas, the oxygen fugacity of the Hawaiian plume, and the role of volcanic gases in Earth's oxygenation. *Proceedings of the National Academy of Sciences*, *114*, 8997–9002. https://doi.org/10.1073/pnas.1619527114

Bryndzia, L. T., Wood, B. J., & Dick, H. J. B. (1989). The oxidation state of the Earth's sub-oceanic mantle from oxygen thermobarometry of abyssal spinel peridotites. *Nature*, *341*, 526–527. https://doi.org/10.1038/341526a0

Canfield, D. E. (2005). The early history of atmospheric oxygen: Homage to Robert A. Garrels. *Annual Review of Earth and Planetary Science*, *33*, 1–36. https://doi.org/10.1146/annurev.earth.33.092203.122711

Canil, D. (1997). Vanadium partitioning and the oxidation state of Archaean komatiite magmas. *Nature*, *389*, 842–845. https://doi.org/10.1038/39860

Cavell, P. A., Wijbrans, J. R., & Baadsgaard, H. (1992). Archean magmatism in the Kaminak Lake area, District of Keewatin, Northwest territories – ages of the carbonatite-bearing alkaline complex and some host granitoid rocks. *Canadian Journal of Earth Sciences*, *29*, 896–908. https://doi.org/10.1139/e92-076

Condie, K. C., & Pease, V. (2008). When did plate tectonics begin on Planet Earth? *Geological Society of America Special Paper,* 440, 294pp.

Creighton, S., Stachel, T., Eichenberg, D., & Luth, R. W. (2010). Oxidation state of the lithospheric mantle beneath Diavik diamond mine, central Slave craton, NWT, Canada. *Contributions to Mineralogy and Petrology*, *159*, 645–657. https://doi.org/10.1007/s00410-009-0446-x

Dalton, J. A., & Presnall, D. C. (1998). The continuum of primary carbonatitic-kimberlitic melt compositions in equilibrium with lherzolite: data from the system CaO-MgO-Al_2O_3-SiO_2-CO_2 at 6 GPa. *Journal of Petrology*, *39*, 1953–1964. https://doi.org/10.1093/petroj/39.11-12.1953

Dasgupta, R., & Hirschmann, M. M. (2006). Melting in the Earth's deep upper mantle caused by carbon dioxide. *Nature*, *440*, 659–662. https://doi.org/10.1038/nature04612

Dasgupta, R., & Hirschmann, M. M. (2007). A modified iterative sandwich method for determination of near-solidus partial melt compositions. II. Application to determination of near-solidus melt compositions of carbonated peridotite. *Contributions to Mineralogy and Petrology*, *154*, 647–661. https://doi.org/10.1007/s00410-007-0214-8

Dasgupta, R., Hirschmann, M. M., & Smith, N. D. (2007). Partial Melting Experiments of Peridotite + CO_2 at 3 GPa and Genesis of Alkalic Ocean Island Basalts. *Journal of Petrology*, *48*, 2093–2124. https://doi.org/10.1093/petrology/egm053

Delano, J. W. (2001). Redox history of the Earth's interior since similar to 3900 Ma: Implications for prebiotic molecules. *Origins of Life and Evolution of the Biosphere*, *31*, 311–341. https://doi.org/10.1023/A:1011895600380

Eggler, D. H., & Lorand, J. L. (1994). Sulfides, diamonds, and mantle f_{O_2}. In: Meyer, H. O. A. & Leonardos, O. H. (eds.) *Diamonds: characterization, genesis and exploration*. Brasilia: C.P.R.M., 160–169.

Eriksson, K. A., Krapez, B., & Fralick, P. W. (1994). Sedimentology of Archean greenstone belts: signature of tectonic evolution. *Earth-Science Reviews*, *37*, 1–88. https://doi.org/10.1016/0012-8252(94)90025-6

Evans, K. A., & Tomkins, A. G. (2011). The relationship between subduction zone redox budget and arc magma fertility. *Earth and Planetary Science Letters*, *308*, 401–409. https://doi.org/10.1016/j.epsl.2011.06.009

Falloon, T. J., & Green, D. H. (1989). The solidus of carbonated, fertile peridotite. *Earth and Planetary Science Letters*, *94*, 364–370. https://doi.org/10.1016/0012-821X(89)90153-2

Ferry, J. M., & Baumgartner, L. P. (1987). Thermodynamic models of molecular fluids at the elevated pressures and temperatures of crustal metamorphism. *Reviews in Mineralogy*, *17*, 323–365. https://doi.org/10.1515/9781501508950-011

Foley, S. F. (1989a). Experimental Constraints on Phlogopite Chemistry in Lamproites .1. the Effect of Water Activity and Oxygen Fugacity. *European Journal of Mineralogy*, 1, 411–426. https://doi.org/10.1127/ejm/1/3/0411

Foley, S. F. (1989b). The genesis of lamproitic magmas in a reduced, flourine-rich mantle. In: Ross, J. (ed.) *Kimberlites and related rocks*. Melbourne: Blackwell, 616–631.

Foley, S. (1991). High-pressure stability of the fluor- and hydroxyl-endmembers of pargasite and K-richterite. *Geochimica et Cosmochimica Acta*, *55*, 2689–2694. https://doi.org/10.1016/0016-7037(91)90386-J

Foley, S. (1992a). Vein-plus-wall-rock melting mechanisms in the lithosphere and the origin of potassic alkaline magmas. *Lithos*, *28*, 435–453. https://doi.org/10.1016/0024-4937(92)90018-T

Foley, S. (1992b). Petrological characterization of the source components of potassic magmas: geochemical and experimental constraints. *Lithos*, *28*, 187–204. https://doi.org/10.1016/0024-4937(92)90006-K

Foley, S. F. (2008). Rejuvenation and erosion of the cratonic lithosphere. *Nature Geoscience*, *1*, 503–510. https://doi.org/10.1038/ngeo261

Foley, S. F. (2011). A reappraisal of redox melting in the Earth's mantle as a function of tectonic setting and time. *Journal of Petrology*, *52*, 1363–1391. https://doi.org/10.1093/petrology/egq061

Foley, S. F., & Fischer, T. P. (2017). An essential role for continental rifts and lithosphere in the deep crabon cycle. *Nature Geoscience*, *10*, 897–902. https://doi.org/10.1038/s41561-017-0002-7

Foley, S. F., & Pintér, Z. (2018). Primary melt composition in the Eath's mantle. In Kono, Y., & Sanloup, C. (eds.) *Magmas under pressure – advances in high-pressure experiments on structure and properties of melts, 1-40*. Amsterdam: Elsevier. https://doi.org/10.1016/B978-0-12-811301-1.00001-0

Foley, S., Musselwhite, D., & Van der Laan, S. (1999). Melt compositions from ultramafic vein assemblages in the lithospheric mantle: a comparison of cratonic and non-cratonic settings, Proceedings of the 7th International Kimberlite Conference, pp. 238–246.

Foley, S. F., Buhre, S., & Jacob, D. E. (2003). Evolution of the Archean crust by shallow subduction and recycling. *Nature*, *421*, 249–252. https://doi.org/10.1038/nature01319

Foley, S. F., Yaxley, G. M., Rosenthal, A., Buhre, S., Kiseeva, E., Rapp, R. P., & Jacob, D. E. (2009). The composition of near-solidus melts of peridotite in the presence of CO_2 and H_2O between 40 and 60 kbar. *Lithos*, *112*, 274–283. https://doi.org/10.1016/j.lithos.2009.03.020

Förster, M. W., Prelevic, D., Buhre, S., Mertz-Kraus, R., & Foley, S. F. (2019). An experimental study of the role of partial melts of sediments versus mantle melts as sources of potassic magmatism. *Journal of Asian Earth Sciences*, *177*, 76–88. https://doi.org/10.1016/j.jseaes.2019.03.014

Frost, B. R. (1979). Mineral equilibria involving mixed-volatiles in a C-O-H fluid phase: the stabilities of graphite and siderite. *American Journal of Science*, *279*, 1035–1059. https://doi.org/10.2475/ajs.279.9.1033

Frost, D. J., & McCammon, C. A. (2008). The redox state of Earth's mantle. *Annual Review of Earth and Planetary Science*, *36*, 389–420. https://doi.org/10.1146/annurev.earth.36.031207.124322

Gaillard, F., Scaillet, B., & Arndt, N. T. (2011). Atmospheric oxygenation caused by a change in volcanic degassing pressure. *Nature*, *478*, 229–232, 2011. https://doi.org/10.1038/nature10460

Gaillard, F., Scaillet, B., Pichavant, M., & Iacano-Marziano, G. (2015). The redox geodynamics linking basalts and their mantle sources through space and time. *Chemical Geology*, *418*, 217–233. https://doi.org/10.1016/j.chemgeo.2015.07.030

Garber, J. M., Maurya, S., Hernandez, J-A., Duncan, M. S., Zeng, L., Zhang, H. L., et al. (2018). Multidisciplinary constraints on the abundance of diamond and eclogite in the cratonic lithosphere. *Geochemistry, Geophysics, Geosystems*, *19*, 2062–2086. https://doi.org/10.1029/2018GC007534

Ghosh, S., Litasov, K. D., & Ohtani, E. (2014). Phase relations and melting of carbonated peridotite between 10 and 20 GPa: a proxy for alkali- and CO_2-rich silicate melts in the deep mantle. *Contributions to Mineralogy and Petrology*, *167*, 964. https://doi.org/10.1007/s00410-014-0964-z

Glaser, S. M., Foley, S. F., & Günther, D. (1999). Trace element compositions of minerals in garnet and spinel peridotite xenoliths from the Vitim volcanic field, Transbaikalia, eastern Siberia. *Lithos*, *48*, 263–285. https://doi.org/10.1016/S0024-4937(99)00032-8

Goncharev, A. G., & Ionov, D. A. (2012). Redox state of deep off-craton lithospheric mantle: new data from garnet and spinel peridotites from Vitim, southern Siberia. *Contributions to Mineralogy and Petrology*, *164*, 731–745. https://doi.org/10.1007/s00410-012-0767-z

Green, D. H. (1973). Experimental melting studies on a model upper mantle composition at high pressure under water-saturated and water-undersaturated conditions. *Earth and Planetary Science Letters*, *19*, 37–53. https://doi.org/10.1016/0012-821X(73)90176-3

Green, D. H., & Falloon, T. J. (1998). Pyrolite: a Ringwood concept and its current expression. In: Jackson, I. (ed.) *The Earth's mantle: composition, structure, and evolution*. Cambridge: Cambridge University Press, pp. 311–378.

Green, D. H., Falloon, T. J., & Taylor, W. R. (1987). Mantle-derived magmas – roles of variable source peridotite and variable C-H-O fluid compositions. In: Mysen, B. O. (ed.)

Magmatic processes: physicochemical principles. Washington: Geochemical Society, 139–154.

Green, D. H., Falloon, T. J., Eggins, S. M., & Yaxley, G. M. (2001). Primary magmas and mantle temperatures. *European Journal of Mineralogy*, *13*, 437–451. https://doi.org/10.1127/0935-1221/2001/0013-0437

Griffin, W. L., O'Reilly, S. Y., Afonso, J. C., & Begg, G. C. (2009). The composition and evolution of lithospheric mantle: a re-evaluation and its tectonic implications. *Journal of Petrology*, *50*, 1185–1204. https://doi.org/10.1093/petrology/egn033

Griffin, W. L., Begg, G. C., & O'Reilly, S. Y. (2013). Continental-root control on the genesis of magmatic ore deposits. *Nature Geoscience*, *6*, 905–910. https://doi.org/10.1038/ngeo1954

Griffin, W. L., Afoso, J. C., Belousova, E. A., Gain, S. E., Hong, X. H., Gonzalez-Jiménez, J. M., et al. (2016). Mantle recycling: transition zone metamorphism of Tibetan ohpiolitic peridotites and its tectonic implications. *Journal of Petrology*, *57*, 655–684. https://doi.org/10.1093/petrology/egw011

Griffin, W. L., Huang, J.-X., Thomassot, E., Gain, S. E. M., Toledo, V., & O'Reilly, S. Y. (2018). Super-reducing conditions in ancient and modern volcanic systems: sources and behaviour of carbon-rich fluids in the lithospheric mantle. *Mineralogy and Petrology*, *2018*. https://doi.org/10.1017/s00710-018-0575-x

Hamilton, D. L., Freestone, I. C., Dawson, J. B., & Donaldson, C. H. (1979). Origin of carbonatites by liquid immiscibility. *Nature*, *279*, 52–54. https://doi.org/10.1038/279052a0

Hermann, J., & Spandler, C. J. (2008). Sediment melts at sub-arc depths: An experimental study. *Journal of Petrology*, *49*, 717–740. https://doi.org/10.1093/petrology/egm073

Hirschmann, M. M., & Stolper, E. M. (1996). A possible role for garnet pyroxenite in the origin of the "garnet signature" in MORB. *Contribution to Mineralogy and Petrology*, *124*, 185–208. https://doi.org/10.1007/s004100050184

Holloway, J. R. (1987). Igneous fluids. *Reviews in Mineralogy*, *17*, 211–233.

Holloway, J. R., & Blank, J. G. (1994). Application of experimental results to C-O-H species in natural melts. *Reviews in Mineralogy*, *30*, 187–230. https://doi.org/10.1515/9781501509674-012

Howarth, G. H., Day, J. M. D., Pernet-Fisher, J. F., Goodrich, C. A., Pearson, D. G., Luo, Y., et al. (2017). Precious metal enrichment at low-redox in terrestrial native Fe-bearing basalts investigated using laser-ablation ICP-MS. *Geochimica et Cosmochimica Acta*, *203*, 343–363. https://doi.org/10.1016/j.gca.2017.01.003

Hutchinson, M. T., Faithfull, J. W., Barford, D. N., Hughes, J. W., & Upton, B. G. J. (2018). The mantle of Scotland viewed through the Glen Gollaidh aillikite. *Mineralogy and Petrology*, *112*, 115–132. doi: 10.1007/s00710-018-0610-y

Ionov, D. A., & Wood, B. J. (1992). The oxidation-state of subcontinental mantle – oxygen thermobraometry of mantle xenoliths from Central-Asia. *Contributions to Mineralogy and Petrology*, *111*, 179–193. https://doi.org/10.1007/BF00348950

Ishimaru, S., Arai, S., & Shukuno, H. (2009). Metal-saturated peridotite in the mantle wedge inferred from metal-bearing peridotite xenoliths from Avacha volcano, Kamchatka. *Earth and Planetary Science Letters*, *284*, 352–360. https://doi.org/10.1016/j.epsl.2009.04.042

Jacob, D. E. (2004). Nature and origin of eclogite xenoliths from kimberlites. *Lithos*, *77*, 295–316. https://doi.org/10.1016/j.lithos.2004.03.038

Jacob, D. E., & Foley, S. F. (1999). Evidence for Archean ocean crust with low high field strength element signature from diamondiferous eclogite xenoliths. *Lithos*, *48*, 317–336. https://doi.org/10.1016/S0024-4937(99)00034-1

Jacob, D., Jagoutz, E., Lowry, D., Mattey, D., & Kudrjatseva, G. (1994). Diamondiferous eclogites from Siberia: remnants of Archean oceanic crust. *Geochimica et Cosmochimica Acta*, *58*, 5191–5207. https://doi.org/10.1016/0016-7037(94)90304-2

Jacob, D. E., Kronz, A., & Viljoen, K. S. (2004). Cohenite, native iron and troilite inclusions in garnets from polycrystalline diamond aggregates. *Contributions to Mineralogy and Petrology*, *146*, 566–576. doi: 10.1007/s00410-003-0518-2

Jacob, D. E., Bizimis, M., & Salters, V. J. M. (2005). Lu-Hf and geochemical systematics of recycled ancient oceanic crust: evidence from Roberts Victor eclogites. *Contributions to Mineralogy and Petrology*, *148*, 707–720. doi: 10.1007/s00410-004-0631-x

Jacob, D. E., Piazolo, S., Schreiber, A., & Trimby, P. (2016). Redox-freezing and nucleation of diamond via magnetite formation in the Earth's mantle. *Nature Communications 7*, 11891. 10.1038/ncomms11891

Jakobsson, S., & Holloway, J. R. (1986). Crystal-liquid experiments in the presence of a C-O-H fluid buffered by graphite +iron+wustite: experimental method and near-liquidus relations in basanite. *Journal of Volcanology and Geothermal Research*, *29*, 265–291. https://doi.org/10.1016/0377-0273(86)90048-X

Jakobsson, S., & Holloway, J. R. (2008). Mantle melting in equilibrium with an iron-wüstite-graphite buffered COH-fluid. *Contributions to Mineralogy and Petrology 155*, 247–256. https://doi.org/10.1007/s00410-007-0240-6

Kadik, A. (1988). Effects of melting on the evolution of fluid and redox conditions in the upper mantle. *Geochemistry International*, *25*, 76–84.

Kadik, A. (1997). Evolution of Earth's redox state during upwelling of carbon-bearing mantle. *Physics of the Earth and Planetary Interiors*, *100*, 157–166. https://doi.org/10.1016/S0031-9201(96)03237-2

Kadik, A. A. (2003). Mantle-derived reduced fluids: Relationship to the chemical differentiation of planetary matter. *Geochemistry International*, *41*, 844–855.

Kasting, J. F., Eggler, D. H., & Raeburn, S. P. (1993). Mantle redox evolution and the oxidation-state of the Archean atmosphere. *Journal of Geology*, *101*, 245–257. https://doi.org/10.1086/648219

Keller, C. B., & Schoene, B. (2012). Statistical geochemistry reveals disruption in secular lithospheric evolution about 2.5 Gyr ago. *Nature*, *485*, 490–493. https://doi.org/10.1038/nature11024

Kjarsgaard, B. A., & Peterson, T. (1991). Nephelinite-carbonatite liquid immiscibility at Shombole volcano, East Africa: petrographic and experimental evidence. *Mineralogy and Petrology*, *43*, 293–314. https://doi.org/10.1007/BF01164532

Kjarsgaard, B. A., Pearson, D. G., Tappe, S., Nowell, G. M., & Dowall, D. P. (2009). Geochemistry of hypabyssal kimberlites from Lac de Gras, Canada: Comparisons to a global database and applications to the parent magma problem. *Lithos, 112S*, 236–248. https://doi.org/10.1016/j.lithos.2009.06.001

Kogiso, T., Hirschmann, M. M., & Pertermann, M. (2004). High-pressure partial melting of mafic lithologies in the mantle. *Journal of Petrology, 45*, 2407–2422. https://doi.org/10.1093/petrology/egh057

Lee, C. T. A. (2006). Geochemical/petrologic constraints on the origin of cratonic mantle. In: Benn, K., Mareschal, J.-C., & Condie, K. C. (eds.) *Archean geodynamics and environments*. Washington, D.C.: American Geophysical Union, 89–114. doi: 10.1029/164GM08

Li, Z. X. A., & Lee, C. T. A. (2004). The constancy of upper mantle f_{O_2} through time inferred from V/Sc ratios in basalts. *Earth and Planetary Science Letters, 228*, 483–493. https://doi.org/10.1016/j.epsl.2004.10.006

Libourel, G., Marty, B., & Humbert, F. (2003). Nitrogen solubility in basaltic melt. *Part I. Effect of oxygen fugacity. Geochimica et Cosmochimica Acta, 67*, 4123–4135. https://doi.org/10.1016/S0016-7037(03)00259-X

Litasov, K.D., Shatskiy, A., & Ohtani, E. (2014). Melting and subsolidus phase relations in peridotite and eclogite systems with reduced C-O-H fluid at 3-16 GPa. *Earth and Planetary Science Letters, 391*, 87–99. https://doi.org/10.1016/j.epsl.2014.01.033

Liu, Z., Rohrbach, A., Klemme, S., Foley, S., & Berndt, J. (2018). The effect of C-O-H fluids on partial melting of eclogite and lherzolite under reducing conditions. International Mineralogical Association, Melbourne, IMA2018-1539.

Luth, R.W., & Stachel, T. (2014). The buffering capacity of lithospheric mantle: implications for diamond formation. *Contributions to Mineralogy and Petrology, 168*, 1083. https://doi.org/10.1007/s00410-014-1083-6

Matveev, S., Ballhaus, C., Fricke, K., Truckenbrodt, J., & Ziegenbein, D. (1997). Volatiles in the Earth's mantle: I. Synthesis of CHO fluids at 1273 K and 2.4 GPa. *Geochimica et Cosmochimica Acta, 61*, 3081–3088. https://doi.org/10.1016/S0016-7037(97)00142-7

McInnes, B. I. A., Gregoire, M., Binns, R. A., Herzig, P. M., & Hannington, M. D. (2001). Hydrous metasomatism of oceanic sub-arc mantle, Lihir, Papua New Guinea: petrology and geochemistry of fluid-metasomatised mantle wedge xenoliths. *Earth and Planetary Science Letters, 188*, 169–183. https://doi.org/10.1016/S0012-821X(01)00306-5

McKenzie, D., & Bickle, M. J. (1988). The volume and composition of melt generated by extension of the lithosphere. *Journal of Petrology, 29*, 625–679. https://doi.org/10.1093/petrology/29.3.625

Millhollen, G. L., Irving, A. J., & Wyllie, P. J. (1974). Melting interval of peridotite with 5.7 per cent water to 30 kilobars. *Journal of Geology, 82*, 575–587. https://doi.org/10.1086/628007

Mironov, N. L., & Portnyagin, M. V. (2018). Coupling of redox conditions of mantle melting and copper and sulfur contents in primary magmas of the Tolbachinsky Dol (Kamchatka) and Juan de Fuca Ridge (Pacific Ocean). *Petrology, 26*, 145–166. https://doi.org/10.1134/S0869591118020030

Mitchell, R. H. (1986). *Kimberlites: mineralogy, geochemistry and petrology*. New York: Plenum Press.

Moore, K. R., & Wood, B. J. (1998). The transition from carbonate to silicate melts in the CaO-MgO-SiO_2-CO_2 system. *Journal of Petrology, 39*, 1943–1951. https://doi.org/10.1093/petroj/39.11-12.1943

Mysen, B. O., Virgo, D., & Seifert, F. (1982). The structure of silicate melts: implications for chemical and physical properties of natural magma. *Reviews of Geophysics Space Physics, 20*, 353–383. https://doi.org/10.1029/RG020i003p00353

Mysen, B. O., Yamashita, S., & Chertkova, N. (2008). Solubility and solution mechanisms of NOH volatiles in silicate melts at high pressure and temperature-amine groups and hydrogen fugacity. *American Mineralogist, 93*, 1760–1770. https://doi.org/10.2138/am.2008.2879

Mysen, B. O., Fogel, M. L., Morrill, P. L., & Cody, G. D. (2009). Solution behavior of reduced C-O-H volatiles in silicate melts at high pressure and temperature. *Geochimica et Cosmochimica Acta, 73*, 1696–1710. https://doi.org/10.1016/j.gca.2008.12.016

Nicklas, R. W., Puchtel, I. S., & Ash, R. D. (2016). High-precision determination of the oxidation state of komatiite lavas using vanadium-liquid mineral partitioning. *Chemical Geology, 433*, 36–45. https://doi.org/10.1016/j.chemgeo.2016.04.011

Nicklas, R. W., Puchtel, I. S., & Ash, R. D. (2018). Redox state of the Archean mantle: evidence from V partitioning in 3.5-2.4 Ga komatiites. *Geochimica et Cosmochimica Acta, 222*, 447–466. https://doi.org/10.1016/j.gca.2017.11.002

O'Neill, H. S. C., & Wall, V. J. (1987). The olivine-orthopyroxene-spinel oxygen geobarometer, the nickel precipitation curve, and the oxygen fugacity of the Earth's upper mantle. *Journal of Petrology, 28*, 1169–1191. https://doi.org/10.1093/petrology/28.6.1169

Olafsson, M., & Eggler, D. H. (1983). Phase relations of amphibole, amphibole-carbonate, and phlogopite-carbonate peridotite: petrologic constraints on the asthenosphere. *Earth and Planetary Science Letters, 64*, 305–315. https://doi.org/10.1016/0012-821X(83)90212-1

O'Reilly, S. Y., & Griffin, W. L. (2013). Mantle metasomatism. In: D.E. Harlov and H. Austrheim (eds.), *Metasomatism and the chemical transformation of rock*. Berlin: Springer-Verlag, Berlin-Heidelberg, pp. 471–533. doi 10.1007/978-3-642-28394-9_12

Palot, M., Pearson, D. G., Stern, R. A., Stachel, T., & Harris, J. W. (2014). Isotopic constraints on the nature and circulation of deep mantle C-H-O-N fluids: Carbon and nitrogen systematics within ultra-deep diamonds from Kankan (Guinea). *Geochimica et Cosmochimica Acta, 139*, 26–46. https://doi.org/10.1016/j.gca.2014.04.027

Pal'yanov, Y. N., Sokol, A. G., Tomilenko, A. A., & Sobolev, N. V. (2005). Conditions of diamond formation through carbonate-silicate interaction. *European Journal of Mineralogy, 17*, 207–214. doi: 10.1127/0935-1221/2005/0017-0207

Pal'yanov, Y.N., Bataleva, Y. V., Sokol, A. G., Borzdov, Y. M., Kupriyanov, I. N., Reutsky, V. N., & Sobolev, N. V. (2013). Mantle-slab interaction and redox mechanism of diamond formation. *Proceedings of the National Academy of Sciences, 110*, 20408–20413. https://doi.org/10.1073/pnas.1313340110

Parkinson, I. J., Arculus, R. J., & Eggins, S. M. (2003). Peridotite xenoliths from Grenada, Lesser Antilles Island Arc.

Contributions to Mineralogy and Petrology, *146*, 241–262. https://doi.org/10.1007/s00410-003-0500-z

Pavlov, A. A., & Kasting, J. F. (2002). Mass-independent fractionation of sulfur isotopes in Archean sediments: Strong evidence for an anoxic Archean atmosphere. *Astrobiology*, *2*, 27–41. doi: 10.1089/15311070275362l321

Perinelli, C., Andreozzi, G. B., Conte, A. M., Oberti, R., & Armieti, P. (2012). Redox state of subcontinental lithospheric mantle and relationships with metasomatism: insights from spinel peridotites from northern Victoria Land (Antarctica). *Contributions to Mineralogy and Petrology*, *164*, 1053–1067. https://doi.org/10.1007/s00410-012-0788-7

Pintér, Z. (2019). Compositions and effects of incipient melts of mantle peridotites in the presence of H_2O and CO_2 – an experimental study. PhD thesis, Macquarie University, Sydney, 132pp.

Pintér, Z., Foley, S.F., Yaxley, G.M., Rosenthal, A., Rapp, R. P., Lanati, A.W., Rushmer, T. (2021). Experimental investigation of the composition of incipient melts in the upper mantle in the presence of CO_2 and H_2O. *Lithos* 396–397, 106224, doi: 10.1016/lithos.2021.106224

Plank, T., Cooper, L. B., & Manning, C. E. (2009). Emerging geothermometers for estimating slab surface temperatures. *Nature Geoscience*, *2*, 611–615. https://doi.org/10.1038/ngeo614

Poli, S., Franzolin, E., Fumagalli, P., & Crottini, A. (2009). The transport of carbon and hydrogen in subducted oceanic crust: An experimental study to 5 GPa. *Earth and Planetary Science Letters*, *278*, 350–360. https://doi.org/10.1016/j.epsl.2008.12.022

Rehfeldt, T., Foley, S. F., Jacob, D. E., Carlson, R. W., & Lowry, D. (2008). Contrasting types of metasomatism in dunite, wehrlite and websterite xenoliths from Kimberley, South Africa. *Geochimica et Cosmochimica Acta*, *72*, 5722–5756. https://doi.org/10.1016/j.gca.2008.08.020

Rohrbach, A., & Schmidt, M. W. (2011). Redox freezing and melting in the Earth's deep mantle resulting from carbon-iron redox coupling. *Nature*, *472*, 209–212. https://doi.org/10.1038/nature09899

Rohrbach, A., Ballhaus, C., Ulmer, P., Golla-Schindler, U., & Schonbohm, D. (2011). Experimental evidence for a reduced metal-saturated upper mantle. *Journal of Petrology*, *52*, 717–731. https://doi.org/10.1093/petrology/egq101

Rollinson, H., Adetunji, J., Kenaz, D., & Szilas, K. (2017). Archaean chromitites show constant $Fe^{3+}/\Sigma Fe$ in Earth's asthenospheric mantle since 3.8 Ga. *Lithos*, *282*, 316–325. https://doi.org/10.1016/j.lithos.2017.03.020

Rosenthal, A., Foley, S. F., Pearson, D. G., Nowell, G. M., & Tappe, S. (2009). Magmatic evolution at the propagating tip of a continental rift – a geochemical study of primitive alkaline volcanic rocks of the western branch of the East African Rift. *Earth and Planetary Science Letters*, *284*, 236–248. https://doi.org/10.1016/j.epsl.2009.04.036

Schidlowski, M. (1981). Uraniferous constituents of the Witwatersrand conglomerates: ore-microscopic observations and implications for the Witwatersrand metallogeny. *Geological Survey Professional Paper 1161*, N1–N29.

Shu, Q., & Brey, G. P. (2015). Ancient mantle metasomatism recorded in subcalcic garnet xenocrysts: temporal links between mantle metasomatism, diamond growth and crustal tectonomagmatism. *Earth and Planetary Science Letters*, *418*, 27–39. https://doi.org/10.1016/j.epsl.2015.02.038

Simon, N. S. C., Carlson, R. W., Pearson, D. G., & Davies, G. R. (2007). The origin and evolution of the Kaapvaal cratonic lithospheric mantle. *Journal of Petrology*, *48*, 589–625. https://doi.org/10.1093/petrology/egl074

Sleep, N.H. (2009). Stagnant lid convection and carbonate metasomatism of the deep continental lithosphere. *Geochemistry, Geophysics, Geosystems*, *10*, Q11010. doi: 10.1029/2009GC002702

Sleep, N. H., Meibom, A., Fridriksson, T., Coleman, R. G., & Bird, D. K. (2004). H_2-rich fluids from serpentinization: Geochemical and biotic implications. *Proceedings of the National Academy of Sciences of the United States of America 101*, 12818–12823. https://doi.org/10.1073/pnas.0405289101

Smart, K. A., Heaman, L. M., Chacko, T., Simonetti, A., Kopylova, M., Mah, D., & Daniels, D. (2009). The origin of high-MgO diamond eclogites from the Jericho Kimberlite, Canada. *Earth and Planetary Science Letters*, *284*, 527–537. https://doi.org/10.1016/j.epsl.2009.05.020

Smart, K. A., Tappe, S., Simonetti, A., Simonetti, S. S., Woodland, A. B., & Harris, C. (2017). Tectonic significance and redox state of Paleopreoterozoic eclogite and pyroxenite components in the Slave cratonic mantle lithosphere, Voyageur kimberlite, Arctic Canada. *Chemical Geology*, *455*, 98–199. https://doi.org/10.1016/j.chemgeo.2016.10.014

Smith, E. M., Shirey, S. B., Nestola, F., Bullock, E. S., Wang, J. H., Richardson, S. H., & Wang, W. Y. (2016). Large gem diamonds from metallic liquid in Earth's deep mantle. *Science*, *354*, 1403–1405. doi: 10.1126/science.aal1303

Sobolev, A. V., Hofmann, A. W., & Nikogosian, I. K. (2000). Recycled oceanic crust observed in 'ghost plagioclase'within the source of Mauna Loa lavas. *Nature*, *404*, 986–990. https://doi.org/10.1038/35010098

Song, S. G., Su, L., Niu, Y. L., Lai, Y., & Zhang, L. F. (2009). CH_4 inclusions in orogenic harzburgite: Evidence for reduced slab fluids and implication for redox melting in mantle wedge. *Geochimica et Cosmochimica Acta 73*, 1737–1754. doi: 10.1016/j.gca.2008.12.008

Stachel, T., & Harris, J. W. (2008). The origin of cratonic diamonds – Constraints from mineral inclusions. *Ore Geology Reviews*, *34*, 5–32. https://doi.org/10.1016/j.oregeorev.2007.05.002

Stachel, T., & Luth, R. W. (2015). Diamond formation – where, when and how? *Lithos*, *220*, 200–220. https://doi.org/10.1016/j.lithos.2015.01.028

Stachel, T., Chacko, T., & Luth, R. W. (2017). Carbon isotope fractionation during diamond growth in depleted peridotite: Counterintuitive insights from modelling water-maximum CHO fluids as multi-component systems. *Earth and Planetary Science Letters*, *473*, 44–51. https://doi.org/10.1016/j.epsl.2017.05.037

Stagno, V., Ojwang, D. O., McCammon, C. A., & Frost, D. J. (2013). The oxidation state of the mantle and the extraction of carbon from Earth's interior. *Nature 493*, 84–88. https://doi.org/10.1038/nature11679

Sweeney, R. J., Green, D. H., & Sie, S. H. (1992). Trace and minor element partitioning between garnet and amphibole and carbonatitic melt. *Earth and Planetary Science Letters*, *113*, 1–14. https://doi.org/10.1016/0012-821X(92)90207-C

Tappe, S., Foley, S. F., Jenner, G. A., & Kjarsgaard, B. A. (2005). Integrating ultramafic lamprophyres into the IUGS classification of igneous rocks: rationale and implications. *Journal of Petrology*, *46*, 1893–1900. https://doi.org/10.1093/petrology/egi039

Tappe, S., Foley, S. F., Jenner, G. A., Heaman, L. M., Kjarsgaard, B. A., Romer, R. L., et al. (2006). Genesis of ultramafic lamprophyres and carbonatites at Aillik Bay, Labrador: A consequence of incipient lithospheric thinning beneath the North Atlantic craton. *Journal of Petrology*, *47*, 1261–1315. https://doi.org/10.1093/petrology/egl008

Tappe, S., Foley, S. F., Stracke, A., Romer, R. L., Heaman, L. M., Kjarsgaard, B. A., & Joyce, N. (2007). Craton reactivation on the Labrador Sea margins: $^{40}Ar/^{39}Ar$ age and Sr-Nd-Hf-Pb isotope constraints from alkaline and carbonatite intrusives. *Earth and Planetary Science Letters*, *256*, 433–454. https://doi.org/10.1016/j.epsl.2007.01.036

Taylor, W. R. (1985). *The role of C-O-H fluids in upper mantle processes - a theoretical, experimental and spectroscopic study*. Ph.D. thesis, Geology Department, University of Tasmania, Hobart.

Taylor, W. R., & Green, D. H. (1987). The petrogenetic role of methane: effect on liquidus phase relations and the solubility mechanism of reduced C-H volatiles. In: Mysen, B. O. (ed.) *Magmatic processes: physicochemical principles*. Washington: Geochemical Society, 121–138.

Taylor, W. R., & Green, D. H. (1988). Measurement of reduced peridotite-C-O-H solidus and implications for redox melting of the mantle. *Nature*, *332*, 349–352. https://doi.org/10.1038/332349a0

Tomkins, A. G., & Evans, K. A. (2015). Separate zones of sulphate and sulphide release from subducted mafic oceanic crust. *Earth and Planetary Science Letters*, *428*, 73–83. https://doi.org/10.1016/j.epsl.2015.07.028

Treiman, A.H., & Schedl, A. (1983). Properties of carbonatite magma and processes in carbonatite magma chambers. *Journal of Geology*, *91*, 437–447. https://doi.org/10.1086/628789

Tumiati, S., Fumagalli, P., Tiraboschi, C., & Poli, S. (2013). An experimental study on COH-bearing peridotite at up to 3.2 GPa and implications for crust-mantle recycling. *Journal of Petrology*, *54*, 453–479. https://doi.org/10.1093/petrology/egs074

Uenver-Thiele, L., Woodland, A. B., Downes, H., & Altherr, R. (2014). Oxidation state of the lithospheric mantle below the Massif Central, France, *Journal of Petrology*, *55*, 2457–2480. https://doi.org/10.1093/petrology/egu063

Uenver-Thiele, L., Woodland, A. B., Seitz, H.-M., Downes, H., & Altherr, R. (2017). Metasomatic processes revealed by trace element and redox signatures of the lithospheric mantle beneath the Massif Central, France, *Journal of Petrology*, *58*, 395–422. https://doi.org/10.1093/petrology/egx020

Villeneuve, M. E., & Relf, C. (1998). Tectonic setting of 2.6 Ga carbonatites in the Slave Province, NW Canada. *Journal of Petrology*, *39*, 1975–1986. https://doi.org/10.1093/petroj/39.11-12.1975

Wallace, M. E., & Green, D. H. (1988). An experimental determination of primary carbonatite magma composition. *Nature*, *335*, 343–346. https://doi.org/10.1038/335343a0

Wallace, M. E., & Green, D. H. (1991). The effect of bulk rock composition on the stability of amphibole in the upper mantle: implications for solidus positions and mantle metasomatism. *Mineralogy and Petrology*, *44*, 1–19. https://doi.org/10.1007/BF01167097

Walter, M. J. (1998). Melting of garnet peridotite and the origin of komatiite and depleted lithosphere. *Journal of Petrology*, *39*, 29–60. https://doi.org/10.1093/petroj/39.1.29

Wang, J., Hattori, K. H., Kilian, R., & Stern, C. R. (2009). Metasomatism of sub-arc mantle peridotites below southern South America: reduction of f_{O_2} by slab-melt. *Contributions to Mineralogy and Petrology*, *153*, 607–624. https://doi.org/10.1007/s00410-006-0166-4

Wang, J., Hattori, K., & Xie, Z. (2013). Oxidation state of lithospheric mantle along the northeastern margin of the North China Craton: implications for geodynamic processes. *International Geology Review*, *55*, 1418–1444. https://doi.org/10.1080/00206814.2013.780722

Weiss, Y., McNeill, J. D., Pearson, D. G., Nowell, G. M., & Ottley, C. J. (2015). Highly saline fluids from a subducting slab as the source for fluid-rich diamonds. *Nature*, *524*, 339–342. https://doi.org/10.1038/nature14857

Weiss, Y., Navon, O., Goldstein, S. L., & Harris, J. W. (2018). Inclusions in diamonds constrain thermo-chemical conditions during Mesozoic metasomatism of the Kaapvaal cratonic mantle. *Earth and Planetary Science Letters*, *491*, 134–147. https://doi.org/10.1016/j.epsl.2018.03.014

Woermann, E., & Rosenhauer, M. (1985). Fluid phases and the redox state of the Earth's mantle – Extrapolations based on experimental, phase-theoretical and petrological data. *Fortschritte der Mineralogie 63*, 263–349.

Woodland, A. B., & Peltonen, P. (1999). Ferric iron contents of garnet and clinopyroxene and estimated oxygen fugacities of peridotite xenoliths from the eastern Finland kimberlite province. In: Gurney, J. J., Gurney, J. L., Pascoe, M. D., & Richardson, S. H. (eds.) *P.H.Nixon Volume*. Cape Town: Red Roof Design, 904–911.

Wyllie, P.J. (1978). Mantle fluid compositions buffered in peridotite-CO_2-H_2O by carbonates, amphibole, and phlogopite. *Journal of Geology*, *86*, 687–713. https://doi.org/10.1086/649737

Wyllie, P.J. (1988). Magma genesis, plate tectonics, and the chemical differentiation of the Earth. *Reviews of Geophysics*, *26*, 370–404. https://doi.org/10.1029/RG026i003p00370

Yaxley, G. M., Berry, A. J., Kamenetsky, V. S., Woodland, A. B., & Golovin, A. V. (2012). An oxygen fugacity profile through the Siberian Craton – Fe K-edge XANES determinations of $Fe^{3+}/\Sigma Fe$ in garnets in peridotite xenoliths from the Udachnaya East kimberlite. *Lithos*, *140–141*, 142–151. https://doi.org/10.1016/j.lithos.2012.01.016

Yaxley, G. M., Berry, A. J., Rosenthal, A., Woodland, A. B., & Paterson, D. (2017). Redox preconditioning deep cratonic lithosphere for kimberlite genesis – evidence from the central Slave Craton. *Scientific Reports* 7, 30. https://doi.org/10.1038/s41598-017-00049-3

Part II

Redox at Work: From Magma Sources to Volcanic Phenomena

6

Ionic Syntax and Equilibrium Approach to Redox Exchanges in Melts: Basic Concepts and the Case of Iron and Sulfur in Degassing Magmas

Roberto Moretti[1,2]

ABSTRACT

The properties of melts and glasses are heavily affected by redox equilibria and these are sensitive to changes in temperature, oxygen fugacity and composition. The compositional control includes the amphoteric behaviour of some components such as iron oxides and demands the understanding of the link between redox, acid-base properties and structural coordination. This goes through a precise and coherent chemical "syntax" for redox and more in general chemical exchanges in melts and magmas. This syntax applies also to the understanding of how bulk composition drives the redox evolution via melt and gas speciation from source depths up to surface. Useful shortcuts and rules of thumbs might be adopted, but we must keep in mind that redox is not a Maxwell's demon or an adjustable parameter. Rather, it is the twin-face of acid-base exchanges that occur in chemically complex mixtures and solutions. In this study we show that recognizing chemically reactive entities allows decrypting the effect of composition on redox patterns of iron and sulfur. Besides, this allows proposing a stoichiometry-based methodology to describe in detail the chemical mechanisms that drive the melt system to get rid of sulfur via decompression, degassing, and precipitation of S-bearing phases. These stoichiometric constraints allow exploring the different features of magma degassing in basaltic and andesitic systems, independently of any model of redox state proposed in the literature.

6.1. INTRODUCTION

In multiphase magmas (melt±crystals±gas), redox state, or better redox potential, is one the most important variables involved in chemical reactions fixing the speciation of fluids discharged by magmatic systems. Redox potential in all hot environments of the Earth's interior is related to oxygen, which controls redox exchanges of elements such as iron and sulfur in their different oxidation states (or oxidation numbers), that is, the (hypothetical) charges that redox-sensitive elements would have if all bonds to other atoms (e.g., oxygen) were 100% ionic, with no covalent bond fraction.

Because iron is orders of magnitude more abundant than other transition elements, the proportion of ferric

[1] *Université de Paris, Institut de Physique du Globe de Paris, Paris, France*

[2] *Observatoire Volcanologique et Sismologique de Guadeloupe, Institut de Physique du Globe de Paris, Gourbeyre, France*

Magma Redox Geochemistry, Geophysical Monograph 266, First Edition.
Edited by Roberto Moretti and Daniel R. Neuville.

DOI: 10.1002/9781119473206.ch6

(Fe^{3+}) and ferrous (Fe^{2+}) iron may control (or buffer) the overall oxygen fugacity (fO_2) of magmas. Therefore, iron redox equilibria and structural conditions in silicate melts and rock mineral assemblages strongly affect magma properties and drive phase equilibria. Among the other multivalent elements, sulfur species play a crucial role in redox exchanges since the oxidation of sulfide to sulfate involves eight electrons. Therefore, any variation of the Fe^{3+}/Fe^{2+} redox ratio implies an eight-fold variation for sulfur species (S^{6+}/S^{2-}; e.g. Douglas & Zaman, 1969; Moretti & Ottonello, 2003a; Nash et al., 2019; Cicconi et al., 2020a; Moretti & Stefansson, 2020). Sulfur in magmas partitions between different phases (gas, solids, such as pyrrhotite and anhydrite, and liquid, such as immiscible Fe–O–S liquids; Haughton et al., 1974; Moretti & Baker, 2008 and references therein) and despite the large electron transfer the effectiveness of the sulfide to sulfate ratio as a buffer of the redox potential is limited by the abundance of sulfur in melts, which is significantly lower than iron in most commonly encountered natural magmas.

Empirical and theoretical formulations have been proposed in literature to assess iron and sulfur oxidation states in natural samples and glasses and establish their relations with fO_2. Such techniques allow the assessment of redox conditions in magmas and their impact on the gas phase speciation on magma rise (especially the H_2S/SO_2 ratio, but also H_2/H_2O and CO/CO_2) and on the phase equilibria involving melt and mineral phases, especially the FeO/Fe_2O_3 ratio (e.g. Frost, 1991 and references therein; Giggenbach et al., 1996; Moretti & Papale, 2004; Behrens & Gaillard, 2006; Burgisser & Scaillet, 2007; Moretti & Stefansson, 2020).

The possibility offered by analytical techniques to measure directly the oxidation state of these redox-sensitive elements in quenched glasses (e.g., sulfur and iron redox ratios in melt inclusions) has offered the opportunity to test these empirical and theoretical formulations to retrieve the evolution of fO_2 for the studied system. Nevertheless, different formulations yield different or even contrasting results for fO_2 and/or FeO/Fe_2O_3 and or S^{6+}/S^{2-} from the same set of data (Oppenheimer et al., 2011; Moussallam et al., 2014). Many of these discrepancies are imputable to how compositional variables are treated, especially in case of sulfur redox state, for which the role of composition has too often been discarded. In any system, bulk composition plays a pivotal role in determining the nature of the ligand (i.e., the solvent structure, such as the melt oxygen network), hence the type and amount of reactive species that it forms by complexing central atoms. A typical example is offered by tetrahedral units in silicate networks and their connectivity (or polymerization) in more complex and continuous structures that are broken by so-called network modifying cations, which include many, but not all, metal cations of transition elements (Mysen & Richet, 2018; Le Losq et al., 2019).

Rather than using the oxide notation typical of mineral chemistry, the adoption of an ionic notation has to be preferred, as it is a straightforward and simpler way to highlight oxidation states and introduce, via charge balance constraints, the stoichiometry of reactions that involve species having a structural significance. Nevertheless, this demands a correct approach to the speciation state of the melt phase, that is, the correct identification of actual chemical reactive entities. The speciation state cannot be simply assumed, but it must explain observed data (e.g. oxidation states and/or solubility behaviors; e.g. Fincham & Richardson, 1954; Moretti et al., 2003; Moretti & Ottonello, 2005) and reconcile with spectroscopic observations, particularly those about oxygen coordination around cations (e.g., Neuville et al., 2020; Le Losq et al. 2020 and references therein).

The aim of this paper is to provide the basic chemical arguments about redox reactivity in melts that are too often overlooked in the Earth Sciences literature, thus allowing the interested readers to orient themselves among the various available models and redox descriptions. In this contribution we then develop on qualitative grounds and discuss by data the premises to more complex theories and models (e.g. the ionic-polymeric one) about the role played by composition via melt/glass structure (see also Moretti & Neuville, 2021). In particular, we show how to deal with the current knowledge of silicate melt structure to set species that describe redox exchanges. To do so, we consider that the melt/glass structure is in turn a reflection of its acid–base behavior, which determines species and ionic complexes that participate in reactions where charge (electrons, cations, and anions) is transferred from each complexing cation to the surrounding ligand, that is, the O^{2-} ions covalently combined in a continuous network. Briefly, we have to assume an understanding of the bonding, structure, and properties of individual molecules in melts, in which the oxidation state plays a basic role for the attribution of formal charges and to "convert" structural findings into speciation hypotheses for chemical reactions. By using data on Fe and S oxidation states, we show how this knowledge is fundamental to describe the redox interactions of magmas and identify possible external factors that shift redox equilibria as a response to acid–base interaction involving gas and melt phases.

6.2. IONIC SYNTAX, SPECIATION STATE AND THE MELT/GLASS NETWORK: STATE OF THE ART AND CONCEPTUAL FRAMEWORK

6.2.1. The Effect of Melt Composition on Iron and Sulfur Oxidation State

To deal with redox exchanges a chemical syntax is needed to describe reactive entities (or moieties). These can be molecules, simple ions, or complexes with a formal charge and which intervene in the process of charge transfer (the reaction). The charge transfer involves donor-acceptor pairs, as defined by Lewis: an acid is any substance *acceptor* of an electron pair and a base is any substance *donor* of an electron pair.

In oxygen-based, non-protonated substances such as molten oxide and particularly silicate melts, the following redox exchange (a so-called half-reaction) occurs

$$1/2\ O_{2(La)} + 2e^- \Leftrightarrow O^{2-}{}_{(Lb)} \tag{6.1}$$

that can be also seen as an electron transfer which identifies a Lewis base (*Lb*), the electron donor, and the corresponding acid (*La*), the electron acceptor. Besides, equilibrium associated with the reaction (also called oxygen electrode; Flood & Forland, 1947; Ottonello et al., 2001; Moretti, 2005: Cicconi et al., 2020b) clearly relates the molecular oxygen activity aO_2, (henceforth replaced with its fugacity, fO_2) to the activity of oxygen ion O^{2-}. The latter can be described as a single, free (unpolarized) oxide ion which is the basic brick that depending on adjacent cations (hence, under different polarization states) builds oxyanion melts and glasses, and particularly the oxygen ligand that frame the silicate network (Flood & Forland, 1947; Duffy & Ingram, 1971). It is straightforward to see that the oxygen electrode is analogous to the hydrogen electrode

$$2H^+{}_{(La)} + 2e^- \Leftrightarrow H_{2(Lb)} \tag{6.2}$$

involved in highly protonated systems, such as aqueous solutions in the hydrosphere. For both Reactions 6.1 and 6.2, the reduced species is a Lewis base and the oxidized one is the Lewis acid.

The *oxygen electrode* (Reaction 6.1) implies that any shift of the (Lewis) acid–base character of the melt is also a redox shift. Variations in composition (i.e. any change of the activity of O^{2-} at constant temperature) will determine a shift of fO_2, the parameter widely accepted to be at the core of redox exchanges in igneous petrology. Vice-versa, changes of fO_2 will affect the acid–base character of the melt and its evolution.

We can already state that acid–base and redox properties are then intimately interrelated and that in a perfectly closed system it is meaningless to think of fO_2 as an independent variable, such as pressure and temperature. Rather, in closed systems, fO_2 is an intensive property related to the composition. Nevertheless, if the mutual redox/acid–base relation appears clearly from the oxygen electrode Reaction 6.1, its parameterization needs to define aO^{2-} in an operational way, that is, with reference to O^{2-} concentration and its variations during the process of silicate network building, which occurs with the change of mean O^{2-} polarization state (Ottonello et al., 2001; Moretti, 2005).

Same as for the electrochemistry of aqueous solution (see also Moretti & Neuville, 2021) based on the hydrogen electrode (Reaction 6.2), in the melt phase equilibria between two species of redox-sensitive elements will result from the sum of two half-reactions: one is the oxygen electrode Reaction 6.1, describing the ligand (or the solvent), and the other involves entities (molecules or ion complexes) in which roman numerals are used to assign the formal oxidation numbers of the central coordinating element, thus avoiding confusion about the distribution of the effective charge in the reactive entity.

Simple half-reactions such as

$$Fe^{3+} + e^- \Leftrightarrow Fe^{2+} \tag{6.3}$$

will then combine with Reaction 6.1 to give

$$Fe^{3+} + \tfrac{1}{2}\,O^{2-} \Leftrightarrow Fe^{2+} + \tfrac{1}{4}\,O_2 \tag{6.4}$$

whose equilibrium constant is:

$$\log K_4 = \log \frac{XFe^{2+}_{melt}}{XFe^{3+}_{melt}} - \frac{1}{2}\log XO^{2-}_{melt} + \frac{1}{4}\log fO_2 \tag{6.5}$$

and where X denotes concentration of the ionic species of interest (Fe^{2+}, Fe^{3+}; O^{2-}). This equation is not very dissimilar from the equilibrium constant that can be computed from the reaction involving macroscopic iron oxide *components*:

$$FeO_{1.5} = FeO + \tfrac{1}{4}\,O_2 \tag{6.6}$$

(note that nothing prevents us from using $FeO_{1.5}$ rather than Fe_2O_3 as macroscopic oxide component) whose equilibrium constant is:

$$\begin{aligned}\log K_6 &= \log \frac{aFeO_{melt}}{aFeO_{1.5,\ melt}} + \frac{1}{4}\log fO_2 \\ &= \log \frac{\gamma FeO_{melt}}{\gamma FeO_{1.5,\ melt}} + \log \frac{XFe^{2+}_{melt}}{XFe^{3+}_{melt}} + \frac{1}{4}\log fO_2\end{aligned} \tag{6.7}$$

with γ representing the activity coefficients of the iron oxides. The important difference is that Equation 6.5 includes a sort of network contribution via aO^{2-}, which is actually a measure of the system basicity, whereas Equation 6.7 relies on the description of activity-composition relation via activity coefficients. Worth noting that both Equations 6.5 and 6.7 show that fO_2 is a thermodynamic quantity that expresses the redox potential and which can be defined also for single-phase melts, in the absence of a gas phase.

These expressions can be tested for a simple binary high-T melt equilibrated in air ($fO_2 = 0.21$), like in glass industry batches. Following equation (6.5), iron oxidation state should decrease if the activity of O^{2-} increases, for example by adding alkalis to the batch. Neverthless, earlier data from glass science (e.g., Paul & Douglas, 1965; Figure 6.1) shows that alkali addition (i.e., increase of the basic R_2O) and increasing aO^{2-} favor Fe^{3+} (Figure 6.1a).

This anomalous behavior cannot be explained by Equation 6.5 or by canonical expressions of the activity coefficients γ for iron oxides in Equation 6.7 ($\log K_6$). These can be calculated by difference of $\log K_6$ (from Ottonello et al., 2001 for melt-relaxed FeO–$FeO_{1.5}$ components or alternatively by adopting the JANAF thermochemical tables;

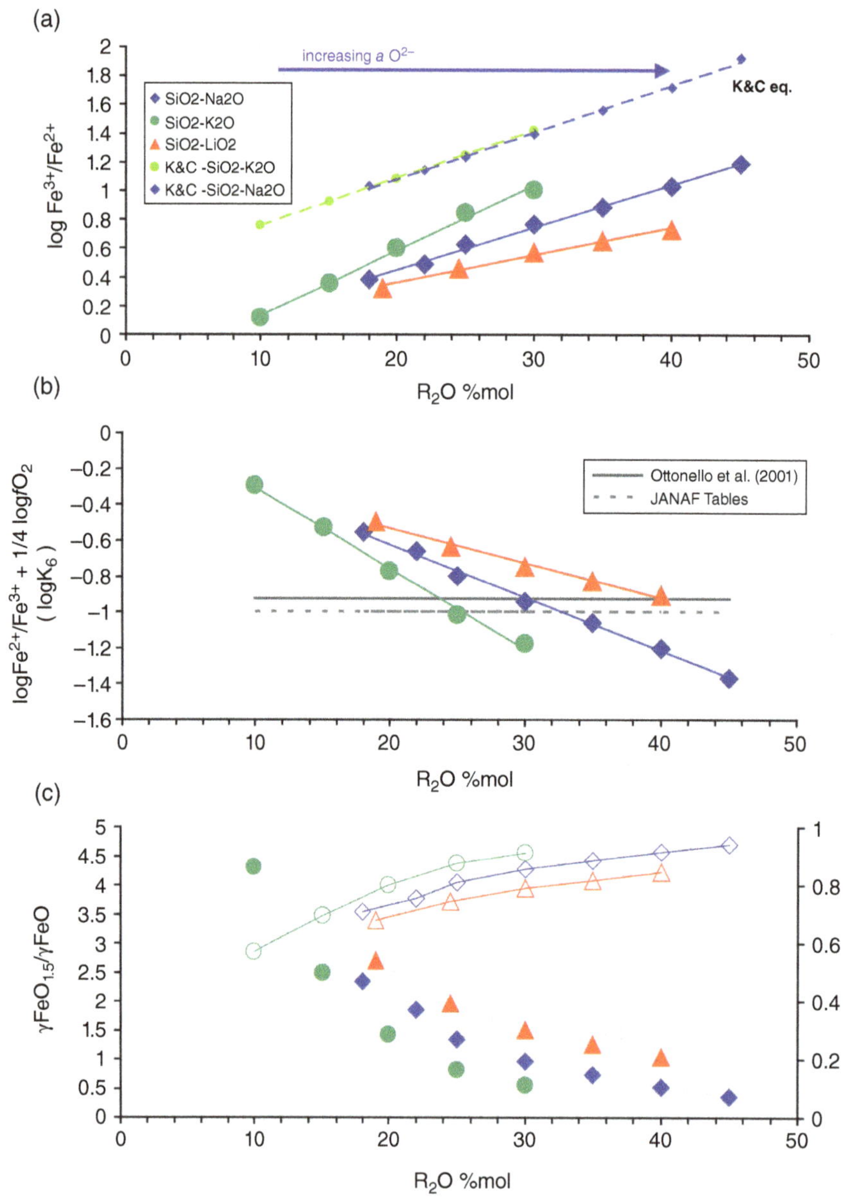

Figure 6.1 (a) Oxidized/reduced iron (log-form) for simple binaries at 1400°C equilibrated on air (Paul & Douglas, 1965). Also shown is the performance of the empirical Kress and Carmichael equation (1991). (b) Evaluation of the equilibrium constant of Reaction 6.6 at 1400°C (grey lines) and of its compositional term. (c) Computation of activity coefficient ratios (filled symbols) reconciling $\log K_6$ (Equation 6.7 and b) with the measured oxidized/reduced iron ratio. Also shown is the relative proportion of oxidized iron (empty symbols; secondary Y-axis).

Chase, 1998) and with the $\log \frac{XFe^{2+}_{melt}}{XFe^{3+}_{melt}} + \frac{1}{4}\log fO_2$ quantity (Figure 6.1b). We see that to fit experimental redox data for increasing alkali content, γ increases for FeO and decreases for $FeO_{1.5}$ (Figure 6.1c), which is then stabilized. This behavior goes contrary to Equation 6.5 predictions and is counterintuitive, because the activity coefficient ratio should approach 1 given the low concentration (0.7–0.8 mol %; i.e. diluted conditions) of total iron in the glass samples of Figure 6.1a.

An empirical and widely used expression somehow related to Reaction 6.6 was derived by Kress and Carmichael (1991) to reproduce a large dataset of experimental Fe^{3+}/Fe^{2+} ratios obtained under controlled fO_2 and including data from Na_2O–FeO–Fe_2O_3–SiO_2 (Lange & Carmichael, 1989) and CaO–FeO–Fe_2O_3–SiO_2 (Kress & Carmichael, 1989) systems, all showing the same "anomalous" behavior displayed in Figure 6.1a (Fe^{3+} favored by alkali addition and aO^{2-} increase). Nevertheless, this formulation cannot catch the differences displayed by binary joins of Figure 6.1a, and largely overestimates oxidized iron. Besides, the Kress and Carmichael equation violates the stoichiometric constraint on $\log fO_2$ imposed by Equations 6.5 and 6.7, presenting a factor 0.2 instead of 0.25. The authors attribute this small but significative difference to the activity coefficient ratio non-linearly deviating from unity in Equation 6.7. Although the complexities in describing the activity coefficient ratio $\gamma FeO/\gamma FeO_{1.5}$, deviations from ideality cannot be reproduced by scaling fugacities, activities, or concentrations with exponential factors inconsistent with reaction stoichiometry. Doing so implies the violation of mass conservation which characterizes any chemical reaction.

Summarizing, the compositional control on the iron oxidation state is very strong, largely non-linear, and displays a dual behavior (particularly relevant for potassium-bearing systems) (Figure 6.2a,b). Such a dual behavior is related to composition and implies the failure of classical solution models involving macroscopic components FeO and Fe_2O_3. It demands a more subtle approach to the speciation state that cannot be approximated by two ionic species only, one for reduced and

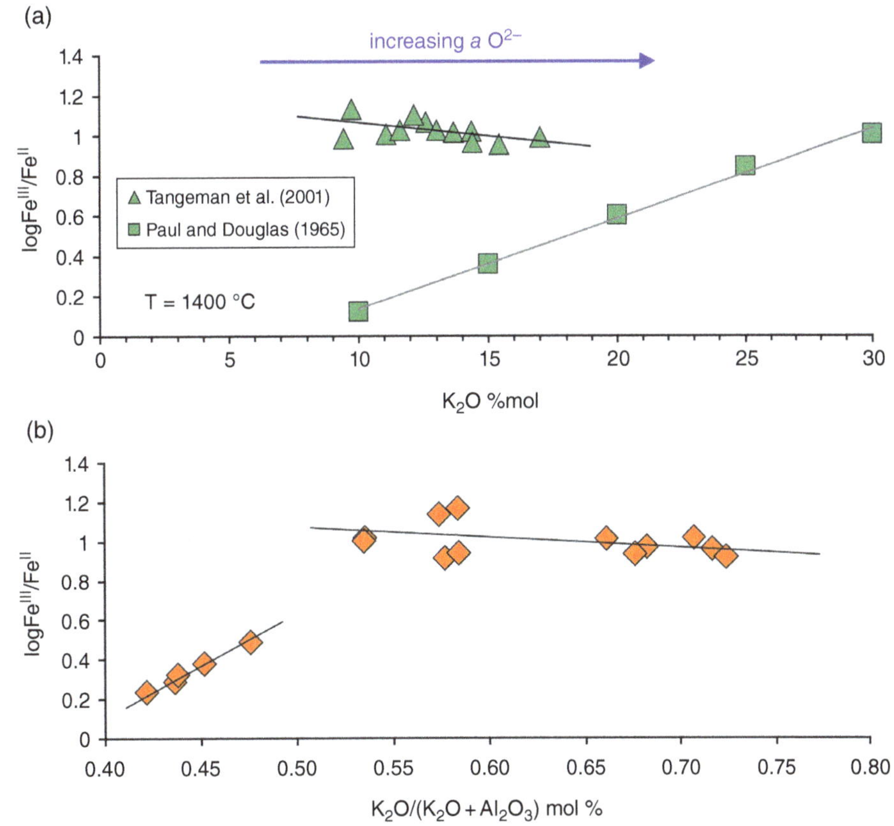

Figure 6.2 (a) Comparison of Paul and Douglas (1965) data with those of Tangeman et al (2001), showing contrasting behavior of iron with increasing K_2O content. (b) Iron redox ratios from Dickenson and Hess (1980), which at low-Fe_{tot} contents shift from oxidation to reduction for increasing $K_2O/(K_2O+Al_2O_3)$ ratio.

one for oxidized iron. Briefly, other chemical mechanisms other than Reaction 6.4 must occur.

Ottonello et al. (2001), Moretti and Ottonello (2001, 2003a), Moretti and Papale (2004), Moretti (2005), and more recently Le Losq et al. (2020) have shown that another chemical mechanism involves trivalent iron, which in coordination IV forms a stable polyhedron, FeO_4^{5-} that heavily contributes to the speciation state of iron in melts and to its redox state via half-reaction

$$FeO_4^{5-} + e^- \Leftrightarrow Fe^{2+} + 4O^{2-} \quad (6.8)$$

which combined with the oxygen electrode 6.1 yields

$$FeO_4^{5-} \Leftrightarrow Fe^{2+} + 7/2O^{2-} + ¼\,O_2 \quad (6.9)$$

It is then clear that the speciation state of iron cannot be treated only considering simple Fe^{2+} and Fe^{3+} cations. These in fact represent network modifying species that do not give rise to a coordination polyhedron in the sense of Pauling, that is characterized by a strong covalency of Fe–O bonds. On the contrary, in FeO_4^{5-} trivalent iron behaves as a network former.

Because Reaction 6.9 displays oxygen ions among the products, increasing aO^{2-} (or XO^{2-}), at fixed T and fO_2, leads to iron oxidation, in agreement with data in Figure 6.1a and as required by its equilibrium constant:

$$\log K_9 = \log \frac{XFe^{2+}_{melt}}{XFeO^{5-}_{4,melt}} + \frac{7}{2}\log XO^{2-}_{melt} + \frac{1}{4}\log fO_2 \quad (6.10)$$

It is then evident that the bulk iron ratio will result from the competitive effect of Reactions 6.4 and 6.9. By introducing henceforth roman numerals to identify oxidation numbers, whereas arabic numbers will be left for the bulk formal charge of the species of interest, we have:

$$\log \frac{Fe^{III}}{Fe^{II}} = \log \frac{NFe^{2+}_{melt}}{NFe^{3+}_{melt} + NFeO^{5-}_{2-,melt}} \quad (6.11)$$

with N indicating number of moles and replacing X, the mol fraction, because molar concentrations and activities in ionic models should be computed over the appropriate matrix (the anionic population for FeO_4^{5-}, the cationic one for Fe^{3+} and Fe^{2+}) for any fused salt (Temkin, 1945), including silicate melts.

Following the guidelines in Fraser (1975), who treated the case of europium (see also Cicconi et al., 2020c), Equation 6.9 can be reduced to (Holmquist, 1966; Ottonello et al., 2001; Moretti, 2005):

$$FeO_2^- \Leftrightarrow Fe^{2+} + 3/2O^{2-} + ¼\,O_2 \quad (6.12)$$

hence

$$\log K_{12} = \log \frac{XFe^{2+}_{melt}}{XFeO^{-}_{2,melt}} + \frac{3}{2}\log XO^{2-}_{melt} + \frac{1}{4}\log fO_2 \quad (6.13)$$

The use of the FeO_2^- species, instead of FeO_4^{5-}, does not modify the conceptual scheme and allows an even more precise calibration of the equilibrium constant because it limits the propagation of errors related to the estimation of XO^2 (or aO^{2-}). It is worth stressing that Reaction 6.12 allows to appreciate that O^{2-} is six times more influential than O_2 in affecting the redox state of Fe in silicate melts (Nesbitt et al., 2020), independently of any coordination issue of trivalent iron (Le Losq et al, 2020b and references therein).

Also the case of sulfur redox equilibria in melts involves two oxidation states, S^{-II} and S^{VI}, but it is well known that for sulfur the redox couple involves the simple anion S^{2-} and the anionic complex SO_4^{2-} (Fincham & Richardson, 1954; Carroll & Rutherford, 1988; Paris et al., 2001; Baker & Moretti, 2011 and references therein), resulting in reaction:

$$SO_4^{2-} + 8e^- \Leftrightarrow S^{2-} + 4O^{2-} \quad (6.14)$$

which combined with the oxygen electrode Reaction 6.1 (taken eight times) yields:

$$SO_4^{2-} \Leftrightarrow S^{2-} + 2O_2 \quad (6.15)$$

Reaction 6.15 may suggest the occurrence of loose and dispersed sulfide and sulfate anions, both speciating over the melt anionic matrix. This idea is also a consequence of the relatively low abundance of sulfur dissolved in natural melts (generally less than 1 wt%; Wallace & Edmonds, 2011 and references therein) and has led to empirical formulations calibrated on compositionally restricted datasets in which the sulfate/sulfide ratio is independent of the melt composition (Wallace & Carmichael, 1994; Matthews et al., 1999; Jugo et al., 2005, 2010) independent of the melt polymerization and structural configuration. For example, based on a small dataset of 22 submarine glasses dominated by reduced sulfur (16 samples with $S^{VI}/S_{tot} < 0.1$; 5 samples with $0.1 < S^{VI}/S_{tot} < 0/9$; 1 sample approaching unity) Wallace and Carmichael (1994) extended the results of the earlier Carroll and Rutherford (1988) electron microprobe study and proposed the following equation:

$$\log \frac{XSO_4^{2-}}{XS^{2-}} = 1.02 \log fO_2 + \frac{25410}{T} - 10 \quad (6.16)$$

in which the factor multiplying $\log fO_2$ is sensibly lower than 2, the stoichiometric coefficient required by Equation 6.15. Again, this low value of the $\log fO_2$ factor cannot be explained by the activity coefficient ratio of sulfide and sulfate deviating from unity. Rather, it must embody compositional variations unexplained by Equation 6.16 (Moretti & Ottonello, 2005).

Figure 6.3a plots S^{VI}/S_{tot} ratios versus logfO_2 (as relative to the fayalite–magnatite–quartz redox buffer, FMQ) from empirical formulas proposed in the literature. All curves show that sulfide to sulfate transition occurs over a narrow fO_2 range (about two log units), much narrower than that of iron, reported for comparison in Figure 6.3a (red solid line). Note the shift to lower relative fO_2 values of Nagashima and Katsura (1973) data, which were questioned and dismissed by Jugo et al. (2005) who stated that

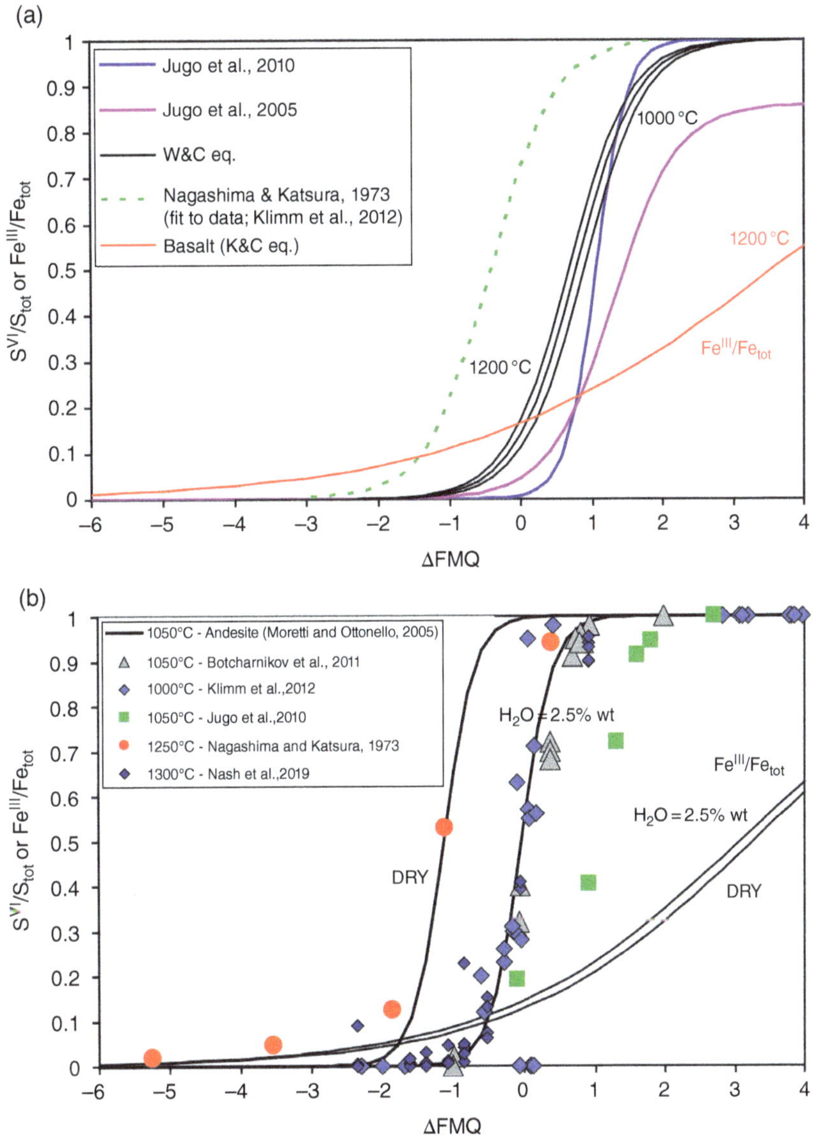

Figure 6.3 (a) S^{VI}/S_{tot} from empirical formulations available in the literature and disregarding the effect of composition. Wallace and Carmichael (1994) equation ("W&C eq."; thin black lines in figure) was used at three different temperatures: 1000°C, 1100°C, and 1200°C. Also reported are the fit to Nagashima and Katsura (1973) data (1300°C) on the silica-soda binary and Fe^{III}/Fe_{tot} for a basalt computed at 1200°C via the Kress and Carmichael (1991) equation. The figure is based on Klimm et al. (2012). (b) XANES-based data of S^{VI}/S_{tot} (Botcharnikov et al., 2011: hydrous andesite at 1050°C; Kimm et al., 2012: soda-lime-silica, potassium-lime-silica, albite and trondhjemite compositions at 1000°C; Nash et al, 2019: model Martian basalts, primitive Fe-rich terrestrial basalts, natural dacites, and rhyolites at 1300°C). Also reported are results of the Moretti and Ottonello (2005) model on andesite (Equation 6.19), for both iron and sulfur redox states and under dry and hydrous conditions.

departure of these data from existing empirical fits (see Figure 6.3a) was due to a wrong fO_2 assessment, based on the pretended outdated and inaccurate fluxing technique.

Figure 6.3b reports XANES experimental datasets, selected for their accuracy, and shows that the S-redox curve shifts in response to compositional variations. The figure also shows the shift of about one log unit in S^{VI}/S_{tot} predicted by the Moretti and Ottonello (2005) model for a dry and a hydrous (2.5 wt%) andesite at 200 MPa and 1050°C. The simulated hydrous andesite compares very well to the andesite investigated at the same P and T by Botcharnikov et al. (2011), to the soda-lime-silica, potassium-lime-silica, albite and trondhjemite compositions in Klimm et al. (2012) at 1000°C, and with model Martian basalts, primitive Fe-rich terrestrial basalts, natural dacites, and rhyolites from Nash et al. (2019) at 1300°C. The dry andesite curve is shifted to lower fO_2 and also overlaps the experimental data (at 1250°C) on the SiO_2–Na_2O (2:1) binary (Nagashima & Katsura, 1973).

Therefore, composition also affects the sulfur oxidation state and among the many compositional variables, concentration of highly soluble water plays a major role. In agreement with the strong impact of water on rheological properties and melt phase equilibria, the effect of dissolved water reflects the shift in acid–base properties hence in the speciation state, that is, in the number, kind, and size of chemically reactive structural groups. As an example of the compositional control of water on the sulfur oxidation state of silicate melts, we wan recall here the simple expressions given in Marini et al. (2011) for Mt. Mazama rhyodacite at 900°C:

$$\log \frac{XSO_4^{2-}}{XS^{2-}} = -0.556C_{H_2O} + 2\log fO_2 + 25.372 \tag{6.17}$$

and for Mt. Etna basalts at 1200°C:

$$\log \frac{XSO_4^{2-}}{XS^{2-}} = -0.374C_{H_2O} + 2\log fO_2 + 16.158 \tag{6.18}$$

C_{H_2O} being the water concentration in wt%. A similar expression, over a wide range of T, can be given for the andesite plotted in Figure 6.3b. By using the Moretti and Ottonello (2005) for different temperatures up to 1300°C, we obtain:

$$\log \frac{XSO_4^{2-}}{XS^{2-}} = 2\log fO_2 - C_{H_2O}\left(\frac{4486.4}{T} - 2.512\right) - 23.125 + \frac{59660}{T} \tag{6.19}$$

6.2.2. Fe–S Mutual Interactions

Figure 6.3b shows that increasing dissolved water at constant fO_2 and T slightly favors Fe^{III}, which however implies a predominance of Reaction 6.9 over Reaction 6.4 for the treated andesitic composition. On the other hand, the sulfate/sulfide ratio decreases, the effect being dramatic for the S^{VI}/S_{tot} ratio between $-2<\Delta FMQ<0$. The figure also shows that for a given Fe^{III}/Fe^{II} ratio, water loss increases oxygen fugacity of only 0.2 log-units at most, whereas for a given SO_4^{2-}/S^{2-} ratio water exsolution decreases oxygen fugacity (see also Equations 6.17–6.19), up to one log-unit when dehydration is achieved. To reconcile these two apparently contradictory effects we have to combine iron and sulfur redox equilibria and plot the oxidized/reduced log-ratios to better appreciate shifts induced by water (Figure 6.4). By combining half-reactions 6.14 and 6.3 (taken eight times) we obtain an expression for mutual Fe–S interactions in melts:

$$8Fe^{3+} + S^{2-} + 4O^{2-} \Leftrightarrow 8Fe^{2+} + SO_4^{2-} \tag{6.20}$$

where O^{2-} species appear on the left side. Therefore, by increasing the aO^{2-} (and in turn the basicity) by *adding* alkalis such Na_2O or K_2O to the melt, the equilibrium is shifted rightward, which implies sulfur oxidation and iron reduction.

If we now consider Reaction 6.8 (always taken eight times) instead of Reaction 6.3, we obtain:

$$8FeO_4^{5-} + S^{2-} \Leftrightarrow 8Fe^{2+} + SO_4^{2-} + 28O^{2-} \tag{6.21}$$

which can be reduced to

$$8FeO_2^{-} + S^{2-} \Leftrightarrow 8Fe^{2+} + SO_4^{2-} + 12O^{2-} \tag{6.22}$$

and which shows that if alkali species are added, sulfur is reduced and iron gets oxidized. In magmatic melts both Reactions 6.20 and 6.22 (or alternatively 6.21) occur, giving rise to the following equilibrium relations

$$\log \frac{XSO_4^{2-}}{XS^{2-}} = 8\log \frac{XFe^{3+}}{XFe^{2+}} + 4\log aO^{2-} + \log K_{20} \tag{6.23}$$

$$\log \frac{XSO_4^{2-}}{XS^{2-}} = 8\log \frac{XFe_2^{-}}{XFe^{2+}} - 12\log aO^{2-} + \log K_{21} \tag{6.24}$$

The simultaneous occurrence of Equations 6.23 and 6.24 allows modeling the mutual interaction of iron and sulfur in melts (Moretti & Ottonello, 2003a), although in most cases one of the two equilibria prevail over the other, which is then negligible. This is the case for the andesite plotted in Figures 6.3b and 6.4, modelled via Moretti and Ottonello (2005) and reflecting essentially Equation 6.24. Log-ratios of iron and sulfur redox state

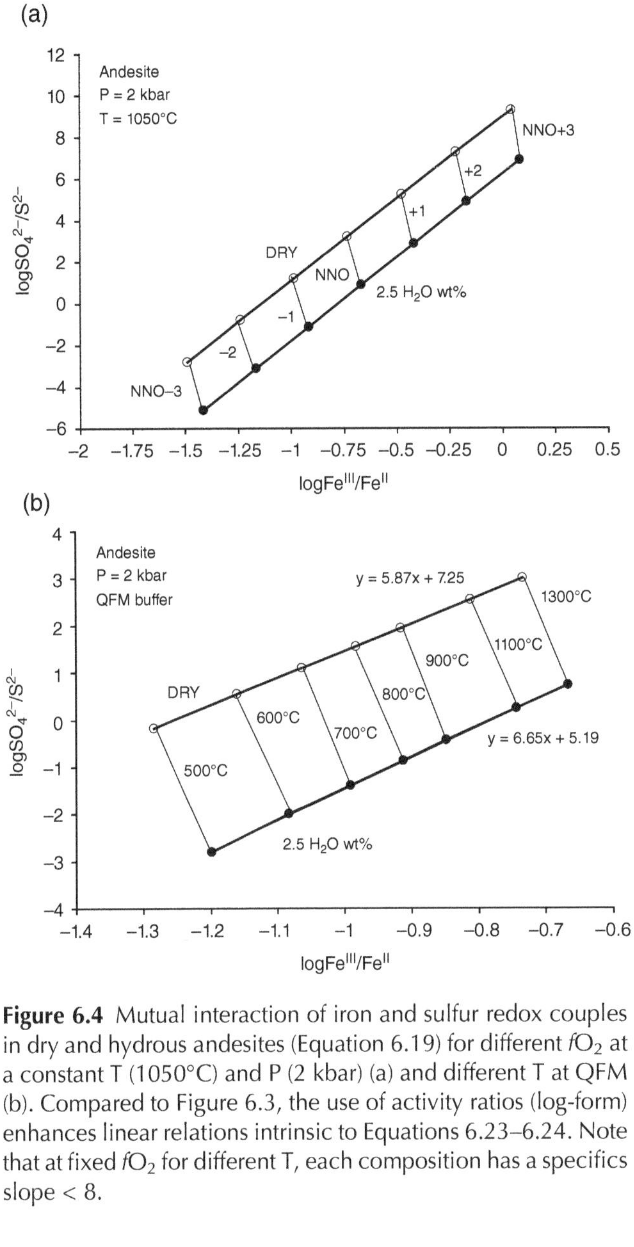

Figure 6.4 Mutual interaction of iron and sulfur redox couples in dry and hydrous andesites (Equation 6.19) for different fO_2 at a constant T (1050°C) and P (2 kbar) (a) and different T at QFM (b). Compared to Figure 6.3, the use of activity ratios (log-form) enhances linear relations intrinsic to Equations 6.23–6.24. Note that at fixed fO_2 for different T, each composition has a specifics slope < 8.

define (at constant T) lines of slope 8, constrained by stoichiometry of Reactions 6.20–6.22 (Figure 6.4a). On the other hand, at a given fO_2 and for varying T, the same composition yields a slope lower that the stoichiometric one (Figure 6.4b).

In a recent paper, Nash et al. (2019) describe mutual Fe–S relations in melt as due to the following electron exchange:

$$S^{2-} + 8Fe^{3+} \Leftrightarrow S^{6+} + 8Fe^{2+} \qquad (6.25)$$

thus assuming a fully ionized, plasma-like, solution which *a priori* cannot account for the complexity of the melt network, hence the role of composition via the speciation state of iron and sulfur in the melt. Moreover, to constrain the equilibrium constant of Reaction 6.25, the authors impose the T-dependence obtained by thermodynamic data on the homogenous *solid phase* reaction

$$FeS + 8FeO_{1.5} \Leftrightarrow 8FeO + FeSO_4 \qquad (6.26)$$

thus assuming that the fusion of solid phases plus their relaxation (dissolution) in the melt phase is uniquely determined by entropic terms. This fact contradicts the thermodynamic findings of Lange and Navrotsky (1992) and Ottonello et al. (2001) for Reaction 6.6 as well as the enthalpy terms for *molten* FeS and $FeSO_4$ components in Moretti and Ottonello (2005), all giving rise to simple equilibrium constants of the type $\log K = a + b/T$ because heat capacities (Cp) of components are constant in the melt phase. Finally, Nash et al. (2019) propose the following equation:

$$\log\frac{XSO_4^{2-}}{XS^{2-}} = 8\log\frac{XFe^{3+}}{XFe^{2+}} + \frac{8.7436{\cdot}10^6}{T^2} - \frac{27703}{T} + 20.273 \qquad (6.27)$$

On the basis of their results, the authors state that composition does not affect the sulfate/sulfide ratio. The Fe^{III}/Fe^{II} ratio in Equation 6.27 can be converted to $\log fO_2$ (e.g., via Kress & Carmichael, 1991), to show that the sulfur redox ratio depends only and dramatically on temperature, at least for selected sets of data about the sulfate/sulfide ratio in Fe-bearing melts and glasses. Nash et al. (2019) explain the fO_2-shift between basaltic datasets (Brounce et al., 2017; see also Figure 6.7) uniquely on the basis of temperature. Based on the Moretti and Ottonello (2005) model, Fe–S mutual interactions (Equations 6.23 and 6.24) show that in a closed single-melt system, at equilibrium, it is trivial to inquire whether the buffering redox-couple is either given by iron or sulfur species. However, only iron fulfills the two basic requirements for an efficient redox couple in natural silicate melts: (i) that a redox-sensitive element must occur in two oxidation states giving rise to species with comparable thermodynamic stabilities; and (ii) that these species must be abundant and pervade their chemical environment.

On the contrary, the sulfate–sulfide pair cannot be candidates for a buffer, given the generally low abundance of sulfur in melts. Besides, the transition from fully oxidized sulfate to fully reduced sulfide occurs in a relatively narrow range (two log-units) whereas the Fe^{III}/Fe^{II} pair can accommodate much larger fO_2 variations (Figures 6.3 and 6.4). Therefore, compared to iron, sulfur has a limited "buffering" redox capacity, but for the same reason it can be a precise indicator of fO_2. Nevertheless, the high sensitivity of the sulfate/sulfide pair with respect to fO_2 challenges the calibration of accurate oxythermometers.

6.3. REDOX EVOLUTION AND MAGMATIC DEGASSING

We are now in an ideal position to more fully discuss redox changes occurring on magmatic degassing. For this purpose, we will base our development on the approach followed by Marini et al (2011) to model the redox-related isotopic evolution of S-isotopes in magmatic systems that get rid of sulfur through multiple chemical mechanisms on their ascent to the surface. These mechanisms are described here with application to the datasets of Brounce et al. (2017) on Mauna Kea lavas and Bénard et al (2019) on arc magmas.

6.3.1. Mechanisms of Sulfur Release on Magma Rise

In a silicate melt, sulfur solubility (or exsolution) is expressed as (Fincham & Richardson, 1954; Haughton et al., 1974; Wallace & Carmichael, 1994; Carroll & Webster, 1994; Moretti et al., 2003; Moretti & Ottonello, 2003b; Moretti & Ottonello, 2005; Baker & Moretti, 2011; Cicconi et al., 2020a):

$$SO_4^{2-} \Leftrightarrow SO_{2(g)} + 1/2O_2 + O^{2-} \qquad (6.28)$$

$$S^{2-} + 3/2O_2 \Leftrightarrow SO_{2(g)} + O^{2-} \qquad (6.29)$$

In natural degassing systems, where fO_2 cannot be imposed or controlled a priori like in experiments, it is better to consider that molecular oxygen is a virtual quantity not detectable in volcanic gases, despite the clear thermodynamic significance of its fugacity. O_2 is better replaced by the H_2/H_2O couple in the gas phase (e.g., Giggenbach, 1987; 1996; Moretti & Stefansson, 2020) by accounting for water dissociation:

$$H_2O_{(g)} \Leftrightarrow H_{2(g)} + 1/2O_{2(g)} \qquad (6.30)$$

Reactions 6.28 and 6.29 can be rewritten as

$$SO_4^{2-} + H_{2(g)} \Leftrightarrow SO_{2(g)} + H_2O_{(g)} + O^{2-} \qquad (6.31)$$

$$S^{2-} + 3H_2O_{(g)} \Leftrightarrow SO_{2(g)} + 3H_{2(g)} + O^{2-} \qquad (6.32)$$

The reactions above are the starting point to write a set of chemical equations whose stoichiometry allows us to identify mechanisms responsible for variations in both sulfide and sulfate species in the melt phase. These can be easily visualized on the nS^{2-} vs nSO_4^{2-} diagram of Figure 6.5, in which analytical data on sulfur oxidation state in the melt can be plotted to give information on processes of S release.

Figure 6.5 shows that a magma can get rid of sulfur by losing:

(ii) S^{2-} via H_2S degassing:

$$H_2O_{(g)} + S^{2-} \Leftrightarrow H_2S_{(g)} + O^{2-} \qquad (6.33)$$

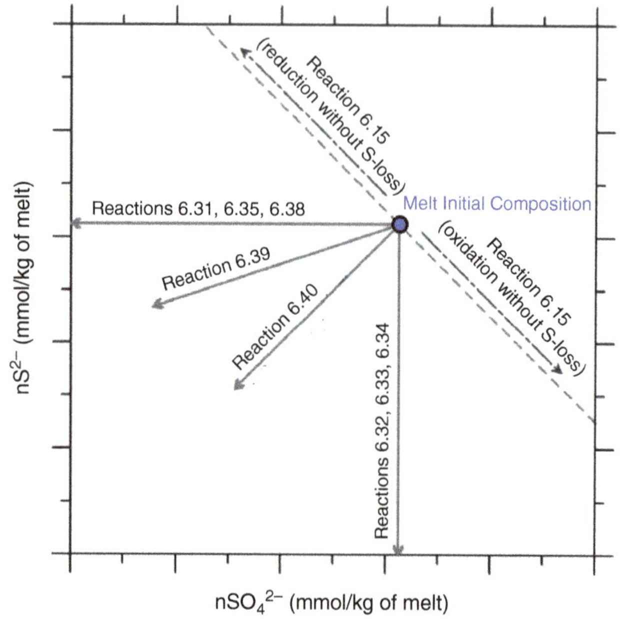

Figure 6.5 Patterns of sulfur release and their effect on melt sulfur distribution (reworked after Marini et al., 2011). Data are plotted as molar amounts (mmol/kg) of both oxidized and reduced sulfur. All trends in this figure have been referred to a unique pole (initial melt composition), although this is not necessary.

(ii) S^{2-} via sulfide separation, either liquid FeS or pyrrhotite (po):

$$S^{2-} + Fe^{2+} \Leftrightarrow > FeS_{(\text{liq or po})} \qquad (6.34)$$

(iii) S^{2-} due to SO_2 via degassing (Reaction 6.32)
(iv) SO_4^{2-} due to SO_2 via degassing (Reaction 6.31)
(v) SO_4^{2-} via H_2S degassing at high pressure (depth) via

$$SO_4^{2-} + 4H_{2(g)} \Leftrightarrow H_2S_{(g)} + 3H_2O_{(g)} + O^{2-} \qquad (6.35)$$

which, at high pressure, is favored with respect to Reaction 6.31 because the net balance between produced (four) and consumed (four) gaseous molecules is zero. Reactions 6.31 and 6.35 are connected via the following homogenous gas reaction

$$H_2S_{(g)} + 2H_2O_{(g)} \Leftrightarrow SO_{2(g)} + 3H_{2(g)} \qquad (6.36)$$

widely adopted also in terms of molecular oxygen (e.g. Aiuppa et al., 2011)

$$H_2S_{(g)} + 3/2O_{2(g)} \Leftrightarrow SO_{2(g)} + H_2O_{(g)} \qquad (6.37)$$

(vi) SO_4^{2-} due to anhydrite separation:

$$SO_4^{2-} + Ca^{2+} \Leftrightarrow CaSO_{4(anh)} \qquad (6.38)$$

(vii) S^{2-} and SO_4^{2-} through disproportionation of sulfate and sulfide ions and degassing of SO_2, as expressed by the reaction:

$$3SO_4^{2-} + S^{2-} \Leftrightarrow 4SO_{2(g)} + 4O^{2-} \quad (6.39)$$

which is obtained by summing up Reactions 6.28 (taken three times) and 6.29 and finally

(viii) S^{2-} and SO_4^{2-} on joint degassing of SO_2 and H_2S in a 1:1 proportion:

$$SO_4^{2-} + S^{2-} + H_{2(g)} \Leftrightarrow H_2S_{(g)} + SO_{2(g)} + 2O^{2-} \quad (6.40)$$

obtained by combining Equations 6.31 and 6.33 to generate a gas phase condition typically observed in andesitic volcanism and suggested to act as a redox buffer (Giggenbach, 1996; see also Section 6.4.2) via Reactions 6.36–6.37.

In Figure 6.5, arrows relate to the above reactions of melt S-loss and their slopes are constrained by the corresponding reaction stoichiometry. Their location depends on measured abundances. Independently of any model of the melt redox state, Figure 6.5 allows putting constraints for interpreting the sulfur oxidation state obtained from melt inclusions in crystals in a magma batch that has evolved from depth to surface.

The dashed gray line in Figure 6.5 identifies the line of equal sulfur content (iso-concentration line) corresponding to the initial abundance. Along this iso-concentration line, oxidation state may shift without any loss due to degassing or crystallization of a S-bearing phase. Examples of iso-concentration lines to be used on this kind of plots are for instance, the sulfur content at sulfide saturation (SCSS) or at anhydrite concentration (SCAS) (e.g., Carroll & Rutherford, 1985; Mavrogenes & O'Neill, 1999; Moretti & Baker, 2008; Li & Ripley, 2009; Baker & Moretti, 2011), or any other significant threshold such the initial content of S in the system.

Along S-isolines sulfide and sulfate proportions may interconvert without any sulfur loss, via fO_2 variations as demanded by Reaction 6.15. Oxygen fugacity variations might be better constrained if we consider that either Reactions 6.20 or 6.21 may shift redox couples of iron and sulfur, depending on bulk composition. Nevertheless, as we have seen, these two reactions have opposite effects, depending on speciation of trivalent iron: if O^{2-} activity increases, Reaction 6.20 shifts rightward and iron reduction and sulfur oxidation are favored. On the other hand, increasing O^{2-} activity produces iron oxidation and sulfur reduction in case of Reaction 6.21. To discriminate which one is effective on the S-isoline, estimates of aO^{2-} or a study of trivalent iron coordination should accompany the S-oxidation data.

The data plotted in Figure 6.5 speak for themselves and show dominant trends and processes. The role of temperature changes is not explicitly accounted for in this representation, which aims at extracting model-independent information about occurring processes and constrain mass variations along S-loss paths.

6.3.2. Application to Real Cases

Data from Bénard et al. (2018) are very useful to understand the mechanisms of S-loss occurring in volatile-rich arc magmas. Figure 6.6 adopts the previously described

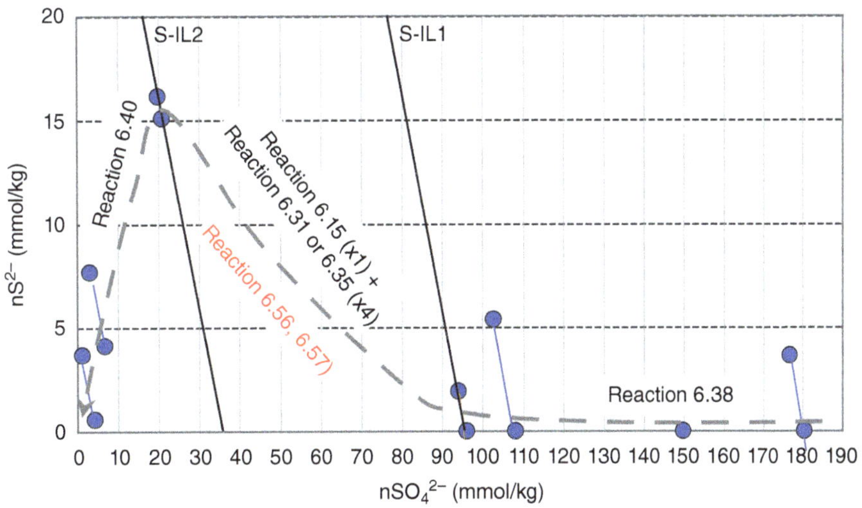

Figure 6.6 Covariation of sulfate and sulfide contents in andesitic to rhyolitic samples from oxidized arc magmas (Bénard et al., 2018). Data doublets (two redox measurements for the same melt inclusion analysis) are connected by thin blue solid segments. S-IL1 and S-IL2 are iso-concentration lines of S that bound the observed interval of reduction. Reported reactions can explain the variations observed along the S-loss pathway (dashed grey line). Reactions 6.56 and 6.57 provide a mechanism complementary to Reactions 6.15, 6.31, 6.34 to explain S-reduction in the system.

chemical semantics and uses data from melt inclusions in Cr-bearing spinel from Kamchatka xenoliths and shows that the richest S-bearing samples of approximately andesitic composition are dominated by SO_4^{2-}, consistent with observed anhydrite saturation (Bénard et al., 2018). The evolutionary path accompanying the S-loss of the whole system (dashed grey line in Figure 6.6) shows that early S-loss occurs essentially via Reaction 6.38 and maintains a very high sulfur oxidation state (S^{VI}/S_{tot} close to 1). When the S-content approaches 3100 ppm (96 mmol/kg), S-loss continues but shifts to more reduced conditions. This shift may occur along one or more S-concentration isolines comprised between two extremes (S-IL1 and S-IL2 isolines in Figure 6.6), and likely resulting in a multi-step process alternating Reaction 6.15 *(reduction without S-loss*; Figure 6.5) and Reactions 6.31 and 6.35 (SO_4^{2-} loss via SO_2 and/or H_2S degassing, depending on depth). We can estimate that for one part of SO_4^{2-} converted to S^{2-} via Reaction 6.15, four parts of SO_4^{2-} are released according to Reactions 6.31 and 6.35. The final evolution sees S-loss following roughly a SO_4^{2-}/S^{2-} slope of about one, in agreement with Reaction 6.40, thus suggesting the occurrence of the $SO_2/H_2S \approx 1$ fluid buffer (e.g., Giggenbach, 1996); which is compatible with extensive upward fluid fluxing in arc settings.

The very different behavior of basaltic systems can be appreciated on the basis of joint measurements of sulfur and iron redox states in Mauna Loa melt inclusion samples (Brounce et al., 2017). Figure 6.7a shows that sulfide is the dominant sulfur species measured in inclusions evolving below the SCSS (low-limit: 1340 ppm; high-limit: 1730 ppm; Brounce et al., 2017). Sulfur is then released via a combination of several degassing mechanisms (Reactions 6.31, 6.32, 6.33, 6.35, 6.39, 6.40) that decrease the sulfur oxidation state. The ensemble of available data can be approximately bracketed by two boundaries (dotted grey lines in Figure 6.7a) of slope 10, indicating that overall, most of the sulfur released is provided by sulfide via Reactions 6.32 and 6.33.

When exploited for mutual Fe–S interactions, pairs of iron and sulfur redox ratio (log-form) tend to form two goups: low-SiO_2 data plot at lower fO_2, whereas high-SiO_2 data at higher fO_2 (Figure 6.7). Besides, but without distinction between the two groups, datapoints plot over several lines of slope 8 rather than a unique one (red lines in Figure 6.7b). This could suggest that a compositional shift takes place, related for example to water content, hence aO^{2-} variations associated with depolymerization (aO^{2-} increases) or polymerization (aO^{2-} increases). However the variation, measured between bounding lines of slope 8 (see Figure 6.4) is mostly contained in 0.07 log units of iron redox ratio (log-form) or 0.4 log unit of sulfur redox ratio. We cannot exclude that these differences may also relate to different equilibrium temperatures, likely due to cooling between groups of samples plotting over different lines. In fact, Figure 6.7b is similar to Figure 6.4b and thus shows that the ensemble of Mauna Kea samples follow a trend of slope < 8 (the black line fitting all data has slope ~4).

Note that by adopting the composition-independent Equation 6.27 as an oxythermometer, Nash et al. (2019) explain the simultaneous variations of Fe^{III}/Fe^{II} and SO_4^{2-}/S^{2-} ratios in Brounce et al. data (Figure 6.7b) via Fe–S secondary electron exchange on cooling from 1200°C to 933°C.

6.4. DISCUSSION

6.4.1. Who Controls What: the Composition/Polymerization/Acid–Base/Redox Connection

Silicate melts are a special category of fused salts in which the interconnected (polymerized) silicate structure is such that cations and anions have actual charges lower than formal ones, and the residual charge distribution from bonding of bridging oxygen to silicon (or other network former) allows oxygen bonding with other cations. This characteristic prevents from readily distinguishing solute and solvent like in aqueous solutions and consequently the speciation state. Therefore, the distinction between solute and solvent becomes blurred in such systems, because speciation is not only complex, but it changes with the marked depolymerization of the silicate framework that obtains from pure SiO_2 to metal-oxide rich compositions (Toop & Samis, 1962a,b; Mysen & Richet, 2018). If in aqueous solutions the formation of ions *sensu stricto* is due to the high dielectric constant of the aqueous medium (efficient shielding of charge and formation of hydration spheres); in contrast, the dielectric properties of silicate melts are very different, no volumetric electrostriction is known, and covalency forces do not exhaust within a single polyhedron or a combination of them, but distribute over the network (Moretti et al., 2014).

Despite the limits summarized here, to assess redox (and acid–base) exchanges in silicate melts we can still calculate the cationic and anionic populations that make up the silicate network and over which concentrations/activities of reactive species can be defined. Consequently, the concentrations of network modifiers (such as Fe^{2+}, or any other network modifier, including the Fe^{3+} cation) will be calculated with reference to the cationic population, whereas the concentration of complexes building the silicate network (*structons*, following the terminology of Fraser, 1977) and formed by tetrahedral units coordinated by Si^{4+}, Al^{3+}, and also by Fe^{3+} must be calculated over the anionic population. The latter includes

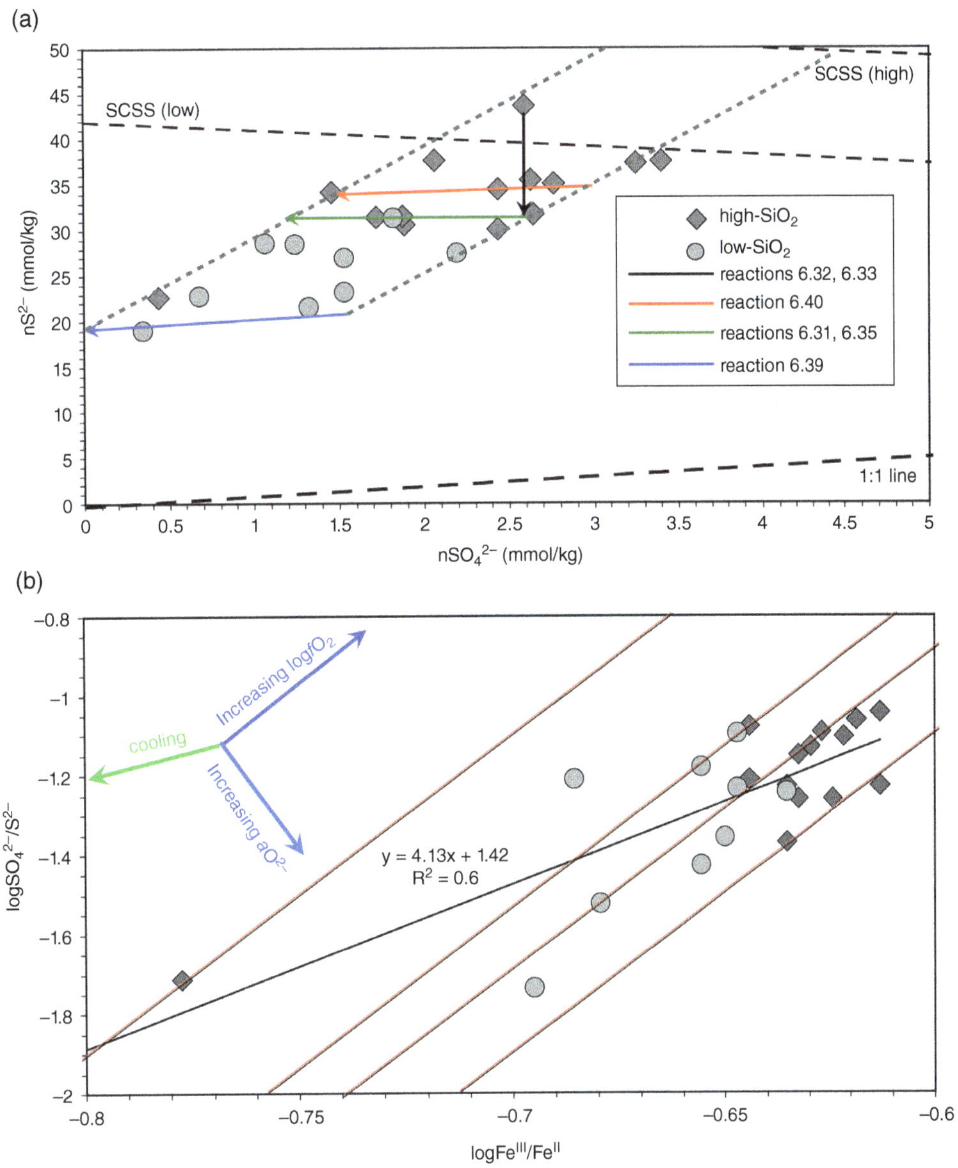

Figure 6.7 (a) Covariation of sulfate and sulfide fractions in melt inclusions from Mauna Kea submarine lavas (Brounce et al., 2017). Sulfur content at sulfide saturation (SCSS) is given a s a range delimited by two S-isolines (high: 54 mmol/kg; low: 42 mmol/kg). (b) Isothermal (1150°C) mutual interactions of Fe and S (red lines are constant composition trendlines with slope 8; see also Figure 6.4a) in Mauna Kea samples; both S and Fe redox ratios (log-form) were from XANES measurements (Brounce et al., 2017); cooling effects (see also Figure 6.4b) may shift datapoints along one or more trendlines with slope lower than 8. Combination of compositional changes and cooling effects is then behind the observed data pattern, approximated by the fitting straight black line of slope 4.13. The data at S^{VI}/S_{tot} detection limit (0.01) were discarded.

a) O^{2-} and *b*) sulfide and sulfate groups, although these are not part of the network (e.g., Moretti and Ottonello, 2003a,b, 2005).

The conceptual basis is the same as the (ideal) Temkin model for fused salt, but in silicate melt the complication arises from the connectivity of tetrahedral units sharing apical oxygens (so-called Bridging Oxygens) and making up a polymerized network without long-range ordering (Mysen & Richet, 2018 and references therein). We keep here only the basics behind melt polymerization, without the details on how to assess it, which instead are described in a wide literature (Moretti, 2005 and references therein). Polymerization is treated as the mixing of three kinds of oxygens: (i) bridging oxygens (BO), denoted as O^0;

(ii) non-bridging oxygens (NBO), which are denoted as O^- and represent unshared (covalently unbounded) apical oxygens resulting from the inter-tetrahedral disruption of the network due to network modifiers; and (iii) the non-network associated or free oxygens, O^{2-}, which are the oxygens not bounded to network formers in the formation of a coordination polyhedron (Fincham & Richardson, 1954; Toop & Samis, 1962a;b; Fraser, 1975; Ottonello et al., 2001; Moretti, 2005; Mysen & Richet, 2018; Nesbitt et al., 2015a,b; Le Losq et al. 2019 and references therein).

Melt polymerization is then the results of the equilibrium reaction of the three oxygen species (Fincham & Richardson, 1954; Toop & Samis, 1962a,b; Fraser, 1977; Ottonello et al., 2001; Ottonello & Moretti, 2004; Moretti, 2005)

$$O^0 + O^{2-} \Leftrightarrow 2O^- \tag{6.41}$$

Reaction equilibrium $\log K_{41}$ (so-called melt polymerization constant; Toop and Samis, 1962a,b) shifts with composition because any silicate network is a solvent *per se* with peculiar dielectric and electrostriction properties; hence with a specific structural framework whose polymerization is the main chemical descriptor. Because of this, the continuous shift of polymerization with composition impacts on redox equilibrium in a non-linear way via aO^{2-} (see also Moretti, 2005 for details). In fact, depending on bulk composition, the polymerization equilibrium $\log K_{41}$ involves a change in the oxidation state of oxygen and hence in the polarization state of the ligand and in network relaxation (Ottonello & Moretti, 2001; Moretti, 2005). This affects the electron pressure of the ligand (the Lewis donor) surrounding the cations (the acceptors), thus the aO^{2-} term entering reactions such as 6.4 or 6.9. Ergo, variations of the activity of O^{2-} reflect changes in the charge transfer operated by the oxygen network in response to compositional variations. That's why any redox change is also a change of the acid–base character of the melt system.

So-called ionic-polymeric models are just a consistent way to treat the intimate connection between network polymerization and acid–base properties of dissolved oxides. They provide the syntax to write chemical exchanges but have no pretension to being a structural description. For example, X-ray Photoelectron Spectroscopic (XPS) results for K-silicate glasses show the presence of at least two types of BO, which can be distinguished by the number of K ions in the moiety (Nesbitt et al., 2017). This yields BO atoms of differing electron density and then affects the charge distribution in a Si tetrahedron and bonding properties. However, ionic-polymeric models accomodate these compositional features by considering the property and energy shift of the melt solvent with composition rather than by increasing the number of oxygen species in the chemical systems and thus the number of reactions of the type of Equation 6.41.

Adopted reactive species In ionic-polymeric models may not explain the whole structural complexity revealed by many spectroscopic techniques (e.g., Henderson et al., 2014 and references therein) but have to be consistent with observations on the short-range ordering, in order to discriminate between (i) cations in true coordination polyhedra making up the network, such as Fe^{III} in Fe_4O^{5-}; (ii) cations acting as network modifiers, such as Fe^{2+} and Fe^{3+}; and (iii) anionic groups not associated with the network, such as sulfide and sulfate. In the case of network modifying behavior, the species under consideration are simple cations (Fe^{2+} or Fe^{3+}) because of bonds with surrounding oxygens are mainly electrostatic, rather than covalent (Moretti, 2005). For this reason it is not necessary to write network-modifying iron species in octahedral or higher coordinations, (i.e., $Fe^{II}O_6^{10-}$ and $Fe^{III}O_6^{9-}$) as these would not correspond to true polyhedra that build the network and speciate over the anionic matrix? Rather, they are simple ionic couplings $Fe^{2+}.6O^{2-}$ and $Fe^{3+}.6O^{2-}$. Therefore reactions:

$$Fe^{3+}\cdot 6O^{2-} + 1/2O^{2-} \Leftrightarrow Fe^{2+}\cdot 6O^{2-} + 1/4\,O_2 \tag{6.42}$$

$$FeO_4^{5-} + 5/2O^{2-} \Leftrightarrow Fe^{2+}\cdot 6O^{2-} + 1/4\,O_2 \tag{6.43}$$

reduce to Reactions 6.4 and 6.9, respectively. In case of network forming behavior, a true polyhedron (in the sense of Pauling, 1960) is formed (FeO_4^{5-} tetrahedron), with Fe^{III}–O bonds essentially covalent. This behavior is confirmed by structural studies and appear to be the normal condition for trivalent iron (Le Losq et al., 2020 and references therein), in agreement with the well-known fact that a strong Lewis bases such as O^{2-} stabilizes high oxidation states in covalent coordination compounds (e.g., Chiorboli, 1980).

If now we want to link the process of ion formation to the dissolution of oxide components, we see that the network-modifying behavior corresponds to dissociation of O^{2-} by the iron macroscopic oxide

$$FeO_{(adc)} \Leftrightarrow Fe^{2+}{}_{(La)} + O^{2-}{}_{(Lb)} \tag{6.44}$$

$$Fe_2O_{3(adc)} \Leftrightarrow Fe^{3+}{}_{(La)} + 3O^{2-}{}_{(Lb)} \tag{6.45}$$

in which iron oxide (which we can see as the Lewis adduct complex, *adc* as subscript) dissolution leads to formation of Lewis acids (the network modifying cation) and the O^{2-} base. On the other hand, network-forming behavior of trivalent iron oxide (the Lewis acid) leads to associate O^{2-} (consumption of the Lewis base) and stabilizes a network structural unit (a coordination polyhedron, that is, the Lewis adduct complex):

$$Fe_2O_{3(La)} + 5O^{2-}{}_{(Lb)} \Leftrightarrow 2Fe^{III}O_4{}^{5-}{}_{(adc)} \tag{6.46}$$

which can also be expressed as

$$Fe_2O_{3(La)} + O^{2-}{}_{(Lb)} \Leftrightarrow 2Fe^{III}O_2{}^{-}{}_{(adc)} \tag{6.47}$$

and provides a better justification for Reaction 6.12.

To realize the analogy between aO^{2-} and pH we must now consider that the above Reactions 6.44 to 6.47 imply the Lux-Flood acid–base formalism (Flood & Forland, 1947) for molten oxides and silicate melts. Schematically:

$$Acid_{(La)} + O^{2-}{}_{(Lb)} \Leftrightarrow Base_{(adc)} \tag{6.48}$$

Reaction 6.48 defines acid–base exchanges on the basis of the transfer of an oxygen ion and recalls the perhaps more familiar acid–base definition of Bronsted-Lowry for aqueous solutions:

$$Acid_{(adc)} \Leftrightarrow Base_{(Lb)} + H^{+}{}_{(La)} \tag{6.49}$$

in which the acid–base character is determined by proton transfer, thus showing the clear analogy between pH and aO^{2-}.

The behavior of trivalent iron, that can be either former or modifier is then related to the dual or *amphoteric* behavior of Fe_2O_3 (Reactions 6.45 and 6.46). Based on Equation 6.48, cations behaving as network modifiers correspond to Lux-Flood basic oxides, whereas cations behaving as network formers correspond to Lux-Flood acidic oxides. The amphoteric behavior can also be seen in terms of Lewis acid–base behavior.

Because in Reaction 6.48 i) the Lux-Flood acid is also a Lewis acid; ii) O^{2-} is the Lewis base; and iii) the Lux-Flood base is the Lewis adduct complex, we see that there is a considerable overlap between Lux-Flood and Lewis acid–base definitions. This overlap is such that Reaction 6.48 represents the obvious mechanism to set a basicity indicator in melts and glasses via aO^{2-}, same as pH as acidity indicator via Reaction 6.49. On the other hand, Reaction 6.48 represents also the transfer of polarizable oxygen ion to an acidic solute (iron cations in Reactions 6.44–6.45; trivalent iron oxide in Reactions 6.46–6.47) leading to formation of an oxyanion species (e.g., iron oxides in Reactions 6.45–6.46; the network structural units in Reactions 6.46–6.47) characterized by covalent bonds. This overlap implies the possibility of setting a quantitative scale of melt basicity unique for all oxide systems including silicate melts of any composition under the more comprehensive and general Lewis definition, provided. Duffy and Ingram (1971, 1975) and Duffy (1993) have shown that this unique scale can be established on the basis of optical basicity, a parameter experimentally observed with the shift of UV absorption bands of metals in glasses (or other media). The observed shift measures the mean polarization state of the various ligands, such as oxide ions in natural silicate melts and glasses, and their ability to transfer fractional electronic charges to the central metal cation and form covalent bonds. Therefore, optical basicity is directly related to aO^{2-} as defined in the Lux-Flood acid–base formalism (Reaction 6.48) The concept of optical basicity arises primarily from the systematic study of the orbital expansion (or "nephelauxetic effect") induced by the increased localized donor pressure on p-block metals (e.g. Tl^{I}, Pb^{II}, Bi^{III}), which are used at trace levels as probes to measure the shift of UV absorption bands. Optical basicity can also be calculated, depending upon the medium stoichiometry, the oxidation numbers of the cations, and upon a "basicity moderating" parameter related to the Pauling electronegativity (Duffy & Ingram, 1975; Duffy, 1993; Moretti, 2005).

Based on the optical basicity contrast of the anionic and cationic matrixes (a quantity related to melt composition), Ottonello et al. (2001) proposed an assessment of the polymerization constant K_{41}, allowing the computation of the three species of oxygen and related activities, particularly aO^{2-}, which measures the Lux-Flood basicity appearing in redox equilibria so far described. This led the Authors to compute the bulk Fe^{III}/Fe^{II} ratio as:

$$\frac{Fe^{III}}{Fe^{II}} = \frac{n_{Fe^{III}}}{n_{Fe^{II}}} = \frac{[FeO_2^-]\sum anions + [Fe^{3+}]\sum cations}{[Fe^{2+}]\sum cations}$$
$$= \frac{f_{O_2}^{1/4}}{K_6} \times \frac{K_{47}^{1/2} a_{O^{2-}}^{2} \sum anions + K_{45}^{1/2} \sum cations}{a_{O^{2-}}^{1/2} K_{44} \sum cations} \tag{6.50}$$

Note that with Equation 6.50 I start using square brackets to define ion activities, meaning that these are computed as the concentration of the ion of interest over the appropriate matrix (either cationic or anionic). Ottonello et al. (2001) have also shown that oxide dissolutions Reactions 6.44, 6.45, 6.47 can be seen as transposition reactions to link thermodynamically macroscopic oxide components and the ionic species that detail the solution behavior of iron oxides. Therefore, their reaction Gibbs energies (or equilibrium constants, i.e., K_{45}, K_{46} an K_{48}) are the energy differences between the standard state for pure molten iron oxides between (i) that at P and T of interest (or 1 bar and T of interest) and (ii) the one of completely dissolved ionic substance. The first is the standard state used to compute reaction equilibria (typically from thermochemical tables, such NIST-JANAF ones; Chase et al., 1998), whereas the second is the standard state in which each ion speciates over the corresponding matrix (fused salts Temkin model). If we take the example of Reaction 6.44, in the latter standard state the activity of FeO in Reaction 6.44 should be one (e.g., Lewis & Randall, 1970, see chapter 22). We then have (Ottonello et al; 2001):

$$\log K_{46} = \frac{[Fe^{2+}][O^{2-}]}{aFeO} \neq 1 = 1.1529 - \frac{1622.4}{T} \quad (6.51)$$

whose value fills the energy gap between the two standard states. Equilibria of the type of Equation 6.51, in which oxide components and the ions derived from their dissociation appear, are the key to reconcile experiments based on the macroscopic oxide notation with a detailed ionic description of the dissociation products (Ottonello et al., 2001). Besides, Equation 6.51 ($\log K_{46}$) is a measure of the strength of the Lux-Flood acid–base pair (Flood & Forland, 1947). This is an important point, as in many studies oxide components and ionic species are mixed, but without any consideration about the different standard states. For example, in the case of joint Fe and S redox variations, it is easy to find the following:

$$S^{2-} + 4Fe_2O_3 \Leftrightarrow SO_4^{2-} + 8FeO \quad (6.52)$$

Because Equation 6.52 mixes up iron *oxide components* and sulfur *ionic species*, it cannot have the pretention to be used to constrain thermodynamically the Fe–S interaction from an analytical database about Fe and S redox ratios. In fact, pseudo-reactions such as 6.52 are solved only empirically by coupling non-thermodynamic expressions of iron and sulfur redox states, such as Kress and Carmichael (1991) and Wallace and Carmichael (1994) or Jugo et al. (2010), in order to constrain possible T-fO_2 solutions. Nevertheless, these approaches lead to wrong results outside the calibration domains. Also the pseudo-equilibrium constant proposed by Nash et al. (2019) (Equation 6.27) illustrates this problem, because it neglects compositional terms related to the activity of O^{2-} and pass on compositional effects to temperature by adopting a T^2 term derived from enthalpy fitting of Fe–S exchange in the solid phase, rather than in pure-liquid or melt.

It is my opinion that these approaches derive from two wrong considerations that have however been deeply rooted in the current Earth Science literature:

1. Reaction 6.6 is considered to be completely addressed by the empirical Kress and Carmichael (1991) equation. However, this equation is not an equilibrium constant and requires compositional fitting parameters that are purely empirical.
2. Based on Reaction 6.15, it is assumed that the sulfide and sulfate exchange is not related to melt composition.

With reference to point 2, we should remark that a direct assessment of the equilibrium constant of Reaction 6.15 ($\log K_{15}$) based on available experiments under controlled fO_2 conditions is actually meaningless because, independent of temperature, SO_4^{2-}/S^{2-} ratios (measurable in the interval 0.01 to 100) do not affect appreciably the constant (K_{15}) value, which basically correspond to fO_2^{2}. This was shown in the compilation reported in by Moretti and Ottonello (2003a; see their figure 3) covering sixteen $\log fO_2$ units and based at that time on available electron microprobe data and older gravimetric analyses of sulfur speciation. Basically, analytical data return a reaction constant ($\log K_{15}$) which is well approximated by one of the terms ($-2\log fO_2$). Compared to the compilation reported by Moretti and Ottonello (2003a), the more recent and accurate XANES data cover a smaller fO_2 interval but display the same behavior, with all the data contained within –3 and +2 log-units (absolute average one log-unit) of deviation from a regression line having slope of ~ –2 (Figure 6.8a). This implies that any operational formula for fO_2 based on SO_4^{2-}/S^{2-} is going to be profoundly affected by the accuracy of the fO_2 estimate, whose uncertainty may encompass the entire range of *measurable* sulfur redox ratios.

The clustering of $\log(SO_4^{2-}/S^{2-})$ values in a small interval around the regression line of Figure 6.8a does not allow $\log K_{15}$ to be solved for temperature and compositional effects. Nevertheless, using datasets grouped for the same temperature, we can plot the $\log(SO_4^{2-}/S^{2-})$ ratios versus $\log fO_2$ to search for the emergence of isothermal trends. However, the isothermal groups of data do not fall completely on straight line of slope 2, as required by stoichiometry of Reaction 6.15: although the clear role of temperature in shifting sulfate/sulfide ratios to higher fO_2, and the departures from a slope of 2 are clearly determined by compositional effects rather than by measurement errors (Figure 6.8b), and these undermine the model of Nash (2019).

To address compositional effects and assess the equilibrium constant of Reaction 6.15, we must make for sulfide and sulfate the same considerations as for O^{2-}, and consider that both S^{2-} and SO_4^{2-} anions are not in an unpolarized state. Instead, they are solvated into the melt structure and thus are more or less polarized depending on whole composition and not only on adjacent cations (Moretti & Ottonello, 2003b; 2005). Based on this consideration, Moretti and Ottonello (2005) modelled sulfur equilibria 26 and 27, rewritten for S_2 via gas phase conversion:

$$1/2S_2 + O_2 \Leftrightarrow SO_2 \quad (6.53)$$

in order to provide a description of S dissolution based on fO_2 and fS_2. as commonly done since Fincham and Richardson (1954).

The two competing solubility equilibria were than modelled by weighting for each experimental composition the metal oxide-metal sulfide and metal oxide-metal sulfate pairs contributing to S^{2-} and SO_4^{2-} dissolution. Weighting factors were established via the Flood-Grjothein thermochemical cycle by using *electrically equivalent ion fractions* (Flood & Grjotheim, 1952; Moretti & Ottonello, 2005):

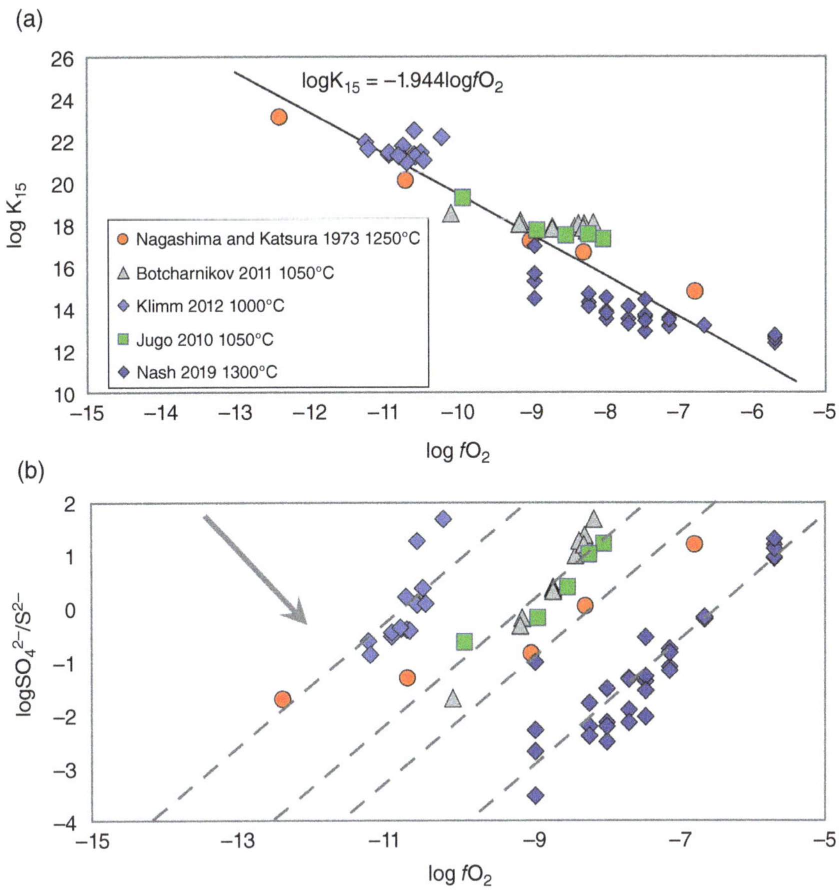

Figure 6.8 (a) Plot of $\log K_{15}$ vs $\log fO_2$, showing the nearly negligible effect of SO_4^{2-}/S^{2-} ratios on log K. (b) SO_4^{2-}/S^{2-} ratios (log-form) vs $\log fO_2$. Data are grouped per author and temperature along isolines of slope 2, as required by Reaction 6.15. Compositions and symbols same as in Figure 6.3.

$$Y_{M_i^{\nu+}} = \frac{\nu_i^+ X_{M_i^{\nu+}}}{\sum \nu_i^+ X_{M_i^{\nu+}}} \qquad (6.54)$$

Assessing the two solubility equilibria for sulfide and sulfate species with respect to a gas phase automatically leads to the assessment of $\log K_{15}$ by subtracting one mechanism from the other. On this basis, Moretti and Ottonello (2005) showed that they could reproduce microprobe and gravimetric SO_4^{2-}/S^{2-} data in the $\log fO_2$ range $-2<QFM<2$, where electron microprobe measurements of both species were shown to be most reliable.

6.4.2. Magma Degassing: Oxygen as a Perfectly Mobile Component and the Role of Gas Redox Buffers

In a volcanic-magmatic system only in a few ideal cases is the redox potential of the whole system imposed uniquely by iron in the melt, that is, by $Fe^{III}/Fe^{II} \approx 1$, such as in the case of volatile-induced re-melting of crystal mush patches (e.g. Moretti et al., 2013). Instead, the buffering of redox potential depends greatly on fluid-rock (crystals)-melt interactions. Modeling and observations show that the oxygen fugacity of magmas can be strongly affected by volatile exsolution and degassing during magma ascent. The fO_2 values measured in volcanic gases and plumes may only partially reflect magma source characteristics, due to magma chemical differentiation including degassing (Moretti & Stefansson 2020 and references therein). For example, Burgisser and Scaillet (2007) applied mass conservation to oxygen in the case of a fast decompressing rhyolite in which gas and melt remain coupled and no oxygen exchange between melt and crystals take place. Oxygen was modeled to be taken up by H_2O, SO_2, and CO_2 in the gas phase and by dissolved FeO, Fe_2O_3, and SO_2 in the melt, without any consideration of how the speciation of volatile species affects mass balances. This simplified approach revealed some features about the evolution of relative fO_2, such as relative oxygen fugacity (ΔFMQ) decreasing from depth to surface, with the exception of the Fe-and S-free calculations. It is also a good example of how in a closed system, without fractionation and phase separation, the question of what controls

fO_2 is trivial, because at equilibrium, as a consequence of the Duhem's principle, either saying that the redox state variations control the system evolution or vice versa that the system evolution fixes the redox state becomes a meaningless statement.

The fO_2 governing buffer arises through exchanges of mobile components between the phases of the magmatic system and the surrounding environment. An unspecified ensemble of reactions, depending on the relative mobility of the fluid phase, the melt, the fractionated crystals and also the wall-rock will then make up multiple redox buffers, each governing a sub-portion of the magmatic system. Basically, we lack a detailed knowledge about the multiple redox reactions involving elements which buffer the fO_2 in a crystallizing and degassing magma. For this reason, Ghiorso and Kelemen (1987) define oxygen as a perfectly mobile component that is continuously added to or removed from the system and modify the Gibbs free energy to model phase equilibria in magmatic systems that are selectively open to oxygen exchange. Therefore, it is not realistic to consider a magmatic system as "closed" to oxygen, unless the full reactivity of the system and its surrounding is entirely described and it is preferred not to apply mass conservation of oxygen (e.g. Moretti & Papale, 2004). Rather than forward modeling, several different examples (Moussallam et al., 2014, 2017; Oppenheimer et al., 2011; Brounce et al., 2017) show that the best way to reconstruct the $\log fO_2$ variations of a magmatic-volcanic system is the backtracking of fO_2 in natural samples (e.g., CO_2/CO, H_2O/H_2, and SO_2/H_2S in volcanic gases, iron and sulfur redox ratios in glasses, fO_2-T pairs from mineral assemblages). Nevertheless, the performance of fO_2 backtracking from measured Fe and S redox states in natural glasses is highly dependent on the choice of adopted oxythermobarometers and may lead to contrasting interpretations (see Moretti & Stefansson, 2020) especially because compositional effects are treated very differently by operational forms proposed in the literature (see Section 6.2). In order to reduced ambiguities in data treatment, I strongly suggest adopting the simple and model-independent procedures described in Section 6.3.1 and constrain main degassing processes behind S-release in volcanic systems (see Figures 6.5 to 6.7).

Figure 6.6 suggests that the gas phase may buffer the redox potential, via SO_2 and H_2S species (Reaction 6.36 or 6.37). Giggenbach (1987; 1996) proposed that in "andesitic" volcanism, where SO_2 and H_2S exist in comparable amounts; a highly mobile gas phase may flux through the magmatic system and buffer the redox potential, before separating to evolve at a post-magmatic stage and buffer the early redox conditions occurring in ore genesis (e.g., Giggenbach, 1987; Einaudi et al., 2003 and references therein).

To investigate more quantitatively the link between the sulfur gas redox buffer and sulfate/sulfide measurements in volcanic glasses, we can then consider the redox properties of the average andesite experimentally investigated by Botcharnikov et al. (2011) and modeled as in Figures 6.3 and 6.4. In a natural degassing system, gas and melt sulfur speciation are conveniently represented by Reaction 6.37 and Reaction 6.15 rewritten to account for water dissociation (reaction 6.30), i.e.

$$S^{2-} + 4H_2O_{(g)} \Leftrightarrow SO_4{}^{2-} + 4H_{2(g)} \tag{6.55}$$

We then have a six component system (S^{2-}, $SO_4{}^{2-}$ SO_2, H_2S, H_2O, H_2) which can be fully described by two reactions, each having five components:

1. by subtracting Reaction 6.37 to Reaction 6.55 to remove *water* we obtain:

$$S^{2-} + 2SO_{2(g)} + 2H_{2(g)} \Leftrightarrow SO_4{}^{2-} + 2H_2S_{(g)} \tag{6.56}$$

2. by subtracting Reaction 6.37 to Reaction 6.55 to remove *hydrogen* we obtain:

$$3S^{2-} + 4SO_{2(g)} + 4H_2O(g) \Leftrightarrow 3SO_4{}^{2-} + 4H_2S_{(g)} \tag{6.57}$$

It is worth noting that either Reactions 6.56 or 6.57 can be used to explain the reduction event in Figure 6.6 via decompression-driven removal of SO_2 and H_2 (Reaction 6.56) or SO_2 and H_2O (Reaction 6.57).

Reaction 6.57 can also be obtained by subtracting Reaction 6.36 from Reaction 6.15 to remove O_2 in the hypothetical five-component system (S^{2-}, $SO_4{}^{2-}$ SO_2, H_2S, H_2O, H_2). Therefore, the equilibrium constant of Reaction 6.37 (see Giggenbach, 1987; Aiuppa et al., 2011):

$$\begin{aligned}\log K_{37} &= \log \frac{XSO_{2(g)}}{XH_2S_{(g)}} + \log fH_2O - \frac{3}{2}\log fO_2 \\ &= -3.986 + \frac{27377}{T}\end{aligned} \tag{6.58}$$

can be combined to the andesite-specific solution for $\log K_{15}$ (Equation 6.19) and give the equilibrium constant of Reaction 6.59:

$$\begin{aligned}\log K_{59} &= \log \frac{XSO_4^{2-}}{XS^{2-}} - \frac{4}{3}\log \frac{XSO_{2(g)}}{XH_2S_{(g)}} - \frac{4}{3}\log fH_2O = \\ &= -17.810 + \frac{23157.3}{T} - C_{H_2O}\left(\frac{4486.4}{T} - 2.512\right)\end{aligned} \tag{6.59}$$

Assuming the sulfur gas buffer ($XSO_2/XH_2S \approx 1$), we can then use Equation 6.59 to compute $SO_4{}^{2-}/S^{2-}$ ratios expected in the andesitic melt for given temperature, dissolved water content, and water fugacity and then

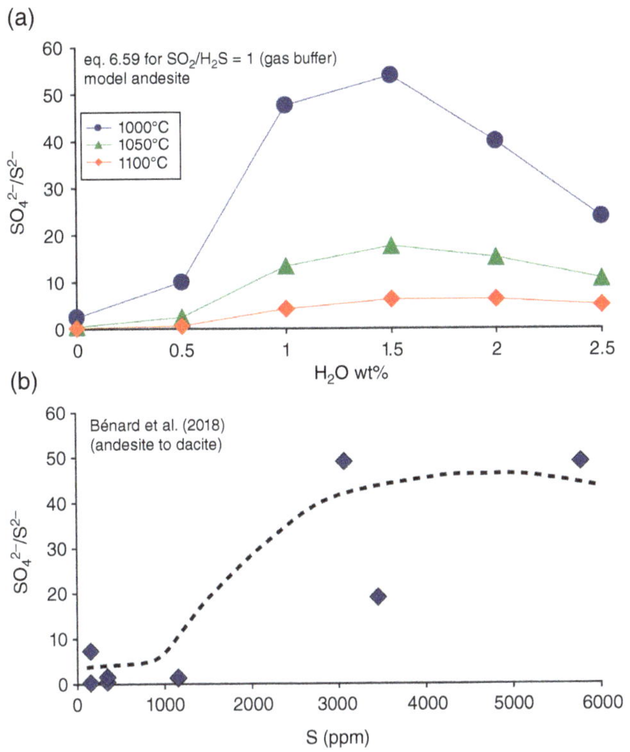

Figure 6.9 (a) Melt sulfate/sulfide ratios vs. dissolved water contents from Equation 6.59 for model andesite with redox state constrained by the sulfur gas buffer, SO_2/H_2S = 1, (Giggenbach, 1987; 1996) at three different temperatures; fH_2O values can be read on the secondary Y-axis. (b) Sulfate/sulfide ratios in andesite to rhyolite melt inclusions measured by Bénard et al. (2018); see also Figure 6.6. Datasets in the two panels compare well, despite the lack of a degassing relation between dissolved H_2O and S to establish the correspondence between computed and measured sulfate/sulfide ratios. The dotted curve is drawn as a guide to the eye.

compare them to data from natural specimens, such as Bénard et al. (2018) data discussed in Section 6.3.1 and Figure 6.6.

Figure 6.9a shows the computed variation of the SO_4^{2-}/S^{2-} ratio with melt water content for the modeled andesite at 1000°C, 1050°C, and 1100°C. For a given water content, fH_2O values have been computed at water saturation pressure by using the Burnham model (Burnham, 1994 and references therein), which for the model andesite yields:

$$fH_2O = 82.12\ C_{H_2O^2} - 29.018\ C_{H_2O} + 1 \qquad (6.60)$$

valid in the 1000°C–1050°C T-range and for $0 < C_{H_2O}$ %wt < 3.5. Calculated SO_4^{2-}/S^{2-} values compare fairly well to measured sulfate/sulfide ratios in andesite to rhyolite melt inclusions (Figure 6.9b; see also Figure 6.6) despite a direct correspondence between datasets not being able to be made due to the lack of data on dissolved H_2O contents.

6.5. CONCLUSIONS

In this study we show that adoption of the ionic notation allows a precise description of redox exchanges of silicate melts, provided that these involve ionic species consistent with the melt acid–base character. The latter is determined by composition and via melt polymerization determines the "free oxygen" (or oxide ion) activity, aO^{2-}. In a melt system, this intensive parameter is like pH in waters and enters redox reactions, representing the basicity of the system, i.e., some kind of network contribution. In silicate melts, the formal link between redox and acid–base exchanges is represented by the "normal oxygen electrode," which relates aO^{2-} and oxygen fugacity, fO_2, the master variable used to quantify the redox potential and its dependence on bulk composition.

By means of the ionic notation we could formally evaluate how both iron and sulfur redox state may be deeply affected by compositional effects. These effects are the larger the compositionally simpler the system. The case of iron is quite emblematic, as its redox parameterization must include the amphoteric behavior of trivalent iron, which can act as either network former of modifier depending on melt basicity. Although natural melts tend to average the effects of composition, the redox state of sulfur too cannot be approximated by operational forms that depend only on temperature and fO_2.

We have also developed a geochemical methodology to investigate redox evolution in degassing magmas. In particular, data from basaltic and andesitic melt inclusions can be used to describe the chemical pathways of sulfur release in these systems and put constraints on the redox state that are independent of models. Depending on residence times and the general conditions behind fluid-melt-rock interactions, the gas phase may buffer the redox potential via SO_2 and H_2S species, which results in useful equations to quantify degassing and S-redox variations in andesitic melts.

The ionic understanding of iron and sulfur redox behaviors in magmas, and particularly the approach to composition and speciation can greatly help to model the redox state of the different magmatic reservoirs on Earth as well as terrestrial planets. In particular, the ionic approach to Fe and S redox equilibria and their compositional dependence offers new possibilities to better discriminate between the primary redox signature inherited from the phase assembly and distinctive mineralogy of the source, and the redox contribution intrinsic to magma evolution on its rise to surface.

CODE AVAILABILITY

Author's numerical routines for Fe and S redox state, as well as S solubility, are made publicly available via github: https://github.com/charlesll.

ACKNOWLEDGMENTS

The Author thanks Charles Le Losq (IPGP) for FORTRAN 90 standardization of numerical routines. The manuscript benefited greatly from the comments of two anonymous reviewers. This study contributes to the IdEx Université de Paris ANR-18-IDEX-0001.

REFERENCES

Aiuppa, A., Shinohara, H., Tamburello, G., Giudice, G., Liuzzo, M., & Moretti, R. (2011). Hydrogen in the gas plume of an open-vent volcano, Mount Etna, Italy. *Journal of Geophysical Research: Solid Earth*, *116*(B10). https://doi.org/10.1029/2011JB008461

Baker, D. R., & Moretti, R. (2011). Modeling the solubility of sulfur in magmas: a 50-year old geochemical challenge. *Reviews in Mineralogy and Geochemistry*, *73*(1), 167–213. https://doi.org/10.2138/rmg.2011.73.7

Behrens, H., & Gaillard, F. (2006). Geochemical aspects of melts: volatiles and redox behavior. *Elements*, *2*(5), 275–280. https://doi.org/10.2113/gselements.2.5.275

Bénard, A., Klimm, K., Woodland, A. B., Arculus, R. J., Wilke, M., Botcharnikov, R. E., et al. (2018). Oxidising agents in sub-arc mantle melts link slab devolatilisation and arc magmas. *Nature Communications*, *9*(1), 1–10. https://doi.org/10.1038/s41467-018-05804-2

Botcharnikov, R. E., Linnen, R. L., Wilke, M., Holtz, F., Jugo, P. J., & Berndt, J. (2011). High gold concentrations in sulphide-bearing magma under oxidizing conditions. *Nature Geoscience*, *4*(2), 112–115. https://doi.org/10.1038/ngeo1042

Brounce, M., Stolper, E., & Eiler, J. (2017). Redox variations in Mauna Kea lavas, the oxygen fugacity of the Hawaiian plume, and the role of volcanic gases in Earth's oxygenation. *Proceedings of the National Academy of Sciences*, *114*(34), 8997–9002. https://doi.org/10.1073/pnas.1619527114

Burgisser, A., & Scaillet, B. (2007). Redox evolution of a degassing magma rising to the surface. *Nature*, *445*(7124), 194–197. https://doi.org/10.1038/nature05509

Burnham, C. W. (1994). Development of the Burnham model for prediction of H_2O solubility in magmas. In: Carroll, M. R., & Holloway, J. R. (eds.) Volatiles in magmas, *Reviews in Mineralogy*, *30*, 123–129. https://doi.org/10.1515/9781501509674-009

Carroll, M. R., & Rutherford, M. J. (1985). Sulfide and sulfate saturation in hydrous silicate melts. *Journal of Geophysical Research: Solid Earth*, *90*(S02), C601–C612. https://doi.org/10.1029/JB090iS02p0C601

Carroll, M. R., & Rutherford, M. J. (1988). Sulfur speciation in hydrous experimental glasses of varying oxidation state; results from measured wavelength shifts of sulfur X-rays. *American Mineralogist*, *73*(7–8), 845–849.

Carroll, M. R., & Webster, J. D. (1994). Reviews in Mineralogy and Geochemistry. *Reviews in Mineralogy and Geochemistry*, *30*(1), 231–279.

Chase, M. W. (1998). NIST-JANAF Thermochemical Tables 2 Volume-Set (Journal of Physical and Chemical Reference Data Monographs), American Institute of Physics.

Chiorboli P. (1980). *Fondamenti di Chimica*, UTET, Torino, 1133 pp.

Cicconi M. R., Le Losq C., Moretti R., & Neuville D. R. (2020a). Magmas are the Largest Repositories and Carriers of Earth's Redox Processes. *Elements* *16*(3), 173–178. https://doi.org/10.2138/gselements.16.3.173

Cicconi M. R., Moretti, R., & Neuville, D. R. (2020b). Earth's Electrodes. *Elements* *16*(3), 157–160. https://doi.org/10.2138/gselements.16.3.157

Cicconi M. R., Le Losq, C., Henderson, G., & Neuville D. R. (2020c). The redox behavior of Rare Earth Elements. In: Moretti, R., & Neuville, D. R. (eds). *Magma Redox Geochemistry*. Geophysical Monograph Series. American Geophysical Union.

Dickenson, M. P., & Hess, P. C. (1982). Redox equilibria and the structural role of iron in alumino-silicate melts. *Contributions to Mineralogy and Petrology*, *78*(3), 352–357. https://doi.org/10.1007/BF00398931

Douglas, R. W., & Zaman, M. S. (1969). Chromophore in iron-sulphur amber glasses. *Physics and Chemistry of Glasses*, *10* (4), 125–132.

Duffy, J. A. (1993). A review of optical basicity and its applications to oxidic systems. *Geochimica et Cosmochimica Acta*, *57*: 3961–3970. https://doi.org/10.1016/0016-7037(93)90346-X

Duffy, J. A., & Ingram, M. D. (1971). Establishment of an optical scale for Lewis basicity in inorganic oxyacids, molten salts, and glasses. *Journal of the American Chemical Society*, *93*(24), 6448–6454. https://doi.org/10.1021/ja00753a019

Duffy, J. A., & Ingram, M. D. (1975). Optical basicity—IV: Influence of electronegativity on the Lewis basicity and solvent properties of molten oxyanion salts and glasses. *Journal of Inorganic and Nuclear Chemistry*, *37*(5), 1203–1206. https://doi.org/10.1016/0022-1902(75)80469-6

Einaudi, M. T., Hedenquist, J. W., & Inan, E. E. (2003). Sulfidation state of fluids in active and extinct hydrothermal systems: Transitions from porphyry to epithermal environments. *Special Publication–Society of Economic Geologists*, *10*, 285–314. https://doi.org/10.5382/SP.10.15

Fincham, C. J. B., & Richardson, F. D. (1954). The behaviour of sulphur in silicate and aluminate melts. *Proceedings of the Royal Society of London. Series A. Mathematical and Physical Sciences*, *223*(1152), 40–62. https://doi.org/10.1098/rspa.1954.0099

Flood, H., & Förland, T. (1947). The acidic and basic properties of oxides. *Acta Chemica Scandinavica*, *1*, 592–604. doi: 10.3891/acta.chem.scand.01-0592

Flood H., & Grjotheim T. (1952). Thermodynamic calculation of slag equilibria. *Journal of the Iron and Steel Institute*, *171*, 64–80.

Fraser, D. G. (1975). Activities of trace elements in silicate melts. *Geochimica et Cosmochimica Acta*, *39*(11), 1525–1530. https://doi.org/10.1016/0016-7037(75)90154-4

Fraser, D. G. (1977). Thermodynamic properties of silicate melts. In: Fraser, D.G. (Ed.), *Thermodynamics in Geology*. D. Reidel Pub., Dortrecht-Holland.

Frost, B. R. (1991). Introduction to oxygen fugacity and its petrologic importance. *Reviews in Mineralogy and Geochemistry*, *25*, 1–9. https://doi.org/10.1515/9781501508684-004

Ghiorso, M. S., & Kelemen, P. B. (1987). Evaluating reaction stoichiometry in magmatic systems evolving under generalized thermodynamic constraints: examples comparing isothermal and isenthalpic assimilation. *Magmatic Processes: Physicochemical Principles*, *1*, 319–336.

Giggenbach, W. F. (1987). Redox processes governing the chemistry of fumarolic gas discharges from White Island, New Zealand. *Applied Geochemistry*, *2*(2), 143–161. https://doi.org/10.1016/0883-2927(87)90030-8

Giggenbach, W. F. (1996). Chemical composition of volcanic gases. In: Scarpa, R., & Tilling, R. I. (eds.) *Monitoring and Mitigation of Volcano Hazards*. Springer, Berlin, Heidelberg, pp. 221–256.

Haughton, D. R., Roeder, P. L., & Skinner, B. J. (1974). Solubility of sulfur in mafic magmas. *Economic Geology*, *69*(4), 451–467. https://doi.org/10.2113/gsecongeo.69.4.451

Henderson, G., Neuville, D.R., & Downs, R. (Eds.). (2014). Spectroscopic methods in mineralogy and material sciences. *Reviews in Mineralogy and Geochemistry*, vol. *78*. Walter de Gruyter GmbH & Co KG.

Holmquist S. B. (1966). Ionic formulation of redox equilibria in glass melts. *Journal of the American Ceramic Society*, *49*, 228–229. https://doi.org/10.1111/j.1151-2916.1966.tb13241.x

Jugo, P. J., Luth, R. W., & Richards, J. P. (2005). An experimental study of the sulfur content in basaltic melts saturated with immiscible sulfide or sulfate liquids at 1300° C and 10 GPa. *Journal of Petrology*, *46*(4), 783–798. https://doi.org/10.1093/petrology/egh097

Jugo, P. J., Wilke, M., & Botcharnikov, R. E. (2010). Sulfur K-edge XANES analysis of natural and synthetic basaltic glasses: Implications for S speciation and S content as function of oxygen fugacity. *Geochimica et Cosmochimica Acta*, *74*(20), 5926–5938. https://doi.org/10.1016/j.gca.2010.07.022

Klimm, K., Kohn, S. C., & Botcharnikov, R. E. (2012). The dissolution mechanism of sulphur in hydrous silicate melts. II: Solubility and speciation of sulphur in hydrous silicate melts as a function of f_{O_2}. *Chemical Geology*, *322*, 250–267. https://doi.org/10.1016/j.chemgeo.2012.04.028

Kress, V. C., & Carmichael, I. S. (1989). The lime-iron-silicate melt system: Redox and volume systematics. *Geochimica et Cosmochimica Acta*, *53*(11), 2883–2892. https://doi.org/10.1016/0016-7037(89)90165-8

Kress, V. C., & Carmichael, I. S. (1991). The compressibility of silicate liquids containing Fe_2O_3 and the effect of composition, temperature, oxygen fugacity and pressure on their redox states. *Contributions to Mineralogy and Petrology*, *108*(1–2), 82–92. https://doi.org/10.1007/BF00307328

Lange, R. A., & Carmichael, I. S. (1989). Ferric-ferrous equilibria in Na_2O-FeO-Fe_2O_3-SiO_2 melts: Effects of analytical techniques on derived partial molar volumes. *Geochimica et Cosmochimica Acta*, *53*(9), 2195–2204. https://doi.org/10.1016/0016-7037(89)90343-8

Lange, R. A., & Navrotsky, A. (1992). Heat capacities of Fe_2O_3-bearing silicate liquids. *Contributions to Mineralogy and Petrology*, *110*(2–3), 311–320. doi: 10.1007/BF00310746

Le Losq, C., Cicconi, M. R., Greaves, G. N., & Neuville, D. R. (2019). Silicate glasses. In: Musgraves, J. D., Hu, J., & Calvez, L. (eds.) *Springer Handbook of Glass*, Springer, Cham. pp. 441–503.

Le Losq, C., Moretti, R., Oppenheimer, C., Baudelet, F., & Neuville D. R. (2020). In situ XANES study of the influence of varying temperature and oxygen fugacity on iron oxidation state and coordination in a phonolitic melt. *Contributions to Mineralogy and Petrology*, *175*, 7, 1–13. https://doi.org/10.1007/s00410-020-01701-4

Lewis, G. N., Randall, M. (1970). *Termodinamica*. Leonardo Edizioni Scientifiche, Roma.

Li, C., & Ripley, E. M. (2009). Sulfur contents at sulfide-liquid or anhydrite saturation in silicate melts: empirical equations and example applications. *Economic Geology*, *104*(3), 405–412. https://doi.org/10.2113/gsecongeo.104.3.405

Marini, L., Moretti, R., & Accornero, M. (2011). Sulfur isotopes in magmatic-hydrothermal systems, melts, and magmas. *Reviews in Mineralogy and Geochemistry*, *73*(1), 423–492. https://doi.org/10.2138/rmg.2011.73.14

Matthews, S. J., Moncrieff, D. H. S., & Carroll, M. R. (1999). Empirical calibration of the sulphur valence oxygen barometer from natural and experimental glasses: method and applications. *Mineralogical Magazine*, *63*(3), 421–431.

Mavrogenes, J. A., & O'Neill, H. S. C. (1999). The relative effects of pressure, temperature and oxygen fugacity on the solubility of sulfide in mafic magmas. *Geochimica et Cosmochimica Acta*, *63*(7–8), 1173–1180. https://doi.org/10.1016/S0016-7037(98)00289-0

Moretti, R. (2005). Polymerisation, basicity, oxidation state and their role in ionic modelling of silicate melts. *Annals of Geophysics*, *48*, 583–608. https://doi.org/10.4401/ag-3221

Moretti, R., & Baker, D. R. (2008). Modeling the interplay of f_{O_2} and fS_2 along the FeS-silicate melt equilibrium. *Chemical Geology*, *256*(3–4), 286–298. https://doi.org/10.1016/j.chemgeo.2008.06.055

Moretti R. & Neuville D. (2021). Redox Equilibria: From Basic Concepts to the Magmatic Realm. In: Moretti, R., and Neuville, D. R. (eds.) Redox Magma Geochemistry. Geophysical Monograph Series, 266. American Geophysical Union.

Moretti R., & Ottonello, G. (2003a). Polymerization and disproportionation of iron and sulfur in silicate melts: insights from an optical basicity-based approach. *Journal of Non-Crystalline Solids*, *323*, 111–119. https://doi.org/10.1016/S0022-3093(03)00297-7

Moretti, R., & Ottonello, G. (2003b). A polymeric approach to the sulfide capacity of silicate slags and melts. *Metallurgical and Materials Transactions B*, *34*(4), 399–410. 10.1007/s11663-003-0066-1

Moretti, R., & Ottonello, G. (2005). Solubility and speciation of sulfur in silicate melts: The Conjugated Toop-Samis-Flood-Grjotheim (CTSFG) model. *Geochimica et Cosmochimica Acta*, *69*(4), 801–823. https://doi.org/10.1016/j.gca.2004.09.006

Moretti, R., & Papale, P. (2004). On the oxidation state and volatile behavior in multicomponent gas–melt equilibria. *Chemical Geology*, *213*(1–3), 265–280. https://doi.org/10.1016/j.chemgeo.2004.08.048

Moretti, R., Le Losq, C., & Neuville, D. R. (2014). The amphoteric behavior of water in silicate melts from the point of view of their ionic-polymeric constitution. *Chemical Geology*, *367*, 23–33. 10.1016/j.chemgeo.2013.12.012

Moretti R., & Stefansson, A. (2020) Volcanic and geothermal redox engines. *Elements*, *16*(3), 179–184. https://doi.org/10.2138/gselements.16.3.179

Moretti, R., Papale, P., & Ottonello, G. (2003). A model for the saturation of COHS fluids in silicate melts. *Geological Society, London, Special Publications*, *213*(1), 81–101. 10.1144/GSL.SP.2003.213.01.06

Moretti, R., Arienzo, I., Orsi, G., Civetta, L., & D'Antonio, M. (2013). The deep plumbing system of Ischia: a physico-chemical window on the fluid-saturated and CO_2-sustained Neapolitan volcanism (southern Italy). *Journal of Petrology*, *54*(5), 951–984. 10.1093/petrology/egt002

Moussallam, Y., Oppenheimer, C., Scaillet, B., Gaillard, F., Kyle, P., Peters, N., et al. (2014). Tracking the changing oxidation state of Erebus magmas, from mantle to surface, driven by magma ascent and degassing. *Earth and Planetary Science Letters*, *393*, 200–209. https://doi.org/10.1016/j.epsl.2014.02.055

Moussallam, Y., Edmonds, M., Scaillet, B., Peters, N., Gennaro, E., Sides, I., & Oppenheimer, C. (2016). The impact of degassing on the oxidation state of basaltic magmas: a case study of K⊠lauea volcano. *Earth and Planetary Science Letters*, *450*, 317–325. https://doi.org/10.1016/j.epsl.2016.06.031

Mysen, B. O., & Richet, P. (2018). *Silicate glasses and melts*. Elsevier, 720 pp.

Nagashima, S., & Katsura, T. (1973). The solubility of sulfur in $Na_2O–SiO_2$ melts under various oxygen partial pressures at 1100° C, 1250° C, and 1300° C. *Bulletin of the Chemical Society of Japan*, *46*(10), 3099–3103. https://doi.org/10.1246/bcsj.46.3099

Nash, W. M., Smythe, D. J., & Wood, B. J. (2019). Compositional and temperature effects on sulfur speciation and solubility in silicate melts. *Earth and Planetary Science Letters*, *507*, 187–198. https://doi.org/10.1016/j.epsl.2018.12.006

Nesbitt, H. W., Henderson, G. S., Bancroft, G. M., & Ho, R. (2015a). Experimental evidence for Na coordination to bridging oxygen in Na-silicate glasses: Implications for spectroscopic studies and for the modified random network model. *Journal of Non-Crystalline Solids*, *409*, 139–148. https://doi.org/10.1016/j.jnoncrysol.2014.11.024

Nesbitt, H. W., Bancroft, G. M., Henderson, G. S., Sawyer, R., & Secco, R. A. (2015b). Direct and indirect evidence for free oxygen (O2−) in MO-silicate glasses and melts (M= Mg, Ca, Pb). *American Mineralogist*, *100*(11–12), 2566–2578. 10.2138/am-2015-5336

Nesbitt H. W., Bancroft G. M., & Henderson G. S. (2020). Polymerization during melting of ortho- and leta-silicates: Effects on Q species stability, heats of fusion, and redox state of mid-ocean range basalts (MORBs). *American Mineralogist*, *105*, 716–726.

Neuville, D. R., Cicconi, M. R., and Le Losq, C. (2021) How to measure the oxidation state of multivalent elements in minerals, glasses, and melts? In: Moretti, R., & Neuville, D. R. (eds.) *Magma Redox Geochemistry*. Geophysical Monograph Series. American Geophysical Union.

Oppenheimer, C., Moretti, R., Kyle, P. R., Eschenbacher, A., Lowenstern, J. B., Hervig, R. L., & Dunbar, N. W. (2011). Mantle to surface degassing of alkalic magmas at Erebus volcano, Antarctica. *Earth and Planetary Science Letters*, *306* (3–4), 261–271. https://doi.org/10.1016/j.epsl.2011.04.005

Ottonello, G., Moretti, R., Marini, L., & Zuccolini, M. V. (2001). Oxidation state of iron in silicate glasses and melts: a thermochemical model. *Chemical Geology*, *174*(1–3), 157–179. https://doi.org/10.1016/S0009-2541(00)00314-4

Ottonello, G., & Moretti, R. (2004). Lux-Flood basicity of binary silicate melts. *Journal of Physics and Chemistry of Solids*, *65*(8–9), 1609–1614. https://doi.org/10.1016/j.jpcs.2004.01.012

Paris, E., Giuli, G., Carroll, M. R., & Davoli, I. (2001). The valence and speciation of sulfur in glasses by X-ray absorption spectroscopy. *The Canadian Mineralogist*, *39*(2), 331–339.

Paul, A., & Douglas, R. W. (1965). Ferric-ferrous equilibrium in binary alakali silicre glasses. *Physics and Chemistry of Glasses*, *6* (1965) 207–211.

Tangeman, J. A., Lange, R., & Forman, L. (2001). Ferric-ferrous equilibria in K_2O-FeO-Fe_2O_3-SiO_2 melts. *Geochimica et Cosmochimica Acta*, *65*(11), 1809–1819. https://doi.org/10.1016/S0016-7037(00)00561-5

Temkin, M. (1945). Mixtures of fused salts as ionic solutions. *Acta Physiochimica,* USSR, *20*, 411.

Toop, G. W., & Samis, C. S. (1962a). Some new ionic concepts of silicate slags. *Canadian Metallurgical Quarterly*, *1*, 129–152. https://doi.org/10.1179/cmq.1962.1.2.129

Toop, G. W., Samis, C. S. (1962b). Activities of ions in silicate melts. *Transactions of the Metallurgical Society of AIME*, *224*, 878–887.

Wallace, P. J., & Carmichael, I. S. (1994). S speciation in submarine basaltic glasses as determined by measurements of S Kα X-ray wavelength shifts. *American Mineralogist*, *79*(1–2), 161–167.

Wallace, P. J., & Edmonds, M. (2011). The sulfur budget in magmas: evidence from melt inclusions, submarine glasses, and volcanic gas emissions. *Reviews in Mineralogy and Geochemistry*, *73*(1), 215–246. https://doi.org/10.2138/rmg.2011.73.8

7

The Petrological Consequences of the Estimated Oxidation State of Primitive MORB Glass

Paul D. Asimow

ABSTRACT

Discrepancies in Fe^{2+}/Fe^{T} in mid-ocean ridge basalt glasses may be an analytical problem with implications principally for the oxidation state of the source mantle, but the Fe^{2+}/Fe^{T} of such glasses also influences their equilibrium with olivine and petrological models of magmatic temperature. Hence Fe^{2+}/Fe^{T} affects the inference of potential temperature and extent of melting in the sub-ridge mantle and models of complementary formation of erupted basalts and cumulate lower crust. Conversely, evaluation of equilibria between olivine phenocrysts and host glasses may provide an independent constraint on oxidation state. This chapter reviews petrological methods for estimating primary basalt composition, potential temperature, and thickness of igneous crust, focusing on the influence of glass redox state. The 1990's consensus around mutually consistent average values of Fe^{2+}/Fe^{T} of MORB glass, thickness of oceanic crust, and mantle potential temperature apparently needs revisiting on several fronts. Values of average MORB glass $Fe^{2+}/Fe^{T} < 0.88$ are apparently inconsistent with olivine-glass data and require primary liquidus temperatures too low for traditional estimates of the average thickness of oceanic crust or of lower crustal cumulates. Critical assessments of the absolute accuracy of Fe^{2+}/Fe^{T} measurements, of electron microprobe measurements of olivine composition, of olivine/liquid Fe/Mg partitioning models, and of thermodynamic models of peridotite melting are essential to resolving this inconsistency. New melting models, source compositions, crustal thickness estimates, or interpretations of the plutonic-volcanic relationship may be needed to reconcile the bulk properties of the oceanic crust and its layers with the measured Fe^{2+}/Fe^{T} of MORB glass.

7.1. INTRODUCTION

Mid-ocean ridge basalts (MORBs), the most abundant magma type on Earth, have become, over the last fifty years, remarkably influential sources of information for development and testing of ideas including principles of igneous petrology (Grove et al., 1992; Hess, 1992; Stolper, 1980; Walker et al., 1979), the workings of plate tectonics and crustal recycling (Hofmann, 2003), and the composition and thermal state of the upper mantle (Behn & Grove, 2015; Dalton et al., 2014; Langmuir et al., 1992; McKenzie & Bickle, 1988; McKenzie & O'Nions, 1991). The influence of MORB is due in part to the fortunate coincidence of the discovery of globally distributed submarine pillow basalts with glassy rinds (Engel et al.,

Division of Geological and Planetary Sciences, California Institute of Technology, Pasadena, CA, USA

Magma Redox Geochemistry, Geophysical Monograph 266, First Edition.
Edited by Roberto Moretti and Daniel R. Neuville.

DOI: 10.1002/9781119473206.ch7

1965) and the widespread adoption of the electron microprobe (EPMA) for routine chemical analysis of minerals and glasses (Bence & Albee, 1968; Langmuir et al., 1977). The presence of glass is critical because it resolves the ambiguities inherent in using crystalline igneous rocks to characterize the liquid line of descent in magmatic evolution (Bowen, 1928), preserves volatile components (Dixon et al., 1988; Michael, 1994), and allows accurate microanalytical methods to substitute for vastly slower whole-rock analyses. Given glassy samples, the electron microprobe enables rapid, precise, and accurate measurement of major and minor elements aided by backscattered electron imaging to exclude phenocrysts and microlites.

One unintended consequence of the great abundance and intellectual influence of electron microprobe analyses of MORB glasses, though, is due to the microprobe's ability to measure only total Fe and lack of sensitivity (in routine work) to the oxidation state of that Fe. At the oxygen fugacity that prevails during terrestrial magmatic processes, most of the total Fe (Fe^T) in basaltic liquids is ferrous (Fe^{2+} or FeO), but a significant fraction is ferric (Fe^{3+} or Fe_2O_3). In whole-rock analysis by wet chemistry, the standard method of petrological analysis until the 1960s, FeO and Fe_2O_3 were independently measured (Presnall, 1966) and considered separate components. In X-ray Fluorescence or EPMA data, on the other hand, only total Fe is measured and expressed as FeO^T or $Fe_2O_3{}^T$. Thus, actual measurements of the Fe^{2+}/Fe^T ratio of basalt glasses are sparse (on the order of 10^2 total published values) compared to total Fe measurements (on the order of 10^4 published values), and it is commonplace to assume a typical value based on those sparse measurements and apply it to all the world's mid-ocean ridges. Furthermore, there have been analytical challenges associated with the measurement of Fe^{2+}/Fe^T in basalt glass that have caused estimates of this number to evolve over time.

The wet chemical analysis of whole rock samples typically yields Fe^{2+}/Fe^T ratios around 0.85 (BVSP, 1981; Carmichael & Ghiorso, 1986) and was taken to imply oxidation states close to the quartz–fayalite–magnetite (QFM) buffer (N.B., for consistency throughout this paper we will report ratios of Fe oxidation states as Fe^{2+}/Fe^T; conversion to other metrics such as Fe^{3+}/Fe^T, Fe^{2+}/Fe^{3+}, FeO/Fe_2O_3, etc., is usually a trivial matter). However, the influential paper of Christie et al. (1986) focused on the glassy rinds of submarine pillows to avoid effects of alteration and seawater interaction during crystallization. Combining wet chemistry and colorimetric titration for FeO with microprobe analyses of FeO^T, thereby obtaining Fe_2O_3 by difference, these authors found a narrow distribution of Fe^{2+}/Fe^T of 0.93±0.02 in a survey of MORB glasses, implying – according to then-current functions relating Fe speciation to oxygen fugacity (fO_2) – typical values of fO_2 that are 1.2±0.44 log units lower than the QFM buffer (Figure 7.1).

This value became popular and widely assumed for about twenty years, a time during which a number of influential models of mantle melting, primary melt accumulation, and crustal fractionation processes were published (Asimow et al., 2001, 2004; Kinzler, 1997; Kinzler & Grove, 1992; Klein & Langmuir, 1987; Langmuir et al., 1992; McKenzie & Bickle, 1988; Niu & Batiza,

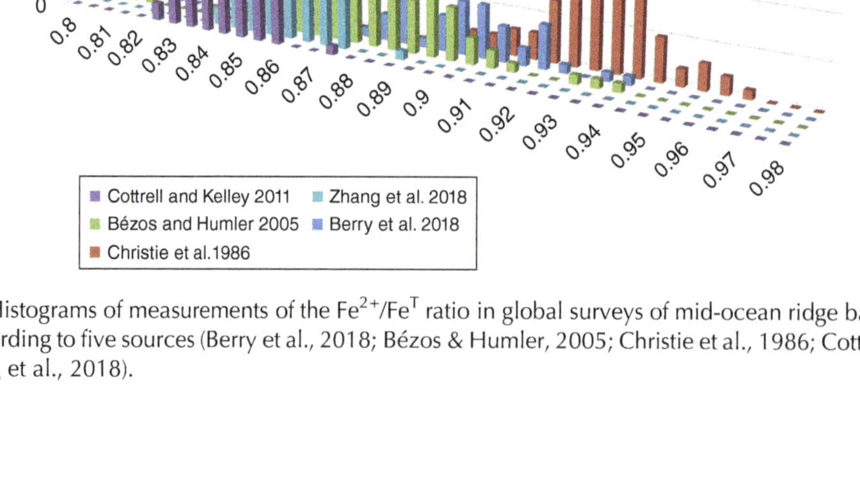

Figure 7.1 Histograms of measurements of the Fe^{2+}/Fe^T ratio in global surveys of mid-ocean ridge basalt (MORB) glasses, according to five sources (Berry et al., 2018; Bézos & Humler, 2005; Christie et al., 1986; Cottrell & Kelley, 2011; Zhang et al., 2018).

1991). Despite differences among these models, many of them were successful in finding that melting of typical, slightly depleted upper mantle peridotite by corner flow and near-fractional decompression melting beneath a ridge could, at a given value of the potential temperature (a measure of the specific entropy of the mantle source and a conserved quantity during adiabatic, reversible pressure changes), produce simultaneously the right amount of partial melt to explain the typical thickness of igneous oceanic crust (then thought to be 7±1 km; White et al., 2001) and a composition of aggregated primary melt able to produce the observed quantity of lower-crustal cumulates while fractionating to yield the observed compositions of seafloor volcanic rocks. Most of these models did not explicitly consider the effect of redox state. Those that did, such as PRIMELT (Herzberg et al., 2007; Herzberg & Asimow, 2008, 2015) and MELTS (Asimow et al., 2001, 2004), assumed Fe^{2+}/Fe^T ratios from Christie et al. (1986).

Such successes supported the widespread adoption of a simple paradigm for the origin of oceanic crust and its layers, in which (1) a mildly depleted (relative to primitive mantle) average upper mantle upwelling due to ridge spreading with (2) a narrow range of potential temperatures underneath normal (non-hotspot-affected) ridge segments underwent (3) progressive near-fractional melting, (4) continuing to near the base of the lithosphere (except at ultra-slow-spreading or highly oblique ridge segments), (5) most of the melt being then extracted and focused to the ridge axis where it was mixed to form a primary aggregate liquid that subsequently experienced fractional crystallization such that (6) the cumulates form the bulk of the lower crust (seismic layer 3) and the complementary fractionated liquid erupts to form the volcanic sequence (layers 2a and 2b).

However, Bézos and Humler (2005) published a new survey of the oxidation state of Fe in MORB glasses, using many of the same samples as Christie et al. (1986) but applying a new titration method for whole-rock analysis of glasses. They obtained systematically lower Fe^{2+}/Fe^T but a similar range within the population, 0.88±0.02, which put the estimated fO_2 of typical MORB glass at 0.41±0.43 log units below QFM. The author is unaware of any published discussion of whether this revised estimate of MORB glass redox state required any reconsideration of the paradigmatic conclusions or quantitative potential temperature estimates listed above.

Recently, the µXANES method was calibrated against wet chemical and Mössbauer analysis (Cottrell et al., 2009) and applied to the problem of determining Fe^{2+}/Fe^T of basaltic glasses including a global survey of MORB glass, again featuring many of the same reference samples as previous studies (Cottrell & Kelley, 2011). This study again reported a similar variance in the Fe^{2+}/Fe^T of the population of MORB glasses but yielded another significant shift in the mean of the distribution, to 0.84±0.01, corresponding to fO_2 ~QFM. A further recalibration of the Mössbauer spectra underlying the µXANES working curve by Zhang et al. (2018) has led to a small revision of the mean absolute Fe^{2+}/Fe^T of MORB, which currently stands by this method at 0.86±0.01, corresponding to fO_2 0.18±0.16 log units below QFM.

Finally, Berry et al. (2018) have proposed an entirely different interpretation of the Mössbauer spectra of basalt glass that yields a MORB range for Fe^{2+}/Fe^T of 0.90 ±0.02. Despite the large difference from the Cottrell et al. (2009) and Zhang et al. (2018) calibrations, there is a very important point of agreement among these studies, which is that the XANES spectra of natural MORB glass match those of synthetic glasses equilibrated close to the QFM buffer (O'Neill et al., 2018). That is, there is currently disagreement about the Fe^{2+}/Fe^T ratio of MORB glass but consensus on the fO_2 recorded by iron speciation in the same samples. However, other redox-sensitive elements such as V may also be pointing to discrepant fO_2 values more oxidizing than QFM (Mallmann & O'Neill, 2013; Nicklas et al., 2019). All this points to a need for a new consensus on an accurate and widely applicable calibration of the relationship between fO_2 and liquid Fe^{2+}/Fe^T (Berry et al., 2018; Borisov et al., 2018; Danyushevsky & Sobolev, 1996; Kilinc et al., 1983; Kress & Carmichael, 1991; O'Neill et al., 2018; Putirka, 2016; Sack et al., 1981). The reader can find an effort to reconcile the various studies of this relationship elsewhere in this book (Cottrell et al., this volume).

An oxidation state close to the QFM buffer in MORB glasses was quickly recognized as apparently discrepant with the oxidation state of abyssal peridotites, thought to be the mantle residues of MORB extraction. These peridotites have mineral assemblages that yield, according to well-calibrated spinel-olivine-orthopyroxene exchange equilibria, more reduced conditions by 1 to 2 log units (Sack et al., 1981). However, the recent survey of Birner et al. (2018), while recognizing considerable heterogeneity in the oxidation state recorded by peridotites, argues that MORB basalts are approximately in redox equilibrium with the average of associated peridotites.

Given the thorny analytical issues, it is not the intent of this chapter to try to settle the actual value of the Fe^{2+}/Fe^T or fO_2 of typical MORB glass. Instead, let us examine the effect of the full plausible range of Fe^{2+}/Fe^T on models of MORB primary melt composition, potential temperature, phenocryst-glass equilibrium, crustal thickness, the origin of cumulate lower oceanic crust, and indeed the entire paradigm for formation of the oceanic crust and its implications for the thermal and compositional state of the upper mantle.

7.2. MODELING METHODS AND SAMPLE SELECTION

7.2.1. Simple Model

The fundamental parameter used to analyze olivine-liquid equilibrium is the distribution coefficient $K_D{}^{Ol/L}$, defined as follows:

$$K_D^{Ol/L} \equiv \frac{D_{FeO}^{Ol/L}}{D_{MgO}^{Ol/L}} = \frac{\left(X_{FeO}^{Ol}/X_{FeO}^{L}\right)}{\left(X_{MgO}^{Ol}/X_{MgO}^{L}\right)} = \frac{\left(X_{FeO}^{Ol}/X_{MgO}^{Ol}\right)}{\left(X_{FeO}^{L}/X_{MgO}^{L}\right)} = \frac{Mg\#^L/\left(1-Mg\#^L\right)}{Fo^{Ol}/\left(1-Fo^{Ol}\right)}, \quad (7.1)$$

where $D_{FeO}^{Ol/L}$ and $D_{MgO}^{Ol/L}$ are the molar partition coefficients between olivine and liquid for FeO and MgO, respectively; X^i_j refers to mole fraction in phase i of component j; $Mg\#^L \equiv X_{MgO}^{L}/\left(X_{MgO}^{L}+X_{FeO}^{L}\right)$; and $Fo^{Ol} \equiv X_{MgO}^{Ol}/\left(X_{MgO}^{Ol}+X_{FeO}^{Ol}\right)$. The first two equalities in Equation 7.1 express that a distribution coefficient is the ratio of two partition coefficients, whereas the last two equalities show the relationship between $K_D{}^{Ol/L}$ and the quantities usually used to express MgO/FeO ratios in silicate liquids and olivine crystals. Roeder and Emslie (1970) showed that $K_D{}^{Ol/L} = 0.30 \pm 0.01$ in the composition, pressure, and temperature range they studied. We will begin with a simple illustrative exercise in which we hold $K_D{}^{Ol/L}$ constant at this value; later we will apply more contemporary models that include variations of this quantity with temperature and melt composition.

The most salient point to make about the definition of $K_D{}^{Ol/L}$ is that it is *not* defined in terms of total Fe (FeO^T) in the liquid: $Fe_2O_3{}^L$ is not counted in any of the quantities in Equation 7.1. This is important. First, there is a thermodynamic logic to this definition. The justification for use of the empirical parameter $K_D{}^{Ol/L}$ arises from its relation to analysis of the chemical exchange reaction

$$Mg_2SiO_4^{Ol} + 2FeO^L \rightleftharpoons Fe_2SiO_4^{Ol} + 2MgO^L. \quad (7.2)$$

In this expression, because the fayalite component of olivine (Fe_2SiO_4) admits only the Fe^{2+} cation, it is only the FeO component of the liquid that participates. Application of the law of mass action to this equation and assumptions about the extent of deviation of the olivine and liquid solutions from ideality lead directly to the prediction that $K_D{}^{Ol/L}$ should display simple and systematic behavior, with small pressure and temperature dependences captured by the small volume and entropy change of reaction 7.2. Second, there is strong experimental justification for the definition and use of $K_D{}^{Ol/L}$ that includes only the FeO component of the liquid. This was classically established by Roeder and Emslie (1970), who studied the composition of olivine coexisting with basaltic liquid across 11 orders of magnitude in fO_2, which at constant temperature and constant FeO^T in the liquid caused variation in the coexisting olivine composition from $<Fo_{75}$ to $>Fo_{95}$.

Now, consider the alternative empirical quantity that we shall call $K'_D{}^{Ol/L}$,

$$K'^{Ol/L}_D \equiv \frac{D_{FeO^T}^{Ol/L}}{D_{MgO}^{Ol/L}} = \frac{\left(X_{FeO}^{Ol}/X_{FeO^T}^{L}\right)}{\left(X_{MgO}^{Ol}/X_{MgO}^{L}\right)} = \frac{\left(X_{FeO}^{Ol}/X_{MgO}^{Ol}\right)}{\left(X_{FeO^T}^{L}/X_{MgO}^{L}\right)} = \frac{Mg\#'^L/\left(1-Mg\#'^L\right)}{Fo^{Ol}/\left(1-Fo^{Ol}\right)}, \quad (7.3)$$

where $Mg\#'^L \equiv X_{MgO}^{L}/\left(X_{MgO}^{L}+X_{FeO^T}^{L}\right)$. This quantity differs from the distribution coefficient defined by Roeder and Emslie (1970), $K_D{}^{Ol/L}$, in being based on total Fe as FeO in the liquid rather than strictly on FeO itself in the liquid. While convenient to calculate when working with electron probe or XRF data that give FeO^T in glass, $K'_D{}^{Ol/L}$ is *not* the true Fe/Mg distribution coefficient for olivine-liquid equilibria. Furthermore, a simple corollary of Roeder and Emslie's result is that the properly defined $K_D{}^{Ol/L}$ (using only FeO^L) *is* constant with respect to Fe^{2+}/Fe^T variations whereas $K'_D{}^{Ol/L}$ is *not* constant with respect to such variations. Any suggestion that $K'_D{}^{Ol/L}$ is effectively reliable or constant with respect to fO_2 must surely be a result of examination of data across a narrow range in Fe^{2+}/Fe^T, insufficient to resolve the dependence in the face of inevitable analytical uncertainty and other sources of error.

In turn, because $K_D{}^{Ol/L}$ is based only on the FeO content rather than the FeO^T of the liquid and models of olivine-liquid equilibria are (or should be) based on $K_D{}^{Ol/L}$, it is not surprising that changes in the estimation of FeO/FeO^T (i.e., Fe^{2+}/Fe^T) will have an impact on models of the compositions of coexisting liquid and olivine and therefore on models of primary magma composition that are linked to either olivine composition or MgO–FeO systematics. We illustrate this first by constructing the simplest possible model of this effect in order to emphasize that what follows is not an artifact of the more complex models to be applied later. Rather we will see that the PRIMELT3 and MELTS models yield a quantitative estimate of an effect that must be qualitatively correct in any model based on a correctly defined $K_D{}^{Ol/L}$.

Figure 7.2 is a plot of liquid FeO vs. liquid MgO in weight percent. Contours of equal liquid FeO/MgO (and $Mg\#^L$) are shown radiating from the origin. For simplicity, at this stage (and only at this stage) we take $K_D{}^{Ol/L} = 0.3$ and treat it as constant. Hence the contours of constant liquid FeO/MgO are also contours of constant

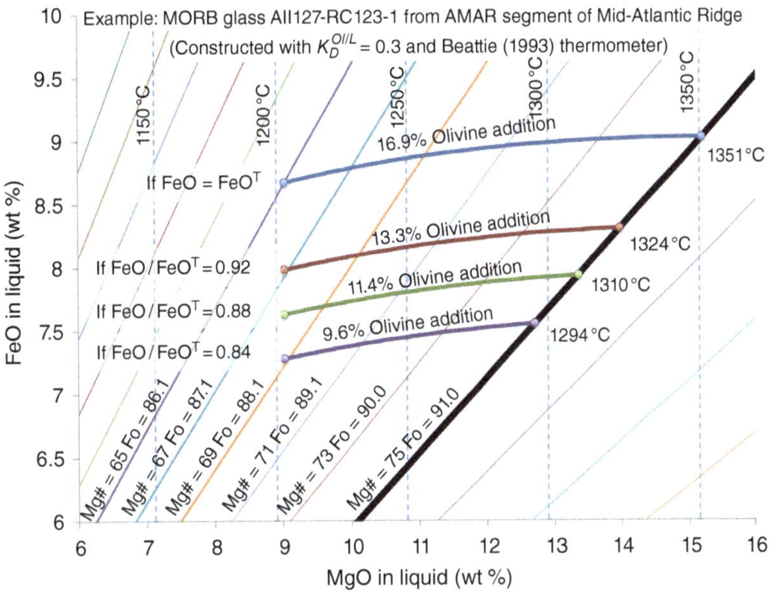

Figure 7.2 A simple model of olivine-liquid equilibria and FeO-MgO systematics to demonstrate the impact of estimates of Fe^{2+}/Fe^T on estimation of primary melt composition and temperature. The figure is constructed with a constant $K_D^{Ol/L}$ of 0.3, the Beattie (1993) thermometer (which is strictly a function of MgO in olivine-saturated liquids), and using the "olivine-targeted" method of correcting melts back to equilibrium with Mg# 91.0 olivine.

olivine FeO/MgO, and hence constant Fo content in coexisting equilibrium olivine. A primitive MORB glass from the Mid-Atlantic Ridge, AII127-RC123-1 (AMAR segment at 36.394°N, 33.653°W, 2560 m water depth), with 9.03 wt.% MgO and 8.67% FeO^T by electron probe (Gale et al., 2013), is plotted at four possible locations in FeO-MgO space. All have 9.03% MgO, but locations are shown for Fe^{2+}/Fe^T = 1.0, 0.92, 0.88, and 0.84. To be clear, although FeO^T is 8.67%, FeO in the glass might be as low as 7.28% for this sample (whose Fe^{2+}/Fe^T has not been measured and so would normally be estimated from the global average MORB value). As the oxidation state increases, the implied Mg# of the glass varies from 65 to 69 and the Fo content of the estimated coexisting olivine composition varies from 86.1 to 88.1.

Although the AMAR glass sample plotted in Figure 7.2 is relatively primitive among the population of MORB glasses, it is not a primary melt; it has experienced olivine fractionation. As long as a glass has experienced only olivine fractionation, it is considered straightforward to attempt to reverse the process by incremental addition of the liquidus olivine composition, which produces an array of parental liquids along an "olivine control line." The only quantity needed to make this incremental olivine addition calculation is $K_D^{Ol/L}$. The additional challenge of this type of primary magma estimation, though, is choosing where along the olivine control line to stop, i.e. to decide which of the family of inferred parental liquids is actually the primary liquid. For this simple exercise we use the widely understood (but probably flawed; see discussion of the PRIMELT3 model), "olivine-targeted" method of inferring primary magma composition by selecting a fixed Fo content of the first olivine to crystallize at the liquidus, chosen to be similar to the observed olivine compositions in residual mantle peridotites, $Fo_{91.0}$. For $K_D^{Ol/L}$ = 0.3, this implies the primary $Mg\#^L$ = 75.0. For each selected value of Fe^{2+}/Fe^T, the olivine control lines representing incremental liquidus olivine addition (with $K_D^{Ol/L}$ = 0.3) are shown, labeled by the mass fraction of olivine added and ending at specific points along the $Mg\#^L$ = 75 contour. The resulting estimated primary liquid ranges from 15.2 to 12.7% MgO as the oxidation state increases and the masses of olivine added to move up the liquidus from the observed composition to the estimated primary state range from 16.9% down to 9.6%.

Finally, using the Beattie (1993) olivine-liquid thermometer (which is strictly a function of liquid MgO content for olivine-saturated liquids), the primary liquids can be associated with an inferred liquidus temperature. Results for the full Fe^{2+}/Fe^T range considered vary from 1351 down to 1294°C. Within the plausible range for MORB, from Fe^{2+}/Fe^T = 0.93 to 0.84, the range of estimates in primary liquid MgO is 2.2 wt.%, the range in liquidus temperatures is 30°C, and the range in mass fraction of olivine added to recover Mg# 91.0 olivine is 4%.

This simplified exercise is presented here with two goals. First, the entire argument can be summarized on one plot and explained completely in a few paragraphs, which will be helpful before proceeding to apply more complex models to refine its conclusions. Second, it shows that in essence the effects highlighted by this chapter are indisputably real; any deviations from the simplifying assumptions of constant $K_D^{Ol/L}$ and constant Fo content in the first liquidus olivine are either small or likely to enhance the MgO differences in the primary liquids, and the Beattie thermometer is unlikely to be in error in this well-calibrated range by more than a few degrees.

7.2.2. PRIMELT3

PRIMELT3 (Herzberg & Asimow, 2015) is the latest refinement of a model originally developed by Herzberg and O'Hara (2002) to address the primary magma problem for peridotite-derived mafic magmas sufficiently primitive to have fractionated only olivine. The detailed assumptions, parameterizations, and limitations of the method have been well-described in the literature (Herzberg et al., 2007; Herzberg & Asimow, 2008, 2015; Herzberg & O'Hara, 2002) and will only be summarized here. The method combines a conventional incremental addition of liquidus olivine ("the inverse calculation") with a novel method of assessing which among the family of parental liquids along the olivine control line could be either a batch melt or an accumulated fractional melt (AFM) of a peridotite source ("the forward calculation"). The incremental olivine addition is in essence quite similar to the simple case presented above, but it is based on the Toplis (2005) model for $K_D^{Ol/L}$, which captures known variations in this quantity as a function of temperature and melt composition; in order to apply the Toplis parameterization and also to enable the forward calculation, all oxides (not just FeO and MgO) are tracked during olivine addition, using appropriate partition coefficients for those elements not perfectly incompatible in olivine (notably Ni and Ca).

The forward model is based on two parameterizations of the extent of batch or accumulated fractional melting of peridotite. Both FeO–MgO space and projections of the CIPW-normative mineralogy of the melt from diopside into the anorthite-olivine-quartz ternary form two-dimensional plots that can be contoured for extent of batch melting of peridotite. Calculation of the contours in FeO-MgO space requires only a parameterization of $K_D^{Ol/L}$ and an assumed MgO and FeO content of the source peridotite (by default, the KR-4003 composition studied by Walter [1998], but this can be changed). Parameterization of the contours in normative mineral space is more complex and requires consideration of separate surfaces for garnet peridotite, spinel peridotite, and harzburgite; the calibration in this case is explicitly tuned to fertile peridotite KR-4003. However, the assumption that liquids derived from various other sources will project on the KR-4003 multiple saturation surface at the correct extent of melting can be tested whenever well-constrained experiments are available. The model includes a further parameterization for the relation between the extent of melting contours of batch and accumulated fractional melting.

The key to understanding how PRIMELT3 locates a unique primary liquid is to observe that, with increasing MgO content, the candidate parental melts along the olivine control line traverse the extent of melting contours in MgO-FeO and in normative mineral space at different rates. A unique solution arises where these two estimates of the extent of melting cross and are equal. This is the only melt within the family of melts parental to an observed rock that is consistent with being a peridotite-derived primary melt. Once this composition is located, its MgO content, liquidus temperature, and liquidus olivine composition can all be obtained directly, and the model has a further parameterization of the mantle potential temperature that yields a primary liquid with a given liquidus temperature. Critically, for current purposes, the model allows a more refined version of the simple calculation given above, both because it allows for variations in $K_D^{Ol/L}$ using the Toplis (2005) model and because it does not assume an arbitrary primary olivine composition but instead outputs a value that is found to be consistent with the primary melt being derived from a certain amount of accumulated fractional melting of fertile peridotite. Higher extents of calculated melting correspond to higher Fo content in the residual and primary liquidus olivine estimates.

PRIMELT3 is realized in the form of a Microsoft Excel spreadsheet (Herzberg & Asimow, 2015) and is equipped for rapid calculation of many input glasses or whole rocks. It is armed with filters that attempt to detect liquids where the model will give incorrect results because they are likely to have fractionated augite, to be derived from pyroxenite sources, or to be derived from volatile-rich sources. Importantly for current purposes, the assumed (or measured) Fe^{2+}/Fe^T of the input rock is a free parameter that the user can set. Hence, we can use the model to explore the consequences of Fe^{2+}/Fe^T estimates for primary magma estimation, much as we did in the simplified model.

7.2.3. pMELTS and rhyoliteMELTS

Although PRIMELT3 provides a means to estimate the primary liquid composition, liquidus temperature, and mantle potential temperature, it is not a complete model of the mid-ocean ridge magmatic process. It does not calculate, for example, the thickness of ocean crust that results from the two-dimensional shape of the melting regime in across-axis cross-section, generated by corner flow beneath spreading plates. It also does not calculate the consequences of fractionation beyond the appearance of plagioclase or augite in the assemblage. For these purposes, we turn to the pMELTS (Ghiorso et al., 2002) and rhyoliteMELTS (Gualda et al., 2010) models. These are updated calibrations based on the original MELTS model (Ghiorso & Sack, 1995). Again, the thermodynamic basis and development of these models and their strengths and shortcomings for peridotite melting at upper mantle pressure are well-described in the literature (Behn & Grove, 2015; Hirschmann et al., 1998, 1999a, 1999b) and we will only briefly summarize the approach here. The use of MELTS to construct a forward model of passively plate-driven, near-fractional melting of peridotite and the aggregation of those liquids to predict both the quantity and composition of primary melt delivered to the crust as functions of source composition and potential temperature was laid out in Asimow et al. (2001). Asimow and Longhi (2004) showed that the updated pMELTS model correctly predicts the compositional range of MORB glasses, and Asimow et al. (2004) explained the method for the inclusion of trace H_2O in the mantle source.

There are several published models that calculate the complete output of passive-flow melting regimes as a function of potential temperature (Asimow et al., 2001; Behn & Grove, 2015; Kinzler, 1997; Langmuir et al., 1992; McKenzie & Bickle, 1988). Although there are offsets between the crustal thicknesses predicted by these models, they all agree (because they have similar entropy of fusion estimates) that – in the range of normal mid-ocean ridge potential temperatures – a 25 °C increase in potential temperature yields an increase in crustal thickness of about 1 km. Given the total range of 5 to 8 km in crustal thickness estimates for non-ultraslow and non-hotspot-affected ridges (Behn & Grove, 2015; White et al., 2001), this is consistent with the bulk of normal ridges sampling a potential temperature range of about 50–75 °C. There are still challenges in making this consistent with inferences of a 100–150 °C global range in the potential temperature sampled by normal ridges (Dalton et al., 2014), but that argument is about the extremes of the data and the models. Here we are interested in the mean. For typical spreading rates, crustal thicknesses, and potential temperatures, the estimate of a potential temperature vs. crustal thickness sensitivity of 25 °C/km appears to be a reasonable consensus.

The oceanic lower crust is thought to represent the accumulated crystals fractionated during cooling of primary aggregate mantle-derived melt, leaving a residual liquid to erupt and form the volcanic sequence. To examine the consequences of a change in the estimate of the composition of the primary melt for the fractionation and crustal construction processes, we turn to the rhyoliteMELTS model, which is more accurate than pMELTS at crustal pressures. We note that the name rhyoliteMELTS is slightly misleading for this purpose; although Gualda et al. (2010) made adjustments to the standard state properties of quartz and sanidine in order to improve the fit of MELTS in felsic systems, this change has no effect in mafic systems, so the results here are the same as would be generated with the original MELTS calibration. For an assumed primary liquid composition (including H_2O content), fractionation pressure, and either fO_2 or Fe^{2+}/Fe^T, the model predicts the liquid line of descent including the masses and compositions of all fractionating solids and the relation between mass of liquid remaining, temperature, and liquid composition.

It is important to note that the Kress and Carmichael (1991) parameterization of Fe^{2+}/Fe^T of silicate liquids as a function of fO_2 was explicitly incorporated in the calibration of all the MELTS models, since their liquid mixing model is based on experiments at known fO_2 and FeO^T without independent measurements of Fe^{2+}/Fe^T (Ghiorso & Sack, 1995). Most often, MELTS models are then run with model fO_2 also defined by Kress and Carmichael (1991), although the alphaMELTS interface (Smith & Asimow, 2005) offers the option to instead use the self-consistent internal thermodynamic definition of fO_2. In either case, though, Davis and Cottrell (2018) have shown that the solid and liquid models in MELTS are inconsistent in their handling of oxygen fugacity. So, MELTS does not provide an independent test of the correct relationship between Fe^{2+}/Fe^T and fO_2 and it is not used here for this purpose. It is used here only to explore the relationship between potential temperature, Fe^{2+}/Fe^T, crustal thickness, and mass of cumulates. It is not obvious how the assumptions and inconsistencies in the handling of fO_2 in MELTS might undermine these exercises, but it is an uncertainty to keep in mind and a reason to consider attempting similar calculations with independent models in the future.

7.2.4. Choice of Reference Samples

We can apply the modeling strategies described above to typical ridges and global averages. However, for the consideration of direct observations of olivine-liquid equilibria and the consequences of redox state on such equilibria, we need to select particular well-characterized suites. For this purpose, if we wish to use basalts to probe the mantle, it is best to work with samples that have escaped magma chamber processing as much as possible. Of all the MORB localities so-far sampled, those that come closest to meeting this requirement are samples from the Siqueiros Transform on the East Pacific Rise (Perfit et al., 1996). Samples recovered from dredging there are distinguished from other MORB locations in containing plagioclase-free and clinopyroxene-free high-MgO glasses with variable amounts of olivine (Hays, 2004; Perfit et al., 1996). Unlike most MORB, the primary magma problem is tractable for these Siqueiros glasses because they differ from their primary magmas by only a small amount of olivine fractionation (Herzberg & Asimow, 2015; Perfit et al., 1996). This great petrological fortune was most likely the outcome of rapid melt transport in transform-related fractures that extend into the mantle (Hebert & Montési, 2011). We will show below a profile of high-precision microprobe analyses across an olivine phenocryst from a well-characterized Siqueiros transform glass; these are unpublished data from the Rutgers University microprobe lab courtesy of C. Herzberg. The high-current olivine microprobe protocol achieves precision and accuracy relative to the San Carlos olivine standard of ±0.01 units in forsterite content (Sobolev et al., 2007). Because this suite of samples is also nearly unique in being vapor-undersaturated at their eruption depth, they have attracted considerable study of their volatile concentrations through detailed analysis of their olivine-hosted melt inclusions (Saal et al., 2002); we use the published melt inclusion data set from Saal et al. (2002), uncorrected for post-entrapment crystallization.

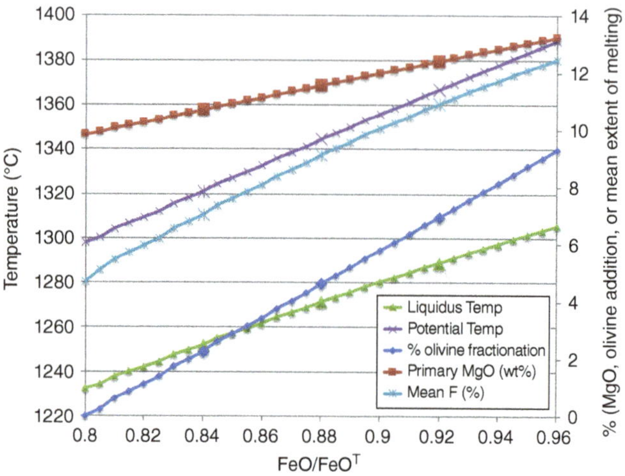

Figure 7.3 PRIMELT3 outputs from calculation of primary magma parental to the average analysis of Siqueiros glass D20-15, as a function of the assumed Fe^{2+}/Fe^T. Liquidus temperature and potential temperature are plotted against the left-hand axis, whereas % olivine fractionation, wt.% MgO in the primary magma, and extent of melting (Mean *F*) are all given on the right-hand axis. The large symbols for each quantity mark the particular output values at Fe^{2+}/Fe^T = 0.84, 0.88, and 0.92.

7.3. RESULTS

7.3.1. PRIMELT3, Potential Temperature, and Host-Phenocryst-Inclusion Equilibrium

We illustrate a refinement of the conceptual calculation of primary melt shown in Figure 7.2 by applying PRIMELT3 to a primitive glass from the Siqueiros transform, D20-15 (Perfit et al., 1996). Figure 7.3 shows key outputs of the PRIMELT3 AFM solution for this glass as a function of the assumed Fe^{2+}/Fe^T in the glass. Consistent with the simple model shown above, as the oxidation state increases and ferrous Fe decreases from Fe^{2+}/Fe^T = 0.92 to Fe^{2+}/Fe^T = 0.84, the percent olivine added to the measured glass to reach the primary melt solution decreases from 7% to 2%; the MgO content of the primary melt decreases from 12.40 to 10.68 wt.%; the inferred mean extent of melting of the mantle source (FAFM) decreases from 10.9% to 7.0%; the liquidus temperature decreases from 1289 to 1252°C (a decrease of 37 °C); and the potential temperature inferred for the underlying melting region decreases from 1367 to 1321 °C (a decrease of 46 °C). We will use this estimated range of potential temperature below to consider the predicted consequence for oceanic crustal thickness using the pMELTS model. First, though, note that over this same range of estimated Fe^{2+}/Fe^T, the composition of the first olivine estimated to crystallize at low pressure from the well-mixed aggregate primary melt decreases from $Fo_{90.54}$ to $Fo_{90.17}$ (Figure 7.4).

Helz et al. (2017) provides an example of using olivine-liquid equilibria (in the well-characterized Kilauea Iki lava lake sequence) and Fe^{2+}/Fe^T measured by XANES to examine the implied values of $K_D^{Ol/L}$. The result, 0.28 (or 0.29 if updated to the Zhang et al. (2018) calibration), is consistent with the temperature and composition dependence of the Toplis (2005) partitioning model. Hence, here we invert this type of calculation and use olivine-liquid pairs and the Toplis (2005) $K_D^{Ol/L}$ function to obtain a bound on allowable Fe^{2+}/Fe^T. Figure 7.4 shows the PRIMELT3 estimates of the most magnesian olivine expected to crystallize at the low-pressure liquidus of the well-mixed primary aggregate fractional melt that eventually fractionated to create the observed Siqueiros D20-15 glass. This is compared directly to the Fo contents measured with a high-precision electron probe profile across an olivine from the same sample, DR20-15-B2-#5 (unpublished data courtesy of C. Herzberg). There is minor reverse zoning at the core of this grain, a thick mantle of constant composition $Fo_{90.4}$, and a narrow normally zoned rim. Apart from the small core area, this grain appears to be a normal phenocryst, grown along the liquid line of descent leading from the primary melt to the erupted glass. For estimates of $Fe^{2+}/Fe^T \geq 0.885$, Figure 7.4 shows that the system is expected to have begun crystallizing olivine at least as magnesian as $Fo_{90.4}$, which

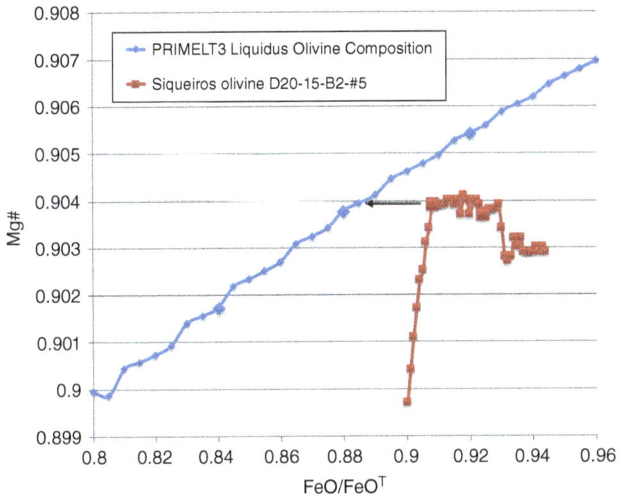

Figure 7.4 PRIMELT3 prediction for the Mg# of the liquidus olivine that first crystallizes at low pressure from the calculated aggregate primary magma parental to the average analysis of Siqueiros glass D20-15, as a function of the assumed Fe^{2+}/Fe^T, compared to measured Mg# profile across olivine grain D20-15-B2-#5 from the same sample (x-axis position arbitrary). The mantle region of this olivine grain, with Mg# = 0.904, is only consistent with being a phenocryst grown during the differentiation path leading to the measured glass for $Fe^{2+}/Fe^T \leq 0.885$. See text for an assessment of the uncertainties on the model and data values shown.

makes this olivine a plausible phenocryst. That is, its Fo content is less than or equal to the maximum Fo of the phenocrysts that would grow during fractionation of the estimated primary melt. On the other hand, for estimates of $Fe^{2+}/Fe^T < 0.885$, we would be forced to seek an alternative origin for this olivine since its Fo content is greater than the maximum Fo content we should expect for a phenocryst. In general, as we raise the estimate of oxidation state of the glass, we shift more of its olivine cargo out of the acceptable phenocryst category.

The analysis shown in Figure 7.4 depends on systematic uncertainty in the measurements of the host glass and olivine phenocryst composition as well as in the parameterization of PRIMELT3. Some of these uncertainties can be assessed. The host glass and olivine phenocryst were measured on the same electron microprobe using the same Fe and Mg standards, but different analytical protocols that might introduce some shift in their relative Fe/Mg ratios; this is hard to quantify. However, if the analyzed forsterite content of the olivine core is shifted up or down by 0.001 units (10 times the reported uncertainty according to Sobolev et al., 2007), then the threshold value of Fe^{2+}/Fe^T shifts up or down by ~0.02. Likewise, if the $K_D^{Ol/L}$ model of Toplis (2005) gives values too large by 0.01, then the inferred primary Fo content of olivine at a given Fe^{2+}/Fe^T shifts upwards by ~0.003 and becomes higher than that measured in the core of the phenocryst at all considered values of Fe^{2+}/Fe^T. Conversely, if the $K_D^{Ol/L}$ model values are too small by 0.01, then the inferred primary Fo content of olivine shifts downwards by ~0.003 and becomes smaller than the core of the measured phenocryst at all considered values of Fe^{2+}/Fe^T. Apparently, the quality of the $K_D^{Ol/L}$ model is the main limiting factor on whether an exercise like this can constrain the Fe^{2+}/Fe^T of the glass.

Figure 7.5 shows a similar exercise based on a large dataset of 107 olivine-hosted melt inclusion analyses from Siqueiros transform MORB samples published by Saal et al. (2002). Using the raw inclusion analyses (not corrected for post-entrapment crystallization), for each inclusion we have plotted the calculated primary liquidus olivine compositions obtained when we assume Fe^{2+}/Fe^T = 0.84, 0.88, and 0.92 against the composition of the actual host olivine crystal containing the inclusion. In this case, petrological expectation is that no inclusion should ever be found in an olivine more magnesian than the first olivine that crystallized along the liquid line of descent leading to the melt inclusion composition, and

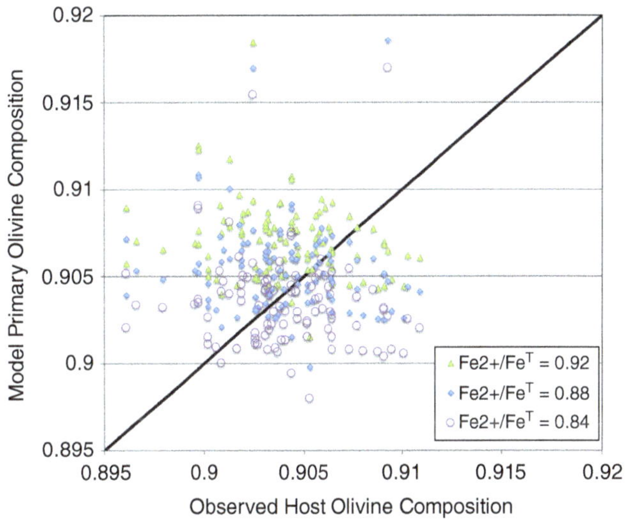

Figure 7.5 PRIMELT3 predictions for the Mg# of the liquidus olivine that first crystallizes at low pressure from the calculated aggregate primary magma parental to each of a family of olivine-hosted melt inclusions from Siqueiros MORB samples (Saal et al., 2002), plotted against the composition of the host olivine of each inclusion. The PRIMELT3 calculation is performed at three different values of Fe^{2+}/Fe^T. Points plotting above the 1:1 line are consistent with the host olivine having grown during fractional crystallization along the liquid line of descent from the model primary magma to the observed inclusion composition. Points below the 1:1 line are not, because the host olivine is too magnesian to be a phenocryst formed from the primary magma or its derivative liquids. See text for an assessment of the uncertainties in this calculation.

this is true whether the crystallization happens before or after entrapment. In Figure 7.5, this condition is expressed by being above the 1:1 line in a plot with most magnesian model olivine composition on the y-axis and observed host olivine composition on the x-axis. For Fe^{2+}/Fe^T = 0.92, 84 out of 107 inclusions plot above the 1:1 line, in the sector where their host olivines could plausibly have crystallized along a path from a well-mixed primary liquid to the observed inclusion composition. For Fe^{2+}/Fe^T = 0.88, 68 out of 107 inclusions plot above the 1:1 line. For Fe^{2+}/Fe^T = 0.84, only 45 out of 107 inclusions plot above the 1:1 line and many are below it by >0.8 units, larger than the combined margin of error in (conventional) olivine analysis, glass analysis, and liquidus olivine reconstruction. One way to express the uncertainty of this analysis is to compare the effect of possible errors in $K_D^{Ol/L}$ to the effect of variations in assumed Fe^{2+}/Fe^T; this works out quite neatly, as increasing all values of $K_D^{Ol/L}$ to 0.01 above the Toplis (2005) model value has the same effect as decreasing Fe^{2+}/Fe^T by 0.04 and decreasing all values of $K_D^{Ol/L}$ to 0.01 below the Toplis (2005) model value has the same effect as increasing Fe^{2+}/Fe^T by 0.04.

In summary, for the particular case of the primitive Siqueiros Transform basalts, the nominal measured compositions of coexisting MORB glass, olivine crystals, and melt inclusions are difficult to reconcile using a popular $K_D^{Ol/L}$ model if Fe^{2+}/Fe^T is significantly smaller than 0.88. Of course, the data, one of our assumptions, or the model may be substantially in error. For example, electron probe analysis may be affected by interlaboratory biases, or – even more pernicious – biases between liquid Mg# and olivine Fo measurements in the same laboratory. Such errors might have affected the calibration of $K_D^{Ol/L}$ models or the analyses of the Siqueiros samples or other natural suites we might observe. Or, the assumption that all increments of melt are perfectly mixed to form a single, well-homogenized primary liquid before it begins to fractionate olivine may be wrong in some or all cases. Later increments of fractional melting will have higher Mg# than the mixture and could have begun fractionating high-Fo olivine that becomes incorporated in liquids produced by later mixing. This might explain an anomaly like that noted in Figure 7.4, but it is difficult to invoke this explanation for the melt inclusion argument in Figure 7.5, since olivines grown from such unmixed increments of high-Mg# fractional melt should not contain normal, well-mixed MORB-like inclusion glasses.

7.3.2. pMELTS, Potential Temperature, and Typical Oceanic Crustal Thickness

One of the principal observables that make mid-ocean ridges amenable to comprehensive modeling of their magmatic systems is the total thickness of igneous crust generated by steady-state spreading. This quantity, based on seismically measured velocity profiles and travel time to the Moho, is converted to depth and subsequently (with an assumption about crustal density) to mass per unit ridge length per increment of spreading. Crustal thickness is thought to describe the total quantity of melt generated and segregated from the mantle at the ridge axis. Classical compilations of seismic observations and of inversions for total melt mass from trace-element patterns led to the widely adopted estimate that normal mid-ocean ridges with full spreading rates faster than 20 mm/yr generate 7±1 km of crust, independent of spreading rate (White et al., 2001). However, this consensus value has recently been undergoing significant downward revision. For example, in a focused study of the segment of the East Pacific Rise between the Siqueiros and Clipperton Fracture Zones thought to have crustal thickness of 6.7–6.8 km based on two-dimensional seismic survey (Canales et al., 2003), analysis of a more recent three-dimensional multi-channel seismic survey (Aghaei et al., 2014) derived an average crustal thickness of 5740 m. Expanded to global surveys, recent results have yielded similar revisions, with global mean crustal thickness of 5.8±1 km (Behn & Grove, 2015; Gale et al., 2013; Le Voyer et al., 2019; Van Avendonk et al., 2016). This revision is critically important because, as we will see in this analysis, the difference between ~7 km and ~6 km of typical MORB crust is in fact roughly equivalent to the difference in model predictions that results from a change of ~0.04 in mean Fe^{2+}/Fe^T.

Here, we take the potential temperatures for various Fe^{2+}/Fe^T values from the PRIMELT3 solutions for the well-defined (though perhaps not entirely typical) Siqueiros suite and use them as inputs to the pMELTS model of melt production from a passive-flow mid-ocean ridge melting regime with the Workman and Hart (2005) source composition, assuming 150 ppm H_2O in the source. As shown in Figure 7.6, the model predicts crustal thickness of 7.2 km at potential temperature (T_P) = 1367 °C and Fe^{2+}/Fe^T = 0.92, 6.1 km at T_P = 1344 °C and Fe^{2+}/Fe^T = 0.88, and 5.1 km at T_P = 1321 °C and Fe^{2+}/Fe^T = 0.84. As expected, since the impact of a change in Fe^{2+}/Fe^T of 0.04 leads to a change in the PRIMELT3 estimate of potential temperature of ~25 °C, it leads in turn to a change of about 1 km in predicted crustal thickness. In short, if the pMELTS model is correct, then as the assumed Fe^{2+}/Fe^T for normal MORB decreases below 0.88, predicted crustal thicknesses begin to decrease below the values consistent with seismic observations. Curiously, the changes in potential temperature and in oxidation state have opposing effects on the total Fe and Na contents of the fractionated melts, such that Fe_8 and Na_8 (Klein & Langmuir, 1987) are insensitive to this mode of variation of pressure and extent of melting (Figure 7.6).

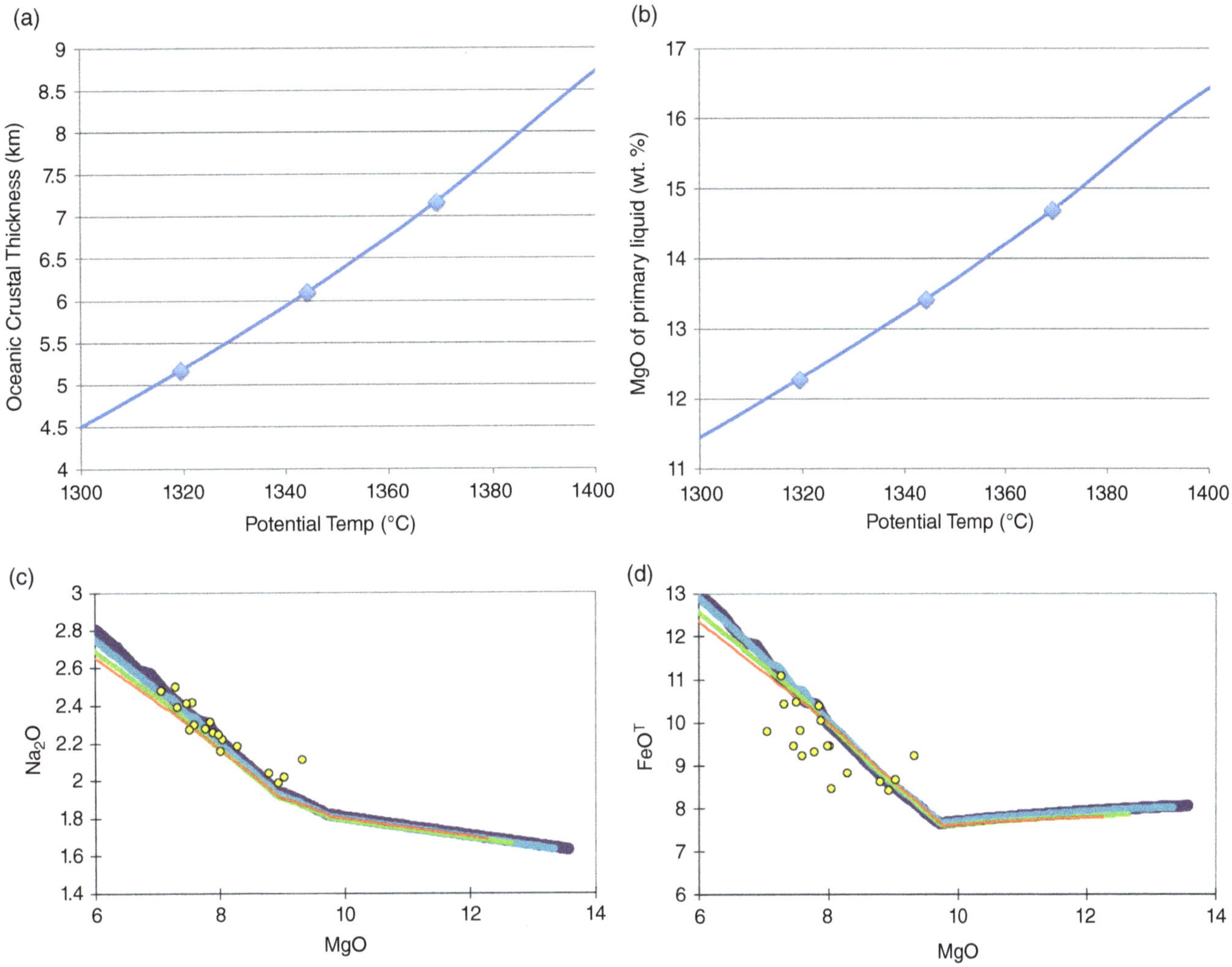

Figure 7.6 Predictions of the pMELTS model of oceanic crustal production for various values of the assumed Fe^{2+}/Fe^T at the potential temperatures indicated by the PRIMELT3 calculations in Figure 7.3. (a) Predicted crustal thickness, using the method of Asimow et al. (2001). As expected from previous calculations of this type, each increment of 0.04 units in Fe^{2+}/Fe^T and 25 °C in potential temperature corresponds to a change in predicted oceanic crustal thickness of ~1 km. (b) Predicted MgO content of the primary liquid for the same cases shown in (a), confirming that the general values and slope of the relationship between MgO of primary liquid and potential temperature is similar to that parameterized in PRIMELT3. (c) and (d) Fractionation of the primary melts from a series of four calculations assuming various values of MORB glass Fe^{2+}/Fe^T: 0.83 (red), 0.87 (green), 0.92 (blue), and 0.94 (purple). Each is tuned in potential temperature to fractionate along the trend of the glass data from the AMAR segment of the Mid-Atlantic Ridge (Gale et al., 2014) when fractionated at the appropriate fO_2. The four models range in potential temperature from 1325 °C to 1362 °C, in crustal thickness from 5.1 km to 6.5 km, and in primary liquid MgO from 12.2 to 13.6 wt.%. Yet, because of cancelling effects of potential temperature and fO_2 on melt chemistry, their Na_8 and Fe_8 (Klein & Langmuir, 1987) values are identical within error.

Again, before jumping to the conclusion that this exercise proves anything about the actual oxidation state of MORB glass, we should consider all other hypotheses that might explain the discrepancy. Some objections do not help, and in fact make the situation worse. For example, stranded melt in the mantle decreases the predicted crustal thickness at a given potential temperature. Fractionation occurring in the mantle lithosphere before the primary melt reaches the crust also decreases the predicted crustal thickness. A more fertile source composition is a possibility, but this has numerous consequences for trace and radiogenic isotope geochemistry that need to be explored; such a solution should not be freely invoked without further testing. The pMELTS model may be systematically in error; this is certainly true to some extent, but it remains true that its sensitivity to crustal thickness as a function of potential temperature is similar to other models such as Langmuir et al. (1992) or Behn and Grove (2015) and these models were developed from piston cylinder experiments in graphite capsules and lack a mechanism for evaluating how their predictions would differ in an oxidizing case.

7.3.3. rhyoliteMELTS and Implications for Origin of Oceanic Lower Crust

A robust constraint from marine seismology and from ophiolite observations is the layered structure of oceanic crust, which is thought to include an extrusive layer of pillow basalts and sheeted dikes (Layer 2) and an intrusive layer of broadly gabbroic cumulates (Layer 3). In general, studies that reported normal oceanic crust about 7 km thick observed that Layer 2 is typically 2–3 km thick and Layer 3 is typically 4–5 km thick (Hooft et al., 2000; Jacobs et al., 2007; White, 1979). Recent compilations that suggest a lower total thickness for average oceanic crust have not included revisions to the relative thickness of the layers, so it is probably still our best estimate that, of the total quantity of melt that escapes the mantle at a normal ridge axis, about 60–70% of its mass crystallizes on the way to forming the erupted sequence. In a simple view where the lower plutonic crust is complementary to the upper volcanic crust in mass and composition, a prediction of the primary aggregate liquid composition and observations of the erupted composition can be tested to see if the fractionation process relating them generates enough cumulate mass to form the lower crust. Recall the difference in estimates of mass of olivine fractionated between the primary melt and the observed melt in both the simple exercise in Figure 7.2 and the PRIMELT3 calculation in Figure 7.3. However, these were both based on an atypically high-MgO glass, and not on the average volcanic composition.

In Figure 7.7, we illustrate fractionation paths computed with the well-calibrated rhyoliteMELTS model that begin from the pMELTS primary melt solutions and pass through the MgO content of the mean MORB composition from Gale et al. (2013), 7.58 wt.%. For the case with potential temperature 1367°C, Fe^{2+}/Fe^{T} = 0.92, and 7.2 km of total crust, fractionation removes 50% of the mass as cumulates along the path from primary to erupted compositions. On the other hand, beginning with the primary melt estimated for potential temperature of 1321°C, Fe^{2+}/Fe^{T} = 0.84, and 5.1 km total crust, fractionation to reach the mean MgO of average erupted MORB removes only 40% of the mass as cumulates. Neither case quite satisfies the expectation of 60–70% fractionation, but the more reducing cases come much closer to this expectation.

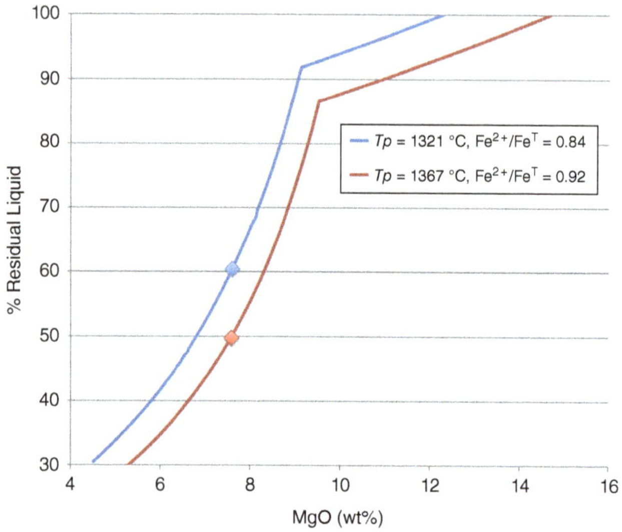

Figure 7.7 rhyoliteMELTS calculations showing the fraction of original liquid mass remaining during fractional crystallization of two representative primary magma compositions to yield residual liquids with the average MgO content of mid-ocean ridge basalts, 7.58 wt.% (Gale et al., 2013). The two cases shown are selected from Figure 7.6, the case with Fe^{2+}/Fe^{T} = 0.84 and potential temperature 1321 °C and the case with Fe^{2+}/Fe^{T} = 0.92 and potential temperature 1367 °C. Fractionation is calculated at constant pressure of 50 MPa in a system closed with respect to oxygen exchange.

As before, we should consider whether the solution to the discrepancy between mass fractionated and Layer 3 thickness for oxidizing cases rests in model errors or paradigmatic assumptions. The mean sampled MORB composition may not describe the mean composition of seismic layer 2, especially if the sheeted dike complex is systematically different from the erupted flows (Stewart et al., 2003). Moreover, if much of the lower crust consists of frozen sills rather than strictly of cumulates (Kelemen et al., 1997), then its mass and thickness would increase relative to what we expect from a pure cumulate calculation. This idea makes strong predictions for the seismic velocity structure of oceanic lower crust, since fully crystallized sills will be more felsic and seismically slower than mafic cumulates.

7.4. SUMMARY AND PROSPECTS

Values of Fe^{2+}/Fe^{T} lower than 0.88, as suggested by some recent calibrations of the μXANES analysis method, call for revision to one or more elements of our understanding of mid-ocean ridge processes. Unless some of the modeling tools we have applied are significantly in error, such oxidizing conditions predict cold, low-MgO primary aggregate magmas that have difficulty crystallizing the most magnesian olivine phenocryst compositions found in MORB, that imply potential temperatures too cold to generate the traditionally assumed typical thickness of oceanic crust, and that are so close to the erupted basalt composition that it becomes difficult to explain the origin of a thick, cumulate lower crust as the complementary product of fractional crystallization. This discrepancy can be interpreted in more than one way. On the one hand, it may suggest that further work on the absolute calibration of the μXANES method is called for. On the other hand, it lends support to the more recent consensus that is emerging around a smaller typical value for the thickness of oceanic crust. The main point of this chapter is not to second-guess the analytical results for Fe^{2+}/Fe^T but rather to make it clear that those results have implications beyond the specific question of oxidation state. In general, assessing whether the paradigm for generation of oceanic crust is internally consistent calls for careful evaluation of the accuracy of our data and our models and the assumptions underlying their application to oceanic crust formation. There are a number of possibilities, and at this time it is not clear which of them may be true.

ACKNOWLEDGMENTS

The high-precision olivine data in Figure 7.4 are shown courtesy of Claude Herzberg. The author wishes to thank Claude Herzberg, Elizabeth Cottrell, Charles Langmuir, Maryjo Brounce, and Esteban Gazel for discussions during the development of these ideas, but the responsibility for any shortcomings in the final product rests entirely on the author. Formal reviews from an anonymous reviewer and Elizabeth Cottrell were very helpful in developing the final version of this chapter.

REFERENCES

Aghaei, O., Nedimović, M. R., Carton, H., Carbotte, S. M., Canales, J. P., & Mutter, J. C. (2014). Crustal thickness and Moho character of the fast-spreading East Pacific Rise from 9°42′N to 9°57′N from poststack-migrated 3-D MCS data. *Geochemistry, Geophysics, Geosystems*, *15*(3), 634–657. https://doi.org/10.1002/2013GC005069

Asimow, P. D., & Longhi, J. (2004). The significance of multiple saturation points in the context of polybaric near-fractional melting. *Journal of Petrology*, *45*(12), 2349–2367. https://doi.org/10.1093/Petrology/Egh043

Asimow, P. D., Hirschmann, M. M., & Stolper, E. M. (2001). Calculation of peridotite partial melting from thermodynamic models of minerals and melts. IV. Adiabatic decompression and the composition and mean properties of mid-ocean ridge basalts. *Journal of Petrology*, *42*(5), 963–998.

Asimow, P. D., Dixon, J. E., & Langmuir, C. H. (2004). A hydrous melting and fractionation model for mid-ocean ridge basalts: Application to the Mid-Atlantic Ridge near the Azores. *Geochemistry, Geophysics, Geosystems*, *5*(1), Q01E16, doi:10.1029/2003GC000568.

Beattie, P. (1993). Olivine-melt and orthopyroxene-melt equilibria. *Contributions to Mineralogy and Petrology*, *115*(1), 103–111. https://doi.org/10.1007/BF00712982

Behn, M. D., & Grove, T. L. (2015). Melting systematics in mid-ocean ridge basalts: Application of a plagioclase-spinel melting model to global variations in major element chemistry and crustal thickness. *Journal of Geophysical Research: Solid Earth*, *120*(7), 4863–4886. https://doi.org/10.1002/2015JB011885

Bence, A. E., & Albee, A. L. (1968). Empirical Correction Factors for the Electron Microanalysis of Silicates and Oxides. *The Journal of Geology*, *76*(4), 382–403. https://doi.org/10.1086/627339

Berry, A. J., Stewart, G. A., O'Neill, H. S. C., Mallmann, G., & Mosselmans, J. F. W. (2018). A re-assessment of the oxidation state of iron in MORB glasses. *Earth and Planetary Science Letters*, *483*, 114–123.

Bézos, A., & Humler, E. (2005). The Fe 3+/ΣFe ratios of MORB glasses and their implications for mantle melting. *Geochimica et Cosmochimica Acta*, *69*(3), 711–725.

Birner, S. K., Cottrell, E., Warren, J. M., Kelley, K. A., & Davis, F. A. (2018). Peridotites and basalts reveal broad congruence between two independent records of mantle f_{O_2} despite local redox heterogeneity. *Earth and Planetary Science Letters*, *494*, 172–189. https://doi.org/10.1016/j.epsl.2018.04.035

Borisov, A., Behrens, H., & Holtz, F. (2018). Ferric/ferrous ratio in silicate melts: a new model for 1 atm data with special emphasis on the effects of melt composition. *Contributions to Mineralogy and Petrology*, *173*(12), 98. https://doi.org/10.1007/s00410-018-1524-8

Bowen, N. L. (1928). *The evolution of the igneous rocks*. Princeton, NJ: Princeton University Press.

Bryndzia, L. T., Wood, B. J., & Dick, H. J. (1989). The oxidation state of the Earth's sub-oceanic mantle from oxygen thermobarometry of abyssal spinel peridotites. *Nature*, *341*(6242), 526–527.

BVSP. (1981). *Basaltic Volcanism on the Terrestrial Planets*. New York: Pergamon.

Canales, J. P., Detrick, R. S., Toomey, D. R., & Wilcock, W. S. D. (2003). Segment-scale variations in the crustal structure of 150–300 kyr old fast spreading oceanic crust (East Pacific Rise, 8°15′N–10°5′N) from wide-angle seismic refraction profiles. *Geophysical Journal International*, *152*(3), 766–794. https://doi.org/10.1046/j.1365-246X.2003.01885.x

Carmichael, I. S. E., & Ghiorso, M. S. (1986). Oxidation-reduction relations in basic magma: a case for homogeneous equilibria. *Earth and Planetary Science Letters*, *78*(2), 200–210. https://doi.org/10.1016/0012-821X(86)90061-0

Christie, D. M., Carmichael, I. S., & Langmuir, C. H. (1986). Oxidation states of mid-ocean ridge basalt glasses. *Earth and Planetary Science Letters*, *79*(3), 397–411. https://doi.org/10.1016/0012-821X(86)90195-0

Cottrell, E., & Kelley, K. A. (2011). The oxidation state of Fe in MORB glasses and the oxygen fugacity of the upper mantle.

Earth and Planetary Science Letters, *305*(3), 270–282. https://doi.org/10.1016/j.epsl.2011.03.014

Cottrell, E., Kelley, K. A., Lanzirotti, A., & Fischer, R. A. (2009). High-precision determination of iron oxidation state in silicate glasses using XANES. *Chemical Geology*, *268*(3), 167–179. https://doi.org/10.1016/j.chemgeo.2009.08.008

Dalton, C. A., Langmuir, C. H., & Gale, A. (2014). Geophysical and Geochemical Evidence for Deep Temperature Variations Beneath Mid-Ocean Ridges. *Science*, *344*(6179), 80–83. https://doi.org/10.1126/science.1249466

Danyushevsky, L. V., & Sobolev, A. V. (1996). Ferric-ferrous ratio and oxygen fugacity calculations for primitive mantle-derived melts: calibration of an empirical technique. *Mineralogy and Petrology*, *57*(3–4), 229–241. https://doi.org/10.1007/BF01162360

Davis, F. A., & Cottrell, E. (2018). Experimental investigation of basalt and peridotite oxybarometers: Implications for spinel thermodynamic models and Fe3+ compatibility during generation of upper mantle melts. *American Mineralogist*, *103*(7), 1056–1067. https://doi.org/10.2138/am-2018-6280

Dixon, J. E., Stolper, E., & Delaney, J. R. (1988). Infrared spectroscopic measurements of CO2 and H2O in Juan de Fuca Ridge basaltic glasses. *Earth and Planetary Science Letters*, *90*, 87–104. https://doi.org/10.1016/0012-821X(88)90114-8

Engel, A. E., Engel, C. G., & Havens, R. G. (1965). Chemical characteristics of oceanic basalts and the upper mantle. *Geological Society of America Bulletin*, *76*, 719–734. https://doi.org/10.1130/0016-7606(1965)76[719:CCOOBA]2.0.CO;2

Gale, A., Dalton, C. A., Langmuir, C. H., Su, Y., & Schilling, J.-G. (2013). The mean composition of ocean ridge basalts. *Geochemistry, Geophysics, Geosystems*, *14*(3), 489–518. https://doi.org/10.1029/2012GC004334

Gale, A., Langmuir, C. H., & Dalton, C. A. (2014). The Global Systematics of Ocean Ridge Basalts and their Origin. *Journal of Petrology*, *55*(6), 1051–1082. https://doi.org/10.1093/petrology/egu017

Ghiorso, M. S., & Sack, R. O. (1995). Chemical mass transfer in magmatic processes IV. A revised and internally consistent thermodynamic model for the interpolation and extrapolation of liquid-solid equilibria in magmatic systems at elevated temperatures and pressures. *Contributions to Mineralogy and Petrology*, *119*, 197–212. https://doi.org/10.1007/BF00307281

Ghiorso, M. S., Hirschmann, M. M., Reiners, P. W., & Kress, V. C. (2002). The pMELTS: A revision of MELTS for improved calculation of phase relations and major element partitioning related to partial melting of the mantle to 3 GPa. *Geochemistry Geophysics Geosystems*, *3*, doi: 10.1029/2001GC000217.

Grove, T. L., Kinzler, R. J., & Bryan, W. B. (1992). Fractionation of mid-ocean ridge basalt (MORB). In: J. Phipps Morgan, D. K. Blackman, & J. M. Sinton (Eds.), *Mantle Flow and Melt Generation at Mid-Ocean Ridges* (Vol. *71*, pp. 281–310). American Geophysical Union

Gualda, G. A., Ghiorso, M. S., Lemons, R. V., & Carley, T. L. (2010). Rhyolite-MELTS: A modified calibration of MELTS optimized for silica-rich, fluid-breaing magmatic systems. *EOS Transactions AGU*, *91*(Fall meeting supplement), abstr. V43A-2352.

Hays, M. R. (2004). *Intra-transform volcanism along the Siqueiros fracture zone 8° 20'N - 8°30'N, East Pacific Rise* (Master of Science). University of Florida, Gainesville, FL.

Hebert, L. B., & Montési, L. G. J. (2011). Melt extraction pathways at segmented oceanic ridges: Application to the East Pacific Rise at the Siqueiros transform. *Geophysical Research Letters*, *38*, L11306. https://doi.org/10.1029/2011GL047206

Helz, R. T., Cottrell, E., Brounce, M. N., & Kelley, K. A. (2017). Olivine-melt relationships and syneruptive redox variations in the 1959 eruption of K⊠lauea Volcano as revealed by XANES. *Journal of Volcanology and Geothermal Research*, *333*, 1–14.

Herzberg, C., & Asimow, P. D. (2008). Petrology of some oceanic island basalts: PRIMELT2.XLS software for primary magma calculation. *Geochemistry Geophysics Geosystems*, *9*. https://doi.org/10.1029/2008gc002057

Herzberg, C., & Asimow, P. D. (2015). PRIMELT3 MEGA. XLSM software for primary magma calculation: Peridotite primary magma MgO contents from the liquidus to the solidus. *Geochemistry Geophysics Geosystems*, *16*(2), 563–578. https://doi.org/10.1002/2014gc005631

Herzberg, C., Asimow, P. D., Arndt, N., Niu, Y. L., Lesher, C. M., Fitton, J. G., et al. (2007). Temperatures in ambient mantle and plumes: Constraints from basalts, picrites, and komatiites. *Geochemistry Geophysics Geosystems*, *8*(2), Q02006. https://doi.org/10.1029/2006gc001390

Herzberg, C. T., & O'Hara, M. J. (2002). Plume-associated ultramafic magmas of Phanerozoic age. *Journal of Petrology*, *43*(10), 1857–1883.

Hess, P. C. (1992). Phase equilibria constraints on the origin of ocean floor basalt. In J. Phipps Morgan, D. K. Blackman, & J. M. Sinton (Eds.), *Mantle Flow and Melt Generation at Mid-Ocean Ridges* (Vol. *71*, pp. 67–102). American Geophysical Union.

Hirschmann, M. M., Ghiorso, M. S., Wasylenki, L. E., Asimow, P. D., & Stolper, E. M. (1998). Calculation of peridotite partial melting from thermodynamic models of minerals and melts. I. Review of methods and comparison with experiments. *Journal of Petrology*, *39*(6), 1091–1115. https://doi.org/10.1093/petroj/39.6.1091

Hirschmann, M. M., Ghiorso, M. S., & Stolper, E. M. (1999a). Calculation of peridotite partial melting from thermodynamic models of minerals and melts. II. Isobaric variations in melts near the solidus and owing to variable source composition. *Journal of Petrology*, *40*(2), 297–313. https://doi.org/10.1093/petroj/40.2.297

Hirschmann, M. M., Asimow, P. D., Ghiorso, M. S., & Stolper, E. M. (1999b). Calculation of peridotite partial melting from thermodynamic models of minerals and melts. III. Controls on isobaric melt production and the effect of water on melt production. *Journal of Petrology*, *40*(5), 831–851. https://doi.org/10.1093/petroj/40.5.831

Hofmann, A. W. (2003). Sampling mantle heterogeneity through oceanic basalts: isotopes and trace elements. In H. D. H. and K. K. Turekian (Eds.), *Treatise on Geochemistry* (Vol. *2*, pp. 1–44). Elsevier. https://doi.org/10.1016/B0-08-043751-6/02123-X

Hooft, E. E. E., Detrick, R. S., Toomey, D. R., Collins, J. A., & Lin, J. (2000). Crustal thickness and structure along three contrasting spreading segments of the Mid-Atlantic Ridge, 33.5°N-35°N. *Journal of Geophysical Research*, *105*(B4), 8205–8226. https://doi.org/10.1029/1999JB900442

Jacobs, A. M., Harding, A. J., & Kent, G. M. (2007). Axial crustal structure of the Lau back-arc basin from velocity modeling of multichannel seismic data. *Earth And Planetary Science Letters*, *259*, 239–255. doi: 10.1016/j.epsl.2007.04.021

Kelemen, P. B., Koga, K., & Shimizu, N. (1997). Geochemistry of gabbro sills in the crust-mantle transition zone of the Oman ophiolite: Implications for the origin of the oceanic lower crust. *Earth and Planetary Science Letters*, *146*(3), 475–488. https://doi.org/10.1016/S0012-821X(96)00235-X

Kilinc, A., Carmichael, I. S. E., Rivers, M. L., & Sack, R. O. (1983). The ferric-ferrous ratio of natural silicate liquids equilibrated in air. *Contributions to Mineralogy and Petrology*, *83* (1–2), 136–140. https://doi.org/10.1007/BF00373086

Kinzler, R. J. (1997). Melting of mantle peridotite at pressures approaching the spinel to garnet transition: Application to mid-ocean ridge basalt petrogenesis. *Journal of Geophysical Research*, *102*(B1), 853–874. https://doi.org/10.1029/96JB00988

Kinzler, R. J., & Grove, T. L. (1992). Primary magmas of mid-ocean ridge basalts 2. Applications. *Journal of Geophysical Research*, *97*(B5), 6907–6926. https://doi.org/10.1029/91JB02841

Klein, E. M., & Langmuir, C. H. (1987). Global correlations of ocean ridge basalt chemistry with axial depth and crustal thickness. *Journal of Geophysical Research*, *92*, 8089–8115. https://doi.org/10.1029/JB092iB08p08089

Kress, V. C., & Carmichael, I. S. E. (1991). The compressibility of silicate liquids containing Fe 2 O 3 and the effect of composition, temperature, oxygen fugacity and pressure on their redox states. *Contributions to Mineralogy and Petrology*, *108*(1), 82–92. https://doi.org/10.1007/BF00307328

Langmuir, C. H., Bender, J. F., Bence, A. E., Hanson, G. N., & Taylor, S. R. (1977). Petrogenesis of basalts from the FAMOUS area: Mid-Atlantic Ridge. *Earth and Planetary Science Letters*, *36*, 133–156. https://doi.org/10.1016/0012-821X(77)90194-7

Langmuir, C. H., Klein, E. M., & Plank, T. (1992). Petrological systematics of mid-ocean ridge basalts: Constraints on melt generation beneath ocean ridges. In J. Phipps Morgan, D. K. Blackman, & J. M. Sinton (Eds.), *Mantle Flow and Melt Generation at Mid-Ocean Ridges* (Vol. *71*, pp. 183–280). American Geophysical Union

Le Voyer, M., Hauri, E. H., Cottrell, E., Kelley, K. A., Salters, V. J. M., Langmuir, C. H., et al. (2019). Carbon Fluxes and Primary Magma CO2 Contents Along the Global Mid-Ocean Ridge System. *Geochemistry, Geophysics, Geosystems*, *20*(3), 1387–1424. https://doi.org/10.1029/2018GC007630

Mallmann, G., & O'Neill, H. S. C. (2013). Calibration of an empirical thermometer and oxybarometer based on the partitioning of Sc, Y and V between olivine and silicate melt. *Journal of Petrology*, *54*(5), 933–949. https://doi.org/10.1093/petrology/egt001

McKenzie, D., & Bickle, M. J. (1988). The volume and composition of melt generated by extension of the lithosphere. *Journal of Petrology*, *29*, 625–679. https://doi.org/10.1093/petrology/29.3.625

McKenzie, D., & O'Nions, R. K. (1991). Partial melt distributions from inversion of rare-earth element concentrations. *Journal of Petrology*, *32*(5), 1021–1091. https://doi.org/10.1093/petrology/32.5.1021

Michael, P. (1994). Regionally distinctive sources of depleted MORB: evidence from trace elements and H2O. *Earth and Planetary Science Letters*, *131*, 301–320. https://doi.org/10.1016/0012-821X(95)00023-6

Nicklas, R. W., Puchtel, I. S., Ash, R. D., Piccoli, P. M., Hanski, E., Nisbet, E. G., et al. (2019). Secular mantle oxidation across the Archean-Proterozoic boundary: Evidence from V partitioning in komatiites and picrites. *Geochimica et Cosmochimica Acta*, *250*, 49–75. https://doi.org/10.1016/j.gca.2019.01.037

Niu, Y., & Batiza, R. (1991). An empirical method for calculating melt compositions produced beneath mid-ocean ridges: application to axis and off-axis (seamounts) melting. *Journal of Geophysical Research*, *96*, 21,753–21,777. https://doi.org/10.1029/91JB01933

O'Neill, H. S. C., Berry, A. J., & Mallmann, G. (2018). The oxidation state of iron in Mid-Ocean Ridge Basaltic (MORB) glasses: Implications for their petrogenesis and oxygen fugacities. *Earth and Planetary Science Letters*, *504*, 152–162. https://doi.org/10.1016/j.epsl.2018.10.002

Perfit, M. R., Fornari, D. J., Ridley, W. I., Kirk, P. D., Casey, J., Kastens, K. A., et al. (1996). Recent volcanism in the Siqueiros transform fault: Picritic basalts and implications for MORB magma genesis. *Earth and Planetary Science Letters*, *141*(1–4), 91–108.

Presnall, D. C. (1966). The join forsterite-diopside-iron oxide and its bearing on the crystallization of basaltic and ultramafic magmas. *American Journal of Science*, *264*(10), 753–809. https://doi.org/10.2475/ajs.264.10.753

Putirka, K. (2016). Rates and styles of planetary cooling on Earth, Moon, Mars, and Vesta, using new models for oxygen fugacity, ferric-ferrous ratios, olivine-liquid Fe-Mg exchange, and mantle potential temperature. *American Mineralogist*, *101*(4), 819–840. https://doi.org/10.2138/am-2016-5402

Roeder, P. L., & Emslie, R. F. (1970). Olivine-Liquid Equilibrium. *Contributions to Mineralogy and Petrology*, *29*, 275–289.

Saal, A. E., Hauri, E. H., Langmuir, C. H., & Perfit, M. R. (2002). Vapor undersaturation in primitive mid-ocean ridge basalt and the volatile content of the Earth's upper mantle. *Nature*, *419*(6906), 451–455.

Sack, R. O., Carmichael, I. S. E., Rivers, M., & Ghiorso, M. S. (1981). Ferric-ferrous equilibria in natural silicate liquids at 1 bar. *Contributions to Mineralogy and Petrology*, *75*(4), 369–376. https://doi.org/10.1007/BF00374720

Smith, P. M., & Asimow, P. D. (2005). Adiabat_1ph: A new public front-end to the MELTS, pMELTS, and pHMELTS models. *Geochemistry Geophysics Geosystems*, *6*, Artn Q02004. https://doi.org/10.1029/2004gc000816

Sobolev, A. V., Hofmann, A. W., Kuzmin, D. V., Yaxley, G. M., Arndt, N. T., Chung, S.-L., et al. (2007). The amount of recycled crust in sources of mantle-derived melts. *Science*, *316*(5823), 412–417.

Stewart, M. A., Klein, E. M., & Karson, J. A. (2003). Geochemical relationships between dikes and lavas at the Hess Deep Rift: Implications for magma eruptibility. *Journal Of Geophysical Research-Solid Earth*, *108*(B4), 2184. https://doi.org/10.1029/2001JB001622

Stolper, E. M. (1980). A phase diagram for mid-ocean ridge basalts. *Contributions to Mineralogy and Petrology*, *74*, 13–27. https://doi.org/10.1007/BF00375485

Toplis, M. J. (2005). The thermodynamics of iron and magnesium partitioning between olivine and liquid: criteria for assessing and predicting equilibrium in natural and experimental systems. *Contributions to Mineralogy and Petrology*, *149*(1), 22–39. https://doi.org/10.1007/s00410-004-0629-4

Van Avendonk, H. J. A., Davis, J. K., Harding, J. L., & Lawver, L. A. (2016). Decrease in oceanic crustal thickness since the breakup of Pangaea. *Nature Geoscience*, *10*, 58. 10.1038/ngeo2849

Walker, D., Shibata, T., & DeLong, S. E. (1979). Abyssal tholeiites from the Oceanographer Fracture Zone. *Contributions to Mineralogy and Petrology*, *70*, 111–125. https://doi.org/10.1007/BF00371875

Walter, M. J. (1998). Melting of garnet peridotite and the origin of komatiite and depleted lithosphere. *Journal of Petrology*, *39*, 20–60. https://doi.org/10.1093/petroj/39.1.29

White, R. S. (1979). Oceanic crustal structure from variable angle seismic reflection-refraction profiles. *Geophysical Journal of the Royal Astronomical Society*, *57*, 683–726. https://doi.org/10.1111/j.1365-246X.1979.tb06785.x

White, R. S., Minshull, T. A., Bickle, M. J., & Robinson, C. J. (2001). Melt generation at very slow-spreading oceanic ridges: Constraints from geochemical and geophysical data. *Journal of Petrology*, *42*(6), 1171–1196. https://doi.org/10.1093/petrology/42.6.1171

Workman, R. K., & Hart, S. R. (2005). Major and trace element composition of the depleted MORB mantle (DMM). *Earth And Planetary Science Letters*, *231*(1–2), 53–72. https://doi.org/10.1016/j.epsl.2004.12.005

Zhang, H. L., Cottrell, E., Solheid, P. A., Kelley, K. A., & Hirschmann, M. M. (2018). Determination of Fe3+/ΣFe of XANES basaltic glass standards by Mössbauer spectroscopy and its application to the oxidation state of iron in MORB. *Chemical Geology*, *479*, 166–175. https://doi.org/10.1016/j.chemgeo.2018.01.006

8

Oxygen Content, Oxygen Fugacity, the Oxidation State of Iron, and Mid-Ocean Ridge Basalts

Andrew J. Berry and Hugh St.C. O'Neill

ABSTRACT

Oxygen is the third most abundant element in the Solar System and the most abundant in the silicate Earth. As the second most electronegative element its concentration in silicate rocks is fixed by the concentrations and oxidation states of the cations to which it bonds as O^{2-}. All of the major cation forming elements in terrestrial silicates have only one oxidation state, except for Fe. Thus, charge balancing with Fe is the only significant reason for variability in the amount of O in silicate rocks. It is difficult to measure differences in the oxygen content of rocks directly and instead it is usually proxied by Fe^{3+}/Fe^{2+}. The Fe^{3+}/Fe^{2+} of a rock is often used to define an oxygen fugacity (fO_2), however, the relationship between Fe^{3+}/Fe^{2+} and fO_2 is not monotonic, as illustrated by the quartz-fayalite-magnetite (QFM) oxygen buffer where at constant fO_2 Fe^{3+}/Fe^{2+} can vary from ~ 0 to 2. Since 1991 the empirical expression of Kress and Carmichael has been used to relate the oxidation state of Fe to fO_2, temperature, composition, and pressure. A feature of this expression is the non-ideal dependence of Fe^{3+}/Fe^{2+} on fO_2, suggesting Fe-Fe interactions. However, recent studies have shown that Fe^{3+}/Fe^{2+} behaves ideally. This has necessitated a refit of the available data. The new expression, for compositions with < 60 wt% SiO_2, is $\log_{10}(Fe^{3+}/Fe^{2+}) = 0.25\,\Delta QFM - 1.35 + 0.023[CaO] + 0.034[Na_2O] + 0.044[K_2O] - 0.018[P_2O_5]$, where fO_2 is in log units relative to QFM (and thus temperature does not need to be specified) and the oxide components are in weight percent. This equation reproduces the data from the Carmichael laboratory better than their original expression and identifies offsets in the literature data that distorted their fit. The average $Fe^{3+}/\Sigma Fe$ (where $\Sigma Fe = Fe^{2+} + Fe^{3+}$) value of mid-ocean ridge basalt (MORB) glasses was found to be 0.10 by XANES spectroscopy, which is consistent with the result of the most recent redox titration study, both for global datasets of samples. The XANES result was based on a new interpretation of the Mössbauer spectra used to determine $Fe^{3+}/\Sigma Fe$ of standards, which removes several long-standing anomalies: (i) the non-ideal dependence of Fe^{3+}/Fe^{2+} on fO_2, (ii) the inability to resolve $Fe^{3+}/\Sigma Fe$ values < ~0.10, and (iii) variation of the isomer shift of Fe^{3+}, but not Fe^{2+}, with fO_2. Using the new expression the average fO_2 of MORB is QFM+0.1. The large dataset of $Fe^{3+}/\Sigma Fe$ values for MORB can be used to show that Fe^{3+} has an incompatibility between that of In and Mn. The bulk partition coefficient of Fe^{3+} is 0.32 and the concentration of Fe_2O_3 in a primitive MORB melt (with 10.4 wt% MgO) is 0.61 wt%. The $Fe^{3+}/\Sigma Fe$ values of MORB are as expected for closed system evolution, with both Fe^{3+} and Fe^{2+} behaving like other moderately incompatible elements during partial melting and fractional crystallisation.

Research School of Earth Sciences, Australian National University, Canberra, ACT, Australia

Magma Redox Geochemistry, Geophysical Monograph 266, First Edition.
Edited by Roberto Moretti and Daniel R. Neuville.

DOI: 10.1002/9781119473206.ch8

8.1. OXYGEN CONTENT, OXYGEN FUGACITY, AND THE OXIDATION STATE OF IRON

Oxygen is the most abundant element in the bulk silicate earth (BSE), whether considered by mass (44.3 wt %) or by atomic fraction (58.4%; Palme & O'Neill, 2014). As a proportion of the anion-forming electronegative elements it figures even more prominently, comprising > 99.95% by weight, mainly as O^{2-} (plus minor contributions from OH^-, CO_3^{2-}, and SO_4^{2-}). The most abundant other anions in the BSE are S^{2-}, Cl^-, and F^-. The high abundance of O reflects, firstly, its ease of nucleosynthesis, which makes it the third-most abundant element in the solar composition after H and He (e.g. Asplund et al., 2009; Palme et al., 2014), and secondly, its fundamental property as the second-most electronegative of all the elements after F, which is not readily formed in standard nucleosynthetic processes. Therefore, unlike the other cosmochemically abundant elements (H, He, C, N, and Ne), some O is calculated to condense at high temperatures from a gas of solar composition as O^{2-} bonded to the refractory lithophile elements (e.g. Al, Ca, Ti). A substantial proportion of the total O available (20%) then condenses with Mg and Si as the components MgO and SiO_2 in magnesian silicates, which, together with Fe-Ni alloys, constitute more than 90% of the rocky bodies of the solar system, including the terrestrial planets. More O then condenses with the moderately volatile elements and by oxidizing metallic Fe to the component $Fe^{2+}O$ (and eventually $Fe^{3+}O_{1.5}$) and S^{2-} to SO_4^{2-}, before the remaining O condenses as OH^- and H_2O in hydrous minerals (e.g. serpentine, epsomite), and eventually as H_2O in planetary ices. Uniquely, O condenses from the solar nebula in significant amounts from start to finish (Table 8.1).

The unique status of O may be emphasized by considering the atomic ratio Mg/O. This ratio in the BSE is 0.33 (Palme & O'Neill, 2014), which reflects a depletion of O relative to Mg compared to the solar ratio of 0.08 (Asplund et al., 2009). This depletion, expressed as $[Mg/O]_{BSE}/[Mg/O]_{solar} = 4.1$, is orders of magnitude less than that of C ($[Mg/C]_{BSE}/[Mg/C]_{solar} = 7400$), N ($[Mg/N]_{BSE}/[Mg/N]_{solar} = 10^5$), and H ($[Mg/H]_{BSE}/[Mg/H]_{solar} = 10^6$). These elements mostly condense as ices and are the main constituents of the giant icy planets. The depletions reflect the relative proportions of each element in the terrestrial and icy planets, with O being the only element that is abundant in both.

The path that O followed during the accretion of the rocky material of the solar system means that its concentration in the silicate rocks of the resulting terrestrial planets was mostly fixed by the concentrations and oxidation states of the electropositive elements (i.e., the cations) to which O^{2-} is bonded. With the exception of Fe, all the major cation-forming elements in the BSE (and hence also in a typical silicate rock), namely Si, Mg, Al, Ca, Ti, Na, K, and P, have only one valence state, as do most of the minor and trace elements. This means that the only significant variability in the amount of O is due to charge balancing with Fe (Figure 8.1). A useful concept for treating the mass balance of variable oxygen is "marginal oxygen," defined as the excess or deficit compared to that bound to the polyvalent elements in their most common oxidation state. For example, the common oxidation state of Fe is 2+, so the marginal oxygen in $Fe^{2+}O$ is zero, but one-third in $Fe^{3+}O_{1.5}$. Of the elements contributing to marginal oxygen in the BSE, Fe is by far the most abundant. Thus, for most purposes, marginal oxygen in rocks is adequately proxied by Fe^{3+}/Fe^{2+}, although a potential complication arises when Fe^{2+} is bonded to S^{2-} in sulfides.

Table 8.1 The percentage of O in different forms, or condensed by various reactions, calculated using the solar abundances in Asplund et al. (2009) and the 50% condensation temperatures at 10^{-4} bar in Wood et al. (2019).

1. High Temperatures (>1250 K)	
(a) with Ca, Al, Ti, and other refractory lithophile elements	1.3%
(b) with Mg and Si	20.2%
2. Mid Temperatures (1250–650 K)	
(a) oxidation of Fe to $Fe^{2+}O$ components*	2.9%
(b) with moderately volatile elements (Na, P, Mn)	0.3%
(c) oxidation of Cr	0.2%
3 Low Temperatures (650–250 K)	
(a) oxidation of Fe^{2+} components	4.8%
(b) oxidation of S^{2-} to SO_4^{2-}	11.5%
(c) oxidation of Ni to Ni^{2+}	0.3%
4 Very Low Temperatures (< 250 K)	
(a) as hydrous minerals (including hydrated salts) } (b) as planetary ices	58.5%

*Only 50% of the available Fe undergoes this reaction; the other 50% combines with S at similar temperatures.

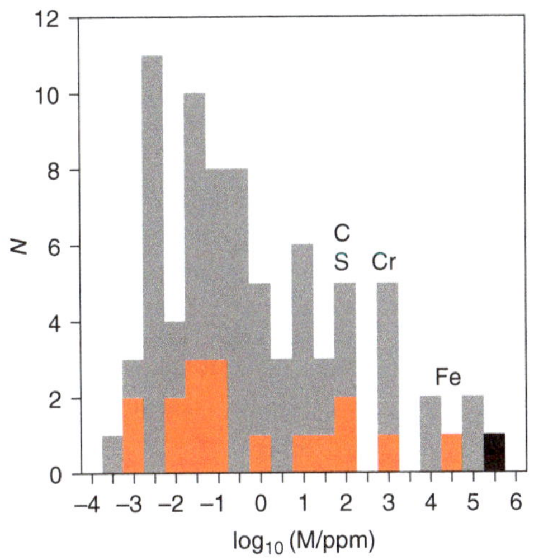

Figure 8.1 Histogram of the log(abundance) of 78 elements (M; bin size 0.5) in the bulk silicate Earth. Oxygen is shown in black, elements known to be redox variable in the upper mantle in red, and other elements in grey. The most abundant redox variable elements are labelled. Data from Palme and O'Neill (2014).

Marginal oxygen is usually such a small proportion of the total oxygen in a rock that it is almost never analyzed directly. Significant variations in marginal oxygen would correspond to small differences between large numbers, and unless a direct oxygen analysis was extraordinarily precise, it would not be useful. For example, in a basalt with 10 wt% FeO_t (where FeO_t is the total Fe, all assigned to a notional oxidation state of 2+), changing $Fe^{3+}/\Sigma Fe$ (where $\Sigma Fe = Fe^{2+} + Fe^{3+}$) from 0 to 1 would increase its oxygen content by only 2.5% relative (by weight). But petrologically, changes in $Fe^{3+}/\Sigma Fe$ of only 0.05 can affect significantly the evolution of basaltic magmas by bringing Fe-Ti oxides onto the liquidus, resulting in sulfide saturation and the consequent removal of chalcophile elements (e.g. Jenner et al., 2010). The change in oxygen content corresponding to a change in $Fe^{3+}/\Sigma Fe$ of 0.05 is 0.125% relative. The precision needed to resolve such a small relative change in oxygen content directly is beyond routine analytical capabilities. Hence, oxygen contents are determined indirectly through the oxidation state of Fe, making allowance for other redox-variable elements, if necessary.

The difficulty of analyzing oxygen contents has led to the redox states of geological systems being defined by their chemical potential of oxygen (μO_2), usually expressed as oxygen fugacity (fO_2). For oxygen, $\mu O_2 = \mu O_2^\circ + RT \log fO_2/pO_2^\circ$, where the superscript ° refers to a standard state of the ideal gas at a pressure of one bar (10^5 Pa) and the temperature of interest. Oxygen fugacity may be thought of as an alternative frame of reference to that of marginal oxygen and its $Fe^{3+}/\Sigma Fe$ proxy. The fugacity frame of reference has advantages, one of which is in comparing natural systems to experiments. In the laboratory, the practical difficulties of keeping systems closed to oxygen has promoted techniques that impose fO_2 instead. It would require unusual circumstances for natural geological processes to occur under controlled fO_2. A similar problem exists with temperature versus heat: in experimental petrology, experiments are done with temperature controlled and heat flows ignored, because it would be too difficult to do otherwise, but when applying laboratory results to large-scale natural processes it is necessary to transfer the frame of reference from constant temperature to one that takes into account the heat budget, for example the adiabatic model for decompression melting of the mantle (e.g. McKenzie & Bickle, 1988).

Problems caused by the fugacity frame of reference are unique to oxygen since no other element is treated geochemically in this way. One of these problems is that there is not a simple relationship between fO_2 and marginal oxygen. Indeed, discontinuities in the relationship occur where there is an isobarically invariant phase assemblage, known as an oxygen buffer. Adding or subtracting oxygen to or from a buffered system does not change its fO_2, i.e., for a perfect buffer $d(\log fO_2)/dO_2 = 0$. An example of such an oxygen buffer is the equilibrium between quartz-fayalite-magnetite (QFM) in the system Fe-Si-O, which allows the fO_2 of natural systems to be compared at different pressures (P) and temperatures (T). This is because, to a first approximation, the change in fO_2 with P or T for a closed system is generally closer to the change in fO_2 of the QFM buffer than the change in absolute fO_2 values relative to the standard state.

This absence of a simple relationship means that there is an asymmetry in the recoverable information when switching between the two frames of reference. For example, in the system Fe–Si–O, knowledge of the oxygen content and the Fe/Si ratio can tell us whether the composition is buffered, and if so, the fO_2; but knowing the Fe/Si ratio and that the fO_2 is that of the QFM buffer tells us only that $Fe^{3+}/\Sigma Fe$ is between 0 ($Fe_2SiO_4 + SiO_2 +$ infinitesimal Fe_3O_4) and 0.67 ($Fe_3O_4 + SiO_2 +$ infinitesimal Fe_2SiO_4). The redox changes through a petrogenetic sequence could in principle be traced using the oxygen content frame of reference without considering fugacity, given partitioning behaviour, but the reverse could not. At some steps in the process, oxygen fugacities would need to be converted to oxygen concentrations (effectively $Fe^{3+}/\Sigma Fe$) in order to determine the mass balance before converting back to oxygen fugacities.

A second problem with the fugacity frame of reference is that fugacity is a concept for a system at equilibrium, and defies definition for systems that are not, such as crystallized basalts (due to the range of temperatures over which they crystallize). The oxygen fugacity frame of reference is irrelevant for many common petrogenetic processes operating far from equilibrium, such as magma mixing, even if the oxygen fugacities of each of the mixing magmas are well known. The application of the oxygen-content frame of reference to magma mixing is as straightforward as the mass balance approach that would be used for any other element.

Nevertheless, the composition of fluids and the speciation of redox-variable minor and trace elements are more easily handled in the fugacity frame of reference, as is open-system behaviour where oxygen is transferred in or out by an equilibrium or near-equilibrium process, such as in degassing of a magma. Clearly both frames of reference are useful, and it is desirable to be able to switch between them. For basaltic melts, this would require knowing the relationship between $Fe^{3+}/\Sigma Fe$ and fO_2 as a function of temperature, pressure, and the composition of the melt.

8.2. MID-OCEAN RIDGE BASALTS

Determining $Fe^{3+}/\Sigma Fe$ in quenched silicate melts, or glasses, is not straightforward. Traditionally this has been attempted by either redox titrations, colloquially known as "wet chemistry," or Mössbauer spectroscopy. In redox titrations a glass sample is dissolved in acid and the Fe, which is assumed to enter the solution without changing its oxidation state, titrated against another element with which it undergoes a redox reaction. When all the Fe has been either oxidized or reduced by a measured amount of the other element (usually indicated by a color change) the amount of Fe^{2+} or Fe^{3+} can be calculated. To convert to $Fe^{3+}/\Sigma Fe$ the total amount of Fe needs to be determined by another technique, often by electron probe microanalysis. There are errors associated with each measurement. The most common methods are the direct and back titrations (see Bézos & Humler, 2005). In the direct method the amount of Fe^{2+} is determined by titration with Cr^{6+} (or Mn^{7+}). In the back method Fe^{2+} is fully oxidized by the addition of V^{5+} and the change in V^{5+} determined. Both methods assume that there is no change in oxidation state during digestion/dissolution, which could occur due to oxidation or reaction of either Fe or the titrant with other redox variable elements. For example, if S dissolved as S^{2-} in basalt was to react with either Fe^{3+} or V^{5+} this would increase the apparent amount of Fe^{2+} and hence reduce $Fe^{3+}/\Sigma Fe$ (Saikkonen & Rautiainen, 1993). Even though the amount of S in mid-ocean ridge basalt (MORB; ~ 1200 ppm as S^{2-}) is far less than the amount of Fe, the potential for each mole of S to reduce eight moles of Fe^{3+} (if oxidized from S^{2-} to S^{6+}) means that there is sufficient S to reduce over one-third of the total Fe. Large errors have also been reported for samples that should only contain Fe^{2+}, such as olivine ($Fe^{3+}/\Sigma Fe$ = 0.07±0.03) and $(NH_4)_2Fe(SO_4)_2.6H_2O$ ($Fe^{3+}/\Sigma Fe$ = 0.05±0.04) (Schüssler et al., 2008). Mössbauer spectroscopy relies on fitting the spectra to components that are attributed to either Fe^{2+} or Fe^{3+}. The fitting is not unique and different models can produce equally good fits. There is thus a degree of subjectivity. $Fe^{3+}/\Sigma Fe$ obtained from Mössbauer spectra are also dependent on the value chosen for the recoil-free fraction, which is a temperature-dependent term that defines how the proportion of Fe^{2+} and Fe^{3+} relate to the spectral areas attributed to these components. For glasses this is usually assumed to be ~ 1 at room temperature (i.e., equal spectral areas for Fe^{2+} and Fe^{3+} correspond to equal amounts of each oxidation state) but for crystalline materials it can vary significantly. For example, for garnet a correction factor of 0.72(6) should be applied to the area of the Fe^{3+} components (Woodland & Ross, 1994). $Fe^{3+}/\Sigma Fe$ values determined by Mössbauer spectroscopy have been compared with those from redox titrations and a 1:1 correlation was found, although the standard deviation of the differences between the values was over 0.075 in $Fe^{3+}/\Sigma Fe$ (Mysen et al., 1985).

Two redox titration studies have investigated large numbers of natural basaltic glasses from global sources and found $Fe^{3+}/\Sigma Fe$ values of 0.07 (n = 78, where n is the number of analyses; Christie et al., 1986) and 0.12 (or 0.11 when corrected for plagioclase phenocrysts; n = 104; Bézos & Humler, 2005). Other studies have reported values from 0.08 to 0.18 (see Table 1 of Bézos & Humler, 2005). Bézos and Humler (2005) compared the direct and back methods and investigated the effects of mechanical grinding time, acid digestion time, and digestion under air and Ar on $Fe^{3+}/\Sigma Fe$. They also analyzed a large number of standards and reference materials and re-analyzed samples from Christie et al. (1986), concluding that this study overestimated Fe^{2+} by ~ 3.5% (relative). The accuracy of the optimized direct titration method used by Bézos and Humler (2005) for determining Fe^{2+} was 2% (relative), with a standard deviation of 0.03 wt%. The $Fe^{3+}/\Sigma Fe$ values they report for MORB glasses would appear to be the most accurate and precise available.

To convert $Fe^{3+}/\Sigma Fe$ to fO_2 requires an understanding of how $Fe^{3+}/\Sigma Fe$ varies not only with fO_2, but also temperature, pressure, and composition. This requires $Fe^{3+}/\Sigma Fe$ to be determined for synthetic samples that explore the effect of these parameters. In 1991 all the available data were fit to an equation of the form:

$$\log(Fe_2O_3/FeO) = 0.25\alpha \log fO_2 + a + b/T + \Sigma c_M X_M \tag{8.1}$$

where α is a scaling factor to account for non-ideal mixing due to interactions between Fe^{2+} and Fe^{3+}, T is temperature in K, X are mole fractions of metal (M) oxides, and *a*, *b*, and *c* are constants, to give an empirical expression (Kress & Carmichael, 1991), which has become the standard equation for calculating $Fe^{3+}/\Sigma Fe$ in silicate melts (cited over 750 times at the time of writing). The $0.25a$ coefficient for fO_2 is 0.196 (and was found to be 0.2071(72), 0.2185(44), and 0.2181(70) in earlier versions of the expression; Kress & Carmichael, 1988; Kilinc et al., 1983; Sack et al., 1980), which differs from the ideal value of 0.25 expected for a one-electron oxidation/reduction reaction (i.e., $Fe^{2+}O + 0.25O_2 = Fe^{3+}O_{1.5}$). This nonideal behavior could be attributed to either interactions between Fe^{2+} and Fe^{3+} (i.e., non-Henry's Law behavior) or a change in melt structure with varying $Fe^{3+}/\Sigma Fe$ that affects the relative stability of either Fe^{2+} or Fe^{3+}. Using the Kress and Carmichael (1991) expression, Bézos and Humler determined the fO_2 of MORB glasses to be QFM-0.41±0.43.

More recently, $Fe^{3+}/\Sigma Fe$ of a global set of MORB glasses was determined by X-ray absorption near edge structure (XANES) spectroscopy to be 0.16 ($n = 103$; Cottrell & Kelley, 2011). This study used a set of basaltic glasses prepared as a function of fO_2, for which $Fe^{3+}/\Sigma Fe$ was determined by Mössbauer spectroscopy, as standards. An almost identical study found $Fe^{3+}/\Sigma Fe$ to be 0.10 ($n = 41$; Berry et al., 2018). In the study of Berry et al. (2018), glass standards were prepared over a larger range of fO_2. In particular, samples were prepared under more reduced conditions (QFM-4.6, or IW-1, where IW is the iron-wüstite buffer; at lower fO_2s Fe metal precipitates). The spectra of samples equilibrated at IW and IW-1 were essentially identical, suggesting that the samples approached $Fe^{3+}/\Sigma Fe = 0$. A redox titration of a similar sample found $Fe^{3+}/\Sigma Fe \approx 0$ (Cottrell & Kelley, 2011). A sample was also prepared at QFM+11, under which conditions $Fe^{3+}/\Sigma Fe \approx 1$ (Berry et al., 2003). The Mössbauer spectra of samples prepared at intermediate fO_2s could be fit by a linear combination of the spectra of these "end-members." The $Fe^{3+}/\Sigma Fe$ values obtained by this approach varied with fO_2 with the ideal coefficient of 0.25. This approach also required no assumptions about the assignment of the Mössbauer spectra. The spectra of the most reduced samples require two components to be well fit. Berry et al. (2018) assigned both of these to Fe^{2+}, whereas the minor component was assigned to Fe^{3+} by Cottrell et al. (2009). While the Mössbauer parameters of this component (isomer shift and quadrupole splitting) are not typical of either Fe^{2+} or Fe^{3+}, it must be emphasized that the fit is simply a model, without physical constraints, and that other models, which yield different Mössbauer parameters, can fit the spectra equally well (Berry et al., 2018). The consequences of assigning the minor component to Fe^{2+} are not only the ideal fO_2 coefficient value of 0.25, but also it is now possible to determine $Fe^{3+}/\Sigma Fe$ values < ~ 0.10, which were previously considered to be below the detection limit, and the isomer shift of Fe^{3+} as a function of fO_2 is now constant, as for Fe^{2+}, the variation of which had seemed anomalous given that the coordination environment of Fe^{2+} is more sensitive to melt structure than that of Fe^{3+} (Guillot & Sator, 2007). The apparent variation in isomer shift was a consequence of different proportions of Fe^{2+} being erroneously attributed to Fe^{3+}. With this simple reinterpretation of the Mössbauer spectra the description of Fe^{3+} in basaltic melts suddenly becomes rather simple. A result of this difference in assignment is that the Fe^{3+} values of the standards used by Cottrell et al. (2009) are overestimated, particularly at low values of $Fe^{3+}/\Sigma Fe$ (Figure 8.2), and with the overestimation of $Fe^{3+}/\Sigma Fe$ in the calibration standards, $Fe^{3+}/\Sigma Fe$ was correspondingly too high for the unknowns. The $Fe^{3+}/\Sigma Fe$ values in Cottrell et al. (2009) can be easily corrected for the calibration error, and when done so agree almost exactly with the results of Berry et al. (2018).

The ideal fO_2 coefficient of 0.25 was also found, within error, in studies of "basaltic" glasses with 1 wt% (Jayasuriya et al., 2004) and ~ 19 wt% (by fitting the data in Righter et al., 2013) ΣFe. For $Fe^{3+}/\Sigma Fe$ to vary with fO_2 with the ideal coefficient for glasses with 1, 10 (Berry et al., 2018), and 19 wt% ΣFe suggests that Fe-Fe

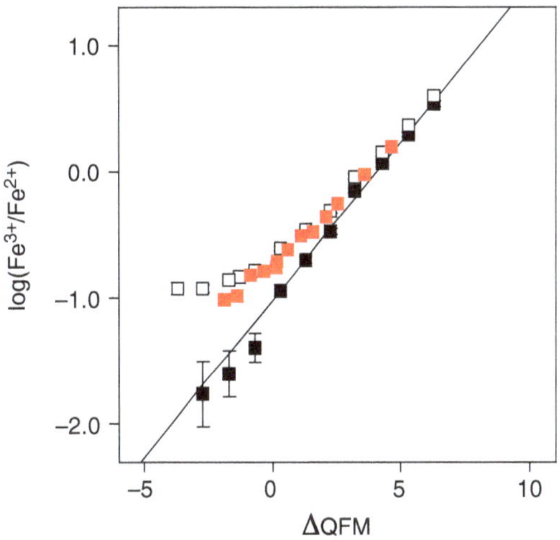

Figure 8.2 $\log(Fe^{3+}/Fe^{2+})$ as a function of fO_2, relative to the quartz-fayalite-magnetite (QFM) buffer, determined by fits to the Mössbauer spectra reported by Berry et al. (2018) when all the components in the spectra of the samples prepared under the most reduced conditions were assigned to Fe^{2+} (solid black squares) and when part of the spectrum was assigned to Fe^{3+} (open black squares), and the results for similar samples reported by Cottrell et al. (2009) (solid red squares). The solid line is the fit to the solid black squares using Equation 8.1 with the ideal value of 0.25 for the fO_2 coefficient.

interactions do not affect $Fe^{3+}/\Sigma Fe$. These studies, in which all variables were held constant except fO_2, allowed the fO_2 coefficient to be determined without distortion by other variables, for which the data may be limited. A coefficient of 0.25 is different to that in the Kress and Carmichael (1991) expression, which suggests that in their fit this coefficient was distorted by incorporating the effects of other parameters, which are in turn probably not optimally described. Further, it has been shown that $\log(Fe^{3+}/Fe^{2+})$ does not vary linearly with mole fractions (as in Equation 8.1) but has a U-shaped dependence on SiO_2 with a minimum at ~ 60 wt% (Borisov & McCammon, 2010). This necessitated a new parameterization of $Fe^{3+}/\Sigma Fe$.

The experimental data in the literature for samples with < 60 wt% SiO_2 (478 data) were fit to Equation 8.1 by statistically weighted least squares fitting. The data were weighted according to their uncertainties, which are 0.005 to 0.02 for the studies of Berry et al. (2018), Righter et al. (2013), Cottrell et al. (2009), and Jayasuriya et al. (2004), and 0.05 was assigned to the $Fe^{3+}/\Sigma Fe$ values determined by redox titrations (407 data from 12 studies; see Table 2 of O'Neill et al., 2018 for details). The fO_2 coefficient was found to be close to 0.25 and was fixed at this value. The inclusion of X_M terms for SiO_2, MgO, Al_2O_3, TiO_2, and ΣFe had only a minor effect on the fit and so were removed. The effect of temperature was found to be parallel to the temperature dependence of the QFM buffer so $\log fO_2$ was replaced with ΔQFM (the difference between the fO_2 of the silicate melt and QFM in log units, $\log fO_2(\text{QFM}) = 8.58 - 25050/T(\text{K})$; O'Neill et al., 2018). The fitting revealed substantial systematic offsets between the predicted and quoted $Fe^{3+}/\Sigma Fe$ values of some studies (Fudali, 1965; Thornber et al., 1980; Moore et al., 1995; Partzsch et al., 2004), however rather than discard and lose information on composition, an offset parameter was introduced for each of these datasets and refined as part of the fit. The final parameterization is:

$$\log_{10}\left(Fe^{3+}/Fe^{2+}\right) = 0.25\Delta\text{QFM} - 1.35 + 0.023[\text{CaO}] + 0.034[\text{Na}_2\text{O}] + 0.044[\text{K}_2\text{O}] - 0.018[\text{P}_2\text{O}_5] \quad (8.2)$$

where $[Na_2O]$, $[K_2O]$, [CaO], and $[P_2O_5]$ are the concentrations of the oxide components in weight percent (O'Neill et al., 2018). The expression allows values of fO_2 relative to QFM to be calculated without having to specify temperature. The reduced χ^2 is 1.0, which suggests that the weighting of the data was appropriate.

Equation 8.2 replaces Kress and Carmichael (1991) as the best available expression relating $Fe^{3+}/\Sigma Fe$ to fO_2 for basaltic to andesitic compositions. Equation 8.2 actually reproduces the Kress and Carmichael data (and all those arising from Carmichael's laboratory) better than their own parameterization, which was degraded by the inclusion of other, less accurate $Fe^{3+}/\Sigma Fe$ values. In particular, the data of Fudali (1965) and Thornber et al. (1980) were influential in producing the low value of the fO_2 coefficient. The difference between the values of QFM predicted by Equations 8.1 and 8.2 for the average composition of over 600 MORB samples from the Smithsonian Institution's Volcanic Glass Project is shown in Figure 8.3. The Kress and Carmichael expression overestimates $Fe^{3+}/\Sigma Fe$ at a given fO_2, with the greatest discrepancy at ~ QFM+1.

For an average $Fe^{3+}/\Sigma Fe$ value for natural MORB of 0.10(2) Equation 8.2 gives QFM+0.2. The value predicted by Kress and Carmichael (1991) is QFM-0.9. Although Bézos and Humler found $Fe^{3+}/\Sigma Fe$ = 0.12 (0.11 if corrected for plagioclase phenocrysts) their value of QFM-0.41(43) was obtained using Kress and Carmichael (1991). The value reported by Cottrell and Kelly (2011) (QFM+0.10(18)) is fortuitously correct since their high values of $Fe^{3+}/\Sigma Fe$ were countered by the underestimation of ΔQFM using Kress and Carmichael (1991).

The large dataset of $Fe^{3+}/\Sigma Fe$ values determined for well-characterized MORB glasses then allowed the petrogenetic behaviour of Fe to be investigated. For a global MORB dataset, the logarithm of the concentration of a trace element, M, and the concentration of MgO are linearly related (O'Neill & Jenner, 2012). In the model of O'Neill and Jenner (2012) the slope of the correlation, or d(log[M])/d([MgO]), is the result of fractional crystallization of olivine, plagioclase, and clinopyroxene in an "average" magma chamber that is replenished, mixed,

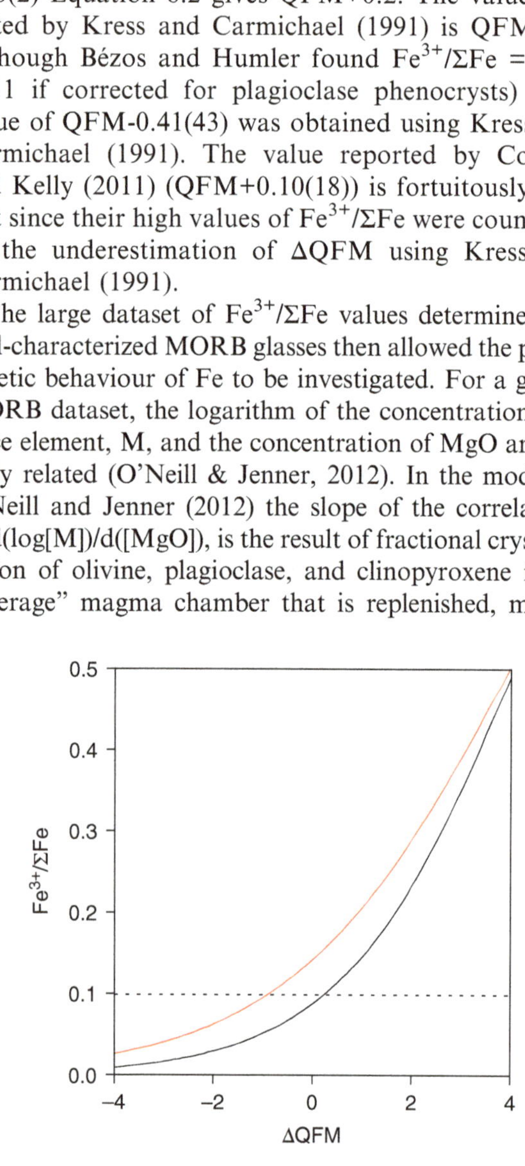

Figure 8.3 $Fe^{3+}/\Sigma Fe$ as a function of fO_2, relative to the QFM buffer, predicted for melts with < 60 wt% SiO_2 using Equation 8.2 (black) and the expression of Kress and Carmichael (1991) (red). The dashed line corresponds to the average $Fe^{3+}/\Sigma Fe$ value of natural MORB.

tapped, and fractionated. The scatter or variability, outside analytical uncertainty, is due to the varying ratios of olivine, plagioclase, and clinopyroxene. The slopes thus reflect the compatibility of M. The correlations for Fe_2O_3 using the data of Berry et al. (2018), Cottrell and Kelley (2011), Bézos and Humler (2005) and Christie et al. (1986), are shown in Figure 8.4a. Only glasses with > 5.5 wt% MgO were included to avoid complications arising from the crystallization of Fe-Ti oxides. Each dataset defines a linear trend with a negative slope (offset from each other due to the different average $Fe^{3+}/\Sigma Fe$ values). The slope obtained from a fit to the data of Berry et al. was -0.079(12). Slopes were also calculated for other elements in the samples of Berry et al. (2018) and compared to those obtained from the 597 samples with > 5.5 wt% MgO reported in Jenner and O'Neill (2012). There is a 1:1 correlation between the results (Figure 8.4b), indicating that the 42 samples studied by Berry et al. (2018) are representative of global MORB compositions (O'Neill et al., 2018). The slopes determined for Fe^{3+} and Fe^{2+} constrain the compatibility of Fe^{3+} to be between that of In and Li, and Fe^{2+} to be between Mn and Na (Figure 8.4b). Fe^{3+} thus behaves like any other moderately incompatible trace element during the evolution of MORB and there is nothing "anomalous" about the limited variability of $Fe^{3+}/\Sigma Fe$ values.

Bulk partition coefficients can be calculated from the slope and these are in excellent agreement with those determined from mineral-melt partitioning (Figure 5 of O'Neill and Jenner, 2012). The bulk partition coefficient calculated in this way for Fe^{3+} is 0.35(6) and the value estimated from mineral-melt partition coefficients is 0.32(6) (O'Neill et al., 2018). The good agreement between the two values further indicates that Fe^{3+} behaves like other incompatible elements and that there is no evidence for external buffering of fO_2, oxidation due to degassing, contamination by hydrothermal fluids, or reactions with other redox variable elements.

The concentration of M in a parental MORB melt can be predicted by extrapolating its correlation with MgO to the assumed MgO content. A parental melt with 10.4 wt% MgO would have 0.61(14) wt% Fe_2O_3, which suggests a MORB source with 0.27(5) wt% Fe_2O_3, assuming 15–20% closed-system partial melting of spinel peridotite (O'Neill et al., 2018). This value agrees well with that of 0.3 estimated from direct measurement of $Fe^{3+}/\Sigma Fe$ in mantle xenoliths (Canil et al., 1994).

The Fe^{3+}/Fe^{2+} ratio defines the fO_2 of a melt. As Fe is the major redox variable element and given that the oxygen content is constant, Fe^{3+}/Fe^{2+} will set the oxidation state ratio of redox variable minor and trace elements, which will need to equilibrate with the imposed redox potential. Given the relative

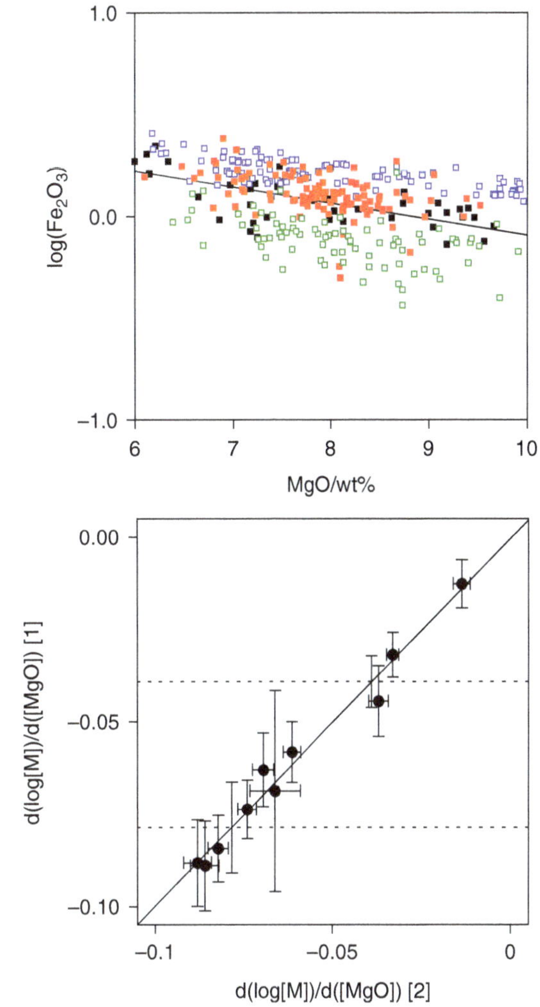

Figure 8.4 (a) log Fe_2O_3 (in wt%) of MORB glasses determined by Berry et al. (2018) (solid black symbols), Bézos and Humler (2005) (solid red), Cottrell and Kelley (2011) (open blue), and Christie et al. (1986) (open green) as a function of MgO (wt %). The solid line is the best fit to the data of Berry et al. (2018). (b) Comparison of d(log[M])/d([MgO]) for elements M (e.g. the slope defined by the line of best fit in (a)) for the samples studied in Berry et al. (2018) ([1]) and O'Neill and Jenner (2012) ([2]). The solid line is the best fit to the data. The dashed lines correspond to the values for Fe^{2+} (-0.039(7)) and Fe^{3+} (-0.079(12)). Error bars are shown for Fe^{2+} and Fe^{3+} at the points where the dashed lines intersect the line of best fit. The data points, in order of increasing d(log[M])/d([MgO]) [2], correspond to Yb, Lu, Li, In, V, Sr, Zn, Na, Mn, and Sc.

concentrations of other redox variable elements this will have a negligible effect on $Fe^{3+}/\Sigma Fe$.

ACKNOWLEDGMENTS

H.O'N. thanks the Australian Research Council for the award of grant FL130100066.

REFERENCES

Asplund, M., Grevesse, N., Sauval, A. J., & Scott, P. (2009). The chemical composition of the Sun. *Annual Review of Astronomy and Astrophysics*, *47*, 481–522. doi: 10.1146/annurev.astro.46.060407.145222

Berry, A. J., O'Neill, H. St.C., Jayasuriya, K. D., Campbell, S. J., & Foran, G. J. (2003). XANES calibrations for the oxidation state of iron in a silicate glass. *American Mineralogist*, *88*, 967–977. https://doi.org/10.2138/am-2003-0704

Berry, A. J., Stewart, G. A., O'Neill, H. St.C., Mallmann, G., & Mosselmans, J. F. W. (2018). A re-assessment of the oxidation state of iron in MORB glasses. *Earth and Planetary Science Letters*, *483*, 114–123. https://doi.org/10.1016/j.epsl.2017.11.032

Bézos, A. & Humler, E. (2005). The $Fe^{3+}/\Sigma Fe$ ratios of MORB glasses and their implications for mantle melting. *Geochimica et Cosmochimica Acta*, *69*, 711–725. https://doi.org/10.1016/j.gca.2004.07.026

Borisov, A., & McCammon, C. (2010). The effect of silica on ferric/ferrous ratio in silicate melts: An experimental study using Mössbauer spectroscopy. *American Mineralogist*, *95*, 545–555. https://doi.org/10.2138/am.2010.3217

Canil, D., O'Neill, H. St.C., Pearson, D. G., Rudnick, R. L., McDonough, W. F., & Carswell, D. A. (1994). Ferric iron in peridotites and mantle oxidation states. *Earth and Planetary Science Letters*, *123*, 205–220. https://doi.org/10.1016/0012-821X(94)90268-2

Christie D. M., Carmichael I. S. E., & Langmuir C. H. (1986). Oxidation states of mid-ocean ridge basalt glasses. *Earth and Planetary Science Letters*, *79*, 379–411. https://doi.org/10.1016/0012-821X(86)90195-0

Cottrell, E., & Kelley, K. A. (2011). The oxidation state of Fe in MORB glasses and the oxygen fugacity of the upper mantle. *Earth and Planetary Science Letters*, *305*, 270–282. https://doi.org/10.1016/j.epsl.2011.03.014

Cottrell, E., Kelley, K. A., Lanzirotti, A., & Fischer, R. A. (2009). High-precision determination of iron oxidation state in silicate glasses using XANES. *Chemical Geology*, *268*, 167–179. https://doi.org/10.1016/j.chemgeo.2009.08.008

Fudali, R. F. (1965). Oxygen fugacities of basaltic and andesitic magmas. *Geochimica et Cosmochimica Acta*, *29*, 1063–1075. https://doi.org/10.1016/0016-7037(65)90103-1

Guillot, B., & Sator, N. (2007). A computer simulation study of natural silicate melts. *Geochimica et Cosmochimica Acta*, *71*, 1249–1265. https://doi.org/10.1016/j.gca.2006.11.015

Jayasuriya, K. D., O'Neill, H. St.C., Berry, A. J., & Campbell, S. J. (2004). A Mössbauer study of the oxidation state of iron in silicate melts. *American Mineralogist*, *89*, 1597–1609. https://doi.org/10.2138/am-2004-11-1203

Jenner, F. E., & O'Neill, H. St.C. (2012). Analysis of 60 elements in 616 ocean floor basaltic glasses. *Geochemistry, Geophysics, Geosystems*, *13*. doi:10.1029/2011GC004009

Jenner, F. E., O'Neill, H. St.C., Arculus, R. J., & Mavrogenes, J. A. (2010). The magnetite crisis in the evolution of arc-related magmas and the initial concentration of Au, Ag and Cu. *Journal of Petrology*, *51*, 2445–2464. https://doi.org/10.1093/petrology/egq063

Kilinc, A., Carmichael, I. S. E., Rivers, M. L., & Sack, R. O. (1983). The ferric-ferrous ratio of natural silicate liquids equilibrated in air. *Contributions to Mineralogy and Petrology*, *83*, 136–140. https://doi.org/10.1007/BF00373086

Kress, V. C., & Carmichael, I. S. E. (1988). Stoichiometry of the iron oxidation reaction in silicate melts. *American Mineralogist*, *73*, 1267–1274.

Kress, V. C., & Carmichael, I. S. E. (1991). The compressibility of silicate liquids containing Fe_2O_3 and the effect of composition, temperature, oxygen fugacity and pressure on their redox states. *Contributions to Mineralogy and Petrology*, *108*, 82–92. https://doi.org/10.1007/BF00307328

McKenzie, D., & Bickle, M. J. (1988). The volume and composition of melt generated by extension of the lithosphere. *Journal of Petrology*, *29*, 625–679. https://doi.org/10.1093/petrology/29.3.625

Moore, G., Righter, K., & Carmichael, I. S. E. (1995). The effect of dissolved water on the oxidation state of iron in natural silicate liquids. *Contributions to Mineralogy and Petrology*, *120*, 170–179. https://doi.org/10.1007/BF00287114

Mysen, B. O., Carmichael, I. S. E., & Virgo, D. (1985). A comparison of iron redox ratios in silicate glasses determined by wet chemical and ^{57}Fe Mössbauer resonant absorption methods. *Contributions to Mineralogy and Petrology*, *90*, 101–106. https://doi.org/10.1007/BF00378253

O'Neill, H. St.C., & Jenner, F. E. (2012). The global pattern of trace-element distributions in ocean floor basalts. *Nature*, *491*, 698–704. https://doi.org/10.1038/nature11678

O'Neill, H. St.C., Berry, A. J., & Mallmann, G. (2018). The oxidation state of iron in mid-ocean ridge basaltic (MORB) glasses: implications for their petrogenesis and oxygen fugacities. *Earth and Planetary Science Letters*, *504*, 152–162.

Palme, H., Lodders, K., & Jones, A. (2014). Solar system abundances of the elements. *Treatise on Geochemistry* (second edition), *2*, 15–36. Elsevier.

Palme, H., & O'Neill, H. St.C. (2014). Cosmochemical estimates of mantle composition. *Treatise on Geochemistry* (second edition), *3*, 1–39. Elsevier.

Partzsch, G. M., Lattard, D., & McCammon, C. (2004). Mössbauer spectroscopic determination of Fe^{3+}/Fe^{2+} in synthetic basaltic glass: a test of empirical f_{O_2} equations under superliquidus and subliquidus conditions. *Contributions to Mineralogy and Petrology*, *147*, 565–580. doi:10.1007/s00410-004-0571-5

Righter, K., Danielson, L. R., Pando, K., Morris, R. V., Graff, T. G., Agresti, D. G., et al. (2013). Redox systematics of martian magmas with implications for magnetite stability. *American Mineralogist*, *98*, 616–628. https://doi.org/10.2138/am.2013.4251

Sack, R. O., Carmichael, I. S. E., Rivers, M., & Ghiorso, M. S. (1980). Ferric-ferrous equilibria in natural silicate liquids at 1 bar. *Contributions to Mineralogy and Petrology*, *75*, 369–376. https://doi.org/10.1007/BF00374720

Saikkonen, R. J., & Rautiainen, I. A. (1993). Determination of ferrous iron in rock and mineral samples by three volumetric methods. *Bulletin of the Geological Society of Finland*, *65*, 59–63.

Schuessler, J. A., Botcharnikov, R. E., Behrens, H., Misiti, V., & Freda, C. (2008). Oxidation state of iron in hydrous phono-tephritic melts. *American Mineralogist*, *93*, 1493–1504. https://doi.org/10.2138/am.2008.2795

Thornber, C. R., Roeder, P. L., & Foster, J. R. (1980). The effect of composition on the ferric-ferrous ratio in basaltic liquids at atmospheric pressure. *Geochimica et Cosmochimica Acta*, *44*, 525–532. https://doi.org/10.1016/0016-7037(80)90048-4

Wood, B. J., Smythe, D. J., & Harrison, T. (2019). The condensation temperatures of the elements: A reappraisal. *American Mineralogist*, *104*, 844–856. https://doi.org/10.2138/am-2019-6852CCBY

Woodland, A. B., & Ross, C. R. (1994). A crystallographic and Mössbauer spectroscopy study of $Fe_3Al_2Si_3O_{12}$-$Fe^{2+}{}_3Fe^{3+}{}_2Si_3O_{12}$ (almandine-"skiagite") and $Ca_3Fe^{3+}{}_2Si_3O_{12}$-$Fe^{2+}{}_3Fe^{3+}{}_2Si_3O_{12}$ (andradite-"skiagite") garnet solid solutions. *Physics and Chemistry of Minerals*, *21*, 117–132. doi: 10.1007/BF00203142

9

Chromium Redox Systematics in Basaltic Liquids and Olivine

Aaron S. Bell

ABSTRACT

Chromium is a minor yet important multivalent element in basaltic melts. Numerous experimental studies have conclusively demonstrated that mixed-valence chromium is present in mafic silicate liquids over a wide range of redox conditions (i.e., IW-1 to FMQ+2). Mixed-valence Cr has also been documented in experimentally grown olivine and pyroxene as well as olivine phenocrysts present in achondritic meteorites. Synchrotron XANES studies $Cr^{2+}/\Sigma Cr$ ratios preserved in olivine and pyroxene phenocrysts can yield a high-fidelity record of Cr valence evolution in their parental liquid, provided that the crystal-liquid partitioning behavior of multivalent Cr is sufficiently well understood. As such, spatially resolved measurements of $Cr^{2+}/\Sigma Cr$ in olivine and pyroxene phenocrysts that record protracted growth histories can provide a powerful, temporally resolved, oxybarometer for tracking the redox evolution of basaltic magmas. This chapter specifically focuses on (i) synchrotron XANES techniques applied to the measurement of $Cr^{2+}/\Sigma Cr$ in geologic materials; (ii) our current understanding of how fO_2, temperature, liquid composition, and quench-induced effects control the $Cr^{2+}/\Sigma Cr$ measured in experimental glasses; and (iii) the relationship between the equilibrium $Cr^{2+}/\Sigma Cr$ in silicate liquids and their olivine daughter crystals. Although the discussion presented in this chapter is primarily devoted to the ongoing development of Cr-based magmatic redox tracers, it should also be noted that Cr valence systematics have implications for how we understand the fractionation of Cr isotopes in igneous systems, as well as the development of thermodynamic models of Cr-spinel crystallization as they relate to the formation of chromite horizons in layered mafic intrusions.

9.1. INTRODUCTION

Oxygen fugacity is a fundamental variable that plays important roles in influencing magmatic phase equilibria and dictating their liquid lines of descent, controlling C–O–H–S volatile speciation and outgassing behavior of magmatic liquids, and controlling the distribution of siderophile elements between silicate liquids and immiscible sulfide liquids and metallic alloys. The evaluation and interpretation of oxygen fugacity (fO_2) signatures carried by silicate liquids and mineral assemblages that comprise basaltic magmas has been an active field of petrologic research for the last 40 years (Fudali, 1965; Anderson & Wright, 1972; Sack et al., 1980; Carmichael & Ghiorso, 1986; Kress & Carmichael, 1991; Carmichael, 1991; Bezos & Humler, 2005; Lee et al., 2005; Kelly & Cottrell, 2009; Shorttle et al., 2015; Gaetani, 2016).

Iron (Fe) is, by far, the most abundant multivalent element in magmatic silicate liquids. Given that Fe is a major constituent of all basaltic magmas, the homogenous equilibrium between ferric and ferrous Fe in silicate

Department of Geological Sciences, University of Colorado Boulder, Boulder, CO, USA

Magma Redox Geochemistry, Geophysical Monograph 266, First Edition.
Edited by Roberto Moretti and Daniel R. Neuville.

DOI: 10.1002/9781119473206.ch9

liquids is largely responsible for controlling the oxygen fugacity of mafic silicate liquids (Sack et al., 1980; Kress & Carmichael, 1991). At fO_2 values more oxidizing than the IW (Fe-wustite) buffer reaction, the ratio of $a\text{FeO}_{1.5}$ to $a\text{FeO}$ in the silicate liquid effectively defines its fO_2 value.

Although the Fe redox equilibrium plays an outsized role with respect to controlling the fO_2 in basaltic liquids, it should be stressed that it is by no means the only multivalent element present in mafic magmatic systems. A plethora of minor and trace elements (both transition metals and REE) are also known to occur in mixed valence states in silicate liquids. The most common of these elements include: Ce, Eu, V, Ti, Mn, Cu, Mo, Sn, W, and Cr. The low abundances of these elements limit the power of their homogenous redox equilibria to dictate magmatic fO_2 values in Fe-rich basaltic liquids. The valence ratios of these elements are controlled by the prevailing magmatic fO_2 value, which is fixed by the homogeneous Fe redox equilibrium in th liquid. In a manner analogous to Fe, the homogenous redox reactions and redox sensitive partitioning behavior of these elements also may serve as useful indicators of magmatic fO_2.

Of the previously mentioned elements, Cr is particularly well suited to serve as a magmatic redox tracer for several reasons. Unlike many of the elements listed above (e.g., V), Cr is typically present at relatively high concentrations in primitive basaltic magma systems. The bulk concentration of Cr in basaltic magmas typically ranges from several hundred ppm on the low end, to several thousand ppm at the high end. From an analytical standpoint, the relatively high concentration of Cr in basaltic liquids and their crystallization products mitigates many difficulties associated with accurately measuring Cr valence ratios. Perhaps more importantly, numerous studies have shown that Cr occurs as a mixture of Cr^{2+} and Cr^{3+} over a range of fO_2 conditions that are pertinent for terrestrial and planetary basaltic magmatism (Kelly & Cottrell 2009; Shorttle et al., 2015; Gaetani, 2016; and Berry et al 2018).

The relative abundances of Cr^{2+} and Cr^{3+} in silicate liquids are formally connected to fO_2 via the homogenous redox equilibrium:

$$\text{CrO}^{\text{liq}} + 0.25\text{O}_2 = \text{CrO}_{1.5}{}^{\text{liq}} \tag{9.1}$$

The equilibrium ratio of Cr^{2+} to Cr^{3+} (hereafter referred as $Cr^{2+}/\Sigma Cr$) shows a first order dependence on oxygen fugacity. In standard thermodynamic notation this reaction can be recast as:

$$\frac{\text{Cr}^{2+\,\text{liq}}}{\Sigma\text{Cr}} = \left[1 + e\left[\ln K + 0.25\ln(fO_2) - \ln\frac{\gamma\text{CrO1.5}}{\gamma\text{CrO}}\right]\right]^{-1} \tag{9.2}$$

where lnK is the equilibrium constant of the reaction in Equation 9.1 and the $\gamma CrO_{1.5}$ and γCrO represent the activity coefficients of the macroscopic oxide components $CrO_{1.5}$ and CrO in the liquid. This expression shows that $Cr^{2+}/\Sigma Cr$ in the liquid is controlled by the value of the equilibrium constant (lnK) of Equation 9.1, oxygen fugacity, and the values of $\gamma CrO_{1.5}$ and γCrO. It should be stressed that the value of lnK is itself intrinsically a function of temperature and pressure, and that the values of $\gamma CrO_{1.5}$ and γCrO are controlled by the liquid composition and its relationship with the thermodynamic mixing properties of $CrO_{1.5}$ and CrO. Our understanding of the variables that control the homogenous Cr valence equilibrium is the foundation that underpins the use of $Cr^{2+}/\Sigma Cr$ systematics in silicate liquids and their crystallization products as petrologic tools that shed new light on magmatic redox processes.

Chromium may also be present as Cr^{6+} under highly oxidizing conditions in some silicate liquids. The relevance of Cr^{6+} in natural magmatic systems is relatively limited, as its occurrence is restricted to fO_2 conditions significantly more oxidizing than the magnetite-hematite buffer. As such, this chapter will exclusively focus on Cr^{2+} and Cr^{3+}.

9.1.1. Some Applications of Cr-Redox Systematics to Petrologic Problems

It has been observed that Cr in olivine and pyroxene can exist in mixtures of Cr^{2+} and Cr^{3+}. The ability of these phases to accommodate mixed valence Cr renders them potentially useful recorders of the $Cr^{2+}/\Sigma Cr$ of their parental magmatic liquids. These observations spawned the idea that measurements of the $Cr^{2+}/\Sigma Cr$ in minerals such as olivine could be used to accurately retrieve the $Cr^{2+}/\Sigma Cr$ of their parental liquid. The $Cr^{2+}/\Sigma Cr$ values of the parental liquid could then be exploited and used in conjunction with an empirical or thermodynamic model to calculate magmatic fO_2 values in a manner analogous to using $Fe^{3+}/\Sigma Fe$ measurements in quenched glasses and the model Fe-redox model of Kress and Carmichael (1991). This approach is predicated on an understanding of the petrologic context of olivine crystallization, particularly the temperature and liquid composition. This approach has several advantages over other traditional oxybarometers. The first advantage is that spatially resolved spot measurements of Cr valence in olivine can be placed into the context of its growth history, potentially revealing a wealth of information about temporal variations in the oxidation state of the magmatic system. The second advantage is that this approach does not require a glassy quenched liquid to be present. This circumvents the issue of quench modification of the $Cr^{2+}/\Sigma Cr$, as well as difficulties associated with finding primitive melt inclusions and deciphering potential post-entrapment modification. The Cr valence in olivine oxy-barometer (CrViO) still requires additional development and calibration before it is a fully operational tool; decoding the $Cr^{2+}/\Sigma Cr$ of natural olivine phenocrysts and exploiting this information as a redox "chronometer" requires comprehensive knowledge of how $Cr^{2+}/\Sigma Cr$ is controlled by fO_2, olivine crystallization temperature, as well as the evolution of melt chemistry that occurs during crystallization.

Knowledge of the Cr-redox systematics in silicate liquids also has important applications to several other fundamental petrologic problems. Perhaps the most direct application of this knowledge is to the modeling of Cr-spinel saturation in basaltic magmas. The crystallization of spinel enriched in the $FeCr_2O_4$ and $MgCr_2O_4$ components critically depends on the activity of trivalent Cr in the liquid (i.e., $CrO_{1.5}$). The activity of $CrO_{1.5}$ in the liquid, as per Equation 8.1, is controlled by the magmatic fO_2, bulk composition, and temperature of the liquid. The collective influence of these variables has important consequences for our understanding of the petrogenesis of chromite horizons in layered mafic complexes, such as the Stillwater (Montana, USA) and the Bushveld (South Africa). The construction of thermodynamic models that can accurately predict the onset of Cr-spinel crystallization in basaltic magmas requires an in-depth understanding of Cr-redox systematics in silicate liquids.

Finally, it should be noted that Cr redox systematics also have important consequences for our understanding of the silicate-metal partitioning behavior of Cr during differentiation and core segregation in the earth and other differentiated planetary bodies. Understanding Cr valence in the ultramafic liquids involved in core formation has a direct relevance for variable fO_2 core formation models. The predominate oxidation state of Cr in the silicate liquid directly influences the slope of the silicate-metal partition coefficient as a function of fO_2 (Wood et al. 2008). Several studies have also concluded that the redox systematics of Cr can significantly impact the fractionation of stable Cr isotopes between silicate liquids, metallic alloys, and oxides (Moynier et al., 2011). The development and future application of stable Cr isotopes to high temperature petrologic problems will almost certainly draw upon our understanding of Cr valence in silicate liquids and minerals such as olivine.

9.1.2. Overview of Chapter Content

The common link among the above applications is that they each require a broad understanding of the Cr redox systematics in silicate liquids. The content of this chapter will focus on several important facets of our current understanding of Cr redox systematics in basaltic magma systems. This chapter will include (i) a basic description of the synchrotron-XANES (X-ray absorption near edge structure) measurements of Cr valence in quenched silicate liquids and minerals; (ii) a review of what is currently known about Cr redox systematics in mafic silicate liquids; and (iii) a review of the olivine-liquid and valence partitioning behavior Cr and its relationship to Cr-redox systematics in basaltic liquids.

9.2. MEASURING CR VALENCE IN GEOLOGIC MATERIALS WITH CR-K EDGE XANES SPECTROSCOPY

Synchrotron-based Cr K-edge XANES spectroscopy is the most widely used analytical method for quantifying Cr valence in geologic materials. There are two major advantages to using synchrotron-based XANES methods over more traditional wet chemical titrations or electron paramagnetic spectroscopy. Synchrotron-based measurements typically have a relatively small analytical footprint – modern Kirkpatrick-Baez X-ray focusing optics can routinely produce beam spot sizes smaller than 10μm by 10μm. XANES-based valence measurements are analogous to electron microprobe point analyses with respect to the dimensions of the beam on the sample surface; it is also important to recognize that the depth from which a typical XANES signal originates can be substantially deeper (on the order of 10s of μm) than the relatively shallow analytical volume associated with typical EMP point analyses. The main point stressed here is that synchrotron XANES analyses afford the analyst an opportunity to gather valence information in a spatial context. The second advantage of XANES is that the high brilliance of the synchrotron X-ray sources allows for the acquisition of spectra in materials with relatively dilute total Cr concentrations – as low as 100 ppm.

The various absorption features in the Cr K-edge XANES spectra are produced by the absorption of incident X-rays that excite 1s core electrons to various bound and continuum states. Numerous discrete electron transitions are observed in the Cr K-edge spectra of silicate glasses; these transitions include the 1s-3d transition that comprises the weak pre-edge feature, the 1s-4s shoulder transition, and the main K- edge absorption (Fig. 9.1). The centroid energies and the intensity of each of these transitions reflect both the formal oxidation state of Cr, as well as its mean coordination geometry and symmetry. The 1s-3d transition is extremely weak to nearly absent for Cr^{2+} and Cr^{3+} bearing materials; this transition shows no discernable systematic variation with changing Cr valence (Berry & O'Neill, 2004). In contrast, the 1s-3d absorption is strongly present and a

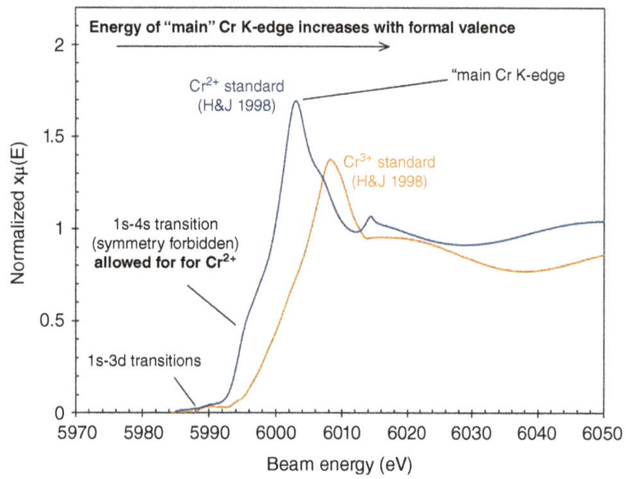

Figure 9.1 XANES spectra from Cr^{2+} and Cr^{3+} standard glasses. There are several notable differences in the spectral features of the two glasses. The most important differences in the spectra are (i) the increase in energy of the rising edge and edge crest as a function of increasing formal Cr valence and (ii) the presence of a salient absorption feature that appears on the shoulder of the rising edge. This feature has been attributed to the 1s-4s transition that is diagnostic of the presence of Cr^{2+}. The intensity of this absorption feature scales with the ratio of Cr^{2+} to Cr^{3+} present in the material of interest.

diagnostic spectral feature for Cr^{6+} bearing oxides and glasses. The intense 1s-3d absorption associated with Cr^{6+} can be attributed to its tetrahedral coordination symmetry which allows for p-d orbital mixing that enhances the intensity of the ordinarily symmetry forbidden 1s-3d dipole transition.

Sutton et al. (1993) and Berry and O'Neill (2004) identified the 1s-4s transition that appears as a shoulder on the main absorption edge as a feature that is diagnostic of the presence of Cr^{2+}. The 1s-4s transition is formally symmetry forbidden by Laporte's rule, therefore this absorption is weak or entirely absent for Cr^{3+} which is known to show a strong preference for high symmetry regular octahedral coordination (Burns, 1976; O'Neill & Berry, 2006). On the other hand, 1s-4s transition is allowed for Cr^{2+} as it is a Jahn-Teller active cation that prefers a tetragonally distorted, reduced symmetry, octahedral site (O'Neill & Berry, 2006; McKeown et al., 2014).

Berry and O'Neill (2004) developed a method for quantifying the $Cr^{2+}/\Sigma Cr$ in silicate quenched silicate glasses by measuring the intensity of the 1s-4s absorption feature in derivative XANES spectra; this study demonstrated that the intensity of the 1s-4s absorption feature is a linear function of the $Cr^{2+}/\Sigma Cr$ in CMAS glasses, and therefore with the appropriate reference standards, measurements of the 1s-4s absorption intensity can be translated into $Cr^{2+}/\Sigma Cr$ values. The precision of $Cr^{2+}/\Sigma Cr$ values determined using this method are on the order of ± 0.025 in silicate glasses (Berry & O'Neill, 2004; Berry et al., 2006; Bell et al., 2014). The linear relationship between the intensity of the 1s-4s absorption and $Cr^{2+}/\Sigma Cr$ has also been observed to hold true for Cr in olivine (Bell et al., 2014; McKeown et al., 2014), and the use of this method has been extended to quantifying Cr valence in olivine. It should be noted that the uncertianty of XANES measured $Cr^{2+}/\Sigma Cr$ values in olivine and other crystalline materials may be much larger as these measurements are often subject to orientation-related absorption anisotropy effects.

9.2.1. Quench Modification of Cr^2+/ΣCr in Fe-bearing Glasses

A major obstacle to quantifying $Cr^{2+}/\Sigma Cr$ in silicate glasses is that the "pristine" Cr valence ratios of natural mafic silicate liquids are substantially modified during quench. In-situ XANES analyses performed on an Fe-bearing liquid at controlled redox conditions (Berry et al., 2006) conclusively demonstrated that Cr^{2+} is present in the liquid prior to quench. These authors concluded that a temperature dependent electron exchange reaction was responsible for altering the equilibrium $Cr^{2+}/\Sigma Cr$ of the liquid as it is cooled through the glass transition. In this exchange reaction, Fe^{3+} is reduced by Cr^{2+} to produce Fe^{2+} and Cr^{3+}. This reaction is strongly favored at low temperatures, such that it runs to near completion, even in rapidly quenched glasses. The proposed electron exchange reaction effectively erases or distorts any "memory" of the Cr^{2+} that may have been in the liquid prior to cooling and quench.

The recognition of this problem has restricted XANES studies of Cr valence to materials that do not contain significant quantities of Fe^{3+}. So far, this limitation has severely hampered further investigations of Cr valence systematics in Fe-bearing liquids.

In contrast to the Cr^{2+} preservation issues that plague silicate liquids, silicate minerals that contain limited quantities of Fe^{3+}, such as olivine, appear to be relatively immune to electron exchange modification effects; Bell et al. (2014) showed that olivine crystals gown from liquids at fO_2 conditions as oxidizing as ΔFMQ+0 (where FMQ represents the fayalite-magnetite-quartz buffer) appear to retain their high temperature $Cr^{2+}/\Sigma Cr$ values. Both the relatively low abundance of Fe^{3+} in olivine, as well as steric and local charge balance constraints, contribute to the inhibition of the Fe^{3+}-Cr^{2+} electron exchange reaction. At present, there are no studies that have investigated Cr valence preservation in other common magmatic silicate minerals that can accommodate significant quantities of Fe^{3+} (e.g., clinopyroxene), therefore it remains unclear if the ability of olivine to resist Cr valence modification on quench is a unique or relatively common trait among minerals that can accommodate mixed valence Cr in their structures.

9.2.2. XANES Anisotropy in Non-Isometric Silicate Minerals

Another significant barrier to obtaining accurate XANES measurements of $Cr^{2+}/\Sigma Cr$ is mitigating the effects of crystallographic orientation on the accuracy of valence measurements. Bell et al. (2017) investigated the effects of crystallographic orientation on the intensity of the 1s-4s absorption peaks in olivine. This study showed that orientation-related absorption anisotropy can significantly distort the apparent $Cr^{2+}/\Sigma Cr$ values of olivine (Fig. 9.2). The dataset generated from this work was used to construct a corrective model that effectively mitigates the effects of absorption anisotropy on the calculated $Cr^{2+}/\Sigma Cr$ values. The inclusion of this correction procedure as a part of spectra processing significantly improves the accuracy of the resultant $Cr^{2+}/\Sigma Cr$ values. It is probable that other Cr^{2+} bearing silicate minerals are likewise affected by orientation anisotropy issues, however there are no available data with which to assess this possibility.

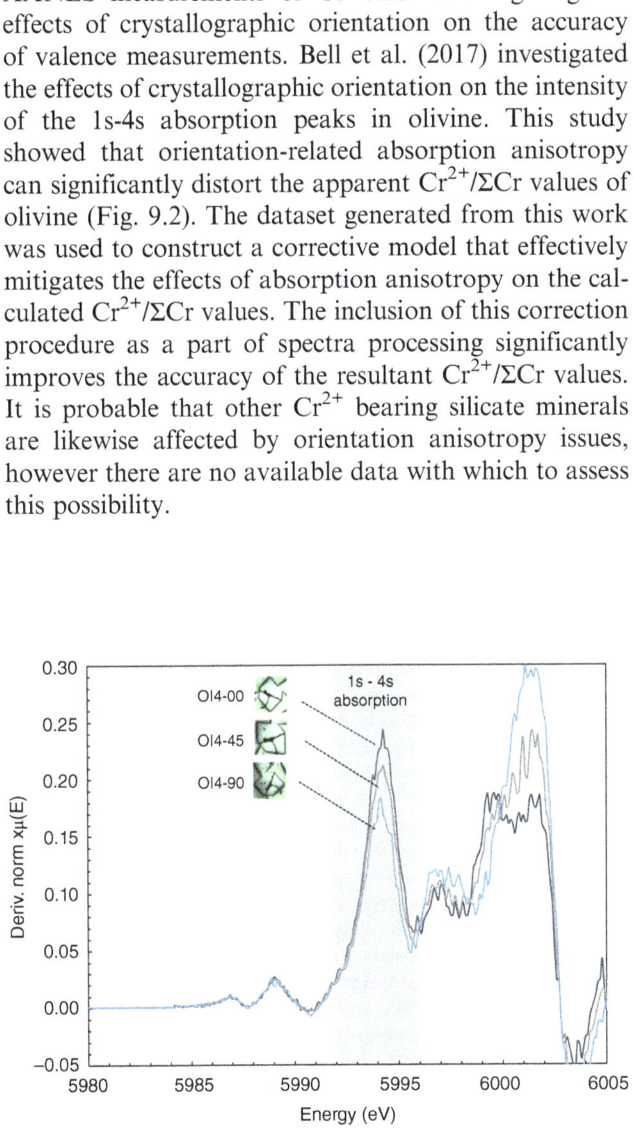

Figure 9.2 Adapted from Bell et al. (2017); illustrates how the crystallographic orientation of olivine relative to the electric field polarization direction of the incident x-ray beam influences the apparent intensity of the 1s-4s absorption feature. The effect of absorption anisotropy in olivine must be corrected to retrieve accurate Cr valence ratios from XANES spectra.

9.3. Cr-REDOX SYSTEMATICS IN SILICATE LIQUIDS: WHAT WE KNOW AND DON'T KNOW

Our current understanding of Cr redox systematics in silicate liquids has been primarily derived from two sources: (i) experimental spectroscopic studies of Cr valence in quenched silicate liquids; and (ii) experimental studies of the redox sensitive spinel-liquid partitioning behavior of Cr. Despite the growing number of experimental studies that have been conducted to elucidate how Cr valence varies in silicate liquids, there are still no thermodynamic or empirical models that can accurately predict $Cr^{2+}/\Sigma Cr$ values as a function of fO_2, temperature, and liquid composition. The lack of a Cr valence model may be largely attributed to the resetting of equilibrium $Cr^{2+}/\Sigma Cr$ that occurs in Fe^{3+} bearing liquids. Although our present knowledge of Cr-redox systematics is incomplete, several existing datasets provide a firm foundation for future Cr valence studies. This section contains a roughly chronological review of the existing Cr valence studies silicate liquids and crystalline phases of geological interest.

9.3.1. The First Direct Measurements of Cr Valence in Quenched Liquids

Schreiber and Haskin (1976) completed the first systematic experimental study of Cr valence in geologically relevant mafic silicate liquids. These experiments were conducted on two iron-free bulk compositions: a forsterite-anorthite-diopside mixture and a forsterite-anorthite silica mixture. The liquids produced in these experiments were equilibrated over a range of geologically pertinent fO_2 values at temperatures from 1300 °C to 1500 °C. The $Cr^{2+}/\Sigma Cr$ of the quenched liquids in the run products were measured by wet chemical titration, electron paramagnetic resonance spectroscopy, and optical spectroscopy.

The Schreiber and Haskin (1976) study produced several important new results, including the first conclusive evidence that Cr^{2+} is a stable valence species in haplobasaltic (i.e., Fe-free) liquids. Furthermore, the Cr valence measurements from their study suggested that the quenched liquids contained binary mixtures of Cr^{2+} and Cr^{3+} under all but the most oxidizing conditions (e.g., log fO_2 = -2 or ΔFMQ+3.4 at 1500°C), and that the $Cr^{2+}/\Sigma Cr$ values of the quenched liquids varied proportionally with the experimentally imposed oxygen fugacity.

Although their experiments were limited to simple iron-free bulk compositions, Schreiber and Haskin (1976) observed that the $Cr^{2+}/\Sigma Cr$ values were broadly correlated with the concentration of network modifying cations (i.e., Ca^{2+} and Mg^{2+}) in their experimental liquids. This observation suggested that the bulk composition of silicate liquids may strongly influence the equilibrium $Cr^{2+}/\Sigma Cr$ ratios. This early conclusion is broadly consistent with our current understanding of how homogenous redox equilibria are connected to theoretical chemical descriptors such as optical basicity and structural descriptors such as the ratio of non-bridging oxygen to tetrahedrally coordinated oxygen (Duffy, 1993; Moretti, 2005).

It should also be mentioned that Schrieber and Haskin (1976) attempted to measure the $Cr^{2+}/\Sigma Cr$ of several Fe-bearing liquid compositions; however, after finding no evidence of the presence of Cr^{2+} in the quenched glasses, they spuriously concluded that Cr^{2+} could not coexist with Fe^{3+} in the liquid; they attributed this result to mutual electron transfer interactions between Cr^{2+} and Fe^{3+} in the liquid at high temperature. Despite this result, Schrieber and Haskin (1976) noted that their Fe-free analog FAD and FAS liquids contained significant quantities of Cr^{2+} even at oxygen fugacity values relevant to terrestrial basaltic magmatism. This observation suggested that the presence of Cr^{2+} in terrestrial basaltic liquids could not be dismissed simply because most terrestrial magma systems evolve under relatively oxidized conditions.

The primary motivation of the Schrieber and Haskin (1976) work was to investigate the pyroxene-liquid and olivine-liquid valence partitioning behavior of Cr in lunar basalts. This focus on relatively low fO_2 lunar magma systems kept the importance of this work from being fully recognized until the completion of later chromite solubility studies that had specific relevance to terrestrial petrologic problems.

9.3.2. Indirect Studies of Cr Valence using Cr-Spinel Buffered Basaltic Liquids

Several more recent experimental studies of Cr-spinel saturated basalts also produced datasets indicating that both Cr^{2+} and Cr^{3+} are present in Fe-bearing basaltic liquids over a wide range of fO_2. Barnes (1986) and Roedder and Reynolds (1991) each investigated Cr-spinel solubility in Fe-bearing basaltic liquids. These studies convincingly showed that the total Cr content of basaltic liquids required to sustain Cr-spinel saturation varied systematically with the experimentally imposed redox conditions. Electron microprobe (EMP) measured Cr concentrations of the Cr-spinel saturated liquids in both studies showed a clear monotonic decrease as a function of increasing experimental fO_2. Both Barnes (1986) and Roedder and Reynolds (1991) reasoned that the stability of Cr-spinel must be inherently controlled by the activity of Cr^{3+} (i.e., $aCrO_{1.5}$) in the liquid, as Cr^{2+} is not readily accommodated in the spinel structure; therefore, the relationship between the measured Cr concentrations and fO_2 reflected underlying changes in the equilibrium $Cr^{2+}/\Sigma Cr$ of the experimental liquids. Liquids with high $Cr^{2+}/\Sigma Cr$ values require high total Cr concentrations to achieve an $aCrO_{1.5}$ value large enough to crystallize Cr-spinel. Even in the absence of direct spectroscopic measurements of Cr valence in the experimental liquids, these chromite solubility datasets provided the first compelling evidence that terrestrial basaltic liquids commonly contain stable Cr^{2+}. Roedder and Reynolds (1991) went on to conclude that the equilibrium $Cr^{2+}/\Sigma Cr$ of MORB liquids at ΔFMQ+0 at 1300°C were approximately 0.30. The reported $Cr^{2+}/\Sigma Cr$ values from the Roedder and Reynolds (1991) experiments on the "401 diabase" basaltic compositions have been re-plotted in Fig. 9.3 as a function of fO_2 and fit using a reformulation of Equation 9.2 that employs an apparent equilibrium constant K'.

The Roedder and Reynolds (1991) study also demonstrated that plots of the $\log(Cr^{3+}/Cr^{2+})$ determined for their experimental liquids vs. $\log(fO_2)$ produced linear arrays with slopes that closely approximated the predicted stoichiometric value of 0.25. Even though it was not explicitly recognized by Roedder and Reynolds (1991), this observation has important implications in that it suggests that the activity coefficient ratio $\gamma CrO_{1.5}/\gamma CrO$ in natural basaltic liquids is relatively invariant with respect to fO_2. On the surface, the Roedder and Reynolds (1991) data suggest that at low concentrations, Cr^{2+} and Cr^{3+} appear to comply with Henry's law. This finding implies that redox behavior of Cr, unlike that of Fe, is not be

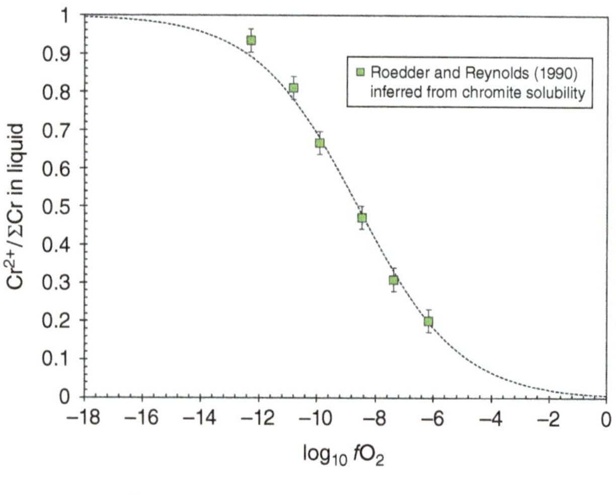

Figure 9.3 $Cr^{2+}/\sum Cr$ ratios plotted vs. experimental fO_2 from Roedder and Reynolds (1991). These $Cr^{2+}/\sum Cr$ data were inferred indirectly from the chromite solubility approach. The sigmoidal curve in the plot was generated by calculating the mean lnK' value from the reported $Cr^{2+}/\sum Cr$ and fO_2 values and using this value with to calculate a sigmoidal curve with a modified version of Equation 9.2 (i.e., the activity coefficient terms dropped as they are implicitly included in the lnK' term is which describes the Cr valence equilibrium for liquids of a single, fixed composition).

complicated by concentration dependent activity coefficients. It has widely been recognized that redox systematics of Fe in silicate liquids may be significantly complicated by fO_2 induced changes in Fe speciation and $\gamma FeO_{1.5}/\gamma FeO$; the complex behavior of Fe has been attributed to potential interactions between Fe^{3+}-Fe^{2+} in the liquid (Kress & Carmichael, 1988; Jayasuriya et al., 2004). The apparent absence of complex activity-concentration relationships for Cr suggests that future thermodynamic models of Cr valence in silicate liquids may justifiably assume that the Cr activities in the liquid are Henrian.

Hanson and Jones (1998) performed a similar series of redox controlled, low-pressure crystallization and chromite solubility experiments. These experiments were conducted on Fe-free haplobasaltic and natural basaltic bulk compositions. Although the primary goal of these experiments was to systematically investigate the valence partitioning behavior of Cr between olivine and mafic liquids as a function of fO_2, Hanson and Jones also calculated $Cr^{2+}/\Sigma Cr$ values of the experimental liquids using a Cr-spinel solubility approach analogous to the one employed by Barnes (1986) and Roedder and Reynolds (1991). Like Roedder and Reynolds (1991), the Hanson and Jones 1998 dataset indicated that (i) the inferred $Cr^{2+}/\Sigma Cr$ values of their Fe bearing liquids showed a systematic relationship to fO_2; (ii) the observed relationship between fO_2 and $Cr^{2+}/\Sigma Cr$ was consistent with the expected stoichiometry of the homogenous Cr redox reaction; and (iii) at temperatures between 1200 °C and 1250 °C, the Fe-bearing MORB liquids contain mixtures of Cr^{2+} and Cr^{3+} at fO_2 values between the IW and FMQ buffers; the regression equations in table 5 of Hanson and Jones (1998) predict $Cr^{2+}/\Sigma Cr$ values at ΔFMQ+0 and 1200 °C for the MORB2 and MORB3 compositions of 0.24 and 0.32, respectively. At these fO_2 and temperature conditions the Kress and Carmichael (1991) equation predicts $Fe^{3+}/\Sigma Fe$ values of 0.15 for the MORB2 liquid and 0.13 for the MORB3 composition, respectively. The results of Hanson and Jones (1998) not only corroborated the chromite solubility results of Roedder and Reynolds (1991) but they also provided an additional piece of evidence indicating that Cr^{2+} does coexist with Fe^{3+} in modestly oxidized mafic liquids.

Like the previous chromite solubility studies, the Hanson and Jones (1998) work made no attempt to explicitly relate the compositional variation among Fe-bearing liquids in their study to either the Cr concentrations at chromite saturation or the inferred and Cr valence ratios of the liquids. Differences in the activity of the chromite component of the experimental Cr-spinel caused by variations in the equilibrium liquid composition may induce changes in the total Cr concentration required to support Cr-spinel saturation in the liquid. In other words, increases or decreases in the chromite activity of the spinel caused by changes in liquid composition may masquerade as changes in $Cr^{2+}/\Sigma Cr$ of the liquid. Changes in chromite solubility caused by variations in the chromite activity of the spinel are difficult to disentangle from genuine changes in the underlying $Cr^{2+}/\Sigma Cr$ of the liquid, thus given this complication, it is difficult to use Cr valence data inferred from chromite solubility experiments to draw concrete conclusions about how liquid compositions influence Cr redox systematics. Perhaps a version of the chromite solubility approach that employs appropriate activity-composition models for Cr-spinel could be applied to existing Cr-spinel solubility datasets to reveal a more nuanced view of the interplay among liquid composition, spinel composition, and $Cr^{2+}/\Sigma Cr$.

9.3.3. XANES Studies of Cr Valence in CMAS and Basaltic Liquids

Berry and O'Neill (2004) and Berry et al. (2006) conducted the first systematic XANES studies of Cr valence in quenched CMAS liquids and produced a set of groundbreaking, high-temperature in situ XANES measurements of $Cr^{2+}/\Sigma Cr$ in a simplified MORB liquid. These XANES studies provide the most comprehensive and robust investigation of $Cr^{2+}/\Sigma Cr$ systematics currently available.

One of the most important advances came from the in-situ XANES measurements on an alkali-free synthetic MORB composition. These experiments were conducted in a specially designed furnace that that allowed XANES measurements to be made up to 1400 °C and under controlled fO_2 values (Fig. 9.4). The results of these in-situ liquid experiments conclusively showed that Cr^{2+} is present at fO_2 conditions as oxidized as the Ni–NiO buffer; the relatively high $Cr^{2+}/\Sigma Cr$ value measured in this oxidized liquid ($\log fO_2 = -6$) clearly demonstrates that Cr^{2+} does in fact coexist with Fe^{+3} at high temperatures, and that $Cr^{2+}/\Sigma Cr$ measured in quenched Fe^{3+} bearing liquid is significantly altered from its equilibrium, high temperature state. The results of the in-situ experiments also showed that the $Cr^{2+}/\Sigma Cr$ decreases systematically with fO_2 in a manner consistent with the experimental results of Roedder and Reynolds (1991) and Hanson and Jones (1998).

Berry et al. (2006) presented the first systematic investigation of how liquid composition and fO_2 collectively control the equilibrium $Cr^{2+}/\Sigma Cr$ of 27 CMAS (CaO-MgO-Al_2O_3-SiO_2) liquids. These experiments were conducted at 1400 °C and at fO_2 values that spanned more than 10 log units. These experiments firmly established that $Cr^{2+}/\Sigma Cr$ values are strongly correlated with changes in liquid composition. For example, the different liquid compositions in the Berry et al. (2006) experiments conducted at 1400 °C and a $\log fO_2$ of -9 (ΔFMQ -2.75)

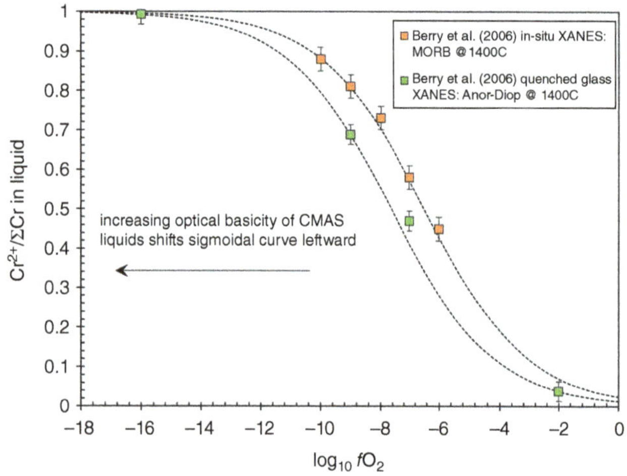

Figure 9.4 Contains plots the XANES measured the $Cr^{2+}/\sum Cr$ ratios from quenched CMAS liquids of Berry et al. (2006) and the in-situ measurements from a simplified MORB liquid. Data from the Berry et al. (2006) study clearly demonstrate that the Cr valence equilibrium shifts with changing melt composition. Increasing the optical basicity of the liquid precipitates a leftward shift in the sigmoidal $Cr^{2+}/\sum Cr$ curve. At constant absolute fO_2 values and temperature, increasing liquid basicity causes a systematic decrease in the equilibrium $Cr^{2+}/\sum Cr$.

produced a broad spectrum of $Cr^{2+}/\Sigma Cr$ values. The compositional variation among these liquids produced a remarkable degree of dispersion in the measured $Cr^{2+}/\Sigma Cr$ values, such that $Cr^{2+}/\Sigma Cr$ ranged from a high of 0.87 in a SiO_2 and Al_2O_3 enriched liquid to a low of 0.48 in a liquid enriched in CaO and depleted in SiO_2. This example illustrates the importance of understanding compositional effects. To put this in perspective, a 40% reduction in $Cr^{2+}/\Sigma Cr$ of a liquid of constant bulk composition requires fO_2 to increase by nearly three orders of magnitude. On the other hand, the range of liquid compositions encountered in natural basaltic liquids is less extreme than the compositional range of the Berry et al. (2006) liquids, therefore it seems unlikely that the compositional evolution brought about by the crystallization of natural basaltic magma systems would induce changes in $Cr^{2+}/\Sigma Cr$ as large as those observed in the Berry et al. (2006) experimental liquids. Assuming the Cr valence behavior in the Fe bearing liquids can be reasonably well predicted by optical basicity as shown in Berry et al. (2006), it appears that wholesale changes in the bulk compositions of basaltic liquids would be required to alter equilibrium $Cr^{2+}/\Sigma Cr$ values by more than 10–20%. As is often the case, additional experimental work is needed to provide more robust constraints on how equilibrium and fractional crystallization might affect $Cr^{2+}/\Sigma Cr$ in natural mafic magmas.

Like Schrieber and Haskin (1976), Berry et al. (2006) observed that $Cr^{2+}/\Sigma Cr$ is linearly correlated with the optical basicity values calculated for the experimental liquids, such that liquids with higher optical basicity values favor lower $Cr^{2+}/\Sigma Cr$ values (Fig. 9.4). The authors also noted that the CaO concentration of the CMAS liquids appeared to be the most important compositional factor with respect to its ability to depress the $Cr^{2+}/\Sigma Cr$ of the liquid; the authors also showed that increasing the mole fraction of CaO in the liquid was generally correlated with substantial decreases in the calculated $\gamma CrO_{1.5}$ values for the investigated suite of CMAS liquids. For example, increasing the mole fraction of CaO in the from 0.10 to 0.40 caused the calculated $\gamma CrO_{1.5}$ values to decrease by roughly a factor of three. In contrast to the significant effects of CaO concentration, the concentrations of MgO and Al_2O_3 showed little significant correlation with the measured $Cr^{2+}/\Sigma Cr$ values in the CMAS liquids. The authors suggested that the observed effects of liquid composition could essentially be explained by two factors: (i) the underlying ability of the liquid to accommodate Cr^{3+} in its preferred high symmetry octahedral site; and (ii) the intrinsic effects of polymerization state or optical basicity of the liquid on stabilizing cations with higher formal valences.

9.3.4. Existing Knowledge Gaps and Future Directions

The Berry et al. (2006) experiments clearly demonstrate the importance of liquid composition with respect to dictating equilibrium $Cr^{2+}/\Sigma Cr$. Despite this importance, there are still considerable gaps in our understanding of Cr valence systematics. The most glaring of these gaps is the fact that relatively little is known about Fe-bearing bulk compositions. The extent to which experimental data from Fe-free haplobasalts can be extrapolated to predict or interpret the $Cr^{2+}/\Sigma Cr$ values of compositionally complex natural mafic silicate liquids has yet to be fully vetted.

The construction of a comprehensive Cr valence model requires a significant experimental investment to quantify how changes in temperature and pressure affect the homogenous Cr redox equilibrium. Additional experimental investigations that are designed to systematically investigate the connection between $Cr^{2+}/\Sigma Cr$ and the concentrations of Fe^{3+}, Fe^{2+}, Na, and K in the liquid are also needed. Given the inherent difficulties in accurately capturing the $Cr^{2+}/\Sigma Cr$ values of Fe^{3+} bearing liquids, future experimental investigations likely will rely on either difficult in-situ XANES studies or exploit XANES measurements of $Cr^{2+}/\Sigma Cr$ minerals such as olivine or pyroxene as proxies for their equilibrium liquids silicate liquids.

9.4. Cr-VALENCE SYSTEMATICS IN EQUILIBRIUM LIQUID-OLIVINE PAIRS

Olivine is unique among most commonly occurring magmatic silicate phases in that it readily accommodates both Cr^{2+} and Cr^{3+} in its structure. Numerous experimental partitioning studies have shown that the olivine-liquid partition coefficient for Cr is nearly invariant with respect to changes in fO_2 (Fig. 9.5). Hanson and Jones (1998) concluded that insensitivity of the "total" Cr partition coefficient to the fO_2 reflects the fact that olivine has a roughly equal affinity for incorporating Cr^{2+} and Cr^{3+} into its structure. This conclusion is strongly supported by the results of an experimental XANES study of $Cr^{2+}/\Sigma Cr$ in olivine performed by Bell et al. (2014). This study performed XANES measurements of $Cr^{2+}/\Sigma Cr$ in olivine that crystallized from basaltic liquids under controlled fO_2 values. The $Cr^{2+}/\Sigma Cr$ of the experimental olivine varied with fO_2 in a manner that mirrored the variation of $Cr^{2+}/\Sigma Cr$ predicted for its parental liquid (Fig. 9.6). This work also confirmed that olivine effectively has no preference for Cr^{2+} over Cr^{3+} or vice-versa. This study provides a strong line of evidence that olivine possesses the unique ability to serve as a high-fidelity recorder of the $Cr^{2+}/\Sigma Cr$ of its conjugate liquid. It is the unique ability of olivine to provide an undistorted view of Cr valence in its parental liquid that forms the rationale behind the Cr valence in olivine oxybarometer proposed in Bell et al. (2014).

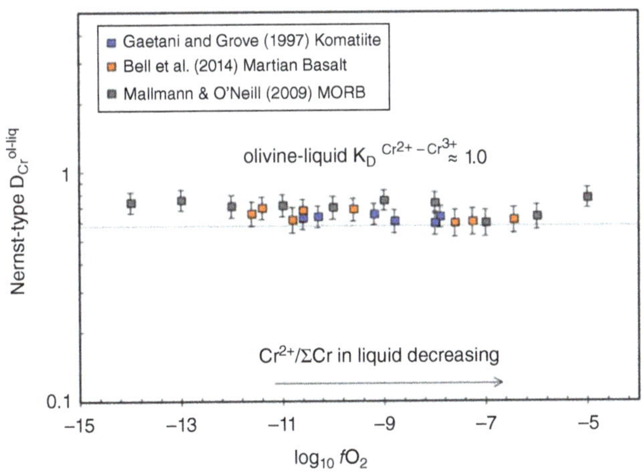

Figure 9.5 Cr partitioning data compiled from several studies plotted vs. experimental fO_2. These data demonstrate that the olivine-liquid partition coefficient for Cr is not influenced by changes in $Cr^{2+}/\sum Cr$ of the liquid. This data suggests olivine has an equal affinity for Cr^{2+} and Cr^{3+}, thus the $Cr^{2+}/\sum Cr$ of olivine should mirror that of its equilibrium liquid. Adapted from Gaetani and Grove (1997), Bell et al. (2014), Mallmann & O'Neill (2009).

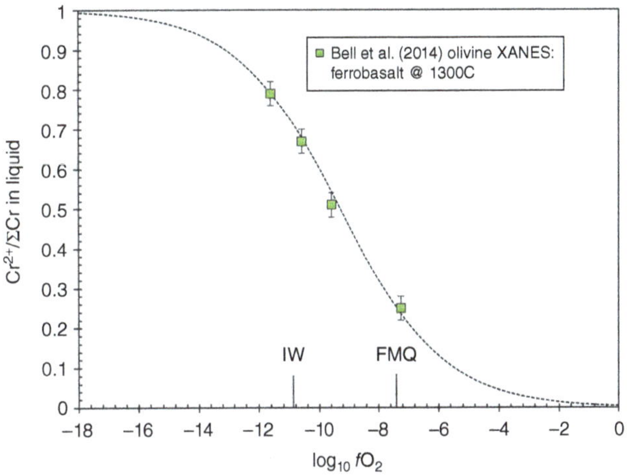

Figure 9.6 XANES measured $Cr^{2+}/\sum Cr$ values of a set of experimentally grown olivine phenocrysts plotted vs. of experimental fO_2. These data demonstrate that the $Cr^{2+}/\sum Cr$ of the olivine phenocrysts are in fact a function of fO_2, and that the Cr valence ratios of olivine that can be described by the same sigmoidal functions that describe Cr valence behavior in silicate liquids. Modified from Bell et al. (2014).

The use of olivine as a Cr valence recorder requires some understanding of the crystal chemical rationale that underpins its ability to accommodate both Cr^{2+} and Cr^{3+}. On the surface, the relatively high partition coefficient values of Cr^{3+} in olivine are somewhat strange considering the difficulty it has in accommodating most trivalent cations. At high fO_2 values, where Cr^{3+} was inferred to be the dominate valence species present in the liquid, Hanson and Jones (1998) calculated ${}^{ol\text{-}liq}D_{Cr3+}$ values that ranged from 0.60–0.70. The partition coefficients from Hanson and Jones (1998) contrast sharply with partition coefficient values for Al^{3+} and Sc^{3+}. Mallmann and O'Neill (2009) reported ${}^{ol\text{-}liq}D_{Al3+}$ values that ranged from 5.5×10^{-3} to 7×10^{-3} and ${}^{ol\text{-}liq}D_{Sc3+}$ values from 0.10 to 0.20.

Several explanations for the relatively high compatibility of Cr^{3+} in olivine have been independently proposed. Burns (1975) pointed out that Cr^{3+} has a strong preference for octahedral coordination; furthermore, Burns (1975) calculated that octahedrally coordinated Cr^{3+} in forsterite has an unusually large crystal field stabilization energy that is on the order of 230kJ (g ion)$^{-1}$. It has also been noted that from a lattice strain perspective, the ionic radius of Cr^{3+} is similar enough to Mg^{2+} and Fe^{2+} that its incorporation into the olivine M-sites is not that energetically taxing.

The incorporation of Cr^{3+} into the olivine structure must also be reconciled with charge balance considerations. Several charge-coupled substitution mechanisms have been proposed to explain how Cr^{3+} can be accommodated in olivine. The most likely substitution involves the presence of an M-site vacancy where

$$(Mg^{2+})^{VI} = \square^{VI} + 2(Cr^{3+})^{VI} \quad (9.3)$$

The existence of this charge couple is supported by an electron paramagnetic resonance study of Cr-doped synthetic forsterite (Shukarov & Tsarov, 2001); this study showed that the Cr^{3+} – vacancy charge couple is realized by a pair of Cr^{3+} cations present in two M1 sites that are separated by an M1 site vacancy (Cr^{3+}M1 – □M1 – Cr^{3+} M1). Hanson and Jones (1998) pointed out that Sc^{3+} adopts a similar vacancy-balanced charge couple, therefore this mechanism may be somewhat common. A second "spinel-structure" charge couple has been proposed by Jollands et al. (2018). This charge couple requires Mg^{+2} to enter the olivine tetrahedral site where it substitutes for Si^{4+}; charge balance is maintained allowing Cr^{3+} to occupy adjacent octahedral sites. At present little is known about the relative importance of these proposed substitution mechanisms or what variables influence which mechanism is more common in natural olivine phenocrysts.

The relatively low compatibility of Cr^{2+} in olivine is also somewhat curious and requires some crystal chemical insight. It is widely known that end-member Cr^{2+} olivine does not exist in nature, which suggests that Cr^{2+} compatibility in the olivine structure is limited. A crystallographic study of synthetic Cr^{2+}-orthosilicate concluded that Cr^{2+} occupies a square planar-like site in the structure (Dollase et al., 1994). Similarly, the EXAFS and XANES study of Cr^{2+} rich olivine from a reduced chondritic meteorite concluded that Cr^{2+} was present in a highly distorted olivine M1 site with a coordination state intermediate between square planar and square pyramidal (McKewon et al., 2014). This site distortion is consistent with the widely known Jahn-Teller distortion commonly observed in the coordination polyhedra of Cr^{2+} in other compounds (Burns, 1975). A simple first order explanation is that the excess local lattice strain caused by the incorporation the Cr^{2+} in distorted coordination is responsible for its limited compatibility in olivine.

9.4.1. Existing Knowledge Gaps and Future Directions

Proxy measurements of $Cr^{2+}/\Sigma Cr$ in olivine show much promise as a method to accurately assess the Cr valence of their parental liquids. There is, however, still a need for additional studies that specifically investigate the potential cause and effect relationships between the bulk composition of olivine (i.e., its fayalite content) and Cr valence partitioning behavior and the dominate Cr^{3+} substitution mechanisms.

9.5. CONCLUDING REMARKS

Perhaps the most powerful application of $Cr^{2+}/\Sigma Cr$ valence systematics is to the assessment of the fO_2 of the mantle sources of primitive basaltic magmas. Using basalts to peer into the oxidation state of the mantle is a notoriously challenging problem; using $Cr^{2+}/\Sigma Cr$ systematics in high Mg# olivine may prove to be an effective method to produce an undistorted view of mantle oxidation state manifested in their partial melts. Furthermore, careful studies of $Cr^{2+}/\Sigma Cr$ systematics in olivine may provide new insights into the importance (or non-importance) of magmatic redox processes and magma mixing in the formation of economic quantities of chromite layered mafic intrusions. The applications of $Cr^{2+}/\Sigma Cr$ valence systematics to Cr isotope fractionation is still in its infancy. Combined XANES studies of Cr valence in minerals and studies of stable Cr isotope fractionation in silicate liquids and magmatic minerals may prove to be a particularly fertile field of future research.

It is also worth mentioning that Cr valence partitioning into other phases such as pyroxenes and amphiboles may also prove to be useful redox proxies. At present there is a paucity of experimental data to assess the possibility. Significant growth in our understanding of Cr valence in magmatic systems seems inevitable given the recent explosion in the number of XANES studies of minerals and silicate glasses appearing the literature.

ACKNOWLEDGMENTS

Andrew Berry and an anonymous reviewer are both thanked for providing through reviews that substantially improved the final version of this chapter. I would also like to acknowledge the financial support of NSF EAR 1550929 which has supported my own work on Cr redox systematics and made the completion of this chapter possible.

REFERENCES

Anderson, A. T., & Wright, T. L. (1972). Phenocrysts and glass inclusions and their bearing on oxidation and mixing of basaltic magmas, Kilauea volcano, Hawaii. *American Mineralogist*, *57*, 188–216.

Ballhaus, C. (1993). Redox states of lithospheric and asthenospheric upper-mantle. *Contributions to Mineralogy and Petrology*, *114*, 331–348. https://doi.org/10.1007/BF01046536

Barnes, S. J. (1986). The distribution of chromium among orthopyroxene, spinel, and silicate liquid at atmospheric pressure. *Geochimica et Cosmochimica Acta, 50*, 1889–1909. https://doi.org/10.1016/0016-7037(86)90246-2

Bell, A. S., Burger, P. V., Le, L., Shearer, C. K., Papike, J. J, Sutton, S. R., et al. (2014). XANES measurements of Cr valence in olivine and their applications to planetary basalts. *American Mineralogist, 99*, 1404–1412. doi: 10.2138/am.2014.4646

Bell, A. S., Shearer, C. K., Burger, P. V., Ren, M., Newville, M., & Lanzirotti, A. (2017). Quantifying and correcting the effects of anisotropy in XANES measurements of chromium valence in olivine: Implications for a new olivine oxybarometer. *American Mineralogist, 102*, 1165–1172. https://doi.org/10.2138/am-2017-5911

Berry, A. J., & O'Neill, H. St.C. (2004). A XANES determination of the oxidation state of chromium in silicate glasses. *American Mineralogist, 89*, 790–798. https://doi.org/10.2138/am-2004-5-613

Berry, A. J., O'Neill, H. St.C., Scott, D. R., Foran, G. J., & Shelley, J. M. G. (2006). The effect of composition on Cr2+/Cr3+ in silicate melts. *American Mineralogist, 91*, 1901–1908. https://doi.org/10.2138/am.2006.2097

Bezos, A., & Humler, E. (2005). The Fe3+/Sigma Fe ratios of MORB glasses and their implications for mantle melting. *Geochimica et Cosmochimica Acta, 69*, 711–725.

Burns, R. G. (1975). Crystal field effects in chromium and its partitioning in the mantle. *Geochimica et Cosmochimica Acta, 39*, 857–864. https://doi.org/10.1016/0016-7037(75)90031-9

Carmichael, I. S. E. (1991). The redox states of basic and silicic magmas—a reflection of their source regions. *Contributions to Mineralogy and Petrology, 106*, 129–141. https://doi.org/10.1007/BF00306429

Carmichael, I. S. E., & Ghiorso, M. S. (1986). Oxidation–reduction relations in basic magma: a case for homogeneous equilibria. *Earth Planetary Science Letters, 78*, 200–210. https://doi.org/10.1016/0012-821X(86)90061-0

Dollase, W. A., Seifert, F., & O'Neill, H. St. C. (1994). Structure of Cr2SiO4 and possible metal-metal interactions in crystals and melt. *Physics and Chemistry of Minerals, 21*, 104–109. doi: 10.1007/BF00205221

Duffy, J. A. (1993). A review of optical basicity and its applications to oxide systems. *Geochimica et Cosmochimica Acta, 57*, 3961–3970. https://doi.org/10.1016/0016-7037(93)90346-X

Fudali, R. F. (1965). Oxygen fugacities of basaltic and andesitic magmas. *Geochimica et Cosmochimica Acta, 29*, 1063–1075. https://doi.org/10.1016/0016-7037(65)90103-1

Gaetani, G. A., & Grove, T. L. (1997). Partitioning of moderately siderophile elements among olivine, silicate melt, and sulfide melt: Constraints on core formation in the Earth and Mars. *Geochimica et Cosmochimica Acta, 61*, 9, 1829–1846. https://doi.org/10.1016/S0016-7037(97)00033-1.

Hanson, B., & Jones, J. H. (1998). The systematics of Cr3+ and Cr2+ partitioning between olivine and liquid in the presence of spinel. *American Mineralogist, 83*, 669–684.

Jayasuriya, K. D., O'Neill, H. St.C., Berry, A., & Campbell, S. J. (2004). A Mössbauer study of the oxidation state of Fe in silicate melts. *American Mineralogist, 89*, 1597–1609. https://doi.org/10.2138/am-2004-11-1203

Jollands, M. C., O'Neill, H. St.C., Van Orman, J., Berry, A. J., Hermann, J., Newville, M., & Lanzirotti, A. (2018). Substitution and diffusion of Cr2+ and Cr3+ in synthetic forsterite and natural olivine at 1200–1500 °C and 1 bar. *Geochimica et Cosmochimica Acta, 220*, 407–428. https://doi.org/10.1016/j.gca.2017.09.030

Kress, V. C., & Carmichael, I. S. E. (1991). The compressibility of silicate liquids containing Fe2O3 and the effect of composition, temperature, oxygen fugacity and pressure on their redox states. *Contributions to Mineralogy and Petrology, 108*, 82–92. https://doi.org/10.1007/BF00307328

Lee, C.-T., Leeman, W. P., Canil, D., & Li, Z.-X.A. (2005). Similar V/Sc systematics in MORB and arc basalts: implications for the oxygen fugacities of their mantle source regions. *Journal of Petrology, 46*, 2313–2336. https://doi.org/10.1093/petrology/egi056

Mallmann, G., & O'Neill, H. S. C. (2009). The crystal/melt partitioning of v during mantle melting as a function of oxygen fugacity compared with some other elements (Al, P, Ca, Sc, Ti, Cr, Fe, Ga, Y, Zr and Nb). *Journal of Petrology, 50*, 1765–1794. https://doi.org/10.1093/petrology/egp053

McKeown, D. A., Buechele, A. C., Tappero, R., McCoy, T. J., & Gardner-Vandy, K. (2014). X-ray absorption characterization of Cr in forsterite within the MacAlpine Hills 88136 EL3 chondritic meteorite. *American Mineralogist, 99*, 190–197. https://doi.org/10.2138/am.2014.4508

Moretti, R. (2005) Polymerisation, basicity, oxidation state and their role in ionic modelling of silicate melts. *Annals of Geophysics, 48*, 583–608. https://doi.org/10.4401/ag-3221

Moynier, F., Yin, Q.-Z., & Schauble, E. (2011). Isotopic evidence of Cr partitioning into Earth's core. *Science, 331*, 1417–1420. doi: 10.1126/science.1199597

Kelley, K. A., & Cottrell, E. (2009). Water and the oxidation state of subduction zone magmas. *Science, 325*, 605–607. doi: 10.1126/science.1174156

Righter K., Danielson L. R., Pando K. M., Shofner G. A., Sutton S. R., Newville M., & Lee C.-T. (2016). Valence and metal/silicate partitioning of Mo: Implications for conditions of Earth accretion and core formation. *Earth and Planetary Science Letters, 437*, 89–100. https://doi.org/10.1016/j.epsl.2015.12.025

Roedder, P. L., & Reynolds, I. (1991). Crystallization of chromite and chromium solubility in basaltic melts. *Journal of Petrology, 32*, 909–934. https://doi.org/10.1093/petrology/32.5.909

Sack, R. O., Carmichael, I. S. E, Rivers, M., & Ghiorso, M. S. (1980). Ferric-ferrous equilibria in natural silicate liquids at 1 bar. *Contributions to Mineralogy and Petrology, 75*, 369–376. https://doi.org/10.1007/BF00374720

Schreiber, H. D., & Haskin, L. A. (1976). Chromium in basalts: Experimental determination of redox states and partitioning among silicate phases. *Proceedings of the Lunar Science Conference*, 7, 513–538. New York: Pergamon Press.

Shakurov, G. S., & Tarasov, F. (2001). High-frequency tunable EPR spectroscopy of Cr3+ in synthetic forsterite. *Applied Magnetic Resonance*, *21*, 597–605. 10.1007/BF03162432

Shorttle, O., Moussallam, Y., Hartley, M. E., Maclennan, J., Edmonds, M., & Murton B. J. (2015). Fe-XANES analyses of Reykjanes Ridge basalts: implications for oceanic crust's role in the solid Earth oxygen cycle. *Earth and Planetary Science Letters*, *427*, 272–285. https://doi.org/10.1016/j.epsl.2015.07.017

Sutton, S. R., Jones, K. W., Gordon, B., Rivers, M. L., Bajt, S., & Smith, J. V. (1993). Reduced chromium in olivine grains from lunar basalt 15555: X-ray absorption near edge structure (XANES). *Geochimica et Cosmochimica Acta*, *57*, 461–468. https://doi.org/10.1016/0016-7037(93)90444-2

Wood, B. J., Wade, J., & Kilburn, M. R. (2008). Core formation and the oxidation state of the Earth: additional constraints from Nb, V and Cr partitioning. *Geochimica et Cosmochimica Acta*, *72*, 1415–1426. https://doi.org/10.1016/j.gca.2007.11.036

10

The Thermodynamic Controls on Sulfide Saturation in Silicate Melts with Application to Ocean Floor Basalts

Hugh St.C. O'Neill

ABSTRACT

A thermodynamic model to calculate the sulfide content at sulfide saturation or SCSS of basaltic and intermediate composition silicate melts has been built from four independently measurable thermodynamic entities, namely the standard state Gibbs free energy of the saturation reaction, the "sulfide capacity", and the activities of FeO in the silicate melt and of FeS in the coexisting sulfide:

$$\ln\left[S^{2-}\right]_{SCSS} = \Delta G^o_{FeO-FeS}/RT + \ln C_{S^{2-}} - \ln a^{sil\ melt}_{FeO} + \ln a^{sulf}_{FeS}$$

The model was calibrated for silicate melts of basic and intermediate composition from published experimental results as a function of temperature, silicate melt composition, and sulfide matte composition in the system Fe-Ni-Cu-S-O at 1 bar. The likely effects of pressure and H_2O content on SCSS were included in an exploratory way. The model was tuned against the large dataset of S contents in OFB glasses of Jenner and O'Neill (2012), giving it a precision comparable to that of the S analyses themselves, which is ~ 5%. All but 3% the OFB glasses were found to be sulfide saturated within uncertainty; these 3% have lost S by devolatization, revealed by their low S/Se. Applying the model to other OFB datasets suggests sulfide saturation is ubiquitous, including olivine-hosted melt inclusions proposed previously to be sulfide undersaturated. The sulfur fugacity (fS_2) of undegassed Ocean Floor Basalts varies proportionally to fO_2, with $\log_{10}fS_2$ typically within the range −0.6 to +0.4.

10.1. INTRODUCTION

In the Earth sciences, quantifying the thermochemistry of sulfur dissolved in magmatic silicate melts is needed for understanding the behavior of chalcophile elements during partial melting and magmatic differentiation, sulfur degassing from volcanic eruptions, core formation in the early solar system, and for modeling the petrogenesis of many kinds of magmatic ore deposits. Meanwhile, in the technologically directed applied sciences, the thermochemical description of sulfur dissolved in the oxide-silicate melts found as slags or fluxes is needed wherever this element is present, one important application being the desulfurization of steel (e.g., Schrama et al., 2017). Accordingly, the thermochemistry of S in silicate melts has received considerable technological attention. While the fundamental principles are obviously of mutual interest, there are aspects to the understanding of S in silicate melts in the geological context that are irrelevant to technological applications, and vice versa. In the technological sector, the compositional space encompassing silicate melts of practical interest is vast, as well shown by the list assembled by Kang and Pelton (2009) of silicate

Research School of Earth Sciences, Australian National University, Canberra, ACT, Australia

Magma Redox Geochemistry, Geophysical Monograph 266, First Edition.
Edited by Roberto Moretti and Daniel R. Neuville.

DOI: 10.1002/9781119473206.ch10

(and aluminate) melts whose "sulfide capacity" (defined below) has been determined experimentally. Yet this dataset includes hardly any realistic magmatic silicate melt compositions. These tend to be richer in SiO_2 (> 40 wt%), with a limited range of alumina contents, and atomic Si/Al between 2.5 and 4.5. Metallurgical slags are often either almost free of alumina, or alternatively have high Al/Si beyond the range found in nature. Natural magmas have modest but non-negligible alkali contents, with most plotting on the standard total alkali-silica diagram used for rock classification (Na_2O+K_2O vs. SiO_2) between 2 and 10 wt% (Na_2O+K_2O), whereas metallurgical slags either have negligible alkalis (e.g., most steel-making slags), or, elsewhere, Na_2O as the main or even sole basic-oxide constituent (e.g., Chan & Fruehan, 1986, 1988). In terms of melt structure, most magmatic melt compositions have NBO/T, the ratio of non-bridging oxygens to tetrahedrally coordinated cations, between 0 and 1.5, whereas few of the compositions listed by Kang and Pelton (2009) have NBO/T < 2. "Water" (H as H_2O) is a major component in many magmas, but is almost absent in slags, because the dissolution of more than trace amounts of H_2O in silicate melts requires geological pressures. And of course such pressures are not relevant to metallurgical processes, which generally occur not far from atmospheric pressure.

There is a further difference. The technologist needs to predict S solubility in a slag perhaps to a factor of two or so, but over the vast range of slag compositions mentioned above. For example, they may need to know how much lime should be added to a slag to reduce S in the steel by a factor of, say, five. To the petrologist, the precision with which sulfide solubilities need to be known is far greater, albeit only within a relatively narrow range of melt compositions. For example, among Ocean Floor Basalts, almost the entire range of recovered glass samples analyzed for S fall within a factor of three (Jenner & O'Neill, 2012), so predicting S solubilities to a factor of two would be useless. From the perspective of basalt petrology, the aspirational target is to predict the SCSS, which is how much S dissolved as S^{2-} there is in a silicate melt in equilibrium with a sulfide phase, to somewhere approaching the analytical accuracy to which S contents can be measured. In basaltic glasses, this appears to be at about ±5%. Formidable though this challenge may seem, the theme of this review is that it can now be met, thanks to the efforts of many experimental petrologists over the last few decades.

The equilibrium that describes SCSS is:

$$\underset{\text{silicate melt}}{\mathrm{FeO}} + 1/2\,\mathrm{S_2} = \underset{\text{sulfide melt}}{\mathrm{FeS}} + 1/2\,\mathrm{O_2} \tag{10.1}$$

As emphasized by O'Neill and Mavrogenes (2002), this equilibrium can be deconstructed into four thermodynamic entities: (i) the standard-state free energy of the reaction, which is a function of temperature and pressure alone; (ii) the "sulfide capacity" of the melt; (iii) the activity of FeO in the silicate melt; and (iv) the activity of FeS in the coexisting sulfide (often molten, when it is appropriately called a matte). Entities (ii), (iii), and (iv) are dependent on composition as well as temperature and pressure. In principle, each one of these entities can be measured experimentally, and independently of each other. In practice there are experimental limitations, but at least at atmospheric pressure all four items have indeed been measured independently of each other. The standard state free energy and activities of FeO in silicate melts and FeS in sulfide mattes are familiar thermodynamic quantities, but the "sulfide capacity" of a silicate melt is not, which has caused some confusion as to its nature, leading to uncertainty as to how it may depend on melt composition. Its relationship with other potentially useful thermodynamic properties of silicate melts, for example their "carbonate capacity," has also been underappreciated.

10.2. SULFIDE CAPACITY

One area in which the technological sciences provided guidance was in establishing the notion of the "sulfide capacity," denoted $C_{S^{2-}}$ (Fincham and Richardson, 1954). These authors studied the solubility of S in some simple-system silicate melts (and one aluminate melt) in equilibrium with SO_2-H_2-CO_2-N_2 gas mixtures, which were used to control the fugacities of sulfur and oxygen (fS_2 and fO_2). While they mention steelmaking as an application of their work, they emphasized that their goal was a broader understanding of silicate melt thermochemistry. They noted that "Previous attempts to elucidate the chemistry of sulfur in molten silicates (e.g., Grant and Chipman, 1946) have been based on the measurements of the equilibrium distribution of sulfur between complex slags and molten iron. Although a considerable amount of quantitative information has been amassed, the results are too complicated to permit the development of a clear picture from them." If this is a reasonable view from within the restricted context of steelmaking, how much more so is it for geological processes? Their call for some judicious reductionism has much influenced the approach of this present review.

Fincham and Richardson (1954) found that, while keeping other compositional parameters and temperature constant, the sulfur concentration in the melt changed proportionally to $(fS_2)^{1/2}$, when fO_2 was low as well as constant. Likewise, when keeping fS_2 constant, they found that S concentrations changed proportionally to $(fO_2)^{-1/2}$. These observations implied that S was dissolving only as the sulfide species, S^{2-}, and did so by replacing O^{2-}. This has sometimes been summarized by writing the reaction:

$$\underset{\text{sil melt}}{O^{2-}} + \tfrac{1}{2} S_2 = \underset{\text{sil melt}}{S^{2-}} + \tfrac{1}{2}O_2 \quad (10.2)$$

There is no implication that O^{2-} is an abundant species in the silicate melt, which it is not in the compositional space of natural magmas (Mysen & Richet, 2005). Rather, O^{2-} should be viewed as a "structural element," by analogy with vacancies or interstitials in the point-defect thermodynamics of crystals (e.g., Schmalzried, 1995). In this formalism, the "structural elements" are treated like thermodynamic components, with chemical potentials that vary monotonically with the logarithms of their concentrations. Whether structural elements can actually have definable chemical potentials is a moot point, but this is the hypothesis. It follows from this hypothesis that a quantity resembling an equilibrium constant may be written for Reaction 10.2, namely $[S^{2-}]/[O^{2-}](fO_2/fS_2)^{1/2}$, where the square brackets denote concentration. It is then assumed that $[O^{2-}]$ is unaffected by $[S^{2-}]$, at least for small amounts of S^{2-} in the melt, which motivates the definition of the sulfide capacity of the melt ($C_{S^{2-}}$) as the constant of proportionality between $[S^{2-}]$ and $(fO_2/fS_2)^{1/2}$:

$$C_{S^{2-}} = [S^{2-}](fO_2/fS_2)^{1/2} \quad (10.3)$$

where $[S^{2-}]$ is the sulfur content of the melt in any convenient units, like parts per million by weight.

O'Neill and Mavrogenes (2002) pointed out that an analogous reaction could be written using orthosilicate and framework structural units of the type actually found in silicate melts:

$$\tfrac{1}{2} \underset{\text{sil melt}}{SiO_4^{4-}} + \tfrac{1}{2} S_2 = \underset{\text{sil melt}}{S^{2-}} + \tfrac{1}{2} \underset{\text{sil melt}}{SiO_2} + \tfrac{1}{2} O_2 \quad (10.4)$$

This reaction has the same dependence on $(fO_2/fS_2)^{1/2}$, also has two "structural elements" (S^{2-} and SiO_4^{4-}), and in addition one definable thermodynamic component (SiO_2). In principle, reactions that conform even more closely to silicate melt structure could be formulated. According to Mysen (1983), "three-dimensional network units are the most common in natural magmatic liquids. These units will also contain most of the aluminum. Consequently, the anionic structure of anhydrous magmatic liquids at or near 1 atm pressure may be described as a combination of nearly pure $Si_2O_6^{4-}$ (NBO/Si = 2) and $Si_2O_5^{2-}$ (NBO/Si = 1) units, together with three-dimensional units in which the aluminum is located." This insight suggests the reaction:

$$\underset{\text{sil melt}}{Si_2O_6^{4-}} + \tfrac{1}{2} S_2 = \underset{\text{sil melt}}{S^{2-}} + \underset{\text{sil melt}}{Si_2O_5^{2-}} + \tfrac{1}{2} O_2 \quad (10.5)$$

All these ways of representing the relationship between fS_2, fO_2 and the S^{2-} concentration of the silicate melt, denoted here as $[S^{2-}]$, come to the same thing, namely $[S^{2-}]$ is proportional to $(fS_2/fO_2)^{1/2}$, provided that $[S^{2-}]$ is sufficiently low that it does not noticeably perturb the concentrations of the other structural elements in the reactions (i.e., O^{2-}, SiO_4^{2-}, etc.). In the case of Reaction 10.5 this implies that $[S^{2-}]$ is too low to affect the ratio $[Si_2O_5^{2-}]/[Si_2O_6^{4-}]$. The proportionality between $[S^{2-}]$ and $(fS_2/fO_2)^{1/2}$ is critical; the concept of a sulfide capacity would not be of much use if this proportionality did not hold.

Fincham and Richardson (1954) went on to show that at high oxygen fugacity (obtained with SO_2-O_2 gas mixes) the proportionality changed, from which they inferred that S was dissolving as sulfate (SO_4^{2-}), hence the definition of the sulfate capacity $C_{SO_4^{2-}}$:

$$C_{SO_4^{2-}} = [SO_4^{2-}](fS_2)^{-1/2}(fO_2)^{-3/2} \quad (10.6)$$

At intermediate fO_2 the levels of S dissolving in the silicate melts dropped below detection limits, implying that S in oxidation states between 2– (in S^{2-}) and 6+ (in SO_4^{2-}) in the melt are unimportant. This precludes a significant role for S dissolving as S^{4+}, or sulfite (SO_3^{2-}), in silicate melts at atmospheric pressure. Whether S^{4+} may play a role at higher pressure, where it is physically possible for pSO_2 to reach values very much greater than 1 bar, remains a grey area – see Métrich et al. (2009) and the discussion on S X-ray spectroscopy below. The fact that their S solubility data could be fitted to the two relationships 10.3 and 10.6 only also meant that experimentally detectable roles could be precluded for S dissolved in silicate melts as elemental S, other molecular S species (e.g., S_2), polysulfide species (e.g., S_3^-), or thiosulfite, thiosulfate, and so on.

The range of melt compositions investigated by Fincham and Richardson (1954) was limited, and they were unable to obtain reproducible results on their Fe-containing composition. They also only looked at a limited range of temperatures. The task of exploring the extent of the compositional space, T and fO_2 over which the Fincham-Richardson sulfide and sulfate solubility laws hold was left for subsequent experimental investigation.

The first attempt to test the Fincham-Richardson relationship on a geologically relevant composition (a komatiite) was by Shima and Naldrett (1975), using fS_2 from 10^{-4} to $10^{-1.4}$ bars and just two different fO_2s at 1450°C. They confirmed that $[S] \propto (fS_2)^{1/2}$ at both fO_2s, but the proportionality between the two fO_2s was only to the power of 0.4, so the relationship looked to fail that part of the test. An even more emphatic lack of agreement was reported later (Buchanan et al. 1983). O'Neill and Mavrogenes (2002) then tested twelve Fe-free compositions in the system CaO-MgO-Al_2O_3-SiO_2 over as wide a range of both fO_2 and fS_2 as their experimental method allowed, but only at one temperature, 1400°C. The low fO_2 and high fS_2 limits were set by the CO_2-CO-SO_2 gas mixes that they used, and the requirement that the partial pressures of all three gases sum to 1 bar. (This combination of input gases does not go to as low fO_2 as the CO_2-H_2-SO_2 mixes of Fincham and Richardson 1954, but has the experimental advantage that achieving the equilibrium gas-phase speciation from the input gas is faster, and there is less susceptibility to thermal segregation in the temperature gradients of conventional gas-mixing muffle furnaces.) The high fO_2 and low fS_2 limits were set by the dissolved S falling below the limit of quantification feasible with the electron microprobe. Contrary to the earlier studies, it was found that the Fincham-Richardson relationship was followed within experimental uncertainty. O'Neill and Mavrogenes (2002) also tested six Fe-containing compositions, and here again the results showed agreement with the Fincham-Richardson hypothesis within experimental uncertainty. The fO_2–fS_2 range that could be accessed experimentally while keeping melt composition constant was not as great as for the Fe-free compositions, because the melt compositions lost some of their FeO towards lower fO_2 and/or higher fS_2, due to exsolution of FeS-rich sulfide melt. Provided the Fincham-Richardson relation can be assumed to hold, this is not disadvantageous, because it throws valuable experimental light on the relationship between $C_{S^{2-}}$ and melt composition (see Section 10.3).

All the compositions examined by O'Neill and Mavrogenes (2002) to test the Fincham-Richardson relationship were broadly similar in terms of melt descriptors like optical basicity or NBO/T (non-bridging oxygens to tetrahedral cations) to natural basalts, *sensu latissimo*, with NBO/T from just above 0 to less than 2. Quantifying solubilities of S as S^{2-} in "acidic" melts with NBO/T ~ 0 was not possible because the solubilities of S in such melts are so low under the experimentally accessible conditions of fO_2 and fS_2 in gas-mixing experiments at atmospheric pressure that adequately precise results cannot be obtained with the electron microprobe. At present there are no unambiguous tests of the Fincham-Richardson relationship for such melts, but this would be feasible with the much better limits of detection afforded by SIMS (< 1 ppm S according to Hauri et al., 2002).

There are currently few published experiments against which to test the Fincham-Richardson relationship for sulfate solubilities (Equation 10.2). An exception is the study of Holmquist (1966), who verified that the relationship of Equation 10.5 generally held well in Na_2O-SiO_2 melts at 1150–1250°C, except at very high concentrations, above several weight percent S. Results mentioned in abstract (O'Neill and Mavrogenes, 2019) confirm the relationship for a wide variety of more magmatically relevant silicate melt compositions.

10.2.1. The Experimental Sulfur Solubility Minimum

One of the main ways that Fincham and Richardson (1954) chose to present their experimental results has subsequently proved unfortunate. The problem that Fincham and Richardson faced was how to display the way that the total amount of S dissolved in a silicate melt changes with fO_2 in going from the sulfide-dominated to the sulfate-dominated regions of fO_2. Clearly connecting the two variables of dissolved S and fO_2 cannot be addressed without taking into account what a third variable, most lucidly fS_2, is doing (Equations 10.2 and 10.5). Illustrating their results under the specification that fS_2 be kept constant was not a practical option for Fincham and Richardson, because to obtain measurable S as S^{2-} in the sulfide field they needed fS_2 to be in the region of 10^{-4} bars, whereas to get measurable S as SO_4^{2-} in the sulfate field, fS_2 needed to be < 10^{-10} bars (depending on temperature). This low fS_2 stems from the requirement for high fO_2. So the quantity that they kept constant for their plots of [$\sum$S] versus log fO_2 was the concentration of SO_2 in their input gas mix, which was either 1% or 2%. Note, this is in the input gas mix, and is not the SO_2 concentration in the gas at equilibrium in the experiments. The resulting plots showed a V-shaped dependence of dissolved S on log fO_2. This same representation was subsequently used by Katsura and Nagashima (1974). Such a variable has no meaning outside the laboratory: for example, nature is not known to meter vapor composed of unequilibrated gas species into magma chambers. Nor is the input gas composition relevant to experiments that do not use gas mixing, such as those under pressure. But somehow the idea that there is a minimum in the solubility of sulfur in silicate melts as a function of oxygen fugacity crept into popular culture. This myth has no substance. Unless a third variable is specified, such as fS_2 (or fSO_2), or the presence and composition of a saturating phase (like a sulfide matte), nothing can be said about the relationship between the two variables of S content and fO_2. In nature, silicate melts with redox speciations towards the reduced end of the terrestrial spectrum, like Ocean Floor basalts, have FeS-rich sulfide matte as their saturating phase, whereas the relevant saturating phase under oxidizing conditions, as for some convergent margin magmatism, might be anhydrite (e.g., Luhr, 2008). There is no continuity in the chemical potentials of sulfur-containing components between the two phases (that is, sulfide matte and anhydrite), hence S contents at phase saturation cannot be conceived as a continuous function of fO_2 between such conditions.

10.3. THE THERMODYNAMIC MEANING OF THE SULFIDE CAPACITY

One way of interpreting the capacities $C_{S^{2-}}$ and $C_{SO_4^{2-}}$ is as the phenomenological constants of proportionality in the relationships of 10.3 and 10.6, as established by experiment. The experiments also show that the capacities vary with the major-element composition of the silicate melt, and temperature. To get a better understanding of these dependences, particularly the functional form of the dependence on melt composition, it is necessary to look beyond the phenomenology, into the thermodynamic foundations. The principle should be the same for both $C_{S^{2-}}$ and $C_{SO_4^{2-}}$ (and indeed for other similar possible "capacities," see Wagner, 1975); here they will be addressed for $C_{S^{2-}}$.

Haughton et al. (1974) suggested that the sulfide capacities measured in their experiments at 1200°C and atmospheric pressure using CO-CO_2-SO_2 gas mixtures could be fitted to an equation with the form:

$$\ln C_{S^{2-}} = A_0^{S^{2-}} + \Sigma X_M A_M^{S^{2-}} \quad (10.7)$$

The justification for this was supplied later by O'Neill and Mavrogenes (2002) from the theory of reciprocal solutions. The thermodynamics of such solutions with mixing on two (or more) sublattices is a well-trod path in physical chemistry (e.g., Flood et al., 1954), and has long been the common choice in the thermochemical modelling of slags (e.g., Hillert, 1998, 2001). Indeed, it forms the basis of most currently used models for calculating $C_{S^{2-}}$ in metallurgical contexts (e.g., Reddy & Blander, 1987; Pelton et al., 1993; Kang & Pelton, 2009).

In these multicomponent ionic solutions, cations mix on one sublattice, anions on the other. Consider the simplest case, the solution $(A^{x+}, B^{x+})(X^{x-}, Y^{x-})$. There exist four end-member compositions, AX, BX, AY, and BY, which together define the limits of the free-energy surface in compositional space, to which all the other contributions to the free energy of the solution are additional, including that from the configurational entropy. The existence of such a free energy surface is not dependent on details of melt structure; it is more fundamental, depending only on the uncontested assumption of charge balance between electropositive elements forming cations (A and B) and electronegative elements forming anions (X and Y). With this assumption, individual ions (A^{x+} etc.) and $A^{x+}B^{x+}$ or $X^{x-}Y^{x-}$ are disregarded as thermodynamic components, because they cannot be independently varied.

The total free energy of the solution may be visualized as being built up from this surface in multicomponent hyperspace by adding the other types of free energies, of which that contributed by the configurational entropy of mixing on each sublattice is fundamental. The free energy of the solution must consist of these two contributions at least; others may or may not be important. These others, collated here under the term "the excess free energy of mixing," include heats of mixing and deviations in the entropies of mixing from whatever model is used for the configurational entropy. The total free energy is therefore given by $G_{melt} = G^o_{end-members} + G_{config} + G_{excess}$.

Assuming that a silicate melt consists of cation and anion sublattices, the three contributions are evaluated as follows:

1. $G^o_{end-members}$ is the sum of the free energies of the pure end-member components, weighted according to the mole fractions of these components. When normalized to the single-anion basis, these components have the general formula $M_{x/z}X$, where M is a cation of valence z, and X are anions of valence x. The fundamental assumption of the reciprocal solution theory is that all possible components that fulfill the criterion of charge balance are present, such that:

$$G^o_{end-members} = \Sigma_{M,X} X_M^{cat} X_X^{an} \mu^o_{M_{x/z}X} \quad (10.8)$$

Here X_M^{cat} is the mole fraction of M^{z+} cations on the cation sublattice, and likewise X_X^{an} is the mole fraction of X^{x-} anions on the anion sublattice. Note that $\Sigma_M X_M^{cat} = \Sigma_X X_X^{an} = \Sigma_{M,X} X_M^{cat} X_X^{an} = 1$. The notation for chemical potential (μ) is used rather than Gibbs free energies for convenience, including avoiding the need to define a reference state, but may be understood to be equivalent in practice (the reference states always cancel out across a balanced chemical reaction). The important point is that the standard state free energy, $\mu^o_{M_{x/z}X}$, of every possible stoichiometric component $M^{z/2}X$ enters into the model. To take an example relevant to the present discussion, the chemical potential of FeS in a multicomponent silicate melt (μ_{FeS}) ineluctably implies something about the value of, say, $\mu^o_{Na_2S}$, unless there is no Na in the melt. The standard-state chemical potential of each possible end-member, $\mu^o_{M_{x/z}X}$, contributes to $G^o_{end-members}$ proportionally to its statistical weight, $X_M^{cat} X_X^{an}$. The logic of this is best understood by noting that the chemical potential of each and every charge-balanced component $M_{x/z}X$ can be in principle evaluated ($\mu_{M_{x/z}X} \equiv \partial G_{melt}/\partial n_{M_{x/z}X}$), and by definition $\mu_{M_{x/z}X} = \mu^o_{M_{x/z}X} + RT \ln a_{M_{x/z}X}$, therefore the free energy of the solution contains information on $\mu^o_{M_{x/z}X}$ for each and every $M_{x/z}X$, so long as some M and some X are present.

2. G_{config} is the ideal configurational entropy that accrues from mixing of cations and anions separately on each sublattice. The usual model is:

$$G_{config} = -RT\left[\Sigma_M n_{cat} X_M^{cat} \ln\left(X_M^{cat}\right) + \Sigma_X n_{an} X_X^{an} \ln\left(X_X^{an}\right)\right] \quad (10.9)$$

here n_{cat} and n_{an} are the fraction of sites available for mixing of the cations and anions, respectively.

3. G_{excess} is the excess free energies of mixing, which accounts for heats of mixing, deviations from the ideal configurational free energy of mixing, and the free energy from any non-configurational entropy of mixing. G_{excess} is conventionally expressed using activity coefficients for cations and anions separately on each sublattice (γ_M^{cat} and γ_X^{an}), by analogy with the configurational free energy:

$$G_{ex} = -\mathrm{RT}\left[\Sigma_M n_{cat} X_M^{cat} \ln\left(\gamma_M^{cat}\right) + \Sigma_X n_{an} X_X^{an} \ln\left(\gamma_X^{an}\right)\right] \tag{10.10}$$

O'Neill and Mavrogenes (2002) presented the next stage of the argument using oxide and sulfide components $M_{z/2}O$ and $M_{z/2}S$, but, as they pointed out, a similar argument could be developed using silicate anion structural elements, such as, for example, orthosilicate unites SiO_4^{2-} and polymerized SiO_2 units, as for Reactions 10.4 and 10.5. Given our present understanding of the structure of silicate melts (e.g., Mysen & Richet, 2005), choosing such structural units might only yield a tractable model between the orthosilicate-equivalent composition at NBO/T = 4, and a fully polymerized melt at NBO/T ~ 0. In actuality, most of the experimental studies of S^{2-} solubility in the geological literature are on melt compositions with NBO/T between 0.05 and 2, and, as noted above, there are few if any precise measurements of $C_{S^{2-}}$ on compositions with NBO/T at or about 0 because the low levels of S^{2-} in such melts are near to the analytical limits of quantification by EMPA under experimentally accessible conditions.

With orthosilicate and silica components, the relationship between [S^{2-}] in the silicate melt and fO_2 and fS_2 is given by:

$$\tfrac{1}{2}\underset{\text{sil melt}}{Mg_2SiO_4} + \tfrac{1}{2}S_2 = \underset{\text{sil melt}}{MgS} + \tfrac{1}{2}\underset{\text{sil melt}}{SiO_2} + 1/2\,O_2 \tag{10.11}$$

The selection of Mg^{2+} as the cation used to describe the equilibrium is arbitrary; for example, Fe^{2+} or Ca would do as well. It is essential to acknowledge this point: the reciprocal theory of mixing on sublattices is not merely a hypothesis, but is mandated by the requirement of internal consistency. This internal consistency means that writing Equation 10.11 using Ca or Fe^{2+} (etc.) rather than Mg should end up giving exactly the same result, which reciprocal solution theory ensures that it does. Other attempts that try to define the activity of FeS in silicate melts without invoking reciprocal solution theory do not meet the requirement of internal consistency and are accordingly wrong. There may be some misconception in petrological thermodynamics, carried over from pioneering days, that an astute choice of components in a multicomponent phase can circumvent having to know the thermodynamic properties of other components in the phase. This has much truth in simple crystalline solutions where there is mixing on only one crystallographic site, but in more complex solutions with mixing on multiple sites, it is a fallacy. Because of the reciprocal solution effect, one cannot choose to define, for example, the activity of FeS in a silicate melt, without implying information about the activity of, say, MgS; this in turn means implying information about the standard state properties of MgS. Hence it is inescapable that the standard state properties of MgS and all other charge-balanced end members must be contained within the model.

All the components in Equation 10.11 are valid thermodynamic components, because they can be independently varied. In formal terms, $(\partial G_{melt}/\partial n_i)_{T,P,n_j}$ ($n_j \neq n_i$) is definable, and, moreover, experimentally measurable. This is not the case for charged species, or "structural elements" in the jargon of point-defect theory (Schmalzried, 1995), like O^{2-} and S^{2-}, because $(\partial G_{melt}/\partial n_{O^{2-}})_{T,P,n_j}$ is not independently variable due to the charge-balance constraint, and is not definable in conventional chemical thermodynamics where only temperature, pressure and composition are the variables. There is no implication in Equation 10.11 that a gas phase need be present, as S_2 and O_2 refer to the components of the system, which may be defined without specifying the phases in the system.

At equilibrium:

$$\mu_{MgS} - \tfrac{1}{2}\mu_{Mg_2SiO_4} = \tfrac{1}{2}\mathrm{RT}\ln(fS_2/fO_2) - \tfrac{1}{2}\mu_{SiO_2} \tag{10.12}$$

The evaluation of μ_{MgS} and $\mu_{Mg_2SiO_4}$ from Equations 10.7 to 10.9 gives:

$$\mu_{MgS} = \mu^o_{MgS} + \Sigma_{N\neq Mg}\left(-X_N^{cat} X_{SiO_4}^{an} \Delta\mu^o_{rec}[S - SiO_4]\right) + \mathrm{RT}\ln\left(X_{Mg}^{cat}\right) + \mathrm{RT}\ln\left(X_S^{an}\right) + \mathrm{RT}\ln\left(\gamma_{Mg}\right) + \mathrm{RT}\ln\left(\gamma_S\right) \tag{10.13}$$

This equation contains a term in X_S^{an}, which is the concentration of S^{2-} in the melt expressed as a mole fraction. The term $\Delta\mu^o_{rec}[S - SiO_4]$ is the change in free energy of the homogeneous equilibrium in the melt; such equilibria are often known as reciprocal reactions:

$$\tfrac{1}{2}Mg_2SiO_4 + N_{2/z}S = MgS + \tfrac{1}{2}N_{4/z}SiO_4 \tag{10.14}$$

$$\Delta\mu^o_{rec}[S - SiO_4] = \left(\mu^o_{MgS} - 1/2\mu^o_{Mg_2SiO_4}\right) - \left(\mu^o_{N_{2/z}S} - 1/2\mu^o_{N_{4/z}SiO_4}\right) \tag{10.15}$$

Likewise for $\mu_{Mg_2SiO_4}$:

$$\mu_{Mg_2SiO_4} = \mu^o_{Mg_2SiO_4} + \Sigma_{N\neq Mg}\left(X_N^{cat} X_S^{an} \Delta\mu^o_{rec}[S - SiO_4]\right) + 2\mathrm{RT}\ln\left(X_{Mg}^{cat}\right) + \mathrm{RT}\ln\left(X_{SiO_4}^{an}\right) + 2\mathrm{RT}\ln\left(\gamma_{Mg}\right) + \mathrm{RT}\ln\left(\gamma_{SiO_4}\right) \tag{10.16}$$

Substituting 10.13 to 10.16 into 10.12, putting $X_{SiO_4}^{an} = 1$ (that is, X_S^{an} is small, and the SiO_2 units do not mix with the charged species S^{2-} and SiO_4^{2-} on the anion sublattice) and rearranging gives:

$$\ln X_s^{an} = \Sigma_M X_M^{cat}\left(\frac{1}{2}\mu^o_{M_{4/z}SiO_4} - \mu^o_{M_{2/z}S}\right)/RT - \tfrac{1}{2}\ln a_{SiO_2} - \ln\gamma_S + \tfrac{1}{2}\ln\gamma_{SiO_4} + \tfrac{1}{2}\ln(fS_2/fO_2) \quad (10.17)$$

This shows that the S content of the melt depends proportionally on $(fS_2/fO_2)^{1/2}$, as in the definition of the sulfide capacity, $C_{S^{2-}}$, in Equation 10.2. Eliminating the term in $(fS_2/fO_2)^{1/2}$ between 10.17 and 10.3 gives:

$$\ln C_{S^{2-}} = B_0^{S^{2-}} + c\ \ln a_{SiO_2} + \Sigma X_M B_M^{S^{2-}} \quad (10.18)$$

where the term $B_0^{S^{2-}}$ also accounts for the conversion from mole fractions on the anion sublattice to weight fractions (conveniently parts per million by weight, while $B_M^{S^{2-}} = k\left(\frac{1}{2}\mu^o_{M_{4/z}SiO_4} - \mu^o_{M_{2/z}S}\right)/RT$, where k is a constant, the same for all M. The activity coefficients, γ_S and γ_{SiO4}, will for simplicity be assumed to vary only with the M^{x+} cation fractions in the melt composition, but this is likely only within a limited range of melt compositions, say with NBO/T between 0.2 and 2, which covers most silicate melt compositions apart from the granite/rhyolite extreme. The average values of γ_S and γ_{SiO4} within this range are implicitly incorporated in the $B_0^{S^{2-}}$ term.

Equation 10.18 also contains a term in the activity of silica, a_{SiO_2}, which may be expected to be a function of the cation mole fractions in the melt. Assume that over a reasonably limited range of silicate melt compositions, it has the functional form $\ln (a_{SiO_2})^c = C_O + X_M C_M$. This permits simplification, to Equation 10.6: $\ln C_{S^{2-}} = A_0^{S^{2-}} + \sum X_M A_M^{S^{2-}}$, where $A_0^{S^{2-}} = B_0^{S^{2-}} + C_O$, and $A_M^{S^{2-}} = B_M^{S^{2-}} + C_M$. In order to achieve a good fit to the data, the cations M include the usual ones that form the basic oxide components important in natural-composition silicate melts (Na^{1+}, Mg^{2+}, K^{1+}, Ca^{2+}, Mn^{2+}, and Fe^{2+}), but also Al^{3+} and Ti^{4+}, whose structural positions in the melts are thought to be different, and in different ways (Mysen & Richet 2005). Si is not included in the sum of these M cations explicitly, because it is implicitly included from the closure condition, $\sum X_M = 1$. The mole fractions are defined on the single-cation basis ($NaO_{0.5}$, MgO, $AlO_{1.5}$, SiO_2, etc.), despite Equations 10.10 to 10.16 being developed with mole fractions on the single-anion basis; for some reason, this way gives a noticeably better fit.

The importance of this derivation of $C_{S^{2-}}$ from solution thermodynamics is that it establishes that the logarithm of the sulfide capacity $C_{S^{2-}}$ should vary linearly with the mole fractions. This theoretical relationship was subjected to detailed experimental scrutiny by O'Neill and Mavrogenes (2002), who investigated nearly 200 different melt compositions covering various proportions of the above cations, with NBO/T ranging from 0 to 2 (the mean NBO/T of the compositions they measured is 0.85, with a standard deviation of 0.3), at 1400°C. Such a high experimental temperature was necessary to access a wide range of melt compositions, and was not meant to mimic magmatic temperatures. The tests included experiments adding progressive amounts of one component to a base composition, the particularly important case being the addition of FeO to a CaO-MgO-Al_2O_3-SiO_2 composition, with results shown in Fig. 13 of O'Neill and Mavrogenes (2012). The reason that this example is so important is that it refutes the notion that there are thermodynamically relevant Fe^{2+}-S^{2-} "complexes" in the silicate melts. Rather, the simple log-linear Equation 10.7 performs well, except at very high FeO contents where FeO loses some of its potency for dissolving S^{2-} (the opposite trend to that which would come from complex formation). This interesting phenomenon will be addressed further below.

The model may be tested from a different perspective, namely the magnitudes of the different $A_M^{S^{2-}}$ terms compared to those expected from standard thermodynamic data. An enabling assumption is that the C_M terms used to parameterize a_{SiO_2} are relatively small, such that neglecting them for the present exposition allows the approximation $A_M^{S^{2-}} \propto \left(\frac{1}{2}\mu^o_{M_{4/z}SiO_4} - \mu^o_{M_{2/z}S}\right)/RT$, where the latter is the difference between the free energies of the orthosilicates and the sulfides. This is an important insight, because if this relationship is valid, it would enable estimates of the effects of melt composition on other "capacities" – sulfate, carbonate, phosphate, etc., providing a thermodynamically based alternative to other concepts like NBO/T or "optical basicity" that have been invoked to predict melt thermodynamic properties. The relationship is tested in Fig. 10.1, using the values of $A_M^{S^{2-}}$ from O'Neill and Mavrogenes (2002) plus that for $A_{Mn}^{S^{2-}}$ from Evans et al. (2008). For the free energies of formation of the orthosilicate and sulfide components, the data for the solid phases at 1400 K were used as a proxy for what is actually required, which is the free energies of the components in the liquid phase in silicate melts extrapolated to infinite dilution at the temperature of interest (here 1673 K). At least this proxy has the merits of consistency for all components. These data are from the JANAF tables (Chase et al., 1998) or Robie and Hemingway (1995).

The reason for not trying to build a model from Reaction 10.5, which may be a more realistic approximation to the structure of most silicate melts of geological interest, is that the analogous thermodynamic data for crystalline

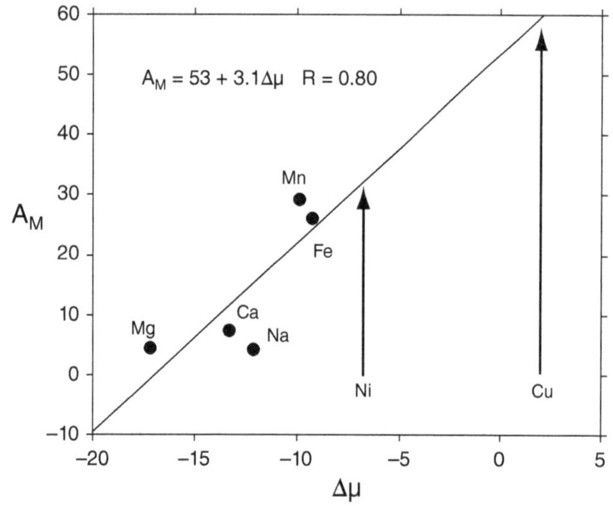

Figure 10.1 Comparison between values of $A_M^{S^{2-}}$ from O'Neill and Mavrogenes (2002) and Evans et al. (2008) with thermodynamic data for the difference in the free energies of the orthosilicate and sulfide solid phases, $\left(\frac{1}{2}\mu^o_{M_{4/z}SiO_4} - \mu^o_{M_{2/z}S}\right)/1400R$. The correlation suggests how approximate values of $A_{Ni}^{S^{2-}}$ and $A_{Cu}^{S^{2-}}$, which are difficult to measure experimentally (Evans et al. 2008), may be estimated.

substances proxying for the $M_{2/z}Si_2O_5$ units are mostly not available, and for the $Si_2O_6^{4-}$ dimer units in the melt (NBO/Si = 2), the equivalent crystalline stoichiometry corresponds to that of the chain silicates (e.g., pyroxenes), which seems dubiously appropriate as structural proxies. The orthosilicate stoichiometry has the same SiO_4^{4-} structural units in crystals as in melts.

There is a fair positive correlation (R = 0.80) between $A_M^{S^{2-}}$ and $\left(\frac{1}{2}\mu^o_{M_{4/z}SiO_4} - \mu^o_{M_{2/z}S}\right)/1400R$. O'Neill and Mavrogenes (2002) pointed out that there was no such correlation between the values of $A_M^{S^{2-}}$ and the analogous terms $\left(\mu^o_{M_{2/z}O} - \mu^o_{M_{2/z}S}\right)/RT$, indicating that the ability of a basic oxide component $M_{2/z}O$ to contribute to the sulfide capacity of a silicate melt is lessened by the strength of the chemical bonding of the M^{z+} cation to the silicate units in the melt. Thus Ca^{2+} is less effective at enhancing $C_{S^{2-}}$ than Fe^{2+} because Ca^{2+}-SiO_4^{4-} interactions are far stronger than Fe^{2+}–SiO_4^{4-} interactions in silicate melts, well known from the thermodynamics of the simple binaries CaO–SiO_2 and FeO-SiO_2 (e.g., Mysen & Richet, 2005).

This correlation also shows that the large magnitude of $A_{Fe}^{S^{2-}}$ relative to the other $A_M^{S^{2-}}$, except $A_{Mn}^{S^{2-}}$, is not an anomaly, but is that expected from the reciprocal solution theory, given the relative magnitudes of the free energies of formation of Fe_2SiO_4 and FeS. There is no reason to postulate complexes between Fe^{2+} and S^{2-} in the silicate melts. The correlation predicts that the effects of Ni^{2+} and Cu^{1+} on $C_{S^{2-}}$ should be even larger than those of Fe^{2+} and Mn^{2+}. Evans et al. (2008) found that $A_{Ni}^{S^{2-}}$ and $A_{Cu}^{S^{2-}}$ could not be measured experimentally with available methods, because at the fS_2–fO_2 conditions needed to achieve measurable S^{2-} in the silicate melt, the Ni and Cu were stripped out into immiscible sulfide melts. It is the same strength of the Ni^{2+}-S^{2-} and Cu^{1+}-S^{2-} chemical bonding in the silicate melts compared to the rather feeble Ni^{2+}-SiO_4^{4-} and Cu^{1+}-SiO_4^{4-} bonding, causing the large values of $A_{Ni}^{S^{2-}}$ and $A_{Cu}^{S^{2-}}$, that also causes the immiscible sulfide melts to form; two sides of the same coin. The correlation in Fig. 10.1 suggests $A_{Ni}^{S^{2-}} \approx 30$ and $A_{Cu}^{S^{2-}} \approx 60$; to estimate the latter the free energy of formation of the nonexistent orthosilicate Cu_4SiO_4 from the oxides was taken to be zero.

This information on the likely values of $A_{Ni}^{S^{2-}}$ and $A_{Cu}^{S^{2-}}$ is somewhat nugatory, as the stripping out of Ni and Cu into a sulfide melt keeps Ni and Cu concentrations in the silicate melt to the 10^2 ppm level, at which they have no discernible effect on $C_{S^{2-}}$ (Evans et al., 2008). Put another way, the atomic ratios of Ni and Cu to Fe in terrestrial basalts are both about 10^{-3}, therefore given these values of $A_{Ni}^{S^{2-}}$ and $A_{Cu}^{S^{2-}}$, neglecting Ni and Cu has an equivalent effect to decreasing the concentration of FeO in the silicate melt by 0.1 and 0.2 % relative, well below the best analytical precision. The effect of Ni and Cu on SCSS is entirely through their obvious influence on the activity of FeS in the sulfide melt. If NiO in a silicate melt has only a slight effect on S^{2-} concentrations, then the corollary is that this S^{2-} will have almost negligible effect on the activity of NiO in the silicate melt, for example, in the context of the partitioning of Ni between olivine and silicate melt. This point is addressed in Tuff and O'Neill (2010).

10.3.1. The Fe^{3+} Problem

Given that the direct measurement of sulfide capacities requires knowing both fO_2 and fS_2, it can be done most easily and also precisely at atmospheric pressure in gas-mixing experiments, where the sum of the partial pressures of all gas species are limited to 1 bar. Unfortunately, increasing fO_2 to where $Fe^{3+}/(Fe^{3+}+Fe^{2+})$ in the silicate melt is more than a couple of percent enters the regime of the experimental S solubility minimum, where S solubilities fall below the limit of quantification of the electron microprobe. Thus only the effect of Fe^{2+} on $C_{S^{2-}}$ is known, and the influence of Fe^{3+} is currently completely mysterious. The correlation used to guestimate $A_{Ni}^{S^{2-}}$ and $A_{Cu}^{S^{2-}}$ (Fig. 10.1) is useless, because thermodynamic data for the non-existent sulfide $Fe^{3+}S^{2-}_{1.5}$ are not available, unsurprisingly. A couple of thoughts arise. One possibility is that the electron configurations of Fe^{3+} is the same as that of Mn^{2+} ($Ar3d^5$), and size differences between cations are less important in the environment of silicate melts where cation-anion interatomic distances are not influenced by crystal symmetries, suggesting that $A_{Fe^{3+}} \sim A_{Mn}$. As the latter is similar to $A_{Fe^{2+}}$ (Evans et al., 2008), changing $Fe^{3+}/(Fe^{3+}+Fe^{2+})$ would have no effect on $C_{S^{2-}}$. Alternatively, on the grounds that Fe^{3+} may play a similar structural role to Ti^{4+} in silicate melts, $A_{Fe^{3+}}$ may be similar to A_{Ti}, and increasing $Fe^{3+}/(Fe^{3+}+Fe^{2+})$ would decrease $C_{S^{2-}}$ accordingly. That changing the oxidation state of a key cationic component of a silicate melt may have a huge effect on $C_{S^{2-}}$ has been claimed for vanadium. Allertz et al. (2016) showed that increasing fO_2 by six orders of magnitude at 1600°C increased $C_{S^{2-}}$ by over two orders of magnitude in CMAS-based melts with 3.6 to 9.5 wt.% V; the increase is presumably due to the V changing from predominantly 2+ to predominantly 3+. Is V a trustworthy analogue for Fe?

10.3.2. Sulfate Capacities, $C_{SO_4^{2-}}$

Following the same logic, the equilibrium between orthosilicate and framework structural units that defines the sulfate capacity, $C_{SO_4^{2-}}$, would be:

$$\tfrac{1}{2}\underset{\text{sil melt}}{SiO_4^{4-}} + \tfrac{1}{2} S_2 + 3/2\, O_2 = \underset{\text{sil melt}}{SO_4^{2-}} + \tfrac{1}{2}\underset{\text{sil melt}}{SiO_2} \quad (10.19)$$

This suggests that the dependence of $C_{SO_4^{2-}}$ on silicate melt composition should take the same form as that for $C_{S^{2-}}$, namely:

$$\ln C_{SO_4^{2-}} = A_0^{SO_4^{2-}} + \Sigma X_M A_M^{SO_4^{2-}} \quad (10.20)$$

where $A_0^{SO_4^{2-}} = B_0^{SO_4^{2-}} + C_0$, $A_M^{SO_4^{2-}} = B_M^{SO_4^{2-}} + C_M$, with $B_M^{SO_4^{2-}} = k\left(\frac{1}{2}\mu^o_{M_{4/z}SiO_4} - \mu^o_{M_{2/z}SO_4}\right)/RT$. The difference between the coefficients for of each M cation for $C_{SO_4^{2-}}$ and $C_{S^{2-}}$ is therefore:

$$A_M^{SO_4^{2-}} - A_M^{S^{2-}} = k\left(\mu^o_{M_{2/z}S} - \mu^o_{M_{2/z}SO_4}\right)/RT \quad (10.21)$$

In taking the difference, most of the approximations used in deriving the individual values of $A_M^{S^{2-}}$ and $A_M^{SO_4^{2-}}$ would cancel out, like the effect of the activity of silica. This then further suggests that the dependence of $C_{SO_4^{2-}}$ on silicate melt composition could be predicted from the experimental measurements on $C_{S^{2-}}$, together with the readily available standard-state thermodynamic data on sulfates and sulfides.

The ratio of sulfide to sulfate dissolved in a silicate melt follows from the definitions of $C_{S^{2-}}$ and $C_{SO_4^{2-}}$:

$$[SO_4^{2-}]/[S^{2-}] = C_{SO_4^{2-}}/C_{S^{2-}}(fO_2)^2 \quad (10.22)$$

The fO_2 at which $[SO_4^{2-}]/[S^{2-}]$ is 1, denoted here as $(fO_2)_{S6+/S2-}$, is therefore $\left(C_{S^{2-}}/C_{SO_4^{2-}}\right)^{1/2}$. For present purposes, the anion complex SO_4^{2-} may be taken as equivalent to S^{6+}. Substituting Equations 10.7 and 10.20 into 10.22 gives:

$$\ln\left([SO_4^{2-}]/[S^{2-}]\right) = +2\ln fO_2 + \left(A_0^{SO_4^{2-}} - A_0^{S^{2-}}\right) + \Sigma X_M\left(A_M^{SO_4^{2-}} - A_M^{S^{2-}}\right) \quad (10.23)$$

where the effect of silicate melt composition is contained in the last term on the right-hand side, whose thermochemical significance is given by Equation 10.21.

From this, it may be predicted that, at a given T, P, and fO_2, increasing FeO in the melt would increase $(fO_2)_{S6+/S2-}$, while increasing CaO and Na_2O would decrease $(fO_2)_{S6+/S2-}$.

The value of $(fO_2)_{S6+/S2-}$ is not well constrained for anhydrous basaltic melts. The inescapable experimental problem is that it falls in the region of the ambient-pressure experimental S solubility minimum. It has been approximately located at about the fO_2 of the quartz–fayalite–magnetite equilibrium (QFM) by Fincham and Richardson (1954) and Katsura and Nagashima (1974) from extrapolation of solubilities measured at higher and lower fO_2. (Oxygen fugacities are conveniently discussed relative to some isobarically invariant "buffer" such as QFM because condensed-phase redox equilibria all tend, to a first approximation, to have similar slopes in 1/T-log fO_2 space. For reference, QFM is given by $\log 10\, fO_{2(QFM)} = 8.58–25050/T$ at 1 bar [O'Neill, 1987], and $\Delta QFM = \log_{10}[fO_2/fO_{2(QFM)}]$).

Recently this approach has been expanded upon by Nash et al. (2019), who exploited the lower limits of detection afforded by SIMS to investigate the solubilities of S in the region of the solubility minimum. Their experiments were at 1300°C, and used CO-CO_2-SO_2 gas mixes. They assumed that total S solubilities at higher fO_2 in excess of those predicted for S^{2-} from the Fincham-Richardson relation at low fO_2, were due to additional S^{6+}. This they corroborated by S K-edge XANES spectroscopy. They found that values of $(fO_2)_{S6+/S2-}$ for six melt compositions lay in the range ΔQFM 0 to +0.6, concordant with the earlier studies. Increasing $(fO_2)_{S6+/S2-}$ correlated qualitatively with increasing FeO in the melt, although the effect was smaller than expected from the sensitivity of $C_{S^{2-}}$ to FeO (O'Neill and Mavrogenes, 2002).

Other information on where $(fO_2)_{S6+/S2-}$ lies in silicate melts comes from X-ray spectroscopic characterization of S^{6+}/S^{2-} in quenched glasses prepared under pressure, and necessarily containing H_2O in order to control fO_2. These measurements have been problematic because of beam-damage issues; see Wilke et al. (2008, 2011), Métrich et al. (2009), and Jugo et al. (2010). The majority view is that S^{4+} (that is, sulfite, SO_3^{2-}) found spectroscopically by XANES is caused by beam damage, at least in glasses of low-pressure provenance (both natural and synthetic). The probability is that most X-ray fluorescence measurements of S oxidation state obtained with the electron microprobe are badly affected by beam damage (Wilke et al., 2008), but because this technique only gives the average oxidation state of S, S^{4+} is not itself distinguished (unlike with XANES), making it difficult to judge exactly what is going on. Because of this knowledge gap, the test for beam damage is whether the measured S^{6+-}/S^{2-} follows the theoretical dependence on fO_2, which is that $[S^{6+-}]/[S^{2-}] \propto (fO_2)^2$ (Equation 10.22), provided that the assumption of negligible S^{4+} is accepted. At present, among studies that cover a sufficiently meaningful range of fO_2, the only experiments to pass this test to date are those of Jugo et al. (2010) and Botcharnikov et al. (2011), who used XANES to quantify $S^{6+}/\sum S$ in a series of glasses quenched from hydrated basaltic and andesitic melts with about 5 wt% H_2O at 1050°C and 0.2 GPa, and those of Nash et al. (2019), described above. Jugo et al. (2010) fitted their results to an equation with the stoichiometry of the speciation constrained at the theoretical value:

$$S^{6+}/\Sigma S = 1/(1 + 10\,(2.1 - 2\Delta QFM)) \qquad (10.24)$$

Hence $(fO_2)_{S6+/S2-}$ is at ΔQFM+1.05. Botcharnikov et al. (2011) then found that this equation described their results on their andesitic composition within uncertainty.

For comparison, the change in oxidation state of the main redox-variable element in a simplified basaltic composition, Fe, given by Berry et al. (2018) is:

$$Fe^{3+}/\Sigma Fe = 1/\left(1 + 10^{(1.0575 - 0.25\Delta QFM)}\right) \qquad (10.25)$$

The relationship between $S^{6+}/\sum S$ and $Fe^{3+}/\sum Fe$ from combining these two equations to eliminate ΔQFM is shown in Fig. 10.2. On the face of it, the range of $Fe^{3+}/\sum Fe$ measured in MORB glasses coincides with $S^{6+}/\sum S$ going from zero to as much as 30%, but this depends on the rather dubious equivalence between hydrated basalts at 1050°C and the natural MORB glasses, which are quenched from ~1200°C and contain little H_2O.

A different view emerges from the recent study of Nash et al. (2019), who calculated Fe^{3+}/Fe^{2+} for the experiments at 1300°C in which they had estimated S^{6+}/S^{2-}

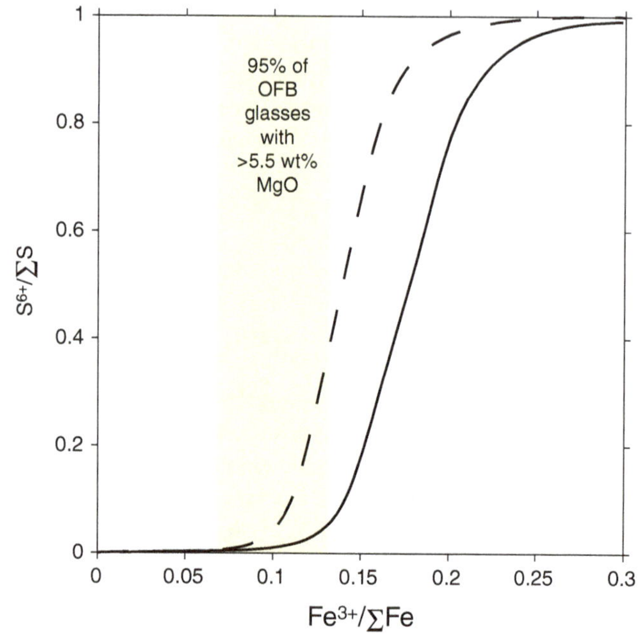

Figure 10.2 The relationship between $S^{6+}/\sum S$ and $Fe^{3+}/\sum Fe$, estimated by combining the results of Jugo et al. (2010) on hydrated basalt at 1050°C and 0.2 GPa, with those of Berry et al. (2018) on MORB glasses (dashed curve), compared to the model of Nash et al. (2019) calculated at 1190°C (solid curve). The ±2s range of measured $Fe^{3+}/\sum Fe$ in MORB glasses from Berry et al. (2018), taken here to correspond to 95% of the population by assuming a normal distribution of $Fe^{3+}/\sum Fe$, is indicated by the shaded region. The model of Nash et al. (2019) predicts only a few percent $S^{6+}/\sum S$ even in the more oxidized MORB glasses.

(as described), and also for those of Jugo et al. (2010) and Botcharnikov et al. (2011) at 1050°C. They then fitted these data to an expression also constrained to the theoretical stoichiometry, which included a dependence on temperature. To constrain this latter variable better (given that the available data are from only two temperatures and those at 1050°C are hydrous melts) they invoked thermochemical data for crystalline-phase analogues of the melt components. As regards the effect of melt composition, their data were too scattered to resolve any (see their Fig. 6). The relationship between $S^{6+}/\sum S$ and $Fe^{3+}/\sum Fe$ from their equation, calculated at 1190° (the average for anhydrous MORB glasses in equilibrium with olivine according to the magmathermometer of Putirka, 2008), is also shown in Fig. 10.2. With their formulation, even the more oxidized MORB glasses would contain only a few percent $S^{6+}/\sum S$. As will be shown, the evidence from the SCSS of MORB glasses is more concordant with this view.

The temperature-dependence of S speciation relative to QFM in silicate melts remains somewhat contentious.

Nash et al. (2019) inferred that decreasing temperature shifts $(fO_2)_{S6+/S2-}$ to higher ΔQFM (i.e., decreasing temperature stabilizes S^{2-} relative to S^{6+} at a given ΔQFM), which is in the opposite direction to that anticipated by Métrich et al. (2009). That S^{6+} forms an anionic complex, SO_4^{2-}, may seem at first sight to endow this speciation with low configurational entropy, but as the effects of entropy are always relative, the entropies of the alternative arrangements need to be considered. Assuming the same orthosilicate and framework species as for modeling of $C_{S^{2-}}$, the two alternatives for redox speciation in a melt containing both Fe and S could be described by the homogeneous equilibrium:

$$S^{2-} + 8\,Fe^{3+} + 2\,SiO_4^{4-} = SO_4^{2-} + 8\,Fe^{2+} + 2\,SiO_2 \tag{10.26}$$

From the point of view of configurational entropy, both sides of this reaction contain the same number of species, but two of the 11 species on the right-hand side (RHS) are polymerized SiO_2 units in the melt, which might be expected to have lower entropy, implying that the speciation might proceed towards the RHS with decreasing temperature.

The experiments of Luhr (1990) on sulfur solubilities in hydrous melts as a function of temperature and fO_2 (at 0.2 and 0.4 GPa) show that at 900°C and above, total S in the oxidized experiments was higher than experiments at QFM (see Fig. 10 of Luhr, 1990), whereas at 800 and 850°C it was the same in the oxidized experiments as in the pyrrhotite-saturated experiments at QFM. This is suggestive that, contrary to the entropy argument, increasing temperature favors the RHS of Reaction 10.26, but it is not definitive as the oxidized experiments did not contain pyrrhotite.

The isochemical redistribution of species according to Reaction 10.26 involves only electron hopping between atoms with coordination changes on the scale of interatomic distances, and is therefore likely very fast, arguably so fast as to be unquenchable from liquidus temperatures (cf. Berry et al., 2003). If the RHS of 10.26 were favored on cooling (the opposite to the inference from the experiments of Luhr, 1990), it might explain the small amounts of sulfate identified by XANES spectroscopy in some natural basaltic glasses (Métrich et al., 2009; Jugo et al., 2010).

The only available indication of the effect of pressure on the S^{2-}/SO_4^{2-} transition is the study of Matjuschkin et al (2016), which indicates that increasing pressure may move $(fO_2)_{6+/2}$ to higher ΔQFM. Measurements of $S^{6+}/\sum S$ and $Fe^{3+}/\sum Fe$ on the same melt compositions over a range of fO_2 (probably from QFM–1 to QFM+2) as a function of temperature and pressure are much needed. But this is a formidable experimental challenge.

10.3.3. Other Capacities for Anions and Anion Complexes in Silicate Melts

The "capacity" idea could be extended to other anionic substitutions in silicate melts, for example, CO_2 dissolving as carbonate, CO_3^{2-}:

$$\underset{\text{sil melt}}{\tfrac{1}{2}\,SiO_4^{4-}} + CO_2 = \underset{\text{sil melt}}{CO_3^{2-}} + \underset{\text{sil melt}}{\tfrac{1}{2}\,SiO_2} \tag{10.27}$$

for which the carbonate capacity would be given by:

$$C_{CO_3^{2-}} = [CO_3^{2-}](fCO_2)^{-1} \tag{10.28}$$

with $A_0^{CO_3^{2-}} - A_M^{S^{2-}} = k\left(\mu^o_{M_{2/x}S} - \mu^o_{M_{2/x}CO_3}\right)/RT$. Likewise, to predict the effect of composition on chlorine in silicate melts, which is difficult to measure experimentally because Cl_2 gas volatilizes most of the oxide components in one-atmosphere experiments, the chloride capacity would be defined as:

$$\underset{\text{sil melt}}{1/4\,SiO_4^{4-}} + \tfrac{1}{2}\,Cl_2 = \underset{\text{sil melt}}{Cl^-} + \underset{\text{sil melt}}{1/4\,SiO_2} + \tfrac{1}{4}\,O_2 \tag{10.29}$$

$$C_{Cl^-} = [Cl^-](fCl_2)^{-1/2}(fO_2)^{1/4} \tag{10.30}$$

and $A_M^{Cl^-} - A_M^{S^{2-}} = k\left(1/2\mu^o_{M_{2/x}S} - \mu^o_{M_{1/x}Cl}\right)/RT$.

A suggestion to relate the various "capacities" of silicate melts to one another was made by Carl Wagner in one of his last contributions (Wagner, 1975). Wagner, who has been called the "Father of Solid-state Chemistry" (Martin, 2001), defined a "basicity" for a silicate melt of a specified composition using carbonate capacities:

$$B_{carb} = C_{CO_3^{2-}}/C^*_{CO_3^{2-}} \tag{10.31}$$

where $C^*_{CO_3^{2-}}$ is the carbonate capacity of a reference melt (he suggested one with composition 0.4 CaO, 0.4 SiO_2, and 0.2 Al_2O_3). Wagner (1975) also discusses how such a definition of "basicity" for a silicate melt provides a more rational measure than the "activity" of O^{2-}, which, despite its appearance as a "structural element" in definitions of capacities like Equation 10.3, is not, as Wagner pointed out, a usefully definable quantity, thermodynamically.

Wagner (1975) then proposed that the capacities of other anionic species $Z^{\nu-}$ (e.g., S^{2-}, SO_4^{2-}, Cl^-, among many others) could be related to this basicity by:

$$C_{Z^{\nu-}} = C^*_{Z^{\nu-}}(B_{carb})^{\nu/2} \tag{10.32}$$

One of the benefits of deriving the thermochemical meaning of these "capacities" from reciprocal solution theory with mixing on sublattices is that it goes straight to the heart of the problem with Wagner's suggestion (which he alluded to himself): it implicitly assumes that the quantities $(\frac{1}{2}\mu^o_{M_{4/x}SiO_4} - \mu^o_{M_{\nu/x}Z})$ are the same for all anionic species Z, and for all relevant elements M. A few minutes perusing a thermochemical database would reveal that this is obviously not the case. The escape clause is that in a majority of basic slag compositions these M cations are dominated by only two, namely Mg and Ca, and it is possible to get away with taking the difference $[(\mu^o_{Ca_{\nu/2}Z} - \mu^o_{Mg_{\nu/2}Z})$ as being constant within limited horizons. Looking a bit further, it is the great stability of the FeS component in silicate melts compared to MgS or CaS (cf. $FeCO_3$ compared to $CaCO_3$ or $MgCO_3$), with the consequence for sulfide capacities, that provides the exception disproving this rule handsomely – although MnS would do as well as FeS.

10.4. A NEW PARAMETERIZATION OF SULFIDE CAPACITY FOR BASALTIC MELTS

From the point of view of reproducing the experimental data, there are nevertheless a few ad-hoc "improvements" that can be made to the simple relationship of Equation 10.7, albeit at the cost of divorcing the parameters $A^{S^{2-}}_M$ from their thermochemical roots. O'Neill and Mavrogenes (2002) showed that adding a cross term in $X_{Fe}X_{Ti}$ is needed to account for Fe^{2+}–Ti interactions in the melt (see Fig. 14 of O'Neill & Mavrogenes, 2002). Smythe et al. (2017) suggested a similar cross-term in $X_{Fe}X_{Si}$, and in the fitting to be described below, it was confirmed that this term was very effective at improving the fit of the experimental data.

A well-known but hitherto unaddressed problem is that the linear relationship between ln $C_{S^{2-}}$ and the cation mole fractions X_M breaks down is at high FeO contents, between 15 and 20 wt.% FeO according to O'Neill and Mavrogenes (2002). This phenomenon has been confirmed by several studies on SCSS (see Section 10.5). The deviation is in the direction of lower S^{2-} in the silicate melt than the linear relationship established at < 15 wt.% FeO would predict, the opposite to that expected from formation of Fe^{2+}-S^{2-} complexes. The deviation becomes noticeable at S^{2-} concentrations that are relatively low compared to those of Fe^{2+}, for example at ~ 3500 ppm S in melt at 20 wt% FeO, which in molar proportions of S to FeO is still only less than 1 to 10 (see Table 13 of O'Neill & Mavrogenes, 2002). Other experiments on Fe-free compositions with high CaO exceed this S^{2-} content with no such deviations apparent (e.g., nearly 6000 ppm dissolve in the AD+Wo composition run 17/2/97 in Table 2 of O'Neill & Mavrogenes, 2002), indicating, importantly, that the Fe problem is not due to the assumption of constant "activities" of "structural elements" like O^{2-} breaking down in the derivation of Equation 10.3. Another curious feature is that the change of $C_{S^{2-}}$ with X_{Fe} is linear within experimental uncertainty up to ~ 15 wt% FeO, but then breaks sharply to a different linear trend of ln $C_{S^{2-}}$ vs X_{Fe} (see Fig. 15 of O'Neill & Mavrogenes, 2002).

This is a peculiar phenomenon, for which there is no explanation at present, and more experimental information would be useful. There seem to be no experiments that have tested the Fincham-Richardson relationship in high FeO melts; for example, it would be interesting to measure $C_{S^{2-}}$ at various fO_2 and fS_2, on a composition with, say, 30 wt% FeO in the melt. The propensity for high FeO, high S^{2-} melts to exsolve FeS-rich sulfide microblobs on quenching despite using the standard rapid-quenching techniques (Métrich et al., 2009; Wykes et al., 2015), has so far prevented obtaining samples suitable for spectroscopy (Métrich et al., 2009). The departure from the simple theory at high FeO has hindered applying sulfide solubility models to many extraterrestrial basalts, which often have FeO > 15 wt%. In view of the lack of theoretical insight but the need for something to be done, it seems expedient to forgo a thermodynamically based model, and find a parameterization of $C_{S^{2-}}$ as a function of melt composition that includes FeO-rich basaltic melts. The shape of the departure from linearity suggests the arctangent function or the error function; the latter was used here. With these three additions to the simple Equation 10.7, all the data from O'Neill and Mavrogenes (2002), including the high FeO compositions, plus those at 1400°C from Tuff and O'Neill (2010), were fitted by weighted least squares regression. The function minimized was $[(\ln(C_{S^{2-}})_{obs} - \ln(C_{S^{2-}})_{calc})/s(\ln C_{S^{2-}})]^2$, where the weight s(ln $C_{S^{2-}}$) was obtained by propagating the uncertainty in measured S, $s([S]_{obs})$ as observed, or 3%, whichever was larger, plus uncertainties in log fO_2 and log fS_2 of 0.05 each. Typically s(ln $C_{S^{2-}}$) is about 0.1 (that is, 10%), except for experiments with very low S contents (< 500 ppm), which may have substantially larger uncertainties, and the 12 CMAS and one CMAS+Ti compositions from Table 1 of O'Neill and Mavrogenes (2002), in which the multiple measurements produce a lower uncertainty of 0.03 to 0.05. The regression gave:

$$\begin{aligned}\ln C_{S^{2-}}(1673\text{K}, 1\text{ bar}) &= -5.330 + 6.7(X_{Na} + X_K) \\ &+ 1.8X_{Al} + 4.9X_{Mg} + 8.1X_{Ca} + 5.0X_{Ti} + 8.9X_{Fe^*} \\ &- 22.2X_{Fe^*}X_{Ti} + 7.2X_{Fe^*}X_{Si} - 2.06\,\text{erf}(-7.2X_{Fe^*})\end{aligned} \tag{10.33}$$

where X_{Fe*} is $(X_{Fe} + X_{Mn})$, and mole fractions are defined on the single cation basis ($NaO_{0.05}$, MgO, $AlO_{1.5}$, SiO_2, etc.). The agreement between the parameterization and the input data (n = 228) is shown in

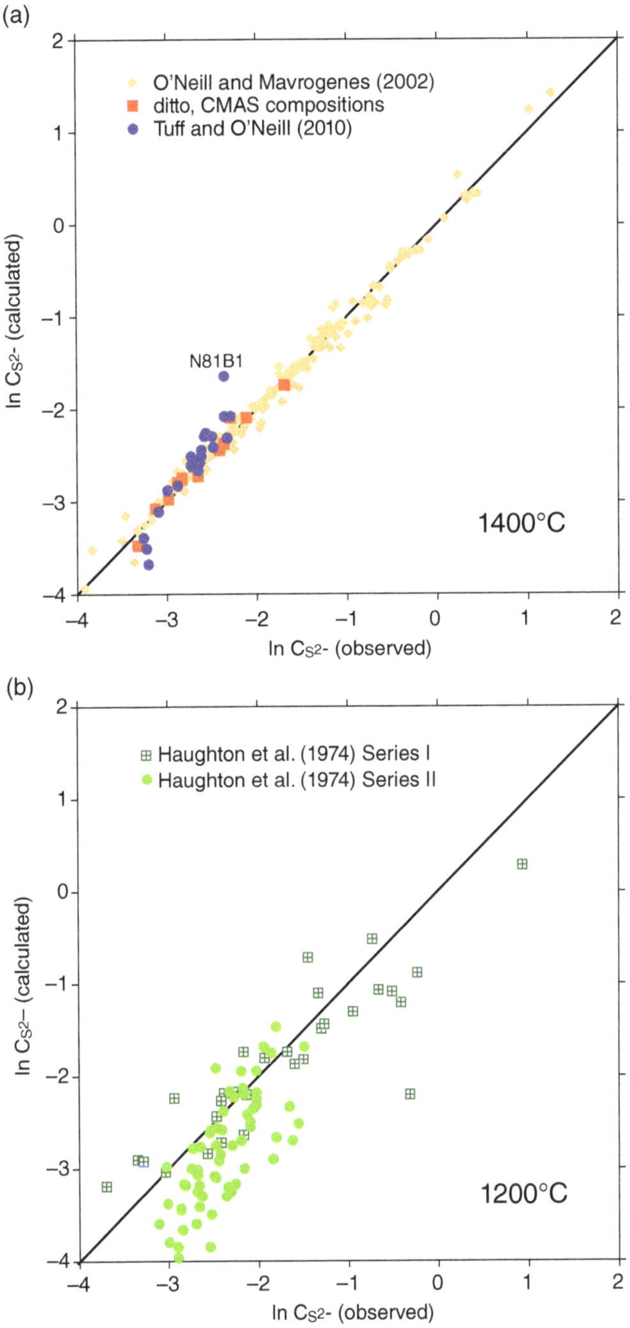

Figure 10.3 (a) Observed values of the sulfide capacity, $C_{S^{2-}}$, at 1400°C of silicate melts at with NBO/T from 0.05 to 2, compared to the parameterization of this work (Equation 10.33). The experimental data are from O'Neill and Mavrogenes (2002) and Tuff and O'Neill (2010). Only the data with $\ln C_{S^{2-}} > -4$ are plotted, as the few data at lower $C_{S^{2-}}$ have NBO/T ~0, and furthermore are imprecise due to low S. (b) Calculated values of $C_{S^{2-}}$ from Equation 10.34 versus the measured values of Haughton et al. (1974) at 1200°C.

Fig. 10.3a. The reduced chi-squared (χ_ν^2) is 1.2, indicating a good fit to the data, apart from two CMAS compositions. One is the CMAS7E composition from Table 1 of O'Neill and Mavrogenes (2002). It is a challenge to model the compositional dependence of $C_{S^{2-}}$ for the twelve CMAS compositions from that Table, because of the small uncertainty in $C_{S^{2-}}$ that comes from the multiple experiments (10 or 12 on each composition). Whilst this challenge is mostly met, the misfit to CMAS7E is a notable failure. The parameterization also fails to fit an anomalous datum from Tuff and O'Neill (2010), N81B1, which was discussed in that paper. Due to the large number of data, these two outliers have negligible effect, but they were removed from the fitting procedure anyway.

10.4.1. Temperature Dependence of $C_{S^{2-}}$

Because the compositional terms come from reciprocal reactions, they are likely to be dominated by enthalpy changes, hence the form of the parameterization for other temperatures is proposed to be:

$$\begin{aligned}
&\ln C_{S^{2-}}(\mathrm{T, 1\ bar}) = \mathrm{A^H/T + A^S} \\
&+ (1673/\mathrm{T})[6.7(\mathrm{X_{Na} + X_K}) + 1.8\mathrm{X_{Al}} + 4.9\mathrm{X_{Mg}} + 8.1\mathrm{X_{Ca}} \\
&+ 5.0\mathrm{X_{Ti}} + 8.9\mathrm{X_{Fe^*}} - 22.2\mathrm{X_{Fe^*}X_{Ti}} + 7.2\mathrm{X_{Fe^*}X_{Si}}] \\
&- 2.06\,\mathrm{erf}(-7.2\mathrm{X_{Fe^*}})
\end{aligned} \tag{10.34}$$

where $A_0 = A^H/T + A^S$. With $(A^H/1673 + A^S)$ constrained to -5.330 to meld with the 1400°C fit, these parameters were evaluated to be $A_H = -23590$ and $A_S = 8.77$, obtained as follows.

The experimental study of Haughton et al. (1974), being ambitious in scope, has had much influence on previous petrologically focused reviews of the behavior of S^{2-} in silicate melts (Poulson & Ohmoto, 1990; Wallace & Carmichael, 1992; Ariskin et al., 2013). The study consisted of 100 experiments at 1200°C, divided into two groups: Series I experiments (n = 37) investigated a single base composition at different fO_2 and fS_2, with the composition varied in its FeO from addition of Fe_2O_3 and/or separation of immiscible sulfide. Their Series II experiments (n = 63) investigated different starting compositions at the same fO_2 and fS_2, with all compositions designed to be saturated in FeS-rich melt. As pointed out in O'Neill and Mavrogenes (2002), their results show a lot of scatter. For example, replicate experiments agree poorly, and the results of the Series I experiments are incommensurate with those of Series II. Two causes are likely: equilibration problems and difficulties analyzing S at relatively low levels by electron microprobe. The silicate melts were held in alumina crucibles, which resulted in continuous reaction with the crucible during the experiment, because the original melt compositions were not saturated in alumina. This effect also biased the melt compositions studied to high Al_2O_3. In addition, some precipitation of FeS during the quenching might be anticipated, given the relatively slow quenching rates possible with the alumina crucibles. The S content in one experiment in Series I, 4F1, which was run at a higher fO_2 than any other experiment in the study, was reported at 270 ppm, which, although low, is far higher than any value that would reasonably be expected for any extrapolation (< 50 ppm), attesting to analytical difficulties. The scatter of the data has meant that attempts to parameterize the results of Haughton et al. (1974) have led to complicated scenarios involving hypothetical Fe-S species or complexes in the silicate melt (Poulson & Ohmoto, 1990; Ariskin et al., 2013), or to the idea that SCSS varies with fO_2 and fS_2 at constant melt composition (Wallace & Carmichael, 1992), which it does not, apart from second order effects like changing Fe^{3+}/Fe^{2+} (Mavrogenes & O'Neill, 1999). These points are discussed further below. Here, the results of Haughton et al. (1974) were used only to obtain an estimate of A_0 at 1200°C. To do this, both series were fitted to Equation 10.34, with the result shown in Fig. 10.3b. Together with some additional information from the author's own preliminary results at 1300°C and 1500°C, these results were used to estimate $A_H = -23590$ and $A_S = 8.77$.

More data over a wider range of temperatures are needed to decrease the considerable uncertainty on these values. The effects of pressure and volatile contents (with H_2O of particular interest) on $C_{S^{2-}}$ cannot at present be measured to useful accuracy by existing experimental methods directly, but their contributions to SCSS are accessible experimentally and will be reviewed.

10.5. SULFIDE CONTENT AT SULFIDE SATURATION (SCSS)

The term "sulfur content at sulfide saturation," meaning the amount of sulfur dissolved in a silicate melt in equilibrium with a sulfide phase (usually but not necessarily a FeS-rich sulfide melt), and acronymed as SCSS, was introduced by Shima and Naldrett (1975). Because some non-negligible proportion of the sulfur dissolved in silicate melts in equilibrium with some sulfides will be SO_4^{2-} towards the high end of the range of oxygen fugacities of natural magmas, and there are possibly other S species present at high pressure or hydrous melts (for example, HS^-), the quantity is less ambiguously labelled "sulfide content at sulfide saturation." This is the meaning of SCSS that will be used here.

The equilibrium between a silicate melt and a sulfide matte (Equation 10.1) gives the relationship:

$$-\Delta G^o_{(FeO-FeS)}/RT = -\ln a^{sil\ melt}_{FeO} + \ln a^{sulf}_{FeS} + 1/2\ln(fO_2/fS_2) \quad (10.35)$$

Substituting Equation 10.3, the definition of $C_{S^{2-}}$, into 10.35 to eliminate (fO_2/fS_2) gives:

$$\ln[S]_{SCSS} = \Delta G^o_{(FeO-FeS)}/RT + \ln C_{S^{2-}} - \ln a^{sil\ melt}_{FeO} + \ln a^{sulf}_{FeS} \quad (10.36)$$

Therefore, SCSS depends on four quantities:

1. The standard state free energy of the Reaction 10.1, $\Delta G^o_{FeO-FeS}$, which is the difference between the two free energies of formation from the elements:

$$\underset{solid}{Fe} + ½\,O_2 = \underset{sil\ melt}{FeO} \quad (10.37)$$

$$\underset{solid}{Fe} + ½\,S_2 = \underset{sulf\ melt}{FeS} \quad (10.38)$$

The free energy of Reaction 10.37, $\Delta G^{o,sil\ melt}_{FeO}$, is taken from Holzheid et al. (1997), who combined the data for crystalline but non-stoichiometric "FeO" in equilibrium with Fe metal from O'Neill and Pownceby (1994) with those for the melting of FeO from Barin (1989), which gave:

$$\Delta G^{o,sil\ melt}_{FeO}(J\ mol^{-1}) = -244118 + 115.559\,T - 8.474\,T\ln(T) \quad (10.39)$$

The data for $\Delta G^{o,sulf\ melt}_{FeS}$ are less accurately known, with different assessments or tabulations of thermodynamic data differing by up to 10 kJ mol^{-1} at relevant temperatures, reflecting different measurements reported in the literature; see Waldner and Pelton (2005) for a review. Here it was assumed that $\Delta G^{o,sulf\ melt}_{FeS} = \Delta H^o_{FeS} - T\,\Delta S^o_{FeS}$, with the values of ΔH^o_{FeS} and ΔS^o_{FeS} to be optimized in the fitting. Whether these optimized values fall within the range of experimentally measured values provides a critical test of the modelling. The effect of pressure on $\Delta G^{o,sulf\ melt}_{FeS}$ could in principle be evaluated from the partial molar volumes of FeO in silicate melts (Lange & Carmichael, 1990) and of FeS in sulfide melts (e.g., Kress et al., 2008). However, for the purpose of calculating SCSS this information is insufficient because the effects of pressure on the other quantities in Equation 10.36 are not known. Henceforward the effects of pressure will be regarded as having to be parameterized empirically from the SCSS experimental data.

2. The sulfide capacity $C_{S^{2-}}$ has been evaluated above as a function of composition and temperature, however, its pressure dependence, and the compositional effects

of H_2O and other volatile species, remain unknown. A hypothesis to explain the currently available experimental data on the effect of H_2O on SCSS will be proposed later, but first only supposedly anhydrous experiments will be considered.

3. The activity of FeO in silicate melts is related to its mole fraction by $a_{FeO}^{sil\ melt} = X_{FeO}^{sil\ melt}\gamma_{FeO}^{sil\ melt}$. The activity coefficients, $\gamma_{FeO}^{sil\ melt}$, have been measured by equilibration with Fe metal according to Reaction 10.37 at atmospheric pressure, using gas-mixing methods, from 1150 to 1500°C. The upper temperature limit is an experimental one, set by the melting point of Fe metal (1538°C). Achieving good accuracy is relatively easy in such measurements compared to most matters in the world of experimental petrology, and studies on silicate melts with compositions of geological relevance are available from Roeder (1974), Doyle and Naldrett (1986), Doyle (1988, 1989), O'Neill and Eggins (2002), Borisov et al. (2004, 2006), and Borisov (2007). In addition, Holzheid et al. (1997) obtained measurements in equilibrium with Fe–Ni–Co alloys. These studies provide over 600 experimental observations, dominated by the Doyle datasets (512 data, all at 1327°C), but the coverage of melt composition is not even. Although the Doyle opus is of great value, it does show a marked predilection for melts high in K, also in Ti, an aversion to Na, and almost a dismissal of Mg; in fact, high MgO compositions are generally lacking in all the studies, and high SiO_2 compositions, such as are relevant to rhyolites (say, $X_{SiO2} > 0.65$), are not represented at all. On the plus side, there is good agreement among all these studies, as noted by O'Neill and Eggins (2002), and by Borisov (2007). When referenced to the standard state of Equation 10.37, the experimental data show that $\gamma_{FeO}^{sil\ melt}$ varies between 0.7 and 2 for most compositions (e.g., Fig. 10.4). Although this is a relatively small variation compared to activity coefficients of other oxide components in silicate melts (O'Neill & Eggins, 2002; Wood & Wade, 2013), this factor of three maps on to a factor of three for $[S]_{scss}$ (Equation 10.36). It therefore needs to be taken into account by parameterizing $\gamma_{FeO}^{sil\ melt}$ as a function of melt composition and temperature. The experimental observations show that $\gamma_{FeO}^{sil\ melt}$ increases with alkali content (Na and K), and decreases with Ti, with a minimum in plots against X_{SiO2} near 0.55 (O'Neill & Berry, 2006; Borisov, 2007). As well as reproducing these features, a parameterization needs to ensure that $\gamma_{FeO}^{sil\ melt} \rightarrow 1$ as $X_{FeO} \rightarrow 1$. To take these factors into account, the experimental data were fitted to the equation:

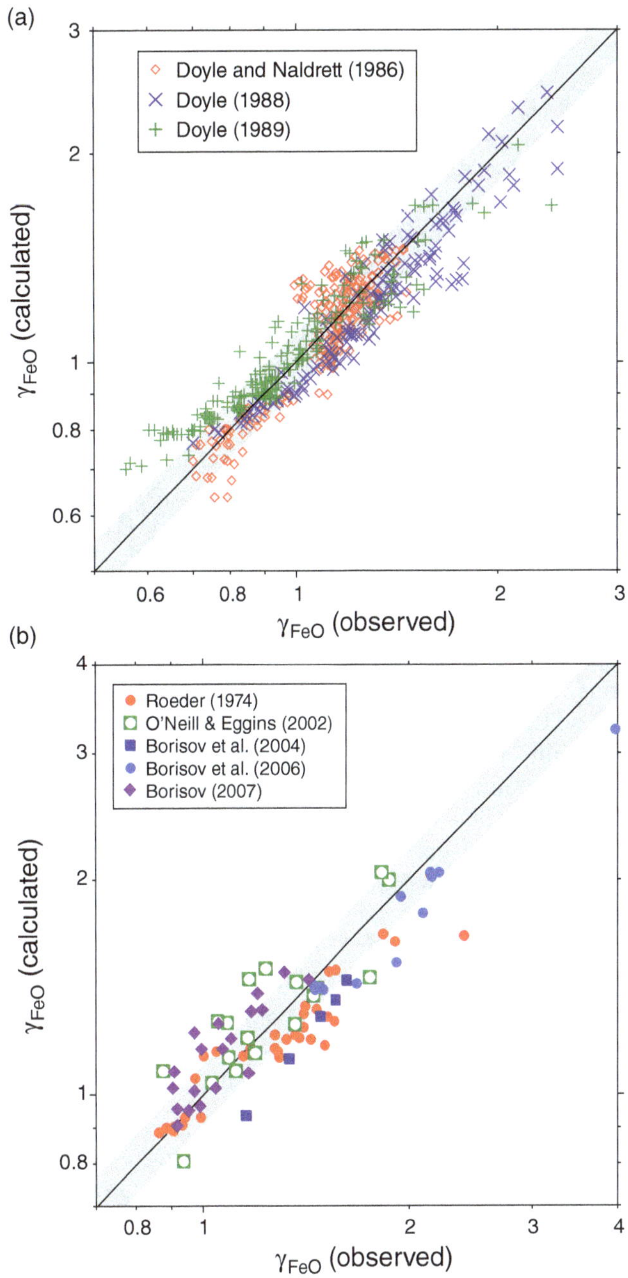

Figure 10.4 Comparison between measured values of the activity coefficient of FeO in silicate melts ($\gamma_{FeO}^{sil\ melt}$) with the parameterization of Equation 10.46. (a) Data of Doyle and Naldrett (1986), and Doyle (1987, 1988), all at 1327°C. (b) Data of Roeder (1974), O'Neill and Eggins (2002), Borisov et al. (2004, 2006), and Borisov (2007), temperatures between 1150 and 1500°C. All experiments equilibrated silicate melt with Fe metal at imposed oxygen fugacity. The gray areas on either side of the 1:1 lines indicates ± 10% relative error.

$$RT \ln \gamma_{FeO}^{sil\ melt} = (1 - X_{FeO})^2 [W_0 + W_{Mg}X_{MgO} + W_{Ca}X_{CaO} + W_{Na}X_{NaO1.5} + W_K X_{KO1.5} + W_{Ti}X_{TiO2}] + (1 - X_{FeO}) \left[U_1 X_{SiO2} + U_2(X_{SiO2})^2\right] \quad (10.40)$$

(the W terms are not Margules coefficients, just ad-hoc parameters). All the experimental data were recalculated to the same standard state using Equation 10.37, with all compositions converted to mole fractions on the single-cation basis. An initial best fit was obtained, but to aid internal consistency a further refinement was done in conjunction with the fitting of the SCSS data, see below.

4. The activity of FeS in sulfide melts presents a somewhat different problem to that of FeO in silicate melts, simply because FeS will be the dominant component in the envisaged applications of Equation 10.36. For example, in ocean floor basalts, X_{FeO} is only about 0.1, whereas a_{FeS}^{sulf} varies from about 0.7 to 0.9. The contributing factors are dilution by O substituting for S, and, in part of the experimental database but ubiquitously in natural systems, dilution by other chalcophile elements, of which Ni and Cu are predominant in nature. There is continuous miscibility between liquid FeS and liquid Fe above the melting point of the latter (e.g., Waldner & Pelton, 2005), such that with decreasing fO_2, and/or decreasing fS_2, Fe-S melts become increasingly metal-rich (Fe/S > 1). However, below Fe/S = 1, the contours of fS_2 vs. mole fraction of FeS on the join Fe–FeS are quite flat (Waldner & Pelton, their Fig. 5), indicating, incidentally, that liquid immiscibility would develop at lower temperatures, were it possible to undercool Fe-S melts. Hence a_{FeS}^{sulf} is rather insensitive to changes to composition in this range. For example, Mavrogenes and O'Neill (1999) calculated that a_{FeS}^{sulf} was ~0.5 in sulfide melts in their experiments in the limiting case of equilibrium with solid Fe metal (see also Ballhaus et al., 2017). Sulfide melts with large metal excesses (Fe/S >> 1) were not included in the experimental database considered here, so this aspect of the a_{FeS}^{sulf} problem was ignored.

The amount of oxygen dissolved in liquid sulfide mattes is a function of fO_2, fS_2, and both Ni and Cu contents, and temperature (Fonseca et al., 2008), indicating considerable complexity. However, switching from system variables (fO_2, fS_2) to focus on the composition of a coexisting silicate melt suggests that the equilibrium between silicate and sulfide melts may be:

$$\underset{sil\ melt}{FeO} = \underset{sulf\ melt}{FeO} \quad (10.41)$$

Hence the amount of the FeO component in the sulfide matte should vary with $a_{FeO}^{sil\ melt}$. Kiseeva and Wood (2013) showed that on a mole fraction basis, $X_{FeO}^{sulf} \simeq X_{FeO}^{sil\ melt}$, which is confirmed here, more or less, from the data used in the SCSS fitting that reported oxygen contents of the sulfide melt, although there is considerable scatter (Fig. 10.5). Measuring the oxygen content of sulfide melts is difficult, because of the complex textures produced on quenching (Kress, 1997; Fonseca et al., 2008), and the task is made harder when silicate melt is present, because this phase provides a refuge for oxygen fleeing the sulfide melt during its crystallization on quenching.

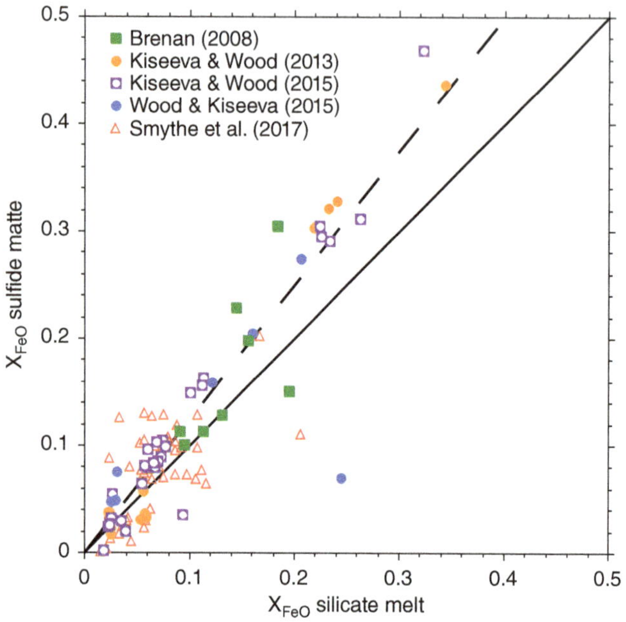

Figure 10.5 The oxygen contents as the component FeO in sulfide mattes coexisting with silicate melt. Data from Brenan (2008), Kiseeva and Wood (2013, 2015), Wood and Kiseeva (2015), and Smythe et al. (2017). The dashed line is the least-squares best fit, $[FeO]_{sulf}$ = 1.20 (±0.04) $[FeO]_{sil\ melt}$, with R^2 = 0.81. The solid line is $[FeO]_{sulf} = [FeO]_{sil\ melt}$, which was used in the SCSS modeling for simplicity.

For Ni and Cu, it has been proposed from available partitioning data that these elements may be treated as inert diluents (Smythe et al., 2017). By this hypothesis, a_{FeS}^{sulf} may be approximated in sulfide mattes that are not metal-rich (that is, their atomic M/[S+O] do not differ greatly from unity) in equilibrium with silicate melts by:

$$a_{FeS}^{sulf} = \left(1 - k_{Ox} X_{FeO}^{sil\ melt}\right) \left[\frac{Fe}{Fe + Ni + Cu + \ldots\ldots}\right]_{sulf} \quad (10.42)$$

Here k_{Ox} is an equilibrium constant for Reaction 10.41, and $\left[\frac{Fe}{Fe + Ni + Cu + \ldots}\right]_{sulf}$ is the atomic ratio of Fe to other cationic diluents in the matte. In natural terrestrial

Table 10.1 Experimental studies used in parameterizing the sulfide content at sulfide saturation (SCSS) in silicate melts.

Reference	T range (°C)	P range (GPa)	No. data used	Chalcophile elements
Anhydrous				
Haughton et al. (1974)	1200	10^{-4}	$C_{S^{2-}}$ only	–
Danckwerth et al. (1979)	1200–1260	10^{-4}	14	–
Peach & Mathez (1993)	1450	0.8	10	Ni
Peach et al. (1994)	1450	0.8	24	–
Gaetani & Grove (1997)	1350	10^{-4}	22	Ni & Cu
Mavrogenes & O'Neill (1999)	1400–1800	0.5–5.5	16	–
Baker et al. (2001)	1225–1450	1.0	6	–
O'Neill & Mavrogenes (2002)	1400	10^{-4}	34	–
Holzheid & Grove (2002)	1350–1600	0.9–2.7	4	Ni
Ripley et al. (2002)	1245	10^{-4}	22	Cu
Sattari et al. (2002)	1330	1.0	5	
Jugo et al. (2005)	1355	1.6	7	–
Liu et al. (2007)	1150–1450	0.5–1.0	41	–
Brenan (2008)	1200–1300	10^{-4}, 0.9–1.6	21	Ni & Cu
Righter et al. (2009)	1200–1500	10^{-4}, 0.8	16	–
Tuff & O'Neill (2010)	1370	10^{-4}	15	minor Ni
Kiseeva & Wood (2013)	1400	1.5	14	Ni & Cu
Ding et al. (2014)	1500–1700	1–4	25	–
Kiseeva & Wood (2015)	1300–1700	1.5	29	Ni & Cu
Wood & Kiseeva (2015)	1400–1460	1–1.5	9	minor Ni
Wykes et al. (2015)	1400	1.5	12	–
Fortin et al. (2015)	1250	1.0	3	–
Smythe et al. (2017)	1400–1800	1.5–5.5	36	Ni & Cu
O'Neill & Mavrogenes (in prep.)	1300	10^{-4}	22	–
Hydrous				
Moune et al. (2009)	1045	0.2–0.3	16	–
Fortin et al. (2015)	1250	1.0	10	–
D'Souza & Canil (2018)	1270	1.0	13	–
This study (Table 10.3)	1400	1.5	6	–

basalts, only Ni and Cu are significant, but other chalcophile elements are present in the sulfide of a few of the experimental studies used as input to the model (Table 10.1), and all were treated similarly. Trials showed that using the simpler $X_{FeO}^{sil\ melt}$ rather than the more logical $a_{FeO}^{sil\ melt}$ in the above expression returned a slightly better result. A linear least-squares best fit found $k_{Ox} = 1.20 \pm 0.04$ (Fig. 10.5), but given the scatter in the data, k_{Ox} was put at 1 following the suggestion of Kiseeva and Wood (2013), which made no perceptible difference to the SCSS fitting.

By splitting the SCSS problem up in this four-fold way, a better understanding is achieved of how SCSS responds to the different variables (T, P, silicate, and sulfide melt compositions). This will enable a quantitative sensitivity analysis. The four-fold way is also useful in divining where model inadequacies may lie, and what further work would be most beneficial.

10.5.1. Fitting the Experimental Data on SCSS on Anhydrous Compositions

The experimental studies from which the data used to model SCSS were taken are listed in Table 10.1. These experiments are all at relatively low fO_2, where SO_4^{2-} should be negligible. Only nominally anhydrous experiments were considered; the effect of H_2O will be addressed briefly in Section 10.5.4. Experiments at > 5 GPa were also not considered, due to the large changes in silicate melt structure expected at higher pressures, and high liquidus temperatures introducing extra experimental uncertainty. Given the lack of experimental information on $\gamma_{FeO}^{sil\ melt}$ at high SiO_2, so were compositions with $X_{SiO2} > 0.62$. The limited information available hints at a large increase in $\gamma_{FeO}^{sil\ melt}$ with X_{SiO2} at higher SiO_2 (O'Neill & Berry, 2006; Borisov, 2007). Data on compositions with < 1 wt% FeO were also excluded for reasons described later.

To account for the effect of pressure in Equation 10.36, two terms were added, the usual one in P/T but modified by a second term in erf(P)/T, to allow for coordination changes in the silicate melt. This extra term will be discussed in detail. The model to which the data were fitted was:

$$\begin{aligned}\ln [S]_{SCSS} = & \left[\Delta G_{FeO}^{sil\ melt} - \Delta H_{FeS}^{o} + T\Delta S_{FeS}^{o}\right]/RT \\ & + \ln C_{S^{2-}} - \ln X_{FeO}^{sil\ melt} - \ln \gamma_{FeO}^{sil\ melt} \\ & + \left(1 - X_{FeO}^{sil\ melt}\right)\left[\frac{Fe}{Fe + Ni + Cu + \ldots}\right]_{sulf} \\ & + (C_1P + C_2\mathrm{erf}(C_3P))/T\end{aligned} \quad (10.43)$$

where the parameters to be determined are ΔH_{FeS}^{o}, ΔS_{FeS}^{o}, C_1, C_2, and C_3 from the SCSS experiments, plus the coefficients in Equation 10.40 for $\gamma_{FeO}^{sil\ melt}$ from the entirely separate $a_{FeO}^{sil\ melt}$ vs. X_{FeO} experiments.

The fitting was by weighted least squares, minimizing the objective function $((\ln [S]_{SCSS(obs)} - \ln[S]_{SCSS(calc)})/s(\ln[S]_{SCSS}))^2$, subject to the constraint that a second objective function for the optimization of $\gamma_{FeO}^{sil\ melt}$, namely $(((RTln\gamma_{FeO}^{sil\ melt})_{obs} - (RTln\gamma_{FeO}^{sil\ melt})_{calc})/s(RTln\gamma_{FeO}^{sil\ melt})^2$, did not increase by more than 2.5% above its best fit value obtained independently of the SCSS data. The 2.5% limit corresponds to less than a percent change in the level of significance. This procedure ensures internal consistency between the different parts of the model. It also has the beneficial effect of bringing more of the relevant compositional space for silicate melts into the parameterization of $\gamma_{FeO}^{sil\ melt}$, and the fact that this can be done within the 2.5% limit chosen shows that, despite the 600 experimental observations, more data are needed on a wider compositional range (particularly towards high MgO, and as pointed out above, high SiO_2). For the weighting, $s(RTln\gamma_{FeO}^{sil\ melt})$ was taken as 800 J mol^{-1} for every datum. This value corresponds approximately to $s(\log fO_2) = 0.03$ and 5% relative uncertainty in $X_{FeO}^{sil\ melt}$. The value of $s([S]_{SCSS})$ was taken as reported, or 5%, whichever was larger. This uncertainty dominates the other uncertainties in the SCSS measurements, namely temperature, pressure, major-element composition, and the minor amounts of H_2O expected to contaminate most high-pressure experiments, but which are usually not measured. The exception is at very low $X_{FeO}^{sil\ melt}$, where $s(X_{FeO}^{sil\ melt})$ becomes dominant as $[S]_{SCSS}$ rises sharply (Wykes et al., 2015). For this reason, and because the composition of the sulfide liquid in equilibrium with such compositions is somewhat uncertain (possibilities include M/S>>1 and the presence of what are normally lithophile components in the sulfide matte), the data with < 1wt.% FeO were not used.

A major problem with combining experimental datasets from different experimental studies performed in different laboratories and addressing different volumes of T-P-compositional space is the way that systematic error can translate into misleading effects. For example, consider a study that reports S contents that are systematically too high, perhaps due to instrument calibration. If that study provided most of the data at, say, high pressure, then this systematic error in S analysis would translate into the modeling as an erroneously enhanced effect of pressure. For analytical aspects, the effects of this type of problem can be ameliorated by reporting analyses of widely available standards. Many of the input studies listed in Table 10.1 have done this using the natural glass VG2 (USNM 111240/52; Jarosewich, 1980) with c. 1400 ppm S, which provides a valuable point of comparison (Table 10.2). The inter-laboratory differences are considerable in some cases. The homogeneity of VG2 is unknown, and the amount of S, although typical of ocean floor basalt glasses, is well below that found in some reduced glasses, let alone in oxidized ones with S as SO_4^{2-}. It would be helpful to have inter-laboratory comparisons available for other standard glasses, perhaps synthetic, with tested homogeneity and higher sulfur contents, one with S as S^{2-} and another with S as SO_4^{2-}.

Of all the initially included studies, only the data of Holzheid and Grove (2002) could not be well fitted, consistently returning calculated values of $[S]_{SCSS}$ higher than observed. This may be due to failure to account for $\gamma_{FeO}^{sil\ melt}$ adequately in these high MgO compositions, alternatively the effect of pressure on depressing $[S]_{SCSS}$ appears higher in these experiments than in others. Although there is no independent reason to do so, the data of Holzheid and Grove (2002) were removed from the fitting, except for their four experiments with high Ni in the sulfide.

The numerical value of C_3 was found to be insignificantly different from 1 (units in GPa^{-1}), so it was fixed at this value, which then gave $C_1 = -291$ K GPa^{-1} and $C_2 = 351$ K. For Reaction 10.38, the optimization produced:

$$\Delta G_{FeS}^{o,sulf\ melt}\left(\mathrm{J\ mol^{-1}}\right) = -106340 + 23.893\,\mathrm{T} \quad (10.44)$$

This gives –70.5 kJ mol^{-1} at 1500 K, in between the values tabulated by Barin (1989) and Robie and Hemingway (1995), namely −70.0 and −73.4 kJ mol^{-1}, respectively. This agreement is a notable affirmation of the model, because the determination of $G_{FeS}^{o,sulf\ melt}$ is independent of anything to do with silicate melts. The free energy for Reaction 10.35 is therefore:

$$\begin{aligned}\Delta G_{FeO-FeS}^{o}\left(\mathrm{J\ mol^{-1}}\right) = & \ 137778 - 91.666\,\mathrm{T} \\ & + 8.474\,\mathrm{T}\ln(\mathrm{T})\end{aligned} \quad (10.45)$$

Table 10.2 Sulfur analyses in natural OFB glass VG2 (USNM 111240/52) by electron microprobe, unless identified otherwise (italics).

S (ppm)	s(S) (ppm)	No. analyses	Reference
1340	80	19	Dixon et al. (1991)
1420	40	6	Wallace & Carmichael (1991)[a]
1320	*50*	*2*	*Wallace & Carmichael (1991)* [a,b]
1400	40	?	Nilsson & Peach (1993) [a]
1348	62	135	Thordarson et al. (1996)[c]
1365	29	139	Thordarson et al. (1996)[c]
1450	30	?	Métrich et al. (1999)
1416	36	10	De Hoog et al. (2001)
1403	31	?	O'Neill & Mavrogenes (2002)
1305	35	203	Thornber et al. (2002)
1310	130	57	Stolper et al. (2004)
1380	60	?	De Hoog et al. (2004)
1549	166	?	Roberge et al. (2004)
1490	60	38	Gurenko et al. (2005)
1414	30	38	Liu et al. (2007)
1405	39	?	Baker (2008)
1362	160	48	Davis et al. (2008)
1410	60	23	Self et al. (2008)
1300	44	?	Bell et al. (2009)
1400	115	37	Kamenetsky et al. (2010)
1415	68	247[d]	Jenner & O'Neill (2012)
1440	30	?	Laurenz et al. (2013)
1610	30	?	Jégo & Dasgupta (2014)
1422	11	17 x 5	Fougeroux (2014)[a]
1392	61	21	Wykes et al. (2015)
1402	28	4	Fortin et al. (2015)
1435	40	43	Vidal et al. (2016)
1486	*27*	*77*	*LeVoyer et al. (2017)*[e]
1521	41	18	D'Souza a&nd Canil (2018)

[a]glass identified as JDFD2 rather than VG2;
[b]wet chemistry;
[c]Two different EMPA routines;
[d]in 19 sessions, spanning two years;
[e]SIMS
Note: Also listed are analyses of JDFD2 (USNM 117354/46), which has often been presumed to be the same sample. The identification of JDFD2 as a different glass (VG7432) is based on the information in Cottrell and Kelley (2011).

The reduced chi-squared for the regression, χ_v^2, was 3.7 (n = 407, four refined parameters). Including the term for pressure in erf(P)/T decreased χ_v^2 from 4.8, a significant improvement since a third of the experiments are at ambient pressure. The rather high value of χ_v^2 is explicable at least in part by inter-laboratory differences in determining S (cf. Table 10.2), pressure, temperature, oxidation state, unanalyzed H_2O in high pressure experiments, and other factors that may differ between studies, and which were not included in the uncertainties used for weighting.

The parameterization of $\gamma_{FeO}^{sil\ melt}$ gave:

$$RT \ln \gamma_{FeO}^{sil\ melt} = \left(1 - X_{FeO}^{sil\ melt}\right)^2 [28870 - 14710X_{MgO}^{sil\ melt} + 1960X_{CaO}^{sil\ melt} + 43300X_{NaO_{0.5}}^{sil\ melt} + 95380X_{KO_{0.5}}^{sil\ melt} - 76880X_{TiO_2}^{sil\ melt}] + \left(1 - X_{FeO}^{sil\ melt}\right)\left[-62190X_{SiO_2}^{sil\ melt} + 31520\left(X_{SiO_2}^{sil\ melt}\right)^2\right] \quad (10.46)$$

with $\chi_v^2 = 2.9$. Note the large effect that the alkalis, Na, and even more so K, have on increasing $\gamma_{FeO}^{sil\ melt}$. The comparison between the measured values of $\gamma_{FeO}^{sil\ melt}$ and those calculated from this parameterization is shown in Fig. 10.4.

$Fe^{3+}/\sum Fe$ in the silicate melts in the experiments that measured $\gamma_{FeO}^{sil\ melt}$ would be insignificantly different from zero, because all these experiments were in equilibrium with metallic Fe. No provision for Fe^{3+} was made in the fitting of the SCSS experiments, either. Because of the constraints on how experiments are done, $Fe^{3+}/\sum Fe$ in the silicate melts of the SCSS database are < 0.02 for atmospheric-pressure experiments (see Tuff & O'Neill, 2010, their Table 5), but likely a little more in many of the higher-pressure experiments, which, because of the prevalence of graphite in the pressure assemblies if not the sample capsules, usually return redox conditions about a log unit in log fO_2 below the graphite-CO_2 buffer, hence two log units below the quartz-fayalite-magnetite equilibrium (Frost & Wood, 1997), corresponding to approximately 0.03 in $Fe^{3+}/\sum Fe$ at atmospheric pressure.

10.5.2. The Effect of Pressure on SCSS

Experimental studies have established that SCSS decreases substantially with increasing pressure at high pressures (Wendlandt, 1982; Mavrogenes & O'Neill, 1999; Holzheid & Grove, 2002; Smythe et al., 2017), as shown in Fig. 10.6. The important consequence is that "primary" magmas should erupt undersaturated in sulfide, if the temperature decrease during their ascent is no more than adiabatic (Mavrogenes & O'Neill, 1999). However, Liu et al. (2007) found no difference in SCSS between 0.5 and 1 GPa in two pairs of experiments, although the uncertainty was large (see their Fig. 2). The present fitting exercise showed that this view has merit. If only the data in the interval from 0 to 1.5 GPa were considered, there would appear to be little or no effect of pressure. The possibility exists that the effect of pressure changes with pressure. This is not surprising;

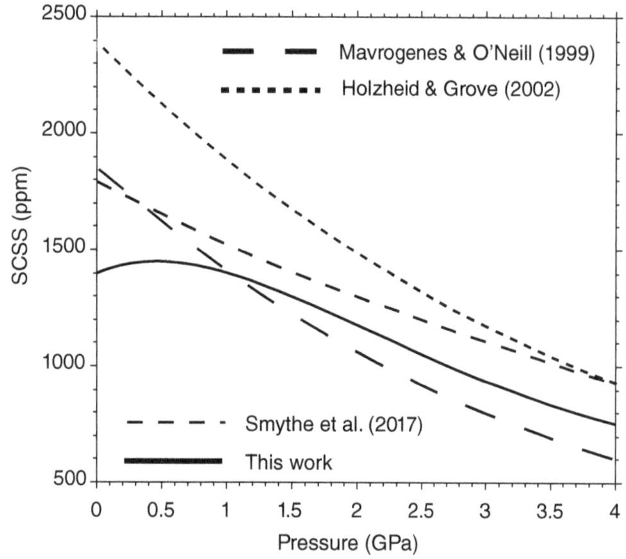

Figure 10.6 The effect of pressure on the SCSS, comparing Equation 10.43 of this study, calculated starting from $[S]_{SCSS}$ = 1400 ppm at ambient pressure, with the equations for SCSS from Mavrogenes and O'Neill (1999), Holzheid and Grove (2002), and Smythe et al. (2017), all at 1400°C. The different calculated values of $[S]_{SCSS}$ at P = 0 are due in part to the different compositions studied. The effect of pressure found by Wendlandt (1982) is similar to that of Mavrogenes and O'Neill (1999) and Holzheid and Grove (2002), and is not shown for clarity. The effect of pressure from Smythe et al. (2017) is smaller than those from Mavrogenes and O'Neill (1997) or Holzheid and Grove (2002), but is similar to the hypothetical parameterization postulated in this study at > 1 GPa. This comparison shows the desirability of integrating experiments at high pressure with those at 1 bar, to constrain the effect of pressure more precisely in the range 1 bar to 1.5 GPa.

previously, it has been assumed in fitting experimental data that the relevant $\int VdP$ terms in the four quantities in Equation 10.36 could be approximated by a constant: that is, ln $[S]_{SCSS}$ varied linearly with P/T (Mavrogenes & O'Neill, 1999). This simplification has long been accepted as a viable first-order approximation in solid-state reactions, where, in essence, it assumes that the effects of the thermal expansivities and bulk moduli of the individual crystalline phases in the reaction cancels out across the reaction. But silicate melts have different P–V characteristics from crystalline phases. With increasing pressure, the main aluminosilicate structural units change, while the average coordinating number of the nearest-neighbor anions forming the coordination polyhedra around the M^{x+} cations tends to increase. Possibly the coordination environment of the S^{2-} anions also changes with pressure. Such changes are more akin to order-disorder phenomena or even phase transitions than to the effects of compressibility on the $\int VdP$ terms in crystals, so carrying over an approximation from experience with crystals may not be such a good idea. Trials fitting the data with polynomials, such that ln $[S]_{SCSS}$ varied with terms in C_n P^n/T, produced a marked hump in $[S]_{SCSS}$ vs. P in the range 0 to 1 GPa, where there are almost no data to constrain reality. As a hump seems implausible, compared to no change in $[S]_{SCSS}$ vs. P in this range, the relationship was described using an error function in Equation 10.43, despite reservations about employing this function twice in one paper. The form of the suggested pressure-dependence is compared to the conventional views in Fig. 10.6. Above ~1 GPa, there is little difference from the conventional assumption.

With increasing pressure above ~5 GPa, SCSS becomes very low (e.g., Li & Agee, 2001), confirming the trend in Fig. 10.6, but precise experimental measurements are difficult because of the high silicate liquidus temperatures, and the fact that the melts do not quench to glasses, making it difficult to know if the tiny sulfide globules commonly encountered were exsolved on quenching or not.

Experimental studies to date have not been designed to cope with the possibility that the pressure dependence of SCSS may be non-linear. For example, the classic study of Wendlandt (1982) investigated pressures only between 1.25 and 3 GPa, completely missing the hypothetical low-pressure inflection shown in Fig. 10.6. A more detailed look at the pressure-dependence of SCSS between 0 and 1 GPa is needed.

10.5.3. The Dependence of SCSS on FeO

SCSS depends on the FeO content in the silicate melt mainly in two ways (Equation 10.36). The $C_{S^{2-}}$ term causes a linear increase of SCSS with increasing $X_{FeO}^{sil\ melt}$, except, as reviewed above, at high FeO (> 15 wt%), whereas the $a_{FeO}^{sil\ melt}$ term causes a logarithmic decrease of SCSS with increasing $X_{FeO}^{sil\ melt}$ (modulated somewhat by changes in $\gamma_{FeO}^{sil\ melt}$ with X_{FeO}). These competing effects may conveniently be studied by measuring SCSS on a compositional pseudo-binary between a FeO-free "matrix" composition and FeO, with temperature and pressure kept constant. Such a series eventuates anyway in 1-atmosphere experiments conducted over a range of fO_2 and fS_2, when FeO is removed from an initial composition by exsolution of an FeS-rich matte in response to decreasing fO_2 and/or increasing fS_2 (Haughton et al., 1974; O'Neill & Mavrogenes, 2002). The result is that graphs of SCSS vs. FeO (wt.%) for these pseudo-binaries should have an asymmetric U shape. Demonstrating this with 1-atmosphere experiments has been unconvincing, because sufficiently high fS_2 cannot be imposed to access much of the left-hand side of the "U" (Haughton et al., 1974; O'Neill & Mavrogenes, 2002). To get around this, Wykes et al.

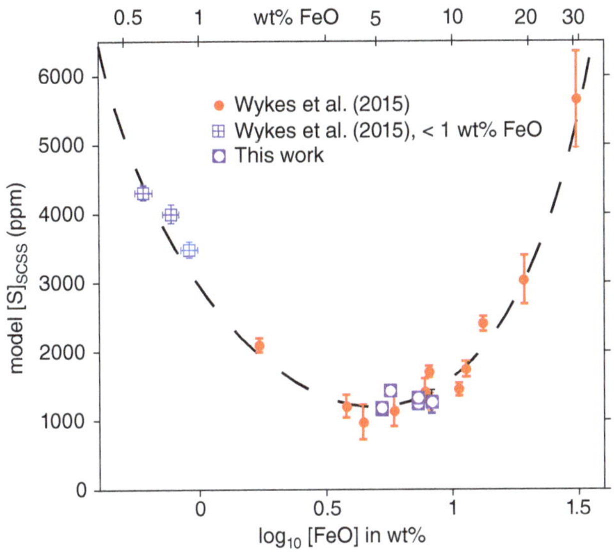

Figure 10.7 SCSS for a binary compositional join between FeO-free "matrix" and FeO. Experimental data from Wykes et al. (2015) and this study (Table 10.3) compared to the SCSS model. The model gives a good fit to the data at < 1 wt.% FeO, which were not used as input, and also to the data at high FeO, which have not been well fitted by previous models; see Fig. 9 of Wykes et al. (2015).

(2015) performed a series of experiments at 1400°C and 1.5 GPa. The results are compared to the present model in Fig. 10.7. The fix of parameterizing $C_{S^{2-}}$ at high FeO with an error-function term (Equations 10.33 and 10.34) works well in describing the data to the highest FeO measured (30 wt%). For a comparison of these data against previous models, see Wykes et al. (2015), their Fig. 9. The model also works well for the data at < 1 wt.% FeO, which were not used in the SCSS fitting. Also plotted on this diagram are some previously unpublished data (Table 10.3), using the same starting composition and Re capsules as described in Wykes et al. (2015), but with water added. The Re capsules did not hold the H_2O quantitatively, but some H_2O was found in the recovered glasses, as determined by FTIR. The concentration of H_2O actually increased from the added amount in one run (G76). The major elements and S analyses were obtained by EMPA, as described in O'Neill and Mavrogenes (2002), but as the major elements were only determined by Energy Dispersive Spectrometry (EDS), they have been recalculated to 100% taking into account the analyzed S and H_2O contents. The latter were measured as described in Liu et al. (2005).

10.5.4. The Effect of H_2O

Naldrett and Richardson (1967) showed that H_2O had no perceptible effect on the melting point of pyrrhotite-magnetite assemblages at 0.2 GPa, implying that this

Table 10.3 SCSS experiments in Re capsules at 1400°C, 1.5 GPa using the starting composition and experimental methods described in Wykes et al. (2015).

run number	G60*	G61	G70	G76*	G93	G94*
added H_2O	1.4	1.8	4	0.2	0.2	5
Na_2O	2.64	2.61	2.46	2.59	2.48	2.41
MgO	9.5	9.5	9.4	9.7	9.3	9.4
Al_2O_3	19.6	19.5	19.3	19.5	19.1	19.2
SiO_2	50.4	50.5	49.8	50.2	49.1	50.1
CaO	10.7	10.7	10.1	10.5	10.3	10.2
TiO_2	1.27	1.28	1.21	1.19	1.23	1.14
FeO(t)	5.2	5.3	7.3	5.6	8.2	7.3
H_2O	0.44	0.43	0.21	0.60	0.17	0.15
S	1176	1193	1332	1434	1268	1256
s(S)	82	90	69	79	169	24

*Sulfur K-edge XANES spectra for these samples were obtained by Métrich et al. (2009)

component has negligibly small solubility in Fe-S-O melts at these conditions. This is supported by the experiments of Moune et al. (2009), who report pyrrhotite, not sulfide melt, at 1045°C, ~0.3 GPa. Thus the effect of H_2O can be isolated to what happens in the silicate melt. In the context of the four-fold way, two variables, $C_{S^{2-}}$ and $a_{FeO}^{sil\ melt}$, will be affected by adding H_2O, but the presence of H in the system could perhaps also result in S dissolving as anionic species in addition to the species that predominate in dry systems (S^{2-} and SO_4^{2-}), for example HS^-, or perhaps, by analogy with OH^-/H_2O speciation, as molecular H_2S in high-SiO_2 melts (Clemente et al., 2004). At present there is no compelling evidence for HS^- in FeO-containing hydrous basic or intermediate melts (Wilke et al., 2011; Wykes et al., 2015), and the following discussion will assume that only S^{2-} is present in the relevant fO_2 range.

Adding H_2O to silicate melts is thought to break up framework silicate units, which decreases the activity of SiO_2. From Reaction 10.10, this should increase S^{2-} in the silicate melt. However, potential additional anion units like OH^- or SiO_3OH^{3-} (the hydroxylated orthosilicate unit), add complexity to the anion sublattice, with input into the reciprocal solution formalism that cannot be predicted at present. But there is a simpler effect that first needs to be addressed. It is dilution. Adding H_2O to a silicate melt adds oxygen anions to the anion sublattice, and when converted into mole fractions, petrologically reasonable amounts translate to large mole fractions: for typical basaltic compositions, 5 wt.% H_2O correspond to about 0.25 in $X_{HO_{0.5}}^{sil\ melt}$ (mole fraction on the single-cation basis). The mole fractions of all the other oxide components $X_{MO_{x/2}}^{sil\ melt}$ are lowered, by a factor of $(1-X_{HO_{0.5}}^{sil\ melt})$. By itself, this dilution effect should influence SCSS in two ways: (i) by lowering $a_{FeO}^{sil\ melt}$; (ii) by lowering all the $X_{MO_{x/2}}^{sil\ melt}$ (M≠H) in the model for $C_{S^{2-}}$ (Equation 10.34).

Potentially, H_2O could change $\gamma_{FeO}^{sil\ melt}$. Not much is known about how H_2O affects individual activity coefficients in silicate melts, but a qualitative judgment can be made from the observation that H_2O has only a minimal influence on the partitioning of Fe^{2+} and Mg between olivine and silicate melt (Toplis, 2005). This indicates that any effect of H_2O cancels out in the ratio $\gamma_{FeO}^{sil\ melt} / \gamma_{MgO}^{sil\ melt}$, which would be unlikely if the effect of H_2O on either activity coefficient were huge. Therefore, it is assumed that $\gamma_{FeO}^{sil\ melt}$ may be evaluated as per Equation 10.46, with no additional terms.

This "dilution" hypothesis was tested against the available experimental database for SCSS in H_2O-containing silicate melts, as follows. Only experiments at sufficiently low fO_2 that S^{6+}/S^{2-} should be near zero were considered: Moune et al. (2009); Fortin et al. (2015); D'Souza and Canil (2018); and Table 10.3. The sulfide capacities of hydrous melts may conceptually be divided into two parts, that due to the anhydrous oxide components, taking account of their dilution by the factor of $(1-X_{HO_{0.5}}^{sil\ melt})$, called here $C_{S^{2-}}^{dil}$, and that due to terms in $X_{HO_{0.5}}^{sil\ melt}$, called here $C_{S^{2-}}^{H}$, such that $C_{S^{2-}} = C_{S^{2-}}^{dil} \cdot C_{S^{2-}}^{H}$. The calculated SCSS due only to dilution is then:

$$\ln [S]_{SCSS}(dil) = \Delta G^{o}_{FeO-FeS}/RT + \ln a_{FeS}^{sulf} + \ln C_{S^{2-}}^{dil} - \ln X_{FeO}^{sil\ melt} - \ln \gamma_{FeO}^{sil\ melt} \quad (10.47)$$

and the difference between the measured S and this calculated $[S]_{SCSS}(dil)$ is given by:

$$\ln C_{S^{2-}}^{H} = \ln [S]_{obs} - \ln [S]_{SCSS}(dil) \quad (10.48)$$

The calculated values of $\ln C_{S^{2-}}^{H}$ from the input experiments are plotted against $X_{HO_{0.5}}^{sil\ melt}$ in Fig. 10.8a, which shows that they scatter around a fairly well defined curve of somewhat parabolic shape. Some scatter would be expected from the effect of H_2O on the activity of SiO_2. D'Souza and Canil (2018) also proposed a strong effect of Na and K, from the difference between the two compositions studied by them. With these considerations, the experimental data were fitted by non-linear least-squares regression assuming uncertainties in both S and H_2O contents, with s([S]) taken as 5% or as reported, whichever is larger, and s(H_2O) as 0.1 wt% + 0.05[H_2O], to parameterize $C_{S^{2-}}^{H}$ as:

$$\ln C_{S^{2-}}^{H} = X_{HO_{0.5}}^{sil\ melt}\left(6.4 + 12.4X_{HO_{0.5}}^{sil\ melt} - 20.3X_{SiO_2}^{sil\ melt} + 73\left(X_{NaO_{0.5}}^{sil\ melt} + X_{KO_{0.5}}^{sil\ melt}\right)\right) \quad (10.49)$$

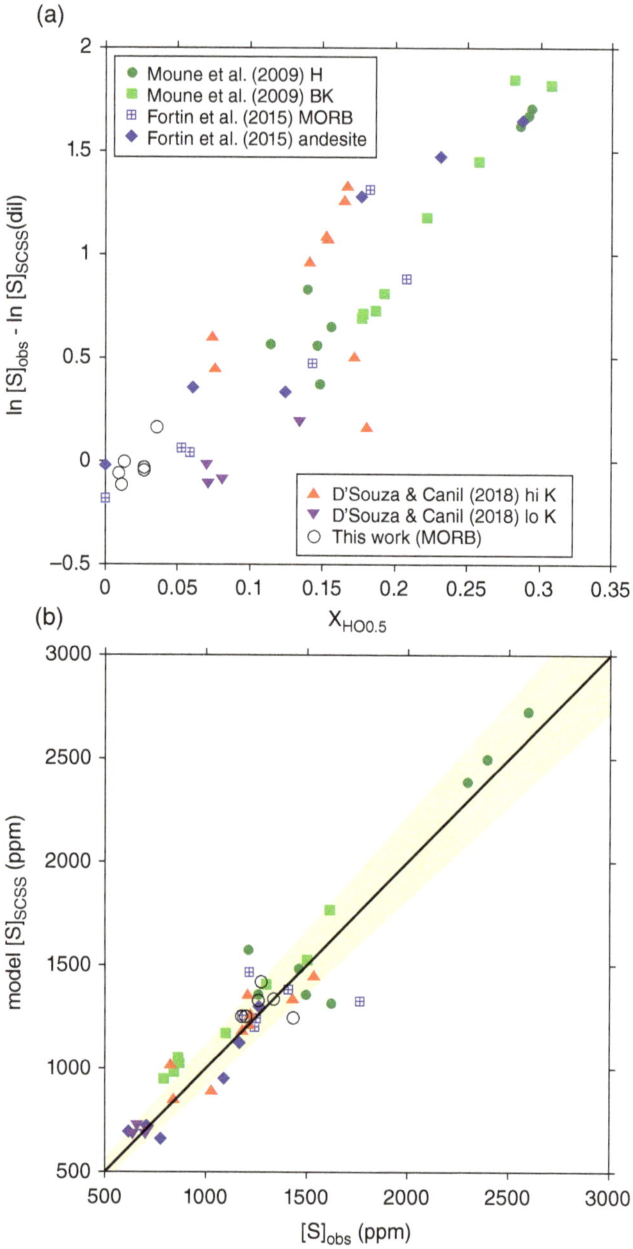

Figure 10.8 The effect of H_2O on SCSS. Data from Moune et al. (2009) at 1045°C, ~ 0.3 GPa, in equilibrium with pyrrhotite; Fortin et al. (2015), 1250°C, 1 GPa; D'Souza and Canil (2018), 1270°C, 1 GPa; and this work, Table 10.3, 1400°C, 1.5 GPa. (a) The difference ln $[S]_{obs}$ – ln $[S]_{SCSS}$(dil), which is equivalent to ln $C_{S^{2-}}^{H}$; see text for further explanation. (b) $[S]_{SCSS}$ calculated from the model vs. $[S]_{obs}$. The shaded area around the 1:1 line indicates ±10%.

The uncertainties in the reported H_2O contents have almost as much influence on the regression as those on the S contents, and cannot be neglected; they must also be propagated through the entire model, because of their effect on calculating the mole fractions of the other components. The reduced chi-squared for the fit was 5.0, indicating that the model is too simplistic, if the weighting of the data is reasonable; but as the four experimental studies address different compositions at different temperatures and pressure, to use more variables could be misleading. The experiments of Moune et al. (2009), all at 1045°C, ~0.3 GPa, were saturated with pyrrhotite, not FeS-rich matte, requiring calculation of the activity of FeS relative to the standard state of liquid FeS. This was done using the enthalpy of melting and the melting temperature of pure FeS of 32.34 kJ mol^{-1} at 1468 K from Robie and Hemingway (1995), to give a_{FeS}^{sulf} = 1.35, with X_{FeO}^{sulf} taken to be zero (no dissolved O in pyrrhotite). No correction was attempted for the stoichiometry of the pyrrhotite (i.e., deviations from atomic Fe/S = 1), as this was not reported. Pyrrhotite in comparable experiments is only a little metal-deficient, with Fe/S ~0.96 (Luhr, 1990). The concentrations of S calculated from the model (Equations 10.47 to 10.49) are compared to the observed values in Fig. 10.8b. Most of the data are reproduced to within ±10%.

The net effect of adding H_2O to a silicate melt is a tradeoff between the dilution effect (where obviously $C_{S^{2-}} > C_{S^{2-}}^{dil}$) and the increase in $C_{S^{2-}}$ due to the addition of the $C_{S^{2-}}^{H}$ term. The dilution effect depends to a large extent on the FeO content of the melt, hence there should be a substantial difference in the net effect of adding H_2O between high and low FeO compositions. For low FeO compositions, adding H_2O should increase $C_{S^{2-}}$, and vice versa. Presumably, for the effect of H_2O on the sulfate capacity, $C_{SO_4^{2-}}$, the same principle would apply ($C_{SO_4^{2-}} = C_{SO_4^{2-}}^{dil} \cdot C_{SO_4^{2-}}^{H}$).

At present, nothing seems to be known about the effects on the solubilities of either S^{2-} or SO_4^{2-} of CO_2, whose own solubility in silicate melts becomes considerable above ~2 GPa, nor of other volatile species.

10.6. APPLICATION TO MID-OCEAN RIDGE AND SIMILAR BASALTS

The geochemistry of ocean floor basaltic (OFB) magmatism has been intensively studied from the perspective of the glassy rinds on erupted pillows. These glasses typically contain ~10^3 ppm total sulfur (Mathez, 1976; Czamanske & Moore, 1977), which is in the range quantifiable by electron microprobe to about 5% precision with a bit of care. Métrich et al. (2009) used XANES spectroscopy to show that the oxidation state of nearly all the S in three representative glasses was S^{2-}, with 4 to 8% as SO_4^{2-}, which they proposed was produced on cooling. Many of the glasses contain globules of quenched sulfide mattes, often nearly spherical in shape, with diameters mostly in the range 10 to 100 μm, with an average of 27 μm according to Patten et al. (2012), although some up to 600 μm have been reported elsewhere (e.g., Czamanske & Moore, 2007). The textures of the phase assemblages in the globules vary, at least in part due to differences in cooling rates (e.g., Patten et al., 2012). Bulk compositions of the globules are more reliably estimated for those with finer-grained textures, and are fairly rich in Ni and Cu. For example, the average Ni and Cu contents of 14 fine-grained globules from Patten et al. (2012) are 6.1 ± 2.9 and 8.5 ± 4.0 wt% respectively, with atomic Fe/(Fe+Ni+Cu) being 0.78 ± 0.05.

The existence of these globules has been attributed to the erupted basalts being at sulfide saturation (Mathez, 1976; Czamanske & Moore, 1977), hence the S contents of their host basalt glasses should correspond to their SCSS at their eruption temperature and pressure. Several attempts have been made to compare their measured S contents against models for the SCSS, but this requires knowing the temperature of the basalt at the time the sulfide matte separated from the silicate liquid, plus correction for its Ni and Cu contents. Only the recent work of Smythe et al. (2017) has addressed this issue quantitatively, although Ariskin et al. (2013) did consider Ni. Another issue is that the $Fe^{3+}/\sum Fe$ in OFB glasses is typically about 10% (Berry et al., 2018), which must lower $a_{FeO}^{sil\ melt}$ by about this amount compared to calculations that assume all Fe in the silicate melt to be Fe^{2+}.

The model for SCSS derived from experimental data in this contribution may be tested against the global dataset of ocean floor basaltic glasses of Jenner and O'Neill (2012), for which S concentrations in 329 samples were determined by EMPA. The advantages of this dataset for present purposes is that it is eclectic in geographical locations and tectonic nuances, and that the determination of S concentrations was standardized against the OFB glass VG2 (Table 10.2). The S concentrations are the means of at least five individual analyses on each sample, with standard deviations giving an indication of sample homogeneity. In order to apply the SCSS model, some enabling assumptions need to be made:

1. Sample temperatures were estimated using the magmathermometer of Putirka (2008), his Equation (15), with H_2O put to zero. This thermometer assumes the presence of olivine, hence is inappropriate for the most evolved OFB glasses, but with the necessary caveat it will be used here in extrapolation for all OFB glasses with > 4 wt% MgO. The equilibrium pressure was taken to be the pressure from which the samples were recovered, or 0.03 GPa if this was not known, but the effects of pressure are trivial with the present model.

2. The H_2O content of OFB glasses is generally < 0.5 wt% (e.g., Michael, 1995; Le Voyer et al., 2015; Shimizu et al., 2016), and in view of the absence of

experimental data on the effect of H_2O at the relevant pressures and temperature, the low sensitivity of SCSS to H_2O in this range (Fig. 10.8), and the lack of direct determination of H_2O in the dataset of Jenner and O'Neill (2012), H_2O has initially been assumed to be zero. This assumption will be examined subsequently.

3. The approximation used in modelling the experimental data, that Fe^{3+} in the silicate melts is low enough to be neglected, is not valid for OFB glasses. During OFB petrogenesis Fe^{3+} behaves as a Lightly InCompatible trace Element (LICE), whose concentration as the oxide component in wt.% is given by ln[Fe_2O_3] = 1.46 – 0.177 [MgO] (Berry et al., 2018; O'Neill et al., 2018). Because Fe^{3+}/Fe^{2+} has been measured in only a subset of the samples analyzed for S by Jenner and O'Neill (2012), it was estimated from the above relationship against MgO. Potentially, Fe^{2+}-Fe^{3+} interactions in the melt could also affect $\gamma_{FeO}^{sil\ melt}$, so an extra term, $W_{Fe^{3+}} X_{Fe^{3+}O_{1.5}}^{sil\ melt}$, was added to Equation 10.46, with $W_{Fe^{3+}}$ to be determined by fitting to the natural data, as described below. As discussed previously, there is no information at all on how Fe^{3+} affects $C_{S^{2-}}$, so for the calculation of this quantity, $Fe^{2+}O$ was put equal to total FeO.

4. Although the composition of the sulfide is not known, it should be specified by the composition of the silicate melt, assuming equilibrium. As diluents of FeS, only Ni, Cu, and O are present in sufficient amounts to affect a_{FeS}^{sulf}. For Ni, the equilibrium between sulfide and silicate melts is:

$$\underset{\text{sil melt}}{\text{FeO}} + \underset{\text{sulf}}{\text{NiS}} = \underset{\text{sulf}}{\text{FeS}} + \underset{\text{sil melt}}{\text{NiO}} \quad (10.50)$$

Hence the atomic ratio Ni/Fe in the sulfide is given by:

$$\text{Ni/Fe(atomic)} = K_{(Ni/Fe)}^{*}[\text{Ni}]/[\text{Fe}^{2+}\text{O}] \quad (10.51)$$

The concentrations of Ni and $Fe^{2+}O$ in the silicate melt are input in ppm and wt% respectively, and [$Fe^{2+}O$] = [FeO_{tot}] – 0.8998 [Fe_2O_3]. The equilibrium constant $K_{Ni/Fe}^{*}$ includes implicitly the ratio of the activity coefficients of NiO and $Fe^{2+}O$ in the silicate melt ($\gamma_{NiO}^{sil\ melt}/\gamma_{Fe^{2+}O}^{sil\ melt}$). The change of $\gamma_{NiO}^{sil\ melt}$ with silicate melt composition parallels that of $\gamma_{Fe^{2+}O}^{sil\ melt}$ (O'Neill & Eggins, 2002), so that this ratio is relatively insensitive to composition and may be approximated as a constant. The stoichiometry of the equivalent Cu-Fe^{2+} exchange reaction is not so convenient, because the oxidation state of Cu is 1+ in the silicate melt and potentially a mixture of $CuS_{0.5}$ and CuS components in the sulfide matte, depending on fS_2 (e.g., Ripley et al., 2002), but to keep things simple, it is proposed that Cu/Fe (atomic) may be approximated in the same way over the limited range of conditions appropriate to OFB. Hence:

$$a_{FeS}^{sulf} = \left(1 - X_{Fe^{2+}O}^{sil\ melt}\right)/\left(1 + K_{(Ni/Fe)}^{*}[\text{Ni}]/[\text{Fe}^{2+}\text{O}] + K_{(Cu/Fe)}^{*}[\text{Cu}]/[\text{Fe}^{2+}\text{O}]\right) \quad (10.52)$$

These enabling assumptions require three unknown parameters, $W_{Fe^{3+}}$, $K_{Ni/Fe}^{*}$ and $K_{Cu/Fe}^{*}$ the values of which were estimated by "tuning" the model to fit the data of Jenner and O'Neill (2012) by least squares, minimizing $(([S]_{obs} - [S]_{SCSS,\ calc})/s[S]_{obs})^2$, where $s[S]_{obs}$ was taken as 5% or as given by Jenner and O'Neill (2012), whichever is larger. Because Ni and Cu decrease, and Fe^{3+} increases monotonically with MgO in OFB glasses, these three parameters are highly correlated with each other.

Of the 329 OFB glasses, 11 were identified as possibly having lost S or being sulfide-undersaturated, and removed from consideration. These samples will be revisited below. Of the remaining 318 glasses, Jenner and O'Neill (2002) identified 17 in which the standard deviation of the S analyses was far above analytical precision, and this heterogeneity was confirmed by replicating the analyses in another session. Although these samples cannot be at equilibrium, their large standard deviations give them little weight in the regression, and they were retained. Initial fitting gave $W_{Fe^{3+}}$ within uncertainty of zero, so it was fixed at this value, giving $K_{Ni/Fe}^{*}$ = 0.013, and $K_{Cu/Fe}^{*}$ = 0.025, with a reduced chi-squared, χ_{ν}^{2}, of 1.45.

The agreement between the measured S content, $[S]_{obs}$, and the calculated SCSS is shown in Fig. 10.9a. Generally, OFB glasses not occurring as melt inclusions have degassed some CO_2, with their remaining CO_2 often postulated to correspond to the CO_2 solubility at the water depth at which they erupted (e.g., Dixon & Stolper, 1995). On the present evidence, this degassing does not usually cause perceptible S loss. Exceptions are 11 data with low $[S]_{obs}$, which are independently distinguishable by their low S/Se ratio (Fig. 10.9b), implying that they have lost S by volatilization, since Se is less volatile than S (Jenner et al., 2010). None of these S-deficient samples (VG numbers 1672, 1673, 1674, 1677, 4750, 4751, 4754, 4755, 6822, 8550, 8555), were erupted at particularly shallow depths. Elsewhere, Nilsson and Peach (1993) demonstrated a correlation between S loss and $Fe^{3+}/\sum Fe$ determined by wet chemistry in back-arc basin glasses from the Lau Basin; see also Jenner et al. (2010). Here the only glass for which Fe^{3+}/Fe^{2+} has been measured is VG6822 (Berry et al., 2018), whose oxidation state is unexceptional (Fig. 10.10), so in this particular case its S loss by degassing was not promoted by an unusually high oxidation state compared to other OFB glasses. Two of the samples (VG8550 and VG8555, from the Tuzo

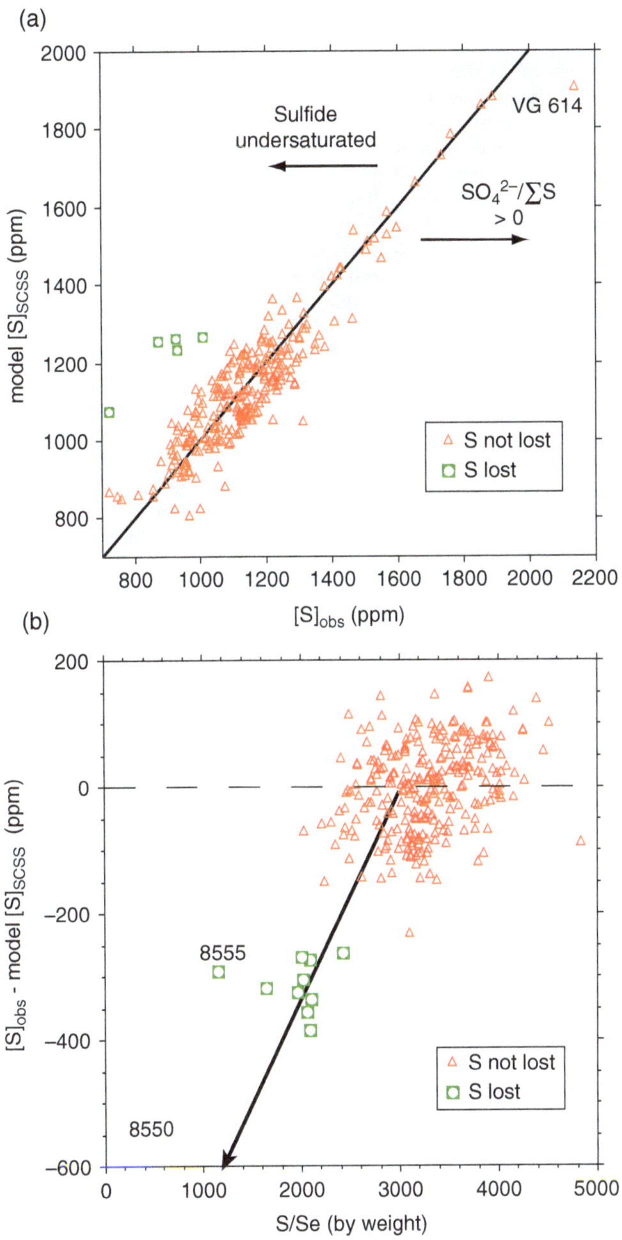

Figure 10.9 Sulfur saturation in ocean floor basaltic (OFB) glasses from Jenner and O'Neill (2012). (a) Calculated SCSS versus S contents analysed by electron microprobe. Of the 329 glasses, only 11 (3%) were identified as probably having lost S. (Two samples with very low S plot off the diagram). (b) This S loss is correlated with decreasing S/Se. The vector for this loss is calculated assuming 1000 ppm S and 0.33 ppm Se, with Se contents of the melt assumed to be conserved (Jenner et al., 2010).

Wilson seamounts) have significantly lower S/Se than that expected from just S loss (Fig. 10.9b), which is intriguing. One possibility is if the devolatilizing magma contained abundant sulfide globules, which were resorbed back into the silicate melt during S loss.

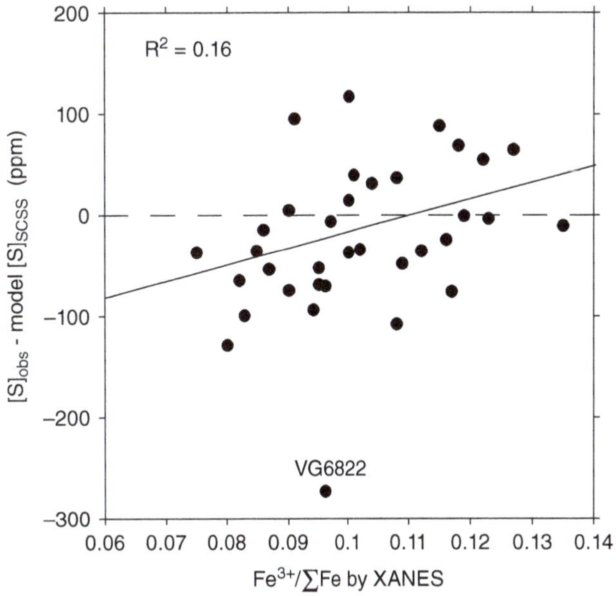

Figure 10.10 The difference between $[S]_{obs}$ and the calculated SCSS from the model for 35 OFB glasses from Jenner and O'Neill (2012) compared against their $Fe^{3+}/\sum Fe$ determined from XANES by Berry et al. (2018). There is almost no correlation (R^2 =0.16). One glass, VG6822, not used in the regression, has low S due to loss by volatility, but retains typical $Fe^{3+}/\sum Fe$.

The calculated compositions of the sulfide mattes that are assumed to be in equilibrium with each glass provide another test of the model. Observed compositions of bulk sulfide globules from OFB are available from Czamanske and Moore (1977) and Patten et al. (2012). As the former study does not give the compositions of the host basalts, only their TiO_2 contents, this variable was used in Fig. 10.11 to compare Ni/Fe and Cu/Fe measured in the real sulfide globules with the calculated ratios in the hypothetical sulfide in equilibrium with each of the glasses from Jenner and O'Neill (2012). For both Ni/Fe and Cu/Fe, the calculated values fall within the range of those observed, indicating that the values of $K^*_{Ni/Fe}$ and $K^*_{Cu/Fe}$ from the model are reasonable. The large scatter in the Ni/Fe and Cu/Fe ratios in the natural sulfide globules evident in these figures reflects a lack of equilibrium between glass and sulfide and/or the difficulties of analyzing the bulk globules on account of their crystallization to polyphase assemblages.

The good agreement between $[S]_{obs}$ and $[S]_{SCSS}$ shown in Fig. 10.9a indicates that S^{6+} is no more than a few percent of total S in these glasses, a possible exception being VG 614, which has the highest $[S]_{obs}$ of any glass in the dataset, and the lowest MgO (4.85 wt%). This finding is compatible with the model of Nash et al. (2019) for $S^{6+}/\sum S$ as a function of $Fe^{3+}/\sum Fe$ (Fig. 10.2). It is reinforced by comparing the values of $Fe^{3+}/\sum Fe$ that were

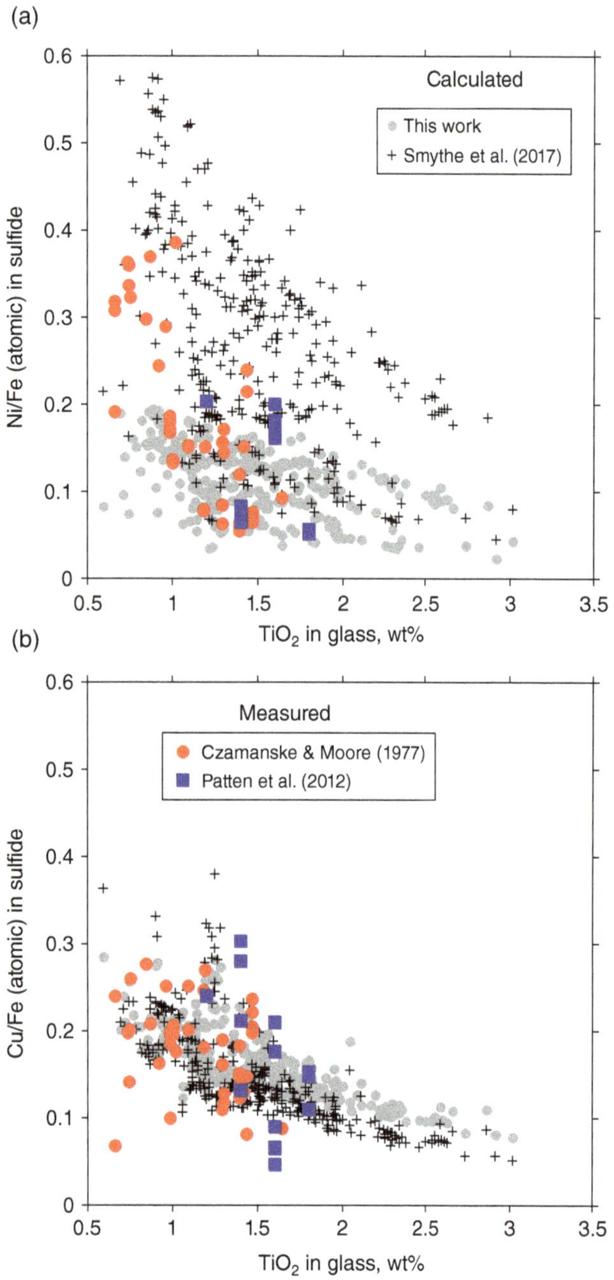

Figure 10.11 Calculated atomic Ni/Fe (a) and Cu/Fe (b) in sulfide liquid in equilibrium with the OFB glasses analyzed by Jenner and O'Neill (2012), compared to observed Ni/Fe and Cu/Fe in sulfide globules from basalts analysed by electron microprobe, by Czamanske and Moore (1977) and Patten et al. (2012). Also plotted are calculated Ni/Fe and Cu/Fe for the same OFB glasses from the model of Smythe et al. (2017).

measured by XANES spectroscopy by Berry et al. (2018) in a subset of 35 of the OFB glasses of Jenner and O'Neill (2012) with the difference between $[S]_{obs}$ and the calculated SCSS (Fig. 10.10). There is almost no correlation

Table 10.4 SCSS model sensitivities (anhydrous model).

Variable (X)	Unit (n_X)	$\partial[S]_{SCSS}/\partial n_X$ (ppm n_X^{-1})
Na_2O	1 wt.%	7
MgO	1 wt.%	60
Al_2O_3	1 wt.%	–37
SiO_2	1 wt.%	–19
CaO	1 wt.%	45
TiO_2	1 wt.%	31
MnO	0.1 wt.%	17.9
FeO_t	1 wt.%	106
Ni	1 ppm	–1.1
Cu	1 ppm	–1.9
$Fe^{3+}/\sum Fe$	0.01	–12
T	1 K	2.25
P	0.01 GPa	1.5

($R^2 = 0.16$), which also implies that the fO_2 of OFB glasses lies below the threshold at which $S^{6+}/\sum S$ becomes more than a few percent (see Fig. 10.2). That the model fits the data with the tuning parameter $W_{Fe^{3+}} \sim 0$ is circumstantial evidence that the effect of changing Fe^{3+}/Fe^{2+} on $C_{S^{2-}}$ may be small. The way that $Fe^{3+}/\sum Fe$ influences SCSS seems to be entirely through its lowering of $X_{FeO}^{sil\ melt}$ from that which would be calculated assuming $Fe^{3+}/\sum Fe = 0$.

An Excel spreadsheet called "SCSS_Calculator.xlsx" for calculating SCSS according to the routine described above is included in Supporting Information.

10.6.1. Sensitivity analysis

For each variable X, the sensitivity may be defined as $\partial[S]_{SCSS}/\partial n_X$, where n_X is a convenient unit of the variable. The values of $\partial[S]_{SCSS}/\partial n_X$ were calculated for representative glass with a typical composition (VG 390), and are given in Table 10.4.

These sensitivities may be used to calculate a representative non-systematic uncertainty (i.e., precision) in $[S]_{SCSS}$, under the approximation of the equation for the propagation of small errors:

$$s([S]_{SCSS}) = \left[\Sigma(s(X)/n_x)^2(\partial[S]_{SCSS}/\partial n_x)^2\right]^{1/2} \qquad (10.53)$$

Taking $s(MO_{x/2}) = 0.1$ wt% (M = Na, Mg, Al, Ca, and Fe as FeO_t), s(MnO), and $s(TiO_2) = 0.03$ wt%, $s(SiO_2) = 0.2$ wt%, s(T) = 20 K, s(Ni) and s(Cu) = 10 ppm, and $s(Fe^{3+}/\sum Fe) = 0.03$, propagates to $s([S]_{SCSS}) = 64$ ppm, or about 5.5% of the average S content of the undegassed OFB glasses of Jenner and O'Neill (2012), which is 1150 ppm. The largest contribution to the uncertainty comes from temperature, followed by $Fe^{3+}/\sum Fe$, then FeO_t, but reasonably accurate knowledge of the Ni and Cu contents of the silicate glass are also needed. Easy to overlook would be the effect of MnO, which contributes to $C_{S^{2-}}$ as much as does FeO on a proportional basis, but does not contribute to $a_{FeO}^{sil\ melt}$. Consequently, ignoring the ~ 0.2 wt% MnO typical of OFB would alone cause an error in SCSS of 3%. The sensitivity of the model to the effect of pressure is speculative only, given the present lack of constraints on the effects of pressure at less than 1 GPa (Fig. 10.6), and was therefore not included.

10.6.2. Is Neglecting H_2O in OFB Glasses Justified?

Although H_2O has not been directly measured in the OFB glasses analyzed by Jenner and O'Neill (2012), their H_2O contents may be estimated approximately, using the observation that H_2O/Ce in OFB glasses is reasonably constant (e.g., Michael, 1995). Assuming H_2O/Ce = 200, the "tuning" procedure was repeated for the 318 undegassed glasses of Jenner and O'Neill (2012), with all mole fractions lowered by the factor $(1-X_{HO_{0.5}}^{sil\ melt})$ to calculate $C_{S^{2-}}^{dil}$, the addition of the term in $C_{S^{2-}}^{H}$, the calculated temperature lowered by 12.83 K per wt% H_2O as specified by the magmathermometer of Putirka (2015), his Equation 15, and $W_{Fe^{3+}}$, $K^{*}_{Ni/Fe}$ and $K^{*}_{Cu/Fe}$ taken as tuning parameters. With this model, it is possible to achieve a statistically better fit (χ^2_ν = 1.16), but at the expense of a very large negative $W_{Fe^{3+}}$, which seems physically unreasonable (although it could be compensating for an effect of Fe^{3+}/Fe^{2+} on $C_{S^{2-}}$), and a larger $K^{*}_{Cu/Fe}$, which translates into higher Cu/Fe ratio in the coexisting sulfide, contrary to the inference from the analyzed sulfides (cf. Fig. 10.11b). On the other hand, constraining $W_{Fe^{3+}}$ = 0 gives χ^2_ν = 1.80, only a little less good than the model assuming zero H_2O, but with lower $K^{*}_{Ni/Fe}$ and $K^{*}_{Cu/Fe}$, which give Ni/Fe and Cu/Fe in good agreement with the available sulfide analyses.

The conclusions are that while the effects of small amounts of H_2O in OFB glasses on SCSS may not be negligible, they cannot be quantified at present. However, the zero-H_2O assumption looks to be a good approximation, and will be used for the following discussions, even where H_2O is known, to ensure unbiased comparisons.

10.6.3. Comparison with Previous Models

With the exception of the recent study of Smythe et al. (2017), previous modeling of SCSS in silicate liquids of geological interest has concentrated on reproducing the available experimental data in systems without Ni and Cu (Moretti & Ottonello, 2005; Moretti & Baker, 2008), and are therefore not directly applicable natural systems containing Fe-Ni-Cu-S-O sulfides.

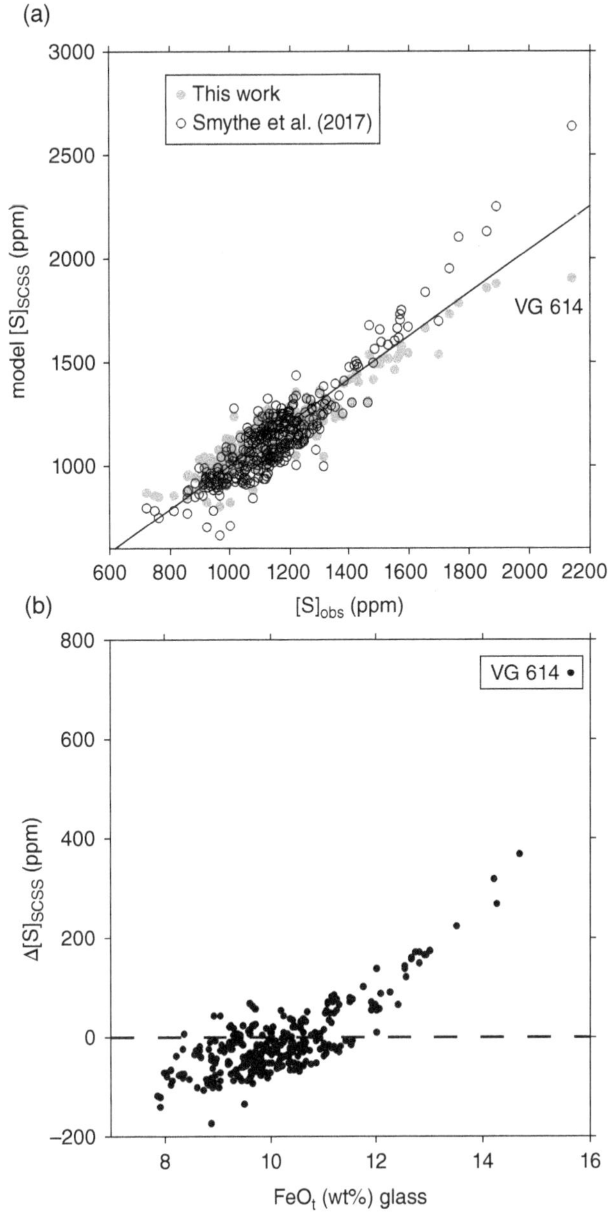

Figure 10.12 Comparison with the SCSS model of Smythe et al. (2017). (a) Calculated $[S]_{SCSS}$ for the OFB glasses of Jenner and O'Neill (2012) from the model of Smythe et al. (2017) vs. $[S]_{obs}$, compared to the model of this work. (b) The difference between the models is a function of the FeO content of the glasses.

The model of Smythe et al. (2017) and that of this work are compared in Fig. 10.12a by applying both to the dataset of Jenner and O'Neill (2012), with liquidus temperatures calculated the same way, that is, with the magmathermometer of Putirka (2008). There is excellent agreement over most of the range of $[S]_{obs}$, from 800 to

1500 ppm, but above 1500 ppm S, the model of Smythe et al. returns higher values of SCSS, which would correspond to such samples being sulfide-undersaturated. Figure 10.12b shows that the difference between the two models is largely due to the different ways that the effects of FeO_t are handled. The model of Smythe et al. does not include the ad-hoc adjustment made in this work for the non-linear decrease of ln $C_{S^{2-}}$ vs. $X_{FeO}^{sil\ melt}$ (Equations 10.33 and 10.34), and does not parameterize $\gamma_{FeO}^{sil\ melt}$ as a function of composition. On the other hand, Smythe et al. (2017) use a more sophisticated treatment of a_{FeS}^{sulf}. Despite this, the calculated values of Ni/Fe in the sulfide from the model of this work appear more realistic when compared to actual sulfide globule analyses (Fig. 10.11a). The calculated Cu/Fe ratios are similar over the range of TiO_2 contents of the glass for which there are actual sulfide globule analyses (Fig. 10.11b).

10.6.4. Application to Other OFB Glass Datasets

Some recent studies have reported sulfur contents in OFB glasses determined by SIMS together with the necessary compositional data including Ni and Cu (Le Voyer et al., 2015, 2017; Shimizu, 2016). The latter study presents new analyses of a number of glasses from Macquarie Island previously analyzed for all relevant elements by Kamenetsky et al. (2000) and Kamenetsky and Eggins (2012), including S by EMPA. This replication is a valuable resource for ground-truthing analytical precision (not just for S), and also provides a test of accuracy, of the "necessary but not sufficient" kind. The average agreement between the S analyses by the two methods is very good (Fig. 10.13a). Non-linear regression of the data assuming that the relative uncertainties are the same in both methods, gives $[S]_{SIMS} = 1.01\ [S]_{EMPA}$, and setting the reduced chi-squared to unity gives the most probable value of this uncertainty as 5.8%. If the SIMS analyses were more precise then the EMPA would be less precise, and vice versa, but 5.8% for EMPA seems reasonable based on the VG2 analyses (Table 10.2). Kendrick et al. (2014) also report replicate measurements of S by both SIMS and EMPA for some Ocean Island basaltic glasses, with fewer samples (n = 18) but over a wider range of S contents, because of variable S loss by degassing. There is less good agreement, with $[S]_{SIMS} = 1.07\ [S]_{EMPA}$ (Fig. 10.13a), largely due to the samples with higher S contents, above the ~ 1400 ppm of the VG2 standard. The most probable uncertainty assuming equal uncertainties is similar, at 5.0%.

In addition to their replication of the S analyses, Shimizu et al. (2016) also report analyses of the major and trace elements in the Macquarie Island glasses that replicate those of Kamenetsky and coworkers, giving some information on the accuracy of calculated SCSS. The values of SCSS calculated from the dataset of Shimizu et al. (2016) are displaced to higher SCSS (Fig. 10.13b). Because of the way the model is assembled, it is easy to find the culprits: they are the Ni and Cu analyses, which

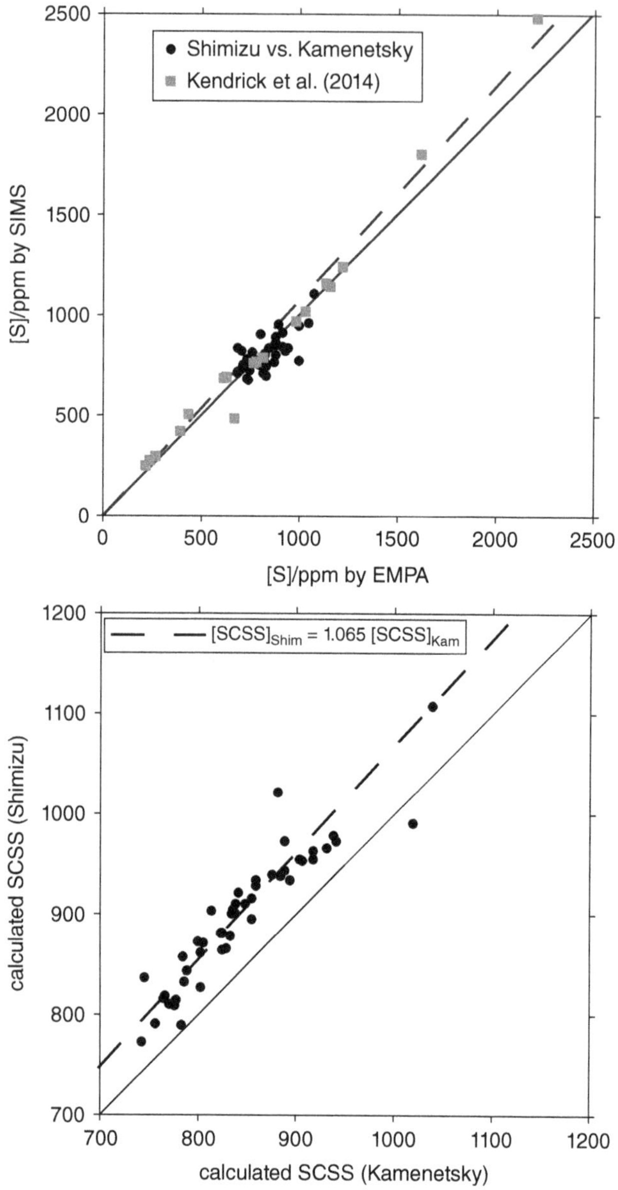

Figure 10.13 (a) Reproducibility of S analyses in basaltic glasses: a comparison of SIMS vs. EMPA. SIMS analyses of Macquarie Island glasses from Shimizu et al. (2016) are compared against EMPA analyses of the same glasses from Kamenetsky et al. (2000), and SIMS and EMPA analyses of some Ocean Island Basalt glasses from Kendrick et al. (2014). (b) Comparison of model $[S]_{SCSS}$ calculated from the analyses of Shimizu et al. (2016) with those calculated for the same glasses from the analyses of Kamenetsky et al. (2000) and Kamenetsky and Eggins (2012).

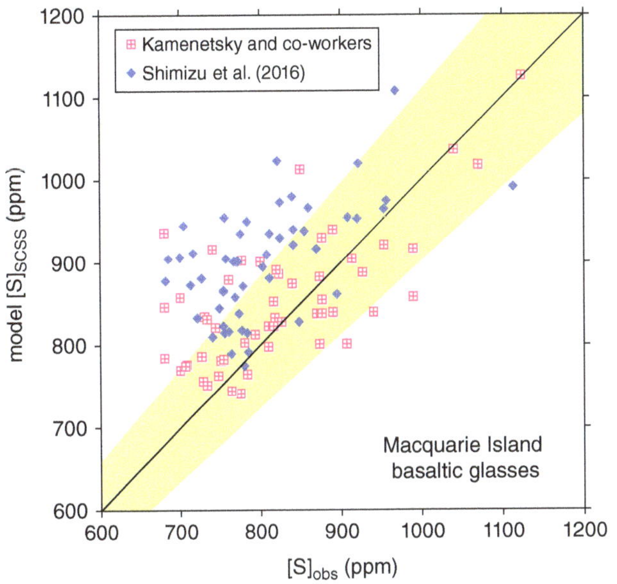

Figure 10.14 Comparison of model $[S]_{SCSS}$ with $[S]_{obs}$ for Macquarie Island basaltic glasses, using the analyses of Shimizu et al. (2016) and those of Kamenetsky et al. (2000) and Kamenetsky and Eggins (2012). The latter scatter around the 1:1 line; the former are displaced to higher model $[S]_{SCSS}$, which, if accurate, could be interpreted as indicating sulfide-undersaturation. The shaded area surrounding the 1:1 line indicates ± 10% uncertainty.

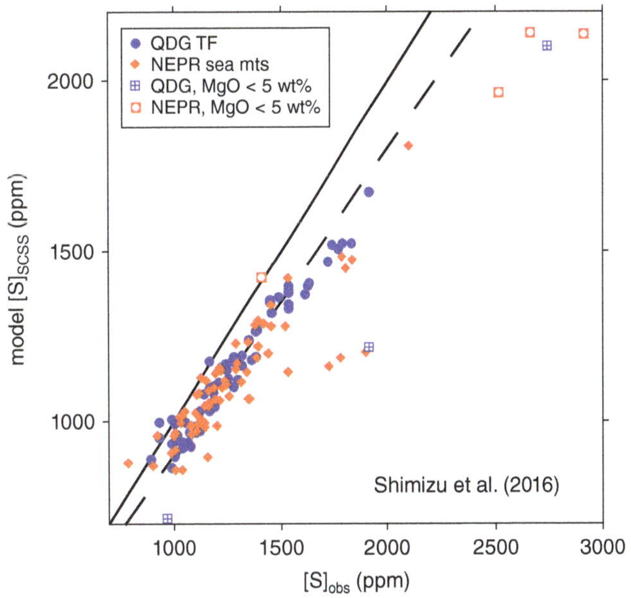

Figure 10.15 Comparison between model $[S]_{SCSS}$ and $[S]_{obs}$ for two sets of glasses from Shimizu et al. (2016) analyzed by SIMS. Most glasses are displaced to high $[S]_{obs}$ by 9% (dashed line) from the 1:1 line, which could be a systematic analytical artifact. However, a few glasses have even higher $[S]_{obs}$, suggesting excess S as SO_4^{2-}. These glasses could be checked for $Fe^{3+}/\sum Fe$ and $S^{6+}/\sum S$ by XANES.

are on average 7% and 14% lower respectively in the dataset of Shimizu et al. (2016). This shows just how important it is to know the Ni and Cu contents of a basalt liquid to judge whether it was sulfide-saturated or not. The comparison of $[S]_{obs}$ with the SCSS model is shown for both datasets in Fig. 10.14. The analyses of Kamenetsky et al. (2000) and Kamenetsky and Eggins (2012) scatter around the model SCSS, although a few are offset to considerably higher SCSS than $[S]_{obs}$, indicating that they may be sulfide-undersaturated. The dataset of Shimizu et al. (2016) is displaced to higher SCSS, as expected from Fig. 10.13b.

Shimizu et al. (2016) report S and all the necessary data for the SCSS model for basaltic glasses from two other geotectonic localities, Northern East Pacific Rise (NEPR) off-axis seamounts, and the Quebrada-Discovery-GoFar (QDG) transform fault system. The comparison between $[S]_{obs}$ and SCSS is shown in Fig. 10.15. There is a systematic offset of 9% for both datasets, with $[S]_{obs}$ being higher than the calculated SCSS. The SCSS models of Smythe et al. (2017) give the same result, almost exactly. Given the complexity of the SCSS models, the only reasonable explanation for why the offset is so systematic is that the calibration of the SIMS for S differs from what has been typically used for EMPA, on which the model is based. Evidence supporting this proposition is: (i) the reported value for VG2 by SIMS (Le Voyer et al., 2017) is higher than most EMPA values by ~ 5% (Table 10.2); and (ii) the 7% offset of SIMS compared to EPMA analyses from Kendrick et al. (2014) shown in Fig. 10.13a. There is, however, the problem that the Macquarie Island glasses do not show the same offset (Fig. 10.13a). The offset is opposite to that expected of undersaturation in sulfide, and towards extra S as SO_4^{2-}. This seems unlikely, because if there was significant SO_4^{2-}, small but plausible variations fO_2 (as monitored by $Fe^{3+}/\sum Fe$) should cause more scatter in $[S]_{obs}$ (cf. Fig. 10.2). A small subset from the NEPR seamounts do seem to contain excess S beyond likely uncertainty (Fig. 10.15), implying non-negligible SO_4^{2-} and inviting further investigation by XANES.

For olivine-hosted melt inclusion glasses from OFB, there is fair agreement between analyzed S and calculated SCSS for those samples in which the necessary Ni and Cu contents have been measured (Le Voyer et al., 2017), shown in Fig. 10.16a, but only if the original analyses, uncorrected for presumed post-entrapment crystallization, are used. The correction used by Le Voyer et al. (2017) hardly changes the Fe, Ni, Cu, or S contents of

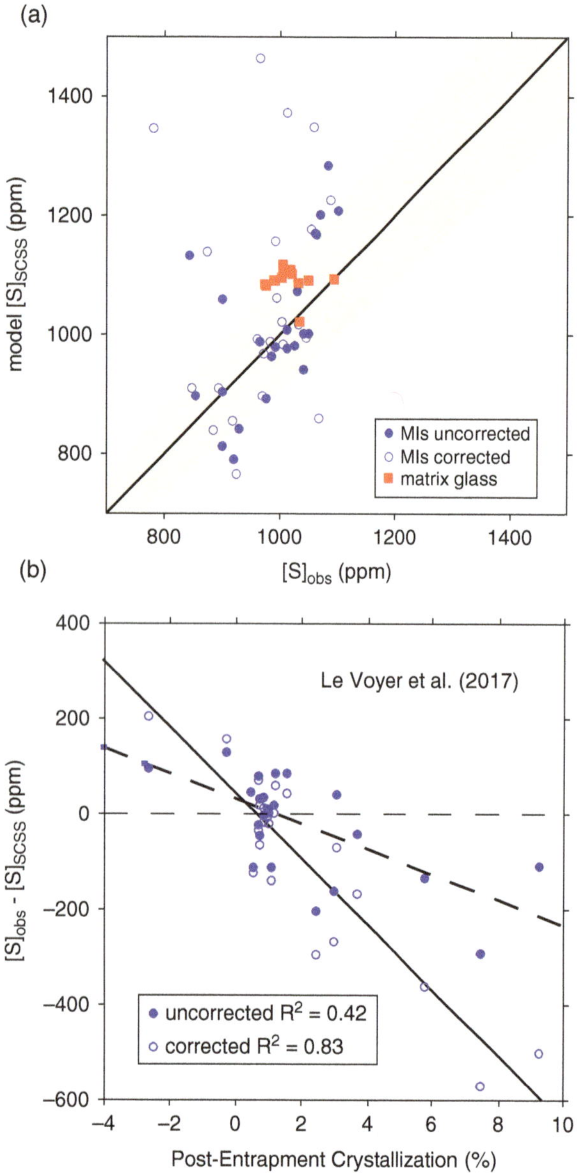

Figure 10.16 Olivine-hosted melt inclusions from Le Voyer et al. (2017). (a) model $[S]_{SCSS}$ compared to $[S]_{obs}$, both uncorrected (i.e., as analyzed), and with silicate melt compositions corrected for post-entrapment crystallization. The agreement for the uncorrected compositions is mostly within ±10% (shaded area). (b) shows that the larger the correction for post-entrapment crystallization, the worse the discrepancy between $[S]_{obs}$ and calculated $[S]_{SCSS}$. However, there is also a significant correlation (R^2 = 0.43) between $[S]_{obs}$ and calculated $[S]_{SCSS}$ for the uncorrected data, indicating that the perceived need for correction may be a sign of disequilibrium between olivine and melt at the time of entrapment.

the trapped melt from their measured values, but mainly affects MgO, hence the calculated temperature from the magmathermometer of Putirka (2008). The result is, the larger the *calculated* degree of post-entrapment crystallization, the greater the difference between $[S]_{obs}$ and the *calculated* SCSS (Fig. 10.16b). This emphasizes the sensitivity of SCSS to temperature. Intriguingly, there is also a correlation between the calculated percentage of post-entrapment crystallization and the uncorrected SCSS (Fig. 10.16b), which can be interpreted as showing that the degree of disequilibrium between the melt inclusion and its olivine host is correlated with undercooling as monitored by the difference between $[S]_{obs}$ and calculated SCSS. Olivine phenocrysts in OFB are often (perhaps mostly) not in Mg-Fe equilibrium with their host melts at any temperature (see, for example, Fig. 6 of Coogan, 2014), and the very existence of a melt inclusion is evidence of rapid crystallization of its host crystal, plausibly caused by undercooling. Melt inclusions have sometimes been regarded as magic bullets to understand basalt petrogenesis, whereas the reality is that a better understanding of basalt petrogenesis is needed to understand melt inclusions.

10.6.5. Siqueiros OFB Olivine-Hosted Melt Inclusions

Saal et al. (2002) studied olivine-hosted melt inclusions in OFB from the Siqueiros Fracture Zone, however they did not measure Ni and Cu, so their claim that some of these melt inclusions were sulfide-undersaturated needs scrutiny. To a first approximation, this may be done if it is assumed that these melt inclusions had the typical Ni and Cu contents for ocean floor basaltic glasses as a function of [MgO]. To implement this idea, the datasets of Jenner and O'Neill (2012) and Gale et al. (2013) were combined, to give, for glasses with [MgO] > 5 wt% (n = 1060):

$$\log_{10}[\text{Ni}](\pm 0.115) = 0.692 + 0.169[\text{MgO}] \quad (10.54)$$

$$\log_{10}[\text{Cu}](\pm 0.103) = 1.620 + 0.037[\text{MgO}] \quad (10.55)$$

as shown in Fig. 10.17. The scatter, which is far outside analytical uncertainty, reflects real geochemical variability, which needs to be taken into account for the SCSS calculation. This may be done using the sensitivities in Table 10.4. The relationship between $[S]_{obs}$ and calculated SCSS enabled by these estimates for Ni and Cu abundances for the Siqueiros inclusions is shown in Fig. 10.18 for both the uncorrected analyses, and with their correction for post-entrapment crystallization. With the exception of five inclusions (out of 107) that have lower $[S]_{obs}$ than calculated SCSS, the inclusions appear to be sulfide-saturated within likely uncertainty. It cannot at present be excluded that the low S in the five anomalous inclusions are due to high Ni and/or Cu or disequilibrium during trapping, as inferred for other melt inclusions (Fig. 10.16).

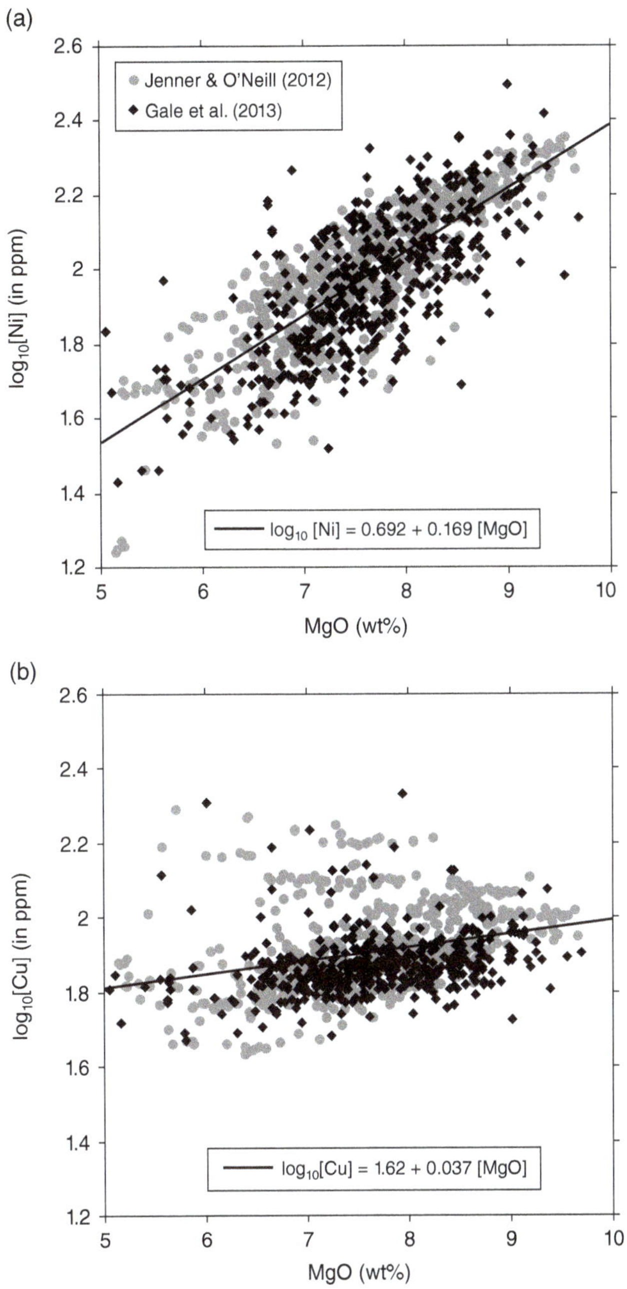

Figure 10.17 Ni and Cu abundances in ocean floor basaltic (OFB) glasses versus MgO. Data from Jenner and O'Neill (2012) and Gale et al. (2013).

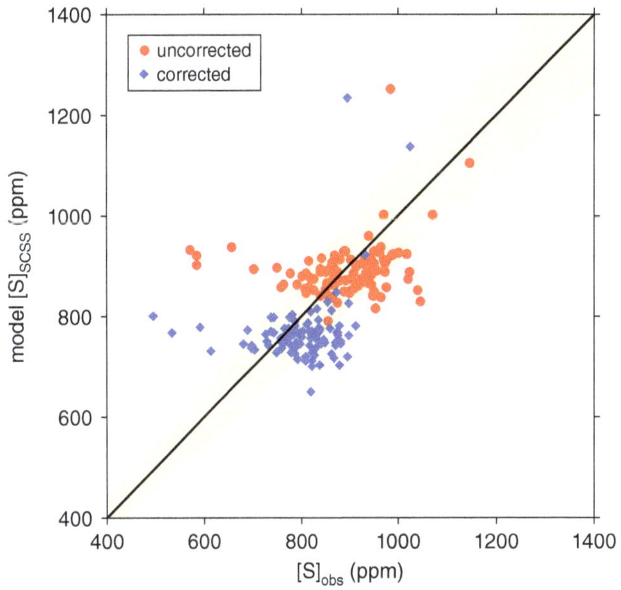

Figure 10.18 Calculated $[S]_{SCSS}$ vs. $[S]_{obs}$ for olivine-hosted melt inclusions from the Siqueiros Fracture Zone (Saal et al. 2002), with Ni and Cu contents estimated using the correlations with MgO shown in Fig. 10.17. Both the uncorrected melt compositions and those corrected for post-entrapment crystallization plot around the 1:1 line, indicating that most inclusions are likely sulfide-saturated. Only five inclusions have significantly lower [S] than expected for saturation.

10.7. THE SULFUR FUGACITY (fS_2) OF OCEAN FLOOR BASALTS

Given T, P, and fO_2, the sulfur fugacity (fS_2) of a basalt at ambient pressure may be calculated from Equation 10.35 if it is sulfide-saturated, but this requires knowing $a_{FeO}^{sil\ melt}$ and a_{FeS}^{sulf}. More generally, at ambient pressure, given T and fO_2, fS_2 may also be calculated from its model $C_{S^{2-}}$ and observed S content, sulfide-saturated or not (Equation 10.3). There is insufficient information to calculate fS_2 at higher pressures by this latter route, because the effect of pressure has been constrained experimentally only on SCSS, and not separately on the four entities that make up SCSS, including $C_{S^{2-}}$. Ambient-pressure values for fS_2 calculated from the $C_{S^{2-}}$ route are plotted against the relative oxygen fugacity ΔQFM in Fig. 10.19a for 35 OFB glasses whose $Fe^{3+}/\sum Fe$ has been determined by Berry et al. (2018) by XANES, and whose S contents were measured by Jenner and O'Neill (2012). The parameterization of O'Neill et al. (2018) was used to convert $Fe^{3+}/\sum Fe$ into ΔQFM. The slope of the correlation between $\log_{10} fS_2$ and ΔQFM is 0.82 ± 0.08, somewhat less than the naïve value of 1 implied by Equation 10.3, due to the effects of silicate melt composition and temperature. This correlation flows on to one of fSO_2 with ΔQFM, since $1/2S_2 + O_2 = SO_2$ (Fig. 10.19b). The large increase in fSO_2 (hence pSO_2) with ΔQFM raises the possibility that some of the rare S-degassed OFB glasses (Fig. 10.9) may have been unusually oxidized (high Fe^{3+}/Fe^{2+}). A high oxidation state has been postulated elsewhere to lead to S degassing during submarine eruptions (Nilsson & Peach, 1993; Jenner et al., 2010). Perversely, degassing of S^{2-} as SO_2 would then reduce

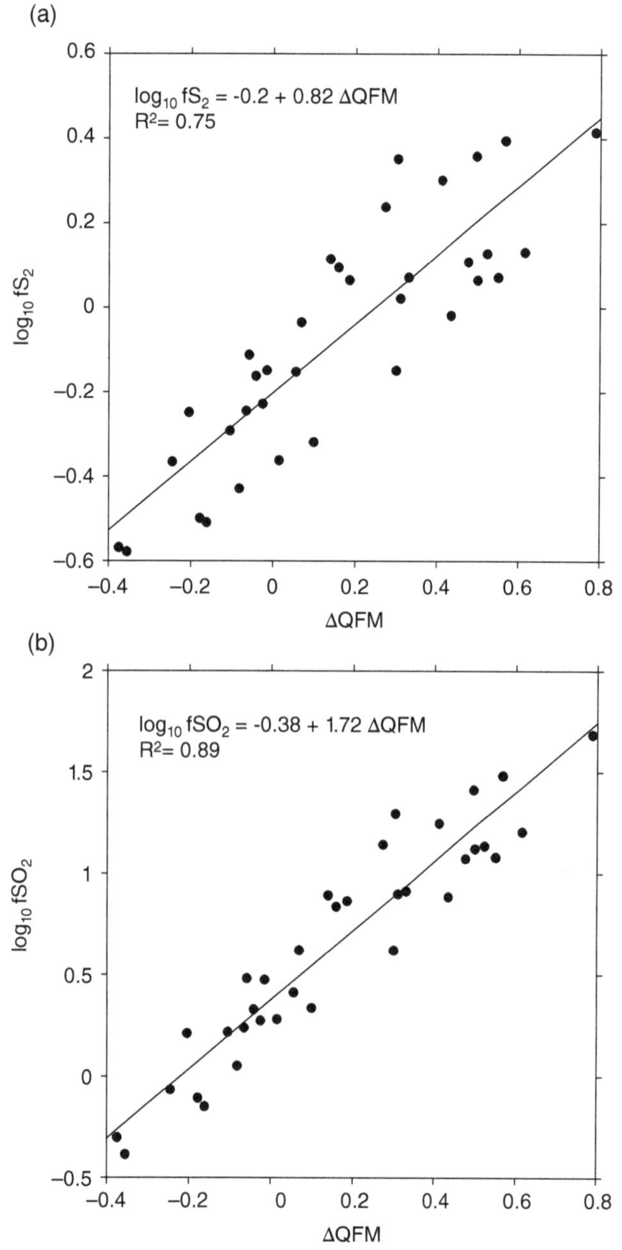

Figure 10.19 (a) $\log_{10} fS_2$ and (b) $\log_{10} fSO_2$ calculated for 35 sulfide-saturated OFB glasses. S analyses from Jenner and O'Neill (2012), $Fe^{3+}/\sum Fe$ from Berry et al. (2018), calculated ΔQFM from O'Neill et al. (2018).

Fe^{3+}/Fe^{2+} (Moussallam et al., 2016). The one glass with low S due to presumed degassing for which $Fe^{3+}/\sum Fe$ was measured, VG6822, shows no evidence of this hypothetical behavior (Fig. 10.10).

10.8. CONCLUSIONS

The "capacity" approach to understanding the thermodynamics of anion species in silicate melts leads to a thermodynamically based model for the SCSS of basic to intermediate magmas. The model has a precision for calculating SCSS of better than 10%, if the composition of the sulfide can be adequately constrained. Application of this model to undegassed basaltic glasses from deep submarine eruptions show that nearly all have S contents consistent with being saturated in a Fe-Ni-Cu-S-O sulfide matte on eruption. The agreement between analyzed S and SCSS indicates that such magmas contained no more than a few percent of S as S^{6+}, which is supported by the results of XANES spectroscopy on their quenched glasses (e.g., Métrich et al., 2009). However, several uncertainties remain. There is a need to better constrain the effects of pressure relevant to the range over which many common basalts are generated and evolve (i.e., 0 to 1.5 GPa). More experimental information is needed on the effects of H_2O, especially confirmation that H-S species (e.g., HS^-, H_2S) are negligible, as assumed here. Nothing is known about the effects of CO_2 or other volatile species. A fascinating problem in physical chemistry yet to be addressed is the kinetics of the electron-exchange interaction between Fe^{3+}/Fe^{2+} and S^{6+}/S^{2-} in silicate melts, which determines to what extent the equilibrium speciation in the silicate melt is preserved when cooled to glass.

ACKNOWLEDGMENTS

Many thanks to John Mavrogenes, my long-term collaborator on matters sulfurous. Thanks also to another long-term colleague, Andrew Berry, for teaching me so much about spectroscopy and for the H_2O determinations by FTIR in Table 10.3. Callum Reekie very kindly worked the Excel spreadsheet into an esthetically pleasing format, and thanks to him and Fran Jenner for their comments and their encouragement. This paper was completed while I was enjoying a visit to the Institut für Mineralogie at the Westfälische Wilhelms-Universität Münster through the generosity of the Alexander-von-Humboldt Stiftung. I also thank two anonymous reviewers, and Roberto Moretti and Daniel Neuville for their editorial handling.

REFERENCES

Allertz, C., Selleby, M., & Sichen, D. (2016). The effect of oxygen potential on the sulfide capacity for slags containing multivalent species. *Metallurgical and Materials Transactions B*, *47*, 3039–3045. https://doi.org/10.1007/s11663-016-0725-7

Ariskin, A. A., Danyushevsky, L. V., Bychkov, K. A., McNeill, A. W., Barmina, G. S., & Nikolaev, G. S. (2013). Modeling solubility of Fe-Ni sulfides in basaltic magmas: The effect of nickel. *Economic Geology*, *108*, 1983–2003. https://doi.org/10.2113/econgeo.108.8.1983

Baker, D. R., Barnes, S. J., Simon, G., & Bernier, F. (2001). Fluid transport of sulfur and metals between sulfide melt and basaltic melt. *The Canadian Mineralogist*, *39*, 537–546.

Baker, D. R. (2008). The fidelity of melt inclusions as records of melt composition. *Contributions to Mineralogy and Petrology*, *156*, 377–395. https://doi.org/10.1007/s00410-008-0291-3

Ballhaus, C., Fonseca, R. O., Münker, C., Rohrbach, A., Nagel, T., Speelmanns, I. M., et al. (2017). The great sulfur depletion of Earth's mantle is not a signature of mantle–core equilibration. *Contributions to Mineralogy and Petrology*, *172*, 68. doi: 10.1007/s00410-017-1388-3

Barin, I., Sauert, F., Schultze-Rhonhof, E., & Sheng, W. S. (1989). *Thermochemical data of pure substances, part I and part II*. Weinheim, Germany: VCH Verlagsgesellschaft, pp. 1–1739.

Bell, A. S., Simon, A., & Guillong, M. (2009). Experimental constraints on Pt, Pd and Au partitioning and fractionation in silicate melt–sulfide–oxide–aqueous fluid systems at 800 C, 150 Mpa and variable sulfur fugacity. *Geochimica et Cosmochimica Acta*, *73*, 5778–5792. https://doi.org/10.1016/j.gca.2009.06.037

Berry, A. J., Shelley, J. M. G., Foran, G. J., O'Neill, H. St.C., & Scott, D. R. (2003). A furnace design for XANES spectroscopy of silicate melts under controlled oxygen fugacities and temperatures to 1773 K. *Journal of Synchrotron Radiation*, *10*, 332–336. doi: 10.1107/s0909049503007556

Berry, A. J., Stewart, G. A., O'Neill, H. St.C., Mallmann, G., & Mosselmans, J. F. W. (2018). A re-assessment of the oxidation state of iron in MORB glasses. *Earth and Planetary Science Letters*, *483*, 114–123. https://doi.org/10.1016/j.epsl.2017.11.032

Borisov, A., Lahaye, Y., & Palme, H. (2004). The effect of TiO_2 on Pd, Ni, and Fe solubilities in silicate melts. *American Mineralogist*, *89*, 564–571. doi: 10.2138/am-2004-0411

Borisov, A., Lahaye, Y., & Palme, H. (2006). The effect of sodium on the solubilities of metals in silicate melts. *American Mineralogist*, *91*, 762–771. https://doi.org/10.2138/am.2006.1977

Borisov, A. A. (2007). Experimental study of the influence of SiO_2 on the solubility of cobalt and iron in silicate melts. *Petrology*, *15*, 523–529. https://doi.org/10.1134/S086959110706001X

Botcharnikov, R. E., Linnen, R. L., Wilke, M., Holtz, F., Jugo, P. J., & Berndt, J. (2011). High gold concentrations in sulphide-bearing magma under oxidizing conditions. *Nature Geoscience*, *4*, 112–115. https://doi.org/10.1038/ngeo1042

Buchanan, D. L., Nolan, J., Wilkinson, N. & de Villiers, J. P. R. (1983). An experimental investigation of sulphur solubility as a function of temperature in synthetic silicate melts. *Geological Society of South Africa Special Publication*, *7*, 383–391.

Chan, A. H., & Fruehan, R. J. (1986). The sulfur partition ratio and the sulfide capacity of Na_2O-SiO_2 slags at 1200° C. *Metallurgical Transactions B*, *17*(3), 491–496. doi: 10.1007/BF02670351

Chan, A. H., & Fruehan, R. J. (1988). The sulfur partition ratio and the sulfide capacity of Na_2O-SiO_2 Slags at 1400° C. *Metallurgical Transactions B*, *19*(2), 334–336. doi: 10.1007/BF02670214

Chase Jr, M. W. (1998). NIST-JANAF thermochemical tables. National Institute of Standards and Technology.

Clémente, B., Scaillet, B., & Pichavant, M. (2004). The solubility of sulphur in hydrous rhyolitic melts. *Journal of Petrology*, *45* (11), 2171–2196.

Coogan, L. A. (2014). The lower oceanic crust. In: Holland, H. D., Turekian, K. K. (Eds.) *Treatise on Geochemistry*. Second Edition, Elsevier, Oxford. Vol. 4. The crust (ed. Rudnick, R. L.) pp. 497–541.

Cottrell, E., & Kelley, K. A. (2011). The oxidation state of Fe in MORB glasses and the oxygen fugacity of the upper mantle. *Earth and Planetary Science Letters*, *305*, 270–282. https://doi.org/10.1016/j.epsl.2011.03.014

Czamanske, G. K., & Moore, J. G. (1977). Composition and phase chemistry of sulfide globules in basalt from the Mid-Atlantic Ridge rift valley near 37 N lat. *Geological Society of America Bulletin*, *88*, 587–599. https://doi.org/10.1130/0016-7606(1977)88<587:CAPCOS>2.0.CO;2

Danckwerth, P. A., Hess, P. C., & Rutherford, M. J. (1979). The solubility of sulfur in high-TiO_2 mare basalts. *Proceedings of the 10th Lunar Planetary Science Conference*, pp. 517–530. New York: Pergamon Press.

Davis, A. S., Clague, D. A., Cousens, B. L., Keaten, R., & Paduan, J. B. (2008). Geochemistry of basalt from the North Gorda segment of the Gorda Ridge: Evolution toward ultraslow spreading ridge lavas due to decreasing magma supply. *Geochemistry, Geophysics, Geosystems*, *9*, 4. https://doi.org/10.1029/2007GC001775

De Hoog, J. C. M., Mason, P. R. D., & Van Bergen, M. J. (2001). Sulfur and chalcophile elements in subduction zones: Constraints from a laser ablation ICP-MS study of melt inclusions from Galunggung Volcano, Indonesia. *Geochimica et Cosmochimica Acta*, *65*, 3147–3164. https://doi.org/10.1016/S0016-7037(01)00634-2

De Hoog, J. C. M., Hattori, K. H., & Hoblitt, R. P. (2004). Oxidized sulfur-rich mafic magma at Mount Pinatubo, Philippines. *Contributions to Mineralogy and Petrology*, *146*(6), 750–761. https://doi.org/10.1007/s00410-003-0532-4

Ding, S., Dasgupta, R., & Tsuno, K. (2014). Sulfur concentration of martian basalts at sulfide saturation at high pressures and temperatures–implications for deep sulfur cycle on Mars. *Geochimica et Cosmochimica Acta*, *131*, 227–246. https://doi.org/10.1016/j.gca.2014.02.003

Dixon, J. E., Clague, D. A., & Stolper, E. M. (1991). Degassing history of water, sulfur, and carbon in submarine lavas from Kilauea Volcano, Hawaii. *Journal of Geology*, *99*, 371–394. https://doi.org/10.1086/629501

Dixon, J. E., & Stolper, E. M. (1995). An experimental study of water and carbon dioxide solubilities in mid-ocean ridge basaltic liquids. Part II: applications to degassing. *Journal of Petrology*, *36*(6), 1633–1646. https://doi.org/10.1093/oxfordjournals.petrology.a037268

Doyle, C. D. (1988). Prediction of the activity of FeO in multicomponent magma from known values in [SiO_2-$KalO_2$-$CaAl_2Si_2O_8$] – FeO liquids. *Geochimica et Cosmochimica Acta*, *52*, 1827–1834. https://doi.org/10.1016/0016-7037(88)90006-3

Doyle, C. D. (1989). The effect of substitution of TiO_2 for SiO_2 on a_{FeO} in magma. *Geochimica et Cosmochimica Acta*, *53*, 2631–2638. https://doi.org/10.1016/0016-7037(89)90134-8

Doyle, C. D., & Naldrett, A. J. (1986). Ideal mixing of divalent cations in mafic magma and its effect on the solution of ferrous oxide. *Geochimica et Cosmochimica Acta*, *5*, 435–443. https://doi.org/10.1016/0016-7037(86)90196-1

D'Souza, R. J., & Canil, D. (2018). Effect of alkalinity on sulfur concentration at sulfide saturation in hydrous basaltic andesite to shoshonite melts at 1270 C and 1 GPa. *American Mineralogist: Journal of Earth and Planetary Materials*, *103*, 1030–1043. https://doi.org/10.2138/am-2018-6404

Evans, K. A., O'Neill, H. St.C., & Mavrogenes, J. A. (2008). Sulphur solubility and sulphide immiscibility in silicate melts as a function of the concentration of manganese, nickel, tungsten and copper at 1 atm and 1400 °C. *Chemical Geology*, *255*, 236–249. https://doi.org/10.1016/j.chemgeo.2008.06.042

Fincham, C. J. B. & Richardson, F. D. (1954). The behaviour of sulphur in silicate and aluminate melts. *Proceedings of the Royal Society of London, Ser. A* 223, 40–62.

Flood, H., Forland, T., & Gzotheim, K. (1954). Ueber den Zusammenhang zwischen Konzentration und Aktivitäten in geschmolzenen Salzmischungen. *Zeitschrift für Anorganische Allgemeine Chemie 276*, 290–315.

Fonseca, R. O., Campbell, I. H., O'Neill, H. St.C., & Fitzgerald, J. D. (2008). Oxygen solubility and speciation in sulphide-rich mattes. *Geochimica et Cosmochimica Acta*, *72*, 2619–2635. https://doi.org/10.1016/j.gca.2008.03.009

Fortin, M. A., Riddle, J., Desjardins-Langlais, Y., & Baker, D. R. (2015). The effect of water on the sulfur concentration at sulfide saturation (SCSS) in natural melts. *Geochimica et Cosmochimica Acta*, *160*, 100–116. https://doi.org/10.1016/j.gca.2015.03.022

Fougeroux, T. (2014). Influence du contexte géodynamique sur l'état d'oxydoréduction du fer dans les magmas. Thèse de doctorat. Nantes.

Frost, D. J., & Wood, B. J. (1997). Experimental measurements of the fugacity of CO_2 and graphite/diamond stability from 35 to 77 kbar at 925 to 1650 C. *Geochimica et Cosmochimica Acta*, *61*(8), 1565–1574. https://doi.org/10.1016/S0016-7037(97)00035-5Gaetani, G. A., & Grove, T. L. (1997). Partitioning of moderately siderophile elements among olivine, silicate melt, and sulfide melt: constraints on core formation in the Earth and Mars. *Geochimica et Cosmochimica Acta*, *61*, 1829–1846. https://doi.org/10.1016/S0016-7037(97)00033-1

Gale, A., Dalton, C. A., Langmuir, C. H., Su, Y., & Schilling, J. G. (2013). The mean composition of ocean ridge basalts. *Geochemistry, Geophysics, Geosystems*, *14*, 489–518. https://doi.org/10.1029/2012GC004334

Grant, N. J., & Chipman, J. (1946). Sulphur equilibria between liquid iron and slags. *Trans. AIME*, *167*, 134–154.

Gurenko, A. A., Belousov, A. B., Trumbull, R. B., & Sobolev, A. V. (2005). Explosive basaltic volcanism of the Chikurachki Volcano (Kurile arc, Russia): Insights on pre-eruptive magmatic conditions and volatile budget revealed from phenocryst-hosted melt inclusions and groundmass glasses. *Journal of Volcanology and Geothermal Research*, *147*, 203–232. https://doi.org/10.1016/j.jvolgeores.2005.04.002

Haughton, D. R., Roeder, P. L. & Skinner, B. J. (1974). Solubility of sulfur in mafic magmas. *Economic Geology*, *69*, 451–467. https://doi.org/10.2113/gsecongeo.69.4.451

Hauri, E., Wang, J., Dixon, J. E., King, P. L., Mandeville, C., & Newman, S. (2002). SIMS analysis of volatiles in silicate glasses: 1. Calibration, matrix effects and comparisons with FTIR. *Chemical Geology*, *183*, 99–114. https://doi.org/10.1016/S0009-2541(01)00375-8

Hillert, M. (1998) Phase equilibria, phase diagrams and phase transformations: Their thermodynamic basis. Cambridge University Press. https://doi.org/10.1017/CBO9780511812781

Hillert, M. (2001). The compound energy formalism. *Journal of Alloys and Compounds*, *320*(2), 161–176. https://doi.org/10.1016/S0925-8388(00)01481-X

Holmquist, S. (1966). Oxygen ion activity and the solubility of sulfur trioxide in sodium silicate melts. *Journal of the American Ceramic Society*, *49*(9), 467–473. https://doi.org/10.1111/j.1151-2916.1966.tb13301.x

Holzheid, A., Palme, H., & Chakraborty, S. (1997). The activities of NiO, CoO and FeO in silicate melts. *Chemical Geology*, *139*(1–4), 21–38. https://doi.org/10.1016/S0009-2541(97)00030-2

Holzheid, A., & Grove, T. L. (2002). Sulfur saturation limits in silicate melts and their implications for core formation scenarios for terrestrial planets. *American Mineralogist*, *87*(2–3), 227–237. https://doi.org/10.2138/am-2002-2-304

Jarosewich, E., Nelen, J. A., & Norberg, J. A. (1979). Electron microprobe reference samples for mineral analyses. *Smithsonian Contributions to Earth Sciences*, *20*, 68–72. https://doi.org/10.1111/j.1751-908X.1980.tb00273.x

Jégo, S., & Dasgupta, R. (2014). The fate of sulfur during fluid-present melting of subducting basaltic crust at variable oxygen fugacity. *Journal of Petrology*, *55*, 1019–1050. https://doi.org/10.1093/petrology/egu016

Jenner, F. E. & O'Neill, H. St.C. (2012). Analysis of 60 Elements in 616 Ocean Floor Basaltic Glasses. *Geochemistry, Geophysics, Geosystems*, *13*, 2. doi: 10.1029/2011GC004009

Jenner, F. E., O'Neill, H. St.C., Arculus, R. J., & Mavrogenes, J. A. (2010). The magnetite crisis in the evolution of arc-related magmas and the initial concentration of Au, Ag, and Cu. *Journal of Petrology*, *51*, 2445–2464. https://doi.org/10.1093/petrology/egq063

Jugo, P. J., Wilke, M., & Botcharnikov, R. E. (2010). Sulfur K-edge XANES analysis of natural and synthetic basaltic glasses: Implications for S speciation and S content as function of oxygen fugacity. *Geochimica et Cosmochimica Acta*, *74*, 5926–5938. https://doi.org/10.1016/j.gca.2010.07.022

Kamenetsky, V. S., Everard, J. L., Crawford, A. J., Varne, R., Eggins, S. M., & Lanyon, R. (2000). Enriched end-member of primitive MORB melts: petrology and geochemistry of glasses from Macquarie Island (SW Pacific). *Journal of Petrology*, *41*, 411–430. https://doi.org/10.1093/petrology/41.3.411

Kamenetsky, V. S. & Eggins, S. M. (2012). Systematics of metals, metalloids, and volatiles in MORB melts: Effects of partial melting, crystal fractionation and degassing (a case study of Macquarie Island glasses). *Chemical Geology*, *302–303*, 76–86. https://doi.org/10.1016/j.chemgeo.2011.04.008

Kamenetsky, V. S., Gurenko, A. A., & Kerr, A. C. (2010). Composition and temperature of komatiite melts from Gorgona Island, Colombia, constrained from olivine-hosted melt inclusions. *Geology*, *38*,11, 1003–1006. https://doi.org/10.1130/G31143.1

Kang, Y. B., & Pelton, A. D. (2009). Thermodynamic model and database for sulfides dissolved in molten oxide slags. *Metallurgical and Materials Transactions B*, *40*(6), 979–994. https://doi.org/10.1007/s11663-009-9283-6

Katsura, T. & Nagashima, S. (1974). Solubility of sulfur in some magmas at 1 atmosphere. *Geochimica et Cosmochimica Acta*, *38*, 517–531. https://doi.org/10.1016/0016-7037(74)90038-6

Kendrick, M. A., Jackson, M. G., Kent, A. J., Hauri, E. H., Wallace, P. J., & Woodhead, J. (2014). Contrasting behaviours of CO_2, S, H_2O and halogens (F, Cl, Br, and I) in enriched-mantle melts from Pitcairn and Society seamounts. *Chemical Geology*, *370*, 69–81. https://doi.org/10.1016/j.chemgeo.2014.01.019

Kiseeva, E. S., & Wood, B. J. (2013). A simple model for chalcophile element partitioning between sulphide and silicate liquids with geochemical applications. *Earth and Planetary Science Letters*, *383*, 68–81. https://doi.org/10.1016/j.epsl.2013.09.034

Kiseeva, E. S., & Wood, B. J. (2015). The effects of composition and temperature on chalcophile and lithophile element partitioning into magmatic sulphides. *Earth and Planetary Science Letters*, *424*, 280–294. https://doi.org/10.1016/j.epsl.2015.05.012

Kress, V. (1997). Thermochemistry of sulfide liquids. I. the system O-S Fe at 1 bar. *Contributions to Mineralogy and Petrology*, *127*, 176–186. https://doi.org/10.1007/s004100050274

Kress, V., Greene, L. E., Ortiz, M. D., & Mioduszewski, L (2008) Thermochemistry of sulfide liquids IV: density measurements and the thermodynamics of O–S–Fe–Ni–Cu liquids at low to moderate pressures. *Contributions to Mineralogy and Petrology*, *156*, 785–797. https://doi.org/10.1007/s00410-008-0315-z

Lange, R. A., Carmichael, I. S. E. (1990). Thermodynamic properties of silicate liquids with an emphasis on density, thermal expansion and compressibility. In: Nicholls, J., Russell, J. K. (eds.) *Modern Methods of Igneous Petrology: Understanding Magmatic Processes*. (Reviews in Mineralogy 24). Washington, D.C.: Mineralogical Society of America. pp 25–64.

Laurenz, V., Fonseca, R. O., Ballhaus, C., Jochum, K. P., Heuser, A., & Sylvester, P. J. (2013). The solubility of palladium and ruthenium in picritic melts: 2. The effect of sulfur. *Geochimica et Cosmochimica Acta*, *108*, 172–183. https://doi.org/10.1016/j.gca.2013.01.013

Le Voyer, M., Cottrell, E., Kelley, K. A., Brounce, M., & Hauri, E. H. (2015). The effect of primary versus secondary processes on the volatile content of MORB glasses: An example from the equatorial Mid-Atlantic Ridge (5°N–3°S). *Journal of Geophysical Research: Solid Earth*, *120*, 125–144. https://doi.org/10.1002/2014JB011160

Le Voyer, M., Kelley, K. A., Cottrell, E., & Hauri, E. H. (2017). Heterogeneity in mantle carbon content from CO_2-undersaturated basalts. *Nature Communications*, *8*, 14062. https://doi.org/10.1038/ncomms14062

Li, J., & Agee, C. B. (2001). The effect of pressure, temperature, oxygen fugacity and composition on partitioning of nickel and cobalt between liquid Fe–Ni–S alloy and liquid silicate: implications for the Earth's core formation. *Geochimica et Cosmochimica Acta*, *65*,1821–1832. https://doi.org/10.1016/S0016-7037(00)00613-X

Li, C., & Naldrett, A. J. (1993). Sulfide capacity of magma: a quantitative model and its application to the formation of sulfide ores at Sudbury, Ontario. *Economic Geology*, *88*, 1253–1260. https://doi.org/10.2113/gsecongeo.88.5.1253

Liu, X., O'Neill, H. St.C., & Berry, A. J. (2005). The effects of small amounts of H_2O, CO_2 and Na_2O on the partial melting of spinel lherzolite in the system $CaO–MgO–Al_2O_3–SiO_2±H_2O±CO_2±Na_2O$ at 1.1 GPa. *Journal of Petrology*, *47*, 409–434. https://doi.org/10.1093/petrology/egi081

Liu, Y., Samaha, N. T., & Baker, D. R. (2007). Sulfur concentration at sulfide saturation (SCSS) in magmatic silicate melts. *Geochimica et Cosmochimica Acta*, *71*, 1783–1799. https://doi.org/10.1016/j.gca.2007.01.004

Luhr, J. F. (1990). Experimental phase relations of water-and sulfur-saturated arc magmas and the 1982 eruptions of El Chichón volcano. *Journal of Petrology*, *31*, 1071–1114. https://doi.org/10.1093/petrology/31.5.1071

Luhr, J. F. (2008). Primary igneous anhydrite: Progress since its recognition in the 1982 El Chichón trachyandesite. *Journal of Volcanology and Geothermal Research*, *175*(4), 394–407. https://doi.org/10.1016/j.jvolgeores.2008.02.016

Martin, M. (2001). Life and achievements of Carl Wagner, 100th birthday. *Solid State Ionics*, *152–153*: 15–17. https://doi.org/10.1016/S0167-2738(02)00318-1

Mathez, E. A. (1976). Sulfur solubility and magmatic sulfides in submarine basalt glass. *Journal of Geophysical Research*, *81*, 4269–4276. https://doi.org/10.1029/JB081i023p04269

Matjuschkin, V., Blundy, J. D., & Brooker, R. A. (2016). The effect of pressure on sulphur speciation in mid-to deep-crustal arc magmas and implications for the formation of porphyry copper deposits. *Contributions to Mineralogy and Petrology*, *171*, 66. https://doi.org/10.1007/s00410-016-1274-4

Mavrogenes, J. A., & O'Neill, H. St.C. (1999). The relative effects of pressure, temperature and oxygen fugacity on the solubility of sulfide in mafic magmas. *Geochimica et Cosmochimica Acta*, *63*, 1173–1180. https://doi.org/10.1016/S0016-7037(98)00289-0

Métrich, N., Joron, J.-L., & Berthier, B. (1998). Occurrence of boron-rich potassic melts in the Vulsini Volcanic District, Italy: Evidence from melt inclusions. *Geochimica et Cosmochimica Acta*, *62*, 507–514. https://doi.org/10.1016/S0016-7037(97)00365-7

Métrich, N., Berry, A. J., O'Neill, H. St.C., & Susini, J. (2009). The oxidation state of sulfur in synthetic and natural glasses determined by X-ray absorption spectroscopy. *Geochimica et Cosmochimica Acta*, *73*, 2382–2399. https://doi.org/10.1016/j.gca.2009.01.025

Michael, P. (1995). Regionally distinctive sources of depleted MORB: Evidence from trace elements and H_2O. *Earth and Planetary Science Letters*, *131*, 301–320. https://doi.org/10.1016/0012-821X(95)00023-6

Moussallam, Y., Edmonds, M., Scaillet, B., Peters, N., Gennaro, E., Sides, I., & Oppenheimer, C. (2016). The impact of degassing on the oxidation state of basaltic magmas: a case study of Kīlauea volcano. *Earth and Planetary Science Letters, 450*, 317–325. https://doi.org/10.1016/j.epsl.2016.06.031

Mungall, J. E., & Brenan, J. M. (2014). Partitioning of platinum-group elements and Au between sulfide liquid and basalt and the origins of mantle-crust fractionation of the chalcophile elements. *Geochimica et Cosmochimica Acta, 125*, 265–289. https://doi.org/10.1016/j.gca.2013.10.002

Mysen, B. O. (1983). The structure of silicate melts. *Annual Review of Earth and Planetary Sciences, 11*, 75–97. https://doi.org/10.1146/annurev.ea.11.050183.000451

Mysen, B. O., & Richet, P. (2005). *Silicate glasses and melts: properties and structure* (Vol. *10*). Elsevier.

Naldrett, A. J., & Richardson, S. W. (1967). Effect of water on the melting of pyrrhotite-magnetite assemblages. Carnegie Institute of Washington, Year Book, *66*, 429–431.

Nash, W. M., Smythe, D. J., & Wood, B. J. (2019). Compositional and temperature effects on sulfur speciation and solubility in silicate melts. *Earth and Planetary Science Letters, 507*, 187–198. https://doi.org/10.1016/j.epsl.2018.12.006

Nilsson, K., & Peach, C. L. (1993). Sulfur speciation, oxidation state, and sulfur concentration in backarc magmas. *Geochimica et Cosmochimica Acta, 57*, 3807–3813. https://doi.org/10.1016/0016-7037(93)90158-S

O'Neill, H. St.C., & Berry, A. J. (2006). Activity coefficients at low dilution of CrO, NiO and CoO in melts in the system CaO–MgO–Al_2O_3–SiO_2 at 1400 C: using the thermodynamic behaviour of transition metal oxides in silicate melts to probe their structure. *Chemical Geology, 231*(1–2), 77–89. https://doi.org/10.1016/j.chemgeo.2006.01.004

O'Neill, H. St.C., Berry, A. J., & Mallmann, G. (2018). The oxidation state of iron in Mid-Ocean Ridge basaltic (MORB) glasses: implications for their petrogenesis and oxygen fugacities. *Earth and Planetary Science Letters, 504*, 152–162. https://doi.org/10.1016/j.epsl.2018.10.002

O'Neill, H. St.C., & Eggins, S. M. (2002). The effect of melt composition on trace element partitioning: an experimental investigation of the activity coefficients of FeO, NiO, CoO, MoO2 and MoO_3 in silicate melts. *Chemical Geology, 186*, 151–181. https://doi.org/10.1016/S0009-2541(01)00414-4

O'Neill, H. St.C., & Mavrogenes, J. A. (2002). The sulfide capacity and the sulfur content at sulfide saturation of silicate melts at 1400 C and 1 bar. *Journal of Petrology, 43*, 1049–1087. https://doi.org/10.1093/petrology/43.6.1049

O'Neill, H. St.C., & Mavrogenes, J. A. (2019). The sulfate capacity of silicate melts at atmospheric pressure. *Goldschmidt Abstracts*, 2019.

O'Neill, H. St.C., & Pownceby, M. I. (1993). Thermodynamic data from redox reactions at high temperatures. I. An experimental and theoretical assessment of the electrochemical method using stabilized zirconia electrolytes, with revised values for the Fe-"FeO", Co-CoO, Ni-NiO and Cu-Cu_2O oxygen buffers, and new data for the W-WO_2 buffer. *Contributions to Mineralogy and Petrology, 114*, 296–314. doi: 10.1007/BF01046533

Patten, C., Barnes, S. J., & Mathez, E. A. (2012). Textural variations in MORB sulfide droplets due to differences in crystallization history. *The Canadian Mineralogist, 50*, 3, 675–692. https://doi.org/10.3749/canmin.50.3.675

Patten, C., Barnes, S. J., Mathez, E. A., & Jenner, F. E. (2013). Partition coefficients of chalcophile elements between sulfide and silicate melts and the early crystallization history of sulfide liquid: LA-ICP-MS analysis of MORB sulfide droplets. *Chemical Geology, 358*, 170–188. https://doi.org/10.1016/j.chemgeo.2013.08.040

Peach, C. L., & Mathez, E. A. (1993). Sulfide melt-silicate melt distribution coefficients for nickel and iron and implications for the distribution of other chalcophile elements. *Geochimica et Cosmochimica Acta, 57*, 13, 3013–3021. https://doi.org/10.1016/0016-7037(93)90290-D

Peach, C. L., Mathez, E. A., Keays, R. R., & Reeves, S. J. (1994). Experimentally determined sulfide melt-silicate melt partition coefficients for iridium and palladium. *Chemical Geology, 117*, 361–377. https://doi.org/10.1016/0009-2541(94)90138-4

Pelton, A. D., Eriksson, G., & Romero-Serrano, A. (1993). Calculation of sulfide capacities of multicomponent slags. *Metallurgical Transactions B, 24B*, 817–825. https://doi.org/10.1007/BF02663142

Poulson, S. R., & Ohmoto, H. (1990). An evaluation of the solubility of sulfide sulfur in silicate melts from experimental data and natural samples. *Chemical Geology, 85*, 57–75. https://doi.org/10.1016/0009-2541(90)90123-O

Putirka, K. D. (2008). Thermometers and barometers for volcanic systems. *Reviews in Mineralogy and Geochemistry, 69*, 61–120. https://doi.org/10.2138/rmg.2008.69.3

Reddy, R. G., & Blander, M. (1987). Modeling of sulfide capacities of silicate melts. *Metallurgical Transactions B, 18B*, 591–596. https://doi.org/10.1007/BF02654272

Righter, K., Pando, K., & Danielson, L. R. (2009). Experimental evidence for sulfur-rich martian magmas: Implications for volcanism and surficial sulfur sources. *Earth and Planetary Science Letters, 288*, 235–243. https://doi.org/10.1016/j.epsl.2009.09.027

Ripley, E. M., Brophy, J. G., & Li, C. (2002). Copper solubility in a basaltic melt and sulfide liquid/silicate melt partition coefficients of Cu and Fe. *Geochimica et Cosmochimica Acta, 66*, 15, 2791–2800. https://doi.org/10.1016/S0016-7037(02)00872-4

Roberge, J., White, R. V., & Wallace, P. J. (2004). Volatiles in submarine basaltic glasses from the Ontong Java Plateau (ODP Leg 192): implications for magmatic processes and source region compositions. *Geological Society, London, Special Publications, 229*, 239–257. https://doi.org/10.1144/GSL.SP.2004.229.01.14

Robie, R.A., & Hemingway, B.S. (1995). Thermodynamic properties of minerals and related substances at 298.15 K and 1 bar (10^5 Pascals) pressure and at higher temperatures. *US Geological Survey Bulletin 2131*, 1–461. https://doi.org/10.3133/b2131

Roeder, P. L. (1974). Activity of iron and olivine solubility in basaltic liquids. *Earth and Planetary Science Letters, 23*, 397–410. https://doi.org/10.1016/0012-821X(74)90129-0

Sattari, P., Brenan, J. M., Horn, I., & McDonough, W. F. (2002). Experimental constraints on the sulfide-and chromite-silicate melt partitioning behavior of rhenium and platinum-group elements. *Economic Geology, 97*, 385–398. https://doi.org/10.2113/gsecongeo.97.2.385

Schrama, F. N. H., Beunder E. M., Van den Berg, B., Yang, Y.-X., & Boom, R. (2017). Sulphur removal in ironmaking and oxygen steelmaking, *Ironmaking & Steelmaking*, *44*, 5, 333–343. https://doi.org/10.1080/03019233.2017.1303914

Schmalzried, H. (1995) *Chemical Kinetics of Solids*, VCH, Weinheim.

Self, S., Blake, S., Sharma, K., Widdowson, M., & Sephton, S. (2008). Sulfur and chlorine in Late Cretaceous Deccan magmas and eruptive gas release. *Science*, *319*, 5870, 1654–1657. doi: 10.1126/science.1152830

Shima, H. & Naldrett, A. J. (1975). Solubility of sulfur in ultramafic melt and the relevance of the system Fe-S-O. *Economic Geology*, *70*, 960–967. https://doi.org/10.2113/gsecongeo.70.5.960

Shimizu, K., Saal, A. E., Myers, C. E., Nagle, A. N., Hauri, E. H., Forsyth, D. W., et al. (2016). Two-component mantle melting-mixing model for the generation of mid-ocean ridge basalts: implications for the volatile content of the Pacific upper mantle. *Geochimica et Cosmochimica Acta*, *176*, 44–80. https://doi.org/10.1016/j.gca.2015.10.033

Stolper, E., Sherman, S., Garcia, M., Baker, M., & Seaman, C. (2004). Glass in the submarine section of the HSDP2 drill core, Hilo, Hawaii. *Geochemistry, Geophysics, Geosystems*, *5*, 7, Q07G15. https://doi.org/10.1029/2003GC000553

Smythe, D. J., Wood, B. J., & Kiseeva, E. S. (2017). The S content of silicate melts at sulfide saturation: new experiments and a model incorporating the effects of sulfide composition. *American Mineralogist*, *102*, 795–803. https://doi.org/10.2138/am-2017-5800CCBY

Thordarson, T., Self, S., Óskarsson, N., & Hulsebosch, T. (1996). Sulfur, chlorine, and fluorine degassing and atmospheric loading by the 1783-1784 AD Laki (Skaftár Fires) eruption in Iceland. *Bulletin of Volcanology*, *58*, 205–225. https://doi.org/10.1007/s004450050136

Thornber, C. R., Sherrod, D. R., Siems, D. F., Heliker, C. C., Meeker, G. P., Oscarson, R. L., & Kauahikaua, J. P. (2002). Whole-rock and glass major-element geochemistry of Kilauea Volcano, Hawaii, near-vent eruptive products: September 1994 through September 2001: U.S. Geological Survey Open-File Report 2002-17, https://pubs.usgs.gov/of/2002/0017/.

Tuff, J., & O'Neill, H. St.C. (2010). The effect of sulfur on the partitioning of Ni and other first-row transition elements between olivine and silicate melt. *Geochimica et Cosmochimica Acta*, *74*, 6180–6205. https://doi.org/10.1016/j.gca.2010.08.014

Vidal, C. M., Métrich, N., Komorowski, J. C., Pratomo, I., Michel, A., Kartadinata, N., et al. (2016). The 1257 Samalas eruption (Lombok, Indonesia): the single greatest stratospheric gas release of the Common Era. *Scientific Reports*, *6*, 34868. https://doi.org/10.1038/srep34868

Wagner, C. (1975). The concept of the basicity of slags. *Metallurgical Transactions B*, 6, 405–409. https://doi.org/10.1007/BF02913825

Wallace, P., & Carmichael, I. S. E. (1992). Sulfur in basaltic magmas. *Geochimica et Cosmochimica Acta*, *56*, 1683–1874. https://doi.org/10.1016/0016-7037(92)90316-B

Waldner, P., & Pelton, A. D. (2005). Thermodynamic modeling of the Fe-S system. *Journal of Phase Equilibria and Diffusion*, *26*, 1, 23–38. https://doi.org/10.1007/s11669-005-0055-y

Wilke, M., Jugo, P. J., Klimm, K., Susini, J., Botcharnikov, R., Kohn, S. C., & Janousch, M. (2008). The origin of S^{4+} detected in silicate glasses by XANES. *American Mineralogist*, *93*, 235–240. https://doi.org/10.2138/am.2008.2765

Wilke, M., Klimm, K., & Kohn, S. C. (2011). Spectroscopic studies on sulfur speciation in synthetic and natural glasses. *Reviews in Mineralogy and Geochemistry*, *73*, 41–78. https://doi.org/10.2138/rmg.2011.73.3

Wood, B. J., & Nicholls, J. (1978). The thermodynamic properties of reciprocal solid solutions. *Contributions to Mineralogy and Petrology*, *66*, 389–400. https://doi.org/10.1007/BF00403424

Wood, B. J., & Wade, J. (2013). Activities and volatilities of trace components in silicate melts: a novel use of metal–silicate partitioning data. *Contributions to Mineralogy and Petrology*, *166*, 3, 911–921. https://doi.org/10.1007/s00410-013-0896-z

11

Redox State of Volatiles and Their Relationships with Iron in Silicate Melts: Implications for Magma Degassing

Nicole Métrich

ABSTRACT

The question of how degassing of H-C-O-S-Cl fluids may change $Fe^{3+}/\Sigma Fe$ ratio and redox conditions in silicate melts has been widely addressed via the experiments and studies of natural systems. After a brief summary of the solubility data and the possible role of the volatile species (CO_2, H_2O, S, Cl) on the redox of magmas during their decompression, the review focuses on the behavior of water and sulfur redox during magma degassing, the relationships between these two species with iron in silicate melts through experimental petrology, geochemical investigations, and modeling of natural systems using cutting-edge methods to determine the redox of S and Fe.

11.1. INTRODUCTION

Carbon, hydrogen, sulfur, and iron are the main redox-sensitive elements in mantle (Frost & McCammon, 2008) and along magma evolution (Mathez, 1984; Berhens & Webster, 2011). The potential role of carbon on the redox of mantle has been recognized early because the oxidation of the reduced C species (graphite or diamond) in the asthenospheric mantle is associated with the reduction of ferric iron in silicate minerals during mantle upwelling (Stagno et al., 2013). A CO_2-rich gas phase may promote magma dehydration and crystallization when it buffers magmatic systems, as it could occur in basaltic systems, and consequently may affect the sulfur behavior and redox (e.g., Moretti et al., 2018). However, the solubility of carbon as CO_3^{2-} species in basalts (Shishkina et al., 2014), and/or CO_2 species in silicic melts (King & Holloway, 2002), is low relative to H_2O at pressure $\leq$ 500 MPa (Moore, 2008; Ni & Keppler, 2013). As an example, Figure 11.1 shows the isobaric evolution of dissolved contents of CO_2 and H_2O in basalt and rhyolite, calculated for different pressures from 50 to 300 MPa (after Burgisser et al., 2015). Water is an abundant volatile species in magmas (Figure 11.1) and most coexisting fluids are partially or prevalently aqueous in composition (Webster et al., 2018). Its key role in the melt properties and magmatic processes has been discussed early (McMillan, 1994), and its effect on the sulfur redox demonstrated (Baker & Moretti, 2011; Berhens & Stelling, 2011). Several wt% of H_2O may be kept dissolved in arc-related silicic melts (Grove et al., 2012), and more broadly in basaltic (Shishkina et al., 2010; Lesne et al., 2011a; Iacono-Marziano et al., 2012) and rhyolitic melts (Moore, 2008).

Iron is the most abundant multivalent element in natural silicate melts and glasses (Kress & Carmichael, 1991). The sulfur concentrations rarely exceed 0.5 wt% in natural hydrous basaltic melts and a few tenths of ppm in rhyolitic melts but sulfur is an essential component, along with H_2O and CO_2, in the volcanic systems (Wallace & Edmonds 2011; Webster & Botcharnikov, 2011).

Université de Paris, Institut de Physique du Globe de Paris, Paris, France

Magma Redox Geochemistry, Geophysical Monograph 266, First Edition.
Edited by Roberto Moretti and Daniel R. Neuville.

DOI: 10.1002/9781119473206.ch11

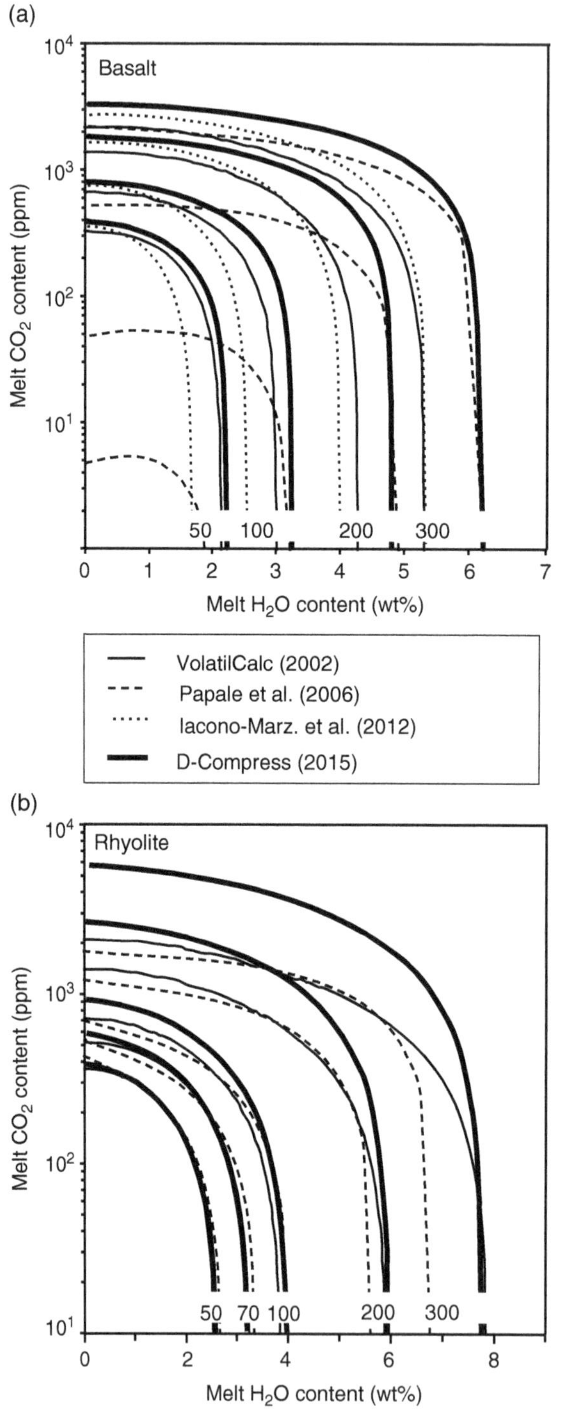

Figure 11.1 Isobaric calculations of H_2O and CO_2 contents in Basaltic melt at 1000 °C and NNO +1 (a) and Rhyolitic melt at 850 °C and NNO +1(b), by four models: D-Compress (Burgisser et al., 2015), VolatilCalc (Newman & Lowenstern 2002), Iacono-Marziano et al. (2012), and Papale et al. (2006). Figure modified from Burgisser et al. (2015). Melt compositions are those proposed by default in D-Compress (Burgisser et al., 2015). In the Papale et al. (2006) model, quantities of Fe_2O_3 and FeO were adjusted using the FeO value and the corresponding molar ratio given by D-Compress. Total fluid pressures are given in MPa. Note that for basalts Shishkina et al. (2010) show that VolatilCalc (Newman & Lowenstern, 2002) tend to overestimate the fluid pressures for P >200 MPa, and to the contrary the Papale et al. (2006) model underestimates the fluid-saturation pressures of tholeiitic melt in the 100–600 MPa. Shishkina et al. (2014) also note the limitations of the Iacono-Marziano et al. (2012) model to predict CO_2 and H_2O solubility in highly depolymerized and alkali-rich mafic melts. Adapted from Burgisser et al. (2015); Newman & Lowenstern (2002); Iacono-Marziano et al. (2012); Papale et al. (2006); Shishkina et al. (2010); and Shishkina et al. (2014).

There is growing evidence that sulfur, because of its multivalence states, plays a dominant role on the redox of silicate melts and vapor phases (see Behrens & Webster, 2011, for a review). Significant redox decoupling between the silicate melt and the coexisting gas phase may occur during the differential transfer, expansion, and closed-system cooling of fast-rising bubbles (Oppenheimer et al., 2018). The oxidation state of the gas emissions is prevalently controlled by their emission temperature (e.g., Giggenbach, 1987; Ohba et al., 1994; Moussallam et al., 2019). Change of the valence state of sulfur and iron, the possible shift of the sulfide–sulfate transition with pressure and thus along magma decompression, the potential remobilization of pre-existing sulfides, and the exsolution of a pre-eruptive gas phase strongly impact the redox evolution of magmas. These processes are particularly efficient in arc volcanic systems where repeated recharges of silicic or intermediate systems by CO_2-rich basaltic batches and magma mixing are ubiquitous processes. As an example, Borisova et al. (2014) propose that a hybrid H_2O-CO_2-S fluid, recorded as fluid-inclusions, formed as the result of the high-pressure mixing between S-rich basaltic and hydrous dacitic magmas prior to the 1991 eruption of Mt Pinatubo in Philippines.

Adding Cl may influence the whole redox state of the systems. Chlorine concentrations widely vary from 5 to 1000 ppm in natural mid-ocean ridge basalts (MORBs) and ocean island basalts (OIBs). They may reach 1000–5000 ppm in arc-related basalts, and 10,000 ppm in alkali-rich silicic melts (Webster et al., 2018). These authors provide a thorough review of Cl and halogens in magmatic systems, their partitioning of between silicate melts, coexisting vapor phase or saline liquids enriched in $H_2O \pm CO_2 \pm SO_2 \pm H_2S$, and of the thermodynamic models that address Cl partitioning in mixed C-O-H-S-Cl fluid. They show that Cl is transported essentially as HCl in high-temperature (T > 500 °C) magmatic gases, and that H_2O-rich gas emissions in arc domain are enriched in Cl relative to the CO_2-rich gases released in within-plate or continental rift contexts. The exsolution of a Cl-bearing volatile phase from mafic melt may cause melt oxidation because of the preferential removal of Fe^{2+} from the melt (Bell & Simon, 2011).

The primary purpose of the present review is to document the considerable progress that has been made, within the last 5–10 years, on the knowledge of the water and sulfur redox behavior during magma degassing, their relationships with iron in silicate melts through experiments on simple and complex systems, geochemical investigations, and modeling of natural systems using cutting-edge methods to determine the redox of S and Fe. I mainly focus here on the redox behavior in magma and coexisting fluid during their crustal transfer and upon eruption. The experimental works reported in literature are conducted under different conditions of fO_2. For a better understanding the oxygen buffers that are used are illustrated in Figure 11.2.

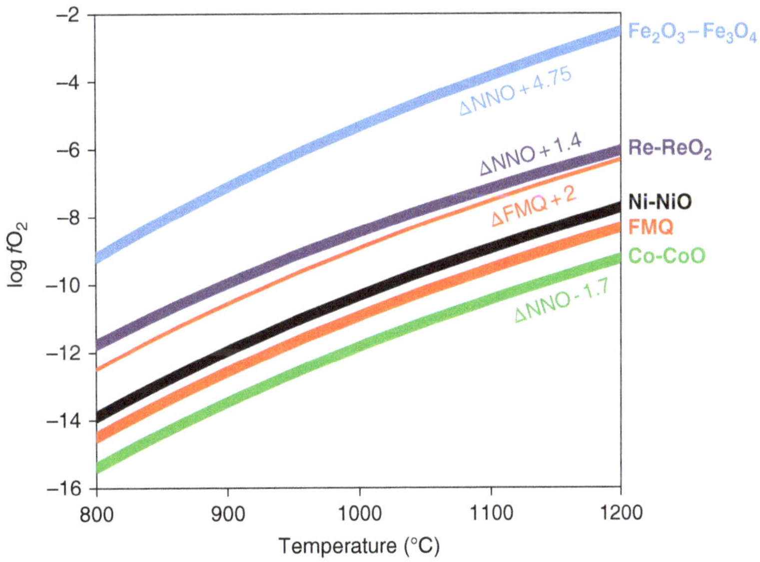

Figure 11.2 Variation of the log fO_2 values as a function of the temperature, for commonly used solid buffers: Co-CoO, Quartz-Fayalite-Magnetite (QFM), Ni-NiO (NNO), Re-ReO, and Hematite-Magnetite. The log fO_2 values relative to the oxygen buffers QFM and Ni-NiO are expressed as ΔFMQ and ΔNNO: Δbuffer = [log fO_2 − log $fO_{2\ buffer}$). As an example, log fO_2 two log units above the FMQ oxygen buffer is noted ΔFMQ +2, and log fO_2 1.7 log units below Ni-NiO is noted ΔNNO -1.7.

11.2. WATER CONCENTRATION IN MELT AND ITS EFFECT ON REDOX

Understanding how water is dissolved in the silicate melts is of interest as water may act as a network-modifier or network-forming complex in polymerized and depolymerized silicate melts, respectively (Moretti et al., 2014; see also Le Losq et al., this book). Whether degassing of H_2O can drive redox change of silicate melts ($Fe^{3+}/\Sigma Fe$ ratio) is a question that is actively debated in literature (e.g., Mathez, 1984; Holloway, 2004), on the basis of experimental data and petrological studies of natural samples (from rhyolite to basalts and phono-tephrites) as summarized by Humphreys et al. (2015) and Waters & Lange (2016).

H_2/H_2O and Fe^{3+}/Fe^{2+} are the redox couples in S and Cl-free, water-bearing melts, following the reaction:

$$H_2O_{melt} + 2FeO_{melt} = H_{2gas} + Fe_2O_{3melt} \quad (11.1)$$

A process of melt oxidation driven by the outward diffusion and loss of hydrogen and the thermal water dissociation was first evoked by Sato and Wright (1966). Such a mechanism has been further debated (e.g., Mathez, 1984), and the question of the melt oxidation through H_2 diffusion explored by Gaillard and co-authors in silicic melts-pure H_2O systems. Gaillard et al. (2001) document the positive effects of water on the redox state of Fe (increasing $Fe^{3+}/\Sigma Fe$) in synthetic and natural hydrous rhyolitic melts (from metaluminous to peralkaline rhyolites; FeO $\approx$0.5 to 4.3 wt%), at intermediate range of fO_2 values (<ΔNNO +1.5). At higher fO_2, this effect vanishes in agreement with previous experiments on peralkaline rhyolites (5.6 wt% total FeO; Moore et al., 1995). Gaillard et al. (2002) experimentally show that the $Fe^{3+}/\Sigma Fe$ ratios re-equilibrate within hours, in hydrous rhyolitic melts at 800 °C and 200 MPa. Because of the limited growth of the oxidized layer, Gaillard et al. (2003) conclude that the redox exchanges between H_2 and iron would have a very limited effect on the iron redox state of bulk magmas during their crustal transfer and possible mixing.

Focusing on experiments on silicic systems, Scaillet and Pichavant (2016) question the common assumption that water-rich magmas are systematically more oxidized than water-poor magmas because of the existence of reduced but H_2O-rich peralkaline rhyolites. They propose that the redox state of the silicate melts would be decoupled from their degree of hydration. Humphreys et al. (2015) have conducted experiments to clarify the relationships between H_2O and iron during hydration and decompression of S-free rhyolitic melts with FeO_{Total} content ranging from 1.0 to 5.7 wt%. In this work, the measurements of the $Fe^{3+}/\Sigma Fe$ ratio (using XANES [X-ray Absorption Near Edge Structure spectroscopy at the S *K*-edge]) and of the H_2O content in the experimental run glasses are combined. In the hydrated samples, the $Fe^{3+}/\Sigma Fe$ ratio (0.1–0.6) correlates with H_2O concentration (from 0.2 wt% in the unhydrated core sample up to ~5.4 wt% in the hydrated parts), over a fO_2 range of ~4 log units. The results of these disequilibrium experiments are in agreement with those of equilibrium experiments on trachybasalts (Botcharnikov et al., 2005). Humphreys et al. (2015) suggest that changes in melt oxidation state are closely linked to variations in melt chemistry and structure, as in their diffusion experiments, any change in $Fe^{3+}/\Sigma Fe$ ratio is shifted toward higher X_{H2O}, at low H_2O contents as melt viscosity is higher. Degassing experiments produce the reverse as the individualization of a free H_2O-vapor phase and the decreasing of the dissolved amount of H_2O (from 7 to ~2wt %) are accompanied by the melt oxidation ($Fe^{3+}/\Sigma Fe$ from 0.53 to 0.65). According to these authors the re-equilibration of fH_2 and fH_2O in the vapor phase would occur following Reaction 11.1.They conclude that in sulfur-free systems, the melt oxidation is consistent with the formation of a free H_2O vapor phase and loss of H_2 and H_2O during closed-system decompression. To the contrary Crabtree & Lange (2012) report no significant difference between the pre-eruptive and post-eruptive FeO (Fe^{2+}) concentrations and the $Fe^{3+}/\Sigma Fe$ ratios of hydrous andesitic and dacitic magmas from the Mexican volcanic arc. The FeO concentrations were measured by titration in the bulk lava samples and also determined from Fe-Ti two-oxide thermobarometry. They thus propose that extensive H_2O degassing (3 to 8 wt %) had negligible effect on oxidation ($Fe^{3+}/\Sigma Fe$ ratio) of these silicic melts. Alternatively, Humphreys et al. (2015) suggest that the degassing of additional multivalent gas species, as sulfur, in the case of the Mexican magmas could have counteracted the melt oxidation driven by the H_2O loss. To test whether H_2O degassing changes the $Fe^{3+}/\Sigma Fe$ ratio in natural rhyolites, Waters & Lange (2016) measured the post-eruptive FeO (Fe^{2+}) concentrations directly on totally degassed obsidian samples, and computed the pre-eruptive Fe^{2+} concentrations of the melt from Fe-Ti oxide pairs. Their results together with compiled Fe-Ti oxide data (Ghiorso & Evans, 2008) support equilibrium degassing of H_2O from magmatic liquids. They conclude that (i) the degassing of up to 6.5 wt% H_2O has no resolvable effect on the ferric–ferrous ratio of the studied obsidian samples, including samples that had magmatic ΔNNO values as low as -0.9 and as high as +1.4; and (ii) the high oxidation state of arc magmas relative to those erupted from mid-ocean spreading ridges and hotspots (e.g., Iceland and Yellowstone) is a magmatic feature, recorded by Fe-Ti oxides during phenocryst growth, and is present prior to extensive H_2O degassing. These authors also propose that arc-related rhyolitic magmas, prior to H_2O degassing, are oxidized relative to silicic magmas from other contexts. Portnyagin et al. (2012) provide the example of silicic magmas from Hekla volcano in Iceland (MIs [melt inclusions] hosted in Fe-rich olivine crystals) that are as rich in H_2O (3.3–6.2 wt%) as oxidized island-arc magmas, but which form at a low fO_2 (ΔQFM from -0.1 to -0.4). The redox state of the silicate melts appear to be decoupled from its water content in agreement with Scaillet and Pichavant (2016).

In summary, the question of the effect of H_2O degassing on the redox requires additional decompression experiments, in Cl-S free silicic systems, with a range of H_2O, FeO, and Fe_2O_3 concentrations, in association with a comprehensive thermodynamic modelling of the H_2O dissolution in silicate melts.

11.3. THE SULFUR SPECIES AND THE REDOX ($Fe^{3+}/\Sigma Fe$ RATIO) OF SILICATE MELTS

11.3.1. Sulfur Speciation in Natural Magmas

It was earlier demonstrated that sulfur is dissolved as sulfate and/or sulfide species in anhydrous silicate melts depending on the fO_2 (Fincham & Richardson, 1954). Since the 1970s, the sulfur concentrations either as sulfide at sulfide saturation (SCSS) or as sulfate at anhydrite saturation (SCAS) in silicate melts have been widely investigated over large ranges of compositions (silicic to mafic melts) in equilibrium with a hydrous fluid or not, of fO_2, of temperatures and pressures, within few experiments a controlled fS_2 as reviewed by Webster and Botcharnikov (2011) and Baker and Moretti (2011). As an example, Figure 11.3 shows the experimental sulfur solubility in basaltic melts as a function of fO_2. The importance of considering the introduction of sulfur in the modelling of the redox state of magma is fully recognized now (Moretti & Papale, 2004, Burgisser & Scaillet, 2007; 2008; Moretti & Baker, 2008; Gaillard et al., 2015). The sulfur concentration, as sulfide (S^{2-}) species, is dominantly associated with Fe in silicate melts, limited by the saturation of Fe-bearing sulfide phases (Mavrogenes & O'Neill, 1999), and proportional to $(fS_2/fO_2)^{1/2}$ (O'Neill & Mavrogenes, 2002; Baker and Moretti, 2011; O'Neill, this book). Under reducing conditions, the oxygen fugacity only affects the composition of the sulfide phase (Fincham & Richardson, 1954; Baker & Moretti, 2011; Fortin et al., 2015). At sulfide saturation, adding sulfur to the system, at constant pressure, temperature, and melt composition, will result in a larger amount of sulfide precipitation (Scaillet et al., 1998). In most MORBs and OIBs, the S and FeO concentrations are positively correlated (e.g.,

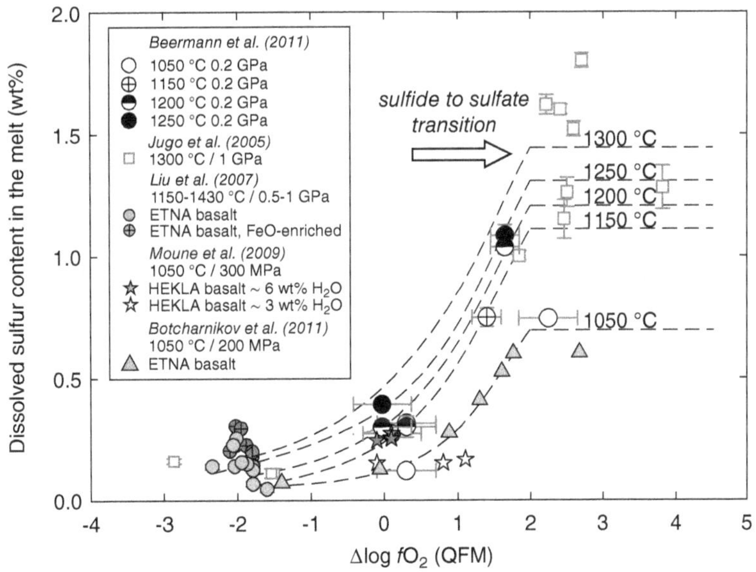

Figure 11.3 Experimentally determined S solubility data of basaltic melts as a function of fO_2 and temperature at 200MPa (Beermann et al., 2011), compared to literature data. Adapted from Beermann et al. (2011). Data for Etna basalt are from Beermann et al. (2011), Liu et al. (2007), and Botcharnikov et al. (2011), for picrobasalt to basalt from Jugo et al. (2005) and for Hekla basalt from Moune et al. (2009). The dashed lines represent interpretative trends for the increase in S solubility with fO_2 at given temperature. The intermediate range of fO_2 values corresponds to the conditions at which both S species are significantly stable (e.g., Jugo et al., 2010). Note that the redox conditions of the sulfide–sulfate transition may change with pressure (see text).

Mathez, 1976; Wallace & Edmonds, 2011). Reekie et al. (2019) also show that most olivine plateau basalts become sulfide-saturated after higher degree of fractional crystallization relative to MORBs, possibly because of their relatively higher total FeO content, an important feature for the behavior of the sulfide-compatible chalcophile and siderophile elements. Smythe et al. (2017) explore the effects of the sulfide composition (Fe-Ni-Cu sulfide melts against pure FeS) on the sulfur solubility in the silicate melt at sulfide saturation and provide new experiments for large ranges of pressure and temperature, from 1.5–7 GPa and of 1400 to 1800 °C, respectively (one experiment at 24 GPa, 1800 °C). They demonstrate that the SCSS in silicate melt depends on the composition of the sulfide liquid, at fixed pressure, temperature, and silicate melt composition. They fitted the data from their experiments and literature to predict the solubility of sulfur at sulfide saturation as a function of the composition of the sulfide liquid, that of the silicate melt (Martian FeO-rich basalt-MORB-komatiite-andesite), the temperature, and pressure. They predict a strong positive dependence of SCSS on the FeO content of the silicate melt and, as an example, a decrease by a factor of ~4 from basalt to an andesite, with 17.90 wt% and 7.65 wt% FeO, respectively, at 1 GPa. The dependence of SCSS in the melt to pressure is predicted to be the reverse. Hydrothermal experiments at 200 MPa confirm the presence of FeS as the dominant reduced S species at ΔNNO -0.5 log units (Zajacz et al., 2012). These experimental results indicate that the sulfur partitioning into the vapor phase is negatively correlated to the FeO concentration of hydrous andesitic glass. Sulfur vapor-melt partition coefficient decreases from 254±25 to 88±6 for melts having 4.0 and 7.5 wt% FeO, respectively. The effect of FeO (Fe^{2+}) content in the melt on the S^{2-} solubility has been experimentally identified in reduced basaltic systems (~QFM to ~QFM+1) (Fiege et al., 2015). More broadly sulfides are common in a wide range of natural volcanic rocks from basalts to rhyolites as reviewed by Parat et al. (2011) and Wallace & Edmonds (2011). Their abundance is often limited as more than 99.5% of the sulfur is dissolved in the basaltic melt (i.e., Pacific–Antarctic ridge basalts). Labidi et al. (2014) propose that MORBs are sulfide-saturated at melting depths, and may remain S-buffered until eruption, in agreement with Smythe et al. (2017), who predict that MORBs are sulfide saturated along their crystallization paths, on the basis of their model for SCSS applied to MORB glasses analyzed by Jenner and O'Neill (2012). OIBs are more variable in fO_2 with variable proportions of S^{2-} and S^{6+} and $Fe^{3+}/\Sigma Fe$ ratios that bracket the MORB values (Jugo et al., 2010; Brounce et al., 2017). Sulfide-bearing submarine basaltic glasses of Samoan Islands (Southwest Pacific) (Labidi et al., 2015) are reported to contain $S^{6+}/\Sigma S$ ratio from 0 to 24±5%% (on average 17±11%), whereas 80% of bulk sulfur is dissolved as sulfide species. The drop of Cu in Samoan basalts (< 6 wt% MgO) is attributed to sulfide fractionation in association with magnetite occurrence. These authors suggest than iron speciation does not buffer the sulfur speciation but that H_2O could act as the oxidizing agent of the dissolved sulfide species in Samoan basalt following the reaction:

$$S^{2-} + 4H_2O = SO_4^{2-} + 4H_2 \qquad (11.2)$$

Undegassed back-arc basin basaltic glasses and arc-related melts (olivine-hosted MIs) are systematically more oxidized than MORBs, indicating that water and fO_2 may correlate in magmas in a wide range of geodynamic settings (Kelley & Cottrell, 2009). This feature is discussed by the authors as a possible consequence of the oxidized conditions of the subducted oceanic lithosphere.

Under oxidizing conditions sulfate anions dominate (e.g., Bonnin-Mosbah et al., 2002; Wallace, 2005; Métrich & Mandeville, 2010; Jugo et al., 2010; Klimm et al., 2012a, b; Zajacz 2015; Debret et al., 2017).

Sulfur as sulfate is an efficient oxidizing agent of iron:

$$S^{2-} + 8Fe^{3+} \Leftrightarrow S^{6+} + 8Fe^{2} \qquad (11.3)$$

There is a consensus to think that arc-related basaltic magmas are, on average, more oxidized than those erupted at mid-ocean ridges in response to the contribution of slab-derived, fluid-rich components to the sub-arc mantle wedge (Wallace & Edmonds, 2011, and references therein; Evans, 2012; Brounce et al., 2014; Humphreys et al., 2015). Similarly, mantle xenoliths in subduction settings are more oxidized with respect to those from continental lithosphere (Frost & McCammon, 2008). There is also growing evidence that sulfate-rich fluids derived from the serpentinite dehydration could act as oxidizing agent of the mantle sources of arc-related magmas (Debret & Sverjensky, 2017; Bénard et al., 2018a). As an example, Bénard et al. (2018a) report the highly oxidized state of sulfur ($S^{6+}/\Sigma S$ ratio close or equal to unit), and the contribution of anhydrite in andesite-dacite melt/glass inclusions, which are regarded as representative of slab-derived melts and are preserved in Cr-bearing spinels of Kamtchacka xenoliths (Figure 11.4). However, sulfate could be released early in the subduction process, in specific conditions, minimizing the contribution of the subducting slab-derived components to the sulfur budgets of arc magmas (Lee et al., 2018). The sharp decrease of Cu along the differentiation of arc-related basalts points to sulfide fractionation (Labidi et al., 2015; Williams et al., 2018), whereas igneous anhydrite crystals have been systematically observed in natural oxidized volcanic rocks (trachyandesites, andesites and dacites) in association with amphibole and apatite

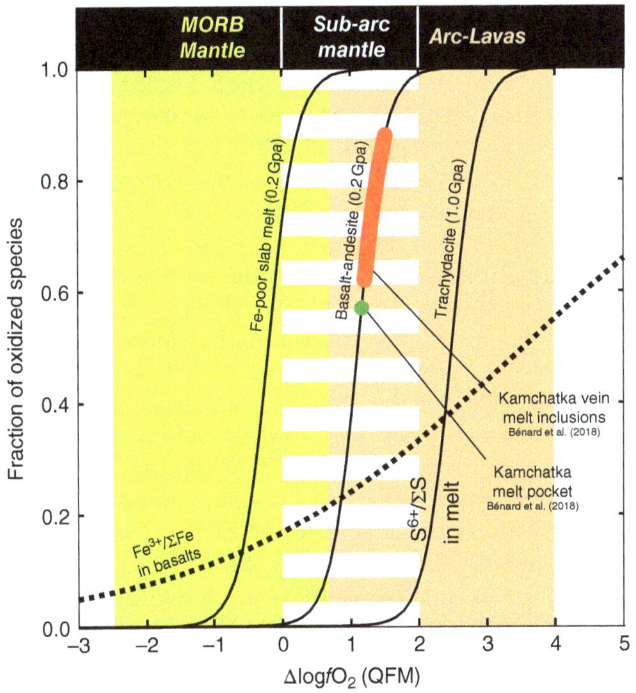

Figure 11.4 Sulfur valence state at fO_2, expressed relative to FMQ. Figure modified from Bénard et al. (2018a). The ranges of $\Delta\log fO_2$ are reported for the abyssal peridotites (MORB mantle) in light green (data from Bryndzia & Wood 1990; Birner et al., 2018); for the sub-arc mantle rocks with white bars (data from Brandon & Draper, 1996, Parkinson & Arculus, 1999; Bénard et al., 2018b); and for the arc lavas in orange (data from Rowe et al., 2009; Jugo et al., 2010; Evans et al., 2012). The $S^{6+}/\Sigma S$ equilibrium curves (black curves) in silicate melts are shown for Fe-poor slab melt and Fe-rich (basalt-andesite) silicate melts at low-pressure ($\leq$ 0.2 GPa) conditions (Jugo et al., 2010; Klimm et al., 2012b), and for trachydacite at 1 GPa (Matjuschkin et al., 2016). The $S^{6+}/\Sigma S$ ratio is determined by μ-XANES and inferred from Raman spectrometry for Kamchatka melt pocket and inclusions. The $Fe^{3+}/\Sigma Fe$ ratio in basalts (dashed line) is calculated for a temperature of 1200 °C, using the model of Kress and Carmichael (1991), according to Klimm et al. (2012b).

(Luhr, 2008; Parat et al., 2011). Vapor-bearing bubbles and anhydrite inclusions described in amphibole of andesite and dacite products of Yanacocha volcano in Peru have been suggested to represent quenched products of $CaSO_4$-rich liquid phase (Chambefort et al., 2008).

11.3.2. Improvement of the Spectroscopic Methods to Probe the Sulfur Valence States in Silicate Melts

During the last 10 years, the spectroscopic methods to identify the sulfur species in silicate glasses have been significantly improved (Wilke et al., 2011; Stelling et al., 2011; Klimm et al., 2012a, b). The $S^{6+}/\Sigma S$ ratios in silicate glasses were widely measured by the electron microprobe analysis (EMPA) using the shift in energy of the S $K\alpha$ X-ray emissions by (Carroll & Rutherford, 1988; Wallace & Carmichael, 1994; Métrich & Clocchiatti 1996; Wilke et al., 2011). It is recognized now that this method tends to underestimate the $S^{6+}/\Sigma S$ ratios of sulfate-dominated run glasses (Jugo et al., 2010). Overestimation and underestimation of the $S^{6+}/\Sigma S$ at reducing and oxidizing conditions, respectively, are also observed by Beermann et al. (2011), even when the electron beam is continuously moved over the sample to minimize the effects of the beam-induced photoreduction of sulfate. The maximum beam damages are expected in Fe-free glasses or glasses with low S-contents that require long acquisition times to increase the detection limit (Klimm et al., 2012b). Longpré et al., (2017) observe that potential beam damages on sulfur speciation, if occur, do not alter their conclusions about the change of the magma redox state upon magma decompression during the 2011–2012 El Hierro eruption. XANES has been intensively developed to probe the sulfur speciation in synthetic and natural glasses (e.g., Bonnin-Mosbah et al., 2002). Avoiding the effects of the photoreduction that significantly affect the sulfur speciation in glasses when using small-sized focused beams of very high energy, XANES spectra have illustrated the stability of the sulfide and/or sulfate species mainly in mafic glasses (Wilke et al., 2008, 2011; Métrich et al., 2009; Jugo et al., 2010). The XANES spectra of serpentinite powders have allowed Debret et al. (2017) to discuss the incorporation of both sulfide and sulfate species during oceanic serpentinization. The XANES spectra at the S *K*-edge of a large range of reduced silicate glasses (Fe-free, CAS/CMAS/ NCMAS to Fe-bearing analogues of natural basalts and andesites) are characterized by a similar absorption edge crest at 2476.4 eV. They exhibit a pronounced absorption edge shoulder in spectra of Fe-bearing compositions (Métrich et al., 2009). Jugo et al. (2010) provided XANES spectra showing the predominance of sulfide species in natural basaltic glasses from spreading centers, whereas ocean island basalts and basanites (i.e., samples from Kilauea, Loihi, Lamont) contained variable proportions of S^{2-} and S^{6+}. However, the acquisition of good XANES spectra of glasses with low S contents (<300 ppm) at intermediate to reducing redox conditions (~QFM + 1.8 to ~QFM) can be difficult, as underlined by Fiege et al. (2014a; 2015), who also questioned the integral method proposed by Jugo et al. (2010) to quantify fO_2. In RAMAN spectra, sulfide species as Fe-S and S-H bonds are identified at 372 cm^{-1} and 2574 cm^{-1}, respectively (Klimm et al., 2012a, b).

XANES spectra at the S *K*-edge of oxidized silicate glasses (with only sulfate species) exhibit an intense peak at around 2482 eV, similar to that of the crystalline sulfate model compounds (e.g., Fleet, 2005; Métrich et al., 2009;

Jugo et al., 2010; Klimm et al., 2012a, b). In RAMAN spectra, the sulfate species are identified by a peak at ca. 1000 cm^{-1} in natural basaltic to rhyolitic glasses that is assigned to S-O bonding (sulfate oxy-anion) in the glass structure (Klimm et al., 2012a, b). Similar features are observed in borosilicate glasses (Lenoir et al., 2009). The development of spectroscopic methods using small-sized beam allows the identification of the sulfur species and the measurement of $Fe^{3+}/\Sigma Fe$ ratios in melt inclusion series and, when combined with the analysis of CO_2, H_2O, trace elements, and sulfur isotopes, provide a new and exciting way for modeling their behavior during magma decompression. To what extents the low pressure degassing induces magma reduction is discussed on the basis of μXANES studies at the S and Fe *K*-edges of natural glassy basaltic clasts of Mauna Kea (Brounce et al., 2017), and olivine-hosted MIs of Kilauea (Moussallam et al., 2014; 2016) as developed in the following sections.

11.3.3. The Sulfur Minimum

Katsura and Nagashima (1974) observed that the sulfur content of the Fe-bearing silicate melts (quenched tholeiite, hawaiite, rhyodacite melts) pass through a minimum with change of fO_2 at 1 atmosphere. Similar minimums near NNO were also reported in hydrous silicic melts, synthetized at 200 MPa, when the amount of S dissolved in the quenched melt is plotted against log fO_2 (Carroll & Rutherford, 1985; Clemente et al., 2004; Botcharnikov et al., 2011). The sulfur concentration at sulfide saturation is independent of fO_2, as demonstrated in basalts (e.g., O'Neill & Mavrogenes, 2002) and trachyandesitic melts (e.g., Carroll & Rutherford, 1987). Matjuschkin et al. (2016) suggest that this minimum results from an unidentified, but relatively insoluble, S-species of intermediate oxidation state. It is established that there is no minimum when the melt is saturated with a sulfide phase, whereas the V-shaped equilibrium solubility curve is seen when the fS_2 is fixed and the fO_2 varies at constant temperature (Baker & Moretti, 2011). This minimum has been suggested to be related to the constraints imposed by standard gas-mixing experimental procedures at one atmosphere (Métrich et al., 2009), as developed by O'Neill (this book).

11.3.4. Sulfur Concentration and Redox as a Function of fO_2

It has been established that the SCAS is about 2 to 5 times higher compared to the SCSS in silicic melts (Carroll & Rutherford, 1987; Luhr, 1990; Masotta & Keppler, 2015), and even 4 to 10 times higher in basaltic (Jugo et al., 2010; Parat et al., 2011; Botcharnikov et al., 2011) and andesitic (Zajacz et al., 2012) melts. As an example, the SCAS measured in the quenched basaltic products equilibrated at 1 GPa and 1300 °C is 1.5±0.2 wt %, and is 10 times higher than the S content in basaltic melts saturated with sulfide liquids (0.14±0.02 wt %; Jugo et al., 2005). Beermann et al. (2011) performed experiments at 200 MPa on basaltic systems covering a wide range of temperature (from 1050 to 1250 °C) and H_2O content (from 1 to 6 wt%). At 1050 °C, the sulfide-saturated run glasses contain 0.12 wt% S ($S^{6+}/\Sigma S$ = 0.16) under reducing conditions (ΔFMQ +0.3), against 0.74 wt% S as sulfate species in the sulfate-saturated glasses ($S^{6+}/\Sigma S$ = 0.95; ΔFMQ +2.2). It is worth noting that, at 1200°C for the same pressure and redox conditions (similar $S^{6+}/\Sigma S$, and oxygen buffer), the SCSS achieves 0.22 wt%. The highest S concentrations, c.a. 1 wt% are measured in sulfate-saturated runs (Figure 11.3). Pyrrhotite is found to be a stable phase coexisting with melt at FMQ - ΔFMQ +0.3, whereas anhydrite is stable at ΔFMQ +1.4 to +2. Similarly, hydrothermal experiments conducted by Botcharnikov et al. (2011), at 1050 °C and 200 MPa, show that the sulfur concentration achieves 0.6 wt% in oxidized basaltic hydrous melt against 0.1 wt% at sulfide saturation (ΔFMQ ≤ -0.5). The amount of dissolved sulfur increases rapidly in the narrow fO_2 range from +0.3 to +1.4 log units above FMQ (Figure 11.3). Fortin et al. (2015) observed a positive correlation between the SCSS and water concentration in experimental runs (basalt-andesite-rhyolite) synthesized at 1 GPa, under reducing conditions (C-CO), in agreement with the model of Moretti and Baker (2008) and the experimental results of Moune et al. (2009). Adding water results in a larger amount of sulfur dissolved in a melt before reaching sulfide saturation (ca. 100 ppm S per wt% of added dissolved water for basalts).

11.3.5. The Sulfide–Sulfate Transition

The sulfide–sulfate transition plays a decisive role in the sulfur behavior during magma evolution as the sulfur solubility in melts and its partitioning into the fluid phase as in the sulfide melts stability. Many works have been dedicated to experimental or natural glass samples. However, one of the key questions of how the measured [$S^{6+}/\Sigma S$] in melts as quenched glasses can be used as proxy of the high temperature redox of magmas has been widely discussed. Electron exchange reaction between sulfur and iron could induce change in the sulfur redox state in natural melts upon cooling (Métrich et al., 2009). However, Jugo et al. (2010) have shown that the redox state of sulfur is quenchable in experimental run glasses. The coupled determination of S and Fe redox state points to the reliability of the Fe-S relationship determined in melt inclusions (Bénard et al., 2018a), and natural submarine glasses (Brounce et al., 2017). Hence, such samples are thought to be suitable to compute the redox state of magmas during their crustal evolution, decompression and eruption, as reviewed here. The condition to fulfill is a rapid cooling during the sample emplacement and, in the case of melt inclusions, minor exchanges between the entrapped melts and their surrounding prior to eruption. Experimental data on sulfur speciation in silicate melts at 200 MPa show that the sulfate stability field is reached at $\Delta NNO \geq 1.5$, and the transition from sulfide to sulfate takes place within 2 log units (Jugo et al., 2010; Botcharnikov et al., 2011; Masotta & Keppler, 2015). The sulfide–sulfate transition is relatively sharp (Figures 11.3, 11.4) compared to the Fe^{2+}–Fe^{3+} transition (Klimm et al., 2012b; O'Neill, this book). Jayasuriya et al. (2004) demonstrated that the $Fe^{3+}/\Sigma Fe$ ratio increases from <0.1 to 0.8 over ca. 7 logfO_2 units in experimental anorthite-diopside eutectic glasses containing 1 wt% FeO_{Total}. The relationship between fO_2 (ΔQFM -1 to 3) and the $Fe^{3+}/\Sigma Fe$ ratio in synthetic MORB glass has been recently quantified by Berry et al. (2018). In Mg-rich basalts synthesized in the range of 1300–1355 °C and 1.0–1.6 GPa, sulfate solid phase is found to be stable at fO_2 as low as ΔFMQ +1.85 (Jugo et al., 2005). Klimm et al. (2012b) combined XANES and RAMAN spectroscopy to study the sulfur behavior in experimental silicic glasses, with low Fe content (0–2.05 wt% expressed as Fe_2O_3), synthesized at 200 MPa under water undersaturated conditions. In the run glasses, sulfur is dissolved as S^{2-} at $fO_2 < \Delta QFM$ -1, and as S^{6+} at $fO_2 > \Delta QFM$ +1. The $S^{6+}/\Sigma S$ ratio increases with fO_2, and the sulfide-sulfate transition in Fe-poor silicate melts at ΔFMQ +1 is shifted towards lower fO_2 (by -1.5 log units) compared to the natural basalts and Fe-rich melts. They confirm the fundamental effect of Fe/S ratio on the S^{6+}/S^{2-} equilibrium. Adding Fe to the system would result in the formation of Fe-S complex in the silicate glasses. They also identify, using RAMAN spectroscopy, the presence of SH^- species that may account for the high S-concentrations in the Fe-poor rhyolitic glasses. Klimm et al. (2012) bring the evidence of the existence of SH^- species, which were early hypothesized by Clemente et al. (2004) in experimental quenched Fe-poor rhyolitic glasses (FeO_{Total} <0.2 wt% and low Fe/S ratios; $fO_2 < \Delta NNO$ +1), following the reaction:

$$H_2O_{gas} + SH^-_{melt} \Leftrightarrow H_2S_{gas} + OH^-_{melt} \quad (11.4)$$

In these experiments the mole fraction of S in the run products was found to be linearly proportional to fH_2S. The fO_2 of the sulfide–sulfate transition has been shown to increase slightly with increasing silica (by ~0.1 log units) for a trachyandesite and by a further 0.5 log units for rhyolite (Masotta & Keppler, 2015).

Thermodynamic modelling of the transition between S^{2-} and S^{6+} show three fundamental results (Moretti & Baker, 2008; Baker & Moretti, 2011): (i) the transition is found to be similar for basaltic melts containing 0.5 wt% H_2O at 1200 °C and hydrous rhyolitic melts at 800 °C investigated, over a range of pressures (100–500 MPa); (ii) the calculated SCSS for silicate melt changes significantly near the NNO buffer; (iii) depending on the pressure, melt composition, temperature, and water content, the fO_2 at which the calculated sulfate species become dominant can change by 2 orders of magnitude. As an example the transition may drift by 1.5 log units towards more oxidizing conditions (ΔNNO +0.5 to +2) as the pressure changes from 100 to 500 MPa in water-saturated rhyolitic melts.

11.3.6. The Sulfide–Sulfate Transition and its Key Role on the Sulfur Behavior During Magma Evolution and Decompression

Recently, Matjuschkin et al. (2016) observed that the sulfide stability field expands towards higher fO_2 with increasing pressure, in agreement with the thermodynamic models (Moretti & Baker, 2008; Baker & Moretti, 2011). These experiments are conducted on water-bearing trachyandesite (57.2 wt% SiO_2; 6.0 wt% H_2O). The quenched silicate glasses display variable SiO_2 concentrations increasing from 60 to 69 wt% with decreasing temperature, and most H_2O concentrations range between 7.0 and 9.5 wt%. Their 0.5 GPa experiment at ΔNNO + 2 (Re-ReO_2 buffer) contains anhydrite. At higher pressure and the same fO_2, the sulfide becomes stable and the stability field of sulfate moves towards higher fO_2 by ~1 log unit at 1.0 GPa (Figure 11.4), and ~1.5 log units at 1.5 GPa. As a result, at ΔNNO +2 and constant temperature of 900 °C, the sulfur content of the quenched trachydacitic glass decreases by a factor of ~3 (from 1032 ppm to 352 ppm) as the pressure increases from 0.5 GPa to 1.5 GPa. Hence, at high pressure a significant fraction of sulfur is possibly locked in sulfide (Cu-bearing sulfide phase) that can become unstable during magma decompression while the stability of S^{6+} species in the melt is prevailing. These authors propose that the record of early sulfide phases (i.e., immiscible sulfide globules) will only be preserved as isolated inclusions in crystals that grew at high pressure in magmas which, in turn, could be oxidized ($fO_2 > \Delta NNO + 2$). Accordingly, it could explain the sulfides reported in the products of the climactic eruption of Mt Pinatubo on 15 June 1991. It would also imply that the mafic magma that intruded the shallow silicic magma was initially oxidized, a feature which has been debated (Wallace & Edmonds, 2011). Sulfide saturation can be triggered by the fractionation of magnetite during arc-related magma evolution, in concordance with a reduction of both Fe and S in the melt (Jenner et al., 2010). A similar process is also invoked along the crystallization path of basalt-andesite suite of Hekla volcano (Moune et al., 2009). Massive sulfide saturation induced by Ti-rich magnetite during the late stage differentiation of Anathan volcanic series, in Marianna island arc, is found to be a viable process to account for the overall $\delta^{57}Fe$ behavior in the andesitic-dacitic melts and the sharp decrease of Cu in the most evolved melts (Williams et al., 2018). Titano-magnetite and Cu-rich sulfide globule are described in association with glass in olivine-hosted melt inclusion representative of shallow, partially degassed magma of Mt Etna, whereas the least differentiated S-rich hydrous magmas are oxidized (Spilliaert et al., 2006; Moretti et al., 2018).

11.3.7. Fluid-Melt Partitioning of Sulfur

It is well established that in the hydrous, oxidized rhyolitic systems, a large part of sulfur is stored into the pre-eruptive vapor phase coexisting with the melt at crustal depth (Scaillet et al., 1998; Wallace, 2001; Scaillet & Pichavant, 2003; Shinohara, 2008; Wallace & Edmonds, 2011). The sulfur partitioning into the H_2O-CO_2 rich gas phase during intrusion of S-rich oxidized basaltic batch into crustal reservoir of silicic magma could be maximum in fO_2 interval of the sulfide-sulfate transition, if the system did not reach saturation with respect to a S-bearing phase (Zajacz et al., 2012). Hence, a reliable quantification of the sulfur budget of volcanic eruptions relies on the good knowledge of the sulfur behavior in magmas and the accurate determination of its partitioning into the volatile phase during their crustal transfer, particularly in the fO_2 range of the sulfide/sulfate transition. Using apatite as proxy for determining S sulfur behavior in the magmatic volatile phase is promising (Parat et al., 2011) as sulfur is incorporated as S^{6+} at high fO_2, but dominantly as S^{2-} under more reducing conditions (FMQ) (Konecke et al., 2017). However, according to Parat et al. (2018), determining the sulfur partitioning between apatite and silicic melts still require experiments to explore the respective effects of the temperature, the fS_2, and P and Ca concentrations of the melt.

Several experiments are available for relatively simple systems. Zajacz (2015) reports experiments at 500 MPa and 1240 °C on the Fe-free, aluminosilicate silicate melts equilibrated with H_2O-S fluids. At oxidizing conditions (Re-ReO_2 buffer) the melt/fluid partition coefficient of sulfur increases exponentially with decreasing degree of the melt polymerization. The experimental $D_S^{melt/fluid}$ value varies from 0.005 to 0.3 in Ca-aluminosilicate melts, as the degree of melt polymerization lowers from the equivalent of a rhyolite to the equivalent of a basalt. Masotta et al., (2016) observe a similar increase of the sulfur partitioning into the fluid phase as the composition of the run glasses – synthesized at 200 MPa under comparable oxidizing conditions (Re-ReO_2 buffer) – changes from trachyte and dacite ($D_S^{fluid/melt}$ 2–20) to rhyolite ($D_S^{fluid/melt}$ 100–120). Binder et al. (2018) performed experiments on Ca and Fe-free haplogranite coexisting with H_2O-S-rich fluid (with NaCl or CO_2) at 750–850 °C and 200 MPa, with oxygen fugacity buffered at Ni-NiO and Re-ReO_2. For metaluminous melt composition they found that, at reducing conditions, $D_S^{fl/melt}$ is nearly one order of magnitude larger (323±14) than under oxidizing conditions (74±5). This effect is ascribed to a major change in sulfur speciation in fluids in response to changing melt composition which was revealed by the Raman spectra of the quenched fluids (H_2S and HS^- under reducing conditions and SO_4^{2-} and HSO_4^- under oxidizing conditions). Under oxidizing conditions, the $D_S^{fl/melt}$ value of 128±06 is independent of the NaCl fraction in the fluid phase. The $D_S^{fl/melt}$ value decreases from 264±25 to 84±12, under reducing conditions, as the fraction of NaCl in the fluid changes from 0.1 to 0.3. CO_2 in the fluid (up to $X_{CO2} = 0.3$) has no effect on the fluid/melt partitioning of sulfur, neither under oxidizing nor under reducing conditions. These experimental results suggest that for typical oxidized, silicic to intermediate magmas of convergent settings, the degassing of sulfur is not influenced by the presence of other volatiles, while under reducing conditions, strong interactions with chlorine are observed.

Experimental systems [melt-CO_2-H_2O-S-Cl] representative of magmatic series (basaltic-dacite-rhyodacite-rhyolites series) and evolved magmas are extensively studied to determine the sulfur behavior in natural silicate magmas, and its partitioning into the coexisting fluids (see Webster & Botcharnikov, 2011, for a review). However the experiments on sulfur-bearing systems are challenging and the question of how the $D_S^{fl/melt}$ varies with fO_2 is still debated as it depends upon multi-parameters such as the activity of fluid components, FeO content in the melt, f_{S2}, temperature, and sulfur speciation (Webster & Bortchanikov, 2011; Baker & Moretti, 2011). The equilibrium partitioning of sulfur between fluid and melt is described as:

$D_S^{fl/melt} = C(S)_{fluid}/C(S)_{melt}$, where C is the S content (in wt%) in the fluid and in the melt, respectively. Sulfur obeys Henry's law in experimental silicic (e.g., Keppler, 2010; Zajacz et al., 2012) and basaltic (Fiege et al., 2015) systems, as established on the basis of the covariation between $D_S^{fl/melt}$ values and bulk S contents of the silicate melt.

At the sulfide–sulfate transition, at 200 MPa, $D_S^{fl/melt}$ drops by a factor 8 (from 171±23 to 21±1) in hydrous andesitic glasses containing 6 wt% FeO (Zajacz et al., 2012). Fiege et al. (2015) report isothermal decompression

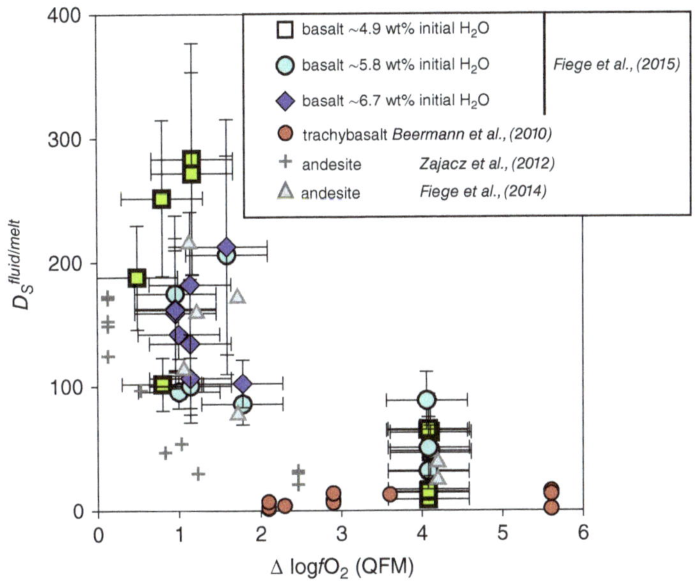

Figure 11.5 Influence of oxygen fugacity on the partitioning of sulfur between the fluid phase and the silicate melt $D_S^{fluid/melt}$. Figure modified from Fiege et al. (2015). Fiege et al. (2015) report the result of decompression experiments on basaltic systems for annealing time of 2 to 5.5 h (near equilibrium conditions), and a range of temperatures from 1150 to 1250 °C. Equilibrium experiments on trachybasalts are conducted at 100 and 200 MPa, 1050 °C (anhydrite- and pyrrhotite-free experiments Beermann et al., 2010), and on andesite at 200 MPa and 1000 °C from Zajacz et al. (2012). Decompression experiments on andesitic system at a rate of ≈0.1 MPa/s, at 1030 °C, 70 MPa, and annealing time ≥ 5 h (near-equilibrium) are from Fiege et al. (2014b).

experiments on water-bearing basaltic melts at ~1050 to ~1250 °C under reducing to oxidizing conditions. They show that (i) $D_S^{fl/melt}$ sharply decreases with increasing fO_2 from 180 at ΔQFM+1 and to 40 at ΔQFM+4 in basaltic melts at ~1200 °C (Figure 11.5); (ii) while limited at oxidizing redox conditions (~ΔQFM +4), the temperature effect is significant at intermediate redox conditions (~ΔQFM +1.5) as $D_S^{fl/melt}$ increases by a factor of ~2 from 94±20 at 1150 °C to 209±80 at 1250 °C. Conversely, such a temperature effect on $D_S^{fl/melt}$ is not observed in Fe-free haplogranitic systems (Keppler, 2010); and (iii) change in melt composition from basaltic to andesitic has a minor effect on $D_S^{fl/melt}$ under oxidizing (ΔQFM +3 to +4) but a major effect on $D_S^{fl/melt}$ under reducing conditions (~QFM). Significant variations of $D_S^{fl/melt}$ at reducing conditions are consistent with the major control of FeO (Fe^{2+}) on sulfur, as sulfide species, in the melt (e.g., Zajacz et al., 2013). Beermann et al. (2011) observe that the $D_S^{fl/melt}$ increases from 27±7 to 200±51 as log fO_2 relative to QFM varies from +3.2 to+0.7 in Cl-bearing trachybasaltic systems (0.05 to ~0.55 wt% Cl) at 200 MPa and 1050 °C. Webster et al. (2009) performed hydrothermal experiments at 200 MPa on alkali-rich trachyte and phonolite-S-H_2O±Cl systems, under relatively high fO_2 (ΔNNO from +0.5 to +1.6). They demonstrate that the dissolved amount of S in the melt increases with the total S content of the system, the CaO content and the $[S^{6+}/(S^{6+}+S^{2-})]$ ratio of the melt, with fO_2 and temperature. In such systems, the $D_S^{fl/melt}$ value increases as the silicate melt becomes polymerized and fO_2 tends to more reducing conditions (Keppler, 2010; Webster & Botcharnikov, 2011). More recently, from a series of hydrothermal experiments on hydrous chloride-rich rhyodacitic melts, Bell and Webster (2015) have documented the role of Cl on the $Fe^{3+}/\Sigma Fe$ ratio, measured in the run glasses using X-ray spectroscopy (XANES). They observe a deviation of the $Fe^{3+}/\Sigma Fe$ ratio – fO_2 evolution relative to the predicted slope that becomes larger as fO_2 increases. They suggest that with increasing fO_2 the change in Cl speciation favors the formation of Na-Cl or Ca-Cl species over the dissolved Fe^{2+}-Cl bonds with in the melt, lowering the stability of ferric-iron-bearing minerals via possible modifications the silicate melt structure. Changes in the $Fe^{3+}/\Sigma Fe$ ratio of silicate melts with addition of Cl indicates strong chemical interactions between Cl and Fe and chloride ions bonding with charge-balancing Fe^{2+} (Webster et al., 2018). These later authors provide a synthetic view of the behavior of Cl, its relationship with Fe and, specifically, its role on the redox of hydrosaline-saturated silicate melts.

11.4. NATURAL SYSTEMS: MAGMA DEGASSING AND REDOX

Sulfur outgassing during magma decompression may induce either reduction or oxidation of the melt depending on the sulfur species (oxidized and/or reduced) dissolved in basaltic melt and in the coexisting gas phase. Sulfur degassing has been widely documented over wide ranges of magma compositions (from silicic to basaltic systems), volcanic activities, and geodynamic contexts (Scaillet et al., 1998; Wallace, 2001; Scaillet & Pichavant, 2003; Métrich & Wallace, 2008; Wallace & Edmonds, 2011; Oppenheimer et al., 2011). A new threshold has been reached in the knowledge of the redox state of a magma while it is losing volatiles, specifically H_2O and S, by carrying out multi-approach studies (i.e., vibrational spectroscopy coupled with elemental and isotopic geochemistry). New software packages are also available to simulate the melt and coexisting vapour compositions in the system C-O-H-S-Cl: SoLeX (Witham et al., 2001) and D-Compress (Burgisser et al., 2015). SolEx applications apply to basaltic systems between 0.5 and 400 MPa. D-Compress software package is calibrated for computing the volatile behavior along the magma compression path for basalt, phonolite, and rhyolite systems between atmospheric pressure and 300 MPa, over a range of temperatures from 790 to 1400 °C. The rapid improvements in probing both the S and Fe redox states in basaltic magmas using cutting-edge methods, and specifically XANES, are illustrated in a series of studies as detailed below, in which the software D-compress (Burgisser et al., 2015), was used to compute C, H, O, and S concentrations and speciation in gas and silicate melt as functions of pressure, temperature, and fO_2. Sulfur exsolution and the S-Fe relationships are expressed via the equations:

$$S^{2-}{}_{(melt)} + 3Fe_2^{3+}O_{3(melt)} \Leftrightarrow S^{4+}O_{2(gas)} + 6Fe^{2+}O_{(melt)} + O^{2-}{}_{(melt)} \quad (11.5)$$

$$S^{6+}O_4^{2-}{}_{(melt)} + 2FeO_{(melt)} \Leftrightarrow S^{4+}O_{2(gas)} + Fe_2^{3+}O_{3(melt)} + O^{2-}{}_{(melt)} \quad (11.6)$$

From the examples dealing with open-condition degassing of mafic magmas, melt reduction following Equation 11.5 appears to be favored relative to melt oxidation. It is a relatively shallow process that involves both S and H_2O loss into the gas phase.

11.4.1. Basaltic Systems

Longpré et al. (2017) investigated olivine-hosted MIs from quenched basanite lava samples from the 2011–2012 submarine eruption at El Hierro, in the Canary Islands. MIs record positive correlations between S and H_2O concentration and the $S^{6+}/\Sigma S$ ratio, whereas S and H_2O are negatively correlated with incompatible elements (alkalis and Nb). Based on the MIs data and simulations using the D-Compress software (Burgisser et al., 2015) starting from melt composition with the highest volatile contents, within a relatively narrow range of fO_2 (ΔFMQ +1.2 and +1.7), they propose a multistage process of an initially oxidized mantle-derived magma, induced by the coupled exsolution of H_2O and S into the fluid phase. Firstly, upon decompression and ascent of the El Hierro basanite the melt is significantly reduced, by at least one log unit of fO_2, following Reaction 11.5. Secondly, there is a significant lowering of the dissolved amount of sulfur to the point of sulfide saturation in El Hierro magma, which is recorded by the late stage precipitation of sulfide globules that are thought to be further (re)dissolved in the silicate melt or removed into a gas phase prior to eruption. Moussallam et al. (2014) studied a series of basanitic to phonolitic melt inclusions from Erebus volcano in Antarctica. MIs recorded magma decompression and evolution from the basanitic magma towards the most differentiated phonolitic magma of Erebus lava lake. At high pressure (~0.8–0.6 GPa) the parent basanite (olivine-hosted MIs) is oxidized with $Fe^{3+}/\Sigma Fe$ ratio up to 0.385, and rich in CO_2 (0.7 wt%) and S (0.25 wt%), whereas its H_2O content does not exceed 1.75 wt%. Sulfide globules are present in MIs and as solid inclusions in minerals. As the degassing proceeds and CO_2, H_2O, and S are exsolved into the gas phase the $Fe^{3+}/\Sigma Fe$ ratio of the melt decreases, predominantly at low pressure, and the computed fO_2 drops by nearly 2 orders of magnitude (from ΔNNO +1 to ΔNNO −1.2). They concluded that the redox evolution of Erebus magmas is prevalently induced by the SO_2 degassing and the $S^{6+}/\Sigma S$ ratio in the melt will re-equilibrate to a lower fO_2, following Reaction 11.5. Moussallam et al. (2016) report XANES studies of olivine-hosted MIs of Kilauea spatter samples from the shallow part of the summit reservoir showing that the magma body was undergoing SO_2 loss prior to eruption. MIs record a clear reduction of iron ($Fe^{3+}/\Sigma Fe$ from 0.30–0.40 to 0.15–0.25) in correlation with the prevalent role of sulfur degassing whereas H_2O and CO_2 would have a negligible effect. The positive correlation observed between the $Fe^{3+}/\Sigma Fe$ ratio and the dissolved amount of H_2O at low pressure is explained by the contemporaneous degassing of H_2O with sulfur. Similarly, the 1959 magma at Kilauea is found to be significantly reduced by late stage degassing of sulfur in course of the eruption (in the rising magma column) (Helz et al., 2017). The $Fe^{3+}/\Sigma Fe$ ratio is correlated to the sulfur content in the melts as it decreases from 0.175–0.160 in the olivine-hosted MIs to ~0.114–0.135 in the degassed matrix glasses. It is in agreement with Carmichael & Ghiorso (1986), who earlier proposed that in Kilauea tholeiites sulfur would act as an oxygen reservoir and the iron and sulfur redox systems as a self-regulating internal oxygen reservoir. A similar relationship between sulfur concentrations and the magma redox state, although not systematic, has been reported in olivine-hosted melt inclusion suites from Marianna arc volcanoes (Brounce et al., 2014). These authors confirm the control of sulfur degassing, as SO_2, on the $Fe^{3+}/\Sigma Fe$ ratios of basaltic melts in agreement with the previous data on olivine-hosted MIs of tephra from Agrigan volcano in Marianna arc (Kelley & Cottrell, 2012). The reduction of both Fe and S in basaltic melts induced by the exsolution of a S-H_2O-CO_2 bearing vapor phase has been recently documented at Mauna Kea volcano in Hawaii, by probing the valence states of Fe and S in basaltic pillow rim glasses using μXANES at Fe and S *K*-edges (Brounce et al., 2017). Sulfur is dominantly dissolved as sulfide species. The $S^{6+}/\Sigma S$ (~0.08–0) and $Fe^{3+}/\Sigma Fe$ (~0.21–0.12) ratios follow the same overall decreasing trend and are correlated with the volatile contents of the glasses in relation with the simultaneous reduction of ferric iron and sulfate species upon degassing. This is a low-pressure process when both S and H_2O are exsolved into the gas phase. The calculated fO_2 (ΔQFM +1 to −0.45) spans ~1.4 orders of magnitude. Using the gas-melt equilibrium model (the software package D-compress; Burgisser et al., 2015), the authors simulate the evolution of the magmatic fO_2 upon degassing starting from a total pressure of 120 bars down to 1 bar. They show that (i) the redox control of sulfur exsolution into the vapor phase, relative to H_2O and CO_2, is prevalent; (ii) fractional degassing of a initially sulfide-saturated basaltic (1.05 wt% H_2O, 1370 ppm S, and 0 ppm CO_2) results in a decrease in fO_2 from ΔQFM from +1.1 to −0.2 that matches 90% of the fO_2 range of the Mauna Kea glasses; and (iii) this range is much more restricted than that determined for the degassed lava samples, hence questioning the use of lava samples to retrieve the initial redox state of the magmas and of their mantle source.

11.4.2. Intermediate to Silicic Systems

The 1835 eruption of Cosigüina volcano in Nicaragua (Longpré et al., 2014) provides a clear example of the distribution of sulfur between the different phases of an andesitic system. According to these authors, the Cosigüina melt was saturated with both a sulfide phase and a hydrous fluid prior to eruption at 100 MPa as revealed by presence of tiny sulfide droplets (5–20 μm) observed in the phenocrysts in association with H_2O-rich fluid inclusions. Sulfur is positively correlated to FeO in the melt. The S repositories in the 1835 Cosigüina system are described to be S dissolved in the andesite melt (1.4–4.5 Tg S), and S exsolved in the hydrous fluid calculated with an amount of exsolved fluid equivalent to 1–2wt % (0.4–8.3 Tg) both being released to the atmosphere at time of eruption, with an addition of the sulfur locked in sulfide droplets earlier formed during magma differentiation from a pre-1835 basaltic andesite (8.4–13 Tg S). In this schema, the breakdown and remobilization of sulfides have significantly contributed (8.4–13 Tg S) to the S global budget of the eruption (a maximum of 10–26 Tg S). The amount of S stored in the fluid phase was calculated to range between 0.9 to 5.6 wt% S using the $D_S^{fl/melt}$ values (between 20 and 80), given for intermediate to silicic magmas, and fO_2 from NNO to ΔNNO +1 (Keppler, 2010; Zajacz et al., 2012). An another example is that of the 1257 CE large plinian eruption of the Samalas volcano in Lesser Sunda arc that produced 158±12 Tg of sulfur dioxide (Vidal et al., 2016). Whereas unstable (Fe, Cu) sulfide globules (up to 10–20 mm in size) are occasionally observed in the titanomagnetite of basaltic scoriae (Métrich et al., 2017), large stable Fe-sulfide bleds are ubiquitous and associated with apatite in the 1257 trachydacitic pumices erupted in the same span of time (Vidal et al., 2016). By combining a detailed geochemistry study of plagioclase-hosted melt inclusions, matrix glasses, and bulk rocks, these authors calculated the respective partitioning of sulfur between the trachydacitic melt, the pre-eruptive H_2O-rich vapor phase, and the poly-metallic sulfides. These results suggest a partition coefficient $D_S^{fl/melt}$ of 13, which fills a gap of experimental constraints on S solubility for these magma conditions. Sulfur is dominantly stored in the pre-eruptive vapor-phase, whereas the amount of sulfur locked in late-stage sulfide globules is limited.

11.5. CONCLUDING REMARKS

• Significant progress in the knowledge of the sulfur behavior, its partitioning into the gas phase, and its importance on the redox state ($Fe^{3+}/\Sigma Fe$ ratio) of magmas have been accomplished. In this regard, multi-approach studies combining vibrational spectroscopy with the geochemistry of volatiles, trace elements, and S and Fe isotopes are highly promising;
• A better understanding of the respective role of S and H_2O, which behave similarly, on the redox during basaltic magma decompression requires further experiments and studies of natural samples;
• Even though very significant progress has been made, further experiments on the fluid/melt partitioning of sulfur, as a function of the melt and fluid composition, are

required for a more accurate quantification of the sulfur budget of the large explosive eruptions, particularly of sulfide-bearing hydrous dacite and trachydacite systems;

• Future efforts will benefit from experiments and modeling of the volatile phase, the fluid/melt partitioning of volatiles to understand the role of volatile on the redox for complex systems melt-CO_2-H_2O-S-Cl-F, in relation with the melts properties.

ACKNOWLEDGMENTS

A. Bénard, K. Klimm, A. Fiege, O. Beermann, and A. Burgisser are kindly thanked for their help with the figures. I am grateful to two anonymous reviewers whose comments helped me to clarify the text.

REFERENCES

Baker, D. R., & Moretti, R. (2011). Modeling the solubility of sulfur in magmas: A 50-year old geochemical challenge. *Reviews in Mineralogy & Geochemistry*, *73*, 167–213. https://doi.org/10.2138/rmg.2011.73.7

Beermann, O., Botcharnikov, R. E., Holtz, F., Diedrich, O., & Nowak, M. (2011). Temperature dependence of sulfide and sulfate solubility in olivine-saturated basaltic magmas. *Geochimica et Cosmochimica Acta*, *75*, 7612–7631. https://doi.org/10.1016/j.gca.2011.09.024.

Behrens, H., & Webster, J. D. (2011). Sulfur in magmas and melts: Its importance for natural and technical processes. *Reviews in Mineralogy & Geochemistry*, *73*, 1–578. https://doi.org//10.2138/rmg.2011.73.0.

Bell, A. S., & Simon, A. (2011). Experimental evidence for the alteration of the Fe^{3+} /ΣFe of silicate melt caused by degassing of chlorine-bearing aqueous volatiles. *Geology*, *39*, 499–502. https://doi.org/10.1130/G31828.

Bell, A. S., & Webster, J. D. (2015). Dissolved Cl, oxygen fugacity, and their effects on Fe behavior in anhydrous rhyodacitic melt. *American Mineralogist*, *100*. 1595–1599. https://doi.org/10.2138/am-2015-5197

Bénard, A., Klimm, K., Woodland, A. B., Arculus, R. J., Wilke, M., Botcharnikov, R. E., et al. (2018a). Oxidising agents in sub-arc mantle melts link slab devolatilisation and arc magmas. *Nature Communications*, *9*, 350. https://doi.org/10.1038/s41467-018-05804-2

Bénard, A., Woodland, A. B., Arculus, R. J., Nebel, O., & McAlpine, S. R. B. (2018b). Variation in sub-arc mantle oxygen fugacity during partial melting recorded in refractory peridotite xenoliths from the West Bismarck Arc. *Chemical Geology*, *486*, 16–30. https://doi.org/10.1016/j.chemgeo.2018.03.004

Berry, A. J., Stewart, G. A., O'Neill, H. S. C., Mallmann, G., & Mosselmans, J. F. W. (2018). A re-assessment of the oxidation state of iron in MORB glasses. *Earth and Planetary Science Letters*, *483*, 114–123. https://doi.org/10.1016/j.epsl.2017.11.032

Binder, B., Wenzel, T., & Keppler, H. (2018). The partitioning of sulfur between multicomponent aqueous fluids and felsic melts. *Contributions to Mineralogy and Petrology*, *173*, 18. https://doi.org/10.1007/s00410-018-1445-6

Birner, S. K., Cottrell, E., Warren, J. M., Kelley, K. A., & Davis, F. A. (2018). Peridotites and basalts reveal broad congruence between two independent records of mantle f_{O_2} and despite local redox heterogeneity. *Earth and Planetary Science Letters*, *494*, 172–189. https://doi.org/10.1016/j.epsl.2018.04.035

Borisova, A. Y., Toutain, J.-O., Dubessy, J., Pallister, J., Zwigk, A., & Salvi, S. (2014). H_2O-CO_2-S fluid triggering the 1991 Mount Pinatubo climactic eruption (Philippines). *Bulletin of Volcanology*, *76*, 800. https://doi.org/10.10007/s00445-014-0800-3

Botcharnikov, R. E., Koepke, J., Holtz, F., McCammon, C., & Wilke, M. (2005). The effect of water activity on the oxidation and structural state of Fe in a ferro-basaltic melt. *Geochimica et Cosmochimica Acta*, *69*, 5071–5085. https://doi.org/10.1016/j.gca.2005.04.023

Botcharnikov, R. E., Linnen, R. L., Wilke, M., Holtz, F., Jugo P. J., & Berndt J. (2011). High gold concentrations in sulphide-bearing magma under oxidizing conditions. *Nature Geoscience*, *4*. https://doi.org/10.1038//NGEO1042

Brandon, A. D., & Draper, D. S. (1996). Constraints on the origin of the oxidation state of mantle overlying subduction zones: an example from Simcoe, Washington, USA. *Geochimica et Cosmochimica Acta*, *60*, 1739–1749.

Brounce, M., Stolper, E., & Eiler, J. (2017). Redox variations in Mauna Kea lavas, the oxygen fugacity of the Hawaiian plume, and the role of volcanic gases in Earth's oxygenation. PNAS 114/ 34, 8997–9002. https://doi.org/10.1073/pnas.1619527114

Bryndzia, L. T., & Wood, B. J. (1990). Oxygen thermobarometry of abyssal spinel peridotites: The redox state and C-O-H volatile composition of the Earth's sub-oceanic upper mantle. *American Journal of Science*, *290*, 1093–1116.

Burgisser, A., & Scaillet, B., (2007). Redox evolution of a degassing magma rising to the surface. *Nature*, *445*, 194–197. https://doi.org/10.1038/nature05509

Burgisser, A., Alletti, M., & Scaillet, B. (2015). Simulating the behavior of volatiles belonging to the C–O–H–S system in silicate melts under magmatic conditions with the software D-Compress. *Computers & Geosciences*, *79*, 1–14.

Burgisser, A., Scaillet, B., & Harshvardhan (2008). Chemical patterns of erupting silicic magmas and their influence on the amount of degassing during ascent. *Journal of Geophysical Research*, *113*, B12204. https://doi.org/10.1029/2008JB00568

Carmichael, I. S. E., & Ghiorso, M. S. (1986). Oxidation-reduction relations in basic magma: a case for homogeneous equilibria). *Earth and Planetary Science Letters*, *78*, 200–210.

Carroll, M. R., & Rutherford, M. J. (1985). Sulfide and sulfate saturation in hydrous silicate melts. *Journal of Geophysical Research, Solid Earth*, *90*, C601–C612.

Carroll, M. R., & Rutherford, M. J. (1987). The stability of igneous anhydrite – experimental results and implications for sulfur behavior in the 1982 El-Chichon trachyandesite and other evolved magmas. *Journal of Petrology*, *28*, 781–801.

Carroll, M.R., & Rutherford, M.J. (1988). Sulfur speciation in hydrous experimental glasses of varying oxidation state: results from measured wavelength shifts of sulfur X-rays. *American Mineralogist*, *73*, 845–849.

Chambefort, I., Dilles, J. H., & Kent A. J. R. (2008). Anhydrite-bearing andesite and dacite as a source of sulfur in magmatic-hydrothermal mineral deposits. *Geology, 36*, 719–722.

Clemente, B., Scaillet, B., & Pichavant, M. (2004). The solubility of sulphur in hydrous rhyolitic melts. *Journal of Petrology, 45*, 2171–2196. https://doi.org/10.1093/petrology/egh052

Crabtree, S. M., & Lange, R. A. (2012). An evaluation of the effect of degassing on the oxidation state of hydrous andesite and dacite magmas: a comparison of pre- and post-eruptive Fe^{2+} concentrations. *Contributions to Mineralogy and Petrology, 163*, 209–224. https://doi.org/10.1007/s00410-011-0667-7

Debret, B., Andreani, M., Delacour, A., Rouméjon, S., Trcera, N., EIMF, & Williams, E. (2017). Assessing sulfur redox state and distribution in abyssal serpentinites using XANES spectroscopy. *Earth and Planetary Science Letters, 466*, 1–11. https://doi.org/10.1016/j.epsl.2017.02.029

Debret, B., & Sverjensky, D. A. (2017). Highly oxidising fluids generated during serpentinite breakdown in subduction zones. *Scientific Reports, 7*, 10351. https://doi.org/10.1038/s41598-017-09626-y

Evans, K. (2012). The redox budget of subduction zones. *Earth Science Reviews, 113*, 11–32. https://doi.org/10.1016/j.earscirev.2012.03.003

Fiege, A., Behrens, H., Holtz, F., & Adams, F. (2014a). Kinetic vs. thermodynamic control of degassing of H_2O-S ± Cl-bearing andesitic melts. *Geochimica and Cosmochimica Acta, 125*, 241–264. https://doi.org/10.1016/j.gca.2013.10.012

Fiege, A., Holtz, F., Behrens, H., Mandeville, C.W., Shimizu, N., Crede, L. S., & Göttlicher, J. (2015). Experimental investigation of the S and S-isotope distribution between H_2O-S ±Cl fluids and basaltic melts during decompression. *Chemical Geology, 393–394*, 36–54. https://doi.org/10.1016/j.chemgeo.2014.11.012

Fiege, A., Holtz, F., Shimizu, N., Mandeville, C. W., Behrens, H., & Knipping, J. L., (2014b). Sulfur isotope fractionation between fluid and andesitic melt: an experimental study. *Geochimica and Cosmochimica Acta, 142*, 501–521. https://doi.org/10.1016/j.gca.2014.07.015

Fincham, C. J. B., & Richardson, F. D. (1954). The behaviour of sulphur in silicate and aluminate melts. *Proceedings of the Royal Society of London, A223*, 40–62.

Fleet, M. E. (2005). XANES spectroscopy of sulfur Earth materials. *Canadian Mineralogist, 43*, 1811–1838.

Fortin, M. A., Riddle, J., Desjardins-Langlais, Y., & Baker, D. R. (2015). The effect of water on the sulfur concentration at sulfide saturation (SCSS) in natural melts. *Geochimica et Cosmochimica Acta, 160*, 100–116. https://doi.org/10.1016/j.gca.2015.03.022

Frost D. J., & McCammon, C. A. (2008). The redox state of Earth's mantle. *Annual Review of Earth and Planetary Sciences, 36*, 389–420. https://doi.org/10.1146/annurev.earth.36.031207.124322

Gaillard, F., Scaillet, B., Pichavant, M., & Bény, J.-M. (2001). The effect of water and f_{O_2} on the ferric–ferrous ratio of silicic melts. *Chemical Geology, 174*, 255–273.

Gaillard, F., Scaillet, B., & Pichavant, M. (2002). Kinetics of iron oxidation–reduction in hydrous silicic melts. *American Mineralogist, 87*, 829–837.

Gaillard, F., Pichavant, M., & Scaillet, B. (2003). Experimental determination of activities of FeO and Fe_2O_3 components in hydrous silicic melts under oxidising conditions. *Geochimica et Cosmochimica Acta, 67*, 4389–4409.

Gaillard, F., Scaillet, B., Pichavant, M., & Iaconno-Marziano, G. (2015). The redox geodynamics linking basalts and their mantle sources through space and time. *Chemical Geology, 418*, 217–233. https://doi.org/10.1016/j.chemgeo.2015.07.030

Ghiorso, M., & Evans, B. (2008). Thermodynamics of rhombohedral oxide solid solutions and a revision of the Fe–Ti two-oxide geothermometer and oxygen barometer. *American Journal of Science, 308*, 957–1039. https://doi.org/10.2475/09.2008.01

Giggenbach, W.F. (1987). Redox processes governing the chemistry of fumarolic gas discharges from White Island, New Zealand. *Applied Geochemistry, 2*, 143–161. https://doi.org/10.1016/0883-2927(87)90030-8

Grove, T. L., Till, C. B., & Krawczynski, M. J. (2012). The role of H_2O in subduction zone magmatism. *Annual Review of Earth and Planetary Sciences, 40*, 413–439. https://doi.org/10.1146/annurev-earth-042711-105310

Helz, R. T., Cottrell, E., Brounce, M. N., & Kelley, K. A. (2017). Olivine-melt relationships and syneruptive redox variations in the 1959 eruption of Kīlauea Volcano as revealed by XANES. *Journal of Volcanology and Geothermal Research, 333–334*, 1–14. https://doi.org/10.1016/j.jvolgeores.2016.12.006

Holloway, J. R. (2004). Redox reaction in seafloor basalts: possible insights into silicic hydrothermal systems. *Chemical Geology, 210*, 225–230. https://doi.org/10.1016/j.chemgeo.2004.06.009

Humphreys, M. C. S., Brooker, R. A., Fraser, D. G., Burgisser, A., Mangan, M. T., & McCammon, C. (2015). Coupled interactions between volatile activity and Fe oxidation state during arc crustal processes. *Journal of Petrology, 56*, 795–814. https://doi.org/0.1093/petrology/egv017

Iacono-Marziano, G., Morizet, Y., Le Trong, E., & Gaillard, F. (2012). New experimental data and semi-empirical parameterization of H_2O-CO_2 solubility in mafic melts. *Geochimica et Cosmochimica Acta, 97*, 1–23. https://doi.org/10.1016/j.gca.2012.08.035

Jayasuriya, K. D., O'Neill, H. St. C., Berry, A. J., & Campbell, S. J. (2004). A Mossbaüer study of the oxidation state of Fe in silicate melts. *American Mineralogist, 89*, 1597–1609.

Jenner, F. E., O'Neill, H. S. C., Arculus, R. J., & Mavrogenes, J. A. (2010). The magnetite crisis in the evolution of arc-related magmas and the initial concentration of Au, Ag and Cu. *Journal of Petrology, 51*, 2445–2464. https://doi.org/10.1093/petrology/egq063

Jugo, P. J., Luth, R.W., & Richards, J. P. (2005). An experimental study of the sulphur content in basaltic melts saturated with immiscible sulfide or sulfate liquids at 1300 °C and 1.0 GPa. *Journal of Petrology, 46*(4), 783–798. https://doi.org/10.1093/petrology/egq097

Jugo, P. J., Wilke, M., & Botcharnikov, R. E. (2010). Sulfur K-edge XANES analysis of natural and synthetic basaltic glasses: implications for S speciation and S content as a function of oxygen fugacity. *Geochimica et Cosmochimica Acta, 74*, 5926–5938. https://doi.org/10.1016/j.gca.2010.07.022

Katsura, T., & Nagashima, S. (1974). Sulfur solubility in some magmas at 1 atmosphere. *Geochimica et Cosmochimica Acta*, *38*, 517–531.

Kelley, K. A., & Cottrell, E. (2012). The influence of magmatic differentiation on the oxidation state of Fe in a basaltic arc magma. *Earth and Planetary Science Letters*, *329–330*, 109–121. https://doi.org/10.1016/j.epsl.2012.02.010

Kelley, K. A., & Cottrell, E. (2009). Water and the oxidation state of subduction zone magmas. *Science*, *325*, 605–607. https://doi.org/10.1126/science.1174156

Keppler, H. (2010). The distribution of sulfur between haplogranitic melts and aqueous fluids. *Geochimica et Cosmochimica Acta*, *74*, 645–660. https://doi.org/10.1016/j.gca.10.010

King, P. L., & Holloway, J. R. (2002). CO_2 solubility and speciation in intermediate (andesitic) melts: The role of H_2O and composition. *Geochimica et Cosmochimica Acta*, *66*, 1627–1640.

Klimm, K., Kohn, S. C., O'Dell, L. A., Botcharnikov, R. E., & Smith, M. E. (2012a). The dissolution mechanism of sulphur in hydrous silicate melts. I: Assessment of analytical techniques in determining the sulphur speciation in iron-free to iron-poor glasses. *Chemical Geology*, *322–323*, 237–249. https://doi.org/10.1016/j.chemgeo.2012.04.027

Klimm, K., Kohn, S. C., & Botcharnikov, R. E. (2012b). The dissolution mechanism of sulfur in hydrous silicate melts. II: solubility and speciation of sulfur in hydrous silicate melts as a function of f_{O_2}. *Chemical Geology*, *322–323*, 250–267. https://doi.org/10.1016/j.chemgeo.2012.04.028

Konecke, B. A., Fiege, A., Simon, A.C., Parat, F., & Stechern A. (2017). Co-variability of S^{6+} S^{4+}, S^{2-} in apatite as a function of oxidation state: Implications for a new oxybarometer. *American Mineralogist*, *102*, 548–557. https://doi.org/10.2138/am-2017-5907

Kress, V. C., & Carmichael, I. S. E. (1991). The compressibility of silicate liquids containing Fe2O3 and the effect of composition, temperature, oxygen fugacity and pressure on their redox states. *Contributions to Mineralogy and Petrology*, *108*, 82–92.

Labidi, J., Cartigny, P., & Jackson, M. G. (2015). Multiple sulfur isotope composition of oxidized Samoan melts and the implications of a sulfur isotope 'mantle array' in chemical geodynamics. *Earth and Planetary Science Letters*, *417* 28–39. https://doi.org/10.1016/j.epsl.2015.02.004

Labidi, J., Cartigny, P., Hamelin, C., Moreira, M., & Dosso, L. (2014). Sulfur isotope bud-get (^{32}S, ^{33}S, ^{34}S and ^{36}S) in Pacific–Antarctic ridge basalts: a record of mantle source heterogeneity and hydrothermal sulfide assimilation. *Geochimica et Cosmochimica Acta*, *133*, 47–67. https://doi.org/10.1016/j.gca.2014.02.023

Lee, C.-T. A., Erdman, M., Yang, W., Ingram, L., Chin, E. J., & DePaolo, D. J. (2018). Sulfur isotopic compositions of deep arc cumulates. *Earth and Planetary Science Letters*, *500*, 76–85. https://doi.org/10.1016/j.epsl.2018.08.017

Lenoir, M., Grandjean, A., Poissonnet, S., & Neuville, D. R. (2009). Quantification of sulfate solubility in borosilicate glasses using Raman spectroscopy. *Journal of Non-Crystalline Solids*, *355*, 1468–1473. https://doi.org/10.1016/j.jnoncrysol.2009.05.015

Lesne, P., Kohn, S. C., Blundy, J., Witham, F., Botcharnikov, R. E., & Behrens, H. (2011b) The experimental simulation of closed-system degassing in the system basalt-H_2O-CO_2-S-Cl. *Journal of Petrology*, *52*, 1737–1762. https://doi.org/10.1093/petrology/egu027

Lesne, P., Scaillet, B., Pichavant, M., Iacono-Marziano, G., & Beny, J. M. (2011a). The H_2O solubility of alkali basaltic melts: an experimental study. *Contributions to Mineralogy and Petrology*, *162*, 133–151.

Longpré, M.-A., Stix, J., Costa, F., Espinoza, E., & Mu⊠oz, A. (2014). Magmatic processes and associated timescales leading to the January 1835 eruption of Cosiguina volcano, Nicaragua. *Journal of Petrology*, *55*, 1173–1208. https://doi.org/10.1093/petrology/egu022

Longpré, M.-A., Stix, J., Klügel, A., & Shimizu, N. (2017). Mantle to surface degassing of carbon- and Sulphur-rich alkaline magma at El Hierro, Canary Islands. *Earth Planetary Science Letters*, *460*, 268–280. https://doi.org/10.1016/j.epsl.2016.11.043

Luhr, J. F. (2008). Primary igneous anhydrite: Progress since its recognition in the 1982 El Chichón trachyandesite. *Journal of Volcanology and Geothermal Research*, *175*, 394–407.

Liu Y., Samaha N.-T., & Baker, D. R. (2007). Sulfur concentration at sulfide saturation (SCSS) in magmatic silicate melts. *Geochimica et Cosmochimica Acta*, *71*, 1783–1799.

Masotta, M., & Keppler, H. (2015). Anhydrite solubility in differentiated arc magmas. *Geochimica et Cosmochimica Acta*, *158*, 79–102. https://doi.org/10.1016/j.gca.2015.02.033

Masotta, M., Keppler, H., & Chaudhari, A. (2016). Fluid-melt partitioning of sulfur in differentiated arc magmas and the sulfur yield of explosive volcanic eruptions. *Geochimica et Cosmochimica Acta*, *176*, 26–43. https://doi.org/10.1016/j.gca.2015.12.014

Mathez, E. A. (1976). Sulfur solubility and magmatic sulfide in submarine basalt glass. *Journal of Geophysical Research*, *81*, 4269–4276.

Mathez, E. A. (1984). Influence of degassing on oxidation states of basaltic magmas. *Nature*, *310*, 371–375.

Matjuschkin, V., Blundy, J. D., & Brooker, R. A. (2016). The effect of pressure on sulphur speciation in mid- to deep-crustal arc magmas and implications for the formation of porphyry copper deposits. *Contributions to Mineralogy and Petrology*, *171*, 66. https://doi.org/10.1007/s00410-016-1274-4

Mavrogenes, J. A., & O'Neill, H. S. (1999). The relative effects of pressure, temperature and oxygen fugacity on the solubility of sulfide in mafic magmas. *Geochimica et Cosmochimica Acta*, *63*, 1173–1180.

McMillan, P. F. (1994). Water solubility and speciation models. *Reviews in Mineralogy and Geochemistry*, *30*, 131–156.

Métrich N., & Wallace P. (2008) Volatile abundances in basaltic magmas and their degassing paths tracked by melt inclusions. *In:* Minerals, Inclusions & Volcanic Processes. K. Putirka & F. Tepley (Eds.), *Reviews in Mineralogy & Geochemistry* 69, pp. 363–402. https://doi.org/10.2138/rmg.2008.69.10

Métrich, N., & Clocchiatti, R. (1996). Sulfur abundance and its speciation in oxidized alkaline melts. *Geochimica et Cosmochimica Acta*, *60*, 4151–4160.

Métrich, N., & Mandeville, C. W. (2010). Sulfur in magmas. *Elements*, *6*, 81–86.

Métrich, N., Berry, A. J., O'Neill, H. S. C., & Susini, J. (2009). The oxidation state of sulfur in synthetic and natural glasses determined by X-ray absorption spectroscopy. *Geochimica et Cosmochimica Acta*, *73*, 2382–2399. https://doi.org/10.1016/j.gca.2009.01.025

Moore, G. (2008). Interpreting H_2O and CO_2 contents in melt inclusions: constraints from solubility experiments and modeling. *Reviews in Mineralogy and Geochemistry*, *69*, 333–362. https://doi.org/10.2138/rmg.2008.69.9

Moore, G., Righter, K., & Carmichael, I. S. E. (1995). The effect of dissolved water on the oxidation state of iron in natural magmas. *Contributions to Mineralogy and Petrology*, *120* 170–179.

Moretti R., Le Losq, C., & Neuville D. R. (2014). The amphoteric behavior of water in silicate melts from the point of view of their ionic-polymeric constitution. *Chemical Geology*, *367*, 23–33. https://doi.org/10.1016/j.chemgeo.2013.12.012

Moretti, R., & Baker, D. R. (2008). Modeling of the interplay of f_{O_2} and fS_2 along the FeS-Silicate Melt equilibrium. *Chemical Geology*, *256*, 286–298. https://doi.org/10.1016/j.chemgeo.2008.06.055

Moretti, R., & Papale, P. (2004). On the oxidation state and volatile behavior in multicomponent gas-melt equilibria. *Chemical Geology*, *213*, 265–280.

Moretti, R., Métrich, N., Arienzo, I., Di Renzo, V., Aiuppa, A., & Allard, P. (2018). Degassing vs. eruptive styles at Mt. Etna volcano (Sicily, Italy). Part I: Volatile stocking, gas fluxing, and the shift from low-energy to highly explosive basaltic eruptions. *Chemical Geology*, *482*, 1–17. https://doi.org/10.1016/j.chemgeo.2017.09.017

Moretti, R., Papale, P., Ottonello, G., Oppenheimer, C., Pyle, D. M., Barclay, J. (2003). A model for the saturation of C-H-O-S fluids in silicate melts. In: *Volcanic Degassing*. 213. *Geological Society of London, Special Publications*, pp. 81–101.

Moune, S., Holtz, F., & Botcharnikov, R. E. (2009). Sulfur solubility in andesitic to basaltic melts: implications for Hekla volcano. *Contributions to Mineralogy and Petrology*, *157*, 691–707.

Moussallam, Y., Edmonds, M., Scaillet, B., Peters, N., Gennaro, E., Sides, I., & Oppenheimer, C. (2016). The impact of degassing on the oxidation state of basaltic magmas: a case study of Kilauea volcano. *Earth Planetary Science Letters*, *450*, 317–325. https://doi.org/10.1016/j.epsl.2016.06.031

Moussallam, Y., Oppenheimer, C., & Scaillet, B. (2019). On the relationship between oxidation state and temperature of volcanic gas emissions. *Earth Planetary Science Letters*, *520*, 260–267. https://doi.org/10.1016/j.epsl.2019.05.036

Moussallam, Y., Oppenheimer, C., Scaillet, B., Gaillard, F., Kyle, P., Peters, N., et al. (2014). Tracking the changing oxidation state of Erebus magmas, from mantle to surface, driven by magma ascent and degassing. *Earth Planetary Science Letters*, *393*, 200–209. https://doi.org/10.1016/j.epsl.2014.02.055

Newman, S., Lowenstern, J. B. (2002). Volatilecalc: a silicate melt-H_2O-CO_2 solution model written in Visual Basic Excel. *Computer Geoscience*, *28*, 597–604.

Ni, H., & Keppler, H. (2013). Carbon in silicate melts. *Reviews in Mineralogy and Geochemistry*, *75*, 251–287. https://doi.org/10.2138/rmg.2013.75.9

Ohba, T., Hirabayashi, J.-I., & Yoshida, M. (1994). Equilibrium temperature and redox state of volcanic gas at Unzen volcano, Japan. *Journal of Volcanology and Geothermal Research*, *60*, 263–272. https://doi.org/10.1016/0377-0273(94)90055-8

O'Neill, H. S. C., & Mavrogenes, J. A. (2002). The sulfide capacity and the sulfur content at sulfide saturation of silicate melts at 1400°C and 1 bar. *Journal of Petrology*, *43*, 1049–1087. https://doi.org/10.1093/petrology/43.6.1049

Oppenheimer, C., Scaillet, B., & Martin, R. S. (2011). Sulfur degassing from volcanoes: Source conditions, surveillance, plume chemistry and Earth system impacts. *Reviews in Mineralogy & Geochemistry*, *73*, 363–421. https://doi.org/10.2138/rmg.2011.73.13

Oppenheimer, C., Scaillet, B., Woods, A., Sutton, A. J., Elias, T., & Moussallam, Y. (2018). Influence of eruptive style on volcanic gas emission chemistry and temperature. *Nature Geoscience*, *678*, 11, 678–681. https://doi.org/10.1038/s41561-018-0194-5

Parat, F., Holtz, F., & Streck, M.J. (2011). Sulfur-bearing magmatic accessory minerals. *Reviews in Mineralogy and Geochemistry*, *73*, 285–314.

Parkinson, I. J., & Arculus, R. J. (1999). The redox state of subduction zones: insights from arc-peridotites. *Chemical Geology*, *160*, 409–423.

Portnyagin, M., Hoernle, K., Storm, S., Mironov, N., van den Bogaard, C., & Botcharnikov, R. (2012). H_2O-rich melt inclusions from Hekla volcano: implications for phase relationships in silicic systems and driving forces of explosive volcanism on Iceland. *Earth Planetary Science Letters*, *357–358*, 337–346. https://doi.org/10.1016/j.epsl.2012.09.047

Reekie, C. D. J., Jenner, F. E., Smythe, D. J., Hauri, E. H., Bullock, E. S., & Williams, H. M. (2019). Sulfide resorption during crustal ascent and degassing of oceanic plateau basalts. *Nature Communications*, *10*, 82. https://doi.org/10.1038/s41467-018-08001-3

Rowe, M. C., Kent, A. J. R., & Nielsen, R. L. (2009). Subduction influence on oxygen fugacity and trace and volatile elements in basalts across the Cascade volcanic arc. *Journal of Petrology*, *50*, 61–91. https://doi.org/10.1093/petrology/egn072

Sato, M., & Wright, T. C. (1966). Oxygen fugacities directly measured in magmatic gases. *Science*, *1*, 53, 1103–1105.

Scaillet, B., & Pichavant, M. (2003). Experimental constraints on volatile abundances in arc magmas and their implications for degassing processes. In Oppenheimer, C., Pyle, D. M., & Barclay, J. (Eds.) *Volcanic degassing*. Geological Society, London, *Special Publications 213*, 23–52.

Scaillet, B., Clemente, B., Evans, B. W., & Pichavant, M. (1998). Redox control of sulfur degassing in silicic magmas. *Journal of Geophysical Research*, *103*(B10), 23937–23949.

Scaillet, B., Holz, F., & Pichavant, M. (2016). Experimental constraints on the formation of silicic magmas. *Elements*, *12*, 109–114. https://doi.org/0.2113/gselements.12.2.109

Shinohara, H. (2008). Excess degassing from volcanoes and its role on eruptive and intrusive activity. *Reviews of Geophysics*, *46*, 1–31.

Schuessler, J. E., Botcharnikov, R., Behrens, H., Misiti, V., & Freda, C. (2008). Amorphous Materials: Properties, structure, and Durability: Oxidation state of iron in hydrous phono-tephritic melts. *American Mineralogist*, *93*. 1493–1504. https://doi.org/0.2113/10.2138/am.2008.2795

Shishkina, T. A., Botcharnikov, R. E., Holtz F., Almeev, R. R. & Portnyagin, M. V. (2010). Solubility of H_2O- and CO_2-bearing fluids in tholeiitic basalts at pressures up to 500 MPa. *Chemical Geology*, *277*, 115–125.

Shishkina, T. A., Botcharnikov, R. E., Holtz, F., Almeev, R. R., Jazwa, A. M., & Jakubiak, A. A. (2014). Compositional and pressure effects on the solubility of H_2O and CO_2 in mafic melts. *Chemical Geology*, *388*, 112–129. https://doi.org/10.1016/j.chemgeo.2014.09.001

Smythe, D. J., Wood, B. J., & Kiseeva, E. S. (2017). The S content of silicate melts at sulfide saturation: new experiments and a model incorporating the effects of sulfide composition. *American Mineralogist*, *102*, 795–803. https://doi.org/10.2138/am-2017-5800CCBY

Spilliaert, N., Allard, P., Métrich, N., & Sobolev, A. V. (2006). Melt inclusion record of the conditions of ascent, degassing, and extrusion of volatile-rich alkali basalt during the powerful 2002 flank eruption of Mount Etna (Italy). *Journal of Geophysical Research*, *111*. https://doi.org/10.1029/2005JB003934.

Stagno, V., Ojwang, D. O., McCammon, C. A., & Frost, D. J. (2013).The oxidation state of the mantle and the extraction of carbon from Earth's interior. *Nature*, *493*, 84. https://doi.org/10.1038/nature11679

Stelling, J., Behrens, H., Wilke, M., Goettlicher, J,. & Chalmin-Aljanabi, E. (2011). Interaction between sulphide and H_2O in silicate melts. *Geochimica et Cosmochimica Acta*, *75*, 3542–3557. https://doi.org/10.1016/j.gca.2011.03.030

Vidal, M. C., Métrich, N., Komorowski, J.-C., Pratomo, I., Michel, A., Kartadinata, N., et al. (2016). The 1257 Samalas eruption (Lombok, Indonesia): the single greatest stratospheric gas release of the Common Era. *Scientific Reports*, *6*, 34868. https://doi.org/10.1038/srep34868

Wallace, P. J. (2001). Volcanic SO_2 emissions and the abundance and distribution of exsolved gas in magma bodies. *Journal of Volcanology and Geothermal Research*, *108*, 85–106.

Wallace, P. J. (2005). Volatiles in subduction zone magmas, concentrations and fluxes based on melt inclusion and volcanic gas data. *Journal of Volcanology and Geothermal Research*, *140*, 217–240.

Wallace, P. J., & Carmichael, I. S. E. (1994). S speciation in submarine basaltic glasses as determined by measurements S $K\alpha$ X-ray wavelength shifts. *American Mineralogist*, *79*, 161–167.

Wallace, P. J., & Edmonds, M. (2011). The sulfur budget in magmas: evidence from melt inclusions, submarine glasses, and volcanic gas emissions. *Reviews in Mineralogy and Geochemistry*, *73*, 215–246. https://doi.org/10.2138/rmg.2011.73.8

Waters, L. A., & Lange, R. A. (2016). No effect of H_2O degassing on the oxidation state of magmatic liquids. *Earth and Planetary Science Letters*, *447*, 48–59. https://doi.org/10.1016/j.epsl.2016.04.030

Webster, J. D., & Botcharnikov, R. E. (2011). Distribution of sulfur between melt and fluid in S–O–H–C–Cl-bearing magmatic system at shallow crustal pressures shallow crustal pressures and temperatures. *Reviews in Mineralogy & Geochemistry*, *73*, 247–283. https://doi.org/10.2138/rmg.2011.73.9

Webster, J. D., Baker, D. R., & Aiuppa, A. (2018). Halogens in mafic and intermediate-silica content magmas. *In The Role of Halogens in Terrestrial and Extraterrestrial Geochemical Processes*, D. E. Harlov & L. Aranovich (Eds.), Springer Geochemistry. https://doi.org/10.1007/978-3-319-61667-4-6

Webster, J. D., Sintoni, M. F., & De Vivo, B. (2009). The partitioning behavior of Cl and S in aqueous fluid- and saline liquid saturated phonolitic and trachytic melts at 200 MPa. *Chemical Geology*, *263*, 19–36. https://doi.org/0.1016/j.chemgeo.2008.10.017

Wilke, M., Jugo, P. J., Klimm, K., Susini, J., Botcharnikov, R., Kohn, S. C., & Janousch, M. (2008). The origin of S^{4+} detected in silicate glasses by XANES. *American Mineralogist*, *93*, 235–240.

Wilke, M., Klimm, K., & Kohn, S. C., (2011). Spectroscopic studies on sulfur speciation in synthetic and natural glasses. *Reviews in Mineralogy & Geochemistry*, *73*, 41–78. https://doi.org/10.2138/rmg.2011.73.3

Williams, H. M., Prytulak, J., Woodhead, J. D., Kelley, K. A., Brounce, M., & Plank, T. (2018). Interplay of crystal fractionation, sulfide saturation and oxygen fugacity on the iron isotope composition of arc lavas: An example from the Marianas. *Geochimica et Cosmochimica Acta*, *226*, 224–243. https://doi.org/10.1016/j.gca.2018.02.008

Witham, F., Blundy, J., Kohn, S. C., Lesne, P., Dixon, J., Chukarov, S. V., & Notcharnikov R. (2012). SolEx: A model for mixed COHSCl-volatile solubilities and exsolved gas compositions in basalt. *Computers & Geosciences*, *45*, 87–97. https://doi.org/10.1016/j.cageo.2011.09.021

Zajacz, Z. (2015). The effect of melt composition on the partitioning of oxidized sulfur between silicate melts and magmatic volatiles. *Geochimica et Cosmochimica Acta*, *158*, 223–244. https://doi.org/10.1016/j.gca.2015.01.036

Zajacz, Z., Candela, P. A., Piccoli, P. M., & Sanchez-Valle, C. (2012). The partitioning of sulfur and chlorine between andesite melts and magmatic volatiles and the exchange coefficients of major cations. *Geochimica et Cosmochimica Acta*, *89*, 81–101. https://doi.org/10.1016/j.gca.2012.04.039

12

Iron in Silicate Glasses and Melts: Implications for Volcanological Processes

Charles Le Losq[1,2], Maria Rita Cicconi[2,3], and Daniel R. Neuville[2]

ABSTRACT

Iron is present in magmas at concentrations ranging from less than 1 wt% to more than 10 wt% in two valence states. In general, Fe^{2+} is a network modifier in the melt structure while Fe^{3+} is a weak network former. The ratio $Fe^{3+}/(Fe^{3+}+Fe^{2+})$ depends on temperature, pressure, oxygen fugacity, and melt composition. Parametric models allow its calculation, but the complex links between melt composition, iron oxidation state, and coordination can be further rationalized using a ionic-polymeric model. Constraining concentration and oxidation state of iron is critical for determining magma density and viscosity, which drive exchanges of matter and heat in the Earth. At high pressures, changes in the coordination of elements, including iron, yield a stiffening and densification of magmas, potentially influencing dynamic and geochemical processes. Near surface, crystallization of Fe-bearing phases changes the residual melt composition, including iron content and oxidation state as well as volatile concentration, ultimately driving large changes in density and viscosity of magmas, and, hence, in the dynamic of fluid flow in volcanic systems. The complex interplay between magma iron content and oxidation state, major element chemistry, crystal and volatile contents thus can play a large role on the dynamic of volcanic systems.

12.1. INTRODUCTION

Iron is among the most important elements in telluric planets. Its proportions in the various different terrestrial reservoirs vary greatly, from ~95 % in Earth core (e.g. Jephcoat & Olson, 1987; Badro et al., 2014) to a few ppb in the Earth atmosphere (e.g. Kopcewicz & Kopcewicz, 1992). In magmas, iron concentration varies from thousands of ppm to 10–15 wt% (Figure 12.1). As a result, iron plays an important role in many igneous processes, through its effects on the phase relationship and physicochemical properties of magmas.

Iron is a transition element that presents two stable valence states: 2+ and 3+. Fe^{2+} has an $[Ar]4s^2\ 3d^6$ electronic configuration when in the divalent form, and Fe^{3+} has half-filled *d* orbital ($[Ar]4s^2\ 3d^5$). The redox ratio of iron, defined hereby as the ratio of ferric to total iron, is determined by temperature, oxygen fugacity, iron concentration, and melt composition. For instance, in first approximation, the higher the iron concentration in a melt,

[1]Research School of Earth Sciences, Australian National University, Canberra, ACT, Australia

[2]Université de Paris, Institut de Physique du Globe de Paris, Paris, France

[3]Department of Materials Science and Engineering, Friedrich-Alexander-Universität Erlangen-Nürnberg, Erlangen, Germany

Magma Redox Geochemistry, Geophysical Monograph 266, First Edition.
Edited by Roberto Moretti and Daniel R. Neuville.

DOI: 10.1002/9781119473206.ch12

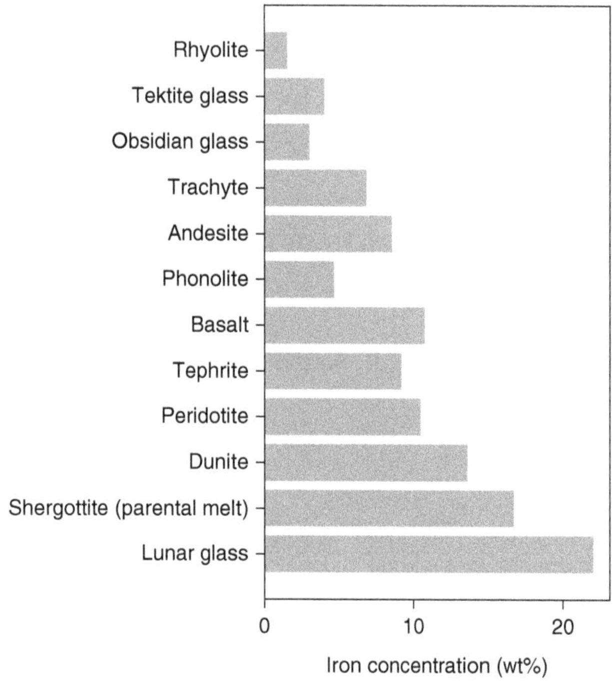

Figure 12.1 Examples of iron concentration ([FeO] equivalent) in various igneous rocks. Data from Geomatériaux - IPGP database, except for the average FeO content of the parental melts of shergottite meteorites estimated by Sossi et al. (2016).

the more iron will be oxidized, i.e., present as Fe^{3+}. Increasing temperature promotes iron reduction, whereas increasing oxygen fugacity promotes its oxidation. Melt composition further plays a critical role on such relationships, leading to a complex link between thermodynamical intensive and extensive parameters and the iron oxidation state in igneous rocks.

In this chapter, we will have a glimpse at how iron is distributed in the disordered geologic matter (glasses and melts) found on Earth. The iron oxidation state in Earth's mantle and core is addressed in other chapters. We will thus essentially focus our attention on describing the proportions, oxidation state, and role of iron in magmas and lavas, as well as their observed amorphous products. We will further discuss the implications of the presence of iron and its oxidation state on the physical properties of magma, such as density and viscosity.

12.2. IRON DISTRIBUTION IN THE DIFFERENT TERRESTRIAL ENVELOPES

While the deepest sections of the Earth (e.g., core) have a very high iron content, its superficial sections still contain moderate but non-negligible fractions of iron. The Earth core is estimated to contain mostly an alloy with a Fe/Ni ratio of ~16 (McDonough & Sun, 1995). Seismological data suggest the presence of ~5 % of light elements in this alloy (e.g. Jephcoat & Olson, 1987; Badro et al., 2014), the best candidates being silicon, sulfur, carbon, oxygen, and hydrogen. Based on the composition of upper mantle xenoliths or on calculations assuming the average solar system element ratios, it is possible to estimate that the Earth mantle contains ~7.2 to 8.1 wt% FeO (Palme & O'Neill, 2014).

Iron in igneous rocks mostly derives from the extraction of the iron from the upper mantle assemblage through partial melting. Mid-ocean ridge basalts, formed from the partial melting of ~10 % of the upper mantle below accretion regions, present an average concentration of FeO of ~8 wt % to 12 wt% (Jenner & O'Neill, 2012). The iron redox ratio, expressed hereby as Fe^{3+}/Fe^{TOT} with $Fe^{TOT} = Fe^{2+} + Fe^{3+}$, in such glasses has been found to be comprised between 0.07 and 0.16, with the latest estimates from X-Ray Absorption Near the Edge Structure (XANES) spectroscopy of 0.10(2) and 0.14(1) (Berry et al., 2018; Zhang et al., 2018), of 0.11(2) from wet-chemistry (Bézos & Humler, 2005), and of 0.09(6) from Raman spectroscopy (Le Losq et al., 2019). Assuming a closed-system condition, such analyses place the average oxidation state of the upper mantle below accretion regions close to the Quartz-Fayalite-Magnetite (QFM) mineral redox buffer. Interestingly, arc basalt magmas present a more oxidized signature (above the QFM buffer), suggesting potentially a more oxidized character of the mantle wedge in subduction regions due to the presence of oxidized subduction fluids (e.g., Brounce et al., 2014). Observing extraterrestrial primitive melts also reveals their enrichement into iron: the FeO content of the parental melts of martian shergottites is estimated at 16.7 wt% in average, and that of lunar glasses reach 22 wt% (Figure 12.1). Iron is thus also a major element in extraterrestrial magmas. Importantly, it most probably played a critical role on the geobiochemistry of planetary surfaces presenting extended solidified lava flows like Mars.

Depending on the crystallization series (tholeiitic or calc-alkaline), the iron concentration of more evolved melts can present different variations, but will ultimately decrease as evolved melts typically present low iron concentrations (Figure 12.1). For instance, the rhyolite magma that was erupted during the 3.1 Ma caldera-forming ignimbrite of the Mt Dore in France contains only 0.7 wt% FeO (Le Losq & Neuville, 2013).

In all cases, iron is prevalent in most igneous rocks, and plays a critical role in determining their properties, such as, for instance, their density and their viscosity. Among major elements in igneous rocks, iron has the particularity to present two different stable valences, Fe^{2+} and Fe^{3+}, which, as we will see, play different roles in the structure of the melt, and, hence, differently influence their properties.

12.3. REDOX EQUILIBRIUM IN MELTS

Reduction-Oxidation (redox) reactions involve the transfer of electrons between chemical species forming the redox couple of a given element, i.e., Fe^{2+} and Fe^{3+} in the case of iron. In silicate melts, oxygen is generally the electron carrier, such that the redox state (ratio of the different species) of iron is driven by changes in the oxygen chemical potential. Oxygen fugacity (fO_2) is conventionally used to describe the oxygen chemical potential. Redox equilibria in glass-forming melts have been exhaustively discussed by the early studies of Schreiber (1986, 1980, 1987). In the case of iron, its redox equilibrium can be described as:

$$4Fe^{3+}_{melt} + 2nO^{2-}_{melt} = 4Fe^{2+}_{melt} + nO_2 \qquad (12.1)$$

with n representing the number of electrons transferred. The relationship described by Equation 12.1 is controlled by several variables: temperature, pressure, oxygen fugacity, and bulk chemical composition. Different models exist to relate the equilibrium constant of Equation 12.1 to those variables. Amongst those, we can cite the models of Sack et al. (1981), Kilinc et al. (1983), Kress and Carmichael (1991), Nikolaev et al. (1996), Jayasuriya et al. (2004), or Borisov et al. (2015). Those models relate melt composition, temperature, and oxygen fugacity through parametric equations of the form:

$$\log\left(\frac{X_{FeO_{1.5}}}{X_{FeO}}\right) = k\log(fO_2) + h/T + \sum d_i X_i + c \qquad (12.2)$$

where T is the temperature in K, and k, h, d_i and c are constants, and X_i are the mole fractions of oxides in the melt. Those models were established using large datasets of melts equilibrated at superliquidus temperatures to constrain the values of the different k, h, d_i, and c constants. While such models are inherently useful, their predictions may be questioned when trying to calculate the iron oxidation state in melts at conditions far from those used in the parameterization. Furthermore, reported lower precisions for predictions of the iron redox ratio in intermediate and silicic compositions (Nikolaev et al., 1996) as well as incomplete considerations of the effect of melt chemistry on iron oxidation state (Borisov et al., 2015) highlight the limits of such models. This pushed, for instance, Borisov et al. (2018) to propose recently a revised equation of the form

$$\log\left(\frac{X_{FeO_{1.5}}}{X_{FeO}}\right) = k\log(fO_2) + h/T + \sum d_i X_i + c + d_{SiAl}X_{SiO_2}X_{Al_2O_3} + d_{SiMg}X_{SiO_2}X_{MgO} \qquad (12.3)$$

where the d_{SiAl} and d_{SiMg} interactive terms allow potentially to better take into account complex compositional effects. Figure 12.2 illustrates variations of the iron oxidation state

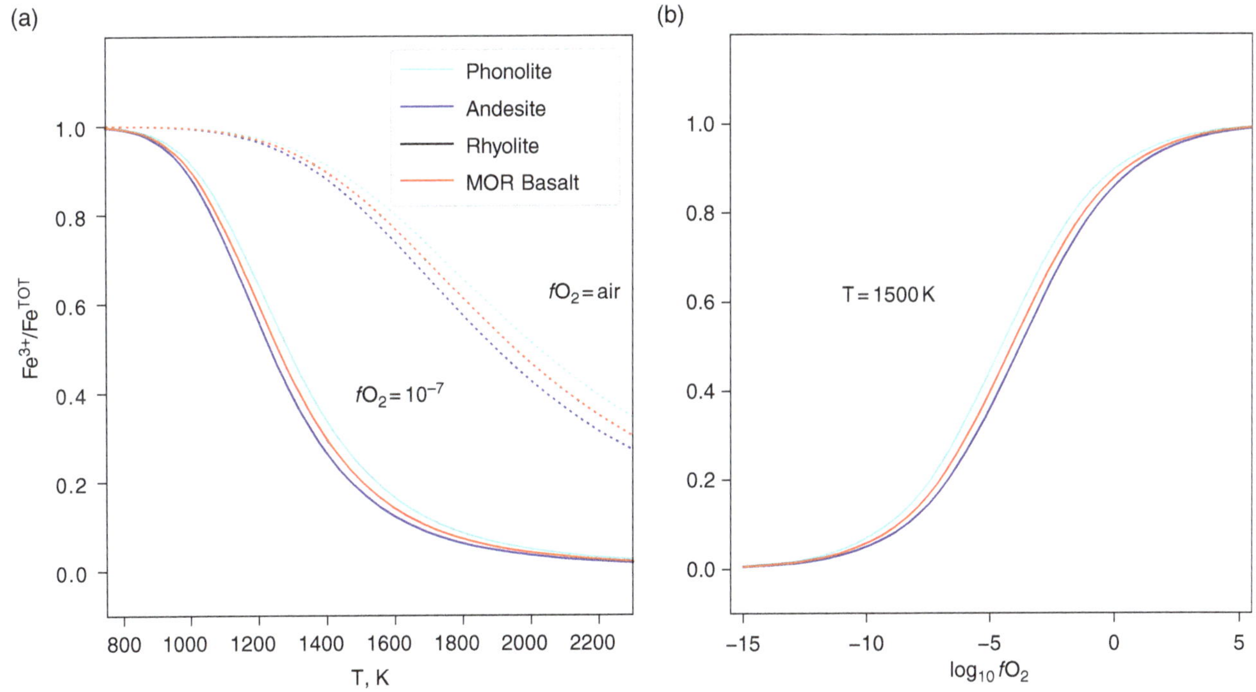

Figure 12.2 Variation of the Fe^{3+}/Fe^{TOT} ($Fe^{TOT} = Fe^{2+} + Fe^{3+}$) redox ratio as a function of (a) temperature and (b) oxygen fugacity, predicted by the model of Borisov et al. (2018) for typical phonolite (Erebus volcano lava bomb, Antartica), andesite (1902 eruption of Mt Pelée, France), rhyolite (3.1 Ma Ignimbrite of Mt Dore, France) and Mid-Ocean Ridge (MOR) basalt (mean composition from Gale et al., 2013) melts.

in different natural melts as a function of temperature and oxygen fugacity calculated using this new model.

Using such a model allows observing that temperature and oxygen fugacity are the main drivers of the iron oxidation state in melts. Chemical composition plays a second order but non-negligible role, particularly if one wants to precisely derive T and fO_2 from measurements of the oxidation state of iron in volcanic glasses for instance (e.g., see Asimow, 2021, chapter 7 from this volume, for an example on basalt glasses). For increased accuracy on such predictions, very precise models are thus needed. This requires the use of models based on thermodynamic and physical frameworks. Aside the problematic highlighted previously, such models further bring a framework to predict the effect of the iron oxidation state on melt properties. This is an important point, as we will see later in this chapter. Therefore, we will not describe further empirical models in the following. Details and discussions can be find in previously cited original publications, such that we refer the reader to them. Rather, we will focus on presenting the effects of extensive and intensive variables on the iron oxidation state and environment in silicate melts, and on constructing a structural background to rationalize such effects.

12.3.1. Temperature, Oxygen Fugacity, and Pressure Effects

Increasing temperature favors the endothermic reaction according to the Van't Hoff law (Schreiber, 1986) and thus shifts the redox equilibrium 12.1 toward reduced species (Figure 12.2a). The log forms of the equilibrium constants of reactions involving Fe^{2+}/Fe^{3+}redox pairs vary linearly with reciprocal temperature (Johnston, 1964; Mysen, 1987; Kress & Carmichael, 1991), with a slope that is equal to the enthalpy of reduction, ΔH, via an approximation of the Van't Hoff relation (Johnston, 1964; Schreiber, 1986):

$$log\left(\frac{Fe^{2+}}{Fe^{3+}}\right) = -\frac{\Delta H}{2.303RT} + b \qquad (12.4)$$

ΔH is endothermic, leading to iron reduction in silicate melts as temperature increases. The effect of the oxygen fugacity, fO_2, an equivalent of the partial pressure of oxygen in the melt corrected for the nonideal character of O_2 (e.g. see Albarède, 2011), is opposite to that of T: increasing fO_2 favors the stabilization of Fe^{3+} species. Similarly to Equation 12.4, this effect can be expressed as:

$$-log\,(fO_2) = 0.25\log\left(\frac{Fe^{2+}}{Fe^{3+}}\right) + k \qquad (12.5)$$

The 0.25 term in Equation 12.5 arises from the fact that this constant usually equals to $\frac{n}{4}$ with n the number of exchanged electrons, which is one in the case of the Fe^{2+}/Fe^{3+}redox pair. For a given composition, and at constant temperature, the relationships described by Equation 12.5 should yield a straight line with slope equal to 0.25. This has been experimentally corroborated by many studies (Fudali, 1965; Schreiber, 1986; Lauer & Morris, 1977; Mysen et al., 1984). However, the melt composition influences this value, such that it may not be ideal in all cases. For example, Mysen et al. (1984) reported a slight decrease of the slope according to the ionic potential of the alkaline-earth cation in alkaline-earth silicate glasses. This trend is corroborated by the recent result of Cicconi et al. (2015), who observed that the mean ionic field strength of the alkali metal cations influences the slope of the Equation 12.5 relationship in multicomponent glasses.

Compared to the effects of oxygen fugacity and temperature, the impact of pressure on the oxidation state of iron has been the subject of few studies and remains not well known. In general, existing studies indicate that higher pressures promote reduced species in many multivalent compounds. For instance, reduction of Fe^{3+}, Cu^{2+}, and Mn^{3+} occurs at high pressure in solid compounds (Drickamer et al., 1970; Wang & Drickamer, 1973; Gibbons et al., 1974). For iron in silicate melts, experimental data confirm such a behavior between 0 and 7 GPa for basaltic and andesitic melt compositions (O'Neill, 2006; Zhang et al., 2017), with a decrease of a few % per GPa of the Fe^{3+}/Fe^{TOT} ratio in this pressure range. Those observations agree with the thermodynamic predictions: as the partial molar volume of Fe_2O_3 (as $FeO_{1.5}$) is higher than that of FeO (see Section 12.4.1), the volume change of Equation 12.1 is negative and reduction of iron is expected with increasing pressure. The study of Zhang et al. (2017) indicates that this behavior remains true up to 7 GPa. However, one should be careful in extrapolating this behavior to higher pressures, because changes in the oxygen coordination number (CN) of Fe are expected (Sanloup, Drewitt, Crépisson, et al., 2013). Such changes in Fe^{2+} and Fe^{3+} CN could affect the dependence of the molar volumes of the Fe_2O_3 and FeO components to pressure, and, in turn, the variations of the Fe^{3+}/Fe^{TOT} ratio in magmas with pressure.

12.3.2. Influence of the Melt Structure and Composition

Introductory Remarks about the Structure of Silicate Melts and Glasses. Before going further in describing the links between iron oxidation state and melt composition, we need to introduce a few basic concepts regarding the role of the different cations in silicate melts, starting from simple binary $M^{x+}_{2/x}O\text{-}SiO_2$ alkali or alkaline-earth silicate compositions. Those compositions contain Si^{4+} *network former* cations, and $M^{x+}_{2/x}$ *metallic* *network modifier* cations, such as Na^+, K^+, Mg^{2+}, or Ca^{2+}. Si is in tetrahedral coordination, forming SiO_4 tetrahedral units that bond with each other through their apical oxygens, which are in this case called bridging oxygens. Tetrahedral units can be discriminated depending on the number n of bridging oxygens they carry, using the Q^n unit notation. The silica glass is mostly composed of Q^4 units, i.e., SiO_4 units with 4 bridging oxygens, and, hence, silica is fully polymerized and its viscosity is the highest known. Modifier cations sever Si-O-Si bonds and form non-bridging oxygen atoms (e.g., see Dupree et al., 1986). For instance, the sodium disilicate glass ($Na_2Si_4O_9$) is composed of a mixture of SiO_4 tetrahedral units with 2 and 3 bridging oxygens, i.e., a mixture of Q^2 and Q^3 units. This profoundly affects the properties of the melts compared to that of pure silica, for instance leading to decrease the melt viscosity of orders of magnitude.

In aluminosilicate glasses, metallic cations (abbrev. M) can also play the role of *compensating* cations. As trivalent Al is present in the network as $[AlO_4]^-$ tetrahedral units, the latter carry a negative charge deficit that is counterbalanced by the compensating cations. The concentrations in modifier and compensating cations depend on the Al/(M+Al) ratio. When Al/(M+Al) is lower than 0.5, modifier cations will be present, and the proportion of compensating cations will be nearly equal to that of $[AlO_4]^-$ tetrahedral units. For $Al/(M+Al) \geq 1$, we can assume at first approximation that all metal cations play a compensating role; such glasses are accordingly referred to as *compensated glasses*. Modifier cations can exist in compensated glasses, but in very low proportions. For further details on such concepts, we invite the reader to refer to Le Losq et al. (2014) and references cited therein. From this brief introduction, we can envision that the chemical composition of silicate melts determines the structural environment of the different cations in it, and that this interplay strongly affects the iron valence and environment in the melts (Borisov & McCammon, 2010; Dickenson & Hess, 1982; Mysen, 2006a; Kress & Carmichael, 1991).

Iron Environment in Melts and Glasses. Ferric iron, Fe^{3+}, is mainly regarded as 4-fold coordinated, but higher coordination numbers have been also reported (Dyar, 1985; Virgo & Mysen, 1985; Hannoyer et al., 1992; Galoisy et al., 2001; Farges et al., 2004; Métrich et al., 2006; Wilke et al., 2007; Rossano et al., 2007; Giuli et al., 2011, 2012; Mysen, 2006b). Interpretations of ^{57}Fe Mössbauer spectroscopy data by Mysen et al. (1985), Virgo and Mysen (1985), and Mysen (2006a) indicate Fe^{3+} in CN 4 at $Fe^{3+}/Fe^{TOT} > \sim 0.5$, and transferring toward a CN 6 at lower Fe^{3+}/Fe^{TOT} ratios. However, analysis of feldspar and haplo-thonalitic glasses by Rossano et al. (2007) suggest a more complex environment, with a wide range of coordinations, and variations with both composition and Fe^{3+}/Fe^{TOT} of the glasses. Interpretations of Fe K-edge X-ray Absorption Spectroscopy (XAS) data seem to support such picture, showing Fe^{3+} present in alumino-silicate glasses mostly in CN 4 and 5 environments (Wilke et al., 2007; Giuli et al., 2011, 2012; Métrich et al., 2006).

Ferrous iron, Fe^{2+}, appears to have a diverse structural environment. The presence of Fe^{2+} in CN 6 has been reported (Calas & Petiau, 1983; Virgo & Mysen, 1985), along with trigonal bipyramidal and tetrahedral coordination (Waychunas et al., 1988; Jackson et al., 1993; Rossano et al., 2007). Fe^{2+} thus appears to have an average coordination close to 5, which could be possibly translated as a coexistence of 4-, 5- and 6- or just the simultaneous presence of 4- and 6- fold Fe^{2+} coordinated (Cicconi et al., 2015).

Melt Composition and Iron Oxidation State. The link between melt composition and iron oxidation state can be understood using different concepts; among those, one of the most successful is that of optical basicity (Moretti & Ottonello, 2003; Duffy & Ingram, 1975; Ottonello et al., 2001; Duffy, 1993). According to Duffy (1993), optical basicity is based on orbital expansion effects reflecting the nature of the chemical bonding between a Lewis acid-base pair; in oxide glasses, oxygens act as bases in the Lewis sense (i.e., as a donor of electron pairs) while metals act as acids. Indeed, the charge balance of the different types of oxygen atoms in the melt can be expressed as (Fincham & Richardson, 1954):

$$2O^- = O^0 + O^{2-} \tag{12.6}$$

with O^-, O^0, and O^{2-} non-bridging, bridging, and "free" oxygens in the melt, respectively. The equilibrium constant K of Equation 12.6 is equal to:

$$K = \frac{[O^{2-}][O^0]}{[O^-]^2} \tag{12.7}$$

assuming that activities of the different oxygen species are equal to their molar concentrations. The free energy of mixing per mole of melt is:

$$\Delta G^{mixing} = \frac{[O^-]}{2} \times R \times T \times ln(K) \tag{12.8}$$

with R the perfect gas constant and T the temperature. An important point is that different metal cations affect K, and thus the fractions of the oxygen species, differently. An important point of considering silicate glasses and melts through Equation 12.6 is that such materials present the characteristics of an acid-base reaction, defined by Flood et al. (1947) as "the transfer of an oxygen ion from a state of polarisation to another." The link between acid-base and redox exchanges in melts can thus be represented by the "normal oxygen electrode":

$$\frac{1}{2}O^2 + 2e^- = O^{2-} \quad (12.9)$$

Melt polymerization is affected by the Lux-Flood acid-base properties of dissolved oxides, with the involvements of free oxygens (Fraser, 1975, 1977):

$$MO + O^{2-} = MO_2^{2-} \quad (12.10)$$

$$MO = M^{2+} + O^{2-} \quad (12.11)$$

Equations 12.10 and 12.11 respectively describe the acidic and basic reactions between the M central cations and the oxygen ligands (O^{2-} and O^-) in silicate melts. The mean polarization state of the oxygen ligands and their ability to transfer fractional electronic charge to the M cation is represented by the optical basicity Λ of the system. It is defined as the ratio of the Jørgensen's function (Jørgensen, 1962) of the ligand in the polarization state of interest over that in an unpolarized state (free O^{2-} ions in an oxidic medium of reference). Different metal cations have different electronic properties and thus different oxide basicity (Table 12.1). It is possible to calculate the glass optical basicity from the oxide optical basicities using the following expression:

$$\Lambda = \sum_i X_{O^i} \Lambda_i \quad (12.12)$$

with X_{O^i} the proportions of oxygen atoms the ith oxide contributes, and Λ_i its oxygen basicity (e.g., see Table 12.1). For instance, for the $CaSiO_3$ glass, obtained from the mixture of 50 mol% CaO and 50 mol% SiO_2, $\Lambda CaSiO_3 = \frac{1}{3} \times 1.00 + \frac{2}{3} \times 0.48 = 0.65$.

Table 12.1 Example of optical basicity Λ of common network formers and modifiers/charge compensators in silicate melts. Values from Ottonello et al. (2001).

Oxide	Λ_i
SiO_2	0.48
Al_2O_3	0.59
TiO_2	0.58
FeO	0.48
Fe_2O_3	0.48
MgO	0.78
CaO	1.00
SrO	1.03
BaO	1.12
H_2O	0.39
Li_2O	1.00
Na_2O	1.15
K_2O	1.36

Optical basicities of silicate glasses correlates well with the oxidation state of iron in the melt equilibrated at given T-fO_2 conditions (Schreiber et al., 1994; Ottonello et al., 2001; Moretti & Ottonello, 2003). Network former cations present close optical basicity in silicate glasses, and thus their influences on the iron oxidation state may be considered as analogous (Mysen et al., 1985; Sack et al., 1981; Kilinc et al., 1983). In detail, changing the melt Al/(Al+Si) leads to small changes in its Fe^{3+}/Fe^{TOT} (e.g. Mysen et al., 1985; Borisov et al., 2015). Metal cations present higher optical basicity than network former cations (Table 12.1), such that increasing their fraction generally promotes oxidation (Schreiber et al., 1994; Mysen et al., 1984) because increasing the melt basicity results in a shift of Equation 12.1 toward the oxidized species. However, different metal cations present different Λ values (Table 12.1), such that the effect of individual cations on the glass oxidation state is complex.

In particular, the effect of the M cation depends on both the Al/(M+Al) ratio of the glass and the M cation electronic properties. The effect of the Al/(Al+M) ratio can be rationalized considering the fact that trivalent iron in polyhedral coordination will present a charge deficit that requires charge balance by metal cations. At Al/(Al+M) < 1, metal cations play a dual role of network modifiers and charge compensators, and, upon addition of trivalent cations, network modifiers can change their role into charge compensators to ensure charge balance of polyhedral units carrying trivalent cations, as shown for Ca (Neuville et al., 2004) or Na (Le Losq et al., 2014). At Al/(Al+M) ≥ 1 and assuming that Fe^{3+} plays a role of network former, introduction of Fe^{3+} in the network will be difficult as metal cations will already be charge compensators of Al, such that a competition between Al and Fe^{3+} for charge compensation will occur. Furthermore, Fe^{2+} will be promoted as it will be able to participate in charge compensating Al-bearing polyhedral units. From those considerations, we expect Fe^{3+} to be stable in peralkaline compositions (Al/(M+Al) < 0.5) and suppressed in peraluminous compositions (Al/(Al+M) ≥ 1). The data of Dickenson and Hess (1982) in K_2O-FeO-Al_2O_3-SiO_2 melts seem to confirm this hypothesis: they clearly show that the Fe^{3+} fraction is not varying much with Al/(Al+M) in peralkaline melts, but decreases with increasing Al/(Al+M) at values higher than ~0.5, highlighting the conversion of Fe^{3+} into Fe^{2+} due to increasing the need for Al charge compensation (Figure 12.3). Borisov et al. (2017) recently questioned such link on the basis of scatter in the data of Dickenson and Hess (1982) at high Fe^{3+}/Fe^{TOT} ratios and on the absence of such a relationship between Al/(Al+K) and iron oxidation state in their study. However, the potassic melts studied by Borisov et al. (2017) also contained

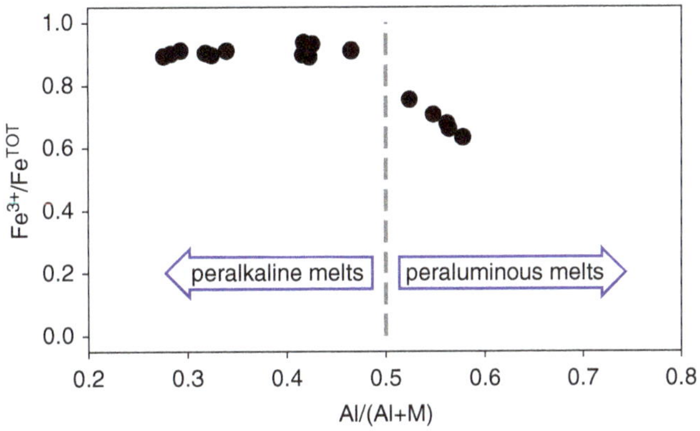

Figure 12.3 Iron oxidation state in K_2O-Al_2O_3-SiO_2-FeO melts at 1673 K in air as a function of the melt Al/(Al+M) molar ratio. The dotted line highlight the tectosilicate join (Al/(Al+M) = 1). Data from Dickenson and Hess (1982).

significant amounts of MgO and CaO, such that Al/(Al +M) (with M=Na+K+Mg+Ca) was never ≥ 0.5 in their samples. Their data are thus not relevant to comment on the links between Al/(Al+M) and Fe^{3+}/Fe^{TOT} in the peraluminous domain.

Despite this, the data of Borisov et al. (2017) shed light on the fact that increasing the fractions of Ca, Na, or K in natural melts yields an increase of the Fe^{3+}/Fe^{TOT} ratio in peralkaline melts. Data from Mysen et al. (1985) in Ca aluminosilicate melts also reveal a decreasing Fe^{3+}/Fe^{TOT} with increasing alkaline-earth metal concentration in aluminosilicate melts, and, hence, with increasing their degree of depolymerization. Those observations agree with earlier data from simple silicate melts of Paul and Douglas (1965), which indicate that increasing the alkali content in binary alkali silicate glasses at fixed T and fO_2 leads to increasing Fe^{3+}/Fe^{TOT}. However, all observations are not unanimous regarding the correlations between metal cation concentrations and iron oxidation state. For instance, Tangeman et al. (2001) reported a decrease in Fe^{3+}/Fe^{TOT} with increasing [K_2O] in K_2O-FeO-SiO_2 glasses, and increases in Fe^{3+}/Fe^{TOT} with increasing [Na_2O] and [CaO] in other binary silicate glasses.

Such complexities indicate that the concentration of metal cations by itself is not sufficient to understand the variations of iron oxidation state with melt composition. Other parameters must be taken into account, such as melt overall degree of polymerization, Al/(Al+Si) and Al/(Al+M) ratios, as well as the ionic field strength of the metal cations in the melt. Indeed, the ionic field strength of metal cations, which plays an important role on the melt structure, correlates with the iron oxidation state (Figure 12.4a). Increasing the ionic field strength of metal cations results in decreasing Fe^{3+}/Fe^{TOT}. According to this, it is expected that mixing metal cations will produce variations in Fe^{3+}/Fe^{TOT} at fixed T and fO_2 conditions. This is confirmed by the study of Cicconi et al. (2015) on phonolite melts, which revealed an increasing Fe^{3+}/Fe^{TOT} with increasing the K/(K+Na) ratio in phonolite magmas, at both oxidized and reduced conditions (Figure 12.4b).

In addition of metal cations, volatile elements can further play a role on the oxidation state of iron. Considering water, the activities of Fe^{2+} and Fe^{3+} change with water content (Gaillard et al., 2003). As a result, the ratio Fe^{3+}/Fe^{TOT} tends to increase in hydrous magmas under reduced conditions, while effects under oxidized conditions are more difficult to distinguish (Gaillard et al., 2003; Moretti, 2005). Sulfur forms a redox couple with iron, such that changes in the redox of one affects that of the other element. Considering S as present mostly as S^{2-} and S^{6+} in magmas, we have:

$$S^{6+} + 8e^- = S^{2-} \tag{12.13}$$

and combining it with the reaction

$$Fe^{3+} + e^- = Fe^{2+} \tag{12.14}$$

we obtain the redox couple

$$S^{2-} + 8Fe^{3+} = S^{6+} + 8Fe^{2+} \tag{12.15}$$

We see there that a small change in S oxidation state can result in larger changes in that of Fe because of the high number of exchanged electrons involved in the redox reaction described by Equation 12.15. As a result, sulfur degassing can affect Fe^{3+}/Fe^{TOT} in magmas (e.g., Moussallam et al., 2014). However, one should remember that in magmas, as in any reacting system, the redox buffer is imposed by a highly abundant component speciating in a reduced and oxidized form. As a result, as Fe is the main multivalent transition element in magmas, their average oxidation state will be generally driven by that of Fe.

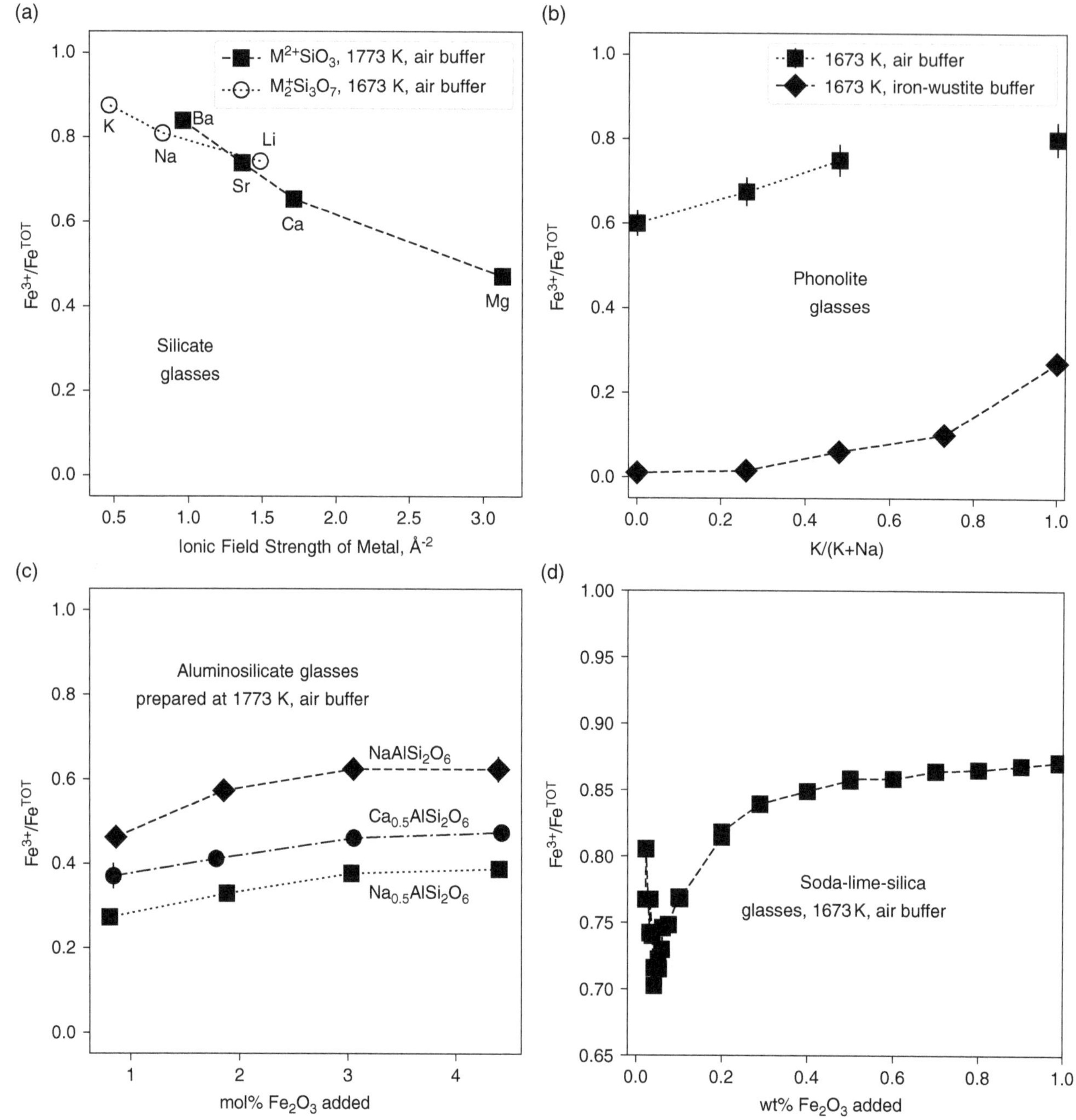

Figure 12.4 Examples of compositional effects on the oxidation state of iron in magmas. (a) Generally, the ionic field strength of metal cations in melts directly influences the iron oxidation state (ionic field strength calculated for a coordination state of 6 with radius from Whittaker & Muntus, 1970). (b) Mixing of different metal cations can lead to linear or non-linear variations of the iron oxidation state at constant T and fO_2. (c) Increasing the concentration of iron in magmas leads to an oxidation of iron. (d) In dilute concentrations, the relationship between iron oxidation state and concentration seem to become complicated. Curves are guides to the eyes. Data are from Densem and Turner (1938); Mysen et al. (1984); Mysen (2006b); Cicconi et al. (2015); Paul and Douglas (1965).

Generally, variations of iron redox ratio with melt composition must further be considered, keeping in mind that Fe^{2+} and Fe^{3+}have different roles in the glass structure. Fe^{2+} is commonly considered as a network modifier element, and Fe^{3+} mainly acts as a network former element, as shown for instance by Extended X-ray Absorption Fine

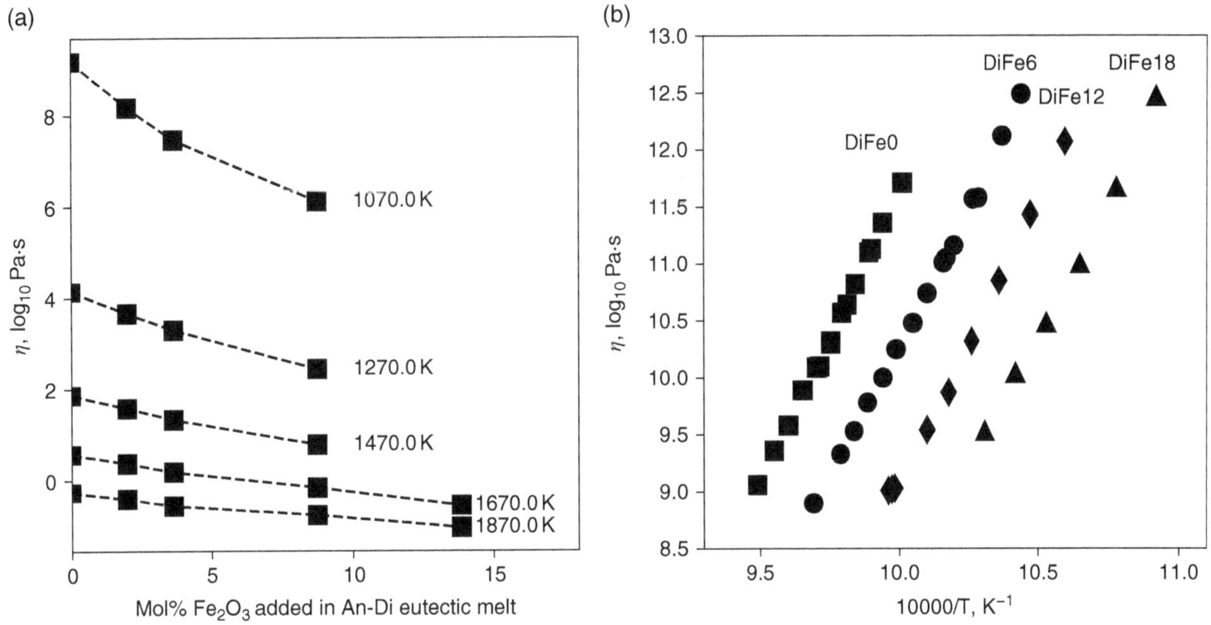

Figure 12.5 Viscosity of (a) anorthite-diopside (An-Di) eutectic melt as a function of added iron, and (b) of diopside melts in which CaO is replaced by FeO. In (a), lines are guides for the eyes; data are from Chevrel et al. (2013). In (b), DiFe0, DiFe6, DiFe12, and DiFe18 melts are compositions where CaO was replaced by FeO in proportions equal to 0, 6, 12, and 18 mol%. $Fe^{3+}/(Fe^{2+}+Fe^{3+})$ redox ratio of those melts is higher than 98% according to the model of Kress and Carmichael (1991); see Table 12.2 for data.

Structure data (Cicconi et al., 2015) and viscosity data (see Section 12.4.2). Interestingly, the change in the melt viscosity at given temperature as a function of the Fe oxidation state further depends on the metal cation in the melt, illustrating the interplay between metal cations and Fe^{3+} for charge compensation (see Section 12.4.2 and Figure 12.5). Furthermore, Fe^{3+} is present as an amphoteric oxide in silicate melts and glasses, i.e. it acts either as an acidic or basic oxide, depending on the bulk system basicity. As a consequence, Fe_2O_3 can either react with bridging or with non-bridging oxygens (Ottonello et al., 2001; Moretti, 2005), and the role of Fe^{3+} in the melt/glass structure is a complex function of the overall system basicity that is dictated by its chemical composition. Another important point of consideration is that, at given T-fO_2 conditions, the concentration of iron affects its oxidation state (Figure 12.4c,d). According to the recent data of Mysen (2006a) or to the older data of Densem and Turner (1938), this effect can shift the oxidation state of iron of ~0.2 Fe^{3+}/Fe^{TOT}. In dilute concentration, it becomes complex as suggested by the data of Densem and Turner (1938). The latter show that, when adding tens to hundreds of ppm of Fe_2O_3 in a soda-lime-silica glass in air at 1400 °C, Fe^{3+}/Fe^{TOT} first decreases of ~0.1, and then increase of ~0.2 with further addition of Fe_2O_3. Such behavior could point to different interactions between Fe^{3+}/Fe^{2+} and the network cations at dilute or high concentrations, but, to conclude on this point, the data of Densem and Turner (1938) need to be corroborated by another study because this contrasting behavior could also arise from an analytical effect.

12.4. PHYSICAL PROPERTIES: HIGHLIGHTS ON DENSITY AND VISCOSITY

The mobility of magmas within the Earth mantle and crust, as well as within superficial volcanic system is determined by two factors: melt density ρ and viscosity η. The former directly determines the melt buoyancy while the latter reflects the internal force resisting uniform melt flow. The importance of ρ and η is illustrated by the expression of the eruption rate Q of a magma ascending into a volcanic conduit of uniform radius a (Jaupart, 1996):

$$Q = \rho \frac{a^4}{8\eta} \left(-\frac{dP}{dz} - \rho g \right) \tag{12.16}$$

with g the gravitational acceleration and $\frac{dP}{dz}$ the vertical pressure gradient. Therefore, the influence of iron on the exchange of matter in igneous geologic systems will be determined by its direct and indirect effects on the density and viscosity of magmas and lavas.

12.4.1. Influence of Iron Content and Redox on the Density of Melts

In general, addition of iron will result in increasing the melt density, due to the large molar mass of iron. Changing the iron oxidation state in the melt will further induce changes in the melt density, because Fe^{2+} and Fe^{3+} present different partial molar volumes. Regarding Fe^{2+}, Lange and Carmichael (1987) reported a partial molar value of 13.6 ± 0.2 $cm^3\ mol^{-1}$ for the oxide component FeO at 1400 °C from density measurements in multicomponent alumino-silicate melts, which agree with the general values of 12.8–14.0 $cm^3\ mol^{-1}$ at 1400 °C reported in silica-rich melts (Bottinga & Weill, 1970; Mo, 1982; Bottinga et al., 1982; Dingwell et al., 1988; Lange & Carmichael, 1989; Liu & Lange, 2006). However, in silica-poor ferrosilicate, the partial molar volume of FeO at 1400 °C approaches 15.8 $cm^3\ mol^{-1}$, as shown by the data of Shiraishi et al. (1978) for instance. Therefore, the partial molar volume of FeO is composition dependent, a fact that can be explained by taking into account variations in the Fe^{2+} local environment with melt composition.

Regarding Fe^{3+}, early reports from Lange and Carmichael (1987) and Dingwell et al. (1988) indicated partial molar volumes of Fe_2O_3 of 42.1 ± 0.3 $cm^3\ mol^{-1}$ and 40.7 ± 0.8 $cm^3\ mol^{-1}$ at 1400 °C, respectively. In Na_2O-FeO-Fe_2O_3-SiO_2 melts, Lange and Carmichael (1989) later reported a value of 41.8 ± 0.4 $cm^3\ mol^{-1}$ at 1400 °C. The recent study of Liu and Lange (2006) reports a similar value of 41.5 ± 0.3 $cm^3\ mol^{-1}$ from high temperature density measurements in Na_2O-FeO-Fe_2O_3-SiO_2 and K_2O-FeO-Fe_2O_3-SiO_2 melts. Those data suggest that the partial molar volume of Fe_2O_3 is independent of temperature and composition in such melt compositions, and further indicate a Fe^{3+} coordination number comprised between 4.5 and 5.0. The results of Liu and Lange (2006) tend to indicate that the partial molar volume of Fe_2O_3 could be independent of composition over an extended range of melt chemical composition (from alkali silicate to magmatic liquids). However, data from Lange and Carmichael (1987, 1989) contradict this, showing a slight dependence of the partial molar volume of Fe_2O_3 on temperature.

Previously reported partial molar values for FeO and $Fe2O_3$ were obtained from density measurements at high temperature in melts, but at room pressure. Increasing pressure changes the overall coordination states of the ions in the melt. The transition of Al from mostly four-fold to mostly six-fold coordination occurs at pressures lower than 15 GPa (Yarger et al., 1995; Allwardt, 2005a; Guillot & Sator, 2007), a phenomenon further enhanced by temperature in aluminium-rich compositions (Allwardt, 2005b; Le Losq et al., 2014). Similarly, Si will undergo a change in its coordination, from a four-fold to a six-fold site, at pressure comprised between 10 and 50 GPa (Guillot & Sator, 2007; Lee et al., 2008; Sanloup, Drewitt, Konôpková, et al., 2013; Wang et al., 2014). This transition to higher coordination numbers with increasing pressure also affects network modifier cations, such as Ca^{2+}, Mg^{2+} and Fe^{2+} (Guillot & Sator, 2007; Sun et al., 2011; Sanloup, Drewitt, Crépisson, et al., 2013). Accordingly, it is expected that Fe^{3+} coordination state also transitions toward higher number, as Al^{3+} does. As a result, while the reported models of magma density may be relevant for sub-crustal pressure, any estimation at pressure higher than a few GPa will be affected by systematic errors due to changes in the coordination state of network former and network modifier cations in melts. Those changes in coordination of all major elements with pressure yield densification of magmas at depth, which can result in controlling the partioning of elements or the mobility of deep melt through density traps (Sanloup, 2016).

12.4.2. Iron and the Viscosity of Silicate Melts

As previously discussed, Fe^{2+} and Fe^{3+} have different roles in the structure of melts, with Fe^{2+} acting as a network modifier element and Fe^{3+} as a weak network former element. This agrees with the observed dependence of melt viscosity on iron concentration and oxidation state. Upon iron addition in a melt with a composition equal to the eutectic anorthite-diopside ($CaAl_2Si_2O_8$-$CaMgSi_2O_6$), the data from Chevrel et al. (2013) show that melt viscosity decreases with increasing iron concentration (Figure 12.5a). We note that this effect is enhanced at undercooled temperatures where entropic effects due to chemical mixing of elements and structural disorder become important. Similarly, substituting CaO by FeO in the diopside melt composition results in a similar decrease in viscosity at undercooled temperatures (Figure 12.5b and Table 12.2) where such melts mostly contain iron in its 3+ valence (calculation from the Kress & Carmichael [1991] model for Ca-Fe-Al-Si-O melts). Therefore, those data together with those in Figure 12.5a indicate that Fe^{3+} appears to act more like a network modifier than like a "traditional" network former as Al^{3+}.

The comparison between the compositions $NaAlSi_3O_8$ and $NaFeSi_3O_8$ further shed light on the role of Fe^{3+} on melt viscosity (Figure 12.6). If, from a structural point of view, Fe^{3+} can be considered as a network former when in tetrahedral state (e.g., Mysen et al., 1985), viscosity data indicate that it does not participate in building a strong network and, hence, acts similarly as a network modifier on melt viscosity (Figures 12.5, 12.6). This probably is due to its ionic field strength that is much weaker than that of Al^{3+}, for instance. In line with such comment, Dingwell and Virgo (1988) noted that the relative viscosity of alumino-, ferro-, and gallio-silicates are inversely

Table 12.2 Viscosity data of DiFe0, DiFe6, DiFe12, and DiFe18 diopside melts, where CaO was replaced by FeO in proportions equal to 0, 6, 12, and 18 mol%. Viscosity was measured following the method of Neuville (2006) between 915 and 1053 K, such that their $Fe^{3+}/(Fe^{2+}+Fe^{3+})$ redox ratio is higher than 98% according to the model of Kress and Carmichael (1991). Errors on viscosity measurements are of 0.03 log unit.

DiFe0		DiFe6		DiFe12		DiFe18	
T, K	log Pa·s	T, K	log Pa·s	T, K	log Pa·s	T, K	log Pa·s
1053.7	10.06	957.9	13.49	982.9	10.87	915.8	13.47
1047.1	10.36	972.4	12.58	943.8	13.07	927.6	12.67
1041.7	10.58	984.7	12.01	954.9	12.43	939.0	12.00
1036.3	10.89	995.3	11.48	965.2	11.85	949.7	11.48
1030.9	11.09	1005.9	11.00	974.8	11.32	959.7	11.04
1029.9	11.10	1016.7	10.53	990.4	10.54	969.9	10.53
1025.6	11.30	964.2	13.12	1001.7	10.03		
1025.6	11.32	974.4	12.57	1002.6	10.02		
1021.5	11.57	983.8	12.05	1003.9	10.01		
1019.4	11.64	990.4	11.74				
1016.3	11.82	1001.2	11.25				
1011.1	12.10	1011.6	10.78				
1010.1	12.13	1021.8	10.33				
1006.0	12.36	1032.2	9.90				
999.0	12.71	981.1	12.16				

correlated with the electronegativity of the trivalent cations, and assigned this to changes in the ionic character of the X^{3+}-O^{2-} bonds (with X any trivalent cation like Al, Ga, or Fe) that ultimately influence melt structure and viscosity. Increasing the ionic character of the X^{3+}-O^{2-} bonds explains the decrease of viscosity observed when replacing Al^{3+} by Fe^{3+} in albitic melts, as shown in Figure 12.6.

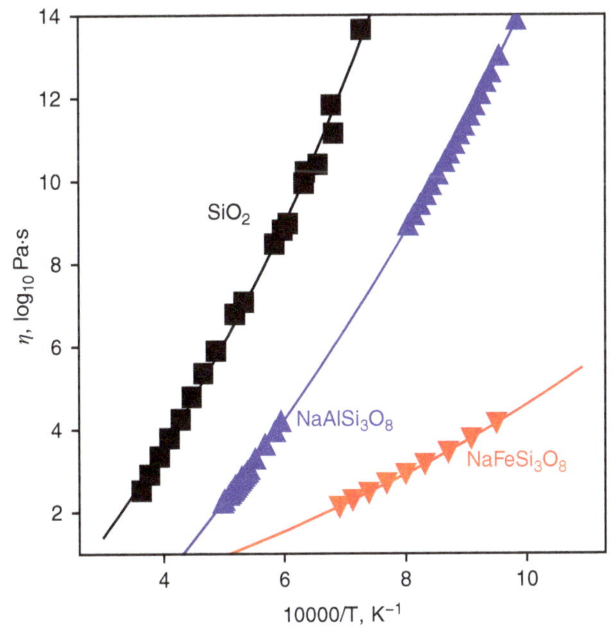

Figure 12.6 Viscosity of SiO_2, $NaAlSi_3O_8$ and $NaFeSi_3O_8$ melts. Data from Hetherington et al. (1964), Urbain et al. (1982), Le Losq & Neuville (2013), and Dingwell et al. (1988).

At fixed melt composition, changing the oxidation state of iron further induces changes in the viscosity of the melt (Figure 12.7). At superliquidus temperatures, the measurements reported by Dingwell and Virgo (1987) and Dingwell (1989, 1991) on ferrosilicate melts indicate that a decrease of the Fe^{3+}/Fe^{TOT} proportion from 0.9 down to 0.2 induces a decrease of viscosity by ~0.5 log units. This is consistent with data on multicomponent basalt and rhyolite melts (Chevrel et al., 2013; Di Genova, Vasseur, et al., 2017). Interestingly, the change of the melt viscosity at given temperature as a function of the iron oxidation state further depends on the metal cation in the melt (Figure 12.7b), illustrating the interplay between metal cations and Fe^{3+} for charge compensation.

Close to the glass transition, data from Liebske et al. (2003) suggest a large viscosity decrease, of nearly 2 orders of magnitude, when Fe^{3+}/Fe^{TOT} decreases from 0.6 down to 0.2 in andesitic melts (Figure 12.7a). Similarly, Chevrel et al. (2013) estimated that the viscosity of depolymerized CaO-Al_2O_3-Fe_2O_3-SiO_2 melts, analogous to basalts, was decreasing of ~1 log unit with decreasing Fe^{3+}/Fe^{TOT} from 0.67 to 0.18 at the calorimetric glass transition temperature. The effect of the iron oxidation state on the melt viscosity is more important near the glass transition, because, as mentioned previously, chemical and structural effects strongly affect the melt configurational entropy, and hence, viscosity at undercooled temperatures (Neuville & Richet, 1991; Richet & Neuville, 1992).

To dive deeper in such concepts, we can use the Adam and Gibbs (1965) theory, which describes viscous flow as a cooperative rearrangement of molecular subunits in the melt. We can write the melt viscosity η as

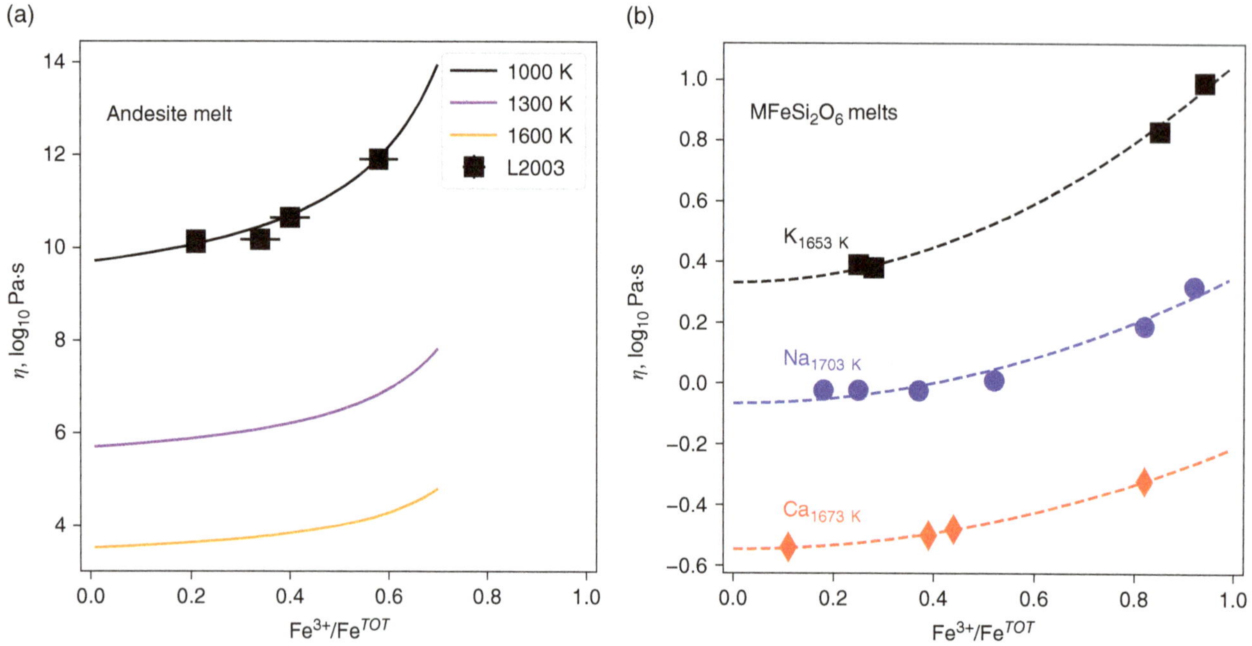

Figure 12.7 Isothermal viscosity as a function of the oxidation state of iron of (a) an andesite melt close to the glass transition and (b) of ferrosilicate melts at superliquidus temperatures. In (a), curves are calculated from the model of Vetere et al. (2008), and symbols are data from Liebske et al. (2003). In (b), curves are polynomial fits to the data from Dingwell and Virgo (1987, 1988); Dingwell (1991).

$$\log_{10}\eta = A_e + \frac{B_e}{T\left(S^{conf}(T_g) + \int_{T_g}^{T} C_p^{conf}/TdT\right)} \quad (12.17)$$

with A_e a constant, B_e a term proportional to the energy barriers opposed to the movement of the molecular subunits, $S^{conf}(T_g)$ the configurational entropy of the melt at the glass transition temperature T_g, and C_p^{conf} the configurational heat capacity of the melt. Actually, the term $S^{conf}(T_g) + \int C_p^{conf}/TdT$ is the configurational entropy of the melt at temperature T, $S^{conf}(T)$. Close to the glass transition, $S^{conf}(T_g)$ is important, implying that melt structure will be particularly influencial on viscous movements (e.g., see Le Losq & Neuville, 2017). As temperature increases, the term $\int C_p^{conf}/TdT$ increases and the melt viscous flow becomes dominated by this term that actually reflect the effect of T on the frequency of T-O bond exchanges (with T network former cations). Adoption of this vision allows understanding the effect of iron oxidation state on melt viscosity: close to the glass transition, in supercooled melts, the different structural environments of Fe^{2+} and Fe^{3+} will influence $S^{conf}(T_g)$, and, hence, will drive large changes in η.

The influence of the iron oxidation state on melt viscosity remains not well constrained, such that the numbers of models that take it into account are limited. Vetere et al. (2008) proposed an equation for andesite melts that allows calculating their viscosity as a function of temperature, water content and iron oxidation state. Their model allows predictions affected by errors lower than 0.2 log units, but is only valid for Fe^{3+}/Fe^{TOT} values lower than 0.7. For a broader range of compositions, we can further cite the parametric model of Duan (2014), which attempts to take into account effects of composition, pressure, water concentration, and iron oxidation state in the modeling of the viscosity of natural magmas. However, predictions from such model should be interpreted with care, because the dataset that includes the effect of iron oxidation state, particularly close to the glass transition, is very limited and may be biased by experimental errors. Indeed, the study of iron-bearing geologic melts always has faced the trouble of crystallization, explaining the scarcity of data on the effect of iron oxidation state on magma viscosity.

12.5. INFLUENCES ON CRYSTALLIZATION AND DEGASSING IN MAGMATIC SYSTEMS

In natural systems, crystallization generally enriches the liquid in Fe^{3+} as olivine and clinopyroxene preferentially incorporate Fe^{2+} in their lattice. Therefore, crystallization of the latter minerals will play an important role in affecting the overall redox condition of the system. Crystallization of iron-bearing oxides can further have a significant influence on the residual melt oxidation state. For instance, superficial magnetite crystallization leading to reduction of the residual melt has been proposed by Oppenheimer et al. (2011) to explain why the lava in the lava lake of Mt Erebus in Antartica is more reduced than the melt inclusions recording the oxidation state of the magma at deeper conditions (conduit/superficial magma chambers). In all cases, crystallization of Fe-bearing phases will change the major element composition of the residual melt, its iron oxidation state, and the magma crystal content. The latter effect induces order of magnitude changes in magma viscosity, as shown in Figure 12.8a. It is affected by the shape and size of crystals, such that crystallization of small phases like spinels may be particularly effective in affecting magma viscosity. In addition to this physical effect, the decrease in the iron content of the residual melt further participates in increasing magma viscosity at isothermal conditions. Building on such idea, Di Genova, Kolzenburg, et al. (2017) proposed, for example, that iron-depletion of magmas by nanolite crystallization could be an important mechanism to drive changes in the rheology of magmas and the dynamic of eruptions of silicic volcanic systems.

In addition to crystallization, changes in volatile content and speciation in magmatic systems needs to be taken into account as they can further modify the oxidation state of silicate melts. Water has been shown to have only a limited effect on iron oxidation state in closed systems (Moretti, 2005; Carmichael, 1991). However, the initial concentration of water in melts is of critical importance, because water dissolved in magmas affects their viscosity by several orders of magnitude regardless of composition (Figure 12.8b). Sulfur and carbon are present in concentrations much lower than that of water, but as they are redox sensitive elements, they play an important role on the magma oxidation state. As a consequence, the relationships between magma oxidation state and volatile speciation should be studied considering a multicomponent C-O-S-H gas-magma system (e.g., Moretti & Papale, 2004). When doing so, significant changes in the magma oxygen fugacity upon magma degassing can be predicted (Moretti & Papale, 2004; Burgisser & Scaillet, 2007). In particular, the interactions between Fe and S are particularly important as the degassing of H_2S and SO_2 can lead to changes in the melt oxidation state, and, hence, Fe^{3+}/Fe^{TOT}. This, in turn, will result in a feedback loop as the melt oxidation state will further influence that of the

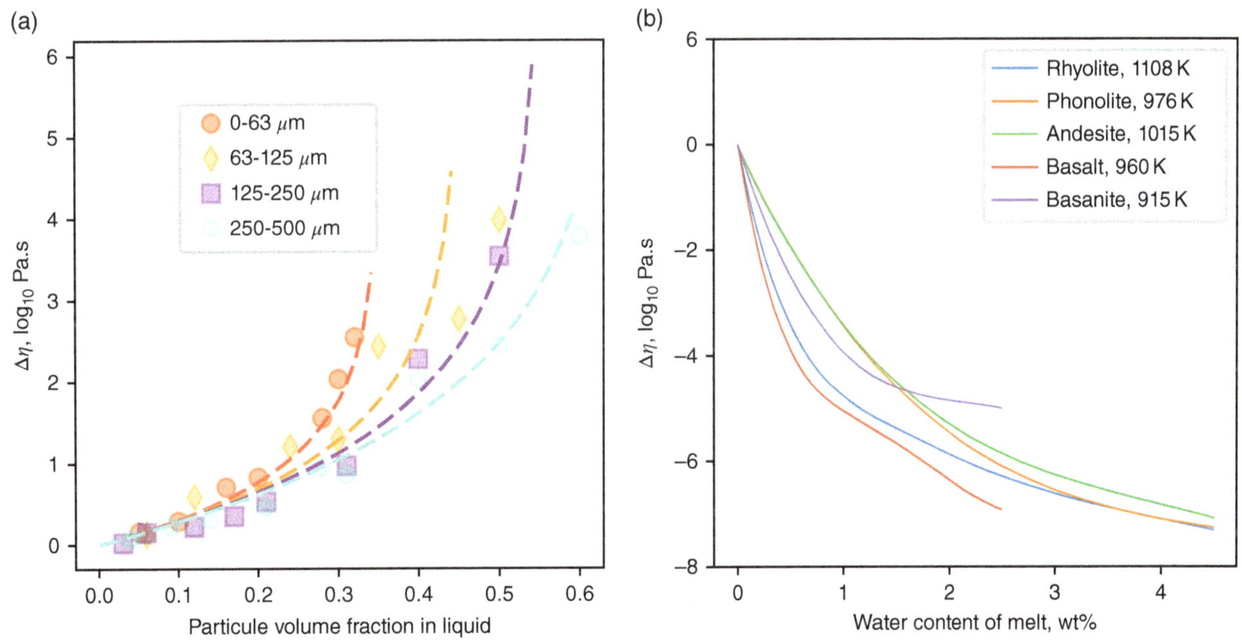

Figure 12.8 Effects of the crystal (a) and water (b) contents on the viscosity of magmas at constant temperature. (a) Experiments from Del Gaudio et al. (2013) on sanidine and pyroxene crystals in silicon oil show a significant effect of the size of crystals on the relative viscosity of the oil ($\Delta\eta = \eta_{crystallized}/\eta_{crystal-free}$). (b) Change in the viscosity of magmas as a function of their water content at supercooled temperatures ($\Delta\eta = \eta_{water-bearing}/\eta_{water-free}$). Curves are calculated for bubble- and crystal-free magmas using equations from Giordano & Dingwell (2003) for a basalt magma of Mt Etna (Italy), Whittington et al. (2000) for a basanite composition, Vetere et al. (2006) for an andesite composition, Le Losq, Neuville, et al. (2015) for the phonolite magma of Mt Erebus (Antartica), and Le Losq (2012) for the rhyolite melt of the 3.1 Ma ingimbritic eruption of the Mt Dore (France).

gas. There is thus a melt-gas equilibria in closed systems, or disequilibra in open systems, that needs to be taken into account for infering variations of the melt and gas oxidation states in degassing volcanic environments. To briefly extend on sulfur in magmas, in addition to interplays of Fe and S during degassing, the complex interplays between the magma fO_2 and its sulfur concentration needs to be mentioned. Indeed, while reduced magmas can contain a few hundred ppm of sulfur as S^{2-} before sulfide saturation, the solubility of sulfur as S^{6+} at oxidized condition is much higher, of a few thousands ppm at anhydrite saturation (see Baker & Moretti, 2011, for a review). Interestingly, magnetite saturation in the melt, which depends on its water concentration, can trigger large changes in the sulfur oxidation state (and thus solubility) as S and Fe form a redox couple in the melt as previously described in Equation 12.15, which can be rewritten in an oxyde form as (Métrich et al., 2009):

$$SO_4^{2-} + 8Fe^{2+}O = S^{2-} + 8Fe^{3+}O_{1.5} \qquad (12.18)$$

Crystallization of magnetite extracting preferentially Fe^{3+} from the melt will drive the melt Fe^{3+}/Fe^{TOT} to a decrease, this resulting in shifting Equation 12.18 to the right, and, hence, will trigger reduction of sulfur, and possibly sulfur saturation and precipitation of sulfide phases (Jenner et al., 2010). Similarly, S degassing will also affect the equilibrium constant of Equation 12.18 and, hence, it may alter the iron oxidation state in the melt.

As a result of the interplays between magma oxidation state, crystallization, and degassing, the influence of the magma oxidation state on the mobility of magmas and the dynamic of volcanic eruptions is plural. Crystallization will affect the major element chemistry of the residual melt, including volatile content, as well as its oxidation state. In addition, it will add a fraction of solid inclusions in the melt. The combination of those effect will affect the effective viscosity of the magma. Degassing can further perturbate the melt's iron redox state, but, maybe more importantly, it will lead to decreasing the melt water content, this being accompanied of significant changes in the rheology of the magma (Figure 12.8b) because of the destructing effects of water on melt polymerization (e.g., see Le Losq et al., 2015). The combination of all those effects is very difficult to study and to model simultaneously (e.g., see Pistone et al., 2012). To highlight possible volcanic outcomes of such effects, we will conclude this part by focusing on the interplay between oxygen fugacity, magma crystallization and rheology.

Experiments in iron-bearing magmatic liquids always faced the problem of crystallization. The presence of iron reduces the undercooled metastable temperature domain in which the melt can be kept free from any crystals. Neuville et al. (1993) noticed this effect while studying the viscosity and heat capacity of andesite and rhyolite melts: they noticed the appearance of magnetite crystals of a scale of 100 nm in the andesite composition after annealing it at 1000 K in air during 24 hours. Nanolites of hercynite ($FeAl_2O_4$) spinels and trydimite also were observed in an andesite composition rich in iron by Linard and Neuville (2000), affecting viscosity measurements (Figure 12.9). TEM observation of the andesite glass after the experiments revealed crystallization of ~25 vol% of nanolites (tablets of 200×100 × 20 nm; Figure 12.9a); the crystallization of hercynite and tridymite phases was metastable, but drove changes in the melt composition toward a rhyolitic one. This effect, combined with that due to crystal addition (Figure 12.8a), induced a large increase in the melt viscosity, for instance of more than 1.5 log unit when starting the dynamic isothermal crystallization-viscosity experiment at 1120 K (Figure 12.9b).

The case of crystallization of nanocrystals in magmatic melts during viscosity experiments was later reported by Liebske et al. (2003) and Villeneuve et al. (2008). Liebske et al. (2003) observed the crystallization of nanolites of magnetites of a size of 10 to 50 nm in andesite melts at undercooled temperatures. Villeneuve et al. (2008) later observed a similar phenomenon when performing viscosity experiments on a basalt composition at undercooled temperatures. Starting from a crystal-free glass, the authors began the viscosity experiments close to the glass transition ($\log(\eta) \sim 10^{12}\ Pa \cdot s$; Figure 12.9c). With gradually increasing temperature and measuring viscosity at the same time, they observed the occurence of crystallization of nanolites of spinel phases at ~975 K. Clinopyroxene crystallization occurred at higher temperature, ~1075 K, causing a sharp increase in melt viscosity related to the presence of crystals as well as changes in residual melt composition, including iron oxidation state as clinopyroxene crystallization preferentially removes Fe^{2+}. As Fe, Ca, and Mg were removed from the residual melt, its degree of polymerization increased, this resulting in increasing its viscosity of orders of magnitudes. Such chemical change magnifies the physical effect of the crystals on the magma viscosity. At higher temperatures, Kolzenburg et al. (2018) recently reported experiments where viscosity variations during the crystallization of basaltic melts upon cooling at constant rate were measured (Figure 12.9d). Those authors varied the fO_2 of the experiments, and results show that the onset of crystallization upon cooling is displaced to lower temperatures under reduced conditions. Those dynamic undercooling experiments are interesting as they are representative of the non-equilibrium conditions encountered during the uprise of magmas in volcanic conduits. In those experiments, the system cools down at cooling rates representative of those in natural systems, such that the degree of undercooling increases continuously, leading to an evolution of the

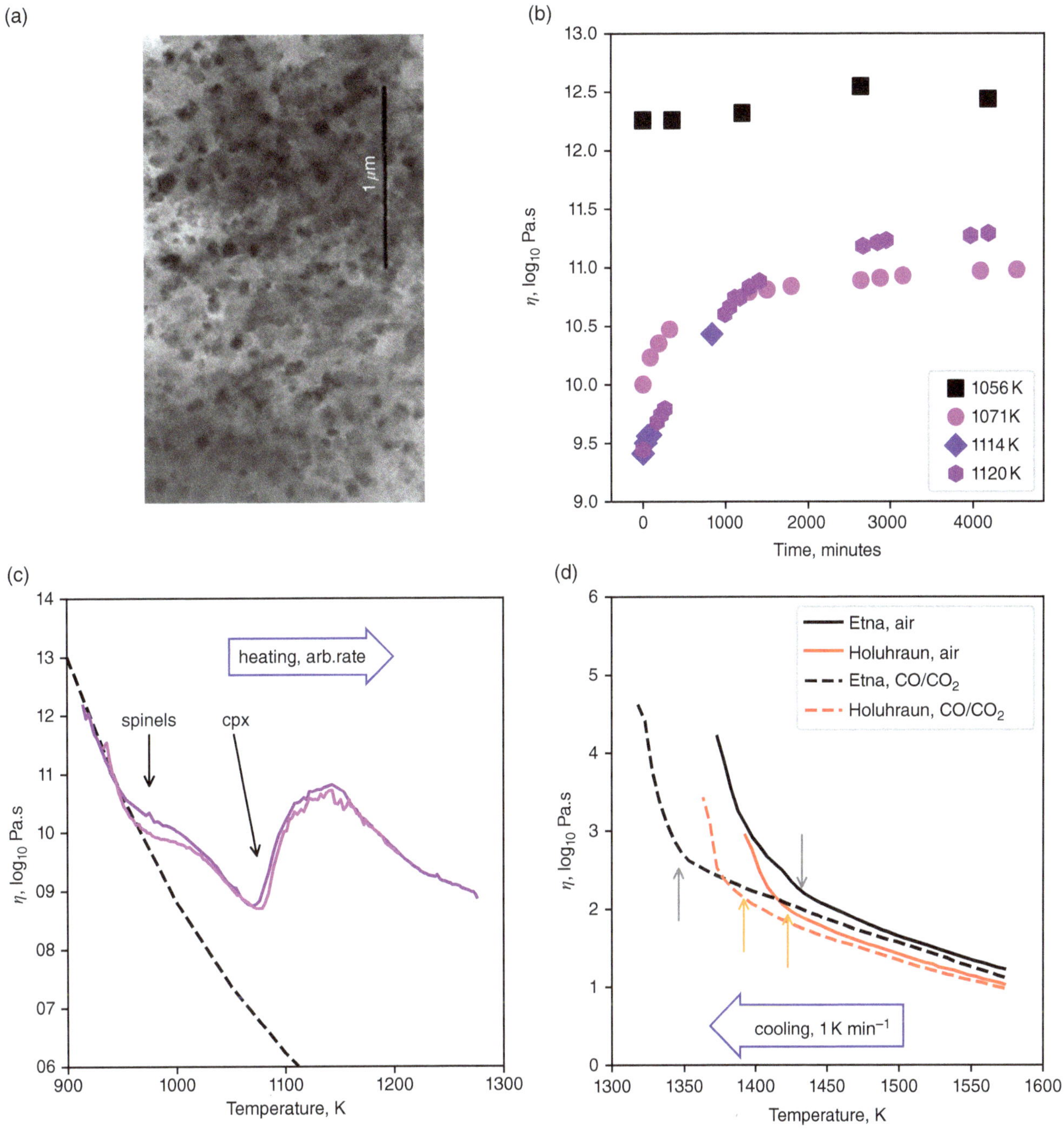

Figure 12.9 Viscosity, crystallinity, and oxidation state of andesite and basalt magmas. (a) Colorized TEM picture of an andesite (ANDE060) glass from Linard and Neuville (2000) after viscosity measurements, containing ~25 vol% of hercynite and trydimite nanocrystals; the size of the tablets are approximatively 200×100×10 nm. (b) Viscosity of the crystallizing andesite (ANDE060) melt from Linard and Neuville (2000) as a function of time. The small changes of the viscosity at 1056 K reflect the absence of extensive crystallization. The two continuous lines are two series of measurements, and the black dotted line is the viscosity of the crystal-free basalt. (c) Viscosity of the Piton de la Fournaise (France) basalt erupted in 1998, measured with increasing temperature at arbitrary rate. Arrows annotated spinels and cpx (clinopyroxene) indicated the onset of the crystallization of those phases while increasing temperature. (d) Viscosity of basalts from Etna (Italy) and Holuhraun (Iceland) measured with decreasing T from 1573 K at a rate of 1 K min^{-1}; arrows highlight the onset of crystallization, which changes with the oxygen fugacity at which the experiments were performed (air, or CO/CO_2 in a 40/60 ratio). Data from Linard and Neuville (2000); Villeneuve et al. (2008); Kolzenburg et al. (2018).

sample crystalline state different from that that would be observed under equilibrium conditions. Such an approach is of particular interest for studying the rheological evolution of lava flows, as those systems are continuously losing heat and, hence, present a degree of undercooling that increases with time. According to Kolzenburg et al. (2018), considering such dynamic evolution of the lava crystallinity is necessary to make representative predictions of the rheology of the lavas, this allowing to improve calculations of the runaway distances of lava flows for instance.

In all cases, the presence of Fe-bearing nanolites in experiments on natural samples highlights (i) the importance of iron and its crystalline phases in natural magmatic systems, and (ii) the difficulty of obtaining crystal-free samples with iron-bearing compositions and to measure their properties at under-cooled temperatures in the laboratory, as the presence of iron enhances crystallization. In particular, one should note that the synthesis of nanolite-free samples in the laboratory requires firing the melts at high temperatures to enhance the reactivity of spinel phases, because those are quite insoluble in silicate melts. Their reaction with the melt can further exhibit a complex behaviour. For instance, the dissolution of high-alumina spinels at 1600 °C in calcium aluminosilicate melts is indirect, with a Ca-Al phase that forms at the melt-crystal interface as an intermediate phase that allows dissolution of alumina into the melt (Sarpoolaky et al., 2003). Therefore, samples fired at too low temperature, even above the liquidus, or for a time too short may still contain nanolite crystals, invisible under the microscope (either binocular or scanning electron microprobe), which can affect *in fine* any data acquired on such products. The presence of small particules is of particular importance for rheology measurements, as the link between the particle concentration and the liquid viscosity is affected by particle size.

12.6. CONCLUDING REMARKS

Iron is a multivalent element present in concentrations ranging from ~1 to more than 15 wt% in magmas. It plays a critical role on the phase diagram as well as on dynamic and thermodynamic properties of magmas, and, as such, is a key component for understanding the evolution of magmatic systems.

The review we performed showed the following key points about iron in silicate melts and glasses:

- In the melt structure, Fe^{3+} is a weak network former and Fe^{2+} is a network modifier.
- Fe^{3+} and Fe^{2+} both participate in decreasing the melt viscosity; Fe^{2+} has a stronger effect on viscosity compared to Fe^{3+}.
- Fe^{3+} and Fe^{2+} have different partial molar volumes, such that magma density depends on iron content and oxidation state.
- Most data suggest that Fe^{3+} is mostly present in CN 4-5 and Fe^{2+} in CN 5–6, but the diversity of results reported in litterature suggest a mixture of environments with CN 4 to 6, and perhaps higher at high pressure.
- The melt composition controls in a complex way the Fe^{3+}/Fe^{2+} ratio, as different metal cations present different oxygen basicities, affecting the equilibrium constant of the redox reaction linking Fe^{3+} and Fe^{2+}.

In magmatic systems, the interplays between crystallization of iron-bearing phases, changes in the composition of the residual melt, Fe^{2+}/Fe^{3+} magma ratio, volatile speciation, sulfur valence and solubility, and overall melt rheology, are very complex. Understanding them is critical to model the evolution of natural magmas. In particular, considering basalt melts in deep magmatic systems, removal of Fe^{2+} by olivine and pyroxene crystallization can be followed by removal of Fe^{3+} by magnetite crystallization, this magnetite crisis leading to change the oxidation state of S in the melt, and, hence, possibly its sulfur content if at saturation. At surface, precipitation of Fe-bearing oxides can lead to further affect the rheology of magmas. Indeed, experiments show that extraction of Fe by magnetite precipitation leads to change the bulk melt composition, driving large increases in the magma viscosity and relaxation time.

Future studies will need to address such issues, particularly considering dynamic systems at meta-equilibrium conditions where crystallization of Fe-bearing phases affects melt properties. In addition to rheological and thermodynamic properties, an interesting additional area of study is iron diffusion in melts. It is important for understanding crystallization and redox exchange phenomena, for instance. Despite this, existing studies do not allow to address the effect of iron oxidation state on the bulk iron diffusion coefficient (e.g., see the review of Zhang et al., 2010). As Fe^{2+} and Fe^{3+} have different environnements, they are expected to diffuse at different rates, the bulk Fe diffusion coefficient being a linear function of those of Fe^{2+} and Fe^{3+} at equilibrium (Zhang et al., 2010). Under disequilibrium conditions as those characterizing dynamic systems (e.g., magma mixing in a reservoir), the differences in Fe^{2+} and Fe^{3+} diffusion rates may result in local variation of iron oxidation state (redox disequilibrium), with various possible implications for crystallization as well as for chemical and isotopic equilibria for example. Such importance justifies further efforts in better defining the diffusion coefficients of Fe^{2+} and Fe^{3+}. In general, there is a clear need for more experimental measurements on controlled systems where oxygen fugacity, temperature and pressure are well constrained. Volatile-bearing systems are of particular interest, as water for instance will influence the diffusivity of iron (e.g., González-García et al., 2017) and, to a lesser extent, its oxidation state. It is only with such data that a complete picture of the complex links between magma composition and properties can be drawn and used for adressing various volcanological and geochemical problems.

ACKNOWLEDGMENTS

CLL acknowledges funding from Chaire d'Excellence IDEX19C627X/FD070/D110 from the IdEX Université de Paris ANR-18-IDEX-0001 and from the Australian Research Council Laureate Fellowship FL1600000 to Prof. Hugh O'Neill (ANU-RSES).

REFERENCES

Adam, G., & Gibbs, J. H. (1965). On the temperature dependence of cooperative relaxation properties in glass-forming liquids. *The Journal of Chemical Physics*, *43*, 1, 139–146. https://doi.org/10.1063/1.1696442

Albarède, F. (2011). Oxygen fugacity. In: M. Gargaud et al. (Eds.), *Encyclopedia of Astrobiology* pp. 1196–1196. Springer Berlin Heidelberg.

Allwardt, J. R., Stebbins, J. F., Schmidt, B. C., Frost, D. J., Withers, A. C., & Hirschmann, M. M. (2005a). Aluminum coordination and the densification of high-pressure aluminosilicate glasses. *American Mineralogist*, *90*, 7, 1218–1222. https://doi.org/10.2138/am.2005.1836

Allwardt, J. R., Poe, B. T., & Stebbins, J. F. (2005b). The effect of fictive temperature on Al coordination in high-pressure (10 GPa) sodium aluminosilicate glasses. *American Mineralogist*, *90*, 8–9, 1453–1457.

Badro, J., Cote, A. S., & Brodholt, J. P. (2014). A seismologically consistent compositional model of Earth's core. *Proceedings of the National Academy of Sciences*, *111*, 21, 7542–7545. https://doi.org/10.1073/pnas.1316708111

Baker, D. R., & Moretti, R. (2011). Modeling the solubility of sulfur in Magmas: A 50-year old geochemical challenge. *Reviews in Mineralogy and Geochemistry*, *73*, 1, 167–213. https://doi.org/10.2138/rmg.2011.73.7

Berry, A. J., Stewart, G. A., O'Neill, H. S. C., Mallmann, G., & Mosselmans, J. F. W. (2018). A re-assessment of the oxidation state of iron in MORB glasses. *Earth and Planetary Science Letters*, *483*, 114–123. https://doi.org/10.1016/j.epsl.2017.11.032

Bézos, A., & Humler, E. (2005). The $Fe^{3+}/\sum Fe$ ratios of MORB glasses and their implications for mantle melting. *Geochimica et Cosmochimica Acta*, *69*, 3, 711–725. doi: 10.1016/j.gca.2004.07.026

Borisov, A., Behrens, H., & Holtz, F. (2015). Effects of melt composition on Fe^{3+}/Fe^{2+} in silicate melts: a step to model ferric/ferrous ratio in multicomponent systems. *Contributions to Mineralogy and Petrology*, *169*, 2. 10.1007/s00410-015-1119-6

Borisov, A., Behrens, H., & Holtz, F. (2017). Effects of strong network modifiers on Fe^{3+}/Fe^{2+} in silicate melts: an experimental study. *Contributions to Mineralogy and Petrology*, *172*, 5. https://doi.org/10.1007/s00410-017-1337-1

Borisov, A., Behrens, H., & Holtz, F. (2018). Ferric/ferrous ratio in silicate melts: a new model for 1 atm data with special emphasis on the effects of melt composition. *Contributions to Mineralogy and Petrology*, *173*, 12, 98. https://doi.org/10.1007/s00410-018-1524-8

Borisov, A., & McCammon, C. (2010). The effect of silica on ferric/ferrous ratio in silicate melts: An experimental study using Mossbauer spectroscopy. *American Mineralogist*, *95*, 4, 545–555. https://doi.org/10.2138/am.2010.3217

Bottinga, Y., & Weill, D. F. (1970). Densities of liquid silicate systems calculated from partial molar volumes of oxide components. *American Journal of Science*, *269*, 2, 169–182. doi: 10.2475/ajs.269.2.169

Bottinga, Y., Weill, D. F., & Richet, P. (1982). Density calculations for silicate liquids. I. Revised method for aluminosilicate compositions. *Geochimica et Cosmochimica Acta*, *46*, 6, 909–919. https://doi.org/10.1016/0016-7037(82)90047-3

Brounce, M. N., Kelley, K. A., & Cottrell, E. (2014). Variations in $Fe^{3+}/\sum Fe$ of Mariana arc basalts and mantle wedge f_{O_2}. *Journal of Petrology*, *55*, 12, 2513–2536. https://doi.org/10.1093/petrology/egu065

Burgisser, A., & Scaillet, B. (2007). Redox evolution of a degassing magma rising to the surface. *Nature*, *445*, 7124, 194–197. https://doi.org/10.1038/nature05509

Calas, G., & Petiau, J. (1983). Coordination of iron in oxide glasses through high-resolution K-edge spectra: Information from the pre-edge. *Solid State Communications*, *48*, 7, 625–629. https://doi.org/10.1016/0038-1098(83)90530-6

Carmichael, I. S. E. (1991). The redox states of basic and silicic magmas: a reflection of their source regions? *Contributions to Mineralogy and Petrology*, *106*, 2, 129–141. https://doi.org/10.1007/BF00306429

Chevrel, M. O., Giordano, D., Potuzak, M., Courtial, P., & Dingwell, D. B. (2013). Physical properties of $CaAl_2Si_2O_8$–$CaMgSi_2O_6$–FeO–Fe_2O_3 melts: Analogues for extra-terrestrial basalt. *Chemical Geology*, *346*, 93–105. https://doi.org/10.1016/j.chemgeo.2012.09.004

Cicconi, M. R., Giuli, G., Ertel-Ingrisch, W., Paris, E., & Dingwell, D. B. (2015). The effect of the [Na/(Na+K)] ratio on Fe speciation in phonolitic glasses. *American Mineralogist*, *100*, 7, 1610–1619. https://doi.org/10.2138/am-2015-5155

Del Gaudio, P., Ventura, G., & Taddeucci, J. (2013). The effect of particle size on the rheology of liquid-solid mixtures with application to lava flows: Results from analogue experiments. *Geochemistry Geophysics, Geosystems*, *14*, 8, 2661–2669. doi: 10.1002/ggge.20172

Densem, N. E., & Turner, W. E. S. (1938). Equilibrium Between Ferrous and Ferrous Oxides in Glasses. *Journal of the Society of Glass Technology*, *22*, 372–389.

Di Genova, D., Kolzenburg, S., Wiesmaier, S., Dallanave, E., Neuville, D. R., Hess, K. U., & Dingwell, D. B. (2017). A compositional tipping point governing the mobilization and eruption style of rhyolitic magma. *Nature*, *552*, 7684, 235–238. https://doi.org/10.1038/nature24488

Di Genova, D., Vasseur, J., Hess, K.-U., Neuville, D. R., & Dingwell, D. B. (2017). Effect of oxygen fugacity on the glass transition viscosity and structure of silica- and iron-rich magmatic melts. *Journal of Non-Crystalline Solids*, *470*, 7885. doi: 10.1016/j.jnoncrysol.2017.05.013

Dickenson, M. P., & Hess, P. C. (1982). Redox equilibria and the structural role of iron in alumino-silicate melts. *Contributions to Mineralogy and Petrology*, *78*, 3, 352–357. https://doi.org/10.1007/BF00398931

Dingwell, D. B., Brearley, M., & Virgo, D. (1988). The dual role of ferric iron in liquid silicates: Effects on density & viscosity. *Chemical Geology*, *70*, 1–2, 86. doi: 10.1016/0009-2541(88)90442-1

Dingwell, D. B. (1989). Shear viscosities of ferrosilicate liquids. *American Mineralogist*, *74*, 9–10, 1038–1044.

Dingwell, D. B. (1991). Redox viscometry of some Fe-bearing silicate melts. *American Mineralogist*, *76*, 9–10, 1560–1562.

Dingwell, D. B., Brearley, M., & Dickinson, J. E. (1988). Melt densities in the Na_2O-FeO-Fe_2O_3-SiO_2 system and the partial molar volume of tetrahedrally-coordinated ferric iron in silicate melts. *Geochimica et Cosmochimica Acta*, *52*, 10, 2467–2475. https://doi.org/10.1016/0016-7037(88)90305-5

Dingwell, D. B., & Virgo, D. (1987). The effect of oxidation state on the viscosity of melts in the system Na_2O-FeO-Fe_2O_3-SiO_2. *Geochimica et Cosmochimica Acta*, *51*, 2, 195–205. https://doi.org/10.1016/0016-7037(87)90231-6

Dingwell, D. B., & Virgo, D. (1988). Viscosities of melts in the Na_2O- FeO- Fe_2O_3- SiO_2 system and factors controlling relative viscosities of fully polymerized silicate melts. *Geochimica et Cosmochimica Acta*, *52*, 2, 395–403. https://doi.org/10.1016/0016-7037(88)90095-6

Drickamer, H. G., Bastron, V. C., Fisher, D. C., & Grenoble, D. C. (1970). The high-pressure chemistry of iron. *Journal of Solid State Chemistry*, *2*, 1, 94–104. https://doi.org/10.1016/0022-4596(70)90038-1

Duan, X. (2014). A model for calculating the viscosity of natural iron-bearing silicate melts over a wide range of temperatures pressures, oxygen fugacites, and compositions. *American Mineralogist*, *99*, 11–12, 2378–2388. https://doi.org/10.2138/am-2014-4841

Duffy, J. A. (1993). A review of optical basicity and its applications to oxidic systems. *Geochimica et Cosmochimica Acta*, *57*, 16, 3961–3970. https://doi.org/10.1016/0016-7037(93)90346-X

Duffy, J. A., & Ingram, M. D. (1975). Optical basicity – IV: Influence of electronegativity on the Lewis basicity and solvent properties of molten oxyanion salts and glasses. *Journal of Inorganic and Nuclear Chemistry*, *37*, 1203–1206. https://doi.org/10.1016/0022-1902(75)80469-6

Dupree, R., Holland, D., & Williams, D. (1986). The structure of binary alkali silicate glasses. *Journal of Non-Crystalline Solids*, *81*, 1–2, 185–200. https://doi.org/10.1016/0022-3093(86)90269-3

Dyar, M. D. (1985). A review of Mössbauer data on inorganic glasses; the effects of composition on iron valency and coordination. *American Mineralogist*, *70*, 3–4, 304–316.

Farges, F., Lefrère, Y., Rossano, S., Berthereau, A., Calas, G., & Brown, G. E. (2004). The effect of redox state on the local structural environment of iron in silicate glasses: a combined XAFS spectroscopy molecular dynamics, and bond valence study. *Journal of Non-Crystalline Solids*, *344*, 3, 176–188. https://doi.org/10.1016/j.jnoncrysol.2004.07.050

Fincham, C. J. B., & Richardson, F. D. (1954). The Behaviour of Sulphur in Silicate and Aluminate Melts. *Proceedings of the Royal Society A: Mathematical Physical and Engineering Sciences*, *223*, 1152, 40–62.

Flood, H., Förland, T., Sillén, L. G., Linnasalmi, A., & Laukkanen, P. (1947). The acidic and basic properties of oxides. *Acta Chemica Scandinavica*, *1*, 592–604. doi: 10.3891/acta.chem.scand.01-0592

Fraser, D. G. (1975). Activities of trace elements in silicate melts. *Geochimica et Cosmochimica Acta*, *39*, 11, 1525–1530. https://doi.org/10.1016/0016-7037(75)90154-4

Fraser, D. G. (1977). Thermodynamic Properties of Silicate Melts. In *Thermodynamics in Geology* (pp. 301–325). Springer Netherlands.

Fudali, R. F. (1965). Oxygen fugacities of basaltic and andesitic magmas. *Geochimica et Cosmochimica Acta*, *29*, 9, 1063–1075. https://doi.org/10.1016/0016-7037(65)90103-1

Gaillard, F., Pichavant, M., & Scaillet, B. (2003). Experimental determination of activities of FeO and Fe_2O_3 components in hydrous silicic melts under oxidizing conditions. *Geochimica et Cosmochimica Acta*, *67*, 22, 4389–4409. https://doi.org/10.1016/S0016-7037(03)00376-4

Gale, A., Dalton, C. A., Langmuir, C. H., Su, Y., & Schilling, J.-G. (2013). The mean composition of ocean ridge basalts. *Geochemistry Geophysics, Geosystems*, n/a–n/a. https://doi.org/10.1029/2012GC004334

Galoisy, L., Calas, G., & Arrio, M. A. (2001). High-resolution XANES spectra of iron in minerals and glasses: structural information from the pre-edge region. *Chemical Geology*, *174*, 1–3, 307–319. https://doi.org/10.1016/S0009-2541(00)00322-3

Gibbons, R. V., Auasus, T. J., & Rossman, G. R. (1974). A spectrographic interpretation of the shock-produced color change in rhodonite ($MnSiO_3$): The shock-induced reduction. *American Mineralogist*, *59*, 177–182.

Giordano, D., & Dingwell, D. (2003). Viscosity of hydrous Etna basalt: implications for Plinian-style basaltic eruptions. *Bulletin of Volcanology*, *65*, 1, 8–14. https://doi.org/10.1007/s00445-002-0233-2

Giuli, G., Alonso-Mori, R., Cicconi, M. R., Paris, E., Glatzel, P., Eeckhout, S. G., & Scaillet, B. (2012). Effect of alkalis on the Fe oxidation state and local environment in peralkaline rhyolitic glasses. *American Mineralogist*, *97*, 2–3, 468–475. https://doi.org/10.2138/am.2012.3888

Giuli, G., Paris, E., Hess, K.-U., Dingwell, D. B., Cicconi, M. R., Eeckhout, S. G., ... Valenti, P. (2011). XAS determination of the Fe local environment and oxidation state in phonolite glasses. *American Mineralogist*, *96*, 4, 631–636. https://doi.org/10.2138/am.2011.3464

Gonzälez-Garcfa, D., Behrens, H., Petrelli, M., Vetere, F., Morgavi, D., Zhang, C., & Perugini, D. (2017). Water-enhanced interdiffusion of major elements between natural shoshonite and high-k rhyolite melts. *Chemical Geology*, *466*, 86–101. https://doi.org/10.1016/j.chemgeo.2017.05.023

Guillot, B., & Sator, N. (2007). A computer simulation study of natural silicate melts. Part II: High pressure properties. *Geochimica et Cosmochimica Acta*, *71*, 18, 4538–4556. https://doi.org/10.1016/j.gca.2007.05.029

Hannoyer, B., Lenglet, M., Dürr, J., & Cortes, R. (1992). Spectroscopic evidence of octahedral iron (III) in soda-lime silicate glasses. *Journal of Non-Crystalline Solids*, *151*, 3, 209–216. https://doi.org/10.1016/0022-3093(92)90031-E

Hetherington, G., Jack, K. H., & Kennedy, J. C. (1964). The viscosity of vitreous silica. *Physic and Chemistry of Glasses*, *5*, 130–136.

Jackson, W. E., de Leon, J. M., Brown, G. E., Waychunas, G. A., Conradson, S. D., & Combes, J. M. (1993). High-temperature XAS study of Fe_2SiO_4 liquid: Reduced coordination of ferrous iron. *Science*, *262*, 5131, 229–233. doi: 10.1126/science.262.5131.229

Jaupart, C. (1996). Physical models of volcanic eruptions. *Chemical Geology*, *128*, 1–4, 217–227. https://doi.org/10.1016/0009-2541(95)00175-1

Jayasuriya, K. D., O'Neill, H. S. C., Berry, A. J., & Campbell, S. J. (2004). A Mössbauer study of the oxidation state of Fe in silicate melts. *American Mineralogist*, *89*, 11–12, 1597–1609. https://doi.org/10.2138/am-2004-11-1203

Jenner, F. E., & O'Neill, H. S. C. (2012). Analysis of 60 elements in 616 ocean floor basaltic glasses. *Geochemistry Geophysics, Geosystems*, *13*, 2, n/a–n/a. https://doi.org/10.1029/2011GC004009

Jenner, F. E., O'Neill, H. S. C., Arculus, R. J., & Mavrogenes, J. A. (2010). The magnetite crisis in the evolution of arc-related magmas and the initial concentration of Au Ag and Cu. *Journal of Petrology*, *51*, 12, 2445–2464. https://doi.org/10.1093/petrology/egq063

Jephcoat, A., & Olson, P. (1987). Is the inner core of the Earth pure iron? *Nature*, *325*, 6102, 332–335. https://doi.org/10.1038/325332a0

Johnston, W. D. (1964). Oxidation-reduction equilibria in iron-containing glass. *Journal of the American Ceramic Society*, *47*, 4, 198–201. https://doi.org/10.1111/j.1151-2916.1964.tb14392.x

Jørgensen, C. (1962). Chemical Bonding. In: *Absorption spectra and chemical bonding in complexes*. pp. 210–243. Elsevier.

Kilinc, A., Carmichael, I. S. E., Rivers, M. L., & Sack, R. O. (1983). The ferric-ferrous ratio of natural silicate liquids equilibrated in air. *Contributions to Mineralogy and Petrology*, *83*, 1–2, 136–140. https://doi.org/10.1007/BF00373086

Kolzenburg, S., Di Genova, D., Giordano, D., Hess, K., & Dingwell, D. (2018). The effect of oxygen fugacity on the rheological evolution of crystallizing basaltic melts. *Earth and Planetary Science Letters*, *487*, 21–32. https://doi.org/10.1016/j.epsl.2018.01.023

Kopcewicz, B., & Kopcewicz, M. (1992). Seasonal variations of iron concentration in atmospheric aerosols. *Hyperfine Interactions*, *71*, 1–4, 1457–1460. https://doi.org/10.1007/BF02397358

Kress, V. C., & Carmichael, I. S. E. (1991). The compressibility of silicate liquids containing Fe_2O_3 and the effect of composition temperature, oxygen fugacity and pressure on their redox states. *Contributions to Mineralogy and Petrology*, *108*, 1–2, 82–92. https://doi.org/10.1007/BF00307328

Lange, R. A., & Carmichael, I. S. (1987). Densities of Na_2O-K_2O-CaO-MgO-FeO-Fe_2O_3-Al_2O_3-TiO_2-SiO_2 liquids: New measurements and derived partial molar properties. *Geochimica et Cosmochimica Acta*, *51*, 11, 2931–2946. https://doi.org/10.1016/0016-7037(87)90368-1

Lange, R. A., & Carmichael, I. S. E. (1989). Ferric-ferrous equilibria in Na_2O-FeO-Fe_2O_3-SiO_2 melts: Effects of analytical techniques on derived partial molar volumes. *Geochimica et Cosmochimica Acta*, *53*, 9, 2195–2204. https://doi.org/10.1016/0016-7037(89)90343-8

Lauer, H. V., & Morris, R. V. (1977). Redox equilibria of multivalent ions in silicate glasses. *Journal of the American Ceramic Society*, *60*, 9–10, 443–451. https://doi.org/10.1111/j.1151-2916.1977.tb15530.x

Le Losq, C., Mysen, B. O., & Cody, G. D. (2015). Water and magmas: insights about the water solution mechanisms in alkali silicate melts from infrared Raman, and ^{29}Si solid-state NMR spectroscopies. *Progress in Earth and Planetary Science*, *2*, 1. https://doi.org/10.1186/s40645-015-0052-7

Le Losq, C., & Neuville, D. R. (2013). Effect of the Na/K mixing on the structure and the rheology of tectosilicate silica- rich melts. *Chemical Geology*, *346*, 57–71. https://doi.org/10.1016/j.chemgeo.2012.09.009

Le Losq, C., Neuville, D. R., Florian, P., Henderson, G. S., & Massiot, D. (2014). The role of Al^{3+} on rheology and structural changes in sodium silicate and aluminosilicate glasses and melts. *Geochimica et Cosmochimica Acta*, *126*, 495–517. https://doi.org/10.1016/j.gca.2013.11.010

Le Losq, C., Neuville, D. R., Moretti, R., Kyle, P. R., & Oppenheimer, C. (2015). Rheology of phonolitic magmas – the case of the Erebus lava lake. *Earth and Planetary Science Letters*, *411*, 53–61. https://doi.org/10.1016/j.epsl.2014.11.042

Lee, S. K., Lin, J.-F., Cai, Y. Q., Hiraoka, N., Eng, P. J., Okuchi, T., … Yoo, C.-S. (2008). X-ray Raman scattering study of $MgSiO_3$ glass at high pressure: Implication for triclustered $MgSiO_3$ melt in Earth's mantle. *Proceedings of the National Academy of Sciences*, *105*, 23, 7925–7929. doi: 10.1073/pnas.0802667105

Le Losq, C. (2012). *Rôle des éléments alcalins et de l'eau sur les propriétés et la structure des aluminosilicates fondus et vitreux: implications volcanologiques* (doctoral dissertation). University Paris VII - Denis Diderot, Paris.

Le Losq, C., Berry, A., Kendrick, M., Neuville, D., & O'Neill, H. (2019). Determination of the oxidation state of iron in mid-ocean ridge basalt glasses by raman spectroscopy. *American Mineralogist*, *104*, 1032–1049. https://doi.org/10.2138/am-2019-6887

Le Losq, C., & Neuville, D. R. (2017). Molecular structure, configurational entropy and viscosity of silicate melts: Link through the adam and gibbs theory of viscous flow. *Journal of Non-Crystalline Solids*, *463*, 175–188. https://doi.org/10.1016/j.jnoncrysol.2017.02.010

Liebske, C., Behrens, H., Holtz, F., & Lange, R. A. (2003). The influence of pressure and composition on the viscosity of andesitic melts. *Geochimica et Cosmochimica Acta*, *67*, 3, 473–485. https://doi.org/10.1016/S0016-7037(02)01139-0

Linard, Y., & Neuville, D. R. (2000). IPGP database. Paris, France.

Liu, Q., & Lange, R. A. (2006). The partial molar volume of Fe_2O_3 in alkali silicate melts: Evidence for an average Fe^{3+} coordination number near five. *American Mineralogist*, *91*, 2–3, 385–393. https://doi.org/10.2138/am.2006.1902

McDonough, W. F., & Sun, S-s. (1995). The composition of the Earth. *Chemical Geology*, *120*, 3–4, 223–253. https://doi.org/10.1016/0009-2541(94)00140-4

Métrich, N., Berry, A. J., O'Neill, H. S. C., & Susini, J. (2009). The oxidation state of sulfur in synthetic and natural glasses determined by X-ray absorption spectroscopy. *Geochimica et Cosmochimica Acta*, *73*, 8, 2382–2399. https://doi.org/10.1016/j.gca.2009.01.025

Métrich, N., Susini, J., Foy, E., Farges, F., Massare, D., Sylla, L., et al. (2006). Redox state of iron in peralkaline rhyolitic glass/melt: X-ray absorption micro-spectroscopy experiments at high temperature. *Chemical Geology*, *231*, 4, 350–363. https://doi.org/10.1016/j.chemgeo.2006.02.001

Mo, X. (1982). The partial molar volume of Fe_2O_3 in multicomponent silicate liquids and the pressure dependence of oxygen fugacity in magmas. *Mineralogical Magazine*, *45*, 337, 237–245.

Moretti, R. (2005). Polymerisation, basicity, oxidation state and their role in ionic modelling of silicate melts. *Annals of Geophysics*, *48*, 4–5. https://doi.org/10.4401/ag-3221

Moretti, R., & Ottonello, G. (2003). Polymerization and disproportionation of iron and sulfur in silicate melts: insights from an optical basicity-based approach. *Journal of Non-Crystalline Solids*, *323*, 1–3, 111–119. https://doi.org/10.1016/S0022-3093(03)00297-7

Moretti, R., & Papale, P. (2004). On the oxidation state and volatile behavior in multicomponent gas–melt equilibria. *Chemical Geology*, *213*, 1–3, 265–280. https://doi.org/10.1016/j.chemgeo.2004.08.048

Moussallam, Y., Oppenheimer, C., Scaillet, B., Gaillard, F., Kyle, P., Peters, N., et al. (2014). Tracking the changing oxidation state of erebus magmas, from mantle to surface, driven by magma ascent and degassing. *Earth and Planetary Science Letters*, *393*, 200–209. https://doi.org/10.1016/j.epsl.2014.02.055

Mysen, B. O. (1987). Redox equilibria and coordination of Fe^{2+} and Fe^{3+} in silicate glasses from ^{57}Fe mossbauer spectroscopy. *Journal of Non-Crystalline Solids*, *95–96*, 247–254. https://doi.org/10.1016/S0022-3093(87)80117-5

Mysen, B. O. (2006a). Redox equilibria of iron and silicate melt structure: Implications for olivine/melt element partitioning. *Geochimica et Cosmochimica Acta*, *70*, 12, 3121–3138. doi: 10.1016/j.gca.2006.03.014

Mysen, B. O. (2006b). The structural behavior of ferric and ferrous iron in aluminosilicate glass near meta-aluminosilicate joins. *Geochimica et Cosmochimica Acta*, *70*, 9, 2337–2353. https://doi.org/10.1016/j.gca.2006.01.026

Mysen, B. O., Virgo, D., Neumann, E.-R., & Seifert, F. A. (1985). Redox equilibria and the structural states of ferric and ferrous iron in melts in the system CaO-MgO-Al_2O_3-SiO_2-Fe-O; relationships between redox equilibria, melt structure and liquidus phase equilibria. *American Mineralogist*, *70*, 3–4, 317–331.

Mysen, B. O., Virgo, D., & Seifert, F. A. (1984). Redox equilibria of iron in alkaline earth silicate melts; relationships between melt structure, oxygen fugacity, temperature and properties of iron-bearing silicate liquids. *American Mineralogist*, *69*, 834–847.

Neuville, D. R. (2006). Viscosity structure and mixing in (Ca, Na) silicate melts. *Chemical Geology*, *229*, 1–3, 28–41. https://doi.org/10.1016/j.chemgeo.2006.01.008

Neuville, D. R., Cormier, L., Flank, A.-M., Briois, V., & Massiot, D. (2004). Al speciation and ca environment in calcium aluminosilicate glasses and crystals by al and ca k-edge x-ray absorption spectroscopy. *Chemical Geology*, *213*, 1–3, 153–163. https://doi.org/10.1016/j.chemgeo.2004.08.039

Neuville, D. R., Courtial, P., Dingwell, D. B., & Richet, P. (1993). Thermodynamic and rheological properties of rhyolite and andesite melts. *Contributions to Mineralogy and Petrology*, *113*, 4, 572–581. https://doi.org/10.1007/BF00698324

Neuville, D. R., & Richet, P. (1991). Viscosity and mixing in molten (Ca Mg) pyroxenes and garnets. *Geochimica et Cosmochimica Acta*, *55*, 4, 1011–1019. https://doi.org/10.1016/0016-7037(91)90159-3

Nikolaev, G. S., Borisov, A. A., & Ariskin, A. A. (1996). Calculation of the ferric–ferrous ratio in magmatic melts: Testing and additional calibration of empirical equations for various magmatic series. *Geochemistry International*, *34*, 8, 9.

O'Neill, H. S. C. (2006). An experimental determination of the effect of pressure on the $Fe^{3+}/\sum Fe$ ratio of an anhydrous silicate melt to 3.0 GPa. *American Mineralogist*, *91*, 2–3, 404–412. https://doi.org/10.2138/am.2005.1929

Oppenheimer, C., Moretti, R., Kyle, P. R., Eschenbacher, A., Lowenstern, J. B., Hervig, R. L., & Dunbar, N. W. (2011). Mantle to surface degassing of alkalic magmas at Erebus volcano Antarctica. *Earth and Planetary Science Letters*, *306*, 3–4, 261–271. https://doi.org/10.1016/j.epsl.2011.04.005

Ottonello, G., Moretti, R., Marini, L., & Zuccolini, M. V. (2001). Oxidation state of iron in silicate glasses and melts: a thermochemical model. *Chemical Geology*, *174*, 1–3, 157–179. https://doi.org/10.1016/S0009-2541(00)00314-4

Palme, H., & O'Neill, H. S. C. (2014). Cosmochemical estimates of mantle composition. In: *Treatise on Geochemistry*, pp. 1–39. Elsevier.

Paul, A., & Douglas, R. W. (1965). Ferrous-ferric equilibrium in binary alkali silicate glasses. *Physics and Chemistry of Glasses*, *6*, 207.

Pistone, M., Caricchi, L., Ulmer, P., Burlini, L., Ardia, P., Reusser, E., et al. (2012). Deformation experiments of bubble- and crystal-bearing magmas: Rheological and microstructural analysis. *Journal of Geophysical Research: Solid Earth*, *117*, B5, n/a-n/a. doi: 10.1029/2011JB008986

Richet, P., & Neuville, D. R. (1992). Thermodynamics of Silicate Melts: Configurational Properties. In: *Thermodynamic Data*. pp. 132–161. Springer New York.

Rossano, S., Behrens, H., & Wilke, M. (2007). Advanced analyses of ^{57}Fe Mössbauer data of alumino-silicate glasses. *Physics and Chemistry of Minerals*, *35*, 2, 77–93. https://doi.org/10.1007/s00269-007-0200-8

Sack, R. O., Carmichael, I. S. E., Rivers, M., & Ghiorso, M. S. (1981). Ferric-ferrous equilibria in natural silicate liquids at 1 bar. *Contributions to Mineralogy and Petrology*, *75*, 4, 369–376. https://doi.org/10.1007/BF00374720

Sanloup, C. (2016). Density of magmas at depth. *Chemical Geology*, *429*, 51–59. https://doi.org/10.1016/j.chemgeo.2016.03.002

Sanloup, C., Drewitt, J. W. E., Crépisson, C., Kono, Y., Park, C., McCammon, C., et al. (2013). Structure and density of molten fayalite at high pressure. *Geochimica et Cosmochimica Acta*, *118*, 118–128. https://doi.org/10.1016/j.gca.2013.05.012

Sanloup, C., Drewitt, J. W. E., Konôpková, Z., Dalladay-Simpson, P., Morton, D. M., Rai, N., et al. (2013). Structural change in molten basalt at deep mantle conditions. *Nature*, *503*, 7474, 104–107. https://doi.org/10.1038/nature12668

Sarpoolaky, H., Zhang, S., & Lee, W. E. (2003). Corrosion of high alumina and near stoichiometric spinels in iron- containing silicate slags. *Journal of the European Ceramic Society*, *23*, 2, 293–300. https://doi.org/10.1016/S0955-2219(02)00170-X

Schreiber, H. D. (1980). Properties of redox ions in glasses: An interdisciplinary perspective. *Journal of Non-Crystalline Solids*, *42*, 1–3, 175–183. https://doi.org/10.1016/0022-3093(80)90019-8

Schreiber, H. D. (1986). Redox processes in glass-forming melts. *Journal of Non-Crystalline Solids*, *84*, 1–3, 129–141. https://doi.org/10.1016/0022-3093(86)90770-2

Schreiber, H. D. (1987). An electrochemical series of redox couples in silicate melts: A review and applications to geochemistry. *Journal of Geophysical Research*, *92*, B9, 9225. https://doi.org/10.1029/JB092iB09p09225

Schreiber, H. D., Kochanowski, B. K., Schreiber, C. W., Morgan, A. B., Coolbaugh, M., & Dunlap, T. G. (1994). Compositional dependence of redox equilibria in sodium silicate glasses. *Journal of Non-Crystalline Solids*, *177*, 340–346. https://doi.org/10.1016/0022-3093(94)90548-7

Shiraishi, Y., Ikeda, K., Tamura, A., & Saito, T. (1978). On the viscosity and density of the molten FeO–SiO_2 system. *Transactions of the Japan Institute of Metals*, *19*, 5, 264–274. https://doi.org/10.2320/matertrans1960.19.264

Sossi, P., Nebel, O., Anand, M., & Poitrasson, F. (2016). On the iron isotope composition of mars and volatile depletion in the terrestrial planets. *Earth and Planetary Science Letters*, *449*, 360–371. https://doi.org/10.1016/j.epsl.2016.05.030

Sun, N., Stixrude, L., de Koker, N., & Karki, B. B. (2011). First principles molecular dynamics simulations of diopside ($CaMgSi_2O_6$) liquid to high pressure. *Geochimica et Cosmochimica Acta*, *75*, 13, 3792–3802. https://doi.org/10.1016/j.gca.2011.04.004

Tangeman, J. A., Lange, R., & Forman, L. (2001). Ferric-ferrous equilibria in K_2O-FeO-Fe_2O_3-SiO_2 melts. *Geochimica et Cosmochimica Acta*, *65*, 11, 1809–1819. https://doi.org/10.1016/S0016-7037(00)00561-5

Urbain, G., Bottinga, Y., & Richet, P. (1982). Viscosity of liquid silica, silicates and alumino-silicates. *Geochimica et Cosmochimica Acta*, *46*, 6, 1061–1072. https://doi.org/10.1016/0016-7037(82)90059-X

Vetere, F., Behrens, H., Holtz, F., & Neuville, D. R. (2006). Viscosity of andesitic melts—new experimental data and a revised calculation model. *Chemical Geology*, *228*, 4, 233–245. https://doi.org/10.1016/j.chemgeo.2005.10.009

Vetere, F., Behrens, H., Schuessler, J. A., Holtz, F., Misiti, V., & Borchers, L. (2008). Viscosity of andesite melts and its implication for magma mixing prior to Unzen 1991–1995 eruption. *Journal of Volcanology and Geothermal Research*, *175*, 1–2, 208–217. https://doi.org/10.1016/j.jvolgeores.2008.03.028

Villeneuve, N., Neuville, D. R., Boivin, P., Bachèlery, P., & Richet, P. (2008). Magma crystallization and viscosity: A study of molten basalts from the Piton de la Fournaise volcano (La Réunion island). *Chemical Geology*, *256*, 3–4, 242–251. https://doi.org/10.1016/j.chemgeo.2008.06.039

Virgo, D., & Mysen, B. O. (1985). The structural state of iron in oxidized vs. reduced glasses at 1 atm: A ^{57}Fe Mossbauer study. *Physics and Chemistry of Minerals*, *12*, 2, 65–76. doi: 10.1007/BF01046829

Wang, P., & Drickamer, H. G. (1973). Reduction of Cu(II) at high pressure. *The Journal of Chemical Physics*, *59*, 2, 713–717. https://doi.org/10.1063/1.1680080

Wang, Y., Sakamaki, T., Skinner, L. B., Jing, Z., Yu, T., Kono, Y., et al. (2014). Atomistic insight into viscosity and density of silicate melts under pressure. *Nature Communications*, *5*, 1. https://doi.org/10.1038/ncomms4241

Waychunas, G. A., Brown, G. E., Ponader, C. W., & Jackson, W. E. (1988). Evidence from X-ray absorption for network-forming Fe^{2+} in molten alkali silicates. *Nature*, *332*, 6161, 251–253. https://doi.org/10.1038/332251a0

Whittaker, E. J. W., & Muntus, R. (1970). Ionic radii for use in geochemistry. *Geochimica et Cosmochimica Acta*, *34*, 9, 945–956. https://doi.org/10.1016/0016-7037(70)90077-3

Whittington, A., Richet, P., & Holtz, F. (2000). Water and the viscosity of depolymerized aluminosilicate melts. *Geochimica et Cosmochimica Acta*, *64*, 21, 3725–3736. https://doi.org/10.1016/S0016-7037(00)00448-8

Wilke, M., Farges, F., Partzsch, G. M., Schmidt, C., & Behrens, H. (2007). Speciation of Fe in silicate glasses and melts by in-situ XANES spectroscopy. *American Mineralogist*, *92*, 1, 44–56. https://doi.org/10.2138/am.2007.1976

Yarger, J. L., Smith, K. H., Nieman, R. A., Diefenbacher, J., Wolf, G. H., Poe, B. T., & McMillan, P. F. (1995). Al Coordination Changes in High-Pressure Aluminosilicate Liquids. *Science*, *270*, 5244, 1964–1967. 10.1126/science.270.5244.1964

Zhang, H., Cottrell, E., Solheid, P. A., Kelley, K. A., & Hirschmann, M. M. (2018). Determination of $Fe^{3+}/\sum Fe$ of XANES basaltic glass standards by Mössbauer spectroscopy and its application to the oxidation state of iron in MORB. *Chemical Geology*, *479*, 166–175. https://doi.org/10.1016/j.chemgeo.2018.01.006

Zhang, H., Hirschmann, M. M., Cottrell, E., & Withers, A. C. (2017). Effect of pressure on $Fe^{3+}/\sum Fe$ Fe ratio in a mafic magma and consequences for magma ocean redox gradients. *Geochimica et Cosmochimica Acta*, *204*, 83–103. https://doi.org/10.1016/j.gca.2017.01.023

Zhang, Y., Ni, H., & Chen, Y. (2010). Diffusion data in silicate melts. *Reviews in Mineralogy and Geochemistry*, *72*, 1, 311–408. doi: 10.2138/rmg.2010.72.8

Part III

Tools and Techniques to Characterize the Redox and its Effect on Isotope Partitioning

13

How to Measure the Oxidation State of Multivalent Elements in Minerals, Glasses, and Melts?

Daniel R. Neuville[1], **Maria Rita Cicconi**[1,2], **and Charles Le Losq**[1,3]

ABSTRACT

All natural materials and most manufactured technologically relevant materials contain a variable but consequent amount of multivalent elements, like iron. Many physicochemical properties of materials highly depend on the type of multivalent cations they contain, as well as on the concentrations and redox states of those multivalent elements. For instance, color is probably the most visible characteristic, resulting from the presence of multivalent transition elements. Other properties like viscosity and density of liquids are affected by those elements, and, as such, they can be seen as important variables for geologic and industrial phenomena. The present contribution focuses on providing an overview of the most common techniques currently in use for the investigation of multivalent elements in natural and industrial materials. It covers both destructive and non-destructive methods, as well as element-specific and bulk techniques.

13.1. INTRODUCTION

Nowadays, several techniques are available for material and elemental characterizations, and, sometimes, it is difficult to decide which method is the most appropriate to obtain the often desired and always necessary information. In most cases, the use of various different methods will provide the necessary information required to understand the connections between macro- and micro-scale properties. In this framework, measuring the redox potential and in turn assessing the redox chemical equilibria of multivalent elements is critical for understanding chemical processes. In this short, non-exhaustive chapter, we will present traditional techniques for measuring the redox ratio of an element, such as wet chemistry analyses, electron microprobe, Mössbauer spectroscopy, and optical spectroscopy. We will also present new developments that now allow performing in situ measurements, like X-ray absorption spectroscopy and Raman spectrometry. We will mention the advantages, disadvantages, and the accuracy of each technique, and provide some advices for their use on this basis. For a detailed review of each of the hereabove mentioned techniques, readers are referred to, e.g., Rossman (1988, 2014), Mottana (2004), Wildner et al. (2004), Henderson et al. (2014), Neuville et al. (2014), and de Ligny and Möncke (2019).

[1] *Université de Paris, Institut de Physique du Globe de Paris, Paris, France*

[2] *Department of Materials Science and Engineering, Friedrich-Alexander-Universität Erlangen-Nürnberg, Erlangen, Germany*

[3] *Research School of Earth Sciences, Australian National University, Canberra, ACT, Australia*

Magma Redox Geochemistry, Geophysical Monograph 266, First Edition.
Edited by Roberto Moretti and Daniel R. Neuville.

DOI: 10.1002/9781119473206.ch13

13.2. WET-CHEMICAL ANALYSES

Wet-chemical analyses are an old and classic method. Almost all classic chemistry textbooks (e.g., Pascal, 1958; Charlot, 1969) mention different recipes for analyzing chemical composition of, and the average oxidation state of, multivalent elements in minerals and bulk rock samples. Wet-chemical analyses can be done on liquids or solids (crystal or glass). They are based on the measurement of the concentration of a species in a solution with oxido-reduction properties. The dosing reaction must be:

- quantitative, i.e., the difference of redox potential between the two pairs must be adequate;
- unique, i.e., there should not be other species able to react with the titrant solution or the species formed;
- and fast, to be sure that no other reactions can occur.

At equivalence, the reagents are mixed in stoichiometric proportions, i.e., the number of electrons that can be captured by the oxidant is equal to the number of electrons that may be transferred from the reducer. The detection of equivalence can be achieved in different ways:

- one of the reagents or one of the products is colored (MnO_4^-, I_2...) and the appearance of the color of the product or the disappearance of the color of the reagent indicates equivalence;
- monitoring the evolution of the potential of oxido-reduction couples, the equivalence of which translates into a potential jump;
- using oxido-reduction indicator, which changes color at a defined potential. We can differentiate two sorts of indicators: (i) general redox indicators like substances that change color when they are in the oxidized or reduced forms; (ii) specific indicators that is a new colored chemical species formed as result of the association between the indicator group and the oxidized or reduced species.

In the rest of this section, we will focus on a method, adapted from Wilson (1960), which is fast and easy to perform, and allows analyzing iron in glasses and minerals with a small amount of starting material. It is important that the analysed material is homogeneous, and is not composed of a mixture of glass and crystal or of different minerals, because otherwise the result will be a convolution of the different redox states of the different phases.

This method is composed of two steps. The first step consists in the determination of Fe^{2+}, while the second one consists in measuring the total quantity of iron.

Step 1. Determination of Fe^{2+} (expressed as FeO) by volumetric measurement. It consists in, first, performing a dissolution of the sample in acids, and then measuring the concentration of FeO in the newly formed solution with a titrated solution of $K_2Cr_2O_7$. The recipe is:

- in a platinum crucible, add 0.5 g of crushed glass with 7 ml of sulfuric acid, H_2SO_4, and 3 ml of hydrofluoric acid, HF;
- place the crucible on a Bunsen burner, and maintain boiling for 3 minutes;
- pour the contents of the crucible and the rinsing water into a beaker containing 1 g of boric acid H_3BO_3, 10 ml of sulfuric acid H_2SO_4 diluted to 50 % by volume, and 5 ml of orthophosphoric acid H_3PO_4. Boric acid will neutralize the HF in the solution, while sulfuric acid makes our solution more acidic. Phosphoric acid will form a complex with iron;
- add a few drops of colored indicator. In the present recipe we use diphenylamine barium sulfonate. The solution is then dosed by gradually adding a titrated solution of dichromate potassium $K_2Cr_2O_7$ and identifying the shift of the color of the solution to purple (provided by the indicator).

The ferrous iron content present in the solution is deduced from the volume, V, of $K_2Cr_2O_7$ required to obtain the purple color shift and the mass, m, of dissolved glass:

$$\%FeO = 100*2V/m*M_{FeO}/M_{Fe} \qquad (13.1)$$

where M_{FeO} and M_{Fe} are the molar mass of FeO and Fe. This gives the final equation:

$$\%FeO = 257.3*V/m. \qquad (13.2)$$

Step 2. Determination of total iron (expressed as Fe_2O_3) by volumetric measurement. This step consists in the reduction and the determination of all the iron in the sample using a titrated solution of $K_2Cr_2O_7$. The recipe is:

- in a platinum crucible, mix 1 g of sample, 40 ml of HF, and 5 ml of H_2SO_4;
- heat the solution on a Bunsen burner until total evaporation;
- add 2 g of boric acid H_3BO_3 and 5 ml of H_2SO_4 to the residues;
- heat the solution again for about 30 minutes;
- pour the contents of the crucible and the rinsing water into a beaker. Keep the solution
- boiling until it clears up;
- reduce the volume of solution to 500 ml with distilled water. Take 50 ml of solution (~ 0.1 g of sample) and pour into a beaker;
- to these 50 ml, add 5 ml of sulfuric acid, 2 or 3 drops of arsenio-tungstate, $(AsO_4\ (WO_3H_3)_9)_2$, a few drops of $TiCl_3$ until the solution's color shifts to blue;
- pour potassium dichromate until the color fades, which corresponds to complete oxidization of $TiCl_3$;
- add 5 ml of acid phosphoric, H_3PO_4, and a few drops of barium diphenylamine sulfonate, $C_{24}H_{20}BaN_2O_6S_2$;
- this solution is dosed with potassium dichromate, $K_2Cr_2O_7$.

The total iron content present is deduced from the volume V of potassium dichromate necessary to observe the color shift, and from the mass m of initial glass/mineral via the equation:

$$Fe_{tot} = 20V/m*M_{Fe2O3}/2M_{Fe}, \quad (13.3)$$

where M_{Fe2O3} and M_{Fe} are the molar mass of Fe_2O_3 and Fe, respectively. This yields the final equation:

$$Fe_{tot} = 2859.3*V/m. \quad (13.4)$$

Thus, knowing the total iron content (expressed as Fe_2O_3), the actual Fe_2O_3 content of the sample can be calculated as:

$$\%Fe_2O_3 = Fe_{tot} - \%FeO*M_{Fe2O3}/2M_{FeO}, \quad (13.5)$$

which corresponds to:

$$\%Fe_2O_3 = Fe_{tot} - 1.1113*\%FeO. \quad (13.6)$$

Advantages, drawbacks, accuracy

Wet chemistry is an old technique with the following advantages:

- it is an absolute method;
- it is very accurate technique, generally more than ±0.05, and in the case of iron redox ± 0.02 on the Fe^{3+}/Fe^{2+};
- it can be used to analyze the concentration and oxidation state of a wide variety of chemical elements.

However, it presents some major drawbacks, preventing its widespread adoption:

- it requires handling potentially dangerous and/or toxic reagents or acids;
- it usually requires a large amount of sample (more than 1 gram);
- it is destructive;
- of course, it is not possible to map the redox ratio of an element in a sample.

13.3. ELECTRONIC MICROPROBE

Electron microprobe, like wet chemistry, allows analyzing a sample chemical composition, with the exception of light elements. In addition to major element composition, several studies also focused on developing this method for the analysis of the oxidation state of iron in minerals and glasses.

The first tests to determine the iron redox ratio with electron microprobe date back to the 1960s. In particular, the works of Fisher (1965) and Albee and Chodos (1970) can be mentioned. Iron has unfilled 3d states ($3d^5$ for Fe^{3+} and $3d^6$ for Fe^{2+} unlike $3d^{10}$ for the filled level). The peaks of the L series emitted from the metals of first order transition are the result of electronic transitions between the 3d orbitals partially filled and the 2p core levels ($L_{2,3}$). Different methods are based on the analysis of the Fe $L_{\alpha,\beta}$ emission spectra of minerals and glasses to measure their Fe^{3+}/Fe^{TOT} (Höfer et al., 1994; Fialin et al., 2001, 2004; Zhang et al., 2018a; Almeev, et al., 2018; Hughes et al., 2018, 2020).

One method for determining Fe^{3+}/Fe^{TOT} is based on the difference in the self-absorption effects on both Fe^{2+} and Fe^{3+} L_α peak positions (Höfer et al., 1994). Self-absorption is the absorption by a chemical species of its own X emissions characteristics. The position, shape, and intensity of the Fe L_α peak all depend strongly on whether Fe in the sample is predominantly Fe^{3+} (low self-absorption) or Fe^{2+} (strong self-absorption). From this point, three approaches allow estimation of the iron redox ratio from the Fe $L_{\alpha,\beta}$ spectra:

- measurement of the energy displacement of the L_α peak (Fialin et al., 2001, 2004);
- measurement of the L_β/L_α intensity ratio (Albee and Chodos, 1970);
- a mixed method developed by Höfer et al. (1994).

Advantages, drawbacks, accuracy

The electron microprobe can be a good tool to measure the Fe redox ratio in glasses because:

- it also allows, in parallel, to perform a full chemical analysis of the material;
- local analysis and mapping are possible;
- it is fast and easy to perform, as in most cases samples will eventually be mounted in epoxy and analyzed by electron microprobe (but this can also be a drawback in some cases, listed here).

However, this method presents various drawbacks:

- accuracy on the Fe^{3+}/Fe^{TOT} redox ratio is of around ± 10 %;
- it requires some sample preparation, including polishing and metallization of epoxy sample mounts;
- samples should contain a maximum of 15 wt% of light elements that tend to migrate under the beam, like sodium or boron;
- good measurements require high iron content, ~3–4 wt% of total iron minimum;
- chemical and redox analyzes of glasses with volatile elements are problematic.

13.4. MÖSSBAUER SPECTROSCOPY

Mössbauer spectroscopy is one of the common tools for determining the redox ratio and the coordination of iron in silicate glasses (Mysen et al., 1980, 1985; Mysen, 1985; Jayasuriya et al., 2004; Zhang et al., 2018b; Cottrell, et al., 2018; Cottrell et al., 2018; Gaborieau et al., 2020). Mössbauer spectroscopy made recently some very important progress through new notable developments on synchrotron light source. Such developments allow measuring the oxidation state of an element *in situ* in a diamond anvil-cell, for example. This can provide very important direct knowledge regarding the iron redox ratio and environment at conditions pertinent for the Earth interior (Potapkin et al., 2012; Cerantola et al., 2017; Maeda et al., 2017; Wei et al. 2017; Cottrell et al., 2018).

Mössbauer spectroscopy is an element-specific technique. It is based on the Mössbauer effect, named after R. L. Mössbauer who received the Nobel prize in 1961 for his discovery. The Mössbauer effect is a phenomenon of nuclear resonance that affects specific isotopes (^{57}Fe, ^{119}Sn, ^{121}Sb, ^{125}Eu). It is apparent in the gamma ray region of the electromagnetic spectrum. A radioactive source emits γ radiation following de-excitation mechanisms of the nucleus between its excited state and its ground state. The gamma radiation emitted is then absorbed by the sample. It probes tiny changes in the energy levels of an atomic nucleus in response to its environment (electric, magnetic, chemical, redox, ligand, etc.). Typically, three types of nuclear interactions may be observed: (i) isomer shift, also called chemical shift in the older literature; (ii) quadrupole splitting; and (iii) magnetic hyperfine splitting. Observation of those phenomena provides information on the charge, spin, and local symmetry of the probed atom, as well as on its internal electric and magnetic fields.

The sample emits a modulated signal, which represents itself as a displacement in energy of the absorption band of the spectrum with respect to the emission energy. The energy displacement of the absorption line with respect to the emission line is named isomer displacement (δ or IS). As it is related to the electronic density of the nucleus, δ is an indicator of its valence. On the other hand, the interaction of the quadrupole moment of the atomic nucleus with the aspherical distribution of electronic charges is presented by a doublet and called quadrupole interaction (Δ or QS). The separation between the two peaks of the doublet is a measure of this interaction. In the case of iron, it is associated with the asymmetry of the environment of the nucleus of iron, and, therefore, informs about the coordination of iron.

An example of a Mössbauer spectrum of iron in a diopside glass with 10 % Fe_2O_3, prepared in slightly reducing conditions at high temperature, is shown in Fig. 13.1. After decomposition of the observed peaks into different components (Lorentzian functions), the relative areas of signals assigned to ferric and ferrous iron are calculated. The redox ratio can be estimated from the ratio of those areas. Spectra of reduced and oxidized samples have an asymmetrical and a symmetrical doublet, respectively. Charts linking isomer displacement and interaction quadrupole provide the coordination of iron.

Advantages, drawbacks, accuracy

Mössbauer spectroscopy has the following advantages:
- it is very accurate, precision is +/– 0.02 Fe^{3+}/Fe^{TOT}
- it is an absolute method;

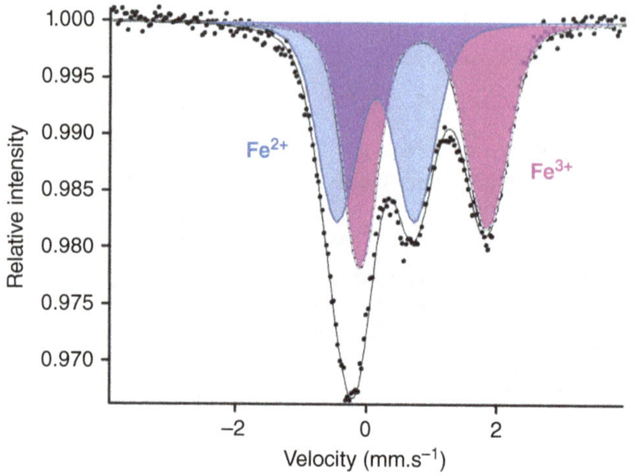

Figure 13.1 Mössbauer spectra of an Fe-bearing diopside glass prepared at 1500 °C, under slightly reducing conditions (log fO_2 = –2.65). Round dots are the experimental measured signal, fitted by the mathematical functions with blue and pink colors, which represent, respectively, the Fe^{2+} and the Fe^{3+} contributions (redraft from Roskosz et al., 2008).

- it does not involve extensive chemical manipulations of the samples;
- thanks to the advances in synchrotron radiation (Potapkin et al. 2012), microbeam are now available and allow in-situ, non-destructive spatially resolved measurements.

However, it also has several drawbacks:
- only a few elements are accessible for redox measurements;
- a large number of samples are needed (excluding synchrotron developments);
- it is a destructive method (excluding synchrotron developments);
- it involves low temperature measurements (excluding synchrotron developments);
- it requires an expensive and time-perishable radioactive source;
- signal decomposition and interpretation may be challenging.

13.5. OPTICAL ABSORPTION SPECTROSCOPY

Optical absorption (OA) techniques allow the identification of electronic transitions in the frequency range 50000 cm^{-1}–2000 cm^{-1}, thus covering the spectral range from ultraviolet (UV) to near-infrared (NIR). It includes the small region that human eyes are able to detect (Visible, VIS) from ~26000 cm^{-1} to ~12800 cm^{-1} (see Fig. 13.2). Specific reviews that describe the basic concepts, the use of this technique in geosciences, or the nature of color in

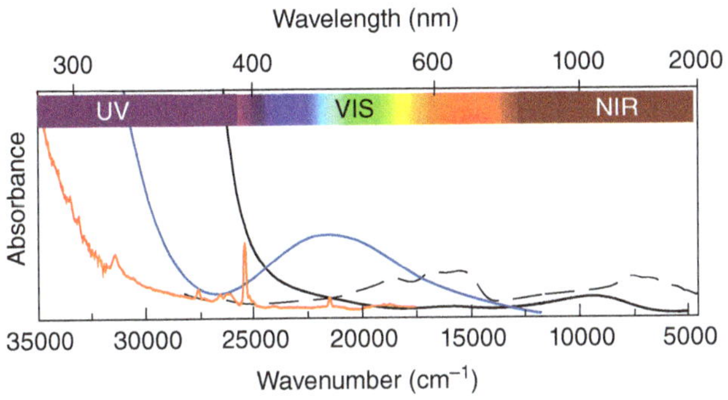

Figure 13.2 Example of Optical Absorption spectra of glasses containing different transition elements: Mn^{3+} doped aluminosilicate glass (blue line); V doped sodium silicate glass (black line); and an archaeological glass containing different multivalent elements (dashed black line). In red, the spectrum of a Eu^{3+}-rich metaphosphate glass (data from Cicconi et al., 2017a). Depending on the absorber(s), bands can be broad or narrow, and occur at different frequencies. On top, the UV–VIS–NIR frequency ranges are reported for reference.

crystalline and amorphous materials are e.g., Wildner et al. (2004), Rossman (2014), and de Ligny and Möncke (2019).

When a radiation interacts with a material, the light can be reflected, scattered, or absorbed. If the outgoing light (with intensity I) has a lower intensity with respect to the incoming beam (with intensity I_0), then the sample has absorbed some of the light. The portion of absorbed light will depend on the thickness of the sample, and on the amount (concentration) of absorbent species. The attenuation of the incident beam intensity I_0, when passing through a sample, is described by the Lambert-Beer equation:

$$I = I_0 \exp(-\mu x), \tag{13.7}$$

with x the thickness of the sample, and μ the linear absorption coefficient that actually depends on the incident wavelength λ. The latter point actually implies that λ intervenes in the Lambert-Beer law described by Equation 13.7 (Wildner et al., 2004).

Using this principle, an UV/VIS or UV/VIS/NIR spectrophotometer measures the intensity I of absorbed or transmitted light as a function of I_0 at each given λ. The ratio $I(\lambda) / I_0(\lambda)$ can be referred to as the transmittance T. In an optically transparent material, $I(\lambda) \sim I_0(\lambda)$ (ignoring reflections), hence the percent transmittance %T will be ~100. On the contrary, a non-transparent material (opaque) will have a %T = 0. The absorbance A is instead defined as:

$$A = \log(I_0(\lambda)/I(\lambda)) = \epsilon C x, \tag{13.8}$$

with C the absorber concentration, x the sample thickness, and ϵ the extinction coefficient. Thus, the amount of light absorbed by a material is directly proportional to the concentration of the absorber, and to the length of the light path (Wildner et al., 2004). We can note that when a photon is absorbed by an atom or molecule it passes from the normally occupied lower energy state, often called the ground state, to an upper or excited state. These transitions are, for example, responsible for colors in minerals and glasses. However, it is important to remark that these electronic transitions are much weaker than charge transfer transitions. For instance, one of the most famous examples is the charge transfer occurring in sapphire, where Ti^{4+} and Fe^{2+} can provide the typical blue color, even if present at ppm levels (Dyar et al., 2008).

An example of optical absorption spectrum, in the range 4000–28500 cm^{-1}, of a blue colored archaeological glass is reported in Fig. 13.3. This glass has many chromophore elements, and the strongest electronic transitions can all be seen in the spectrum, with Co, Mn, Cu, and Fe providing clear absorption bands. Other transitions may be too weak to be visible, such as those associated with 6-fold coordinated Mn^{2+} ions. Indeed, the pinkish/reddish color deriving from $^{[6]}Mn^{2+}$ transitions can be really detected only in the mineral rhodochrosite because of its high Mn^{2+} concentration (Rossman, 1988, 2014). Nevertheless, by combining other optical techniques, i.e., photoluminescence spectroscopy, it could be possible to observe both absorption and emission of Mn^{2+} ions, without interference from the oxidized Mn^{3+} species. Another useful technique that could be associated with the optical ones is EPR spectroscopy (Electron Paramagnetic Resonance). EPR reveals the signal only of species having unpaired electrons (e.g., V^{4+}, Mn^{2+}, Fe^{3+}, Cu^{2+}), thus it allows discriminating many different transition elements (Möncke et al. 2014).

Despite all chromophore elements present in this example, the resulting glass has the typical deep blue color of

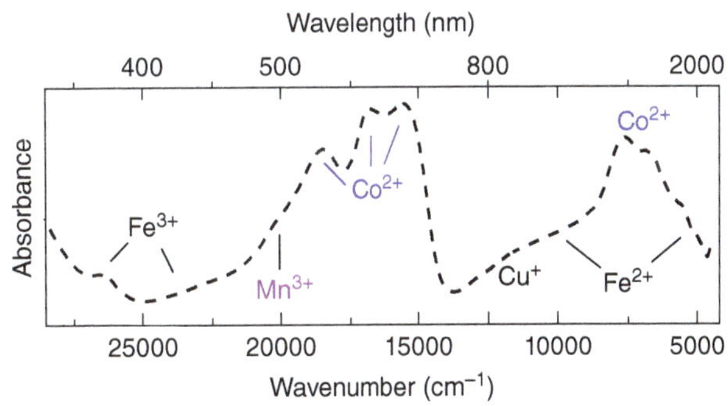

Figure 13.3 Optical absorption spectrum, in the range 4000–28500 cm^{-1}, of a deep blue colored glass. The color is mainly provided by $^{[4]}Co^{2+}$ ions, but several other broad bands are visible as evidence of the presence of other transition elements.

Co^{2+} 4-fold coordinated. All *d–d* transitions are Laporte- or parity-forbidden (Laporte parity selection rule), which reduces the transition probabilities and, in turn, the absorption intensity (intensity of the transitions). However, the Laporte rule can be relaxed (weakened) for non-centrosymmetric geometries, and as a result, high-absorption intensities are associated with the absorption bands of $^{[4]}Co^{2+}$ ions (Möncke et al. 2014; Moncke 2018).

As observed in the previous examples, optical absorption techniques are very sensitive and allow to detect the presence of many elements, even at very dilute levels. However, the quantification of the average oxidation state is very challenging for most of the chemical elements. For example, in the archaeological glass reported in Fig. 13.3, the presence of various broad transitions associated with the different elements introduced in the glass hampers any significative quantification of the redox state (Cicconi et al., 2020). Ergo, often many other techniques are coupled to optical absorption measurements, in order to obtain complementary information (XAS, photoluminescence, XPS, voltammetry, and titration are some examples) (e.g., Leister et al., 1999: Bingham et al., 2002; Möncke et al., 2014; Hunault et al., 2016). Furthermore, even in the case of a simplified bulk composition having a single transition element, the extraction of useful information is not always straightforward, since the appropriate extinction coefficients must be available.

An example of this complexity is reported in Fig. 13.4, where the absorption spectrum of an aluminosilicate glass doped with vanadium is reported. Vanadium in glasses can be stable with three possible valences, V^{5+}, V^{4+}, and V^{3+}, that provide a variety of responses in terms of optical properties (electronic configuration varies from $3d^0$ to $3d^2$) and several bands in the UV-VIS-NIR region. Moreover, V^{5+} species do not provide electronic transitions because of the $3d^0$ electronic configuration. Thus, only charge transfer bands occur that overlap the UV-edge. In order to evaluate the distinct extinction coefficients that are needed to quantify the proportions of the different V species from the absorption spectra, Leister et al. (1999) performed the signal deconvolution of several glasses (an example of signal deconvolution is reported in Fig. 13.4) but also required other techniques. For example, the extinction coefficients of reduced V^{3+} can be obtained from the deconvoluted optical absorption spectra of glasses synthesized at very reducing conditions, where V^{3+} are the only (or predominant) ions present. However, in order to determine the concentration of V^{4+} species and, in turn, to calculate the extinction coefficients for V^{4+}, the authors used other techniques, such as Square-wave voltammetry and Electron Spin Resonance (Leister et al. 1999).

Thus, by using optical absorption, it is possible to quantitatively obtain information on an element oxidation state, along with information on coordination, bonding character and site geometry. Moreover, the high sensitivity of this technique allows detecting species at very low concentrations. Despite that, the use of other spectroscopic or voltammetry techniques is often required for an accurate determination of the proportions of the different redox species.

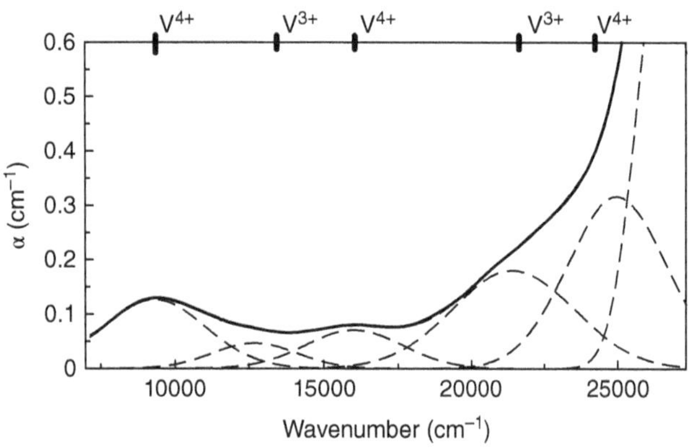

Figure 13.4 Optical Absorption spectrum of a V-doped sodium aluminosilicate glass melted at 1400 °C, in air (1.8 mol% V_2O_5) in the frequency range 7000–27300 cm^{-1}. Several broad bands associated with electronic transitions of V^{4+} and V^{3+} species are visible. V^{5+} charge transfer band(s) are not discernible from the UV-edge. Dashed lines represent the mathematical functions used for signal deconvolution. The proportions of the different redox species can be calculated by using the appropriate extinction coefficient and the area of the deconvoluted bands. Modified after Cicconi et al. (2020).

Advantages, drawbacks, accuracy

Optical Absorption spectroscopy is regularly used in, e.g., analytical chemistry for a qualitative and quantitative determination of different analytes, either in liquid, solid or gas form. In Geosciences, it is widely used to study crystalline and amorphous solids containing transition elements and lanthanides. This technique has several advantages:

• OA has high accuracy and it is a cost-effective method;
• it can provide information not only on the type of species, but also on their local environment (oxidation state, bonding character, coordination);
• OA can be also a quantitative technique, and, in simplified systems, absorber concentration can be derived from the Lambert-Beer's equation;
• it has high sensitivity.

On the other hand, it is important to point out that the measured signals can be very weak, and the signal/noise ratio of the collected signals may be very low. Thus, a good calibration of the instrument is required in order to remove the noise characteristic of the source, and the electronic noise from the detector. Furthermore, for classical transmission measurements, the sample preparation can be time consuming (e.g., optical-grade quality double-sided polishing), and opaque samples must be measured in reflectance mode. Finally, the analysis of natural samples can sometimes be very tricky because of the overlap of many different electronic and charge transfer transitions that may occur in a single sample.

13.6. X-RAY ABSORPTION SPECTROSCOPY

X-ray Absorption Spectroscopy (XAS) is an element-specific spectroscopic technique that allows studying many elements of the periodic table, in many types of phases (crystalline, amorphous, liquids, gases). It provides information on the electronic, structural, and magnetic properties of the sample. Detailed information about the theory as well as data acquisition and reduction protocols for materials can be found in these contributions: Brown et al. (1988), Mottana (2004), Henderson et al. (2014), and Willmott (2011). Moreover, the readers are encouraged to read the recent paper by Abe et al. (2018) to have an overview of the different XAS-based techniques, applications (catalysis, high-pressure), and sources of errors.

XAS is a measurement of the variation of the linear X-ray absorption coefficient $\mu(x)$ as a function of the X-ray energy E. The incident x-ray beam has an energy $E = h\nu$, (E in keV, h is Planck's constant, ν is the frequency) and an intensity I_0. A typical detection mode of the XAS signal is based on transmission measurements, hence on the attenuation of the incident beam intensity when passing through a sample. It follows the Lambert-Beer's law (see Section 13.5, Equation 13.7). However, depending on the X-ray energy range selected, on the element investigated (photoabsorber), on the concentration of the photoabsorber, and on the matrix, there are other detection methods that can be used, i.e., fluorescence and total-electron-yield (TEY). For instance, thin concentrated samples should be collected in transmission mode, whereas dilute, low-energy edges are better detected with fluorescence measurements. Details on the instrumentations, and on the different acquisition modes, are provided in Van Bokhoven and Lamberti (2016).

The X-Ray absorption spectrum of an element, usually displayed as a function of X-rays energy, can be divided into three regions with increasing energy, as shown in Fig. 13.5 (e.g., Natoli and Benfatto, 1986; Mottana, 2004; Bunker, 2010; Willmott, 2011):

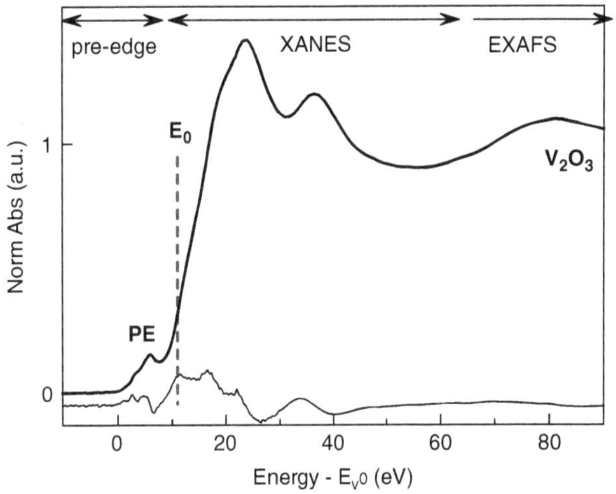

Figure 13.5 Energy scaled K-edge XAS spectrum of vanadium in divanadium trioxide (V_2O_3). The main sub-regions are reported: the pre-edge region before the absorption edge (E_0) that contains the small feature PE (pre-edge peak); the multiple scattering region (XANES) that extends from several eV above the pre-edge up to ca. 50 to 70 eV above the edge; and the single scattering region or extended region (EXAFS). On the bottom, the first derivative of the spectrum is reported. E_0 identifies the edge position, and is defined as the first maximum of the first derivative function. E_{V0} = metallic V K-edge = 5465 eV.

1. a pre-edge region occurring at energies lower than the absorption peak (2–20 eV before the absorption edge), i.e., where the incident X-ray energy E is lower than the binding energy E_b, of an electron in the orbital, $E < E_b$, and electronic transitions have low probabilities (Geiger, 2004; Van Bokhoven and Lamberti, 2016). A feature (pre-edge peak, PE) may be present. It is linked to the crystal field transition, and, in the case of K-edge transitions of *3d* transition elements, it derives from the *1s* to the empty *3d* levels more or less *4p* hybridized by the ligands (*1s* ➔ *3d*-like levels transition);
2. when the X-ray incoming energy approaches the binding energy ($h\nu \sim E_b$) there is a higher probability for an electronic transition to occur, resulting in an absorption edge. This region is called the X-ray Absorption Near Edge Structure or XANES region and is characterized by sharp absorption structures arising from strong multiple scattering of photoelectrons by atoms surrounding the absorber. This scattering process involves multiple scattering centers that often extend beyond the first nearest-neighbor shell of ligands (Geiger, 2004; Bunker, 2010; Willmott, 2011; Van Bokhoven and Lamberti, 2016);
3. The third region of the spectra, extending from about 50 to 1000 eV above the edge, is called the Extended X-ray Absorption Fine Structure, namely EXAFS region and it is characterized by weak oscillations of the spectrum due to backscattering of the photoelectrons (Geiger, 2004).

Since each region is characterized by distinct phenomena, particular care must be taken when using acronyms.

Indeed, many authors misuse the terms XAS, XANES, and EXAFS, and sometimes they are erroneously considered synonymous. From a redox point of view, the first two regions are the ones providing key information regarding an element oxidation state.

XANES covers the range between the threshold (E_0) and the point at which the EXAFS region begins. Because of its origin, XANES is particularly sensitive to the geometrical arrangement of the first neighbors around the central absorber. As a consequence, it provides information on site geometry, symmetry, and bond angles, as well as knowledge about the oxidation state of the absorber. Beyond the absorption edge, the EXAFS region is used to ascertain geometrical parameters such as the interatomic distances between the investigated absorbing atoms and the neighbors in its immediate surrounding, the coordination number with the nearest and next-nearest neighbors and their nature (Galoisy, 2004, and references therein; Henderson et al. 2014).

13.6.1. K-edge XANES Spectra of *3d* Transition Elements

The analysis of the XANES region is widely used in many scientific fields. It provides information about the average valence and coordination number of an elements, as well as the type of ligands and site symmetry. Since the 1980s, many mineralogical, geochemical, and environmental studies are based on XANES spectra at the K-edge (*1s*). These early studies favored the first-raw transition elements since it was possible to obtain spectra with good signal/noise ratio and high energy resolution in air, despite moderately powerful second-generation synchrotron radiation (Mottana, 2004). In particular, K-edge XANES spectra of *3d* transition elements present spectral features and energy shifts that can provide excellent structural information on the photoabsorber in glasses and minerals. Indeed, the pre-edge peak and the edge structures can be used to derive information on the photoabsorber speciation without an exhaustive theoretical work. Information can be extracted using empirical correlations, built from the study of XANES spectra acquired on reference materials with well-known local structure (see Section 13.6.2).

To briefly illustrate the variations occurring in the XANES region depending on element valence and ligand geometry, spectra for some of the first-row transition metal oxides are reported in Fig. 13.6, in the energy range covering the pre-edge peak region up to ca. 50 eV above the absorption edge (E_0). The spectra show many variations depending on the element speciation. For example, the vanadium oxides XANES spectra reported in Fig. 13.5 present a wide range of oxidation state (0 to +5); these yields absorption edge shifts to higher energy with increasing oxidation state, from ~5465 eV for V^0 to ~5480 eV for V^{5+}. Besides oxidation state information, V coordination geometries can be deduced as well from the spectra and Wong et al. (1984) reported a very detailed study of Vanadium oxides. For instance, the V site shows regular octahedral symmetry Oh in VO, but distortion increases from C3 in V_2O_3 to C_s in V_2O_5 (Wong et al. 1984); such changes go along with the pre-edge peak rising in intensity following the loss of symmetry. In detail, VO has a NaCl cubic face-centered structure with regular octahedral VO_6 units. In this structure, the electronic transition *1s* → *3d* is dipole-forbidden, resulting in a weak (or virtually zero) pre-edge peak. On the contrary, when the inversion center is broken, due to the loss of symmetry, as in the case of the distorted 6-fold coordinated VO_6 units in V_2O_3, the pre-edge peak increases its intensity. V_2O_5 displays V in 5-fold coordination in highly distorted tetragonal pyramids, and presents a very strong pre-edge peak (Wong et al. 1984). Thus, the crystal field theory could be used to provide an interpretation of the pre-edge peak characteristics (de

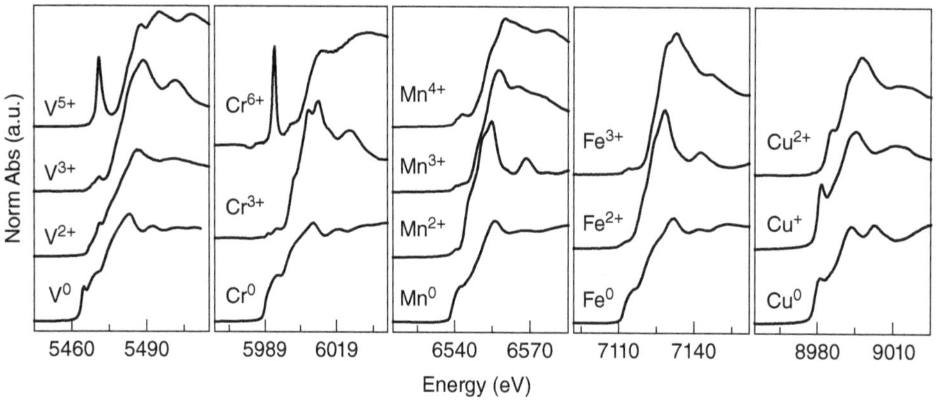

Figure 13.6 XANES spectra of first- row multivalent transition metal oxides (except Cr^{6+} compound: $CaCrO_4$). The spectra follow similar trend with the shift of the absorption edge to higher energies at increased average oxidation state.

Groot et al. 2009) and, in theory, the highest pre-edge peak intensity is reached in case of tetrahedral coordination (T_d).

The other transition element oxides shown in Fig. 13.6 follow similar trends, showing a shift of E_0 to higher energy at increased oxidation state, and a correlation between pre-edge peak intensity and local site symmetry. Cr K-edge XANES spectra are discussed in detail in Bell et al. (2020, this volume).

13.6.2. Extracting Information from XANES Spectra

In order to extract information on elements average oxidation state and local environment from the XANES region, several approaches can be adopted depending on the element studied as well as on the electronic transition considered (e.g., K- or L-edge). Here we will report a few examples on: (i) absorption edge position; (ii) absorption edge and pre-edge peak; (iii) signal deconvolution; and (iv) linear combination. The first two methods are mainly based on the empirical approaches, and thus, based on the study of reference materials with well-known structure.

A basic procedure to determine the average oxidation state of a multivalent element in materials is to calculate the absorption edge position (hereby referred as the inflection point, E_0) of several well-characterized reference materials. Since there is a shift of E_0 to higher values with increasing the oxidation state (Fig. 13.6), a calibration might be established, and then used to estimate the average valence of unknown samples. This method has often shown to be extremely reliable, as in the case of V oxides (Wong et al., 1984) or Mn oxides (Fig. 13.6; Belli et al., 1980). However, in other cases, the correlation between the formal oxidation state and the absorption edge position is not accurate enough (e.g., Farges, 2005). This is especially the case when the element in the investigated material is present with different coordination geometries and site occupancies. Manganese compounds represent a good example to visualize such complexities. Mn has valences ranging from 0 to +7. The Mn K-edge spectra of the crystalline compounds shown in Fig. 13.7a are representative of the most stable oxidation states in geological systems (2+, 3+, and 4+), and of the different Mn coordination geometries (see Table 13.1). Mn K-edge absorption edge shifts to higher energies upon increasing Mn oxidation state. However, an energy shift can be observed as well when changing the local symmetry around Mn, even at constant oxidation state (e.g., Rodocrosite and MnO). Consequently, in this specific case the simple observation of the edge position is not enough to precisely characterize the samples and supplementary analysis are necessary, such as the detailed study of the pre-edge peaks, or the use of a complementary technique.

XANES spectra of some Fe minerals (Table 13.1) and Fe-bearing silicate glasses are shown in Figs. 13.7b and c (experimental details provided in Giuli et al. 2011, 2012a, b; Cicconi et al. 2015a). The spectra of the Fe crystalline compounds reported here represents the most common stable oxidation states (2+ and 3+) and coordination geometries of Fe in geologic materials. In the case of iron, qualitative and quantitative information on $1s \rightarrow 3d/4p$

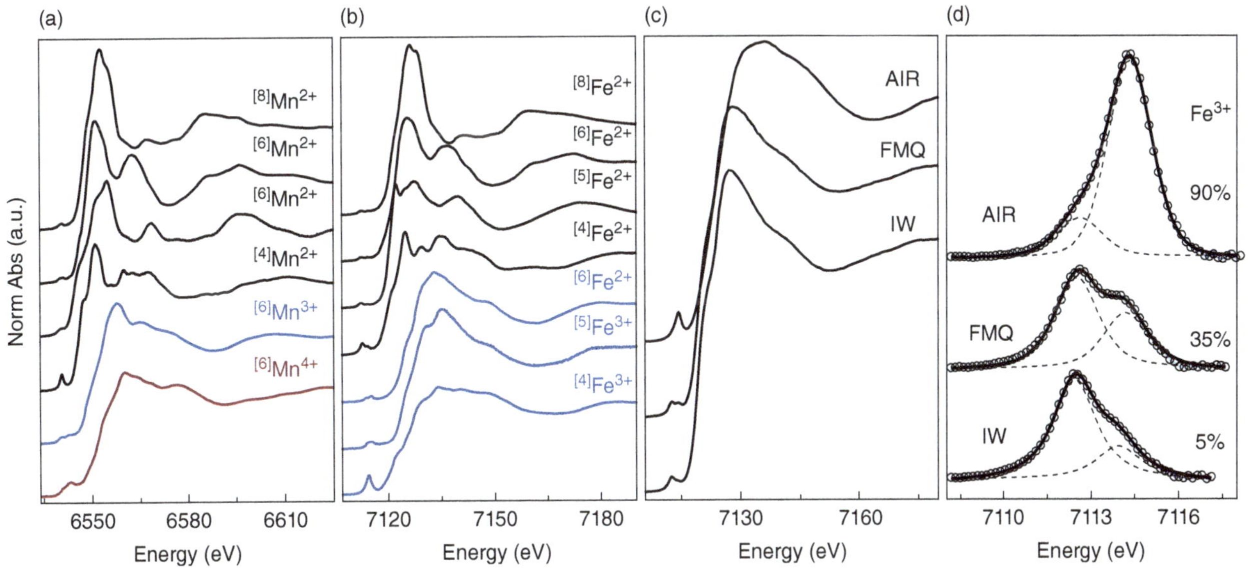

Figure 13.7 K-edge XANES spectra of Mn (a) and Fe (b) crystalline compounds (names listed in Table 13.1). (c) Fe K-edge XANES spectra of silicate glasses equilibrated under different fO_2 conditions. (d) background subtracted pre-edge peaks, and their deconvolution; estimated %Fe^{3+} are reported (±5 %).

Table 13.1 Fe and Mn crystalline compounds reported in Fig. 13.6. Experimental details for Fe and Mn minerals are reported in Giuli et al. (2011, 2012a, b), and Cicconi et al. (submitted), respectively.

Crystalline compound	Valence	Coordination
Almandine	Fe^{2+}	[6]
Siderite	Fe^{2+}	[6]
Ericaite	Fe^{2+}	[5]
Akermanite	Fe^{2+}	[4]
Aegirine	Fe^{3+}	[6]
Yoderite	Fe^{3+}	[5]
Kimzeyite	Fe^{3+}	[4]
Spessartine	Mn^{2+}	[8]
Rodocrosite	Mn^{2+}	[6]
MnO	Mn^{2+}	[6]
Helvite	Mn^{2+}	[4]
Mn_2O_3	Mn^{3+}	[6]
MnO_2	Mn^{4+}	[6]

transitions were obtained for Fe-bearing crystalline and amorphous materials by e.g., Waychunas et al. (1983), Calas and Petiau (1983), Petit et al., (2001), Berry et al., (2003, 2010), Bajt et al. (1994), Galoisy et al. (2001), and Wilke et al. (2001). Since these studies, several others have shown that it is possible to correlate Fe pre-edge peak features (i.e., intensity and barycenter) with both the average site geometry (and symmetry) and iron average valence state.

In-depth pre-edge analysis requires several steps. After the first step that consists of background subtraction, the pre-edge peak can be decomposed in different peaks with given mathematical functions (e.g., pseudo-Voigt). This allows one to estimate, for example, the PE weighted energy position (a.k.a. centroid or barycenter) and the PE total integrated area (Fig. 13.6d; see Wilke et al., 2001, 2004). For Fe, in known crystalline reference materials, with Fe present with a single valence and in a single crystallographic site, the PE analysis has shown that the centroid position of Fe^{3+} compounds is ~2.4 eV above the edge of metallic Fe (7112 eV), whereas the centroid of Fe^{2+} is at ~0.9 eV above the metallic Fe one. For more complex Fe site geometry, or for mixed valences, there is a nonlinear response between pre-edge peak centroid and integrated intensity (see variogram in e.g., Wilke et al., 2001; Giuli et al. 2012a; Henderson et al., 2014). Fortunately, it is possible to estimate both Fe average oxidation state and coordination number with a good accuracy, by using mixing lines between the end-members (e.g., Wilke et al., 2004). As a result, the determination of the average Fe redox ratio and coordination via the study of the Fe pre-edge peak is very common nowadays.

XAS spectra of glasses present less structures and broader features compared to crystalline materials because of the lack of periodicity in amorphous materials (Fig. 13.7c). Nevertheless, the considerations done for the crystalline compounds are still valid for glasses or liquids. Figs. 13.7c and 13.7d show the Fe K-edge spectra of phonolitic glasses, equilibrated over a range of oxygen fugacity (fO_2) conditions [$\log_{10}(fO_2)$ from –0.68 to –11; from air to FMQ, to IW[1]] (Cicconi et al., 2015a). The absorption edge shifts from 7123.5 eV to 7119 eV with decreasing fO_2 conditions, indicating higher amount of reduced Fe species. In addition, the pre-edge peaks strongly change depending on the fO_2 conditions. The PE peak can be fitted with two pseudo-Voigt functions, whose energies at ~7112.65 and ~7114.4 eV are consistent with those of Fe^{3+} reference materials. For samples prepared in air, the dominant peak at higher energies, is that assigned to Fe^{3+}, while for samples equilibrated under reducing conditions (close to the IW buffer) the dominant one is the peak at lower energies, assigned to Fe^{2+}. The estimated centroids, from air to IW, were: 2.34 eV and 1.04 eV (above the Fe metallic energy), respectively providing estimated Fe^{3+} content of ~90 % and 5 % (±5 %), in agreement with the values quantified via wet-chemistry (Cicconi et al., 2015a). Hence, the deconvolution of the pre-edge peak is a very accurate way to quantify the Fe redox ratio in many materials and in some cases, it has reached an extremely high accuracy (Berry et al., 2018; Zhang, Cottrell, et al., 2018). The same approach can be used for signal collected *in-situ* at high temperatures, and/or under different imposed fO_2 conditions (Magnien et al., 2004, 2008; Cochain et al., 2012; Cicconi et al., 2015b).

We have shown that both the absorption edge position and the analysis of the pre-edge peak may provide a satisfactory assessment of an element valence and local environment. However, natural materials, either crystalline or amorphous, usually present a higher complexity in the transition element local environment, making the interpretation of the different XANES spectral features difficult. Furthermore, for many multivalent elements, it is not always possible to perform studies based on crystalline reference materials, simply because they may not be available (e.g., no natural Ti^{3+} or Eu minerals are available). Moreover, the detailed study of edge and pre-edge features require the use of a monochromator that provides high energy resolution. This induces some drawbacks in term of acquisition setup for the acquisition of very good spectra in the extended region.

Other ways to gather information from the XAS data imply peak fitting of the spectral features, and/or the linear combination of two end-members. Spectral peak fitting is a time-consuming procedure, which involves the knowledge of the electronic transitions responsible for the XANES

[1] Respectively, fayalite-magnetite-quartz, and iron-wüstite buffers

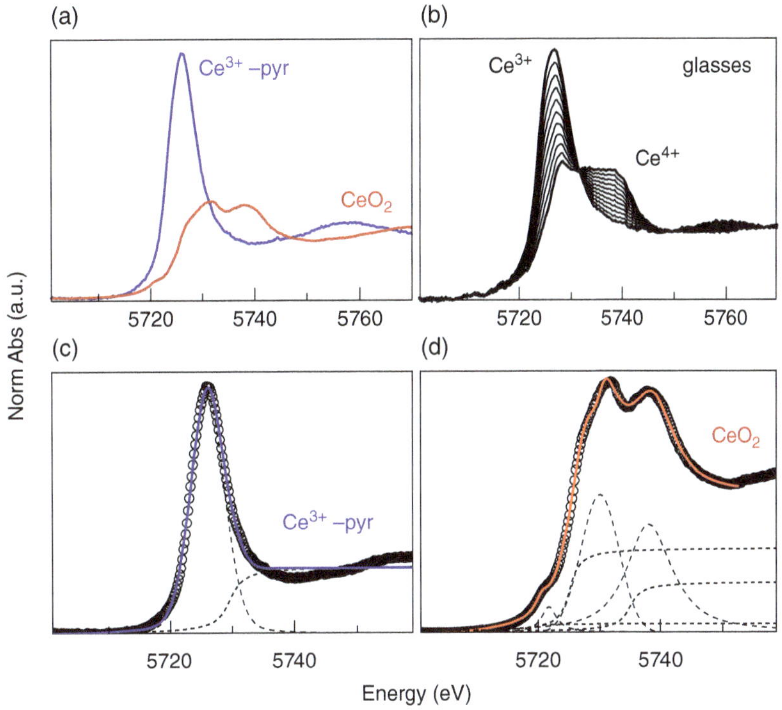

Figure 13.8 Cerium L_{III}-edge XANES spectra for two Ce crystalline materials (a), and examples of linear combination (b) and deconvolution of the spectral features (c and d). See text for details.

features. However, in the absence of proper reference compounds, it may provide means to identify mixed valence states. The analysis of the XANES spectra has been usually performed by fitting a model composed of a combination of several mathematical functions (Lorentzian, Gaussian, or pseudo-Voigt, to represent transitions to bound states) and arctangent functions (transitions to the continuum) to the data (Kaindl et al., 1987; Lytle et al., 1990; Takahashi et al., 2000a, 2005; Cicconi et al., 2012). XANES peak fitting has been used for decades in physic and materials sciences, especially for lanthanides compounds (Kaindl et al., 1987; Lytle et al., 1990; Bianconi et al., 1987). In Geosciences, one of the first examples was the determination of Ce^{3+}/Ce^{4+} by XANES at the Ce L_{III}-edge in ferromanganese nodules, granite and cherts (Takahashi et al., 2000a,b). The authors used Ce salts as reference compounds and performed both the linear combination of end-member spectra from trivalent and tetravalent chemicals, and the spectral fitting of the XANES region. Ce^{3+} and Ce^{4+} XANES spectra, at the L_{III}-edge, show very distinctive shape (Fig. 13.8). Trivalent compounds have a single sharp resonance peak, representing excitation into the empty *5d* band, producing a fully screened $4f^1$ *5d* state (Ce^{3+}-pyrochlore in Fig. 13.8a). Tetravalent Ce compounds (e.g., CeO_2 in Fig. 13.8a) shows at least three peaks: a small band around 5720 eV ($2p \rightarrow 4f^1$) and two strong features located at ~5730 eV and at ~5738 eV, respectively assigned to $(4f^1\ \underline{L})5d$ and $4f^0\ 5d$ configurations[2] (Kaindl, 1987; Takahashi et al., 2000a,b). According to Takahashi et al. (2000a), the sharp peak of trivalent Ce compounds can be fitted by the combination of a Lorentzian and an arctangent function, whereas in Ce^{4+} compounds, one Lorentzian function and one arctangent function are assigned to each peak. Besides, Ce^{4+} compounds show spectral shapes that depends on the chemical environment of Ce (Bianconi et al., 1987; Soldatov et al., 1994). Examples of such fits are reported in Fig. 13.8c,d. The deconvolution of spectra related to mixed valences is very challenging because of the several electronic transitions embedded in the XAS signals.

Based on the linear combination of XANES spectra for Ce^{3+} and Ce^{4+} reference materials, Takahashi et al. (2000a) estimated the Ce^{3+}/Ce^{4+} ratio in different natural materials, with an error of ~10 %. Since the shape of the Ce^{4+} XANES spectra changes depending on the element local environment, the selection of the appropriate reference compound is critical, particularly when studying silicate melts or glasses. Indeed, as previously mentioned, XAS spectra of amorphous materials present less features compared to crystalline compounds, making their interpretation even more challenging. In case of Ce, while

[2] L indicates a hole in the ligand shell

the Ce^{3+} XANES spectrum in a glass does look like that of a crystalline compound, the spectra of Ce^{4+} in glasses do not (e.g., Burnham & Berry, 2014). Thus, in order to estimate the proportions of Ce^{3+} and Ce^{4+} in glasses or melts, it is possible to linearly combine the two end-members spectra of Ce^{3+} and Ce^{4+}, acquired in the same matrix, to fit that of the sample. This linear combination method provides then a ratio that yields the Ce^{3+}/Ce^{4+} ratio in the analyzed material. This is one of the most accurate methods to determine Ce oxidation state in given materials (e.g., Takahashi et al., 2000a; Burnham & Berry, 2014). An example of this method is reported in Fig. 13.8b. In-situ dispersive XANES data were acquired on a Ce-doped sodium disilicate melt (Cicconi et al., 2017b); the trivalent and tetravalent cerium end-members were obtained at T > 1100 K, under oxygen fugacity conditions controlled by O_2 and Ar/H_2 gases, respectively. This approach allows an estimation of Ce redox state with higher accuracy (errors between 2 and 5 %) with respect the deconvolution method, and does not require several well-known model compounds. However, it requires the acquisition of the Ce^{3+} and Ce^{4+} signals in the same matrix, under very reduced and oxidized conditions respectively. This step can represent a challenge, making this method difficult to apply in some situations.

Advantages, drawbacks, accuracy

X-ray Absorption Spectroscopy is a powerful element-selective technique that allows studying a photoabsorber element, even at very dilute conditions, and in any physical state (crystalline, amorphous, liquids, gases). Hence, its use for either standard experiments, sample mapping, or for in-situ experiments at extreme conditions of temperature, pressure, or oxygen fugacity has increased in the last decade. XANES spectroscopy works for all transition elements (like Cu, Cr, Ti, or Eu; for example, see Berry et al., 2006a,b; Doyle et al., 2016; Burham et al., 2015) up to U for stable element (Caisso et al., 2014) and also for transuranian elements like Americium (Caisso et al., 2015; Prieur et al., 2018). Moreover, for some elements, it is also possible to acquire data at different edges. For example, several elements like Fe or V can be investigated at the K-edge (Calas and Petiau, 1983a,b; Wilke et al., 2001; Berry et al., 2010) and the L-edge (Bourdelle et al., 2013; Hoche et al., 2013).

Despite its attractiveness, XAS presents several drawbacks. The first challenge is that XAS experiments usually require the use of a synchrotron source. This typically involves the submission of a scientific proposal to large-scale facilities (synchrotron light sources), and thus it is necessary to provide a clear explanation of why it is critical to obtain data from synchrotron radiation (this statement is valid for all synchrotron-based technique), and most importantly, a prior deep knowledge of the investigated samples is required. Once access to a light source is granted, then several other things need to be carefully prepared. Particular attention to detail is required when planning an experiment as well as during data collection. The quality of the signals and the accuracy of collected data will depend on the probed photoabsorber element, as well as on its concentration.

The selection of the detection mode might be critical, because secondary effects may interfere with the signals. Examples of such unwanted effects are self-absorption or induced beam damage (i.e., see for sulfur Metrich et al., 2002, 2003, 2010; and for iron Goncalves et al., 2013, and Cottrell et al., 2018; Hughes et al. 2018, 2020). Finally, the accuracy of the experiment will depend on the standard used and on how the sample is prepared. Before performing an experiment, it is necessary to select the appropriate reference compounds. All samples should further be prepared according to the selected detection mode chosen and to the type of analyses performed (e.g., XANES vs EXAFS). Particular attention should be placed upon analyzing crystalline materials because of the dependence of the XAS features on polarization. A detailed list of practical issues on sample preparation, errors, and type of experiments is reported in Bunker (2010).

To conclude this part, we want to bring to the attention of the reader problems arising when analyzing complex samples, formed by mixtures of minerals, with possibly additional amorphous phases. For example, Fig. 13.9 represents a schematic drawing of the texture of a lava sample containing glass and crystal, from which XAS spectra at the Fe K-edge and Raman spectra were collected. Such an example is fairly representative of many natural samples, as they usually consist of a mixture of several minerals and possibly glass/amorphous phases. Depending on the beam size and of the experimental settings, it is nowadays easy to analyze small areas, typically of micron size, with a good spatial resolution and precision. Now, it seems obvious that, if the analyzed zone contains several phases, results will represent a convolution of the signals from the different phases (Fig. 13.9). It is quite difficult to find the end-members forming the analyzed mixture by numerical methods. Some studies tried to perform such calculations in order to introduce correcting terms to take into account such mixing effects, but this should be discouraged for several reasons. First, such correction implies a linear mixture of the original signals, and this may be very far from the reality due to, e.g., different refractive index and absorption coefficients of the analyzed phases that result in the beam sampling differently the different phases. Then, such correction has no physical meaning, and, as the ground for any good scientific study is the robustness of the reported data, avoiding mathematical tricks to try gathering information from discussable datasets is best practice.

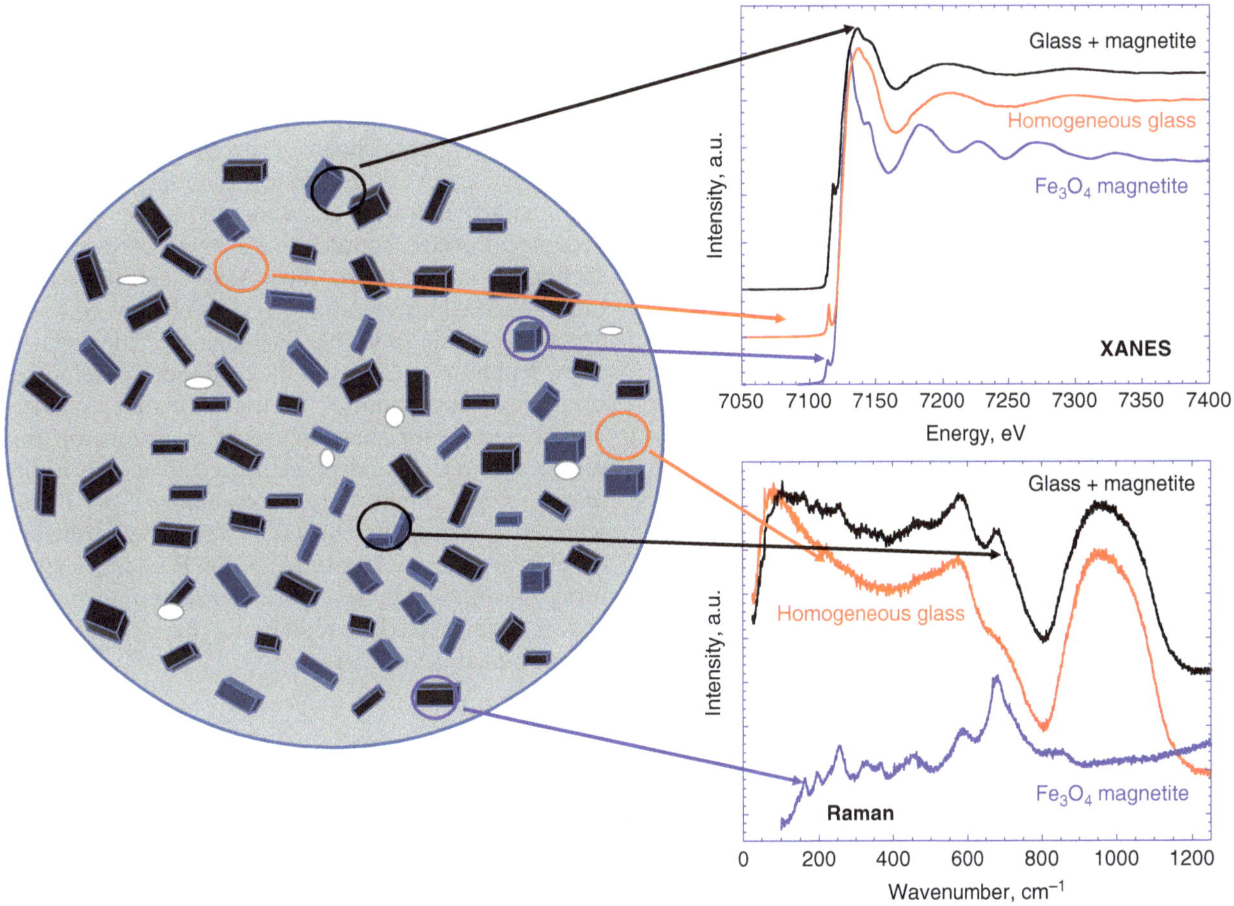

Figure 13.9 Schematic drawing of a lava, and XAS spectra at the Fe K-edge and Raman spectra of magnetite crystal, basalt glass, and their mixture. The mixed spectra are very different from the end-members in both XAS and Raman. In some cases, the end-members spectra can be used to find the proportions of the end-members, but this requires perfect knowledge of the two end-members spectra and of their mixing behavior. For example, De Ligny and Neuville (2017) have shown that it is possible to determine the rate of nucleation and growth in well-characterized, reference silicate melts from the principal component analysis of their Raman spectra acquired in situ at high temperature. However, for geologic products, such approach is very challenging because end-members spectral components as well as their mixing behavior are often unknown.

13.7. RAMAN SPECTROSCOPY

Vibrational spectroscopies, such as Raman spectroscopy, have provided information on the structure of materials in many fields. Numerous Raman studies have been carried out on silicate glasses and have made it possible, among other things, to determine the spectral treatment procedure (Mysen et al., 1982; Seifert et al., 1982), to acquire structural information on different alumino-, alkaline, and alkaline-earth silicate glasses (Mysen et al., 1980, 1981, 1985; McMillan et al., 1992; Mysen and Frantz, 1992; Neuville and Mysen, 1996; Neuville, 2006) and on iron-bearing glasses (Mysen et al., 1984, 1985; Cooney and Sharma, 1990; Wang et al., 1995; Magnien et al., 2004, 2006, 2008; Roskosz et al., 2008; Cochain et al., 2012; Di Genova et al., 2015, 2016, 2017; Le Losq et al., 2019). Despite such interest, few authors have attempted to highlight the influence of redox on Raman spectra or to perform a calibration aimed at the determination of the redox state of iron in glasses. Such analysis could benefit from the advantages of Raman spectroscopy, being a cheap, easy to access, of micrometer-scale resolution, non-destructive method that requires minimal sample preparation.

Materials, whatever their form (liquid, gas, or solid), are an arrangement of atoms linked by chemical bonds, forming molecules that can be assimilated to harmonic oscillators. Vibrational (Infrared, Brillouin, Raman) spectroscopy analyzes the excitation and the vibrations of such molecular vibrators, following the interaction between the material and an incident electromagnetic radiation.

When a monochromatic beam passes through a transparent material, most of the incident electromagnetic beam is transmitted through the material without interacting with it. This process is an elastic scattering or Rayleigh scattering, for which the chemical bond returns to its original vibrational state. In this case, the frequency of the transmitted electromagnetic wave is that of the incident radiation. However, a tiny portion of the incident wave undergoes inelastic scattering: this is the Raman effect. The frequency of scattered radiation differs from that of the incident beam because the excited molecular vibrators did not return to its energy initial state; in other terms, they exchanged energy with the incident photons. Two inelastic scattering scenarios are possible: Raman Stokes and Raman anti-Stokes. For Stokes scattering, the photons gave energy to the vibrator that returns to a higher energy state, such that the scattered photons' vibrational frequency is less than that of the incident radiation. Conversely, in the case of the anti-Stokes scattering, the photons absorbed energy from the molecular vibrators that return to a lower energy level, and, hence, scattered photons have a higher frequency.

As a result, Raman spectra present bands that are signatures of the chemical bonds and molecular groups in a material, in two regions (Stokes and anti-Stokes) which frequencies are located symmetrically on both sides of the exciter frequency. Generally, only the Stokes part of Raman spectra is exploited. Indeed, at room temperature, anti-Stokes lines are less intense than Stokes lines because Boltzmann's statistic indicates that molecules are close to their ground states and thus are more likely to go to a higher energy state. An exhaustive review on Raman spectroscopy applied to Earth and Material Sciences was made by Neuville et al. (2014), and another one by de Ligny and Neuville (2017) on its application to the study of the crystal nucleation and growth phenomenon.

13.7.1. Spectrometer, Technical Designs, Acquisition Parameters

A Raman spectrometer generally consists of a laser source (argon or krypton), which can emit monochromatic light at several different wavelengths. The incident beam is focused on the sample through a confocal system that consists of diaphragms, lenses, and an optical microscope. The confocal system also collects and filters the backscattered beam, then transmits it to a pre-monochromator and spectrometer for analysis. Through a set of networks and mirrors, the beam is filtered and diffracted to arrive on a multichannel CCD detector which converts it into a spectrum I(ω), where I(ω) is the number of photons at the wavenumber ω, usually expressed in cm^{-1}. As Raman spectroscopy measures photons that present a frequency shift from those constituting the incident beam, ω should actually be referred as the Raman shift.

Currently, a large choice of spectrometers is available. Their price can vary from a few thousand euros to over two hundred thousand euros, depending on the spectral accuracy, the laser sources, and the proposed equipment such as a confocal microscope. Aside the traditional laboratory spectrometers, it also is possible now to find relatively accurate portable Raman spectrometers that allow real *in situ* analysis in the field. Such developments bring new opportunities to use Raman spectroscopy directly *in situ*, for analyzing rocks, pigments, or fragile samples for various purposes related to studies in fields like geology, archeology, volcanology, or material sciences. Such attractiveness is reinforced by forthcoming developments in the treatment of the Raman signal with using machine learning algorithms in conjunction with spectral databases. This allows automatic recognition of minerals phases for instance.

13.7.2. Raman Spectra

As previously said, Raman spectra show bands characteristic of specific chemical bonds and molecular groups. More detailed about band vibration, selection rules, molecule vibrations can be found in several books like Hersberg (1945a,b), Wilson et al.(1955), Long (1977), or Harris and Bertolucci (1978). To summarize, the length and the strength of chemical bonds vary according to the cations, their coordination numbers, as well as the strengths of the bonds involved in the molecule. As a result, Raman bands are representative of the coexisting structural units in silicate glasses and fuses, making Raman spectroscopy a powerful tool for probing the structure and state of polymerization of silicates (Brawer, 1975; Brawer & White, 1977; Furakawa & White, 1980; Mysen et al., 1980, 1981).

13.7.3. Spectral Pre-Treatment: Background Subtraction and Normalization

To obtain quantitative information from a Raman spectrum, it is necessary to (i) correct for temperature and excitation line effects; (ii) remove any remaining background; and (iii) normalize the spectra.

The measured intensity depends on the wavelength and temperature (effect of the Bose-Einstein distribution of excited phonons) of acquisition (Long, 1977). Those effects can be removed using the equation

$$I = Iobs\left[\nu o^3 \nu/(\nu o - \nu)^4\right]\left[1 - \exp\left(-hc\nu/kT\right)\right] \quad (13.9)$$

with h the Planck constant, k the Boltzmann constant, c the speed of light in m s^{-1}, T the temperature in K, νo the wavenumber of the incident laser light in m^{-1}, and ν the measured Raman shift in m^{-1}. This correction is mostly important if one wants to observed peaks at low frequency (lower than ~500 cm^{-1}; see Neuville et al., 2014). Its origin and variants can be found in Shuker and Gammon (1970), Galeener and Sen (1978), Brooker et al. (1988), and Hehlen et al. (2010). Some software offers the different variants, see the function *tlcorrection* from the *rampy* library in Python for instance.

Often spectra are affected by a background. Part of it can be removed prior to the temperature and excitation line effect correction (Equation 13.9), but some background can still persist after this correction. Spectral background can arise from many different sources, including adsorbed OH groups on sample surface, black-body radiation in high temperature experiments, and incident light scattering interference in the sample (the latter is particularly visible in multiphasic samples). To remove such background, one can either acquire a spectrum of the spectrometer and setup without the sample. This allows, for instance, the removal of the signal of diamond during diamond-cell experiments. Note that, in this case, the correction described by Equation 13.9 is applied (if applied) after the subtraction of the known background. However, such approach will not allow taking into account any effect due to the interaction between the sample and the laser. Therefore, a common practice is to use a flexible mathematical function fit to the minima in the spectra. This function represents an empirical best approximation of the true background, and after being subtracted from the raw spectrum, it allows obtaining a background corrected spectrum. Several functions can be used, like polynomial functions or splines, but the most promising developments nowadays reside in automatic methods like the ALS (Eilers and Boelens 2005) and arPLS (Baek et al., 2015) background subtraction techniques. Those methods are fairly good for Raman spectra of minerals as they present peaks very distinct from the baseline. The treatment of glass Raman spectra require more effort because minima often contain the tails of the broad peaks arising from the vibration of the disordered molecular entities composing amorphous materials. Therefore, background is often manually fit to specific areas using user-defined functions. In any case, many libraries allow one to perform background subtraction in an interactive way in many programming languages (see for instance *rampy* in Python, *hyperSpec* in R, *Baseline Fit* in Matlab/Octave, *Spectra* in Julia). As different choice of background can severely influence the obtained result, and as a manual fit using a graphical interface is often non-reproducible even by the same user, we advocate the implementation of systematic iterative computer codes for background fitting and subtraction, such that those codes can be provided to the community for their use and accuracy evaluation.

After background fitting, normalization of the intensity is often desired. Depending on the goal of the study, different choices can be made. At this stage, we should recall that Raman spectra represent a glimpse into the vibrational density of state (VDoS) of materials. The integral of the VDoS is equal to 1 by definition, as it is that of the intensities of all the observed vibrations. Therefore, in case of doubt regarding normalization, a best practice is to simply normalize the observed intensities by their integral, i.e., the area under the Raman spectral signal. However, sometimes other choices can be made for representation purpose of easiness of analysis. For instance, it is also possible to normalize the observed intensities to that of the maximum of a given peak, or to a value measured at given Raman shift. In any case, different choices can be considered depending on the goal of the Raman analysis: (i) either treat the data to obtain information relevant to our study, so in this case we can choose any protocol as long as we stick with it; or (ii) process the spectra for comparison with the literature. In the latter case, it is necessary to use the same protocols as described in previous studies. We must conclude by stressing that, if several pieces of information are to be extracted from the same spectrum, normalization to the total area of the spectrum will be necessary. This will provide information comparable to that obtained by other techniques (NMR and XANES, cf. Cochain et al., 2012), as well as volatile content (Le Losq et al., 2012).

13.7.4. Details on Raman Spectra of Silicate Glasses

Fig. 13.10 show a normalized Raman spectrum of pyroxene glass composition (50 mol% SiO_2 – 5 mol% Na_2O – 22.5 mol% MgO – 22.5 mol% CaO). This spectrum is composed by several frequency regions: (i) a low frequency region at Raman shifts < ~600 cm^{-1}; (ii) an intermediate region comprised between ~600 and ~800 cm^{-1}; (iii) a high frequency region between ~800 and ~1200 cm^{-1}. More detail can be found in Neuville et al. (2014). The bands observed were associated with the modes of vibration and various silicate structural units (Mysen et al., 1980; McMillan, 1981; McMillan et al., 1982).

The bands composing the spectra of amorphous silicates are relatively broad and unresolved compared to those of the crystals. The width of the bands results from the disorder of the amorphous structure, marked for instance by the large pseudo-Gaussian distributions of bonding forces and angles inherited from the Brownian motions of ionic entities at high temperature in the liquid state (Mysen et al., 1982), from which the glasses were

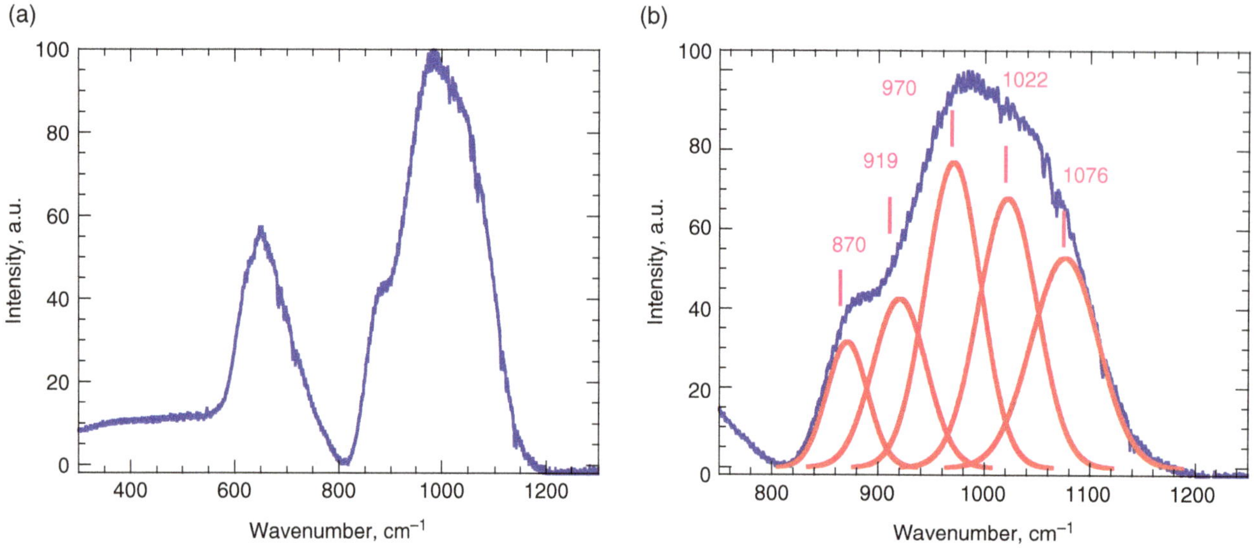

Figure 13.10 Raman spectrum of an iron-free pyroxene glass (CaO-SiO_2-Na_2O-MgO) at room temperature. (a) Standardized whole spectrum (b) high-frequency envelope to highlight Q species. The 870 cm^{-1} band can be attributed to Q^1 species, 919 cm^{-1} to Q^2, 970 cm^{-1} to T_{2s}, 1022 cm^{-1} to Q^3 species, and 1076 cm^{-1} to Q^4 species (based on Magnien et al., 2006).

formed and are an image of, as well as to replacement of Si by other tetrahedral cations and to the variety of modifying cations (Brawer, 1975).

In any solid, the number of vibrational modes and the density of states are complex functions of the frequency. Generally, the analogy between a spring and the chemical bonds leads to the application of the Hooke relation:

$$\nu = ½\, \pi * (k/\mu)^{1/2}. \tag{13.10}$$

with ν the vibration frequency in Hz, μ the mass of the atoms involved in the bond in kg, and k the binding force in $N.m^{-1}$. In a silicate, the lattice vibrations involve a large number of atoms and are mainly stretching and bending vibrations of the M-O bonds accompanied by the movement of the TO_4 tetrahedra (M: modifying or charge compensator cations, and T network formers). Therefore, Raman bands and their frequencies are highly dependent on the state of polymerization of the network. In addition, band shape will depend on the competition between the lifetime of the virtual excited state and the coherence lifetime. Depending on such lifetime effects, bands tend to be either Gaussian, or Lorentzian, or a mixture of both. In a silicate crystal, bands usually tend to be Lorentzian while in glasses, the large bands actually tend to a Gaussian lineshape reflecting the inherent Brownian disorder of the structure of the liquids they mirror. Network modifying cations have large masses and generate weak links (high μ, low k). The frequency region at Raman shifts < 600 cm^{-1} represents the stretching vibration modes of the M-O bonds. With smaller masses and larger force constants, the vibrations associated with tetrahedral cations are localized at a higher frequency. The bands between 600 and 800 cm^{-1} are due to the bending vibrations of the O-T-O bonds and the TO_4 tetrahedra, more particularly to the Si movement in its tetrahedral cage accompanied by a slight movement of the oxygen atoms. Their intensity and frequency decrease with increasing polymerization (Sharma et al., 1984; McMillan, 1994). Finally, the high frequency bands in the 800–1200 cm^{-1} envelope represent the symmetric stretching vibrations of the T-O-T bonds. Their intensity and frequency increase with increasing polymerization (Mysen et al., 1980; McMillan, 1984).

The bending and stretching vibration are essentially in two frequency domains 500–700 cm^{-1} and 800–1200 cm^{-1} and are strongly correlated with the various tetrahedral Q species and thus with non-bridging oxygens (Q^n, where n = number of bridging oxygens) (Mysen et al., 1980; McMillan, 1984). In particular, the relatively high-frequency envelope has several contributions. As a result, it has undergone numerous deconvolution and modeling treatments to determine the number, position, and intensity of contributions for various glass compositions (Mysen et al., 1980, 1981; Seifert et al., 1982, among others). Since Mysen et al., (1980), it is commonly accepted that:

- Q^0 species (SiO_4^{4-}) are between 850 and 880 cm^{-1};
- Q^1 ($Si_2O_7^{6-}$) between 900 and 920 cm^{-1};
- Q^2 ($Si_2O_6^{4-}$) between 950 and 980 cm^{-1};
- Q^3 ($Si_2O_5^{2-}$) between 1050 and 1100 cm^{-1};

• Q^4 (SiO_2) between 1060 and 1190 cm^{-1}.

Furthermore, another band is generally used, called T_{2s}, between 950 and 1050 cm^{-1}, and corresponds to a stretching vibrational mode of TO_4 tetrahedra (see details in Le Losq et al., 2014). Fig. 13.10a,b shows a spectrum of an iron-free silicate glass composition and the deconvolution into five tetrahedral units of the high-frequency envelope.

Raman studies have shown that, in alkali and alkaline-earth silicate glasses, the abundances of the Q species are linked by the disproportionation reactions:

$$2\,Q^n \Leftrightarrow Q^{n+1} + Q^{n-1}, \qquad (13.11)$$

with 4 > n > 0. The equilibrium constants of those reactions depend on temperature, aluminium concentration as well as the mean ionic field strength of the non-network former cations. More details can be found in, e.g., Mysen (1988, 1990, 1996) and Mysen et al. (2003).

13.7.5. Raman Spectra of Iron Silicate Glass

Silicate compositions containing iron have been studied by Raman spectroscopy, mainly to interpret the structural role of iron and to try to link the achieved redox equilibrium to the silicate glass/melt structure (Mysen et al., 1980, 1984; Fox et al., 1982; Cooney & Sharma, 1990; Wang et al, 1995). More recently, studies also focused on the determination of Fe redox ratio in glasses from their Raman spectra (Magnien et al., 2006; Roskosz et al., 2008; Di Muro et al., 2009; Cochain et al., 2012; Neuville et al., 2014a; Di Genova et al., 2016; Le Losq et al., 2019). The vibrations of the bonds depending on the substitution of silicon by other forming cations (Fe^{3+} in particular), the Raman bands, and in particular the high-frequency envelope can be considered as tools for characterizing the structural role of iron.

Magnien et al. (2006) performed an analysis of spectra of iron-pyroxene glasses (Fig. 13.11), revealing that the intensity of bands at ~910 and ~550 cm^{-1} increases with increasing the oxidation of those glasses from 0.09 to 0.99. Despite clear changes, the analysis of Raman spectra with the redox state remained purely qualitative in this study. It corroborated results from earlier works (Mysen et al., 1980; Virgo et al., 1983; Wang et al., 1995) that showed that Fe^{3+}-O stretching vibrations give rise to a signal between 900 and 990 cm^{-1} in Raman spectra of iron-bearing silicate and aluminosilicate glasses. Fe^{2+} in the octahedral position does not yield any signal, or possibly at low frequencies (< 500 cm^{-1}) mixed with stronger signals from T-O-T bonds in silicate glasses.

Fig. 13.12 a,b show Raman decompositions of the 0.09 and 0.99 spectra shown in the previous figure. The intensity of the 910 cm^{-1} band increases clearly with the Fe^{3+} content in the iron-pyroxene glass. This 910 cm^{-1} band

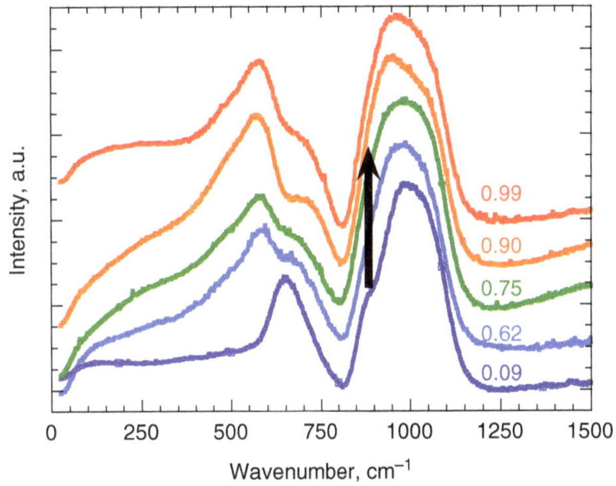

Figure 13.11 Raman spectra at room temperature for a series of iron-pyroxene glasses. The Fe^{3+}/Fe^{tot} redox ratio is indicated to the right of the spectra and the curves have been shifted to facilitate the observation of the changes. These spectra were normalized at the maximum intensity of the spectra corrected from the excitation lines (see Neuville et al., 2014). Figure adapted from Magnien et al., 2006.

shifts in frequency as a function of the chemical composition of the glass. As a consequence, in aluminosilicates, borosilicates, or highly polymerized compositions, this band will be rather be located around 990 cm^{-1}. In borates, this band is around 1300 cm^{-1} whereas in aluminates it is at 650 cm^{-1} (El Hayek, 2016). In any case, in silicate glasses, this Raman band at ~ 910 cm^{-1} correlates perfectly with the variations of the Fe pre-edge peak observed in Fe K-edge XANES spectra.

As a result, it is possible to calibrate the variations of intensity of this peak with the iron redox state in the glass. For instance, the iron-pyroxene glass was investigated by Magnien et al. (2004, 2006, 2008) with different techniques, like Mössbauer spectroscopy, wet chemical analyses, XANES at the Fe K-edge, and Raman spectroscopy. The comparison of all these techniques are plotted in Fig. 13.13 and show a good agreement of the Fe redox states measured by the different techniques. However, it should be remembered that XANES and wet-chemical analyses are absolute measurements, whereas electronic microprobe and Raman spectroscopy need calibrations.

The measurement of the iron redox ratio by Raman spectrometry is based on the use of a set of the Raman spectra of glass standards with known redox ratio, and the establishment of a calibration linking changes in the spectra to variations in the glasses' iron redox ratios. After acquisition, several methods can be used:

• calibrate the changes in intensity of the peak around 910 cm^{-1} to the iron redox state in the glass (Magnien et al., 2006; Roskosz et al., 2008; Le Losq et al., 2019);

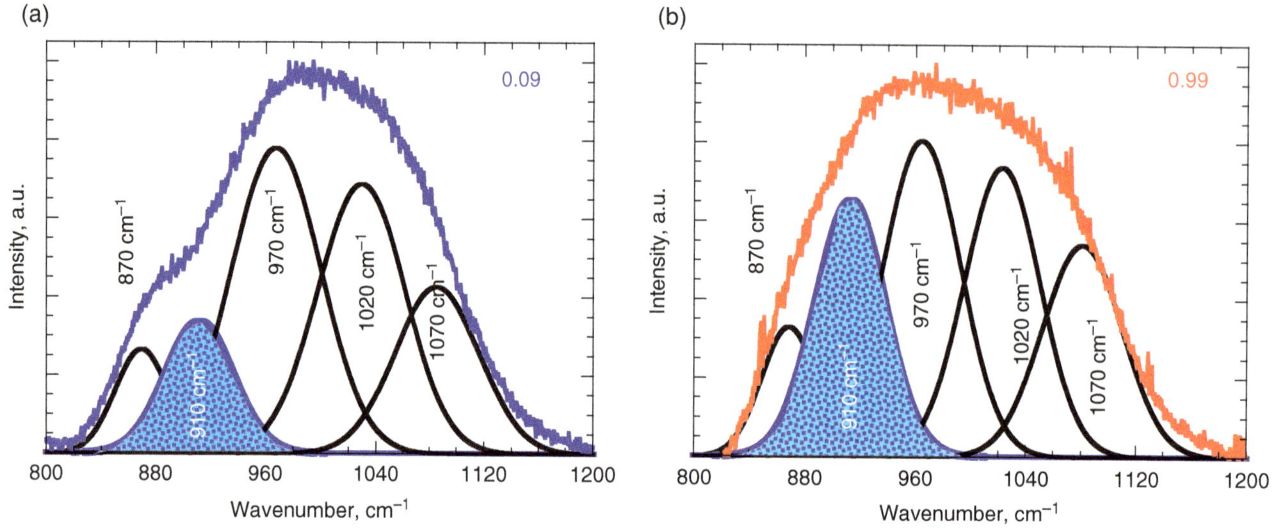

Figure 13.12 a,b: Deconvolution of the 0.09 and 0.99 Raman spectra (based on Magnien et al., 2006).

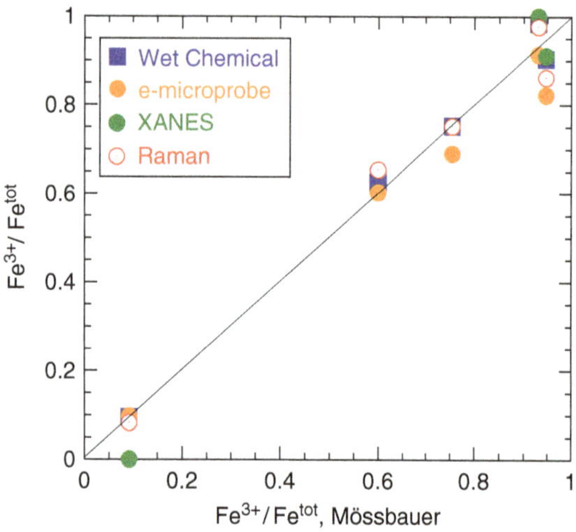

Figure 13.13 Comparison of iron ratios analyzed by different techniques, wet chemical, e-microprobe, XANES at the Fe K-edge, and Raman spectroscopy (redrafted from Magnien et al., 2004, 2006).

• use a linear mixture of end-member spectra (Di Genova et al., 2016);
• peak-fitting (Di Muro et al., 2009; Welsh et al., 2017);
• machine learning methods (Le Losq et al., 2019).

Le Losq et al. (2019) recently compared the use of six methods (intensity, linear mixture, and machine learning algorithms like neural networks or support vector machines) on a set of basalt glass standards synthesized in the lab at well know oxygen fugacity conditions, and analyzed by Mossbauer spectroscopy. They demonstrated that, on a given glass composition, an accuracy of 0.02 Fe^{3+}/Fe^{tot} can be reached with Raman spectroscopy. However, they also showed that, as small changes in glass compositions affect quite strongly the glass structure and, hence, Raman spectra, Fe redox Raman determination is very sensitive to matrix effects. With using a correction for this effect, they were able to measure the redox state of mid-ocean ridge basalts at 0.09 ± 0.06 Fe^{3+}/Fe^{TOT}.

Advantages, drawbacks, accuracy

Raman spectroscopy benefits from several advantages:
• micrometric resolution, allowing mapping and precise spatial measurements;
• non-destructive (but… see below);
• good precision can be achieve with simple technique like following the intensity of a peak (down to 0.02 Fe^{3+}/Fe^{TOT});
• possibility of in situ experiments, at high temperature and high pressure or directly in the field with a portable Raman spectrometer.

However, several drawbacks also affect this method:
• it is a relative method, requiring glass standards…
• … and chemical effects are strong as Raman spectra are a picture of the entire vibrational landscape of the samples;
• at elevated laser power, beam damage can be a problem (e.g., see Di Genova et al., 2017), but this can be avoided with the use of filters to decrease laser power;
• consistent, systematic data treatment protocol are necessary to obtain reproducible results; consequently, data processing via computer programming is strongly advised (spectrum-by-spectrum hand treatment should be avoided at all cost);
• strong sensitivity to the structure, hence composition of glasses, such that changes in major element composition and redox-related changes are mixed and difficult to desantangle;
• fluorescence and luminescence are often important problems when analyzing natural samples;
• it is rare to have pure glasses in natural samples, free from any crystals even nanolites, making the use of Raman challenging as soon as crystal peaks are detected.

In the future, it may be possible to study other pairs of redox than Fe^{3+}/Fe^{tot} in glasses, but for the moment it is the only redox couple well characterized. However, couples such as Ce^{4+}/Ce^{3+} may be considered. Such study requires that the two elements be in different roles depending on their redox, network former or network modifier. A priori Raman spectroscopy should not work to study Cr redox as Cr structural role in silicate glass is independent of its redox. One may note that Raman spectrometry can also be used to investigate redox volatile elements as recently demonstrated by Cicconi et al. (2019a,b) for the study of iodine dissolved in vitreous silicate.

13.8. IN SITU REDOX DETERMINATION AT HIGH TEMPERATURE OR AT HIGH PRESSURE

Recently Mössbauer spectroscopy was used at high pressure to measure the redox state in a diamond anvil cell at 100 GPa. This experiment is very fast, around 10 minutes, whereas such an experiment with a radioactive point source would take more than one week and the data quality would be considerably less (Potapkin et al., 2012). These new technical developments carried out at the ESRF are very promising and should allow in the coming years to better answer the questions of redox of the mantle phases, and more generally to improve the knowledge of redox equilibria at high pressure.

High temperature experiment: several setups were developed to conduct high-temperature experiment to investigate redox state: a HT furnace to made XANES (Berry et al., 2003b); the levitation device (Alderman et al., 2017), and the micro heating device originally developed by Mysen and Frantz (1992) and modify to use for XANES, WAXS, IR, Raman spectroscopies (Neuville, et al, 2014b).

In the next part, two examples will be shown using a micro heating device to measure Raman spectroscopy and XANES at the Fe K-edge on the same pyroxene (50%SiO_2–5%Na_2O–20% MgO–20%CaO-5%FeO) composition at high temperature (Fig. 13.14).

Fe^{3+}/Fe_{Tot} redox ratio plotted in Fig. 13.15 is calculated from the XANES treatment of the pre-edge and Raman deconvolution following procedure given in

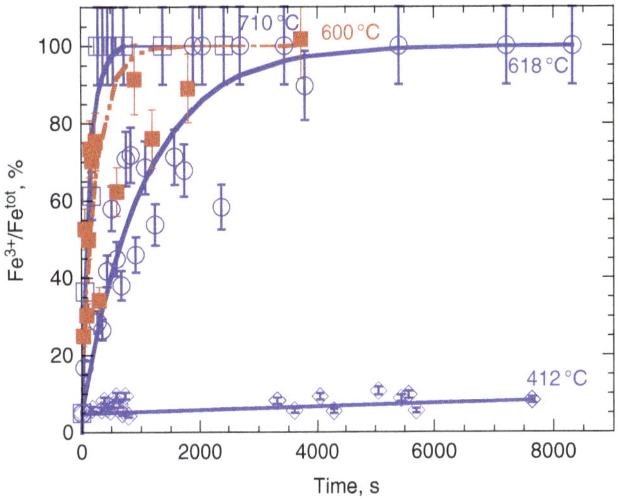

Figure 13.15 Fe^{3+}/Fe_{Tot} redox as different temperatures, 412, 618, 710 °C from XANES measurements at the Fe K-edge and at 600 °C from Raman spectroscopy (redrafted from Magnien et al., 2006, 2008).

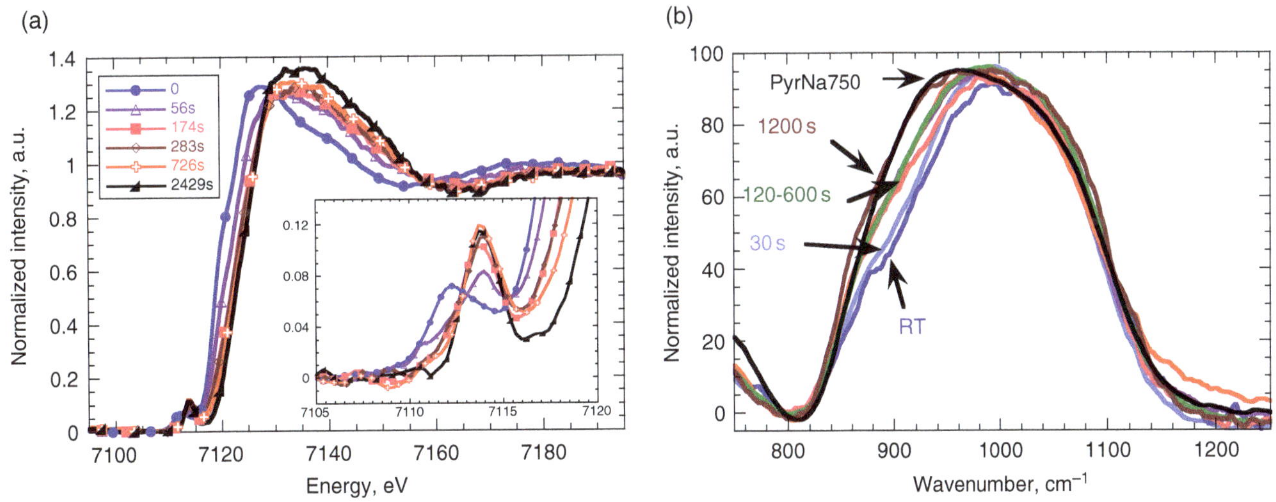

Figure 13.14 (a) Evolution of the normalized XANES spectra at the Fe K-edge and a zoom on the associated pre-edge of PyrNa17R at 710 °C (T g + 100 °C) as a function of time. The times are indicated in the legend. (b) Raman spectra as a function of time for PyrNa17R at 600 °C (redrafted from Magnien, 2005).

Table 13.2 Summarize of analytical techniques and their advantages.

	Wet chemical	Electron microprobe	Mossbauer	XANES	Optical absorption	Raman
All elements	X			X	X	
Specific		Fe	Fe, Sb, I, Ni...			Fe
Destructive	X	X	X	X	X	
Not preparation						X
Short preparation	X	X				
Long preparation			X	X	X	
Easy access	X	X			X	X
Request access			X	X		

Sections 13.6 and 13.7 or more detailed in Magnien et al. (2006, 2008).

The variation of redox ratio as a function of time at different temperature observed by Raman spectroscopy and by XANES at the Fe K-edge show similar variation, which proves a good agreement between the two techniques.

From this redox variation at high temperature observed by Raman spectroscopy or XANES at the Fe K-edge it is possible to determine redox diffusion coefficient as shown by Magnien et al. (2008), for example. This was also recently used by Le Losq et al. (2020) to follow redox disequilibrium at high temperature on the andesite of Erebus volcanoes.

13.9. CONCLUSION

In this chapter, we briefly reviewed six techniques that allow the quantification of the oxidation state of an element in different samples (mineral, glasses, rocks) and in different physical states (solid, melts). Each method presents different advantages and drawbacks. Some will be suited for absolute measurements like wet chemistry, while others require standards but may allow mapping (electron microprobe or Raman spectroscopy). Table 13.2 summarize the six techniques and their advantages and access.

Future developments in term of technique, like synchrotron Mössbauer, or data treatment with, in particular, machine learning, may allow pushing further the boundaries of possibilities and seeking more and more information from experiments and naturals samples. Raman spectroscopy and XANES at the different M, L, or K-edge can be also measured directly at high temperature which can provide the possibility to measure and follow redox kinetics of melts directly at high temperature (Magnien et al., 2008; Neuville et al., 2014; Prieur et al., 2018; Caisso et al., 2017).

ACKNOWLEDGMENTS

MRC thanks the Naturhistorische Gesellschaft Nürnberg, Prof. Dr. Mischka and Ms. Horlacher for the archeological glass.

REFERENCES

Abe, H., Aquilanti, G., Boada, R., Bunker, B., Glatzel, P., Nachtegaal, M., & Pascarelli, S. (2018). Improving the quality of XAFS data. *Journal of Synchrotron Radiation*, *25*(Pt 4), 972–980. doi:10.1107/S1600577518006021

Albee, A. L., & Chodos, A. A. (1970). Semiquantitative electron microprobe determination of Fe 2+/Fe 3+ and Mn2+/Mn3+ in oxides and silicates and its application to petrologic problems. *American Mineralogist*, *55*, 491–501.

Alderman O., Wilding M. C., Tamalonis A., Sendelbach S., Heald S. M., Benmore C. J., et al. (2017). Iron K-edge X-ray absorption near-edge structure spectroscopy of aerodynamically levitated silicate melts and glasses. *Chemical Geology*, *453*, 169–85. doi: 10.1016/j.chemgeo.2017.01.020

Baek S.-J., Park A., Ahn Y.-J., & Choo J. (2015) Baseline correction using asymmetrically reweighted penalized least squares smoothing. *Analyst*, *140*, 250–257. doi: 10.1039/C4AN01061B

Bajt S., Sutton S. R., & Delaney J. S. (1994). X-ray microprobe analysis of iron oxidation states in silicates and oxides using X-ray absorption near edge structure (XANES). *Geochimica et Cosmoschimica Acta*, *58*, 5209–5214. https://doi.org/10.1016/0016-7037(94)90305-0

Bancroft, M., Nesbitt, H. W., Henderson, G. S., O'Shaughnessy, C., Withers, A. C., & Neuville, D. R. (2018). Lorentzian dominated lineshapes and linewidths for Raman symmetric stretch peaks (800–1200 cm−1) in Qn (n=1–3) species of alkali silicate glasses/melts. *Journal of Non-Crystalline Solids*. *484*, 72–83. https://doi.org/10.1016/j.jnoncrysol.2018.01.018

Berry A. J., & O'Neill H.St.C. (2004). A XANES determination of the oxidation state of chromium in silicate glasses. *American Mineralogist*, *89*, 790–798. https://doi.org/10.2138/am-2004-5-613

Berry, A. J., O'Neill, H.St.C., Jayasuriya, K. D., Campbell, S. J., & Foran, G. J. (2003a). XANES calibration for the oxidation state of iron in a silicate glass. *American Mineralogist*, *88*, 967–977. https://doi.org/10.2138/am-2003-0704

Berry, A. J., Shelley, J. M. G., Foran, G. J., O'Neill, H.St.C., & Scott, D. R. (2003b). A furnace design for XANES spectroscopy of silicate melts under controlled oxygen fugacities and temperatures to 1773 K. *Journal of Synchrotron Radiation*, *10*, 332–336. 10.1107/s0909049503007556

Berry, A., O'Neill, H.St.C., Scott, D. R., Foran, G. J., & Shelley, J. M. G. (2006a). The effect of composition on Cr 2+ /Cr 3+ in silicate melts. *American Mineralogist*, *91*, 1901–1908. https://doi.org/10.2138/am.2006.2097

Berry, A., Hack, A. C., Mavrogenes, J. A., Newville, M., & Sutton, S. R. (2006b). A XANES study of Cu speciation in high-temperature brines using synthetic fluid inclusions. *American Mineralogist*, *91*, 1773–1782. https://doi.org/10.2138/am.2006.1940

Berry, A. J., Yaxley, G. M., Woodland, A. B., & Foran, G. J. (2010). A XANES calibration for determining the oxidation state of iron in mantle garnet. *Chemical Geology*, *278*, 31–37. https://doi.org/10.1016/j.chemgeo.2010.08.019

Berry, A. J., Stewart, G. A., O'Neill, H.St.C., Mallmann, G., & Mosselmans, J. F. W. (2018). A re-assessment of the oxidation state of iron in MORB glasses. *Earth and Planetary Science Letters*, *483*, 114–123. https://doi.org/10.1016/j.epsl.2017.11.032

Bianconi, A., Marcelli, A., Dexpert, H., Karnatak, R., Kotani, A., Jo, T., & Petiau, J. (1987). Specific intermediate-valence state of insulating 4*f* compounds detected by L3 x-ray absorption. *Physical Review B*, *35*, 806–812. doi: 10.1103/physrevb.35.806

Bingham, P. A., Parker, J. M., Searle, T., Williams, J. M., & Smith, I. (2002). Novel structural behaviour of iron in alkali-alkaline-earth-silica glasses. *Comptes Rendus Chimie*, *5*, 787–796. https://doi.org/10.1016/S1631-0748(02)01444-3

Bourdelle, F., Benzerara, K., Beyssac, O., Cosmidis, J., Neuville, D. R., Brown, G. E., Jr, & Paineau, E. (2013). Quantification of the ferric/ferrous iron ratio in silicates by scanning transmission X-ray microscopy at the Fe L2,3 edges. *Contribution to Mineralogy and Petrology*, *166*, 423–434. https://doi.org/10.1007/s00410-013-0883-4

Burnham, A. J. Berry, Halse, H. R., Schofield, P. F., Cibin, G., & and Mosselmans, J. F. W. (2015). The oxidation state of europium in silicate melts as a function of oxygen fugacity, composition and temperature. *Chemical Geology*, *411*, 248–259. https://doi.org/10.1016/j.chemgeo.2015.07.002

Brawer, S. (1975). Theory of the vibrational spectra of some network and molecular glasses. *Physical Review B*, *11*, 3173–3194. https://doi.org/10.1103/PhysRevB.11.3173

Brawer, S., & White, W. B. (1977). Raman spectroscopic investigation of the structure of silicate glasses (II). Soda-alkaline earth-alumina ternary and quaternary glasses. *Journal of Non-Crystalline Solids*, *23*, 261–278. https://doi.org/10.1016/0022-3093(77)90009-6

Brown, G. E., Calas, G., Waychunas, G. A., & Petiau, J. (1988). X-ray absorption spectroscopy; applications in mineralogy and geochemistry. *Reviews in Mineralogy and Geochemistry*, *18*, 1, 431–512.

Bunker, G. (2010). *Introduction to XAFS: A Practical Guide to X-ray Absorption Fine Structure Spectroscopy*. Cambridge: Cambridge University Press.

Burnham, A. D. D., & Berry, A. J. (2014). The effect of oxygen fugacity, melt composition, temperature and pressure on the oxidation state of cerium in silicate melts. *Chemical Geology*, *366*, 52–60. https://doi.org/10.1016/j.chemgeo.2013.12.015

Caisso, M., Lebreton, F., Horlait, D., Picart, S., Martin, P. M., Bès, R., et al. (2014). Nanostructured gadolinium-doped ceria microsphere synthesis from ion exchange resin: Multi-scale in-situ studies of solid solution formation. *Journal of Solid State Chemistry*, *218*, 155–163. https://doi.org/10.1016/j.jssc.2014.06.028

Caisso, M., Picart, S., Belin, R., Lebreton, F., Martin, P. Dardenne, K., et al. (2015). In-situ characterization of uranium and americium solid solution formation for CRMP process: first combination of in-situ XRD and XANES measurements. *Dalton Transactions*, *44*, 6391–6399. https://doi.org/10.1039/C4DT03515A

Calas G., & Petiau, J. (1983a). Structure of oxides glasses. Spectroscopic studies of local order and crystallochemistry. Geochemical implications. *Bulletin de Mineralogie*, *106*, 33–55.

Calas, G., & Petiau, J. (1983b). Coordination of iron in oxide glasses through high resolution K-edge spectra: information from pre-edge. *Solid State Communications*, *48*, 625–629. https://doi.org/10.1016/0038-1098(83)90530-6

Cerantola, V., Bykova, E., Kupenko, I., Merlini, M., Ismailova, L., et al. (2017). Stability of iron-bearing carbonates in the deep Earth's interior. *Nature Communications*, *8*, 15960. doi: 10.1038/ncomms15960

Cicconi, M. R. R., Giuli, G., Paris, E., Ertel-Ingrisch, W., Ulmer, P., & Dingwell, D. B. (2012). Europium oxidation state and local structure in silicate glasses. *American Mineralogist*, *97*(5–6), 918–929. https://doi.org/10.2138/am.2012.4041

Cicconi, M. R., Giuli, G., Ertel-Ingrisch, W., Paris, E., & Dingwell, D. B. (2015a). The effect of the [Na/(Na+K)] ratio on Fe speciation in phonolitic glasses. *American Mineralogist*, *100* (7), 1610–1619. https://doi.org/10.2138/am-2015-5155

Cicconi, M. R., Neuville, D. R., Tannou, I., Baudelet, F., Floury, P., Paris, E., & Giuli, G. (2015b). Competition between two redox states in silicate melts: an in-situ experiment at the Fe K-edge and Eu L3-edge. *American Mineralogist*. *100*, 1013–1016. https://doi.org/10.2138/am-2015-5172

Cicconi, M. R., Veber, A., de Ligny, D., Rocherullé, J., Lebullenger, R., & Tessier, F. (2017a). Chemical tunability of europium emission in phosphate glasses. *Journal of Luminescence*, *183*, 53–61. https://doi.org/10.1016/j.jlumin.2016.11.019

Cicconi, M. R., Neuville, D. R., Blanc, W., Lupi, J.-F., Vermillac, M., & de Ligny, D. (2017b). Cerium/aluminum correlation in aluminosilicate glasses and optical silica fiber preforms. *Journal of Non-Crystalline Solids*, *475*, 85–95. doi:10.1016/j.jnoncrysol.2017.08.035

Cicconi, M. R., Pili, E., Grousset, L., & Neuville, D. R. (2019a). The Influence of Glass Composition on Iodine Solubility. MRS Advances. pp. 1–9 https://doi.org/10.1557/adv.2018.665

Cicconi, M. R., Pili, E., Grousset, L., Florian, P. Bouillard, J. C., Vantelon, D., & Neuville, D.R. (2019b). Iodine dissolution and speciation in glasses. *Scientific Report*. https://doi.org/10.1038/s41598-019-44274-4

Cicconi, M. R., Lu, Z., Uesbeck, T., van Wüllen, L., Brauer, D. S., & de Ligny, D. (2020). Influence of Vanadium on optical and mechanical properties of aluminosilicate glasses. *Frontiers in Materials*. https://doi.org/10.3389/fmats.2020.00161

Charlot, G. (1969). *Methods of Analytical Chemistry. Quantitative Analysis of Inorganic Compounds*. 2nd Edition, Chem., Moscow, 668pp.

Cochain, B., Neuville, D. R., de Ligny, D., Roux, J., Baudelet, F., Strukelj, E., & Richet, P. (2009). Kinetics of iron redox reaction in silicate melts: A high-temperature Xanes study on an alkali basalt. *Journal of Physics: IV*, *190*, 12182–12186.

Cochain, B., Neuville, D. R., Henderson, G. S., McCammon, C., Pinet, O., & Richet, P. (2012). Iron content, redox state and structure of sodium borosilicate glasses: A Raman, Mössbauer and boron K-edge XANES spectroscopy study. *Journal of the American Ceramics Society*, *94*, 1–12. https://doi.org/10.1111/j.1551-2916.2011.05020.x

Cook, G.B., & Cooper, R.F. (2000). Iron concentration and the physical processes of dynamic oxidation in alkaline earth aluminosilicate glass. *American Mineralogist*, *85*, 397–406. https://doi.org/10.2138/am-2000-0401

Cook, G. B., Cooper, R. F., & Wu, T. (1990). Chemical diffusion and crystalline nucleation during oxidation of ferrous iron-bearing magnesium aluminosilicate glass. *Journal of Non-Crystalline Solids*, *120*, 207–222. https://doi.org/10.1016/0022-3093(90)90205-Z

Cooney, T. F., & Sharma, S. K. (1990). Structure of glasses in the systems Mg2SiO4-Fe2SiO4, Mn2SiO4-Fe2SiO4, Mg2SiO4-CaMgSiO4 and Mn2SiO4-CaMnSiO4. *Journal of Non-Crystalline Solids*, *122*, 10–32. https://doi.org/10.1016/0022-3093(90)90220-G

Cooper, R. F., Fanselow, J. B., & Poker, D. B. (1996). The mechanism of oxidation of a basaltic glass: chemical diffusion of network-modifying cations. *Geochimica et Cosmochimica Acta*, *60*, 17, 3253–3265. https://doi.org/10.1016/0016-7037(96)00160-3

Cooper, R. F., Fanselow, J. B., Weber, J. K. R., Merkley, D. R., & Poker, D. B. (1996). Dynamics of oxidation of a Fe2+-bearing aluminosilicate (basaltic) melt. *Science*, *274*, 1173–1176. doi: 10.1126/science.274.5290.1173

Cottrell, E., Lanzirotti, A., Mysen, B. O., Birner, S., Kelley, K. A., Botcharnikov, R., et al. (2018). A Mössbauer-based XANES calibration for hydrous basalt glasses reveals radiation-induced oxidation of Fe. *American Mineralogist*, *103*, 489–501. https://doi.org/10.2138/am-2018-6268

de Groot, F. M. F., Vankó, G., & Glatzel, P. (2009). The 1s x-ray absorption pre-edge structures in transition metal oxides. *Journal of Physics: Condensed Matter*, *21*, 10, 104207. https://doi.org/10.1088/0953-8984/21/10/104207

de Ligny, D., & Möncke, D. (2019). Colors in Glasses. Chapter 9 In: *Springer Handbook of Glass*. Musgraves, J. D., Hu, J., Calvez, L. (Eds). doi: 10.1007/978-3-319-93728-1

de Ligny, D., & Neuville, D. R. (2017). Raman spectroscopy: a nice tools to understand nucleation and growth mechanism. In: From glass to crystal: nucleation, growth and phase separation, from research to applications. Neuville, D. R., Cormier, L., Caurant, D., & Montagne, L. (Eds.) EDP-Sciences.

Di Genova, D., Morgavi, D., Hess, K.-U., Neuville, D. R., Borovkov, N., Perugini D., & Dingwell, D. B. (2015). Approximate chemical analysis of volcanic glasses using Raman spectroscopy. *Journal of Raman Spectroscopy*. doi: 10.1002/jrs.4751

Di Genova, D., Kolzenburg, S., Vona, A., Chevrel, M. A., Hess, K.-U., Neuville, D. R., et al. (2016) Raman spectra of Martian glass analogues: a tool to approximate their chemical composition. *Journal of Geophysical Research Planet*, *121*, 740–752. https://doi.org/10.1002/2016JE005010

Di Genova, D., Borovkov, N., Vasseur, J., Hess, K.-U., Neuville, D. R., & Dingwell, D. B. (2017). Structure, glass transition and viscosity of alkali- and iron-rich glasses and melts with different oxygen fugacity. *Journal of Non-Crystalline Solids*, *470*, 78–85. doi: 10.1016/j.jnoncrysol.2017.05.013

Doyle, P. M., Berry, A. J., Schofield, P. F., & Mosselmans, J. F. W. (2016). The effect of site geometry, Ti content and Ti oxidation state on the Ti K-edge XANES spectrum of synthetic hibonite. *Geochimica et Cosmochimica Acta*, *187*, 294–310. https://doi.org/10.1016/j.gca.2016.05.001

Dyar, M. D., Gunter, M. E., & Tasa, D. (2008). *Mineralogy and optical mineralogy* (p. 708). Chantilly, VA: Mineralogical Society of America.

Farges, F. (2005). Ab initio and experimental pre-edge investigations of the Mn K -edge XANES in oxide-type materials. *Physical Review B*, *71*, 15, 155109. https://doi.org/10.1103/PhysRevB.71.155109

Farges F., Lefrère Y., Rossano, S., Berthereau, A., Calas, G., & Brown, Jr., G. E. (2004). The effect of redox state on the local structural environment of iron in silicate glasses: a combined XAFS spectroscopy, molecular dynamics and bond valence study. *Journal of Non-Crystalline Solids*, *344*, 176–188. https://doi.org/10.1016/j.jnoncrysol.2004.07.050

Fialin, M., Bezos, A., Wagner, C., Magnien, V., & Humler, E. (2004). Quantitative electron microprobe analysis of Fe3+/ΣFe: Basic concepts and experimental protocol for glasses. *American Mineralogist*, *89*, 654–662. https://doi.org/10.2138/am-2004-0421

Fialin, M., Wagner, C., Metrich, N., Humler, E., Galoisy, L., & Bezos, A. (2001). Fe3+/ ΣFe vs. FeLα peak energy for minerals and glasses: recent advances with the electron microprobe. *American Mineralogist*, *86*, 456–465. https://doi.org/10.2138/am-2001-0409

Fisher, D. W. (1965). Change in the soft x-ray L emission spectra with oxidation of the first series transition metals. *Journal of Applied Physics*, *36*, 2048–2053. https://doi.org/10.1063/1.1714400

Fox, K. E., Furakawa, Y., & White, W. B. (1982). Transition metals ions in silicate melts. Part 2. Iron in sodium silicate glasses. *Physics and Chemistry of Glasses*, *23*, 169–178.

Fudali F. (1965). Oxygen fugacity of basaltic and andesitic magmas. *Geochimica et Cosmochimica Acta*, *29*, 1063–1075. https://doi.org/10.1016/0016-7037(65)90103-1

Fukumi, K., Hayakawa J., & Komiyama, T. (1990). Intensity of raman band in silicate glasses. *Journal of Non-Crystalline Solids*, *119*, 297–302. https://doi.org/10.1016/0022-3093(90)90302-3

Furakawa, T., & White, W. B. (1980). Vibrational spectra and glass structure. *Journal of Non-Crystalline Solids*, *38/39*, 87–92. https://doi.org/10.1016/0022-3093(80)90399-3

Gaborieau, M., Laubier, M., Bolfan-Casanova, N., McCammon, C., Vantelon, D., Chumakov A. I., et al. (2020). Determination of Fe3+/ΣFe of olivine-hosted melt inclusions using Mössbauer and XANES spectroscopy. *Chemical Geology*. doi: 10.1016/j.chemgeo.2020.119646

Galoisy, L. (2014). X-ray absorption spectroscopy in geosciences. In: *Spectroscopic methods in mineralogy* (pp. 553–587). EMU notes in Mineralogy 6. https://doi.org/10.1180/EMU-notes.6.13

Galoisy, L., Calas, G., & Arrio, M. A. (2001). High-resolution XANES spectra of iron in minerals and glasses: structural information from the pre-edge region. *Chemical Geology, 174*, 307–319. https://doi.org/10.1016/S0009-2541(00)00322-3

Geiger, C. A. (2004). An introduction to spectroscopic methods in the mineral sciences and geochemistry. In: *Spectroscopic Methods in Mineralogy*. Germany: Mineralogical Society of Great Britain and Ireland; 1–42. doi:10.1180/EMU-notes.6.1

Giuli, G., Paris, E., Hess, K.-U., Dingwell, D. B., Cicconi, M. R., Eeckhout, S. G., et al. (2011). XAS determination of the Fe local environment and oxidation state in phonolite glasses. *American Mineralogist, 96*, 4. https://doi.org/10.2138/am.2011.3464

Giuli, G., Alonso-Mori, R., Cicconi, M. R., Paris, E., Glatzel, P., Eeckhout, S. G., & Scaillet, B. (2012a). Effect of alkalis on the Fe oxidation state and local environment in peralkaline rhyolitic glasses. *American Mineralogist, 97*, 2–3. https://doi.org/10.2138/am.2012.3888

Giuli, G., Cicconi, M. R., & Paris, E. (2012b). The $^{[4]}Fe^{3+}$–O distance in synthetic kimzeyite garnet, $Ca_3Zr_2[Fe_2SiO_{12}]$. *European Journal of Mineralogy, 24*, 5, 783–790. https://doi.org/10.1127/0935-1221/2012/0024-2206

Gonçalves Ferreira, P., de Ligny, D., Lazzari, O., Jean, A., Cintora Gonzalez, O., & Neuville, D. R. (2013). Photoreduction of iron by a synchrotron X-ray beam in low iron content soda-lime silicate glasses. *Chemical Geology, 346*, 106–112. doi: 10.1016/j.chemgeo.2012.10.029

Harris, D. C., & Bertolucci, M. D. (1978). *Symmetry and Spectroscopy: An Introduction to Vibrational and Electronic Spectroscopy*. Oxford University Press, New York

Henderson, G. S., De Groot, F. M., & Moulton, B. J. (2014). X-ray absorption near-edge structure (XANES) spectroscopy. *Reviews in Mineralogy and Geochemistry, 78*, 1, 75–138. doi: 10.2138/rmg.2014.78.3

Herzberg, G. (1945a). Molecular spectra and molecular structure. I. Spectra of diatomic molecule. Princeton Press.

Herzberg, G. (1945b). Molecular spectra and molecular structure. II. Infrared and Raman spectra of polyatomic molecules. Princeton Press.

Höche, T., Ikeno, H., Mäder, M., Blyth, R., Sales, B. C., & Tanaka, I. (2013). Vanadium L2,3 XANES experiments and first-principles multielectron calculations: Impact of second-nearest neighboring cations on vanadium-bearing fresnoites. *American Mineralogist, 98*, 665–670. https://doi.org/10.2138/am.2013.4335

Höfer, H. E., Brey, G. P., Schulz-Dobrick, B., & Oberhänsly, R. (1994). The determination of the oxidation state of iron by the electron microprobe. *European Journal of Mineralogy, 6*, 407–418. doi: 10.1127/ejm/6/3/0407

Hughes, E., Buse, B., Kearns, S., Blundy, J, Kilgour, G., Mader, H., et al. (2018). High spatial resolution analysis of the iron oxidation state in silicate glasses using the electron probe. *American Mineralogist, 103*, 1473–1486. https://doi.org/10.2138/am-2018-6546CCBY

Hughes, E. C., Buse, B., Kearns, S. L., Brooker, R. A., Di Genova, D., Kilgour, G., et al. (2020). The microanalysis of iron and sulphur oxidation states in silicate glass - Understanding the effects of beam damage. IOP Conf. Series: Materials Science and Engineering 891, 012014 IOP Publishing - doi:10.1088/1757-899X/891/1/012014

Hunault, M. O. J. Y., Galoisy, L., Lelong, G., Newville, M., & Calas, G. (2016). Effect of cation field strength on Co2+ speciation in alkali-borate glasses. *Journal of Non-Crystalline Solids, 451*, 101–110. https://doi.org/10.1016/j.jnoncrysol.2016.06.025

Jayasuriya, K. D., O'Neill, H.St.C., Berry, A. J., & Campbell, S. J. (2004). A Mössbauer study of the oxidation state of iron in silicate melts. *American Mineralogist, 89*, 1597–1609. https://doi.org/10.2138/am-2004-11-1203

Kaindl, G., Wertheim, G. K., Schmiester, G., & Sampathkumaran, E. V. (1987). Mixed valency versus covalency in rare-earth core-electron spectroscopy. *Physical Review Letters, 58*, 606–609. doi: 10.1103/PhysRevLett.58.606

Leister, M., Ehrt, D., Von der Goenna, G., Ruessel, C., & Breitbarth, F. (1999). Redox states and coordination of vanadium in sodium silicates melted at high temperatures. *Physics and Chemistry of Glasses, 40*, 319–325.

Long, D. A. (1977). *Raman spectroscopy*. Great Britain, McGraw-Hill, Inc. 275 pp.

Le Losq, C., Neuville, D. R., Moretti, R., & Roux, J. (2012). Water quantification and speciation in silicate melt using Raman spectroscopy. *American Mineralogist, 97*, 779–791.

Le Losq, C., Neuville, D. R., Moretti, R., Oppenheimer, C., & Kyle, P. (2015). Rheology of phonolitic magmas – the case of the Erebus lava lake. *Earth and Planetary Science Letters, 411*, 53–61. https://doi.org/10.1016/j.epsl.2014.11.042

Le Losq, C., Berry, A. J., Kendrick, M. A., Neuville, D. R., & O'Neill, HSt. C. (2019). Determination of the oxidation state of iron in basaltic glasses by Raman spectroscopy. *American Mineralogist, 104*, 1032–1042.

Le Losq, C., Moretti, R., Oppenheimer, C., Neuville, D. R. (2020). In situ XANES study of the influence of varying temperature and oxygen fugacity on iron oxidation state and coordination in a phonolitic melt. *Contributions to Mineralogy and Petrology, 175*, 64–77. doi: 10.1007/s00410-020-01701-4

Lytle, F. W., van der Laan, G., Greegor, R. B., Larson, E. M., Violet, C. E., & Wong, J. (1990). Determination of the valence of Pr, Gd, and Ho in YBa2Cu3O7 by x- ray absorption spectroscopy. *Physical Review B, 41*, 8955–8963. doi: 10.1103/physrevb.41.8955

Maeda, F., Kamada, S., Ohtani, E., Hirao, N., Mitsui, T., Masuda, R., et al. (2017). Spin state and electronic environment of iron in basaltic glass in the lower mantle. *American Mineralogist, 102*, 2106–2112. https://doi.org/10.2138/am-2017-6035

Magnien, V. (2005). Etude cinétique des réactions d'oxydoréduction des silicates. PhD Institut de Physique du Globe de Paris. 284pp.

Magnien, V., Neuville, D. R., Cormier, L., Mysen, B. O., & Richet, P. (2004). Kinetics of iron oxidation in silicate melts: A preliminary XANES study. *Chemical Geology, 213*, 253–263. https://doi.org/10.1016/j.chemgeo.2004.08.047

Magnien, V., Neuville, D. R., Cormier, L., Roux, J., Pinet, O., & Richet, P. (2006). Kinetics of iron redox reactions: A high-temperature XANES and Raman spectroscopy study. *Journal of Nuclear Materials*, *352*, 190–195. doi: 10.1016/j.jnucmat.2006.02.053

Magnien, V., Neuville, D. R., Cormier, L., Roux, J., Hazemann, J.-L., de Ligny, D., et al. (2008). Kinetics and mechanisms of iron redox reactions in silicate melts: The effects of temperature and alkali cations. *Geochimica et Cosmochimica Acta*, *72*, 2157–2168. https://doi.org/10.1016/j.gca.2008.02.007

McMillan, P. F. (1984). A Raman spectroscopy study of glasses in the system CaO-MgO-SiO2. *American Mineralogist*, *69*, 645–659.

McMillan P. F. (1984). Structural studies of silicate glasses end melts – applications and limitations of Raman spectroscopy. *American Mineralogist*, *69*, 622–644.

McMillan, P. F., Piriou, B., & Navrotsky, A. (1982). A Raman spectroscopic study of glasses along the joins silica-calcium aluminate, silica-sodium aluminate and silica-potassium aluminate. *Geochimica et Cosmochimica Acta*, *46*, 2021–2037. https://doi.org/10.1016/0016-7037(82)90182-X

McMillan, P. F., Wolf, G. H., & Poe, B. T. (1992). Vibrational spectroscopy of silicate liquids and glasses. *Chemical Geology*, *96*, 351–366. https://doi.org/10.1016/0009-2541(92)90064-C

Metrich, N., Bonnin-Mosbah, M., Susini, J., Menez, B., & Galoisy, L. (2002). Presence of sulfite (S IV) in arc magmas: implications for volcanic sulfur emissions. *Geophysical Research Letters*, *129*, 11, 33-1–33-4. doi: 101029/2001GL014607

Metrich, N., Susini, J., Galoisy, L., Calas, G., Bonnin-Mosbah, M., & Menez, B. (2003). X-ray microspectroscopy of sulfur in basaltic glass inclusions. Influence on the volcanic sulfur emission. *Journal de Physique IV*, *104*, 393–397. doi: 10.1051/jp4:20030107

Möncke, D. (2018). Optical Spectroscopies. In: *The Encyclopedia of Archaeological Sciences* 1–7. John Wiley & Sons, Inc. https://doi.org/10.1002/9781119188230.saseas0423

Möncke, D., Papageorgiou, M., Winterstein-Beckmann, A., & Zacharias, N. (2014). Roman glasses coloured by dissolved transition metal ions: redox-reactions, optical spectroscopy and ligand field theory. *Journal of Archaeological Science*, *46*, 23–36. https://doi.org/10.1016/j.jas.2014.03.007

Mottana, A. (2014). X-ray absorption spectroscopy in mineralogy. In: Spectroscopic methods in mineralogy (pp. 465–552). EMU notes in Mineralogy 6. https://doi.org/10.1180/EMU-notes.6.12

Mysen, B. O. (1988). *Structure and properties of silicate melts*. Elsevier, Amsterdam. p. 354.

Mysen, B. O. (1990). Relationships between silicate melt structure and petrologic processes. *Earth-Science Reviews*, *27*, 281–365. https://doi.org/10.1016/0012-8252(90)90055-Z

Mysen, B. O. (1991). Relations between structure, redox equilibria of iron and properties of magmatic liquids. In: L. L. Perchuk and I. Kushiro (Eds.) *Physical chemistry of magmas, Advances in Physical Geochemistry*, *9*, pp. 41–98. New York: Springer-Verlag.

Mysen, B. O., & Frantz, J. D. (1992). Raman spectroscopy of silicate melts at magmatic temperatures: Na2O-SiO2, K2O-SiO2 and Li2O-SiO2 binary compositions in the temperature range 25-1475°C. *Chemical Geology*, *96*, 321–332. https://doi.org/10.1016/0009-2541(92)90062-A

Mysen, B. O., & Frantz, J. D. (1994). Silicate melts at magmatic temperatures: in-situ determination to 1651°C and effect of temperature and bulk composition on the mixing behavior of structural units. *Contributions to Mineralogy and Petrology*, *117*, 1–14. https://doi.org/10.1007/BF00307725

Mysen, B. O., & Virgo, D. (1978). Influence of pressure, temperature and bulk composition on melt structures in the system NaAlSi2O6-NaFe3+Si2O6. *American Journal of Science*, *278*, 1307–1322.

Mysen, B. O., & Virgo, D. (1985). Iron-bearing silicate melts: relations between pressure and redox equilibria. *Physics and Chemistry of Minerals*, *12*, 191–200. https://doi.org/10.1007/BF00311288

Mysen, B. O., & Virgo, D. (1989). Redox equilibria, structure and properties of Fe-bearing alulminosilicate melts: relationships among temperature, composition and oxygen fugacity in system Na2O-Al2O3-SiO2-FeO. *American Mineralogist*, *74*, 58–76.

Mysen, B. O., Seifert, F. A., & Virgo, D. (1980). Structure and redox equilibria of iron-bearing silicate melts. *American Mineralogist*, *65*, 867–884.

Mysen, B. O., Virgo, D., & Kushiro, I. (1981). The structural role of aluminium in silicate melts – a Raman spectroscopy study at 1 atmosphere. *American Mineralogist*, *65*, 678–701.

Mysen, B. O., Virgo, D., & Seifert, F. A. (1982). Curve-fitting of Raman spectra of silicate glasses. *American Mineralogist*, *67*, 686–695.

Mysen, B. O., Virgo, D., & Seifert, F. A. (1982). The structure of silicate melts: implications for chemical and physical properties of natural magmas. *Reviews of Geophysics*, *20*, 353–383. https://doi.org/10.1029/RG020i003p00353

Mysen, B. O., Virgo, D., & Seifert, F. A. (1984). Redox equilibria of iron in alkaline earth silicate melts: relationships between melt structure, oxygen fugacity, temperature and properties of iron-bearing silicate liquids. *American Mineralogist*, *69*, 834–848.

Mysen, B. O., Virgo, D., & Seifert, F. A. (1985). Relationships between properties and structure of aluminosilicate melts. *American Mineralogist*, *70*, 88–105.

Mysen, B. O., Virgo, D., Neumann, E.-R., & Seifert, F. A. (1985). Redox equilibria and the structural states of ferric and ferrous iron in melts in the system CaO-MgO-SiO2-Fe-O: relationships between redox equilibria, melt structure and liquidus phase equilibria. *American Mineralogist*, *70*, 317–331.

Natoli, C. R. (1983). Near edge absorption structure in the framework of the multiple scattering model. Potential resonance or barrier effects? In: Bianconi, A., Incoccia, L., & Stilpich, S. (Eds.) *EXAFS and Near Edge Structure*, Springer Series in Chemistry and Physics, *27*, 43–56. Berlin: Springer.

Natoli, C. R., & Benfatto, M. (1986). A unifying scheme of interpretation of X-ray absorption spectra based on the multiple scattering theory. *Le Journal de Physique Colloques*, *47*, C8, C8-11-C8-23. https://doi.org/10.1051/jphyscol:1986802

Nesbitt, H. W., O'Shaughnessy, C., Bancroft, G. M., Henderson, G. S., & Neuville, D. R. (2018). Factors affecting line shapes and intensities of Q3 and Q4 Raman bands of Cs silicate glasses. *Chemical Geology*, *5*, 1–11. doi: 10.1016/j.chemgeo.2018.12.009

Neuville, D. R., & Mysen, B. O. (1996). Role of aluminium in the silicate network: in-situ, high temperature study of glasses and melts on the join SiO2-NaAlO2. *Geochimica et Cosmochimica Acta, 60*, 1727–1737.

Neuville, D. R. (2006). Viscosity, structure and mixing in (Ca, Na) silicate melts. *Chemical Geology, 229*, 28–42. https://doi.org/10.1016/j.chemgeo.2006.01.008

Neuville, D. R., Henderson, G. S., & de Ligny, D. (2014a). Advances in Raman spectroscopy applied to earth and material sciences. In: Henderson G. S., Neuville, D. R., Down, B. (Eds.) *Spectroscopic methods in Mineralogy and Material Sciences*, Review in Mineralogy and Geochemistry, Vol *78*, 509–541.

Neuville, D. R., Hennet, L., Florian, P., & de Ligny, D. (2014b). In situ high temperature experiment. In: Henderson, G. S., Neuville, D. R., & Down, B. (Eds.) *Spectroscopic methods in Mineralogy and Material Sciences*, Review in Mineralogy and Geochemistry, Vol *78*, 779–800.

Petit, P. E., Farges, F., Wilke, M., & Solé, V. A. (2001). Determination of the iron oxidation state in Earth materials using XANES pre-edge information. *Journal of Synchrotron Radiation, 8*, 2, 952–954. doi: 10.1107/s0909049500021063

Pascal, P. (1958). *Nouveau traite de chimie minérale*. Masson edt.

Prieur, D., Epifano, E., Dardenne, K., Rothe, J., Hennig, C., Scheinost, A. C., et al. (2018). Negative thermal expansion of the UO2 fluorite local structure. *Inorganic Chemistry*. doi: 10.1021/acs.inorgchem.8b02657

Roskosz, M., Toplis, M. J., Neuville, D. R., & Mysen, B. O. (2008). Quantification of the kinetics of iron oxidation in silicate melts using Raman spectroscopy and assessment of the role of oxygen diffusion. *American Mineralogist, 93*, 1749–1759. https://doi.org/10.2138/am.2008.2861

Rossman, G. R. (1988). Optical spectroscopy. *Reviews in Mineralogy, 18*, 207–254.

Rossman, G. R. (2014). Optical spectroscopy. *Reviews in Mineralogy and Geochemistry, 78*, 1, 371–398.

Soldatov, A. V., Ivanchenko, T. S., Della Longa, S., Kotani, A., Iwamoto, Y., & Bianconi, A. (1994). Crystal-structure effects in the Ce L3-edge x-ray-absorption spectrum of CeO2: Multiple-scattering resonances and many-body final states. *Physical Review B, 50*, 8, 5074–5080. https://doi.org/10.1103/PhysRevB.50.5074

Seifert, F. A., Mysen, B. O., & Virgo, D. (1982). Three dimensional structure of quenched melts (glass) in the system SiO2-NaAlO2, SiO2-CaAl2O4 and SiO2-MgAl2O4. *American Mineralogist, 67*, 696–717.

Takahashi, Y., Shimizu, H., Kagi, H., Yoshida, H., Usui, A., & Nomura, M. (2000a). A new method for the determination of CeIII/CeIV ratios in geological materials; application for weathering, sedimentary and diagenetic processes. *Earth and Planetary Science Letters, 182*, 3–4, 201–207. https://doi.org/10.1016/S0012-821X(00)00250-8

Takahashi, Y., Shimizu, H., Usui, A., Kagi, H., & Nomura, M. (2000b). Direct observation of tetravalent cerium in ferromanganese nodules and crusts by X-ray-absorption near-edge structure (XANES). *Geochimica et Cosmochimica Acta, 64*, 2929–2935.

Takahashi, Y. Kolonin, G. R., Shironosova, G. P., Kupriyanova, I. I., Uruga, T., & Shimizu, H. (2005). Determination of the Eu(II)/Eu(III) ratios in minerals by X-ray absorption near-edge structure (XANES) and its application to hydrothermal deposits. *Mineralogical Magazine 69*, 179–190. https://doi.org/10.1180/0026461056920245

Van Bokhoven, J. A., & Lamberti, C. (Eds.). (2016). *X-ray absorption and X-ray emission spectroscopy: theory and applications* (Vol. *1*). John Wiley & Sons.

Yaxley A., Berry, A. J., Rosenthal, A., Woodland, A. B., & Paterson, D. (2017). Redox preconditioning deep cratonic lithosphere for kimberlite genesis – evidence from the central Slave Craton. *Scientific Reports, 7* 30. 10.1038/s41598-017-00049-3

Waychunas, G. A., Apted, M. J., & Brown, G. E. Jr. (1983). X-ray K-edge absorption spectra of Fe minerals and model compounds: near edge structure. *Physics and Chemistry of Minerals, 10*, 1–9. https://doi.org/10.1007/BF01204319

Wang, Z., Cooney, T. F., & Sharma, S. K. (1995). In situ structural investigation of iron-containing silicate liquids and glasses. *Geochimica et Cosmochimica Acta, 59*, 8, 1571–1577. https://doi.org/10.1016/0016-7037(95)00063-6

Wei, Q., McCammon, C., & Gilder, S.A. (2017). High-pressure phase transition of iron: A combined magnetic remanence and Mössbauer study. *Geochemistry, Geophysics, Geosystems, 18*. doi: 10.1002/2017GC007143

Wildner, M., Andrut, M., & Rudowicz, C. Z. (2014). Optical absorption spectroscopy in geosciences. In: *Spectroscopic methods in mineralogy* (pp. 93–143). Germany: Mineralogical Society of Great Britain and Ireland. https://doi.org/10.1180/EMU-Notes.6.3

Wilke, M., Farges, F., Petit, P.-E., Brown, Jr G.E., & Martin, F. (2001). Oxidation state and coordination of Fe in minerals: An Fe K-XANES spectroscopic study. *American Mineralogist, 86*, 714–730. https://doi.org/10.2138/am-2001-5-612

Wilke, M., Partzsch, G. M., Bernhardt, R., & Lattard, D. (2004). Determination of the iron oxidation state in basaltic glasses using XANES at the K-edge. *Chemical Geology, 213*, 71–87. https://doi.org/10.1016/j.chemgeo.2005.03.004

Willmott, P. R. (2011). *An Introduction to Synchrotron Radiation: Techniques and Applications*. New York: John Wiley and Sons.

Wilson A. D. (1960). The micro-determination of ferrous iron in silicate minerals by a volumetric and colorimetric method. *Analyst, 85*, 823–827. https://doi.org/10.1039/AN9608500823

Wilson, E. B., Decius, J. C., & Cross, P.C. (1955) *Molecular Vibrations: The Theory of infrared and Raman spectra*. New York: McGraw-Hill.

Wong, J., Lytle, F. W., Messmer, R. P., & Maylotte, D. H. (1984). K-edge absorption spectra of selected vanadium compounds. *Physical Review B, 30*, 10, 5596–5610. https://doi.org/10.1103/PhysRevB.30.5596

Zhang, C., Almeev, R., Hughes, E. C., Borisov, A., Wolff, A., Höfer, H., et al. (2018a). Electron microprobe technique for the determination of iron oxidation state in silicate glasses. *American Mineralogist, 103*, 1445–1454. https://doi.org/10.2138/am-2018-6437

Zhang, H. L., Cottrell, E., Peat, S., Kelley, A. K., & Hirschmann, M. M. (2018b). Determination of Fe3 +/ΣFe of XANES basaltic glass standards by Mössbauer spectroscopy and its application to the oxidation state of iron in MORB. *Chemical Geology, 479*,166–175. https://doi.org/10.1016/j.chemgeo.2018.01.006

14

Oxidation State, Coordination, and Covalency Controls on Iron Isotopic Fractionation in Earth's Mantle and Crust: Insights from First-Principles Calculations and NRIXS Spectroscopy

Marc Blanchard[1] **and Nicolas Dauphas**[2]

ABSTRACT

Isotopes provide valuable insights into the complex geochemical behavior of iron. To put the wealth of Fe isotopic data measured in natural samples into a quantitative framework, it is important to know how iron isotopes are fractionated at equilibrium between co-existing iron-bearing phases or species. These isotopic equilibrium fractionation factors can be derived from isotopic exchange experiments and the study of natural samples, but can also be calculated from partition functions, whose main contribution is the vibrational energy. This approach relies on first-principles calculations (atomistic modeling based on quantum mechanics) and vibrational spectroscopies (Mössbauer and Nuclear Resonant Inelastic X-ray Scattering – NRIXS). Comparison of the results obtained from these techniques provides confidence in their reliability and improves our understanding of the parameters controlling iron isotopic fractionation among coexisting phases. After an introduction to the theory and methods applied in this field, the chapter will review how NRIXS and first-principles calculations help interpret iron isotopic variations in natural rocks and minerals. At equilibrium, the heavy isotopes of iron will concentrate in the phases where the interatomic force constants are the greatest, meaning in the phases where iron bonds are the stiffest. Higher oxidation state, higher covalency, and lower coordination (shorter bond length) tend to be associated with stronger bonds and heavy iron isotope enrichments.

14.1. INTRODUCTION

Iron is a ubiquitous element that is abundant in terrestrial rocks, fluids, and magmas. It displays a rich and complex chemistry based on its multiple valence. Its three oxidation states are 0 (metallic iron), +2 (ferrous iron), and +3 (ferric iron). On Earth, these oxidation states are stratified in a manner that follows the structural and chemical layers characterizing our planet. At the center, the core is made of metallic iron alloyed with some lighter elements. The mantle is the main repository for ferrous iron under the form of silicate and oxide minerals. Ferric iron is mainly found in crustal rocks and surface environments due to the presence nowadays of an oxygen-rich atmosphere.

Iron is often involved in oxidation-reduction (redox) equilibria: chemical reactions that involve a transfer of electrons between species having different valence states. This electron transfer implies a readjustment of chemical bonds. As we will see, the modification of the interatomic

[1] *Géosciences Environnement Toulouse (GET), Université de Toulouse, CNRS, IRD, Toulouse, France*

[2] *Origins Laboratory, Department of the Geophysical Sciences and Enrico Fermi Institute, The University of Chicago, Chicago, IL, USA*

Magma Redox Geochemistry, Geophysical Monograph 266, First Edition.
Edited by Roberto Moretti and Daniel R. Neuville.

DOI: 10.1002/9781119473206.ch14

bonds affects in turn the isotopic properties. The ubiquity and rich chemistry of iron explains why variations in its four stable isotopes (^{54}Fe, ^{56}Fe, ^{57}Fe, and ^{58}Fe) have become essential tools to investigate various cosmochemical, geochemical, and biochemical questions. In Earth sciences, iron isotopic analyses have brought new insights into planetary accretion and mantle-core differentiation, crust formation by partial melting, redox evolution of the Earth's mantle, redox evolution of the ocean in relation with the oxygenation of the atmosphere, and magmatic differentiation. An overview of the most recent results on these topics can be found in Dauphas and Rouxel (2006), Johnson et al. (2008), and Dauphas et al. (2017).

The relative enrichment in heavy isotopes of one phase relative to another referred to as isotope fractionation, can be due to kinetic processes or can occur in conditions of thermodynamic equilibrium. Equilibrium isotopic fractionation factors represent reference values that must be known in order to interpret the iron isotope variations measured on natural samples. These key data and their temperature dependence can be obtained through several approaches. The most traditional way consists in designing isotope exchange experiments whose goal is to attain or to point to the equilibrium state (e.g., Schuessler et al., 2007; Shahar et al., 2008, 2017; Saunier et al., 2011; Syverson et al., 2013). Alternative approaches consist of deriving theoretically the equilibrium isotopic fractionation factors from the measurement or the computation of vibrational properties. Electronic structure calculations based on the resolution of quantum mechanics' laws can be used for determining the equilibrium fractionation for any traditional or non-traditional stable isotopes (e.g., Schauble, 2004; Blanchard et al., 2017). This theoretical approach is also of great interest for investigating what crystal chemical parameters govern the isotopic fractionation. Alternatively, for Mössbauer-active elements (like iron), the vibrational properties of the target element in solids can be investigated by Mössbauer spectroscopy through the measurement of the temperature dependence of the isomer shift (e.g., Polyakov & Mineev, 2000) or using nuclear resonance inelastic X-ray scattering, NRIXS (e.g., Polyakov et al., 2005; Dauphas et al., 2012, 2018). While sound theoretically, determining the second-order Doppler shift via the temperature dependence of the isomer shift in Mössbauer spectroscopy is fraught with difficulties. The NRIXS technique provides the vibrational density of states for the resonant atom and the database of equilibrium fractionation factors calculated from NRIXS measurements is rapidly growing.

In this chapter, we will give a brief introduction to the theory of mass-dependent stable isotope fractionation. We will define the quantity that is actually calculated (i.e., the reduced partition function ratio β) and how it relates with the commonly used fractionation factor α. We will explain how the β-factor is derived from atomistic modeling or NRIXS measurements. Iron isotope studies based on these two techniques will be reviewed, focusing on minerals and melts. The reliability of the derived equilibrium fractionation factors will be assessed by comparing the results from the various techniques. The results will be analyzed to identify what are the crystal parameters that control the equilibrium fractionation factors. The final section of this chapter will discuss some examples of igneous systems where equilibrium fractionation factors help assess how iron geochemistry and redox behavior affect iron isotopic variations.

14.2. THEORY: EQUILIBRIUM ISOTOPIC FRACTIONATION FROM VIBRATIONAL PROPERTIES

The theory of equilibrium mass-dependent stable isotope fractionation has been established by Bigeleisen and Mayer (1947) and Urey (1947). Here we define briefly the fractionation factors α and β, and we introduce the expressions that are used for determining their theoretical values. More details can be found in the review article of Blanchard et al. (2017).

Let's consider a reaction during which the X' (light) and X (heavy) isotopes are exchanged between two phases A and B:

$$AX' + BX \rightleftarrows AX + BX' \tag{14.1}$$

Even in conditions of thermodynamics equilibrium, the isotopic exchange leads to an isotopic enrichment of one phase relative to the other. This phenomenon of equilibrium isotopic fractionation has some temperature dependence that can be calculated. Indeed, as any chemical reaction, the isotopic exchange reaction is characterized by an equilibrium constant, K_{eq}, related to the variation of the standard free energy between reactants and products, or to the partition functions, Q, of the reactants and products; the two quantities (free energy and partition function) being linked. The equilibrium constant of Reaction 14.1 is thus given by the following expression:

$$K_{eq} = \frac{Q_{products}}{Q_{reactants}} = \frac{Q(AX) \times Q(BX')}{Q(AX') \times Q(BX)} \tag{14.2}$$

The commonly used fractionation factor, α, between two phases can be defined as the ratio of quantum and classical equilibrium constants (i.e., equilibrium constants obtained considering either quantum or classical mechanics). This is justified by the fact that classical contributions to the partition functions only play a bookkeeping role in the isotopic fractionation, and therefore it is convenient to define a reduced partition function, by ratioing the quantum partition function to its classical counterpart (Q_{qm}/Q_{cl}). Thus using Equation 14.2, the α fractionation factor between phases AX and BX can be expressed as:

$$\alpha_{AX-BX} = \frac{(K_{eq})_{qm}}{(K_{eq})_{cl}} = \frac{Q_{qm}(AX) \times Q_{cl}(AX')}{Q_{qm}(AX') \times Q_{cl}(AX)} \times \frac{Q_{qm}(BX') \times Q_{cl}(BX)}{Q_{qm}(BX) \times Q_{cl}(BX')} = \frac{\beta_{AX}}{\beta_{BX}} \quad (14.3)$$

This expression allows introducing the concept of reduced partition function ratio, also called β. The reduced partition function ratio β_{AX} can be seen as the isotopic fractionation factor between the phase AX and an ideal atomic gas of X. Contrary to the fractionation factor α that can be measured directly from isotope exchange experiments where two or more phases are coexisting, the fractionation factor β is associated to one phase, and in most cases, cannot be retrieved from isotopic compositions. However, β is the quantity that is calculated from the vibrational properties, as described below. As shown in Figure 14.1, β-factors can be conveniently compared and combined in order to derive the equilibrium fractionation factors α between any coexisting phases. Isotopic fractionations are relatively small, on the order of parts per thousand for iron isotopes, so it is common to use the notation 1000 ln α or 1000 ln β, reporting the result in permil (‰).

From Equation 14.3, we understand that the β-factor of a given phase can be calculated from the classical and quantum partition functions. Their expressions are explained and detailed in several articles (e.g., Bigeleisen & Mayer, 1947; Richet et al., 1977). Briefly, the ratio of the classical partition functions of two isotopologues will only depend on the atomic mass of isotopes and symmetry considerations, while the main contribution to the quantum partition function is the vibrational energy. If we consider a crystal undergoing a complete isotopic substitution and if interatomic vibrations are treated as harmonic, we obtain the following β-factor expression:

$$\beta_{AX} = \left[\prod_{i=1}^{3N} \prod_{\{q\}} \frac{\nu_{q,i}}{\nu'_{q,i}} \times \frac{e^{-h\nu_{q,i}/(2kT)}}{1-e^{-h\nu_{q,i}/(kT)}} \times \frac{1-e^{-h\nu'_{q,i}/(kT)}}{e^{-h\nu'_{q,i}/(2kT)}}\right]^{1/nN_q} \quad (14.4)$$

where $\nu_{q,i}$ and $\nu'_{q,i}$ are the vibrational frequencies of the vibrational mode i = 1,3N along wavevector q for the two isotopologues. N is the number of atoms in the crystal unit-cell with n sites for the element of interest. T is the temperature, h the Planck constant and k the Boltzmann constant. The second product is performed on a sufficiently large grid of N_q q-points in the first Brillouin zone. Crystal lattices have some periodicity that can be captured by their Fourier transform, which is also a lattice known as reciprocal space. The first Brillouin zone is a primitive cell in the reciprocal space and this zone is sampled using a uniform grid of q-points. β-factor value converges toward the final value while the number of q-points used is increased. The expression 14.4 takes into account the "rule of the high-temperature product" also called "Redlich-Teller rule" (Kieffer, 1982), which accounts for the fact that there should be no isotopic fractionation in the high-temperature limit. Similar β-factor expressions can be derived for a molecule where rotational and translational motions also contribute to the partition function or for the case of a "site by site" isotopic substitution (Blanchard et al., 2017).

In an equivalent way, one can write the equilibrium fractionation factor β as a function of the kinetic energy K of the element (Polyakov, 1997):

$$ln\,\beta_{AX} = \frac{m-m'}{m}\left(\frac{K}{RT} - \frac{3}{2}\right) \quad (14.5)$$

Here this kinetic energy will be calculated from the vibrational density of states of the element $g(e)$:

$$K = \frac{3}{2}\int_0^{e_{max}} E(e/kT)g(e)de \quad (14.6)$$

where $E(e/kT)$ is the Einstein function for the vibrational energy of a single harmonic oscillator at frequency $\nu = e/h$, and e_{max} corresponds to the maximal energy of the vibrational spectrum. Using a series expansion of the kinetic energy above (Polyakov et al., 2005, 2007) or a Bernoulli

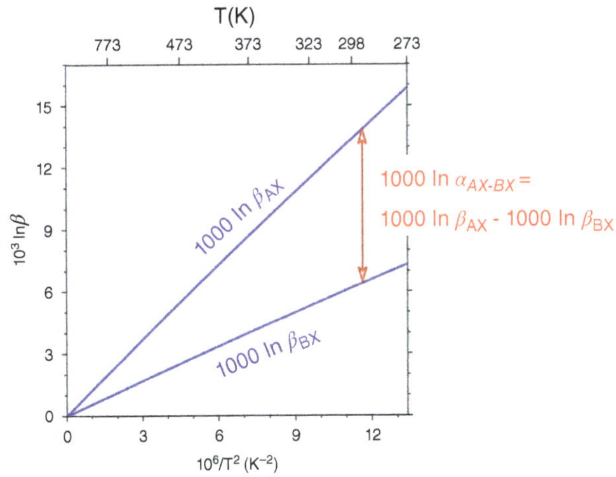

Figure 14.1 Generic example of the temperature dependence of the β-factors of phases AX and BX, expressed in ‰. β-factors can then be combined in order to obtain the equilibrium fractionation factor α between these two phases; a quantity that can be compared to measured isotopic fractionations between co-existing phases.

expansion of the β-factor (Dauphas et al., 2012), the β-factor can be expressed as,

$$ln\,\beta_{AX} = \left(\frac{m'}{m} - 1\right)\left(\frac{m_2^g}{8k^2T^2} - \frac{m_4^g}{480k^4T^4} + \frac{m_6^g}{20160k^6T^6}\right) \quad (14.7)$$

where m_i^g is the ith moment of the vibrational density of states $g(e)$: $m_i^g = \int_0^{+\infty} g(e)e^i de$

Finally, Bigeleisen and Mayer (1947) also derived an approximate formula, which relates the β-factor to the force constants (the equivalent of a spring constant for a mechanical spring) acting on the element of interest:

$$\beta_{AX} = 1 + \frac{m-m'}{mm'}\frac{\hbar^2 F}{8(kT)^2} \quad (14.8)$$

with $\hbar = h/2\pi$. The mean force constant F is related to the second moment of the phonon density of states through Lipkin's sum rule (Lipkin, 1995, 1999), such that Equation 14.8 is equivalent to truncating Equation 14.7 to the first order (Dauphas et al., 2012). F is the sum of force constants in three orthogonal directions opposing displacement of the atom X from its equilibrium position. This expression is a valid approximation of Equation 14.4 when $h\upsilon/kT < 2$; in other words at high-temperature or for a system with relatively low relevant vibrations. For iron, Equation 14.7 is well approximated by Dauphas et al. (2012, 2017),

$$ln\,\beta_{AX} \approx 2.9\frac{F}{T^2} - 38\frac{F^2}{T^4} \quad (14.9)$$

Accordingly, truncating the expansion to the first term in the expansion, which is known as the high temperature approximation, gives rise to a relative error on $ln\beta_{AX}$ of ~$13F/T^2$, meaning that it will be larger for solids that have iron in stiff bonds and it will be smaller at high temperature (see Figure 14.2; Dauphas et al., 2012).

In conclusion of this theoretical section, one has to remember that the equilibrium fractionation factor α can be predicted by combining the relevant β-factors, which in turn are obtained from the vibrational properties. In practice, the variation of vibrational frequencies associated with the isotopic substitution, the vibrational density of states or the force constants of the element of interest can be determined either from first-principles calculations or by NRIXS spectroscopy, as we will see in the next section. Alternatively, although less accurate, it is possible to estimate the force constants using an ionic model based on Pauling's rules, mainly from the valences and ionic radii of the targeted element and its first neighbors. A derivation of this approach and an application to iron isotopes can be found in Young et al. (2015), Macris et al. (2015), and in Sossi and O'Neill (2017).

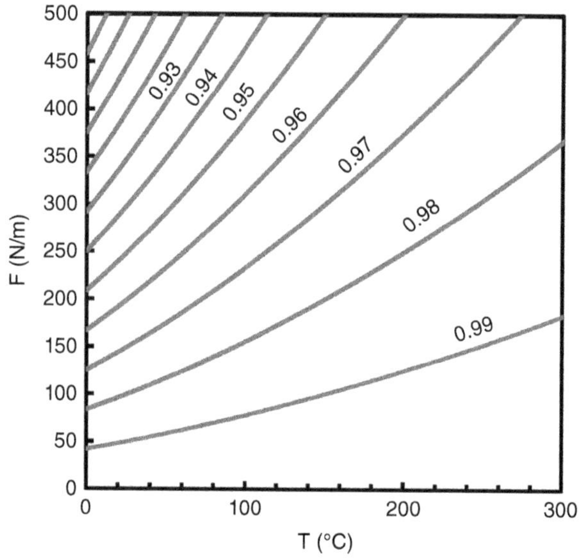

Figure 14.2 Relative departure (numbers on the curves) from Equation 14.8 introduced by the addition of the higher order term, for the $^{56}Fe/^{54}Fe$ ratio. For high-temperatures and relatively low force constants, equation (22.8) is a good approximation (Source: Dauphas et al., 2012. Reproduced with permission of Elsevier).

14.3. CALCULATION OF VIBRATIONAL PROPERTIES

14.3.1. First-Principles Calculations

The vibrational properties (vibrational frequencies, vibrational densities of states) can be calculated from atomistic modeling, which describes the behavior of a physicochemical system from its components at the atomic level. One can consider the ions and treat their interactions using classical mechanics using empirical force fields but most studies investigating non-traditional stable isotopes, and in particular iron isotopes, employ first-principles calculations. This means that the laws of quantum mechanics are used for describing the interactions between nuclei and electrons. In that case, the electronic structure is obtained by solving the Schrödinger equation requiring some approximations but no empirical parameters. A widespread class of first-principles calculations is based on the density functional theory (DFT), which considers that the ground state energy is a functional of the electronic density (Hohenberg & Kohn, 1964). More details about the computational techniques can be found in many books like for instance in the review by Cygan and Kubicki (2001).

There are typically three steps in first-principles calculations for obtaining the vibrational properties needed for the determination of equilibrium isotope fractionation factors using either Equation 14.4, 14.5, or 14.8. In the first step, the minimum-energy structure is determined via geometric relaxation. From an initial guess geometry, often the experimentally determined structure, the forces on each atom, and the stress over the cell are calculated, and a refined guess structure is determined. This procedure continues iteratively until the residual forces and stress are sufficiently small. This relaxation is needed as otherwise the structure is unstable and calculation of vibrational frequencies yields spurious results. Once the minimum-energy configuration has been calculated, the second step is the determination of interatomic force constants for displacements of the atomic nuclei from their equilibrium positions. Finally, the vibrational frequencies or densities of states are determined from force constants and appropriate isotopic masses (Baroni et al., 2001). Uncertainties in calculated frequencies are typically the main factor limiting the accuracy of calculated fractionation factors. Méheut et al. (2009) showed that a systematic correction of n% on the frequencies induces a relative systematic correction on the logarithmic β-factors ($\ln\beta$) between n% (at low temperatures) to 2n% (at high temperatures). The commonly used generalized gradient approximation is for instance associated with a systematic underestimation of ~5% of the harmonic vibrational frequencies. This would lead to a relative uncertainty of ~0.25‰ on a β-factor of 5‰. A virtue of the theoretical approach is that it enables to go beyond the average isotopic properties of one phase by providing the contribution of each crystallographic site. These data allow to identify the crystal chemical factors governing the isotope fractionations.

14.3.2. NRIXS Spectroscopy

Iron-57 being a Mössbauer-active isotope with a low-lying nuclear excited state at 14.4125 keV, one can use the synchrotron radiation technique of nuclear resonant inelastic X-ray scattering (NRIXS) to probe the vibrational properties of iron contained in a solid. The energy of the incoming X-ray photon beam is changed in small steps within a typical interval of –120 to +150 meV (the exact energy range depends on the temperature of the measurements and the nature of the material analyzed) around the nominal resonance energy (i.e., 14.4125 keV). At each energy, the scattered X-rays induced by the nuclear transition are analyzed. When the pulse hits the sample, X-rays are scattered by electrons. This electronic contribution, almost instantaneous, can be discarded by imposing some time discrimination, while the targeted nuclear scattering is delayed due to the finite lifetime (141 ns) of the excited ^{57}Fe nucleus.

When the incoming X-ray's energy is lower than the nominal resonance energy, some nuclear transitions can still occur because lattice vibrations provide the small energy deficit (a phenomenon known as phonon annihilation). When the incoming energy is higher than the nominal resonance energy, the excess energy can be absorbed by lattice vibrations (phonon creation) and the nuclear transitions can still take place. By measuring the flux of the scattered X-rays for various values of the incident X-ray energy, one can thus probe the vibrational properties of the sample, which in turn are used to determine equilibrium fractionation factors. In practice, two approaches exist to retrieve β-factors from NRIXS spectra. Polyakov et al. (2007) used the kinetic energy and first order perturbation theory to calculate the β-factor from the vibrational density of state of the iron sublattice $g(e)$ (Equations 14.5 and 14.6). The vibrational density of states is calculated using a Fourier-Log transformation of the raw spectrum $S(e)$. Dauphas et al. (2012) used a different approach based on the determination of the force constant of iron bonds and other higher order moments of the raw spectrum $S(e)$. The two approaches are mathematically equivalent (Dauphas et al., 2012). A virtue of the second approach is that the error bars are easy to calculate and one knows how data processing affects the results. The following uncertainties are propagated and details on the error propagation calculation are provided in Dauphas et al. (2014) and Hu (2016): counting statistics, uncertainties in the baseline definition, offset in energy scaling, overall energy scaling, and a bin-to-bin energy jitter. In order to retrieve reliable β-factors, special attention must be paid to the energy range that must be extended enough (as mentioned) and to the baseline definition (Blanchard et al., 2015). The reason is that the mean force constant of the iron bonds, which is the main driver for the isotopic fractionation, is calculated from the third moment of the raw spectrum $S(e)$ or the second moment of the phonon density of states [integrals of $S(e)\ e^3$ and $g(e)\ e^2$] such that small errors at the energy tails of the spectra, where is signal is lowest, have a disproportionate influence on the calculated force constant (Dauphas et al., 2012; Blanchard et al., 2015).

14.4. IRON ISOTOPE STUDIES BASED ON NRIXS OR DFT

Two main factors can influence the force constant of the iron bonding environment in condensed phases or fluids: (i) the oxidation state of the metal center; and (ii) the coordination environment, including number and identity of the coordinating ligands. Several DFT studies focused on the magnitude of ligand effect relative to redox effect on Fe isotope fractionation by considering various Fe species relevant of aqueous solutions, such as aquo complexes (Anbar et al., 2005; Ottonello & Zuccolini, 2009) as well as inorganic and organic complexes (Domagal-Goldman & Kubicki, 2008; Hill & Schauble, 2008; Ottonello & Zuccolini, 2008; Hill et al., 2010; Fujii et al., 2014). Other DFT studies focused on iron-bearing minerals such as hematite, goethite, pyrite, and siderite (Blanchard et al., 2009, 2010, 2015). In particular, Blanchard et al. (2010) tackled the question of the effect of the chemical composition by evaluating the Fe β-factor of hematite (Fe_2O_3) as a function of the Al content. Rustad and collaborators (Rustad & Dixon, 2009; Rustad et al., 2010) studied specifically the mineral-solution isotopic fractionation between aqueous iron and hematite, siderite, or goethite (mainly combining their well-tested β-factors for aqueous iron with previous mineral β-factors). In this case, determining accurate values requires special care because of the fundamentally different nature of the iron-bearing species considered, i.e., a periodic crystal versus a disordered solvated species. Rustad and Yin (2009) investigated the isotopic properties of ferropericlase and ferro-perovskite (bridgmanite) in lower-mantle conditions to discuss Earth's accretion and differentiation.

Ab initio studies can most directly be compared with NRIXS studies. Indeed, this synchrotron radiation technique can measure the phonon density of states, which can be compared with theoretical estimates to assess what vibration modes control the isotopic fractionation. Polyakov et al. (2007) and Polyakov and Soultanov (2011) estimated the Fe β-factors of oxide and sulfide minerals from the vibrational densities of states collected in earlier studies whose aims were to derive seismic velocities. Dauphas et al. (2012) also used previously published data in order to document the equilibrium fractionation factors of various compounds among which mainly silicate, oxide, sulfide minerals, and the different phases of metallic iron. They also introduced a new approach to derive equilibrium fractionation factors for iron that involves using the moments of phonon excitation probability density function (as well as the phonon density of states) to calculate the coefficients in the $1/T^2$ polynomial expansion of 1000 lnβ. Dauphas et al. (2012) reported the first NRIXS measurements of goethite and jarosite, which were subsequently reevaluated in Blanchard et al. (2015). This study highlighted the need to collect data over a wide energy range and implemented new data reduction approaches to retrieve accurate β-factors from the NRIXS measurements. Significant improvements have been made to the acquisition and reduction of NRIXS data that have been implemented in a data reduction software called SciPhon (Dauphas et al., 2018).

Roskosz et al. (2015) analyzed by NRIXS a series of spinels with varying compositions and redox states. These Fe β-factors combined with NRIXS data on olivine (Dauphas et al., 2014) and pyroxenes (Jackson et al., 2009) allowed the authors to shed new light on the isotopic signatures of natural peridotites. Several studies addressed the question of metal-silicate fractionation in order to improve our understanding of the conditions of Earth's core formation. Polyakov (2009) calculated the Fe β-factors for metallic iron, ferropericlase, perovskite, and post-perovskite in high-pressure conditions. Some of these NRIXS spectra were acquired for the determination of seismic velocities, which are derived by fitting the PDOS by a Debye model at low energy. The signal-to-noise ratio at those low energies is high and little attention was paid in those studies to the high energy end of the spectra where the signal-to-noise is low. The experiments are done in Diamond Anvil Cells (DACs) with small sample sizes, resulting in long acquisition times. To cut acquisition times, many spectra were truncated in energy before the counts reached background level. As mentioned earlier, this can lead to erroneous estimates of Fe β-factors. More recently, Shahar et al. (2016) investigated the pressure effect (from 2 to 40 GPa) on the iron isotopic properties of iron alloys with hydrogen, carbon, or oxygen. Liu et al. (2017) focused on the fractionation between basaltic glasses, metallic iron and ion-rich alloys of Fe-Ni-Si, Fe-Si, and Fe-S up to 206 GPa. Yang et al. (2018) studied Fe isotopic fractionation in lower mantle minerals bridgmanite and ferropericlase to evaluate whether terrestrial magma ocean crystallization could have imparted some iron isotopic fractionation to Earth's upper mantle. Finally, Dauphas et al. (2014) was interested in the fractional crystallization of magmas and the trend toward heavy iron isotopic compositions of silicic magmas. To do so, these authors determined the Fe β-factors in basaltic, andesitic, dacitic and rhyolitic glasses (taken as proxies for silicate melts) at magmatic temperatures, as well as in olivine. Williams et al. (2016) studied the iron isotopic properties of silicate glasses but this time with compositions relevant to lunar petrology, which includes mare basalts that are richer in titanium than their terrestrial equivalents.

14.5. COMPARISON OF EQUILIBRIUM FRACTIONATION FACTORS DERIVED FROM VARIOUS TECHNIQUES

Table 14.1 displays a comparison of the iron β-factors obtained from NRIXS, Mössbauer measurements, and first-principles calculations (DFT) for various iron-bearing minerals, i.e., sulfides, oxides, and one carbonate. With the exception of goethite, the β-factors derived from the various techniques are in good agreement. With a maximum difference of 0.5‰ at 298 K, the values are the same within the error bars. In the case of pyrite, the apparent discrepancy between DFT and Mössbauer results that were reported in Blanchard et al. (2009) and

Table 14.1 Iron β-factors ($^{56/54}$Fe) derived from first-principles calculations (DFT), NRIXS, and Mössbauer measurements for some minerals. Error bars due to uncertainties on calculated frequencies or on the data processing for NRIXS and Mössbauer measurements are of the order of ±0.2–0.4‰ at 298 K and ±0.01–0.02‰ at 1573 K.

	DFT		NRIXS		Mössbauer	
Mineral	298 K	1573 K	298 K	1573 K	298 K	1573 K
Pyrite, FeS_2	9.0[a]	0.33[a]	–	–	9.3[b]	0.35[b]
Chalcopyrite, $CuFeS_2$	5.5[c]	0.20[c]	5.4[c]	0.20[c]	–	–
Siderite, $FeCO_3$	4.1[a]	0.15[a]	–	–	4.1[d]	0.15[d]
Hematite, Fe_2O_3	7.1[a]	0.27[a]	7.4[e]	0.28[e]	7.6[f]	0.28[f]
Goethite, FeOOH	7.3[g]	0.28[g]	8.1[g]	0.31[g]	5.7[d]	0.21[d]
Ferropericlase, $(Mg_{0.75}Fe_{0.25})O$	4.7[h]	–	5.2[i]	–	–	–

[a] Blanchard et al. (2009).
[b] Blanchard et al. (2012).
[c] Polyakov and Soultanov (2011).
[d] Polyakov and Mineev (2000).
[e] Dauphas et al. (2012).
[f] Polyakov et al. (2001).
[g] Blanchard et al. (2015).
[h] Rustad and Yin (2009).
[i] Polyakov (2009).

in Polyakov and Soultanov (2011) could be resolved by using a better constrained temperature dependence of the Mössbauer spectra (Blanchard et al., 2012). Unlike other minerals, a significant discrepancy remains for goethite, where the DFT result is loosely bracketed by Mössbauer and NRIXS data. Calculations would suggest a possible overestimation of the NRIXS value possibly related to the baseline definition between the lattice modes and OH bending modes, which might be enough to obtain this 0.8‰ difference. The intrinsic features of the samples studied by NRIXS and Mössbauer spectroscopy may also contribute to the discrepancy (Blanchard et al., 2015). The comparison of results from these independent techniques provides a means of assessing the reliability of equilibrium isotope fractionation factors. A virtue of NRIXS and *ab initio* studies is that they provide β-factors for individual minerals as opposed to α-factors between minerals that equilibration experiments can measure. The former are more widely applicable than the later as the equilibrium fractionation between any two pair of minerals can be calculated provided that their β-factors are known, while equilibration experiments are only relevant to the mineral pair that has been equilibrated. Nevertheless, equilibration experiments and measurements of minerals in natural samples that are thought to have achieved equilibrium provide some ground-truthing of *ab initio* or NRIXS determinations and these more empirical approaches can tackle problems that are presently not easily handled by the vibrational approaches, notably with regard to isotopic fractionation involving liquids.

Inter-mineral equilibrium iron isotopic fractionation factors have been thoroughly discussed in the literature. We report here only a few examples of comparison between theoretical fractionation factors and experimental data. Many experiments are designed to allow measurement of the isotopic fractionation between iron aqueous species and a coexisting mineral. DFT β-factors of Fe aqueous species are quite variable in the literature, depending on the computational methods and parameters used. For instance, values for $Fe(III)_{aq}$ range from 8.3 to 10.1‰ at 295 K (Anbar et al., 2005; Hill & Schauble, 2008; Ottonello & Zuccolini, 2009; Rustad & Dixon, 2009; Rustad et al., 2010; Fujii et al., 2014). Details about the complications related to the modeling of aqueous species can be found in Blanchard et al. (2017). Among these studies, Rustad et al. (2010) performed a meticulous investigation of the effect of the solvation and basis set size on the isotopic properties of aqueous species. By doing so, these authors could reconcile some of the theoretical and experimental fractionation results. The experimental value of Wiesli et al. (2004) for $Fe(II)_{aq}$-siderite (+0.48±0.22‰ at 293 K) is in excellent agreement with the DFT value (+0.54‰) that one obtains by combining the β-factors of Rustad et al. (2010) and Blanchard et al. (2009) for $Fe(II)_{aq}$ and siderite, respectively. DFT also predicts that pyrite should be enriched in Fe heavy isotopes relative to $Fe(II)_{aq}$ (+1.1‰ at 623 K), which is confirmed by the experimental value of Syverson et al. (2013), i.e. +0.99±0.29‰. For hematite, the agreement is less remarkable. The experimental fractionation of −0.10±0.20‰ at 370 K between $Fe(III)_{aq}$ and hematite (Skulan et al., 2002) is consistent within the error bars with the fractionation of +0.22‰ found by considering the NRIXS value of hematite β-factor (Dauphas et al.,

2012), while the DFT value (Blanchard et al., 2009) is slightly larger, i.e. +0.43‰ at 370 K. For goethite, there is a disagreement between DFT, NRIXS, and equilibration experiments (Dauphas et al., 2017), as DFT and NRIXS are off by 2‰ from experimental measurements. The reason for the discrepancy is unclear but we note that the experimentally measured equilibrium isotopic fractionation between $Fe(II)_{aq}$ and ferrihydrite is in line with model predictions for goethite. More comparisons, for low- and high-temperature systems, are reviewed in Dauphas et al. (2017).

14.6. PARAMETERS CONTROLLING EQUILIBRIUM FRACTIONATION FACTORS

Provided that the harmonic approximation is valid, theory predicts that isotopic fractionation should scale with the stiffness of the bonds. The stiffness of the interatomic bonds is quantified by the interatomic force constants F. The β-factors increase proportionally with interatomic force constant (Figure 14.3). In other words, understanding fractionation means understanding the parameters that control the force constants of the bonds involving the element of interest, here iron. Among the available data, the iron sulfide troilite displays the lowest Fe force constant while the highest force constant is found in the ferric end-member of a rhyolitic glass. Some studies (e.g., Young et al., 2015; Macris et al., 2015; Sossi & O'Neill, 2017) employed an ionic model for estimating interatomic force constants and from those, equilibrium fractionation factors. This approach has the advantage of helping identify the crystal chemical parameters governing isotopic fractionation. These analytical expressions of F show that the bond strength is mainly controlled by the charge of the two atoms involved in the bonding and the distance between them.

Figure 14.4a shows the temperature dependence of Fe β-factors of minerals where iron is bonded to oxygen atoms. These minerals belong to various mineral families: oxides, carbonates, silicates, and sulfates. In this figure, the three colors allow to separate the minerals with respect to the oxidation state of iron, i.e., presence of only Fe^{2+}, only Fe^{3+}, or a mixed oxidation state with both Fe^{2+} and Fe^{3+}. Minerals with divalent or trivalent iron form two domains that do not overlap. Minerals where iron is divalent display the smallest β-factors, bracketed by siderite at the bottom and olivine, enstatite at the top. Minerals with trivalent iron show larger β-factors, ranging from hematite to jarosite. In this context, the β-factors of minerals with mixed oxidation states (magnetite and spinels) fall in the lower part of the Fe^{3+} domain. Consistently, Macris et al. (2015) measured inter-mineral iron isotopic fractionation from five distinct mantle xenolith lithologies from San Carlos (Arizona) and found that, in all cases where spinel and olivine coexist, the isotopic composition of spinel is heavier than that of the corresponding olivine. The overall picture is that iron oxidation state plays a great role, even if, as we will see, the interatomic force constant is governed by an interplay of several parameters defining the coordination environment (charge, coordination number, bond lengths).

Expectedly, the correlation "β-factor versus force constant" (Figure 14.3) holds for all minerals and melts where iron is bonded to oxygen as well as sulfur. However, the iron sulfides do not follow the trend "β-factor versus

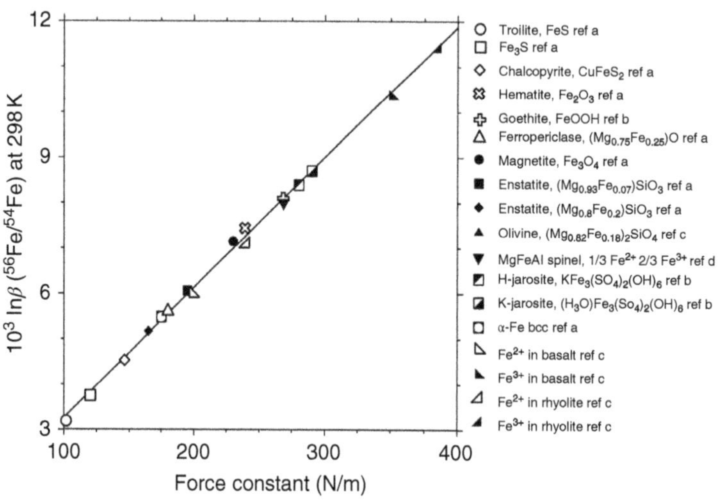

Figure 14.3 Iron β-factors ($^{56/54}Fe$) at 298 K as a function of the iron force constant in various minerals and melts. Data are from the following NRIXS studies: a = Dauphas et al. (2012); b = Blanchard et al. (2015); c = Dauphas et al. (2014); d= Roskosz et al. (2015). Error bars are of the order of ±10 N/m and ±0.2–0.4‰ at that temperature.

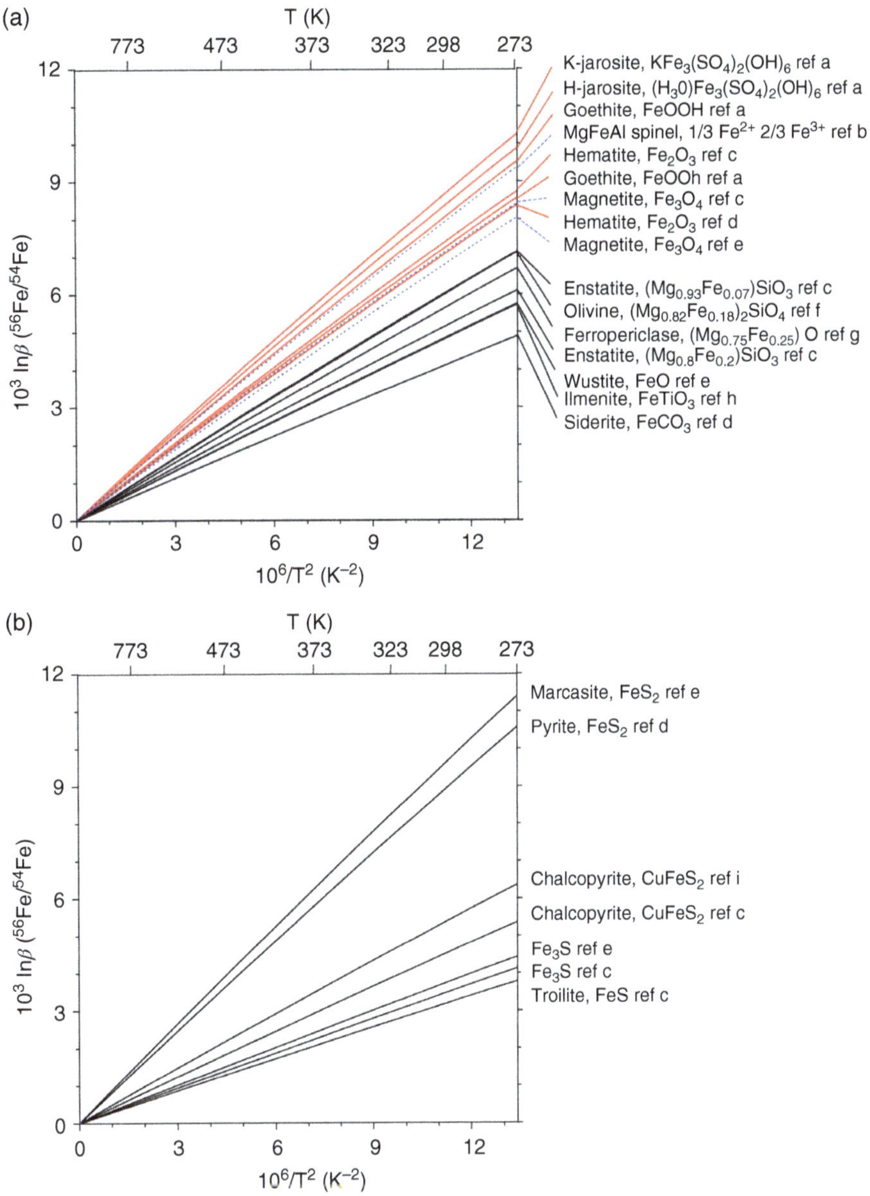

Figure 14.4 Temperature dependence of iron β-factors ($^{56/54}Fe$) for minerals where iron is bonded to oxygen atoms (a) or to sulfur atoms (b). Black curves are for minerals with only Fe^{2+}, red curves for minerals with only Fe^{3+}, and dotted blue curves for minerals with both Fe^{2+} and Fe^{3+}. Data are derived from either first-principles calculations or NRIXS measurements, and are taken from the following references: a = Blanchard et al. (2015); b= Roskosz et al. (2015); c = Dauphas et al. (2012); d = Blanchard et al. (2009); e = Polyakov et al. (2007); f = Dauphas et al. (2014); g = Polyakov (2009); h = Williams et al. (2016); i = Polyakov and Soultanov (2011).

redox state of iron" highlighted for minerals with Fe-O bonds (Figure 14.4) and are also out of the trend "β-factor versus bond length" shown in Figure 14.5. For instance, the two polymorphs pyrite and marcasite with divalent iron ions display β-factors higher than those of all Fe^{3+}-rich oxides and sulfates. This is explained by differing chemical properties of Fe-S bond with respect to Fe-O bond. Fe-S bonds are more covalent than Fe-O bonds as shown by the smaller electronegativy difference (Pauling electronegativies: Fe = 1.83, O = 3.44, S = 2.58). This example shows that at a given redox state, the value of Fe β-factors increase with the degree of covalence of interatomic bonds. We also observe in Figure 14.4 that β-factors of iron sulfides vary over a range of values exceeding that of minerals with Fe-O bonds. Troilite possesses iron as Fe^{2+} in octahedral site like pyrite and

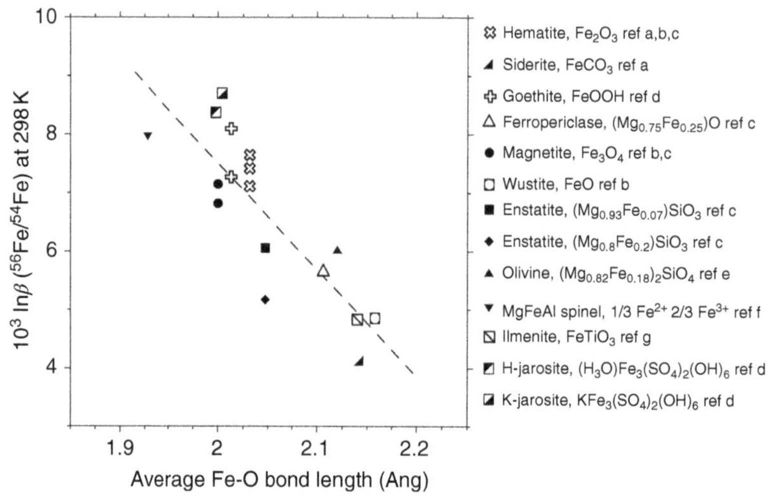

Figure 14.5 Iron β-factors ($^{56/54}$Fe) at 298 K as a function of the average Fe-O bond length for various minerals. Data are derived from either first-principles calculations or NRIXS measurements, and are taken from the following references: a = Blanchard et al. (2009); b = Polyakov et al. (2007); c = Dauphas et al. (2012); d = Blanchard et al. (2015); e = Dauphas et al. (2014); f= Roskosz et al. (2015); g = Williams et al. (2016). The dashed line is a guide for the eye showing the negative correlation trend.

marcasite but in troilite, FeS_6 octahedra are much more distorted and with longer Fe-S bonds, which explains the very low Fe β-factor of troilite. In chalcopyrite, iron is in tetrahedral site but with Fe-S bonds of similar lengths as in marcasite. Chalcopyrite mainly differs by its complex charge distribution between Fe, Cu, and S atoms, where iron would be in high-spin Fe^{3+} state compared to low-spin Fe^{2+} of pyrite and marcasite (Kobayashi et al., 2004). Finally, Fe_3S is the most iron-rich iron sulfide known to date in the iron-sulfur system (Lin et al., 2004). Iron sits in three nonequivalent crystallographic sites with both S and Fe atoms in the first coordination sphere.

Figure 14.5 gives the dependence of Fe β-factors over the average Fe-O bond lengths in minerals. We observe a general decrease of fractionation factors when the average bond length increases, which is in agreement with the general rule of short bonds concentrating heavy isotopes. Bond lengths increase with the coordination number. In this series of minerals, only magnetite and spinels have iron in tetrahedral sites in addition to octahedral sites. The presence of these tetrahedral sites leads to a shortening of the average Fe-O bond length to values lower than 2 Å. Bond lengths are also affected by iron oxidation state. In Figure 14.5, data with an average Fe-O bond length longer than 2.04 Å correspond to minerals containing only divalent iron. Ionic radii are negatively correlated with charge. This plot has been drawn from the crystallographic data usually derived from X-ray diffraction. When Fe presents a partial occupancy of its crystallographic site, like for instance in olivine where Fe shares the metal site with Mg, X-ray diffraction data give an average bond length for this site and without distinguishing Fe from Mg. Therefore, the Fe-O bond lengths are not known accurately. This uncertainty adds to the intrinsic scatter of this relationship. Note that in Figure 14.5, the two enstatite samples differing by their iron content have distinct β-factors. The sample with 7 mol% of Fe displays at 298 K a Fe β-factor 0.9‰ larger than the sample with 20 mol% of Fe. This difference exceeds the error bars of these NRIXS data (Dauphas et al., 2012). Although additional data are required, this effect of the composition may be real. The actual Fe-O bond lengths and thus the force constant and equilibrium isotope fractionation may vary with the Fe content. Similarly, Wang et al. (2017) found using DFT that equilibrium Ca isotope fractionation between pyroxenes depends on the Ca content.

The conclusions regarding the parameters controlling isotopic fractionation, discussed here mainly from mineral data, also hold for magmas. By combining NRIXS and X-ray absorption spectroscopy (XANES) on a series of magmatic glasses, Dauphas et al. (2014) could demonstrate the controls exerted by redox and structural (e.g., polymerization) conditions in magmas. In particular, a structural change of Fe^{2+} in the most silicic magma explains the shift towards higher force constant for Fe^{2+} of the trend defined by rhyolite glasses relative to the single trend defined for basalt, andesite, and dacite glasses.

14.7. SELECTED APPLICATIONS TO THE INTERPRETATION OF IRON ISOTOPIC VARIATIONS IN IGNEOUS ROCKS

Equilibrium fractionation factors have found many applications in low-temperature geochemistry, but kinetic effects are often confounding factors in trying to interpret iron isotopic variations in aqueous systems, products of alteration, or chemical sediments. In that respect, igneous rocks are potentially simpler to interpret. The reason is that in those settings, equilibrium isotopic fractionation between coexisting phases is more easily achieved because diffusion between phases is rapid at high temperature. Kinetic effects are also present but these are mostly due to diffusion, which can be identified by correlating the isotopic variations of one element with another. For example, during diffusion in olivine, Mg and Fe diffuse in opposite directions, resulting in a negative correlation between Mg and Fe isotopic variations (Teng et al., 2011). *In situ* isotopic analyses can also reveal whether kinetic effects are present as the presence of isotopic variations within minerals indicates that the rock is not in isotopic equilibrium (Sio et al., 2013; Oeser et al., 2015). For all these reasons, igneous isotopic geochemistry is a field where determination of iron β-factors is of broad relevance. This is a rapidly expanding field (see Young et al., 2015; Sossi et al., 2016; Dauphas et al., 2017; Bourdon et al., 2018, for some recent reviews on the topic) and we focus on three geological settings where iron isotope geochemistry and redox behavior may play some roles.

14.7.1. Mineral-Melt Fractionation and the Heavy Fe Isotopic Composition of the Crust

The isotopic reference material used for reporting Fe isotopic analyses is IRMM-014. Measurements of chondritic (undifferentiated) meteorites showed that they have similar Fe isotopic composition in bulk and that this composition is indistinguishable from the IRMM-014 value (Craddock & Dauphas, 2011). Mid-ocean ridge basalts have approximately uniform Fe isotopic composition (average δ^{56}Fe = +0.105±0.006‰; Teng et al., 2013), which is markedly heavier than chondrites. MORBs are the most abundant constituent of the oceanic crust, meaning that the bulk oceanic crust as a whole must have a heavy Fe isotopic composition. Continental crust also has heavy Fe isotopic composition (Telus et al., 2012).

The question of the Fe isotopic composition of the mantle is more difficult to tackle. Two categories of samples are best suited for that purpose. One is products of high degree partial melting such as komatiites (Dauphas et al., 2010; Hibbert et al., 2012). The reason is that any melt-solid fractionation would be mitigated if a large fraction of the rock melts (in the limit that the mantle is fully molten, then the magma composition must be equal to that of the source rock). A potential issue with komatiites is that they have been weathered and serpentinized but Craddock et al. (2013) showed that serpentinization had little effect on the Fe isotopic composition of abyssal peridotites. Even well-preserved komatiite samples show Fe isotopic variations resulting from partial melting and fractional crystallization (Hibbert et al., 2012), which complicates estimation of the mantle δ^{56}Fe value. With these caveats in mind, Dauphas et al. (2010) used Alexo komatiites to estimate that the Fe isotopic composition of the mantle must be +0.044±0.030‰. Magnesium isotopes show little variations in the komatiites measured so far, defining a mantle δ^{26}Mg value of -0.275±0.042‰, which is in line with measurements of peridotites (Handler et al., 2009; Teng et al., 2010). Another category of samples is mantle peridotites (Weyer & Ionov, 2007; Poitrasson et al., 2013; Craddock et al., 2013). The problem with these samples is that many samples have been affected by various forms of metasomatism, which fractionates iron isotopes, possibly due to kinetic effects such as diffusive processes in reactions involving melts and minerals (Weyer & Ionov, 2007; Poitrasson et al., 2013). Nevertheless, the distribution of δ^{56}Fe values of peridotites seems to peak at the chondritic value (Craddock et al., 2013; Poitrasson et al., 2013). In particular, abyssal peridotites, which may sample the mantle source of MORBs have Fe isotopic compositions that are close to chondritic (Craddock et al., 2013). Those data suggest that Earth's mantle has a chondritic Fe isotopic composition of +0.025±0.025‰ (however, see Poitrasson et al., 2013, for a different viewpoint). Why then does the oceanic crust have heavy Fe isotopic composition?

Three studies have been performed recently to measure the force constant of iron at high pressure by NRIXS (Shahar et al., 2016; Liu et al., 2017, Yang et al., 2018). These studies focused on iron alloys and they found that given the high equilibration temperature involved, partitioning between mantle and core (with O, Si, or Ni as the most likely alloying elements), or terrestrial magma ocean crystallization did not fractionate iron isotopes by more than ~0.02‰. This is consistent with the view that Earth's mantle has a chondritic or slightly elevated δ^{56}Fe value.

Available evidence suggests that melting the mantle to make the oceanic crust can fractionate iron isotopes. The effect is subtle (~0.07‰) but given that iron is a major constituent of the Earth, understanding what causes this fractionation may shed some light on the processes governing Earth's magmatic evolution. Two approaches have been used to address this question. One is to use estimated equilibrium fractionation between minerals and melts based on measured isotopic variations in peridotites (Williams et al., 2005). This is a proper first-order approach but

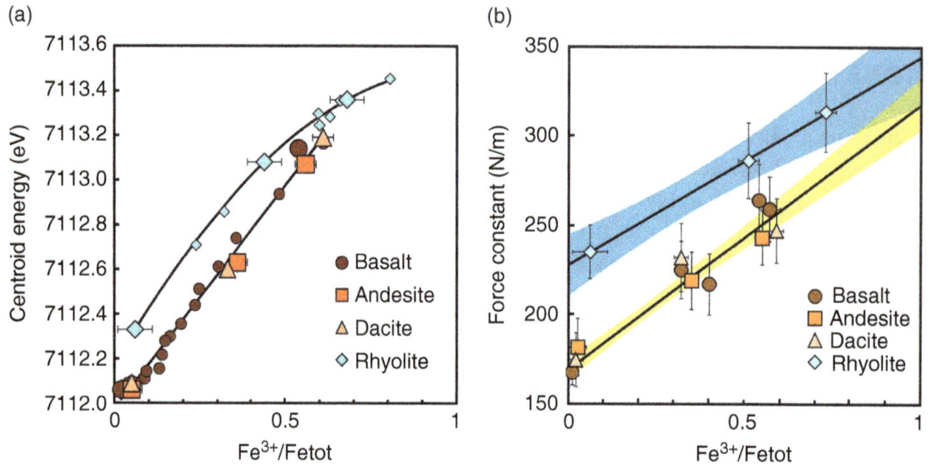

Figure 14.6 XANES centroid energy (left) and mean force constant of iron measured by NRIXS (right) against iron redox state in a variety of synthetic glasses (from Dauphas et al., 2014). The Fe^{3+}/Fe^{2+} ratio has a very strong influence on the stiffness of the iron bonds but superposed on this effect is a control of the silica or alkali content of the melt, as illustrated by the fact that the rhyolite evolves on a different trend than less silicic magma (Source: Dauphas et al., 2014. Reproduced with permission of Elsevier).

to gain a more mechanistic understanding of iron isotopic fractionation during mantle melting, one needs to independently determine the equilibrium fractionation factors between the phases involved.

The main mineral phases relevant to mantle melting to form MORBs are olivine, orthopyroxene, clinopyroxene spinel, and of course basaltic melt. MORBs form by approximately 10% partial melting. Iron in the upper mantle is mostly present as Fe^{2+} (only ~3.5% of iron is in its most oxidized form, Fe^{3+}). The effective melt/solid partition coefficient of iron during melting is ~1, meaning that when 10% of the mantle melts, 10% of iron is partitioned into the melt. This means that to explain the +0.07‰ difference in Fe isotopic composition between mantle and MORBs, a fractionation of approximately +0.07‰ between melt and residue is needed. To our knowledge, no *ab initio* calculations have been performed for iron-bearing minerals relevant to the upper-mantle. There is a more extensive data set available from NRIXS studies (Dauphas et al., 2014; Roskosz et al., 2015; Jackson et al., 2009). Jackson et al. (2009) examined the force constant of iron in orthoenstatite (an orthopyroxene). No data is available for clinopyroxene. Roskosz et al. (2015) studied Fe-bearing spinels characterized by a range of iron contents and oxidation states. Dauphas et al. (2014) measured the force constants of olivine as well as several glasses of basalt, andesite, dacite, and rhyolite containing variable amounts of Fe^{3+}. Prissel et al. (2018) measured the force constant of iron bonds in Fe^{2+}-bearing basaltic glasses doped with various amounts of Ti. The precision of these NRIXS measurements is limited to ~±15 N/m, which translates into an error on the β-factor of ~±0.017‰ at 1300 °C. As discussed in Section 14.6, what emerges from those studies and other experimental studies is that there is a very strong influence of the redox state of iron and a subtler effect of the coordination environment (Figure 14.6). Most phases relevant to mantle are silicates (pyroxenes, olivine, and melt). All Fe^{2+}-bearing phases (minerals and glasses) have Fe force constants that are similar, meaning that taken at face value, little isotopic fractionation is expected between Fe^{2+} in olivine or pyroxene, and Fe^{2+} in melt but higher precision measurements would be needed to tell exactly if any effect is present at the ~ ±0.03‰ level. Prissel et al. (2018) recently determined experimentally the equilibrium Fe isotopic fraction between silicate melts and olivine and found no measurable fractionation between the two.

A significant effect can be driven by redox conditions as Fe^{3+} tends to be enriched in the heavy Fe isotopes relative to Fe^{2+} and during melting Fe^{3+} behaves as a mildly incompatible element, meaning that the melt will have a higher Fe^{3+}/Fe^{2+} than the solid residue. Dauphas et al. (2009, 2014) assessed the degree of iron isotopic associated with this phenomenon. The result of this analysis is that the NRIXS measurements can only explain approximately half of the isotopic fractionation between melt and mantle (assuming that the mantle has near chondritic Fe isotopic composition). It is not clear at the present time what the cause for the discrepancy is. It could be that clinopyroxenes are enriched in the heavy isotopes of iron, that the harmonic approximation breaks down near or above the solidus, or that kinetic isotope effects are present. Clearly, more work is needed.

14.7.2. Isotopic Fractionation During Magmatic Differentiation

Several studies have shown that rocks related by trends of magmatic differentiation can have fractionated iron isotopic compositions (Teng et al., 2008; Schuessler et al., 2009). In some instances, those trends are not due to equilibrium isotopic fractionation but rather to kinetic isotopic fractionation associated with diffusion in olivine (Dauphas et al., 2010; Teng et al., 2011; Sio et al., 2013, 2018; Sio & Dauphas, 2017; Oeser et al., 2015, 2018; Collinet et al., 2017). As olivine crystallizes, the Mg/Fe of the magma decreases (olivine is relatively Mg-rich). The core of the olivine crystals are therefore out of thermodynamic equilibrium and Fe diffuses in while Mg diffuses out of olivine to correct the existing chemical imbalance between olivine rim and core. Because light Mg and Fe isotopes tend to diffuse faster than heavy isotopes, and Mg and Fe diffuse in opposite directions (Richter et al., 2009), olivine acquires light Fe and heavy Mg isotopic compositions. Removal of such olivine drives the residual magma towards heavy Fe isotopic composition. Diffusive fractionation will only play a role when the timescale for intra-mineral diffusive equilibration is commensurate with the timescale of cooling/crystallization. If the diffusive timescale is much shorter than the cooling/crystallization timescale, then bulk crystals will be in bulk chemical and isotopic equilibrium with the host lava. If the diffusive timescale is much longer than the cooling/crystallization timescale, then the crystal rim will be in equilibrium with the host lava and the crystal interior will be unaffected by diffusive re-equilibration. In the latter case, the bulk of the crystals will not be in equilibrium with the host lava but the relevant isotopic fractionation factors will still be the equilibrium ones, the difference being that the process will be modelled as a distillation.

Equilibrium isotopic fractionation factors are relevant to intra-crustal differentiation of mafic and felsic rocks. One aspect that has a dramatic effect on iron isotopic fractionation is that iron tends to form stronger bonds in more silicic and alkali-rich magmas such as rhyolite than in less silicic and alkali-rich magmas such as basalt (Dauphas et al., 2014). Changes in the coordination environment of iron in these silicic magmas is also seen in XANES (Figure 14.6). During fractional crystallization, the melt-mineral equilibrium isotopic fractionation for iron will thus increase as the crystallization proceeds. This process can produce silica-rich melts that have very heavy Fe isotopic compositions (Dauphas et al., 2014), which are found in natural systems (Poitrasson & Freydier, 2005; Telus et al., 2012). Redox conditions could also influence indirectly how iron isotopes are fractionated during differentiation of mafic and felsic rocks. The reason is that the redox state and whether or not the system is buffered influences the Fe^{3+}/Fe^{2+} ratio in the melt and the onset of magnetite crystallization (Sossi et al., 2012; Foden et al., 2015) and sulfide saturation (Williams et al., 2018; Peters et al., 2018).

14.7.3. Isotopic Fractionation in Arc Lavas

The systematics of iron isotope variations in subduction-related rocks has been the focus of significant work (Dauphas et al., 2009; Foden et al., 2018; Nebel et al., 2013, 2015; Williams et al., 2004, 2005, 2018; Debret et al., 2016). One reason is that these variations are presumably related to the mantle redox conditions and prior history of melting in the source of arc lavas. Oceanic arc lavas are produced by melting of the mantle wedge bounded above by the crust above and below by the subducting slab. This process of flux melting involves the transfer of fluids from the cold subducting plate into the hotter mantle wedge above, where melting takes place primarily due to lowering of the solidus temperature by fluid addition. Arc lavas are more oxidized than MORBs and an outstanding question is whether this is a feature of the source or whether it is an artifact from shallow level degassing and crystallization (Kelley & Cottrell, 2009; Lee et al., 2005).

Williams et al. (2004, 2005) first showed that redox conditions could influence iron isotopic fractionation in arc systems. They measured the Fe isotopic of bulk rocks and mineral separates of sub-arc peridotites and found that the Fe isotopic composition was correlated with the redox state. Specifically, spinels with higher Fe^{3+}/Fe_{tot} ratios had more negative $\delta^{56}Fe$ values than the samples with lower Fe^{3+}/Fe_{tot}. This result is counter intuitive because as discussed in Section 14.6, ferric iron tends to concentrate heavy iron isotopes in spinels and melts (Dauphas et al., 2014; Roskosz et al., 2015). A possible explanation for the trend seen by Williams et al. (2004) is that the peridotites record a prior episode of melt extraction under higher oxygen fugacity. Because Fe^{3+} is incompatible, the Fe^{3+}/Fe^{2+} ratio is higher in the melt than in the solid. Elevated oxygen fugacity is associated with a greater difference in Fe^{3+}/Fe^{2+} between melt and solid, and a larger net equilibrium isotopic fractionation between melt and solid. Melt extraction under high oxygen fugacity can lead to a negative shift in the $\delta^{56}Fe$ value of the residue because the melt is enriched in the heavy isotopes of iron. Negative $\delta^{56}Fe$ values could thus be associated with high Fe^{3+}/Fe_{tot} ratios.

Significant work has also been done on arc lavas (Dauphas et al., 2009; Foden et al., 2018; Nebel et al., 2015; Williams et al., 2018). As discussed, measurements of island arc lavas show that they are more oxidized than MORBs. To first order, one would therefore expect those arc lavas to be enriched in the heavy isotopes of iron, which is opposite to what is found. Arc lavas tend to have lower $\delta^{56}Fe$ values than MORBs. In the New-Britain island arc, the lavas formed by the largest amount of fluid fluxing and highest degree of partial melting are the ones that have $\delta^{56}Fe$ values closer to chondritic, while the ones with the lowest degree of partial melting have $\delta^{56}Fe$ values closer to MORB (Dauphas et al., 2009). The $\delta^{56}Fe$ measurements of island arc basalts from Dauphas et al. (2009) could be interpreted in two ways. One is that island arc basalts actually form under rather low oxygen fugacity so that the isotopic fractionation during melting is small and their oxygen fugacity increases by shallow processes such as degassing and fractional crystallization. The other is that the source of island arc basalts experienced previous episodes of melting, which drove the residue towards lower $\delta^{56}Fe$ values (Nebel et al., 2015; Foden et al., 2018). Those low $\delta^{56}Fe$ mantle regions could be re-melted to produce island arc basalts. An observation that speaks against the second scenario is that boninites, which are thought to form by high degree flux melting of depleted sources, have remarkably uniform Fe isotopic composition (Dauphas et al., 2009). Furthermore, their $\delta^{56}Fe$ value is close to that inferred for the mantle from peridotite measurements (Dauphas et al., 2009). It would take a remarkable coincidence for boninites to originate from a low $\delta^{56}Fe$ mantle region and to have then their Fe isotopic compositions shifted by partial melting by just the right amount to end up with such homogeneous and chondritic-like Fe isotopic composition. This is a circumstantial argument, and the second scenario is still viable (Nebel et al., 2015; Foden et al., 2018). Foden et al. (2018) measured the Fe isotopic compositions of mafic lavas from 15 arcs. Like Dauphas et al. (2009), they found that contrary to expectations given their oxidized nature, arc lavas have light Fe isotopic composition relative to MORBs. They also found that across various arcs, the Fe isotopic composition tends to correlate inversely with indirect proxies for the slab transport of oxidants. Cold and wetter subduction systems tend to have the lightest Fe isotopic composition and vice versa. This is more consistent with a scenario that involves large extents of prior melt depletion under buffered oxygen fugacity conditions. Williams et al. (2018) studied extensively the Fe isotopic composition of lavas from the Marianas, focusing in particular on the effect of magmatic differentiation on iron isotopic fractionation in arc settings. In more evolved samples (MgO<~4 wt%), they found significant Fe isotopic fractionation that could be explained by sequestration of magnetite crystals and sulfidic melts. To summarize, redox conditions in the mantle wedge seem to influence the isotopic composition of arc-related lavas but the relationship may be more complex than equating oxidizing conditions in the source to heavier Fe isotopic composition in the lava.

14.8. CONCLUSIONS AND PERSPECTIVES

Iron has received a great deal of attention due to its abundance and importance in all terrestrial reservoirs. Among non-traditional stable isotopes, iron is the most extensively studied element. First-principles calculations and NRIXS measurements are contributing significantly to that field, and fully complement laboratory exchange experiments. Inter-comparisons between all available techniques demonstrate that calculated equilibrium isotopic fractionation factors are reliable enough for interpreting iron isotope variations measured on natural samples. First-principles calculations and NRIXS measurements on well-characterized samples represent efficient tools to understand how crystal chemical parameters control iron isotopic fractionations. Alternatively, an ionic model based on Pauling's rules can be used to get a first-order estimate of the iron β-factor. This approach is satisfactory (at least qualitatively) when the interatomic bonds of interest are ionic but must be taken with care in other cases.

The central parameter controlling the equilibrium fractionation is the interatomic force constant, which reflects the strength of the bonds involving iron. For compounds (minerals, glasses) with relatively ionic Fe-O bonds, the iron oxidation state as well as the Fe-O bond length (in relation with the iron coordination number) will strongly affect the force constant. Whereas the higher degree of covalency of Fe-S bonds explains the singularity of iron sulfides with respect to the trend defined for other compounds involving Fe-O bonds.

Given the richness and complexity of iron chemistry, an extensive set of highly precise and accurate equilibrium fractionation factors is needed to properly interpret natural iron isotopic variations. Special efforts should be made to investigate the effect on isotopic fractionation of the chemical composition (e.g., solid-solutions, chemical impurities) and pressure. To our knowledge, no first-principles calculations have been done to calculate the β-factor of iron in silicate melts. This would be an important direction of investigation even if the quite complex relevant chemistry and redox state of silicate melts, and their high viscosity compared to aqueous solutions represent tremendous challenges for molecular dynamics simulations.

Studies of igneous rocks show that the redox state of iron influences the isotopic composition of iron. This is most clearly seen in arc-related lavas, where recycling of oxidants at subduction zones imparts a specific Fe isotopic signature to the erupted lavas. However, the relationship between redox state in the source and iron isotopic composition of the lavas is complex, reflecting long-term redox conditions in the mantle wedge during prior episodes of melt extraction. The bulk of Earth's oceanic crust has heavy Fe isotopic composition relative to the inferred mantle composition. While significant progress has been made on that front, the underlying mechanism for this enrichment is still uncertain. Silicic rocks can show heavy Fe isotope enrichments reflecting the fact that iron tends to form stiffer bonds in more silicic and alkali rich magmas. In those systems, redox conditions can also influence the pattern of iron isotopic fractionation by controlling the timing of magnetite crystallization. To summarize, there are many lines of evidence showing that iron isotopes are influenced by redox conditions in igneous rock and significant progress has already been made towards understanding these controls through *ab initio* and NRIXS studies of equilibrium fractionation factors.

ACKNOWLEDGMENTS

This work was supported by grants from NSF (EAR1444951, EAR1502591, EAR2001098) and NASA (NNX17AE86G, NNX17AE87G, NNX15AJ25G, 80NSSC17K0744, 80NSSC20K0821) to ND. We are grateful to Paolo Sossi and an anonymous reviewer for their constructive comments.

REFERENCES

Anbar, A. D., Jarzecki, A. A., & Spiro, T. G. (2005). Theoretical investigation of iron isotope fractionation between $Fe(H_2O)_6^{3+}$ and $Fe(H_2O)_6^{2+}$: implications for iron stable isotope geochemistry. *Geochimica et Cosmochimica Acta*, *69*, 825–837. https://doi.org/10.1016/j.gca.2004.06.012

Baroni, S., de Gironcoli, S., Dal Corso, A., & Giannozzi, P. (2001). Phonons and related crystal properties from density functional perturbation theory. *Reviews of Modern Physics*, *73*, 515–561.

Bigeleisen, J., & Mayer, M. G. (1947). Calculation of equilibrium constants for isotopic exchange reactions. *Journal of Chemical Physics*, *15*, 261–267. https://doi.org/10.1063/1.1746492

Blanchard, M., Poitrasson, F., Méheut, M., Lazzeri, M., Mauri, F., & Balan, E. (2009). Iron isotope fractionation between pyrite (FeS_2), hematite (Fe_2O_3) and siderite ($FeCO_3$): A first-principles density functional theory study. *Geochimica et Cosmochimica Acta*, *73*, 6565–6578. https://doi.org/10.1016/j.gca.2009.07.034

Blanchard, M., Morin, G., Lazzeri, M., & Balan, E. (2010). First-principles study of the structural and isotopic properties of Al- and OH-bearing hematite. *Geochimica et Cosmochimica Acta*, *74*, 3948–3962. https://doi.org/10.1016/j.gca.2010.04.018

Blanchard, M., Poitrasson, F., Méheut, M., Lazzeri, M., Mauri, F., & Balan, E. (2012). Comment on "New data on equilibrium iron isotope fractionation among sulfides: Constraints on mechanisms of sulfide formation in hydrothermal and igneous systems" by V. B. Polyakov and D. M. Soultanov. *Geochimica et Cosmochimica Acta*, *86*, 182–195. doi: 10.1016/j.gca.2012.01.048

Blanchard, M., Dauphas, N., Hu, M., Roskosz, M., Alp, E., Golden, D., et al. (2015). Reduced partition function ratios of iron and oxygen in goethite. *Geochimica et Cosmochimica Acta*, *151*, 19–33. https://doi.org/10.1016/j.gca.2014.12.006

Blanchard, M., Balan, E., & Schauble, E. A. (2017). Equilibrium fractionation of non-traditional isotopes: A molecular modeling perspective. *Reviews in Mineralogy and Geochemistry "Non-traditional stable isotopes,"* *82*, 27–63. https://doi.org/10.2138/rmg.2017.82.2

Bourdon, B., Roskosz, M., & Hin, R. C. (2018). Isotope tracers of core formation. *Earth-Science Reviews*, *181*, 61–81. https://doi.org/10.1016/j.earscirev.2018.04.006

Collinet, M., Charlier, B., Namur, O., Oeser, M., Médard, E., & Weyer, S. (2017). Crystallization history of enriched shergottites from Fe and Mg isotope fractionation in olivine megacrysts. *Geochimica et Cosmochimica Acta*, *207*, 277–297. https://doi.org/10.1016/j.gca.2017.03.029

Craddock, P. R., & Dauphas, N. (2011). Iron isotopic compositions of geological reference materials and chondrites. *Geostandards and Geoanalytical Research*, *35*, 101–123. https://doi.org/10.1111/j.1751-908X.2010.00085.x

Craddock, P. R., Warren, J. M., & Dauphas, N. (2013). Abyssal peridotites reveal the near-chondritic Fe isotopic composition of the Earth. *Earth and Planetary Science Letters*, *365*, 63–76. https://doi.org/10.1016/j.epsl.2013.01.011

Cygan, R. T., & Kubicki, J. D. (2001). Molecular modeling theory: Applications in the geosciences. *Reviews in Mineralogy and Geochemistry*, *42*, 531.

Dauphas, N., & Rouxel, O. (2006). Mass spectrometry and natural variations of iron isotopes. *Mass Spectrometry Reviews*, *25*, 515–550. https://doi.org/10.1002/mas.20078

Dauphas, N., Craddock, P. R., Asimow, P. D., Bennett, V. C., Nutman, A. P., & Ohnenstetter, D. (2009). Iron isotopes may reveal the redox conditions of mantle melting from Archean to Present. *Earth and Planetary Science Letters*, *288*, 255–267. https://doi.org/10.1016/j.epsl.2009.09.029

Dauphas, N., Teng, F.-Z., & Arndt, N. T. (2010). Magnesium and iron isotopes in 2.7 Ga Alexo komatiites: Mantle signatures, no evidence for Soret diffusion, identification of diffusive transport in zoned olivine. *Geochimica et Cosmochimica Acta*, *74*, 3274–3291. https://doi.org/10.1016/j.gca.2010.02.031

Dauphas, N., Roskosz, M., Alp, E., Golden, D., Sio, C., Tissot, F., et al. (2012). A general moment NRIXS approach to the determination of equilibrium Fe isotopic fractionation factors: application to goethite and jarosite. *Geochimica et Cosmochimica Acta*, *94*, 254–275. https://doi.org/10.1016/j.gca.2012.06.013

Dauphas, N., Roskosz, M., Alp, E. E., Neuville, D., Hu, M., Sio, C. K., et al. (2014). Magma redox and structural controls on iron isotope variations in Earth's mantle and crust. *Earth and Planetary Science Letters*, *398*, 127–140. https://doi.org/10.1016/j.epsl.2014.04.033

Dauphas, N., John, S. G., & Rouxel, O. (2017). Iron isotope systematics. *Reviews in Mineralogy and Geochemistry "Non-traditional stable isotopes,"* *82*, 415–510. https://doi.org/10.2138/rmg.2017.82.11

Dauphas, N., Hu, M. Y., Baker, E. M., Hu, J., Tissot, F. L. H., Alp, E. E., et al. (2018). SciPhon: a Data Analysis Software for Nuclear Resonant Inelastic X-ray Scattering with Applications to Fe, Kr, Sn, Eu and Dy. *Journal of Synchrotron*

Radiation, *25*, 5, 1581–1599. https://doi.org/10.1107/S1600577518009487

Debret, B., Millet, M.-A., Pons, M.-L., Bouihol, P., Inglis, E., & Williams, H. (2016). Isotopic evidence for iron mobility during subduction. *Geology*, *44*, 215–218. https://doi.org/10.1130/G37565.1

Domagal-Goldman, S. D., & Kubicki, J. D. (2008). Density functional theory predictions of equilibrium isotope fractionation of iron due to redox changes and organic complexation. *Geochimica et Cosmochimica Acta*, *72*, 5201–5216. https://doi.org/10.1016/j.gca.2008.05.066

Foden, J., Sossi, P. A., & Wawryk, C. M. (2015). Fe isotopes and the contrasting petrogenesis of A-, I- and S-type granite. *Lithos*, *212*, 32–44. https://doi.org/10.1016/j.lithos.2014.10.015

Foden, J., Sossi, P. A., & Nebel, O. (2018). Controls on the iron isotopic composition of global arc magmas. *Earth and Planetary Science Letters*, *494*, 190–201. https://doi.org/10.1016/j.epsl.2018.04.039

Fujii, T., Moynier, F., Blichert-Toft, J., & Albarède, F. (2014). Density functional theory estimation of isotope fractionation of Fe, Ni, Cu, and Zn among species relevant to geochemical and biological environments. *Geochimica et Cosmochimica Acta*, *140*, 553–576. https://doi.org/10.1016/j.gca.2014.05.051

Handler, M. R., Baker, J. A., Schiller, M., Bennett, V. C., & Yaxley, G. M. (2009). Magnesium stable isotope composition of Earth's upper mantle. *Earth and Planetary Science Letters*, *282*, 306–313. https://doi.org/10.1016/j.epsl.2009.03.031

Hibbert, K., Williams, H., Kerr, A. C., & Puchtel, I. (2012). Iron isotopes in ancient and modern komatiites: evidence in support of an oxidised mantle from Archean to present. *Earth and Planetary Science Letters*, *321*, 198–207. https://doi.org/10.1016/j.epsl.2012.01.011

Hill, P. S., & Schauble, E. A. (2008). Modeling the effects of bond environment on equilibrium iron isotope fractionation in ferric aquo-chloro complexes. *Geochimica et Cosmochimica Acta*, *72*, 1939–1958. doi: 10.1016/j.gca.2007.12.023

Hill, P. S., Schauble, E. A., & Young, E. D. (2010). Effects of changing solution chemistry on Fe^{3+}/Fe^{2+} isotope fractionation in aqueous Fe–Cl solutions. *Geochimica et Cosmochimica Acta*, *74*, 6669–6689. https://doi.org/10.1016/j.gca.2010.08.038

Hohenberg, P., & Kohn, W. (1964). Inhomogeneous electron gas. *Physical Review*, *136*, B864–B871. https://doi.org/10.1103/PhysRev.136.B864

Hu, M. Y. (2016). Some notes on data analysis for nuclear resonant inelastic x-ray scattering. *Hyperfine Interact*, *237*, 64. https://doi.org/10.1007/s10751-016-1284-7

Jackson, J. M., Hamecher, E. A., & Sturhahn, W. (2009). Nuclear resonant X-ray spectroscopy of $(Mg,Fe)SiO_3$ orthoenstatites. *European Journal of Mineralogy*, *21*, 551–560. https://doi.org/10.1127/0935-1221/2009/0021-1932

Johnson, C. M., Beard, B. L., Roden, E. E. (2008). The iron isotope fingerprints of redox and biogeochemical cycling in modern and ancient Earth. *Annual Review of Earth and Planetary Sciences*, *36*, 457–493. https://doi.org/10.1146/annurev.earth.36.031207.124139

Kelley, K. A., & Cottrell, E. (2009). Water and the oxidation state of subduction zone magmas. *Science*, *325*, 605–607. doi: 10.1126/science.1174156

Kieffer, S. W. (1982). Thermodynamics and lattice vibrations of minerals: 5. Applications to phase equilibria, isotopic fractionation, and high-pressure thermodynamic properties. *Reviews of Geophysics and Space Physics*, *20*, 827–849. https://doi.org/10.1029/RG020i004p00827

Kobayashi, H., Kamimura, T., Alfé, D., Sturhahn, W., Zhao, J., & Alp, E. E. (2004). Phonon density of states and compression behavior in iron sulfide under pressure. *Physical Review Letters*, *93*, 195503. doi: 10.1103/PhysRevLett.93.195503

Lee, C.T., Leeman, W. P., Canil, D., & Li, Z.-X. A. (2005). Similar V/Sc systematics in MORB and arc basalts: Implications for the oxygen fugacities of their mantle source regions. *Journal of Petrology*, *46*, 2313–2336. https://doi.org/10.1093/petrology/egi056

Lin, J.-F., Fei, Y., Sturhahn, W., Zhao, J., Mao, H., & Hemley, R. J. (2004). Magnetic transition and sound velocities of Fe_3S at high pressure: implications for Earth and planetary cores. *Earth and Planetary Science Letters*, *226*, 33–40. https://doi.org/10.1016/j.epsl.2004.07.018

Lipkin, H. J. (1995). Mössbauer sum rules for use with synchrotron sources. *Physical Review B*, *52*, 10073. doi: 10.1103/physrevb.52.10073

Lipkin, H. J. (1999). Mössbauer sum rules for use with synchrotron sources. *Hyperfine Interactions*, *123–124*, 349–366. https://doi.org/10.1023/A:1017019822825

Liu, J., Dauphas, N., Roskosz, M., Hu, M. Y., Yang, H., Bi W., et al. (2017). Iron isotopic fractionation between silicate mantle and metallic core at high pressure. *Nature Communications*, *8*, 14377. https://doi.org/10.1038/ncomms14377

Macris, C. A., Manning, C. E., & Young, E. D. (2015). Crystal chemical constraints on inter-mineral Fe isotope fractionation and implications for Fe isotope disequilibrium in San Carlos mantle xenoliths. *Geochimica et Cosmochimica Acta*, *154*, 168–185. https://doi.org/10.1016/j.gca.2015.01.024

Méheut, M., Lazzeri, M., Balan, E., & Mauri, F. (2009). Structural control over equilibrium silicon and oxygen isotopic fractionation: a first-principles density-functional theory study. *Chemical Geology*, *258*, 28–37. https://doi.org/10.1016/j.chemgeo.2008.06.051

Moynier, F., Fujii, T., Wang, K., & Foriel, J. (2013). Ab initio calculations of the Fe(II) and Fe(III) isotopic effects in citrates, nicotianamine, and phytosiderophore, and new Fe isotopic measurements in higher plants. *Comptes Rendus Géoscience*, *245*, 230–240. https://doi.org/10.1016/j.crte.2013.05.003

Nebel, O., Arculus, R., Sossi, P., Jenner, F., & Whan, T. (2013). Iron isotopic evidence for convective resurfacing of recycled arc-front mantle beneath back-arc basins. *Geophysical Research Letters*, *40*, 5849–5853. https://doi.org/10.1002/2013GL057976

Nebel, O., Sossi, P., Bénard, A., Wille, M., Vroon, P., & Arculus, R. (2015). Redox-variability and controls in subduction zones from an iron-isotope perspective. *Earth and Planetary Science Letters*, *432*, 142–151. https://doi.org/10.1016/j.epsl.2015.09.036

Oeser, M., Dohmen, R., Horn, I., Schuth, S., & Weyer, S. (2015). Processes and time scales of magmatic evolution as revealed by Fe–Mg chemical and isotopic zoning in natural olivines. *Geochimica et Cosmochimica Acta*, *154*, 130−150. https://doi.org/10.1016/j.gca.2015.01.025

Oeser, M., Ruprecht, P., & Weyer, S. (2018). Combined Fe-Mg chemical and isotopic zoning in olivine constraining magma mixing-to-eruption timescales for the continental arc volcano Irazú (Costa Rica) and Cr diffusion in olivine. *American Mineralogist*, *103*, 4, 582–599. https://doi.org/10.2138/am-2018-6258

Ottonello, G., & Zuccolini, M. V. (2008). The iron-isotope fractionation dictated by the carboxylic functional: An ab-initio investigation. *Geochimica et Cosmochimica Acta*, *72*, 5920–5934. https://doi.org/10.1016/j.gca.2008.09.027

Ottonello, G., & Zuccolini, M. V. (2009). Ab-initio structure, energy and stable Fe isotope equilibrium fractionation of some geochemically relevant H–O–Fe complexes. *Geochimica et Cosmochimica Acta*, *73*, 6447–6469. https://doi.org/10.1016/j.gca.2009.06.034

Peters, B. J., Shahar, A., Carlson, R. W., Day, J. M. D., & Mock, T. D. (2019). A sulfide perspective on iron isotope fractionation during ocean island basalt petrogenesis. *Geochimica et Cosmochimica Acta*, *245*, 59–78. https://doi.org/10.1016/j.gca.2018.10.015

Poitrasson, F., Halliday, A. N., Lee, D.-C., Levasseur, S., & Teutsch, N. (2004). Iron isotope differences between Earth, Moon, Mars and Vesta as possible records of contrasted accretion mechanisms. *Earth and Planetary Science Letters*, *223*, 253–266. https://doi.org/10.1016/j.epsl.2004.04.032

Poitrasson, F., & Freydier, R. (2005). Heavy iron isotope composition of granites determined by high resolution MC-ICP-MS. *Chemical Geology*, *222*, 132−147. https://doi.org/10.1016/j.chemgeo.2005.07.005

Poitrasson, F., Delpech, G., & Grégoire, M. (2013). On the iron isotope heterogeneity of lithospheric mantle xenoliths: implications for mantle metasomatism, the origin of basalts and the iron isotope composition of the Earth. *Contributions to Mineralogy and Petrology*, *165*, 1243−1258. https://doi.org/10.1007/s00410-013-0856-7

Polyakov, V. B. (1997). Equilibrium fractionation of the iron isotopes: Estimation from Mössbauer spectroscopy data. *Geochimica et Cosmochimica Acta*, *61*, 4213–4217. https://doi.org/10.1016/S0016-7037(97)00204-4

Polyakov, V. B. (2009). Equilibrium iron isotope fractionation at core-mantle boundary conditions. *Science*, *323*, 912–914. doi: 10.1126/science.1166329

Polyakov, V. B., & Mineev, S. D. (2000). The use of Mössbauer spectroscopy in stable isotope geochemistry. *Geochimica et Cosmochimica Acta*, *64*, 849–865. https://doi.org/10.1016/S0016-7037(99)00329-4

Polyakov, V. B., Mineev, S. D., Gurevich, V. M., Khramov, D. A., Gavrichev, K. S., Gorbunov, V. E., & Golushina, L. N. (2001). The use of Mössbauer spectroscopy and calorimetry for determining isotopic equilibrium constants. *Hematite. Russian Journal of Physical Chemistry*, *75*, 912–916.

Polyakov, V. B., Mineev, S. D., Clayton, R. N., Hu, G., & Mineev, K. S. (2005). Determination of tin equilibrium isotope fractionation factors from synchrotron radiation experiments. *Geochimica et Cosmochimica Acta*, *69*, 5531–5536. https://doi.org/10.1016/j.gca.2005.07.010

Polyakov, V., Clayton, R., Horita, J., & Mineev, S. (2007). Equilibrium iron isotope fractionation factors of minerals: reevaluation from the data of nuclear inelastic resonant X-ray scattering and Mössbauer spectroscopy. *Geochimica et Cosmochimica Acta*, *71*, 3833−3846. https://doi.org/10.1016/j.gca.2007.05.019

Polyakov, V. B., & Soultanov, D. M. (2011). New data on equilibrium iron isotope fractionation among sulfides: constraints on mechanisms of sulfide formation in hydrothermal and igneous systems. *Geochimica et Cosmochimica Acta*, *75*, 1957–1974. https://doi.org/10.1016/j.gca.2011.01.019

Prissel, K. B., Krawczynski, M. J., Nie, N. X., Dauphas, N., Couvy, H., Hu, M. Y., et al. (2018). Experimentally determined effects of olivine crystallization and melt titanium content on iron isotopic fractionation in planetary basalts. *Geochimica et Cosmochimica Acta*, *238*, 580–598. https://doi.org/10.1016/j.gca.2018.07.028

Richet, P., Bottinga, Y., & Javoy, M. (1977). A review of hydrogen, carbon, nitrogen, oxygen, sulphur, and chlorine stable isotope fractionation among gaseous molecules. *Annual Reviews of Earth and Planetary Sciences*, *5*, 65–110.

Richter, F. M., Watson, E. B., Mendybaev, R., Dauphas, N., Georg, B., Watkins, J., & Valley, J. (2009). Isotopic fractionation of the major elements of molten basalt by chemical and thermal diffusion. *Geochimica et Cosmochimica Acta*, *73*, 4250−4263. https://doi.org/10.1016/j.gca.2009.04.011

Roskosz, M., Sio, C. K. I., Dauphas, N., Bi, W., Tissot, F. H. L., Hu, M. Y., et al. (2015). Spinel-olivine-pyroxene equilibrium iron isotopic fractionation and applications to natural peridotites. *Geochimica et Cosmochimica Acta*, *169*, 184–199. https://doi.org/10.1016/j.gca.2015.07.035

Rustad, J. R., & Dixon, D. A. (2009). Prediction of iron-isotope fractionation between hematite (α-Fe_2O_3) and ferric and ferrous iron in aqueous solution from density functional theory. *The Journal of Physical Chemistry A*, *113*, 12249–12255. doi: 10.1021/jp9065373

Rustad, J. R., & Yin, Q.-Z. (2009). Iron isotope fractionation in the Earth's lower mantle. *Nature Geoscience*, *2*, 514–518. https://doi.org/10.1038/ngeo546

Rustad, J. R., Casey, W. H., Yin, Q.-Z., Bylaska, E. J., Felmy, A. R., Bogatko, S. A., et al. (2010). Isotopic fractionation of Mg^{2+}, Ca^{2+}, and Fe^{2+} with carbonate minerals. *Geochimica et Cosmochimica Acta*, *74*, 6301–6323. doi: 10.1016/j.gca.2010.08.018

Saunier, G., Pokrovski, G. S., & Poitrasson, F. (2011). First experimental determination of iron isotope fractionation between hematite and aqueous solution at hydrothermal conditions. *Geochimica et Cosmochimica Acta*, *75*, 6629–6654. https://doi.org/10.1016/j.gca.2011.08.028

Schauble, E. A. (2004). Applying stable isotope fractionation theory to new systems. *Reviews in Mineralogy and Geochemistry*, *55*, 65–111. doi: 10.2138/gsrmg.55.1.65

Schuessler, J. A., Schoenberg, R., Behrens, H., & von Blanckenburg, F. (2007). The experimental calibration of the iron isotope fractionation factor between pyrrhotite and peralkaline rhyolitic melt. *Geochimica et Cosmochimica Acta*, *71*, 417–433. https://doi.org/10.1016/j.gca.2006.09.012

Schuessler, J. A., Schoenberg, R., & Sigmarsson, O. (2009). Iron and lithium isotope systematics of the Hekla volcano, Iceland—evidence for Fe isotope fractionation during magma differentiation. *Chemical Geology*, *258*, 78–91. https://doi.org/10.1016/j.chemgeo.2008.06.021

Shahar, A., Young, E. D., & Manning, C. E. (2008). Equilibrium high-temperature Fe isotope fractionation between fayalite and magnetite: an experimental calibration. *Earth and Planetary Science Letters*, *268*, 330–338. https://doi.org/10.1016/j.epsl.2008.01.026

Shahar, A., Schauble, E. A., Caracas, R., Gleason, A. E., Reagan, M. M., Xiao, Y., et al. (2016). Pressure-dependent isotopic composition of iron alloys. *Science*, *352*, 580–582. 10.1126/science.aad9945

Shahar, A., Elardo S. M., & Macris, C. A. (2017). Equilibrium fractionation of non-traditional stable isotopes: an experimental perspective. *Reviews in Mineralogy and Geochemistry "Non-traditional stable isotopes,"* *82*, 65–83. https://doi.org/10.2138/rmg.2017.82.3

Sio, C. K. I., & Dauphas, N. (2017). Thermal histories of magmatic bodies by Monte Carlo inversion of Mg--Fe isotopic profiles in olivine. *Geology*, *45*, 67–70. https://doi.org/10.1130/G38056.1

Sio, C. K. I., Dauphas, N., Teng, F.-Z., Chaussidon, M., Helz, R. T., & Roskosz, M. (2013). Discerning crystal growth from diffusion profiles in zoned olivine by in situ Mg–Fe isotopic analyses. *Geochimica et Cosmochimica Acta*, *123*, 302−321. https://doi.org/10.1016/j.gca.2013.06.008

Sio, C. K., Roskosz, M., Dauphas, N., Bennett, N. R., Mock, T., & Shahar, A. (2018). The isotope effect for Mg-Fe interdiffusion in olivine and its dependence on crystal orientation, composition and temperature. *Geochimica et Cosmochimica Acta*, *239*, 463–480. https://doi.org/10.1016/j.gca.2018.06.024

Skulan, J. L., Beard, B. L., & Johnson C. M. (2002). Kinetic and equilibrium Fe isotope fractionation between aqueous Fe(III) and hematite. *Geochimica et Cosmochimica Acta*, *66*, 2995–3015. https://doi.org/10.1016/S0016-7037(02)00902-X

Sossi, P. A., Foden, J. D., & Halverson, G. P. (2012). Redox-controlled iron isotope fractionation during magmatic differentiation: an example from the Red Hill intrusion, S Tasmania. *Contributions to Mineralogy and Petrology*, *164*, 757−772. https://doi.org/10.1007/s00410-012-0769-x

Sossi, P. A., & O'Neill, H.St.C. (2015). The effect of bonding environment on iron isotope fractionation between minerals at high temperature. *Geochimica et Cosmochimica Acta*, *196*, 121–143. https://doi.org/10.1016/j.gca.2016.09.017

Sossi, P. A., Nebel, O., & Foden, J. (2016). Iron isotope systematics in planetary reservoirs. *Earth and Planetary Science Letters*, *452*, 295–308. https://doi.org/10.1016/j.epsl.2016.07.032

Syverson, D. D., Borrok, D. M., & Seyfried, Jr. W. E. (2013). Experimental determination of equilibrium Fe isotopic fractionation between pyrite and dissolved Fe under hydrothermal conditions. *Geochimica et Cosmochimica Acta*, *122*, 170–183. https://doi.org/10.1016/j.gca.2013.08.027

Telus, M., Dauphas, N., Moynier, F., Tissot, F. L., Teng, F.-Z., Nabelek, P. I., et al. (2012). Iron, zinc, magnesium and uranium isotopic fractionation during continental crust differentiation: The tale from migmatites, granitoids, pegmatites. *Geochimica et Cosmochimica Acta*, *97*, 247−265. https://doi.org/10.1016/j.gca.2012.08.024

Teng, F.-Z., Dauphas, N., & Helz, R. T. (2008). Iron isotope fractionation during magmatic differentiation in Kilauea Iki Lava Lake. *Science*, *320*, 1620−1622. doi: 10.1126/science.1157166 *Article*

Teng, F.-Z., Li, W.-Y., Ke, S., Marty, B., Dauphas, N., Huang, S., et al. (2010). Magnesium isotopic composition of the Earth and chondrites. *Geochimica et Cosmochimica Acta*, *74*, 14, 4150–4166. https://doi.org/10.1016/j.gca.2010.04.019

Teng, F.-Z., Dauphas, N., Helz, R. T., Gao, S., & Huang, S. (2011). Diffusion-driven magnesium and iron isotope fractionation in Hawaiian olivine. *Earth and Planetary Science Letters*, *308*, 317−324. https://doi.org/10.1016/j.epsl.2011.06.003

Teng, F.-Z., Dauphas, N., Huang, S., & Marty, B. (2013). Iron isotopic systematics of oceanic basalts. *Geochimica et Cosmochimica Acta*, *107*, 12−26. https://doi.org/10.1016/j.gca.2012.12.027

Urey, H. C. (1947). The thermodynamic properties of isotopic substances. *Journal of the Chemical Society* (London) 562–581.

Wang, W., Zhou, C., Qin, T., Kang, J.-T., Huang, S., Wua, Z., & Huang F. (2017). Effect of Ca content on equilibrium Ca isotope fractionation between orthopyroxene and clinopyroxene. *Geochimica et Cosmochimica Acta*, *219*, 44–56. https://doi.org/10.1016/j.gca.2017.09.022

Weyer, S., & Ionov, D. A. (2007). Partial melting and melt percolation in the mantle: The message from Fe isotopes. *Earth and Planetary Science Letters*, *259*, 119−133. https://doi.org/10.1016/j.epsl.2007.04.033

Wiesli, R. A., Beard, B. L., & Johnson, C. M. (2004). Experimental determination of Fe isotope fractionation between aqueous Fe(II), siderite and "green rust" in abiotic systems. *Chemical Geology*, *211*, 343–362. https://doi.org/10.1016/j.chemgeo.2004.07.002

Williams, H. M., McCammon, C. A., Peslier, A. H., Halliday, A. N., Teutsch, N., Levasseur, S., & Burg, J.-P. (2004). Iron isotope fractionation and the oxygen fugacity of the mantle. *Science*, *304*, 1656−1659. doi: 10.1126/science.1095679

Williams, H. M., Peslier, A. H., McCammon, C., Halliday, A. N., Levasseur, S., Teutsch, N., & Burg, J. P. (2005). Systematic iron isotope variations in mantle rocks and minerals: The effects of partial melting and oxygen fugacity. *Earth and Planetary Science Letters*, *235*, 435−452. https://doi.org/10.1016/j.epsl.2005.04.020

Williams, K. B., Krawczynski, M. J., Nie, N. X., Dauphas, N., Couvy, H., Hu, M. Y. and Alp, E. E. (2016) 'The role of differentiation processes in mare basalt iron isotope signatures', *Lunar Planetary Science Conference*, *47*, pp. 2779

Williams, H. M., Prytulak, J., Woodhead, J. D., Kelley, K. A., Brounce, M., & Plank, T. (2018). Interplay of crystal fractionation, sulfide saturation and oxygen fugacity on the iron isotope composition of arc lavas: An example from the Marianas. *Geochimica et Cosmochimica Acta*, *226*, 224–243. https://doi.org/10.1016/j.gca.2018.02.008

Yang, H., Lin, J.-F., Hu, M. Y., Roskosz, M., Bi, W., Zhao, J., et al. (2018). Iron isotopic fractionation in mineral phases from Earth's lower mantle: Did terrestrial magma ocean crystallization fractionate iron isotopes? *Earth and Planetary Science Letters*, *506*, 113–122. https://doi.org/10.1016/j.epsl.2018.10.034

Young, E. D., Manning, C. E., Schauble, E. A., Shahar, A., Macris, C. A., Lazar, C., & Jordan, M. (2015). High-temperature equilibrium isotope fractionation of non-traditional stable isotopes: Experiments, theory, and applications. *Chemical Geology*, *395*, 176–195. https://doi.org/10.1016/j.chemgeo.2014.12.013

15

The Role of Redox Processes in Determining the Iron Isotope Compositions of Minerals, Melts, and Fluids

Paolo A. Sossi[1,2] and Baptiste Debret[2,3]

ABSTRACT

Stable isotope fractionation is a response to the minimization of free energy associated with differences in vibrational frequencies during substitution of isotopic masses between two phases. Site properties are dictated by the bonding environment of iron, and influence its "force constant", a descriptor for bond strength. Because iron readily transitions between its ferrous (Fe^{2+}) and ferric (Fe^{3+}) state over the oxygen fugacities (fO_2) of terrestrial magmas, redox processes are presumed to control iron isotope fractionation. In fact, knowledge of both coordination and redox state are key to interpreting iron isotope variations in high-temperature, high-pressure geological systems. Determinations of iron force constants by experimental, theoretical, and spectroscopic means in minerals, melts, and fluids are reviewed, emphasizing the effect of composition in influencing isotopic fractionation. Within this framework, the application of numerical models of partial melting to iron isotope variations in oceanic basalts and subduction-zone (or arc) magmas has shed light on their genesis. The influence of melt composition, mineral mode, and fluid exsolution in producing iron isotope fractionation in evolving magmas is explored. Iron isotopic signatures in magmatic rocks are linked to that of their sources, and are influenced by the iron isotopic evolution of subducting slabs and transfer of fluids to the mantle wedge.

15.1. INTRODUCTION

The existence of iron in three predominant oxidation states in high temperature processes (Fe^0, Fe^{2+}, Fe^{3+}) provides considerable scope for the fractionation of its isotopes. This property, together with the fact that iron is often the most abundant polyvalent element in natural systems, has made its isotopic composition the most well-studied of the non-traditional systems (see Dauphas et al., 2017, for a recent comprehensive review).

Owing to the role that iron plays in determining the oxygen fugacity (fO_2) of a geological system, variations in its isotopic composition have often been ascribed to changes in its redox state (e.g., Williams et al., 2004; Dauphas et al., 2009). This approach does not explicitly account for the coordination of iron, which can have an equally important role in engendering isotopic fractionation (see Young et al., 2015; Sossi & O'Neill, 2017). Furthermore, the oxygen fugacity of a system may have an indirect effect

[1]*Institute of Geochemistry and Petrology, ETH Zürich, Zürich, Switzerland*

[2]*Université de Paris, Institut de Physique du Globe de Paris, Paris, France*

[3]*Laboratoire G-Time, DGES, Université Libre de Bruxelles, Brussels, Belgium*

Magma Redox Geochemistry, Geophysical Monograph 266, First Edition.
Edited by Roberto Moretti and Daniel R. Neuville.

DOI: 10.1002/9781119473206.ch15

on the iron isotope composition by modifying the stability of different phases and compositions of solid solutions (e.g., the precipitation of pyrite from hydrothermal fluids, Wawryk & Foden, 2015; or spinel compositions in solid solution series, Roskosz et al., 2015). Sossi et al. (2016) and Fujii et al. (2014) emphasized the role of coordinating ligand field strength (e.g., –S, –Cl, and –OH) in affecting iron isotopic fractionation in minerals and fluids, respectively, by way of changing bond covalence and hence bond stiffness (Hill et al., 2010). Although this effect is limited in terrestrial mafic magmas in which iron is almost exclusively coordinated by oxygen, it becomes an important consideration in intermediate- and silicic magmas that can exsolve hydrothermal fluids (see Heimann et al., 2008) and in fluid loss during mineral breakdown and subsequent dehydration of subducting slabs (see Debret & Sverjensky, 2017).

Deconvolving these controls is not straightforward because they tend to be interrelated; for example, simply varying temperature will affect the relative stability of minerals that precipitate from a fluid and these differing assemblages would be reflective of a different fO_2. Similarly, because the Fe^{3+} cation has an ionic radius smaller than that of its ferrous counterpart, it prefers to partition into smaller, tetrahedrally coordinated sites, whereas Fe^{2+} has a greater preference for octahedral sites in melts and minerals (e.g., Wilke, 2005). As such, these compounding effects on the magnitude of stable isotope fractionation require careful theoretical (e.g., Schauble et al., 2001), experimental (e.g., Shahar et al., 2008) and spectroscopic (e.g., Dauphas et al., 2012) approaches in order to tease out the individual contribution of each.

Rather than tackling iron isotopic variation in natural systems by an inverse approach (i.e., through sample measurement and interpretation), in this review, the role of iron oxidation state and coordination number in controlling iron isotope fractionation is critically examined from a first-principles perspective (Section 15.2). In order to do so, the methods by which equilibrium iron isotope fractionation factors can be determined, and a compilation of existing data are briefly reviewed (Section 15.3). A general introduction to the ways in which iron isotope fractionation between minerals, melts and fluids depend on redox processes is presented in Section 15.4. Finally, how these determinations have been applied to interpret variation in natural systems, comprising partial melting, magmatic differentiation, and fluid transfer in the crust and mantle is detailed (Section 15.5).

15.2. PRINCIPLES AND NOMENCLATURE

The ratio of two isotopes, n and d of an element, E, is generally expressed relative to the same ratio in a standard in delta notation, which is given by:

$$\delta^{(n/d)}E = \left(\frac{\left({}^{n}E/{}^{d}E\right)_{sample}}{\left({}^{n}E/{}^{d}E\right)_{standard}} - 1 \right) \times 1000. \quad (15.1)$$

The fractionation factor between two phases A and B is:

$$\Delta^{n/d}E_{A-B} = \delta^{n/d}E_A - \delta^{n/d}E_B, \quad (15.2)$$

and is related to alpha notation by the equivalence:

$$\Delta^{n/d}E_{A-B} = 10^3 \ln {}^{n/d}\alpha_{A/B} \quad (15.3)$$

The exchange of isotopes between two phases, A and B, can be written:

$${}^{d}E_A + {}^{n}E_B = {}^{d}E_B + {}^{n}E_A. \quad (15.4)$$

The equilibrium distribution of isotopes between the two phases is described by the fractionation factor, α:

$$\alpha^{\frac{n}{d}}E_{A/B} = \frac{\left({}^{n}E/{}^{d}E\right)_A}{\left({}^{n}E/{}^{d}E\right)_B}. \quad (15.5)$$

When the B phase consists of dissociated atoms of element E (as in an ideal, monatomic gas), the fractionation factor is given the name 'beta factor' (or β factor) and represents a convenient way of normalising fractionation factors to a "0" value (as there is no energy associated with isotopic substitution in a dissociated monatomic gas, E^0). For cases in which the stoichiometric coefficients are unity, Equation 15.5 is equivalent to an equilibrium constant. Therefore, the free energy of the isotopic substitution at equilibrium can be described:

$$\Delta G^{o}_{(5)} = -RT \ln\left(\alpha^{\frac{n}{d}}E_{A-B}\right). \quad (15.6)$$

At the conditions (temperature, T; pressure, P) relevant to magmatic processes, classical thermodynamic terms cancel and isotopic fractionation is broadly onset by differences in the energy associated with vibrational frequencies in molecules and in condensed phases among one another. In the harmonic approximation, these vibrational energies ($E_{vib,\,i}$) are quantized into n equally spaced levels, where the finite minimum half quantum of energy that arises from the uncertainty in the positions and momenta that a particle, i, could occupy in its ground state ($n = 0$), corresponds to the "zero-point" energy:

$$E_{vib,i} = \left(n_i + \frac{1}{2}\right)h\nu_i. \quad (15.7)$$

Here, h is Planck's constant and ν_i is the vibrational frequency. Because, even at 1000 K, as given by the Boltzmann equation, atoms occupy vanishingly small amounts of their excited states ($n > 0$), the differences in vibrational frequencies between two isotopes is solely controlled by differences in zero-point energies. For changes in pressure, the relationship between the mode Gruneisen parameter ($\gamma_{G,i}$), the vibrational frequency, and molar volume (V) at constant temperature in the quasi-harmonic approximation (Polyakov & Kharlashina, 1994), leads to a predictable increase in frequency with pressure due to contracting molar volumes at constant temperature:

$$\left(\frac{\partial \ln \nu_i}{\partial \ln V}\right) = -\gamma_{G,i}. \tag{15.8}$$

As volume changes induced by isotopic substitutions are typically small, correspondingly large pressure variations are required to significantly influence isotopic fractionation. Frequencies for a simple diatomic molecule (or, approximately, a mean completely symmetric case) are given by:

$$\nu_i = \frac{1}{2\pi}\sqrt{\frac{K}{\mu}}, \tag{15.9}$$

Where K is the force constant of vibration and μ the reduced mass. The derivation of the dependence of isotopic fractionation on the vibrational frequencies of molecules was first expressed by Urey (1947) and Bigeleisen and Mayer (1947) in the harmonic approximation in terms of the vibrational isotopic partition function:

$$Q_{vib,i} = \prod_i \frac{e^{\left(-\frac{h\nu_i}{2kT}\right)}}{1 - e^{\left(-\frac{h\nu_i}{kT}\right)}}. \tag{15.10}$$

Equation 15.10 is the product across all vibrational frequencies ν_i for a given isotope in a given phase, where k is Boltzmann's constant and T absolute temperature. Ratios of vibrational partition functions may be equated to fractionation factors in a simplified manner for $h\nu_i/kT < 5$ (Urey, 1947; Bigeleisen & Mayer, 1947):

$$10^3 \ln \alpha^{\frac{n}{d}}E_{A-B} = \Delta^{\frac{n}{d}}E_{A-B} = \frac{10^3 3N}{96}\left(\frac{h}{\pi kT}\right)^2\left(\frac{1}{m_d} - \frac{1}{m_n}\right)[K_A - K_B]. \tag{15.11}$$

Here, K_A and K_B refer to the iron force constants in phases A and B, respectively, m_d and m_n the atomic masses of the denominator (d) and numerator (n) isotopes, respectively, and the $3N$ term (Avogadro's Number) comes from summation over all vibrations in condensed matter (Bigeleisen & Mayer, 1947). Equation 15.11 states that the isotopic fractionation at the high temperature limit ($T \rightarrow \infty$) will approach zero. For a particular element and isotope ratio, most of the equation reduces to a constant term, yielding, for $^{57/54}$Fe:

$$\Delta^{\frac{57}{54}}Fe_{A-B} = 4284\frac{[K_A - K_B]}{T^2}. \tag{15.12}$$

Equally, for $^{56/54}$Fe;

$$\Delta^{\frac{56}{54}}Fe_{A-B} = 2904\frac{[K_A - K_B]}{T^2}. \tag{15.13}$$

In order to minimize the higher free energy of bonds with higher vibrational frequencies (Equation 15.9), the heavier isotope is favoured over the lighter one.

In general, higher force constants occur in bonding environments with (see Schauble, 2004; Young et al., 2015; Sossi et al., 2016):

(i) higher oxidation state
(ii) lower coordination number
(iii) higher bond covalence
(iv) low spin *vs.* high spin (for Fe and other transition metals)

To a first order, the first three controls may be estimated empirically to derive values of K (see Young et al., 2015; Sossi & O'Neill, 2017), considering that the interactions between cations and anions in a lattice site can be described quantitatively with ionic interactions. Variations of K arise due to a change in the charge and coordination of the cation, which are described by the bond valence principle (Pauling, 1929):

$$\bar{s} = \frac{\overline{Z}}{\overline{C_N}}. \tag{15.14}$$

In this equation, $\bar{s}$ is the bond valence, $\overline{Z}$ is the average charge on the cation, and $\overline{C_N}$ the mean coordination number in the mineral. The force constant is then approximated by solving the Born-Landé equation, after Young et al. (2009), which describes the elastic energy of an ionic lattice:

$$K_{f,Fe-O} = -\frac{\bar{s}_{Fe}\bar{s}_O q^2(1-B)}{4\pi\varepsilon_o r^3}, \tag{15.15}$$

where q is the charge on an electron (1.602×10^{-19} C), B is the Born exponent, dependent on lattice compressibility (taken to be 12 for minerals), $1/4\pi\varepsilon_o$ is Coulomb's constant (9×10^9 Nm^2/C^2), and r is the bond length in metres. More quantitative methods for estimating for K are described in Section 15.3.

15.3. METHODS FOR THE CALIBRATION OF IRON ISOTOPE FRACTIONATION FACTORS

The quantity that drives equilibrium isotope fractionation, the value of the force constant (cf. Equation 15.15), may be calibrated via a variety of methods. These include calculating the structure of the phase (theoretical approaches) or by measuring inelastic scattering properties of Mössbauer-sensitive ^{57}Fe (spectroscopic methods). Alternatively, methods that involve measurement of the isotopic composition of two or more phases in equilibrium with one another to determine fractionation factors (experimental approaches) are also widely used. All three are described briefly in the next sections.

15.3.1. Theoretical

Ab-initio simulations are the most common way of determining vibrational frequencies, typically with density functional theory (DFT; e.g., Schauble, 2004), one such class of these computations. DFT models the electronic structure and density of electrons in individual atoms to find the most probable equilibrium ground state electronic configuration. Once this state is defined for a system consisting of x number of atoms, the interatomic distances between them define a molecular structure from which the force constants are calculated. The interested reader is referred to Blanchard and Dauphas (Chapter 14, this volume) for a fuller description of the method. A simplified calculation based on the bond valence principle can also yield accurate force constants (Equation 15.15; Young et al., 2015, 2009; Sossi & O'Neill, 2017), though it uses the atomic structure of the phase as an input parameter.

15.3.2. Experimental

Here, the phases of interest are synthesised by subjecting a known composition to a specific pressure and temperature. The isotopic composition of one of the resultant phases is measured with respect to that of a second phase, yielding the fractionation factor, $\alpha^{n}_{d}E_{A/B}$. That the two phases grown experimentally are in equilibrium with one another may be determined by two different means: (i) chemical and isotopic equilibrium is demonstrated to have been obtained between the phases of interest (e.g., by time-series and reversal experiments) and (ii) by the three-isotope method (Shahar et al., 2008). Here, a series of experiments that approach equilibrium is performed, and the true equilibrium composition is calculated by extrapolation of the experiments to the known bulk composition. Considerations in designing and interpreting the results of experimentally determined isotopic fractionation are summarised in Shahar et al. (2017) and in Bourdon et al. (2018).

15.3.3. Spectroscopic

Measurement of force constants relies upon the excitation of the vibrational modes of motion of the mineral lattice by incoming X-rays which collide with a ^{57}Fe nucleus (the only iron isotope sensitive to this effect). Such information can be determined by Mössbauer spectroscopy (Polyakov & Mineev, 2000) or by Nuclear-Inelastic Resonant X-Ray Scattering, NRIXS (Polyakov et al., 2007; Dauphas et al., 2012). The latter method is preferred because the characteristics of the impacting photons can be tuned and scanned for purpose at a synchrotron. Both of these methods are dealt with in more detail in Blanchard and Dauphas (Chapter 14, this volume).

15.4. FUNDAMENTAL CONTROLS ON ISOTOPIC FRACTIONATION BETWEEN MINERALS, MELTS, AND FLUIDS

15.4.1. Minerals

The $Fe^{3+}/\sum Fe$ ratio in minerals is controlled by the stoichiometry of the phase. Some series display complete solid solution between compositional end-members, and common igneous examples include:

(i) Magnetite-Ulvöspinel: $^{IV}Fe^{3+}\ ^{VI}(Fe^{2+}Fe^{3+})O_4$ – $^{IV}Fe^{2+VI}(Fe^{2+}Ti^{4+})O_4$

Magnetite shows complete solid solution with ulvöspinel, an exchange that forms the basis for the Fe-Ti exchange oxybarometer of Buddington and Lindsley (1964) (see also Arató & Audétat, Chapter 18, this volume). Both are spinel group minerals *sensu lato*, with the general formula AB_2O_4, where, if the B cations occupy both the tetrahedral and octahedral sites, the structure is said to be "inverse." Magnetite is a 2-3 spinel (one 2+ cation with two 3+) and has an inverse cation distribution at room temperatures, though it tends to the random distribution at magmatic temperatures around 1000 °C (Wu & Mason, 1981). Ulvöspinel, a 4-2 spinel, also has an inverse structure with Ti exclusively occupying the octahedral site, while ferrous iron is both VI- and IV-fold coordinated. Intermediate compositions have site distributions that are temperature-dependent and are controlled by electron exchange between the two valence states of Fe, with the random distribution being favoured at higher temperature (O'Neill & Navrotsky, 1984; Wechsler et al., 1984; Bosi et al., 2009).

Shahar et al. (2008) determined the fractionation factor between magnetite and fayalite experimentally in a piston cylinder to be $\Delta^{57}Fe_{mag\text{-}fay} = +0.30 \times 10^6/T^2$, reflecting the preference of heavier iron for the stronger, Fe^{3+}-bearing IV- and VI-fold sites of magnetite relative to the exclusively octahedral ferrous iron in fayalite. Sossi and O'Neill (2017) measured the fractionation factor between magnetite and $FeCl_2.4H_2O$ fluid in piston cylinder experiments in Ag capsules at 800 °C. They estimated the magnetite-ulvöspinel fractionation factor from fits to their experimental data using bond valence theory, giving $\Delta^{57}Fe_{mag\text{-}usp} = +0.32 \times 10^6/T^2$ (Figure 15.1), very similar to that measured between magnetite and fayalite. There is considerable disagreement as to the force constant of Fe bonds in magnetite; Polyakov et al. (2007) report a value of 228 N/m calculated from NRIXS spectra, similar to that given thereafter by Dauphas et al. (2012), 230 N/

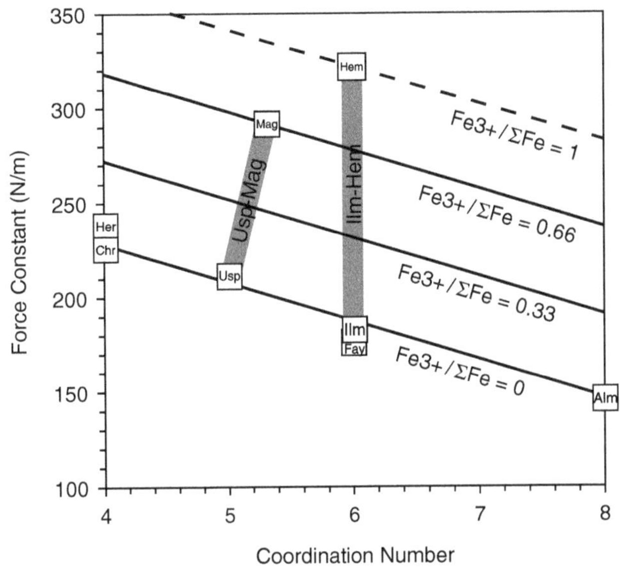

Figure 15.1 After Sossi and O'Neill (2017). Dependence of the calculated force constants on the coordination number and oxidation state of iron. Lines of constant $Fe^{3+}/\sum Fe$ are drawn on in black. Effects of two common solid solutions on the force constant (magnetite–ulvöspinel and hematite–ilmenite) are shown in grey. Minerals are shown in white squares; Her = Hercynite, Chr = Chromite, Usp = Ulvöspinel, Mag = Magnetite, Hem = Hematite, Ilm = Ilmenite, Fay = Fayalite, Alm = Almandine.

m, using the same technique. Roskosz et al. (2015) report a higher value of 264 N/m using NRIXS via extrapolation of trends based on solid solutions of Mg-Fe-Al-Cr spinels, and they note that discrepancies may be due to the quality of the spectra at high energies and the energy range over which spectra are collected. Importantly, though the absolute force constants may vary, the magnitude of mineral-mineral fractionation factors in these studies are in good agreement.

ii) Ilmenite-Hematite: $^{VI}Fe^{2+}TiO_3$ – $^{VI}Fe^{3+}{}_2O_3$

Iron passes from entirely ferrous to entirely ferric along the solid solution series, whilst remaining in octahedral coordination (e.g., Wechsler & Prewitt, 1984). Ilmenite shows limited solid solution with hematite at the fO_2s typical of igneous systems, where X_{Ilm} varies from 0.98 at 2 log units below the Fayalite-Magnetite-Quartz (FMQ) buffer to 0.90 at ΔFMQ+1 during crystallisation of a dry tholeiitic ferrobasalt (see Toplis & Carroll, 1995).

Determinations of force constants in hematite via Mössbauer spectroscopy (Polyakov & Mineev, 2000) ab-initio calculations (Blanchard et al., 2009) and NRIXS (Polyakov et al., 2007; Dauphas et al., 2012) are all in good agreement and yield mean Fe-O force constants of between 235 and 252 N/m. Note that the spinel trend measured in Roskosz et al. (2015) yields higher values for the purely ferric iron-bearing end-members, 303 ± 6 N/m (n = 3). Polyakov and Mineev (2000) report a force constant for ilmenite of 90 N/m, and hence a fractionation factor of $\Delta^{57/54}Fe_{hem\text{-}ilm} = 4284(235 - 90)/T^2 = 0.61 \times 10^6/T^2$. Sossi and O'Neill (2017) measured the isotopic fractionation between ilmenite and $FeCl_2.4H_2O$ fluid at 800 °C and estimated the ilmenite-hematite fractionation factor from fits to their experimental data using bond valence theory to give $0.58 \times 10^6/T^2$ (Figure 15.1). Both fractionation factors are in very good agreement and highlight the importance of mineral composition within a solid solution in determining the isotopic fractionation factor.

iii) Almandine-Pyrope: $^{VIII}Fe^{2+}{}_3{}^{VI}Al_2[SiO_4]_3$ – $^{VIII}Mg_3{}^{VIAl_2}[SiO_4]_3$; Grossular-Andradite: $^{VIII}Ca_3$-$^{VIAl3+}{}_2[SiO_4]_3$ – $^{VIII}Ca_3{}^{VI}Fe^{3+}{}_2[SiO_4]_3$

Ferrous iron is dodecahedrally coordinated by oxygen in almandine garnet, a rarity among Fe-bearing minerals, in which it typically occupies VI-fold sites. Ferric iron in andradite garnet is in octahedral coordination.

Sossi and O'Neill (2017) reported fractionation factors between almandine and other Fe-bearing minerals and showed that its iron isotope composition was invariably lighter than all other phases investigated (including $FeCl_2$ fluid) by virtue of its longer Fe-O bonds (average of 2.37 Å). They calculated a force constant of 147 N/m for almandine (Figure 15.1). No measurements of force constants or fractionation factors involving andradite garnet are presently available. However, natural andradite-bearing garnets show heavier iron isotope compositions relative to their ferrous iron-bearing counterparts (Urosevic et al., 2018) while garnet is almost always the isotopically lightest phase in eclogitic rocks (Williams et al., 2009; Huang et al., 2019), reflecting their generally low $Fe^{3+}/\Sigma Fe$ in mafic and ultramafic rocks (<0.1; Williams et al., 2009; Berry et al., 2010).

iv) Aegirine-Hedenbergite: $Na^{VI}Fe^{3+}Si_2O_6$ – $Ca^{VI}Fe^{2+}Si_2O_6$

The coupled substitution $NaFe^{3+}(CaFe^{2+})^{-1}$ across the aegirine (acmite)-hedenbergite series involves a change in ferric to ferrous iron in the same octahedral crystallographic site (the M1 site). This transition is associated with an increase in mean Fe-O bond length from 2.05 Å in aegirine to 2.15 Å in hedenbergite at igneous temperatures of ≈1000°C (see Cameron et al., 1973).

Both end-members of the series were analyzed by Mössbauer spectroscopy and force constants calculated according to the second-order Doppler shift by Polyakov and Mineev (2000), yielding 274 N/m for aegirine and 112 N/m for hedenbergite, the latter being similar to other $^{VI}Fe^{2+}$-bearing silicates determined in the same study (olivine, diopside and enstatite; Polyakov & Mineev, 2000). More importantly, the fractionation factor between the two is then $\Delta^{57/54}Fe_{aeg\text{-}hed} = 0.68 \times 10^6/T^2$, similar in magnitude to that for the exchange of octahedral Fe^{2+} and Fe^{3+} in the ilmenite-hematite series (Figure 15.1).

v) Riebeckite-Glaucophane: $Na_2{}^{VI}(Mg,Fe^{2+})_3{}^{VI}Fe^{3+}{}_2Si_8O_{22}(OH)_2$-$Na_2{}^{VI}(Mg,Fe^{2+})_3Al^{3+}{}_2Si_8O_{22}(OH)_2$

Riebeckite-group amphiboles are typically found in alkaline magmatic rocks (cf. Schoenberg et al., 2009). They can contain between 100% and 40% of their Fe as Fe^{3+}, in which ferric iron is ordered on the M2 site and small amounts on the M1 (Hawthorne, 1978) whereas glaucophane hosts ferrous iron exclusively and is distributed across its M1 and M3 sites.

From Mössbauer spectroscopy data, Schoenberg et al. (2009) calculated the beta factors for two riebeckitic amphiboles with $Fe^{3+}/\sum Fe$ of 0.46 and 0.44, and found $10^3\ln\beta^{57/54}Fe_{rbk}$ = 0.90 and 0.74 × $10^6/T^2$, respectively, which correspond to force constants of 214 and 175 N/m.

vi) Spinel group; hercynite – ferrochromite – magnetite $^{IV}Fe^{2+}Al_2O_4$ – $^{IV}Fe^{2+}Cr_2O_4$ – $^{IV}Fe^{3+}\ ^{VI}[Fe^{2+}Fe^{3+}]O_4$

Spinel-group minerals are ubiquitous in terrestrial mafic and ultramafic lithologies (e.g., Barnes & Roeder, 2001) and show extensive solid solutions with first-row transition metals representing the A and B cations (O'Neill & Navrotsky, 1984). Natural spinels tend to exhibit Cr-Al or Fe^{3+}-Cr exchange, with a paucity of Fe^{3+}-Al spinels caused by positive deviations from ideality and hence a solvus in the binary system (e.g., Petric et al., 1981; Nell et al., 1989). Chromium, owing to its high octahedral site preference energy $(6/5\Delta_{oct})$[1], is strongly ordered onto the octahedral site, conferring a normal cation distribution on the spinel structure, meaning Fe^{2+} is found exclusively on the tetrahedral site. Similarly, Fe^{3+} is displaced from the octahedral site in magnetite to the tetrahedral site with increasing ferrochromite component (O'Neill & Navrotsky, 1984).

Spinels are generally found in association with olivine, for example in mantle peridotites and in komatiitic magmas, and are used together with orthopyroxene to determine oxygen fugacities (e.g., Woodland et al., 1992). In reciprocal solid solution with Fe-Mg olivine, Fe^{2+} substitution is favoured such that Cr/(Cr+Al) increases with increasing $Fe^{2+}/(Fe^{2+}+Mg)$ (e.g., Poustovetov & Roeder, 2000). At high Fe^{3+} contents, the spinel structure inverts and extensive exchange with Ti occurs as discussed.

In a comprehensive study of the effect of solid solution on the mean strength of the Fe-O bonds, Roskosz et al. (2015) determined the force constants for eight spinel-group minerals. They focused largely on Mg-Al-Fe spinels with variable $Fe^{3+}/\sum Fe$ ratios, from 0 to 1, and one magnesiochromite composition. Their results show that the Fe-O force constant is linearly correlated with

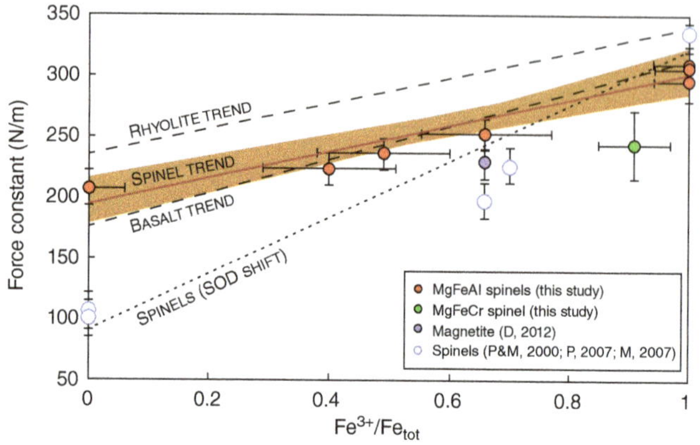

Figure 15.2 Iron-oxygen force constants in spinel group minerals (after Roskosz et al. 2015). The regression and associated uncertainty refers to Al-bearing spinels only, and the rhyolite and basalt trend are from the data of Dauphas et al. (2014). P&M = Polyakov and Mineev 2000; P = Polyakov et al., 2007; M = Mineev et al. 2007; D = Dauphas et al. 2012.

$Fe^{3+}/\sum Fe$ with only subsidiary influence from other structural parameters, ranging from 207 ± 14 in Fe^{2+}-bearing spinels to 303 ± 6 N/m (n = 3) for solely ferric iron-bearing spinels (Figure 15.2). These values compare with Fe-O force constants of 197 – 165 N/m for $^{VI}Fe^{2+}$-bearing silicates (average = 181 ± 15 N/m, n = 4), namely olivine and orthopyroxene (Roskosz et al., 2015), and hence $\Delta^{57/54}Fe_{sp(FeII)\text{-}sil(FeII)}$ = 0.11±0.08 × $10^6/T^2$. In attempting to characterize the influence of coordination environment on iron isotopic fractionation, Sossi and O'Neill (2017) determined the fractionation factor between two spinels; hercynite ($FeAl_2O_4$) and chromite ($FeCr_2O_4$) and $FeCl_2.4H_2O$ fluid at 800 °C. They calculated force constants of 238 N/m and 225 N/m, respectively, higher than that of fayalite (178 N/m), yielding $\Delta^{57/54}Fe_{sp(FeII)\text{-}sil(FeII)}$ = 0.21 × $10^6/T^2$, reflecting the influence of IV-fold *vs.* VI-fold coordination of Fe in spinel and olivine, respectively (Figure 15.1). Compilation of spinel-olivine fractionation factors measured in peridotitic rocks show generally positive $\Delta^{57/54}Fe_{sp\text{-}ol}$ at low Cr# (molar $Cr/(Cr+Al+Fe^{3+})$) compared with negligible or slightly negative $\Delta^{57/54}Fe_{sp\text{-}ol}$ at high Cr# (Zhang et al. 2019), in qualitative agreement with lowering of the force constant in chromite with respect to hercynite (Sossi & O'Neill, 2017) though a change in sign is not predicted.

vii) Bridgmanite – Ferropericlase $^{VIII}(Fe^{2+},Mg^{2+},Fe^{3+})^{VI}(Fe^{3+},Al^{3+},Si^{4+})O_3$ – $^{VI}Fe_x(Mg_{1-x})O$

Though not a solid solution, it is instructive to assess the body of work for these two minerals given their importance for controlling iron isotopic fractionation in the Earth's lower mantle. Iron is present primarily as ferrous iron in the octahedral site of ferropericlase (cubic) and as Fe^{2+}

[1] In crystal field theory, Δ_{oct} relates to the energy difference between the d-shell electrons that are split into the e_g and t_{2g} orbitals. For an octahedral field, t_{2g} orbitals are stabilised with respect to the barycentre by $2/5\Delta_{oct}$ and e_g by $-3/5\ \Delta_{oct}$. Thus, the minimum Δ_{oct} is 0 (e.g., Zn^{2+}, Cu^+) and the maximum is $6/5\Delta_{oct}$ (e.g., V^{2+}, Cr^{3+}).

and Fe^{3+} in the dodecahedral A and octahedral B sites of bridgmanite (orthorhombic). The substitution of ferric iron into bridgmanite is strongly facilitated through its coupled substitution with Al^{3+} (McCammon, 1997), which forces Fe^{3+} onto the dodecahedral site by the mechanism $^{VIII}(Fe^{2+}, Mg^{2+})^{VI}Si^{4+} \leftrightarrow {}^{VIII}Fe^{3+VI}Al^{3+}$ (Vanpeteghem et al., 2006).

Further, the variety of iron spin transitions in lower mantle minerals leads to additional electronic configurations that modify <Fe-O> bond lengths and therefore influence isotopic fractionation. At low pressures, the crystal field splitting energy (see footnote 1) is lower than that required to pair electrons, meaning iron exists in its high spin (HS) configuration. However, compression at lower mantle pressures reduces the interatomic distance between Fe and O, destabilizing the e_g orbitals and making electron spin pairing energetically more favourable, resulting in a low spin (LS) configuration. This effect is most prevalent in the octahedral sites of ferropericlase and bridgmanite for Fe^{2+} and Fe^{3+}, respectively (see Lin et al., 2013, for a review). Along a mantle adiabat, $^{VI}Fe^{2+}$ in ferropericlase begins the transition from HS to LS near 60 GPa (Badro et al., 2003; Wentzcovitch et al., 2009) and is complete by 120 GPa. The transformation is displaced to lower pressures for $^{VI}Fe^{3+}$ in bridgmanite, occurring over a range from 30–70 GPa (Jackson et al., 2005a; Fujino et al., 2012), while $^{VIII}Fe^{2+}$ experiences a spin transition only under very high pressures (> 120 GPa; Badro et al., 2004; McCammon et al., 2010).

Rustad and Yin (2009), using DFT simulations, recorded a systematic increase in the force constant for ferropericlase ($10^3\ln\beta^{57/54}Fe_{fp} = 0.145$ at 4000 K for low spin at 120 GPa and 0.098 for high spin), corresponding to force constants of ≈540 N/m and ≈365 N/m, respectively. By contrast, no such shift was observed for purely Fe^{2+}-bearing bridgmanite due to the asymmetric distortion of the iron-oxygen bonds and it has $10^3\ln\beta^{57/54}Fe_{bg}$ identical to that of ferropericlase in its LS state. Force constants derived from Polyakov (2009) for $Fe_{0.25}Mg_{0.75}O$ (P = 0–109 GPa) and from (Shahar et al., 2016) for FeO (P = 5–39 GPa) using NRIXS yield very similar pressure dependencies; $K_{Fe\text{-}O} = (2.39\pm0.17)P$ (GPa) + 164±10 N/m and $(2.56\pm0.16)P$ (GPa) + 169±4 N/m, respectively. If extrapolated linearly to 120 GPa, these calibrations yield $K_{Fe\text{-}O}$ between 450 and 475 N/m, lower than that calculated by Rustad and Yin (2009) for LS ferropericlase. Williams et al. (2012) reported the first direct measurement of the iron isotope composition of experimentally synthesized Al-bearing (≈0.23 per formula unit, pfu bridgmanite, with $Fe^{3+}/\sum Fe$ ~0.37 in equilibrium with iron metal at 24 GPa and 1850 °C. Their preferred $\Delta^{57/54}Fe_{bg\text{-}met}$ is +1.35‰, whose sense is in accord with recent NRIXS determinations of Yang et al. (2019), though the magnitude is much larger ($\Delta^{57/54}Fe_{bg\text{-}met}$ by NRIXS is +0.2‰ at 2000 K and 30 GPa). These authors collected an extensive set of in-situ high-pressure data for ferropericlase ($Fe_{0.25}Mg_{0.75}O$) and bridgmanites (both Al-free and Al-bearing at 0.11 pfu). Bridgmanite reaches a maximum $K_{Fe\text{-}O}$ of 449±46 N/m (46 GPa), and is isotopically heavier than ferropericlase below 60 GPa, after which the iron in the latter transitions to its LS state that produces a drastic increase in force constant with a slope of dK_{Fe-O}/dP of ~ 3 N/m/GPa, reaching 626±31 N/m at 94 GPa. Above 60 GPa, the force constant of bridgmanite is unchanging around 335±15 N/m up to 104 GPa, independent of $Fe^{3+}/\sum Fe$ and Al content.

15.4.2. Melts

Because of the lack of long-range order in silicate melts (or fluids), the variation of Fe^{3+}/Fe^{2+} is unbound by stoichiometry and can vary freely. The ratio of ferrous to ferric iron in silicate liquids can be understood quantitatively by considering the homogeneous equilibrium:

$$Fe^{3+}O_{1.5}(l) = Fe^{2+}O(l) + \frac{1}{4}O_2. \tag{15.16}$$

The equilibrium constant of this reaction is:

$$K^* = \frac{a\left(Fe^{2+}O\right)f(O_2)^{1/4}}{a\left(Fe^{3+}O_{1.5}\right)}. \tag{15.17}$$

Where a refers to the activity of the iron oxide component dissolved in the liquid and K^* accounts for their possible deviations from the ideal case (K). Written in terms of $Fe^{3+}/\sum Fe$:

$$\frac{XFe^{3+}O_{1.5}}{\sum Fe} = \frac{1}{1 + 10^{\left(-\frac{1}{4}\log f(O_2) + \log K - \log\frac{\gamma FeO}{\gamma FeO_{1.5}}\right)}}. \tag{15.18}$$

The γFeO and $\gamma FeO_{1.5}$ refer to the activity coefficients of these species in the silicate melt referenced to the standard states of the pure liquid oxides at the conditions of interest. Using thermodynamic data for Reaction 15.16, Jayasuriya et al. (2004) calculated values of γFeO = 1.37 and $\gamma FeO_{1.5} = 1.08\pm0.33$ based on Fe^{3+}/Fe^{2+} ratios determined by Mössbauer spectroscopy. These determinations are valid at infinite dilution (which they inferred at 1 wt.% FeO^T) in an Anorthite-Diopside eutectic composition at 1400 °C and 1 bar, meaning that iron dissolution in silicate melts is close to ideal. Furthermore, their data showed that $\log(Fe^{3+}/Fe^{2+})$ is proportional to $(fO_2)^{0.245}$, within uncertainty of the theoretical dependence expected from Equation 15.16, suggestive of no interactions between Fe^{2+}-Fe^{3+} in silicate liquids as a function of fO_2. This hypothesis, that $\frac{\gamma FeO}{\gamma FeO_{1.5}}$ is independent of fO_2 for a fixed composition, was later confirmed by Berry et al. (2018) for a natural MORB composition with higher quantities of iron (10 *wt.%* FeO^T).

By contrast, $\frac{\gamma FeO}{\gamma FeO_{1.5}}$ is readily affected by silicate composition. Dickenson and Hess (1986) show that potassium stabilizes tetrahedrally-coordinated ferric iron by charge balance. Recent work by Borisov et al. (2017) illustrates that this stabilization occurs inversely proportional to their bond valence in silicate melts (e.g., Farges et al., 2004) whereas Al competes with Fe^{3+} for tetrahedral sites. This effect is manifest in the higher $Fe^{3+}/\sum Fe$ in $(Na,K)_2Si_3O_7$ silicate melts (Knipping et al., 2015) relative to alkali-poor MORB compositions (Berry et al., 2018) at the same fO_2.

The effect of composition on $Fe^{3+}/\sum Fe$ is evident in the study of Dauphas et al. (2014), the first to combine quantification of X-Ray Absorption Near-Edge Structure (XANES) spectra to determine the redox ratio of glasses with NRIXS spectra to calculate mean Fe-O force constants. They analyzed glass of four different compositions corresponding to basaltic, andesitic, dacitic, and rhyolitic liquids melted at 1550 °C and 1 bar. Notably, the rhyolitic glass has relatively higher $Fe^{3+}/\sum Fe$ than the more mafic compositions at the same fO_2. While tempting to correlate this shift with silica content, Borisov and McCammon (2010) show that increasing silica results in decreases in $Fe^{3+}/\sum Fe$ up to 60% SiO_2, all else being equal, suggestive instead of the role of alkali metals in stabilizing tetrahedrally coordinated Fe^{3+}.

Dauphas et al. (2014) report an increase in the Fe-O force constant from 199±15 N/m in Fe^{2+}-bearing mafic-intermediate compositions, which they extrapolate to 351 ± 29 N/m for $Fe^{3+}/\sum Fe = 1$ (Figure 15.3). Independently of fO_2, the rhyolitic composition is displaced to higher Fe-O force constants for a given $Fe^{3+}/\sum Fe$, 240±13 N/m ($Fe^{3+}/\sum Fe = 0$) and 385±24 ($Fe^{3+}/\sum Fe = 1$). This distinction therefore likely reflects the greater proportion of IV-fold coordinated Fe^{3+} in rhyolitic compositions relative to more basic melts. Although ferrous and ferric iron in basaltic glasses have average coordination numbers close to V (though this may be through a combination of IV- and VI-fold rather than V-fold itself), ferric iron tends to be in slightly lower coordination environments than does ferrous iron (Wilke et al., 2007; Knipping et al., 2015). In both cases, the addition of alkalis stabilizes tetrahedrally coordinated iron, the effect being more pronounced for Fe^{3+} (Jackson et al., 2005b; Knipping et al., 2015; Borisov et al., 2017), again inversely proportional to their bond valence ($Rb^+ > K^+ > Na^+ > Li^+$), resulting in bond stiffening.

The relative compressibility of oxide components in silicate melts leads to structural changes as a function of pressure (e.g., Kress & Carmichael, 1991) that can affect isotopic fractionation (Section 15.2). At equilibrium, differences in molar volumes ($\overline{V}$) of the FeO and $FeO_{1.5}$ components with pressure (P) impact upon their stability, with the resultant change in Gibbs Free Energy of formation (ΔG^o) described by:

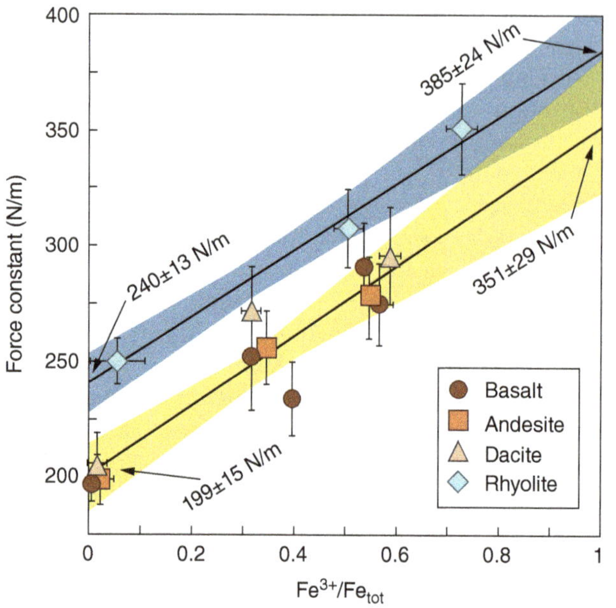

Figure 15.3 After Dauphas et al. (2014). Mean force constants of Fe-O bonds in silicate glasses for basalt (circle), andesite (square), dacite (triangle), and rhyolite (diamond) compositions as a function of their $Fe^{3+}/\sum Fe$ content measured by XANES. The linear trends defined by the data permit extrapolation of Fe-O force constants to $Fe^{3+}/\sum Fe = 1$. Uncertainties on the error envelopes are 95% confidence intervals.

$$\Delta G^o_{(15.16),T,P} = \Delta G^o_{(15.16),T,\ 1\ bar} + \int_{1\ bar}^{P} \Delta \overline{V}_{T,P} dP, \tag{15.19}$$

and, with the relation $\Delta G = -RT\ln K$, the ferric-ferrous ratio is given by:

$$\ln\left(\frac{XFeO_{1.5}}{XFeO}\right) = -\left(\frac{\Delta G^o_{(15.16),T,\ 1\ bar} + \int_{1\ bar}^{P} \Delta \overline{V}_{T,P} dP}{RT}\right) - \ln\left(\frac{\gamma FeO_{1.5}}{\gamma FeO}\right) + \frac{1}{4}\ln fO_2. \tag{15.20}$$

The volume change for Equation 15.16 at 1 bar is negative, such that FeO would be expected to be stabilized with respect to $FeO_{1.5}$ with increasing pressure at constant relative fO_2, assuming no change in their molar volumes. Calculations of iron redox state by Mössbauer and XANES spectra for iron in andesitic melts at the Ru-RuO_2 oxygen buffer as a function of pressure bear out this prediction, and show increasing Fe^{2+}/Fe^{3+} ratios up to 7 GPa (O'Neill et al., 2006; Zhang et al., 2017). This tendency was modelled using a Murnaghan equation of state with molar volumes from Lange and Carmichael (1987) and bulk moduli of 30.3 GPa and 16.6 GPa for FeO and $FeO_{1.5}$, respectively (Kress & Carmichael, 1991). Zhang et al. (2017) recognized a densification of the

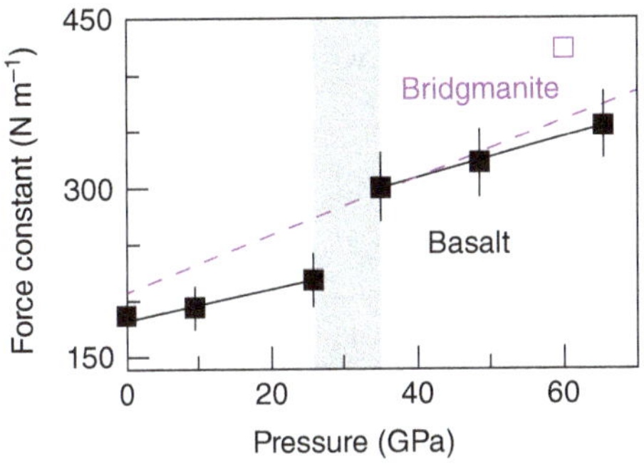

Figure 15.4 After Liu et al. (2017). Force constants (N/m) of iron bonds in Fe^{3+}-free basaltic glass (black squares) with their associated 95% confidence intervals as a function of pressure (GPa) at room temperature (300 K). Black lines are linear fits to the data. Also shown are iron force constants in bridgmanite (Rustad and Yin, 2009, dashed line, and Shahar et al., 2016, purple square).

FeO component by a progressive shift in average iron coordination number from 5.5 to 6 over this pressure range, whereas that of $FeO_{1.5}$ remained constant at ≈5.3. However, higher pressure experiments show that the trend of increasing Fe^{2+}/Fe^{3+} is reversed between 7 and 15 GPa (Armstrong et al., 2019) with Fe^{2+}/Fe^{3+} decreasing thereafter up to 23 GPa, suggesting ferric iron also transitions to higher coordination numbers above ~7 GPa.

Although coordination number increases with pressure, the associated bond stiffening (recalling the inverse proportionality of volume and vibrational frequency, Equation 15.8) results in considerable increases in <Fe-O> force constants in silicate glasses (Liu et al., 2017). At ambient pressure, $K_{Fe\text{-}O}$ in Fe^{3+}-free basaltic glass is ≈190 N/m (Figure 15.3; Dauphas et al., 2014) whereas it reaches 355 N/m at 65.4 GPa (Figure 15.4; Liu et al., 2017); equivalent to the effect of quantitative substitution of ferrous for ferric iron at 1 bar. The determined gradient, dK_{Fe-O}/dP, shows two regimes, 0–25.8 GPa; dK_{Fe-O}/dP = 1.22 ±0.25 N/m/GPa and 35.1–65.4 GPa; dK_{Fe-O}/dP = 1.81±0.10 N/m/GPa, with a discontinuous jump between them (Figure 15.4). Therefore, despite evidence for a gradual increase in FeO coordination number in silicate glasses under conditions that are relevant to partial melting in the upper mantle (< 7 GPa, Zhang et al., 2017), this structural change induces little or no increase in $K_{Fe\text{-}O}$. By contrast, the step-wise increase in $K_{Fe\text{-}O}$ from ≈220 N/m to ≈300 N/m between 25–35 GPa coincides with pronounced densification in basaltic melts, inferred to be caused by the passage of Si from tetrahedral to octahedral coordination (Sanloup et al., 2013). Therefore, bulk structural changes in the melt may be more effective in modifying Fe-O force constants than are changes in local bonding environment.

15.4.3. Fluids

The compositional parameter space for fluids is larger than that for silicate melts, in which most components are dissolved as oxide species. By contrast, metals can be complexed by Cl-, F-, C-, S-, and OH ligands in geologically relevant fluids (see Burnham, 1979; Candela, 1989; Manning, 2004; Richards, 2011, for summaries of fluid compositions in arc magmatic and hydrothermal systems). Magmatic fluids are predominantly aqueous, with dissolved components ranging from 17–83% (Manning, 2004), which are broadly correlative with the *P-T* conditions at which the fluid equilibrated. Low temperature fluids have metal-bearing species dominated by alkali-chloride complexes (NaCl, KCl), in addition to HCl and SO_2 as other major volatiles (e.g., Richards, 2011). The dissolution of Si becomes important at higher temperatures as the composition of the fluid-rock system approaches supercritical behaviour, which occurs at 1.5 GPa and 700 °C in the system albite-H_2O (Manning, 2004).

Iron, like the alkali metals, likely dissolves as Cl-bearing species in hydrothermal fluids, as evidenced by its strong solubility in Cl-rich fluids (e.g., Kwak et al., 1986). It is expected to be predominantly ferrous; the stability of ferric chloride species is much lower (Testemale et al., 2009; Saunier et al., 2011) except during dissolution under very oxidizing conditions (e.g., in equilibrium with hematite; Brugger et al., 2016). Spectroscopic techniques that have been employed to probe speciation in acidic chloride solutions for a range of Cl molalities from 0.5 to 12 up to 450 °C suggest the presence of higher-order complexes, with $FeCl_4^{2-}$ as the dominant Fe-bearing species (Testemale et al., 2009). By contrast, speciation of Fe in fluids at relevant *P-T* conditions (up to 650 °C, 2 kbar; Chou & Eugster, 1977; Boctor et al., 1980; Fein et al., 1992) is largely informed by the stoichiometry of mineral dissolution reactions. Iron solubility in fluids is expected to be influenced by temperature, pressure, pH and by the presence of chlorine. The data of Chou and Eugster (1977) on magnetite solubility in HCl solutions conform to a stoichiometry with $FeCl_2$ as the stable fluid species, whose solubility is given as logm($FeCl_2$) = 6887/T(K)-8.44.[2] Ding and Seyfried (1992) experiments showed that $(FeCl)^+$ and particularly $FeCl_2$ are the dominant Fe species in aqueous fluids at the hematite-magnetite buffer and up to 450 °C and 0.3 to 0.5 kbar. The solubility of Fe increases drastically with increasing NaCl concentrations, passing from 0.369 mmolal of Fe at 350°C and 29.1 mmolal of Na to 1.285 mmolal of Fe at 500°C and 1233 mmolal of Na (see also Figure 12 in Testemale et al., 2009). These authors also observed $FeCl_2$ as the dominant species with

[2] m refers to the molality, in mol/kg; whereas mmolal refers to mmol/kg

increasing temperatures. Indeed, thermodynamic calculations at high pressure and temperature (630–660 °C and 2 GPa) show that Fe is mobile in slab-derived fluids as $FeCl_2$ during antigorite breakdown (Debret & Sverjensky, 2017). It should be noted that the complexation of Fe^{2+} with sulfate, carbonate, or silica was not considered in these thermodynamic models due to the absence of thermodynamic data; although this is not proof that these species do not exist. Furthermore, even though Fe^{3+}-complexes were included, they were found to be unimportant at the investigated conditions.

These studies all point to the predominance of $FeCl_2$, particularly at high temperatures. Nevertheless, there is a dearth of experimental studies at conditions relevant to that of subduction zone magmatism or high-temperature ore-forming hydrothermal fluids (> 800 °C). One of the few are the magnetite solubility experiments of Simon et al. (2004) who also presumed $FeCl_2$ as the stable species and measured up to 7.3 wt% iron in the fluid phase at high NaCl molalities. Sanchez-Valle et al. (2017) recorded XANES spectra of chlorinated-aqueous fluids equilibrated with Fe-bearing minerals in a diamond anvil cell at 2.5 GPa and 800 °C. Their results imply the transport of Fe with $(Fe^{2+}Cl_x)^{2-x}$ stoichiometries, with the stability of Fe^{3+}-bearing chloride species decreasing with increasing temperature.

In a series of studies, Hill and coworkers calculated and measured experimentally the isotopic fractionation between a variety of iron chlor-aquo species at ambient temperatures (Hill & Schauble, 2008; Hill et al., 2009, 2010). These studies show several trends at 298 K:

1. The iron isotope fractionation factor and calculated force constants decrease with increasing chlorinity. Specifically, $10^3\ln\beta^{56/54}Fe\ (FeCl_x(H_2O)_{6-x})^{3-x}$ is 9.41, 8.54, 8.10, 7.13, and 7.14 for x = 0, 1, 2, 3, and 4, respectively.
2. In the equivalent ferrous-iron bearing system, the species have $10^3\ln\beta^{56/54}Fe(FeCl_x(H_2O)_{6-x})^{2-x}$ of 6.39, 5.98, and 5.91 for x = 0, 1, and 2, respectively.
3. Regardless of the basis set used for the ab-initio calculations, the fractionation factor between ferric- and ferrous iron-bearing species, $10^3\ln\alpha_{3+-2+} \approx 3‰$.

Fujii et al. (2014) calculated β factors for a host of aqueous, ferric-, and ferrous iron-bearing species (Figure 15.5). Though there are minor differences in $10^3\ln\beta^{56/54}Fe$ among species of a given Fe oxidation state, these differences become small (< 0.3‰) by 573 K. Invariably, species with chloride ligands are the lightest among other species studied, particularly compared to PO_4^{3-}, OH^-, and SO_4^{2-} which are stronger-field ligands and therefore concentrate the heavier isotopes (Figure 15.5). A similar order can be observed among Fe^{3+}-bearing compounds (Figure 15.5). As per the study of Hill et al. (2010), the isotopic fractionation between Fe^{2+} and Fe^{3+}-bearing species is most marked. At 573 K, the value of $10^3\ln\beta^{56/54}Fe_{3+-2+} \approx 0.90‰$, suggesting iron isotopes should be a powerful discriminant of the oxidation state of iron in fluids, even at high temperatures, but less so between different ligands for the same oxidation state of iron.

Iron isotope compositions of chlorine-bearing fluids relevant to hydrothermal systems (300–350 °C) were explored for the pairs fluid-pyrite (Syverson et al., 2013) and fluid-chalcopyrite (Syverson et al., 2017). Solutions

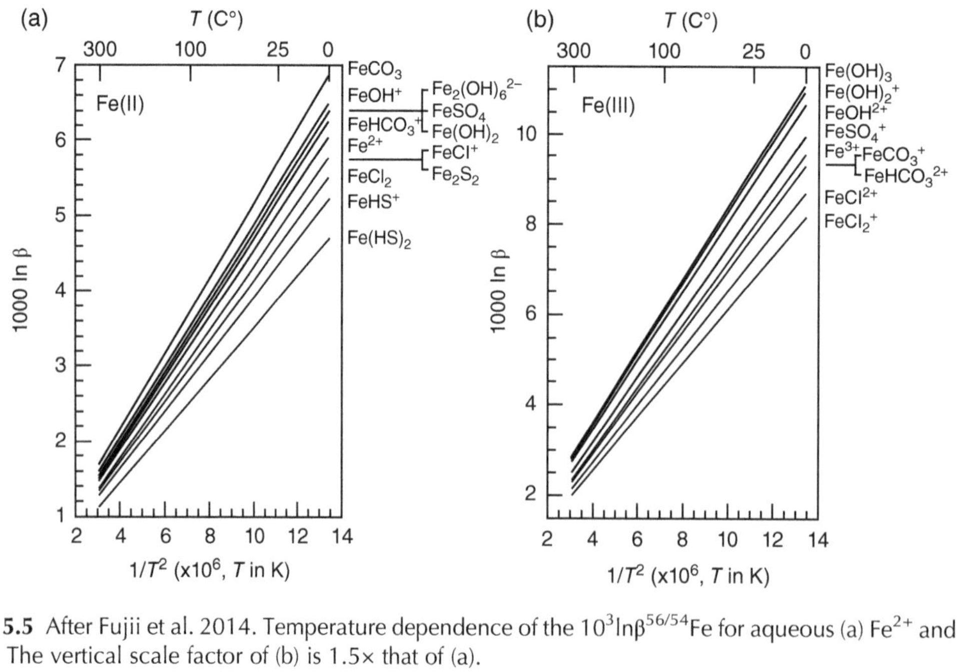

Figure 15.5 After Fujii et al. 2014. Temperature dependence of the $10^3\ln\beta^{56/54}Fe$ for aqueous (a) Fe^{2+} and (b) Fe^{3+} species. The vertical scale factor of (b) is 1.5× that of (a).

equilibrated with pyrite contained ≈10 mmol/kg Fe^{2+} and ≈1000 mmol/kg Na, and, at low pH (≈3 at 350°C) and 500 bar, $Fe(H_2O)^{2+}$ was calculated to have been stable, and defined a fractionation factor $10^3\ln\alpha(^{57/54}Fe_{pyr\text{-}fluid}) = +1.49\pm0.44‰$. This gives a force constant for $Fe(H_2O)^{2+}$ of ≈150 N/m (given that of pyrite is ≈280 N/m; Blanchard et al., 2009). Under the same conditions, the fractionation between chalcopyrite and fluid, $10^3\ln\alpha(^{57/54}Fe_{ccp\text{-}fluid}) = +0.09\pm0.17‰$ is negligible within uncertainty (Syverson et al., 2017). At 300 °C and NaCl-HCl±$FeCl_2$ fluids, dissolution/precipitation of hematite yielded $10^3\ln\alpha(^{57/54}Fe_{hem\text{-}fluid}) = +0.54\pm0.15‰$ in which the fluid species was likely $FeCl_2^0$(aq) or $FeCl^+$, however no fractionation was observed at 450 °C and 600 bar, nor at 200 °C, where the fluid phase was inferred to be dominated by Fe^{3+}-bearing aqueous species (Saunier et al., 2011). Nevertheless, this sequence of isotopic fractionation ($\delta^{57}Fe_{pyr} > \delta^{57}Fe_{hem} > \delta^{57}Fe_{mag} > \delta^{57}Fe_{ccp} \geq \delta^{57}Fe_{fluid}$) is borne out in measurements of hydrothermally precipitated ore phases (Wawryk & Foden, 2015, 2017; Li et al., 2018). Higher temperature experiments carried out at 800 °C and 10 kbar by Sossi and O'Neill (2017) found strong isotopic fractionation between $FeCl_2.4H_2O$ fluid and Fe^{3+}-rich minerals (namely magnetite), where $10^3\ln\alpha(^{57/54}Fe_{mag\text{-}fluid}) = +0.44\pm0.01‰$. By contrast, $^{VI}Fe^{2+}$-bearing minerals (fayalite, ilmenite) had similar iron isotope compositions to the fluid, and almandine garnet was lighter still, at $10^3\ln\alpha(^{57/54}Fe_{alm\text{-}fluid}) = -0.14\pm0.03‰$, which gives the iron force constant for $FeCl_2^0$ in the fluid of 175 N/m. This compares well with force constants of 153–186 N/m for $FeCl_2(H_2O)_4$ calculated in Hill et al. (2010) depending on the DFT basis set used, their Table A1.

15.5. EFFECT OF REDOX PROCESSES IN INFLUENCING IRON ISOTOPE FRACTIONATION IN NATURAL SYSTEMS

15.5.1. Magmatic Processes

Partial Melting. Partial melting is the process that describes the incomplete transformation of solid to liquid phase(s). In terrestrial magmatic settings, this process can occur in both the crust and mantle. Should liquid-solid segregation occur, partial melting has the capacity to impart isotopic fractionation on its constituent phases.

Though early studies considered that the compositions of mantle peridotites and basaltic rocks were indistinguishable (e.g., Zhu et al., 2001; Beard et al., 2003a) leading to the development of the concept of a "mean mafic Earth" with $\delta^{57}Fe = +0.10\pm0.03‰$, later, higher-precision studies soon overturned this perception. In particular, two studies, Williams et al. (2005) and Weyer and Ionov (2007) presented an extensive number of analyses of whole rock peridotites and, in the case of the former, their constituent minerals. The conclusion from both of these works was that peridotitic rocks were isotopically lighter ($\delta^{57}Fe = +0.03\pm0.05‰$ at Mg# = 0.894, that of fertile mantle) than their basaltic counterparts and secondly, that partial melting of the mantle could induce resolvable isotopic fractionation in which the residual peridotite composition would tend to lighter isotopic values with progressive extraction of heavier basaltic melt. Weyer and Ionov (2007) assigned bulk peridotite-liquid fractionation factors that fit arrays of $\delta^{57}Fe$ vs. Mg# (molar Mg/(Mg+Fe^{2+})) to quantify this effect, finding $10^3\ln\alpha(^{57/54}Fe_{mantle\text{-}melt}) = -0.1$ to $-0.3‰$. These fractionation factors were also used to fit the observation of the isotopically heavy ($\delta^{57}Fe = +0.15‰$) compositions of basaltic rocks measured to that stage.

As recognized by Williams et al. (2005) and Weyer and Ionov (2007), iron isotope fractionation arises mechanistically from the incompatibility of Fe^{3+} relative to Fe^{2+}, reflected in the higher $Fe^{3+}/\sum Fe$ of MORB glasses (0.10±0.02; Berry et al., 2018) relative to peridotites (0.037, Canil et al., 1994). Clinopyroxene is the major contributor to partial melting at low pressures (< 3 GPa) (e.g., Hirschmann et al., 1999), and $D(Fe^{2+})/D(Fe^{3+})$[3] ≈10, where $D(Fe^{2+})$ ≈1 (e.g., Mallmann & O'Neill, 2009; Davis et al. 2013). Therefore, during melt extraction, the modal olivine content increases and, as it contains less ferric iron than ortho- and particularly clinopyroxene or spinel (Canil & O'Neill, 1996), tends to concentrate the lighter isotopes. Dauphas et al. (2009) developed a melting model predicated on the difference of Fe^{3+}/Fe^{2+} ratios of the mantle source and the partial melt. As ferric iron behaves as an incompatible element, the extent of isotopic fractionation is highest in the melt at low degrees of melting, a feature also attributable to the mass balance considerations (i.e., the "lever rule"). However, this treatment implicitly neglected the effect of coordination environment, whose isotopic effect was unconstrained until the study of Dauphas et al. (2014), whose measurement of Fe-O force constants in magmatic glasses revealed a preference for the heavier isotopes in lower coordination environments (Section 15.4.2). With this knowledge, combined with $K_{Fe\text{-}O}$ of mantle minerals, Sossi and O'Neill (2017) were able to build self-consistent modal and non-modal melting equations based on the mass balance equations of Shaw (1970), re-cast in terms of isotopic fractionation between phases 1 and 2

[3] The elemental partition coefficient, denoted "D", is given by the ratio of the concentration of the element in a solid phase with respect to that in the liquid phase.

$$\frac{C_l^i}{C_o^i} = \frac{1}{\left(\left(D_{1-2}^i \alpha^i E_{1-2}\right) + F\left(1 - P_{1-2}^i \alpha^i E_{1-2}\right)\right)}, \tag{15.21}$$

where C_l^i and C_o^i denote the concentration of isotope i of element E in the liquid and bulk, respectively, F is the melt fraction, and P the product of a phase's melting reaction coefficient and elemental D summed over each phase. Solutions are found for the concentration of each isotope, and hence the resulting isotopic composition of the liquid is:

$$10^3 \ln\left(\frac{C_l^i}{C_l^j}\right) = \delta^i E. \tag{15.22}$$

The above equations can be applied to any isotope system. For iron, the crucial free variable, D_{1-2}^i, is controlled by the temperature and difference in iron force constants between phases 1 and 2 (mineral and melt):

$$D_{Min-Melt}^{54+nFe} = \left(D_{Min-Melt}^{54Fe}\right) e^{\left(\frac{1435n\left(K_{Fe-O}^{Min} - K_{Fe-O}^{Melt}\right)}{T^2}\right)}. \tag{15.23}$$

Here, $n = 2$ ($^{56/54}Fe$) or 3 ($^{57/54}Fe$). The force constant for each phase is given in the form:

$$K_{Fe-O} = a \times \frac{Fe^{3+}}{\Sigma Fe} + b \tag{15.24}$$

where a describes the slope of the dependence of the force constant on $Fe^{3+}/\sum Fe$, and b the y-intercept, which depends on coordination (VIII < VI < IV). Values of constant a and b can be calculated from Figs. 15.1–15.3, and yield $a = 125$ and $b_{VIII} = 145$ N/m, $b_{VI} = 185$ N/m, and $b_{IV} = 225$ N/m.

Systematics of iron isotope fractionation applying this partial melting model are summarised in Figs. 15.6a and b. As in the models presented in Dauphas et al. (2009), iron isotope fractionation between the melt and its source is inversely proportional to the melt fraction (Figure 15.6a), as the $Fe^{3+}/\sum Fe$ of the melt is maximized at low F due to the relative incompatibility of ferric iron. Independent of this effect is the modal mineralogy of the source; as the modal abundance of garnet increases at the expense of olivine (e.g., via increasing pressure or in a bimineralic eclogite), the fractionation factor increases due to the retention of the light isotopes in its VIII-fold sites.

This mineralogical control on the extent of iron isotope fractionation can account for the isotopic differences between mantle peridotites, MORBs, and some OIBs. The iron isotope variations observed in mantle peridotites are more scattered than those in basaltic rocks. Studies reporting analyses of peridotitic or ultramafic lithologies include (Poitrasson et al., 2004, 2013; Williams et al.,

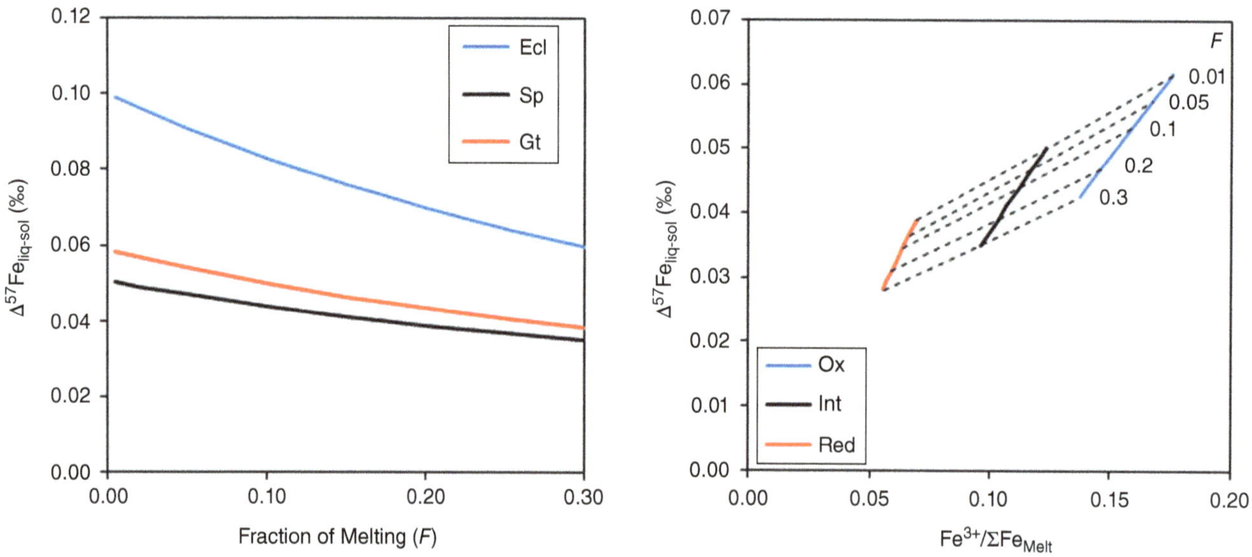

Figure 15.6 Melting models from Sossi & O'Neill (2017) (see Equations 15.21–15.24). (a) Variation of the fractionation factor between liquid and solid as a function of source lithology and degree of melting. Peridotites in the spinel (sp) and garnet (gt) fields show similar fractionation, whereas partial melts of eclogites (ecl) have a more pronounced enrichment in heavy isotopes. (b) Variation of the fractionation factor between solid and liquid as a function of $Fe^{3+}/\sum Fe_{Melt}$ for three difference source compositions; "Ox" = oxidized ($Fe^{3+}/\sum Fe_{Source}$ = 0.051), "Int" = intermediate ($Fe^{3+}/\sum Fe_{Source}$ = 0.036), "Red" = reduced ($Fe^{3+}/\sum Fe_{Source}$ = 0.021). Dashed lines are fraction of melting (F) contours, listed to their right. Melting of more oxidized sources is associated with greater isotopic fractionation.

2004, 2005, 2009; Weyer et al., 2005; Weyer & Ionov, 2007; Zhao et al., 2010, 2012, 2017; Huang et al., 2011; Craddock et al., 2013; Sossi et al., 2016) and were summarized by Sossi et al. (2016), who filtered the data for unmetasomatized, unaltered, and fertile peridotites. This procedure greatly reduced data variability enabling a 2SE uncertainty to be taken of 60 peridotite analyses that yielded $\delta^{57}Fe = +0.05\pm0.01‰$ for the value of fertile mantle. These authors noted that significant iron isotope fractionation in fresh, unmetasomatized peridotites was not evident until reaching Mg#s > 0.91, reflecting the retention of the bulk of the iron budget in the peridotite.

Relative to ultramafic rocks, oceanic basalts almost exclusively preserve heavier iron isotope compositions (Weyer & Ionov, 2007). Nevertheless, data for oceanic basalts remained sporadic until the study of Teng et al. (2013), in which 93 iron isotope analyses of basaltic rocks were presented, comprising 43 MORBs, while the remainder were OIBs from the Society Islands, the Cook-Austral chain, and Hawai'i. The isotopic composition of MORB gave a mean value of $\delta^{57}Fe = 0.15\pm0.01‰$ (2SE, $n = 43$), some 0.1‰ heavier than fertile mantle. Considering that the maximum iron isotope fractionation permitted during melting of peridotite in the spinel facies is 0.05‰ at FMQ (Figure 15.6a), this poses a conundrum for the derivation of MORB from peridotitic upper mantle (see discussion in Dauphas et al., 2014). However, as noted by Sossi et al. (2016), MORBs have Mg#s that are too low (0.54 < Mg# < 0.68) to have been in equilibrium with Fo_{90} mantle olivine, which, since $K_D^{Fe\text{-}Mg}$ ($[Fe^{2+}/Mg]_{Ol}/[Fe^{2+}/Mg]_{Melt}$) is equal to ≈0.3 (Toplis, 2005), should produce melts with Mg# ≈0.74 (compositional effects on $K_D^{Fe\text{-}Mg}$ broadly cancel such that the use of a single value introduces minimal uncertainty). In this scenario, the low-pressure evolution of MORB magmas during decompression expands the olivine phase volume (Kinzler & Grove, 1992) and light iron isotopes are preferentially sequestered into olivine, resulting in heavier values than could be accounted for by single-stage melt extraction. Correction of the iron isotope composition of MORB to a fictive primary melt composition with Mg# ≈0.74 yields $\delta^{57}Fe = 0.10‰$ (Sossi et al., 2016) that can be reconciled with partial melting. Additional possibilities, including diffusive effects during partial melting (e.g., Weyer & Ionov, 2007; Poitrasson et al., 2013; Su et al., 2015), may also become important in fractional melting regimes where melt parcels and residual mantle equilibrate only at infinitesimally small increments (see Asimow et al., 1997). Here, preferential transport of the lighter iron isotopes into peridotite minerals during ascent, due to its faster diffusivity, would lead to heavier isotopic compositions than predicted by equilibrium fractionation factors.

With respect to MORBs, OIBs are frequently (though not always; Peters et al., 2019) higher in $\delta^{57}Fe$ (Teng et al., 2013), a tendency that was highlighted in the data of Williams and Bizimis (2014) and Konter et al. (2016) on MgO-rich pyroxenites and OIBs from Hawai'i and Samoa, respectively. In these samples, whose Mg# reaches 0.69, $\delta^{57}Fe$ can extend to +0.45‰, up to 0.4‰ heavier than recent estimates for fertile mantle peridotite. All things being equal, OIBs should be heavier than MORBs by virtue of the lower degrees of partial melting involved in their genesis (Section 15.5.1.1, Figure 15.6). Nevertheless, this offset is greater than can be produced by degree of melting alone, and the combination of isotopically heavy iron and high Mg# requires additional source-related melting controls. Both studies argue for the presence of pyroxenite in the source, which has the dual effect of (i) increasing the $\delta^{57}Fe$ of the source (pyroxenites from Oahu, Hawai'i have $\delta^{57}Fe$ between +0.10 and +0.27‰, Williams and Bizimis, 2014); and (ii) enhancing iron isotope fractionation during partial melting. This second point was quantified in Sossi and O'Neill (2017), who demonstrated that such heavy values were difficult to create during melting of mantle peridotite, and instead required a high fraction of garnet in the source that simultaneously acted as a sink for light iron isotopes (see Figure 15.6a) and increased the Gd/Yb ratio of the basalts (Figure 15.7).

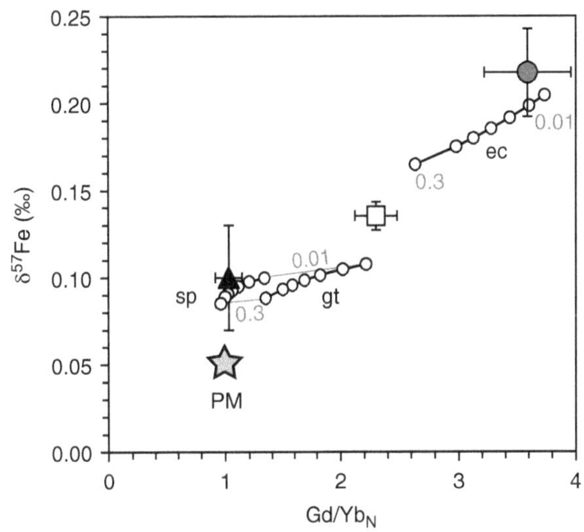

Figure 15.7 After Sossi and O'Neill (2017). Iron isotope composition of mafic rocks (black triangle, MORB; white square, Hawai'i; grey circle, Tahiti) from the ocean basins (Teng et al., 2013), corrected to their primary magma compositions, plotted against Gd/Yb_N. Black lines show the iron isotope compositions of melts modelled by batch melting (Equation 15.21), at melt fractions given in grey, in both the garnet (gt) and spinel (sp) fields from a primitive mantle source (grey star), and of eclogite partial melts (ec), from an N-MORB source ($\delta^{57}Fe = +0.1‰$).

The expected effect of source oxidation is to increase the iron isotope fractionation observed in products of partial melting (Dauphas et al., 2009; Figure 15.6b). Although the consensus is that arc magmas are more oxidized than those from oceanic settings (MORBs and OIBs; see Carmichael, 1991; Kelley & Cottrell, 2009; Cottrell et al., 2021 [Chapter 3, this volume]), conjecture still abounds as to whether the sources of arc magmas are more oxidized than their equivalents under mid-ocean ridges or ocean islands (Lee et al., 2005; Mallmann & O'Neill, 2009). In this scenario, the higher oxygen fugacity of arc magmas could be acquired after segregation from their mantle sources, by fractional crystallisation or degassing of reduced species, such as H_2 or CH_4 (Mathez, 1984). The sensitivity of iron isotopes to the redox state of iron may in turn be able to distinguish between these two hypotheses.

Should the sources of arc magmas be more oxidized, then the mechanics of iron isotope fractionation would predict an iron isotope composition heavier than oceanic basalts. Analyses of arc magmas were, until recently, surprisingly sparse, restricted to studies of back-arc basalts (Nebel et al., 2013, 2018) and two arcs: New Britain (Dauphas et al., 2009) and the sediment-rich Banda (Nebel et al., 2015) prior to the extensive dataset comprising 130 arc magmas of Foden et al. (2018). These authors discovered, contrary to expectations, that arc magmas have systematically lighter iron isotope composition relative to MORBs, extending to values even lighter than fertile mantle peridotites, around -0.15‰ in Avachinsky, Klyuchevskoy, and Ploskiy volcanoes of the Kamchatka arc (Foden et al., 2018). This result has two important implications. Firstly, if the sources of arc magmas are indeed more oxidized than ambient mantle, then iron itself cannot be the cause of this oxidation. Rather, the role of S^{6+} as the oxidising agent for arc magmas is supported by quantitative calculations of the redox budget of subduction zones (Evans & Tomkins, 2011), largely due to its efficiency in oxidising iron via an 8-electron transition from sulfate to sulfide:

$$S^{6+} + 8Fe^{2+} = S^{2-} + 8Fe^{3+}. \tag{15.25}$$

This means 8 moles of Fe are oxidized for every mole of S, which equates to 1.8 wt% FeO per 1000 ppm S; the S content of MORB, though those of arc magmas are even higher (Wallace & Edmonds, 2011). This comes about due to the higher solubility of sulfur in silicate melts at anhydrite saturation with respect to sulfide saturation (O'Neill & Mavrogenes, 2002; Li & Ripley, 2009; Smythe et al., 2017). Alternatively, the idea that the oxygen fugacities of arc magma sources are similar to ambient mantle stands. In either case, iron isotopes show no evidence for the oxidation of arc magmas sources relative to their oceanic counterparts.

Secondly, the observation that arc magmas are *lighter* than oceanic basalts begs explanation. Nebel et al. (2015) argued that isotopically light sources could be generated by the prior extraction of melt from sub-arc mantle peridotite, which is known to be more refractory than ambient mantle on the basis of lower abundances of incompatible high field strength elements (Nb, Ti) in arc magmas. These elements behave conservatively during slab fluid transfer to the mantle wedge and reflect the composition of mantle wedge peridotite (Woodhead et al., 1993). Degrees of partial melting needed to explain the isotopic fractionation are close to 30%, which are observed in xenoliths from the Bismarck arc (Bénard et al., 2018). However, Foden et al. (2018) asserted that values as low as -0.15‰ cannot be produced by melt depletion alone and considered additional scenarios. In particular, time-integrated diffusive stripping of heavier isotopes in preference to their lighter counterparts was cited as a mechanism to generate light isotope compositions in peridotite sources, and, by extension, partial melts derived from them. As discussed in Section 15.4.3, the iron isotope composition of aqueous fluids are light, provided the speciation of iron is dominated by Fe^{2+}-bearing species, particularly $FeCl_2$, as inferred from the measurements of Sanchez-Valle et al. (2017). However, it remains uncertain as to whether the Fe content of these fluids is high enough to shift the iron isotope composition of the mantle wedge (Ayers et al., 1997; see also Section 15.5.2.3). Indeed, supra-subduction zone ophiolitic peridotites, despite their high degree of depletion (≈25–35%) and re-enrichment with subduction zone fluids, resulting in a distinctive "U-shape" REE pattern, are not systematically lighter than abyssal peridotites (Su et al., 2015; Zhang et al., 2019). Although certain samples extend to the requisite light values ($\delta^{57}Fe \approx$ -0.15‰), this variation is, at best, weakly dependent on proxies for the degree of melt extraction (Mg#, Cr# in spinel or clinopyroxene) or fluid addition (Ba/Th, Ba/La). By contrast, Turner et al. (2018) observe a positive correlation between $\delta^{57}Fe$ and $\delta^{7}Li$ to very low values (-0.46‰ and -1.78‰, respectively), pointing to fluid addition and/or diffusive processes, though the primary mechanism for inducing these variations remains uncertain. Further studies of sub-arc mantle peridotites are required to evaluate the degree to which these processes are liable to affect their iron isotopes compositions locally.

On a global scale, Foden et al. (2018) recognized a negative correlation between the iron isotope composition of global arc magmas and the thermal parameter (φ):

$$\phi = age \times \sin\theta u, \tag{15.26}$$

the product of slab age and vertical descent velocity, where θ is the dip angle and u the convergence rate

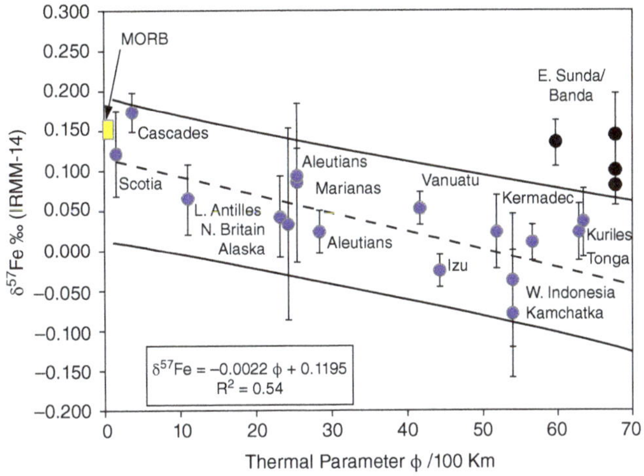

Figure 15.8 After Foden et al. (2018). $\delta^{57}Fe_{(IRMM\text{-}14)}$ *vs.* Slab Thermal Parameter (φ)/100km for population means for each arc or arc segment. The error bars are 2SE and the 95% uncertainty envelope of the linear regression is indicated as the black curves. Value for MORB is from Teng et al. (2013).

(Kirby et al. 1996) (Figure 15.8). Subducting slabs that are old and cold tend to descend at faster rates, such that they do not thermally equilibrate as much with surrounding mantle, and slab geotherms are too cold to permit slab melting. By contrast, hotter slabs descend more slowly due to their greater thermal buoyancy, and hence heat up to greater extents during their protracted residence time in the mantle, allowing partial melting of the slab. Foden et al. (2018) also suggested that the mantle wedge peridotite in longer-lived and therefore colder subduction zones remained trapped in the sub-arc mantle by convective "Richter rolls" (Davies et al., 2016), subjecting the same mantle to continual melt-extraction and fluid ingress, which may have been partially kinetic in nature (see also Zhao et al., 2012), resulting in isotopically light arc magma sources. Alternatively, on the basis of the similarity in iron isotope compositions of sub-arc and convecting mantle, Zhang et al. (2019) propose that redox-unbuffered melting of arc sources may account for their lighter composition than MORB, yet this does not address the very light isotope signatures observed in magmas from high φ arcs such as Kamchatka and Samoa.

Differentiation. Magmatic differentiation comprises the range of processes involving, but not limited to, fractional crystallization, crustal assimilation, fluid exsolution, Soret diffusion, and vapor-phase degassing, all of which have the potential to fractionate iron isotopes by mass transfer.

The earliest concerted effort to study the effect of magmatic differentiation on iron isotope fractionation was that of Poitrasson and Freydier (2005), who analyzed a collection of granites including peralkaline, metaluminous, and peraluminous varieties[4], broadly corresponding to A-, I-, and S-type magmas, respectively (Chappell & White, 1974; Frost & Frost, 2010). These granitic rocks record a near-exponential increase in $\delta^{57}Fe$ with falling MgO (below 1 wt%) or rising SiO_2 (above 70 wt%) contents (their Figs. 5 and 6) relative to the "mean mafic earth" value of $\delta^{57}Fe = +0.10 \pm 0.03$‰ reaching $+0.59 \pm 0.03$‰ (Poitrasson & Freydier, 2005). In order to explain this monotonic increase, Poitrasson and Freydier (2005) appealed to the exsolution of Fe^{2+}-rich fluids that transported isotopically light iron. A similar trend was observed by Schoenberg and von Blanckenburg (2006), who cited that additional constraints were required to explain the isotopic relationship with SiO_2 content. A fluid-exsolution hypothesis was extended further by Heimann et al (2008), who, compiling much of the data available at the time (Figure 15.9), confirmed the trend of Poitrasson and Freydier (2005) in which silicic igneous rocks above 70 wt% SiO_2 exhibited strong positive deviations from mafic igneous rocks. These authors considered that the lack of correlation of iron isotope composition with incompatible element concentrations (which can be inverted to yield the fraction of melt remaining) argued against the role of fractional crystallization in determining iron isotope compositions. The observed inverse correlation between Zr/Hf and $\delta^{57}Fe$ was taken as evidence of fluid exsolution, as Heimann et al. (2008) asserted that these trace

[4] This classification is based on the relative molar quantities of components required to form feldspars, where peralkaline rocks have ($Na_2O + K_2O$) > Al_2O_3, metaluminous rocks have $Al_2O_3 < Na_2O+K_2O+CaO$, but $Al_2O_3 > Na_2O+K_2O$ and peraluminous rocks have $Al_2O_3 >$ ($Na_2O + K_2O + CaO$).

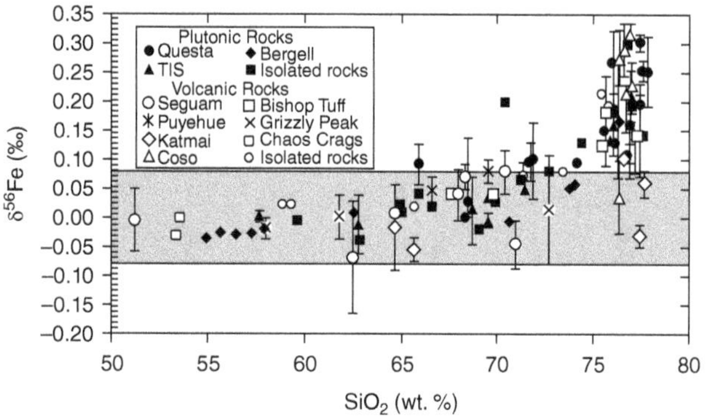

Figure 15.9 After Heimann et al. 2008. An early compilation of iron isotope analyses in igneous rocks.

elements remain little affected by crystal fractionation (but see Linnen & Keppler, 2002). In these highly differentiated rocks, most of the iron is contained in Fe-Ti oxides (specifically magnetite in oxidized granites and rhyolites), and therefore should retain the heavy isotopes, where Heimann et al. (2008) modeled $\Delta^{57}Fe_{mag\text{-}fluid}$ between ≈+0.45‰ at 700 °C and ≈+0.75‰ at 500 °C. Because the solubility of $FeCl_2$ decreases with temperature and isotopic fractionation is proportional to $1/T^2$, the more fractionated granites should preserve the heaviest iron isotope compositions.

Contemporaneously, Teng et al. (2008) and Schuessler et al. (2009) published two studies showing unequivocal evidence for the role of crystal-liquid fractionation in controlling iron isotope variations in differentiating magmatic systems. Whole rocks from the Kilauea Iki lava lake, Hawai'i, record progressive enrichment of the heavy isotopes of iron with decreasing MgO, from values close to 0‰ in olivine cumulates to +0.3‰ in evolved basaltic samples (Teng et al. 2008). Importantly, isotopic fractionation is observed below 70 wt% SiO_2, in contrast to other differentiating magmatic systems. This variation broadly correlates with the $Fe^{3+}/\sum Fe$ content of the whole rock, and, more precisely, reflects the incorporation of isotopically light iron into the $^{VI}Fe^{2+}$ sites of olivine relative to the residual melt, fit with a fractionation factor $10^3 \ln^{57/54}Fe\alpha_{ol\text{-}melt}$ = -0.20±0.10‰. Unlike at Kilaeua Iki, isotopic fractionation at the Hekla volcano, Iceland, occurs only around 70 wt% SiO_2, in line with earlier studies (Schuessler et al., 2009). These authors rejected the hypothesis of fluid exsolution as the cause for this variation, based on the constancy of the isotopes of Li, an element even more fluid-mobile than iron, in the same rocks. Instead, they argued for crystallization of Ti-rich titanomagnetite, which crystallizes as Mag_{35} at Hekla (Baldridge et al., 1973), in elevating the iron isotope composition in evolved rocks.

The absence of iron isotope fractionation below 70 wt% SiO_2, except for the differentiating sequence of basalts at Kilauea Iki behoved an explanation. The Red Hill intrusion, southern Tasmania, exhibits a vertical continuum of compositions created by in situ fractional crystallization of a single injection of magma in a closed system (Sossi et al., 2012). Much like the Kilauea Iki lava lake, the iron isotope composition of dolerites from Red Hill increased in $\delta^{57}Fe$ with decreasing MgO contents, before reaching a maximum near +0.30‰ and declining following the crystallization of magnetite (Mag_{95}), an effect that had been expected, but never observed, on the basis of experimental constraints (Shahar et al., 2008). Thus, iron isotopic evolution for the tholeiitic series mirrored that followed by the liquid line of descent on an AFM (Alkalis-FeO-MgO) diagram, and Sossi et al. (2012) were able to derive fractionation factors; $\Delta^{57}Fe_{cpx\text{-}melt}$ = $-0.25\times10^6/T^2$ and $\Delta^{57}Fe_{mag\text{-}melt}$ = $+0.20\times10^6/T^2$. These authors also pointed out that the differing behaviour of Kilaeua Iki and Red Hill on one hand and Hekla on the other could be attributed to the extent to which the differentiating system remained open to oxygen exchange (Figure 15.10). Systems that are closed to oxygen exchange, which may be imposed by the formation of chilled margins that surround the magma body, have a finite oxygen budget. Therefore, the early crystallization Fe^{2+}-rich phases increases the $Fe^{3+}/\sum Fe$ and $\delta^{57}Fe$ of the melt, prior to the crystallization of magnetite-rich titanomagnetite, whose $Fe^{3+}/\sum Fe$ ratio and hence $\delta^{57}Fe$ is higher than that of the melt. By contrast, systems open to O_2 exchange are buffered at values near FMQ, potentially by chemical and mass exchange with surrounding wall rocks (Carmichael, 1967). Here, the $Fe^{3+}/\sum Fe$ of the melt and minerals remain similar, and they precipitate Ti-rich titanomagnetite, with potentially lower $Fe^{3+}/\sum Fe$ than the melt, causing an increase in the iron isotope composition of the residual melt.

Studies of layered intrusions reveal more complex histories that involve the introduction of several pulses of magma and crystal accumulation that blurs the isotopic signal of the evolving melt(s). However, the conclusion from these studies is that magmatic differentiation does not induce significant isotopic fractionation in bulk rocks (Chen et al., 2014; Liu et al., 2014; Bilenker et al., 2017), notwithstanding the fact that the individual minerals comprising these rocks can be strongly fractionated. Isotopic fractionation generally occurs in the sequence magnetite > clinopyroxene > olivine > ilmenite, in broad agreement with the hierarchy depicted in Section 15.4.1. Equilibration temperatures of ilmenite-magnetite pairs suggest slow cooling as anticipated for intrusive magmas, and point to sub-solidus re-equilibration as the cause for these disparate isotopic signatures, thereby leaving the composition of the whole rock unaffected.

The uniformly heavy iron isotope composition of silicic melts also has its origin in the increased bond strength of iron in these highly polymerized, alkali-rich compositions

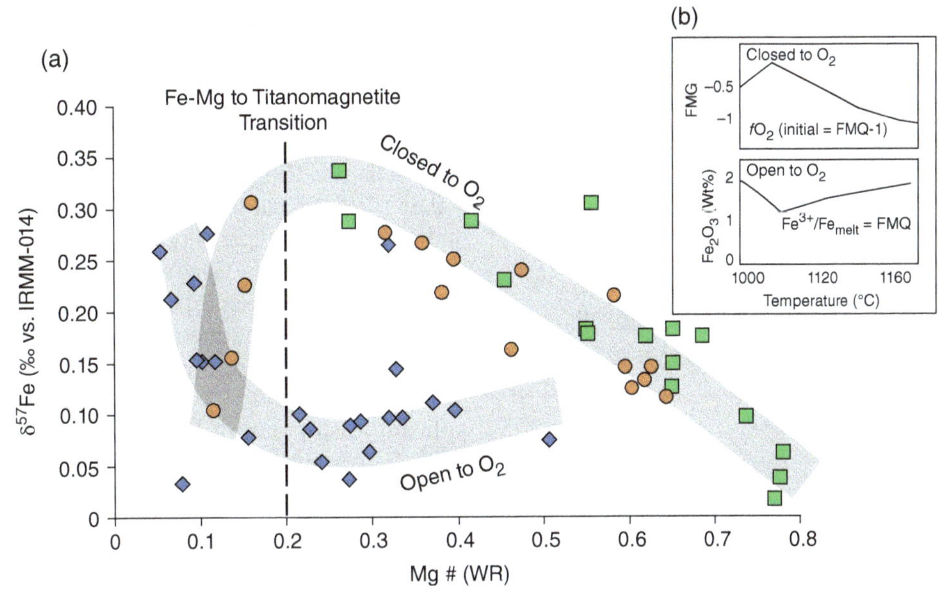

Figure 15.10 δ^{57}Fe of whole rocks against their Mg#, characterizing the change in iron isotope composition with differentiation. (a) Analyses of co-genetic whole-rock samples from tholeiitic suites: green squares = Kilauea Iki (Teng et al. 2008), blue diamonds = Hekla (Schuessler et al. 2009), orange circles = Red Hill (Sossi et al. 2012) display disparate behavior depending on whether the system evolved open (Hekla) or closed (Red Hill, Kilauea Iki) to oxygen exchange. (b) Simplified versions of Figs. 4 and 7 from Toplis and Carroll (1996), highlighting the contrasting change in fO_2 relative to the FMQ buffer and Fe_2O_3 (wt%) for conditions open and closed to oxygen exchange, respectively. These trends directly mirror those seen in (a).

(Section 15.4.2), as suggested by Schuessler et al. (2009) and Sossi et al. (2012). This behaviour was conclusively illustrated in the combined NRIXS-XANES study of Dauphas et al. (2014), which revealed an increase in Fe-O force constants of felsic melts relative to mafic and intermediate compositions at a given $Fe^{3+}/\sum Fe$ (see Section 15.4.2). Foden et al. (2015), upon analyzing 42 A-, I-, and S-type granites, and compiling literature data, found that only the A-type granites, especially the peralkaline ones, that form from protracted crystallization of basaltic magma (Turner et al., 1992) exhibit heavy iron isotope compositions. As these granites are typically reduced and H_2O- and volatile-poor, their heavy compositions likely reflect extensive crystal-liquid segregation rather than fluid exsolution. More reduced, ilmenite-bearing S-type granites also display a similar isotopic evolution, whereas more oxidized, magnetite-series I-type granites never reach heavy iron isotope compositions, likely due to the combined effect of isotopically light olivine and heavy magnetite in canceling each other out. This behavior is evident in evolving magmatic rocks from the Anatahan volcano of the Marianas arc (Williams et al., 2018), in which the effect of magnetite crystallization is less pronounced than at Red Hill (a change in δ^{57}Fe of ≈0.15‰) after displaying mildly rising iron isotopic values prior to this point. The iron isotopic evolution then increases towards heavier values at very low MgO contents below 2 wt%, as per other siliceous igneous rocks.

Within this framework, recent years have seen a flurry of studies focusing on better understanding the nature of inter-mineral and mineral-melt fractionation in evolved magmatic systems (Du et al., 2017; He et al., 2017; Wu et al., 2017; Xia et al., 2017; Li et al., 2019). These studies argue for the predominant role of fractional crystallization in producing the heavy δ^{57}Fe values of high silica igneous rocks, by a combination of the crystallization of Ti-rich titanomagnetite and increasing Fe-O bond strength in highly silicic melts with high total alkali/alkaline earth cation ratios (Du et al., 2017; Wu et al., 2017), with residual garnet contributing in peraluminous compositions (He et al., 2017). Noteworthy is the very heavy isotopic composition of plagioclase (Figure 15.11), which attests to the preferential incorporation of ferric iron into its structure, as expected on thermodynamic grounds (Lundgaard & Tegner, 2004). However, the proportion of total iron hosted in plagioclase is so low so as to be unimportant.

15.5.2. Fluid Transfer from Mantle to Crust

High-pressure (HP) metamorphic ophiolites (meta-ophiolites) returned from former subduction zones provide information on deep geochemical processes

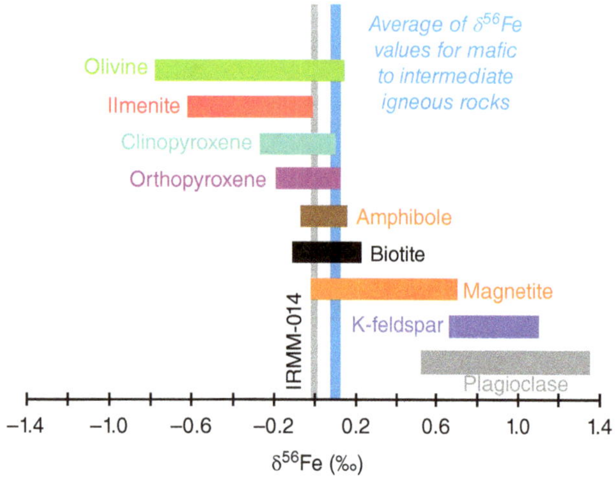

Figure 15.11 After Du et al. 2017. Compilation of Fe isotope mineral compositions in igneous rocks.

occurring in subduction zones (Poli & Schmidt, 2002). They constitute an alternative and complementary way to arc magmas to study the isotopic composition and redox state of the fluids released during slab dehydration and devolatilization at depth. This is of particular importance for establishing the mass balance of key redox-sensitive elements, such as iron, in subduction zones. However, obtaining direct constraints on the redox state and isotopic evolution of Fe in oceanic lithosphere during subduction remains challenging.

Altered Oceanic Lithosphere. The subducting slab consists of ultramafic oceanic mantle, the mafic oceanic crust, and their sedimentary cover, all of which undergo seafloor hydrothermal alteration prior to subduction. In the ultramafic parts, this alteration is manifest in the oxidation of primary Fe^{2+}-bearing phases, such as olivine, pyroxenes, to form Fe^{3+}-bearing phases, namely magnetite at high temperature or Fe^{3+}-bearing serpentine at low temperatures (Andreani et al., 2013; Klein et al., 2014; Bonnemains et al., 2016). All abyssal peridotites are variably to highly serpentinized (Mével, 2003) and therefore display high $Fe^{3+}/\Sigma Fe$ (0.58±0.03, 2SE, n = 111; Andreani et al., 2013, 2014; Paulick et al., 2006). However, their iron contents and isotopic compositions are relatively constant, with the exception of peridotites affected by hydrothermal activity (e.g., the Rainbow hydrothermal site) where Fe mobility and therefore isotopic fractionation was enhanced by the circulation of sulfur-bearing fluids (e.g., Severmann et al., 2004; Debret et al., 2018a). They range from −0.18±0.03‰ to 0.10±0.06‰ (Craddock et al., 2013 – unweathered samples; Debret et al., 2016; Debret et al., 2018a – non-mineralized samples) with a mean value, −0.03±0.03‰ (2SE, n = 32), indistinguishable from that of orogenic lherzolites (Section 15.5.1). Although some variations have been observed among different sites; for example, the South West Indian (0.00±0.03‰) and Gakkel ridge (0.01 ±0.02‰) samples display higher δ^{57}Fe values relative to Mid-Atlantic Ridge Kane (-0.13±0.02‰) and Rainbow (-0.05±0.05‰) samples, this difference has been attributed to melting processes occurring prior to oceanic lithospheric alteration (Debret et al., 2018a). Despite large equilibrium isotope fractionation between Fe^{3+} and Fe^{2+}-bearing aqueous species and minerals (see Section 15.4.3), these observations suggest that the conversion of peridotites to serpentinites occurs without significant loss or gain of Fe, producing little net change in their Fe isotope compositions.

Relative to fresh oceanic basalts, seafloor-altered basalts record larger variations of δ^{57}Fe (−2.49‰ ≤ δ^{57}Fe ≤ +2.05‰; Rouxel et al., 2003), attributed to crystallization of secondary Fe-bearing minerals at low temperature (Rouxel et al., 2003). Indeed, the interaction of mafic rocks with seawater results in oxidation of Fe^{2+} to Fe^{3+} through seawater sulfate reduction at high temperature (200–350 °C; Shanks et al., 1981) and hydrolysis at low temperature (< 150 °C; Bach & Edwards, 2003). The limiting factor for the progress of these reactions is the amount of available reacting seawater (water–rock ratio). Their $Fe^{3+}/\Sigma Fe$ is therefore expected to be highest in the uppermost mafic crust (from seafloor to depth of about 200 m) where permeability is highest. However, the relatively low mean $Fe^{3+}/\Sigma Fe$ (0.22±0.08) of hydrothermally altered mafic rocks suggests they largely escape oxidation (Lécuyer & Ricard, 1999). Similarly, highly fractionated δ^{57}Fe values in mafic rocks may therefore be restricted to the uppermost crust. Indeed, oceanic mafic rocks show a limited range of Fe isotope compositions; in ophiolite sections (+0.18±0.08 ≤ δ^{57}Fe (‰) ≤ +0.43 ±0.11; Inglis et al., 2017) or at mid-oceanic ridges (−0.07 ≤ δ^{57}Fe (‰) ≤ 0.33; Beard et al., 2003b, Nebel et al., 2013, Teng et al., 2013, Weyer & Ionov, 2007; Williams et al., 2009). Thus, there is no clear evidence of any significant or consistent leaching or addition of iron in metabasites during hydrothermal alteration at depth.

Subducted Oceanic Lithosphere. The subduction of oceanic lithosphere and overlying sediments initiates a continuum of metamorphic reactions and dehydration/rehydration events between the downgoing slab (Schmidt & Poli, 1998), the slab-mantle wedge interface, and/or the mantle wedge, which are strongly controlled by intensive variables, such as temperature, pressure, and oxygen fugacity, and the initial composition and mineralogy of the rocks (e.g., Debret & Sverjensky, 2017; Evans et al., 2017). For a given lithology, the metamorphic evolution is therefore not unique and may vary according to the considered setting. We describe some of the main observations on Fe mobility and redox state in metamorphic lithologies composing meta-ophiolites.

Metamorphic ultramafic rocks outcropping in mountain ranges are often highly deformed and recrystallized into antigorite (high P-T variety of serpentine) and metamorphic olivine during prograde metamorphism (e.g., Trommsdorff et al., 1998; Hermann et al., 2000; Padron-Navarta et al., 2011; Scambelluri et al., 2014; Kempf & Hermann, 2018). The recrystallization of slab serpentinites during subduction results in a net decrease of magnetite mode and $Fe^{3+}/\Sigma Fe$ (Debret et al., 2014, 2015; Merkulova et al., 2017). In addition, observations of hematite-magnetite assemblages in dehydrated serpentinites (Debret et al., 2015) suggest high fO_2 (ΔFMQ+1 to +5), during antigorite breakdown in subduction zones, compatible with the release of oxidized S and C in slab-derived fluids (Debret & Sverjensky, 2017). Fluid loss is manifest as an increase in subducted serpentinite $\delta^{57}Fe$ values, which range from 0.00±0.10‰ to 0.21±0.10‰ (Debret et al., 2016). The negative correlation between $\delta^{57}Fe$ and $Fe^{3+}/\Sigma Fe$ of bulk serpentinites provides direct evidence for the release of Fe-bearing fluids during serpentinite devolatilization in subduction zones (Figure 15.12; Debret et al., 2016). Such a correlation is consistent with the release of sulfate-rich and/or hypersaline fluids, which

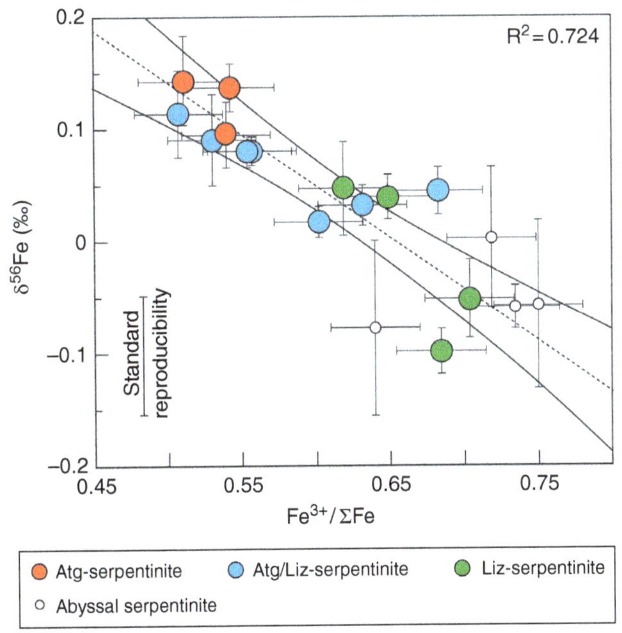

Figure 15.12 After Debret et al. (2016). Increase in $\delta^{56}Fe$ and decrease of $Fe^{3+}/\Sigma Fe$ ratio in serpentinites during subduction. Bulk sample iron isotope composition versus $Fe^{3+}/\Sigma Fe$ in subducted Alpine lizardite (Liz-) and antigorite (Atg/Liz- and Atg-) bearing serpentinites are compared to the one of abyssal serpentinites.

preferentially complex isotopically light Fe in the form of Fe^{2+}-SO_x or Fe^{2+}-Cl_2 species, during ultramafic rock devolatilization. No such correlation is observed in eclogitic rocks from meta-ophiolites, suggesting limited Fe mobility during their dehydration, and their compositions likely reflect protolith variations instead (El Korh et al., 2017; Inglis et al., 2017). In accord with these observations, Huang et al. (2019) report typical basaltic values for eclogites from the Dabie Sulu orogeny, with -0.01±0.10‰ to + 0.15±0.04‰ (n = 9), however, they are associated with oxidized, omphacite-epidote veins inferred to have been formed from the breakdown of lawsonite (±phengite). These veins have elevated $Fe^{3+}/\Sigma Fe$ (≈0.6–0.9) with respect to the adjacent eclogite (≈0.3–0.5) and, consequently higher $\delta^{57}Fe$, from 0.09±0.06‰ to 0.32±0.05‰ (Huang et al. 2019). Though these veins are concluded to have been locally derived from the proximal eclogites on the strength of their similar Sr- and Nd-isotope compositions, they could only have constituted a volumetrically minor fraction of the eclogite source rocks in order to leave their iron isotope compositions unperturbed.

Analyses of metasedimentary rocks in meta-ophiolites are rather sporadic in the literature and show a limited range of $\delta^{57}Fe$, from 0.08±0.03‰ to 0.17±0.02‰ (Inglis et al., 2017). Nebel et al. (2015) analyzed a series of currently subducting sediments sampled from the Banda arc whose $\delta^{57}Fe$ is between +0.10±0.07 to +0.25±0.02‰ with a mean of 0.17±0.03‰ (2SE, n = 9), without any co-variation between $\delta^{57}Fe$ and detrital element proxies or radiogenic isotope compositions or total organic carbon contents. Siderite ($FeCO_3$) and ankerite ($Ca(Fe,Mg,Mn)(CO_3)_2$) are the most common Fe-bearing carbonates occurring in seafloor sediments. Both phases are expected to preferentially incorporate light iron relative to olivine (Polyakov & Mineev, 2000) and can display highly fractionated $\delta^{57}Fe$ values down to −2.94±0.03‰ in sediments (e.g., Johnson et al., 2008). These minerals are also particularly soluble in fluids emanating at temperatures ranging from 200 to 300 °C during shallow subduction (Milesi et al., 2015) and their dissolution would contribute to the formation of carbonate-bearing fluids with a light Fe isotope signature. Consequently, investigations of carbonated serpentinites from blueschist terrains show that the transfer of carbon-bearing fluids from metasediments to serpentinite can contribute to a strong decrease of $\delta^{57}Fe$ of serpentinite, down to −0.40‰ (Debret et al., 2018b). Although the exact nature of Fe complexation in sediment-derived fluids is poorly constrained, this observation further corroborates the release of isotopically light fluids during sediment decarbonation and dehydration in subduction zones.

Iron Bearing Fluids Transferred to the Mantle Wedge. Most studies of meta-ophiolites point to the release of isotopically light Fe in slab-derived fluids during subduction, consistent with the transport of iron bound to chlorine, sulfate, and/or carbonate, as these ligands preferentially complex light Fe with respect to Fe^{3+}-rich antigorite and magnetite (e.g., see Section 15.4.3, Fujii et al., 2014; Hill et al., 2010). Although the vector of fractionation is correct, it remains unknown whether or not enough Fe can be mobilized in slab-derived fluids to significantly modify the δ^{57}Fe of mantle wedge peridotites to produce the light iron isotope composition of arc magmas (Foden et al., 2018).

Polyphase inclusions trapped in high pressure minerals represent the fluid released at high pressure and temperature during the devolatilisation of the subducting lithosphere in subduction zones. Scambelluri et al. (2015) showed that these olivine-hosted HP-fluids have variable FeO contents from 0.17 to 2.13 wt% (assuming $\sum$Fe as FeO), in which Cl is often associated with Fe, as inferred from solubility measurements and thermodynamic calculations (see Section 15.4.3). In addition, the presence of sulfide phases that are not stable in pristine or altered mantle rocks indicates the mobilization of sulfur concomitant with chlorine and iron. These authors also report the existence of magnesite in inclusions from chlorite harzburgite, providing compelling evidence of carbon mobilisation in the meta-peridotite dehydration fluid.

The interaction of slab-derived fluids during their migration through the slab/mantle interface and the mantle wedge may modify their isotopic signature. In order to test how faithfully this fluid signature is transferred to the mantle wedge (i.e., during subduction), Debret et al. (2018b) analyzed metamorphic gem olivines in veins from the Sapat complex, part of the sub-arc mantle of the Kohistan-Ladakh Arc. The vein-forming minerals crystallised from a fluid that migrated through the fore-arc mantle, and display light δ^{57}Fe values (-0.13±0.13‰ to -0.54 ±0.14‰) relative to that of mantle olivine (e.g., San Carlos olivine δ^{57}Fe has ≈+0.02±0.05‰; Sossi et al., 2015) supporting the hypothesis that they crystallised through reaction with an isotopically fractionated fluid. Similarly, Turner et al. (2018) reported sub-arc mantle xenoliths with light δ^{57}Fe (from +0.03 to -0.46‰) relative to mantle peridotites. Such values are too fractionated to result solely from partial melting and require slab fluid addition and/or diffusive effects that produce light values. This is confirmed by the correlation between δ^{57}Fe and U/Th ratios in these sub-arc xenoliths. Although not all sub-arc peridotites show such isotopically light values (Section 15.5.1.1), these observations provide empirical evidence for the transfer of isotopically light Fe across subduction zones, in which Fe can be complexed by chlorine or sulfate or carbonate anions. Existing models (Huang et al., 2019) suggest, given the composition of polyphase inclusions (Scambelluri et al., 2015) and inferred iron isotope fractionation factors between olivine and fluids (Debret et al., 2018), that 30–40% fluid addition to the mantle wedge would be required to shift their iron isotope compositions to sufficiently light values observed in some arc basalts (-0.15‰; Foden et al., 2018), but this increases to unrealistically high values for sub-arc mantle xenoliths with compositions as low as -0.46‰ (Turner et al., 2018). The predominant uncertainties with such models remain (i) the fraction of fluid permitted to metasomatize the mantle wedge; (ii) its iron content and iron isotope composition; (iii) the degree of disequilibrium during fluid-rock interaction; and (iv) the nature and identity of the Fe-bearing fluid species. Determining the dominant Fe-bearing species will require experimental work that can produce thermodynamic data for these species at the pressures, temperatures, and compositions of interest.

15.6. CONCLUSION

At the high temperature conditions relevant to the formation of igneous and metamorphic rocks, the differences in the ground-state vibrational frequencies of isotopic substitutions, a quantum mechanical phenomenon, control isotope partitioning between phases. These frequencies are in turn determined by the bonding environment of iron; where higher redox state and lower coordination number tend to favour the heavier isotopes. Similarly, a reduction in partial molar volume of the iron-bearing site in a condensed phase due to either increasing pressure and/or a transition to a low spin electronic configuration engenders bond contraction and strengthening, also concentrating the heavier isotopes. The role of changing redox state and coordination environment in modifying iron force constants is key, particularly along mineral solid solutions, magmatic liquid compositional series (basalt to rhyolite), and for different ligand identities and complexation stoichiometries in aqueous fluids.

Partial melting in the Earth's upper mantle results in small (≈+0.05‰) iron isotopic enrichment in the liquid relative to its source because isotopically heavy, ferric iron is preferentially concentrated in the melt with respect to lighter ferrous iron, in addition to its slightly lower mean coordination number in silicate liquids relative to mantle minerals. The difference between the melt and its residue decreases with higher degrees of melting and in more reduced sources. Contrastingly, the degree of iron isotopic enrichment in the melt is enhanced in some ocean island basalts due to the presence of heavy isotope-enriched mafic, garnet-bearing components with respect to mid-ocean ridge basalts. Arc magmas are, on average, lighter still, and have iron isotope compositions that are inversely correlative with the (i) age of the subducting slab and (ii) its downgoing velocity. The mechanisms that contribute to this isotopic signature may include continued episodes of light, fluid-fluxed melting in stationary sub-arc mantle that are depleted with time in mature subduction zones. However, understanding the relative contribution of each process requires concerted experimental, natural and spectroscopic work to determine the budget, speciation, and transport of iron in subduction zones.

Magmatic rocks show a tendency towards heavy isotope-enriched compositions with increasing differentiation; however, the conditions under which these liquids cool and crystallize controls their subsequent evolution. Isotopic equilibria between minerals and melt play a predominant role in determining their liquid line of descent; conditions closed to O_2 exchange result in early heavy isotope enrichment in mafic tholeiitic systems, whereas systems open to O_2 exchange (typically calc-alkaline magmas in arc environments) exhibit little change in their iron isotope composition prior to $\approx$65–70% SiO_2. The uniformly heavy iron isotope composition of granites above this value stems from a strengthening of Fe-O bonds in alkali-rich siliceous melts in tandem with crystallization of Fe^{2+}-bearing phases. Because A-type granites undergo more extensive crystal fractionation than do S- and particularly I-types, they tend to have the heaviest isotopic compositions, while the Fe^{3+}-rich mineral assemblage of more oxidized I-type granites limits heavy isotopic enrichment.

Metamorphism accompanying cooling of the oceanic lithosphere and its hydration at mid-oceanic ridges imparts surprisingly little- to no iron isotopic fractionation at the lithospheric scale, whereas local reactions in the uppermost mafic crust or nearby hydrothermal areas (e.g., Rainbow hydrothermal site) provide evidence for iron mobility relative to fresh basalts or peridotites. Analyses of mafic rocks in high pressure meta-ophiolites appear to record primary magmatic signatures and have imprints inherited from prior partial melting episodes. In contrast, sedimentary and ultramafic rocks show progressively heavier iron isotope compositions with increasing metamorphic grade, consistent with the loss of Fe^{2+}-bearing species with an isotopically light signature, in which iron is likely complexed by chlorides or oxidised ligands (i.e., sulfate or carbonate). Whether iron is sufficiently concentrated in these fluids to influence the iron isotope composition of the sub-arc mantle remains an active area of research.

ACKNOWLEDGMENTS

PAS was supported by the European Research Council under the H2020 framework program/ERC grant agreement #637503 (PRISTINE) at IPGP and by an SNF Ambizione Fellowship (#180025) at ETH Zürich. BD acknowledges financial support by LabEx UnivEarthS (ANR-10-LABX-0023 and ANR-18-IDEX-0001). We thank Marc Blanchard and an anonymous reviewer for their thoughtful and constructive reviews. The authors wish to note that this represents an up-to-date survey of the literature at the time of writing in mid-2019.

REFERENCES

Andreani, M., Escartin, J., Delacour, A., Ildefonse, B., Godard, M., Dyment, J., et al (2014). Tectonic structure, lithology, and hydrothermal signature of the rainbow massif (Mid-Atlantic Ridge 36° 14'N). *Geochemistry, Geophysics, Geosystems*, *15*, 3543–3571. doi: 10.1002/2014GC005269

Andreani, M., Muñoz, M., Marcaillou, C., & Delacour, A. (2013). μXANES study of iron redox state in serpentine during oceanic serpentinization. *Lithos*, *178*, 70–83. doi: 10.1016/j.lithos.2013.04.008

Arató, R., & Audétat, A. (2021). Titanomagnetite – silicate melt oxybarometry. In: Moretti, R., & Neuville, D. R. *Magma Redox Geochemistry*. Geophysical Monograph 266. American Geophysical Union. (This volume).

Armstrong, K., Frost, D. J., McCammon, C. A., Rubie, D. C., & Boffa Ballaran, T. (2019). Deep magma ocean formation set the oxidation state of Earth's mantle. *Science*, *365*, 903–906. doi: 10.1126/science.aax8376

Asimow, P. D., Hirschmann, M. M., & Stolper, E. M. (1997). An analysis of variations in isentropic melt productivity. *Philosophical Transactions of the Royal Society A: Mathematical, Physical, and Engineering Sciences*, *355*, 255–281. doi: 10.1098/rsta.1997.0009

Ayers, J. C., Dittmer, S. K., & Layne, G. D. (1997). Partitioning of elements between peridotite and H_2O at 2.0–3.0 GPa and 900–1100 °C, and application to models of subduction zone processes. *Earth and Planetary Science Letters*, *150*, 381–398. doi: 10.2134/jeq2005.0109

Bach, W., & Edwards, K. J. (2003). Iron and sulfide oxidation within the basaltic ocean crust: Implications for chemolithoautotrophic microbial biomass production. *Geochimica et Cosmochimica Acta*, *67*, 3871–3887. doi: 10.1016/S0016-7037(03)00304-1

Badro, J., Fiquet, G., Guyot, F., Rueff, J.-P., Struzhkin, V. V., Vankó, G., & Monaco, G. (2003). Iron partitioning in Earth's mantle: toward a deep lower mantle discontinuity. *Science* *300*, 789–791. doi: 10.1126/science.1081311

Badro, J., Rueff, J.-P., Vankó, G., Monaco, G., Fiquet, G., & Guyot, F. (2004). Electronic Transitions in Perovskite: Possible Nonconvecting Layers in the Lower Mantle. *Science 305*, 383–386. doi: 10.1126/science.1098840

Baldridge, W. S., McGetchin, T. R., Frey, F. A., & Jarosewich, E. (1973). Magmatic evolution of Hekla, Iceland. *Contributions to Mineralogy and Petrology*, *42*, 245–258. doi: 10.1007/BF00371589

Barnes, S. J., & Roeder, P. L. (2001). The Range of Spinel Compositions in Terrestrial Mafic and Ultramafic Rocks. *Journal of Petrology*, *42*, 2279–2302.

Beard, B. L., Johnson, C. M., Skulan, J. L., Nealson, K. H., Cox, L., & Sun, H. (2003a). Application of Fe isotopes to tracing the geochemical and biological cycling of Fe. *Chemical Geology, 195*, 87–117. doi: 10.1016/S0009-2541(02)00390-X

Beard, B. L., Johnson, C. M., Von Damm, K. L., & Poulson, R. L. (2003b). Iron isotope constraints on Fe cycling and mass balance in oxygenated Earth oceans. *Geology, 31*, 629–632. doi: 10.1130/0091-7613(2003)031<0629:IICOFC>2.0.CO;2

Bénard, A., Woodland, A. B., Arculus, R. J., Nebel, O., & McAlpine, S. R. B. (2018). Variation in sub-arc mantle oxygen fugacity during partial melting recorded in refractory peridotite xenoliths from the West Bismarck Arc. *Chemical Geology, 486*, 16–30. doi: 10.1016/j.chemgeo.2018.03.004

Berry, A. J., Yaxley, G. M., Woodland, A. B., & Foran, G. J. (2010). A XANES calibration for determining the oxidation state of iron in mantle garnet. *Chemical Geology, 278*, 31–37. doi: 10.1016/j.chemgeo.2010.08.019

Berry, A. J., Stewart, G. A., O'Neill, H.S.C., Mallmann, G., & Mosselmans, J. F. W. (2018). A re-assessment of the oxidation state of iron in MORB glasses. *Earth and Planetary Science Letters, 483*, 114–123. doi: 10.1016/j.epsl.2017.11.032

Bigeleisen, J., & Mayer, M. G. (1947). Calculation of Equilibrium Constants for Isotopic Exchange Reactions. *The Journal of Chemical Physics, 15*, 261–267. doi: 10.1063/1.1746492

Bilenker, L. D., VanTongeren, J. A., Lundstrom, C. C., & Simon, A. C. (2017). Iron isotopic evolution during fractional crystallization of the uppermost Bushveld Complex layered mafic intrusion. *Geochemistry, Geophysics, Geosystems, 18*, 2016GC006660. doi: 10.1002/2016GC006679.Received

Blanchard, M., Poitrasson, F., Méheut, M., Lazzeri, M., Mauri, F., & Balan, E. (2009). Iron isotope fractionation between pyrite (FeS_2), hematite (Fe_2O_3) and siderite ($FeCO_3$): A first-principles density functional theory study. *Geochimica et Cosmochimica Acta, 73*, 6565–6578. doi: 10.1016/j.gca.2009.07.034

Blanchard, M., & Dauphas, N. (2019). Oxidation state, coordination, and covalency controls on iron isotopic fractionation in Earth's mantle and crust: Insights from first-principles calculations and NRIXS spectroscopy. In: Moretti, R., & Neuville, D. R. *Magma Redox Geochemistry*. Geophysical Monograph 266. American Geophysical Union. (this volume)

Boctor, N. Z., Popp, R. K., & Frantz, J. D. (1980). Mineral-solution equilibria-IV. Solubilities and the thermodynamic properties of $FeCl_2$ in the system Fe_2O_3-H_2-H_2O-HCl. *Geochimica et Cosmochimica Acta, 44*, 1509–1518.

Bonnemains, D., Carlut, J., Escartin, J., et al (2016) Magnetic signatures of serpentinization at ophiolite complexes. *Geochemistry, Geophysics, Geosystems, 17*:2969–2986. doi: 10.1002/2016GC006321

Borisov, A., Behrens, H., & Holtz, F. (2017). Effects of strong network modifiers on Fe^{3+}/Fe^{2+} in silicate melts: an experimental study. *Contributions to Mineralogy and Petrology, 172*, 34. doi: 10.1007/s00410-017-1337-1

Borisov, A., & McCammon, C. (2010). The effect of silica on ferric/ferrous ratio in silicate melts: An experimental study using Mössbauer spectroscopy. *American Mineralogist, 95*, 545–555. doi: 10.2138/am.2010.3217

Bosi, F., Halenius, U., & Skogby, H. (2009). Crystal chemistry of the magnetite-ulvospinel series. *American Mineralogist, 94*, 181–189. doi: 10.2138/am.2009.3002

Bourdon, B., Roskosz, M., & Hin, R. C. (2018). Isotope tracers of core formation. *Earth-Science Reviews, 181*, 61–81. doi: 10.1016/J.EARSCIREV.2018.04.006

Brugger, J., Liu, W., Etschmann, B., Mei, Y., Sherman, D. M., & Testemale, D. (2016). A review of the coordination chemistry of hydrothermal systems, or do coordination changes make ore deposits? *Chemical Geology, 447*, 219–253. doi: 10.1016/j.chemgeo.2016.10.021

Buddington, A. F., & Lindsley, D. H. (1964). Iron-titanium oxide minerals and synthetic equivalents. *Journal of Petrology, 5*, 310–357. doi: 10.1093/petrology/5.2.310

Burnham, C. W. (1979). Magmas and hydrothermal fluids. In: Barnes, H. L. (ed.) *Geochemistry of Hydrothermal Ore Deposits*, 2nd edn. New York: John Wiley & Sons. pp 71–136.

Cameron, M., Sueno, S., Prewitt, C. T., & Papike, J. J. (1973). High-temperature crystal chemistry of acmite, diopside, hedenbergite, jadeite, spodumene and ureyite. *American Mineralogist, 58*, 594–618.

Candela, P. A. (1989). Magmatic ore-forming fluids: thermodynamic and mass transfer calculations of metal concentrations. In: Whitney, J. A., & Naldrett, A. J. (eds.) *Ore Deposition Associated with Magmas*. Reviews in Economic Geology. Society of Economic Geologists. pp. 203–221.

Canil, D., & O'Neill, H. S.C. (1996). Distribution of ferric iron in some upper-mantle assemblages. *Journal of Petrology, 37*, 609–635. https://doi.org/10.1093/petrology/37.3.609

Canil, D., O'Neill, H. S. C., Pearson, D. G., Rudnick, R. L., McDonough, W. F., & Carswell, D. A. (1994). Ferric iron in peridotites and mantle oxidation states. *Earth and Planetary Science Letters, 123*, 205–220. doi: 10.1016/0012-821X(94)90268-2

Carmichael, I. S. E. (1967). The mineralogy of Thingmuli, a Tertiary volcano in eastern Iceland. *American Journal of Science, 276*, 309–329.

Carmichael, I. S. E. (1991). The redox states of basic and silicic magmas: a reflection of their source regions? *Contributions to Mineralogy and Petrology, 106*, 129–141. doi: 10.1007/BF00306429

Chappell, B. W., & White, A. J. R. (1974). Two contrasting granite types. *Pacific Geology, 8*, 173–174.

Chen, L.-M., Song, X.-Y., Zhu, X.-K., Zhang, X.-Q., Yu, S.-Y., & Yi, J. N. (2014). Iron isotope fractionation during crystallization and sub-solidus re-equilibration: Constraints from the Baima mafic layered intrusion, SW China. *Chemical Geology, 380*, 97–109. doi: 10.1016/j.chemgeo.2014.04.020

Chou, I.-M., & Eugster, H. P. (1977). Solubility of Magnetite in Supercritical Chloride Solutions. *American Journal of Science, 277*, 1296–1314.

Cottrell, E., Birner, S. K., Brounce, M., Davis, F. A., Waters, L. E., & Kelley, K. A. (2021). Oxygen Fugacity Across Tectonic Settings. In: Moretti, R., & Neuville, D. R. (Eds.) *Magma Redox Geochemistry*. Geophysical Monograph Series. American Geophysical Union.

Craddock, P. R., Warren, J. M., & Dauphas, N. (2013). Abyssal peridotites reveal the near-chondritic Fe isotopic composition of the Earth. *Earth and Planetary Science Letters, 365*, 63–76.

Dauphas, N., Craddock, P. R., Asimow, P. D., Bennett, V. C., Nutman, A. P., & Ohnenstetter, D. (2009). Iron isotopes may reveal the redox conditions of mantle melting from Archean to Present. *Earth and Planetary Science Letters, 288*, 255–267.

Dauphas, N., Roskosz, M., Alp, E. E., Golden, D. C., Sio, C. K., Tissot, F. L. H., et al (2012). A general moment NRIXS approach to the determination of equilibrium Fe isotopic fractionation factors: Application to goethite and jarosite. *Geochimica et Cosmochimica Acta*, *94*, 254–275. doi: 10.1016/j.gca.2012.06.013

Dauphas, N., Roskosz, M., Alp, E. E., Neuville, D. R., Hu, M. Y., Sio, C. K., et al (2014). Magma redox and structural controls on iron isotope variations in Earth's mantle and crust. *Earth and Planetary Science Letters*, *398*, 127–140. doi: 10.1016/j.epsl.2014.04.033

Davies, D. R., Le Voci, G., Goes, S., Kramer, S. C., & Wilson, C. R. (2016). The mantle wedge's transient 3-D flow regime and thermal structure. *Geochemistry, Geophysics, Geosystems*, *17*, 78–100.

Davis, F. A., Humayun, M., Hirschmann, M. M., & Cooper, R. S. (2013). Experimentally determined mineral/melt partitioning of first-row transition elements (FRTE) during partial melting of peridotite at 3 GPa. *Geochimica et Cosmochimica Acta*, *104*, 232–260. doi: 10.1016/j.gca.2012.11.009

Debret, B., Andreani, M., Muñoz, M., Bolfan-Casanova, N., Carlut, J., Nicollet, C., et al. (2014). Evolution of Fe redox state in serpentine during subduction. *Earth and Planetary Science Letters*, *400*, 206–218. doi: 10.1016/j.epsl.2014.05.038

Debret, B., Beunon, H., Mattielli, N., Andreani, M., Ribeiro Da Costa, I., & Escartin, J. (2018a). Ore component mobility, transport and mineralization at mid-oceanic ridges: a stable isotopes (Zn, Cu and Fe) study of the Rainbow massif (Mid-Atlantic Ridge 36°14'N). *Earth and Planetary Science Letters*, *503*, 170–180. doi:10.1016/j.epsl.2018.09.009

Debret, B., Bolfan-Casanova, N., Padrón-Navarta, J. A., Martin-Hernandex, F., Andreani, M., Garrido, C. J., et al. (2015). Redox state of iron during high-pressure serpentinite dehydration. *Contributions to Mineralogy and Petrology*, *169*, 1–18. doi: 10.1007/s00410-015-1130-y

Debret, B., Bouilhol, P., Pons, M. L., & Williams, H. (2018b). Carbonate transfer during the onset of slab devolatilization: New insights from Fe and Zn stable isotopes. *Journal of Petrology*, *59*, 1145–1166. doi: 10.1093/petrology/egy057

Debret, B., Millet, M. A., Pons, M. L., Bouilhol, P., Inglis, E., & Williams, H. (2016). Isotopic evidence for iron mobility during subduction. *Geology*, *44*, 215–218.

Debret, B., & Sverjensky, D. A. (2017). Highly oxidising fluids generated during serpentinite breakdown in subduction zones. *Scientific Reports*, *7*, 1–6. doi: 10.1038/s41598-017-09626-y

Dickenson, M. P., & Hess, P. C. (1986). The structural role and homogeneous redox equilibria of iron in peraluminous, metaluminous and peralkaline silicate melts. *Contributions to Mineralogy and Petrology*, *92*, 207–217.

Ding, K., & Seyfried, W. E. (1992). Determination of Fe-Cl complexing in the low pressure supercritical region (NaCl fluid): Iron solubility constraints on pH of subseafloor hydrothermal fluids. *Geochimica et Cosmochimica Acta*, *56*, 3681–3692. doi: 10.1016/0016-7037(92)90161-B

Du, D. H., Wang, X. L., Yang, T., Chen, X., Li, J.-Y., & Li, W. (2017). Origin of heavy Fe isotope compositions in high-silica igneous rocks: A rhyolite perspective. *Geochimica et Cosmochimica Acta*, *218*, 58–72. doi: 10.1016/j.gca.2017.09.014

Evans, K. A., Reddy, S. M., Tomkins, A. G., Crossley, R. J., & Frost, B. R. (2017). Effects of geodynamic setting on the redox state of fluids released by subducted mantle lithosphere. *Lithos*, *278–281*:26–42. doi: 10.1016/j.lithos.2016.12.023

Evans, K. A., & Tomkins, A. G. (2011). The relationship between subduction zone redox budget and arc magma fertility. *Earth and Planetary Science Letters*, *308*, 401–409. doi: 10.1016/j.epsl.2011.06.009

Farges, F., Lefrère, Y., Rossano, S., Betherau, A., Calas, G., Brown, Jr., G. E. (2004). The effect of redox state on the local structural environment of iron in silicate glasses: a combined XAFS spectroscopy, molecular dynamics, and bond valence study. *Journal of Non-Crystalline Solids*, *344*, 176–188. doi: 10.1016/j.jnoncrysol.2004.07.050

Fein, J. B., Hemley, J. J., Angelo, W. M. D., Komminou, A., & Sverjensky, D. A. (1992). Experimental study of iron-chloride complexing in hydrothermal fluids. *Geochimica et Cosmochimica Acta*, *56*, 3179–3190. doi: 10.1016/0016-7037(92)90296-U

Foden, J., Sossi, P. A., & Nebel, O. (2018). Controls on the iron isotopic composition of global arc magmas. *Earth and Planetary Science Letters*,*494*, 190–201. doi: 10.1016/j.epsl.2018.04.039

Foden, J., Sossi, P. A., & Wawryk, C. M. (2015). Fe isotopes and the contrasting petrogenesis of A-, I- and S-type granite. *Lithos*, *212–215*, 32–44. doi: 10.1016/j.lithos.2014.10.015

Frost, C. D., & Frost, B. R. (2010). On ferroan (A-type) granitoids: Their compositional variability and modes of origin. *Journal of Petrology*, *52*, 39–53. doi: 10.1093/petrology/egq070

Fujii, T., Moynier, F., Blichert-Toft, J., & Albarède, F. (2014). Density functional theory estimation of isotope fractionation of Fe, Ni, Cu, and Zn among species relevant to geochemical and biological environments. *Geochimica et Cosmochimica Acta*, *140*, 553–576. doi: 10.1016/j.gca.2014.05.051

Fujino, K., Nishio-Hamane, D., Seto, Y., Sata, N., Nagai, T., Shinmei, T., et al (2012). Spin transition of ferric iron in Al-bearing Mg–perovskite up to 200 GPa and its implication for the lower mantle. *Earth and Planetary Science Letters*, *317–318*, 407–412. doi: 10.1016/J.EPSL.2011.12.006

Hawthorne, F. C. (1978). The crystal chemistry of the amphiboles. VIII. The crystal structure and site chemistry of fluor-riebeckite. *The Canadian Mineralogist*, *16*, 187–194.

He, Y., Wu, H., Ke, S., Liu, S.-A., & Wang, Q. (2017). Iron isotopic compositions of adakitic and non-adakitic granitic magmas: Magma compositional control and subtle residual garnet effect. *Geochimica et Cosmochimica Acta*, *203*, 89–102. doi: 10.1016/j.gca.2017.01.005

Heimann, A., Beard, B. L., & Johnson, C. M. (2008). The role of volatile exsolution and sub-solidus fluid/rock interactions in producing high $^{56}Fe/^{54}Fe$ ratios in siliceous igneous rocks. *Geochimica et Cosmochimica Acta*, *72*, 4379–4396. doi: 10.1016/j.gca.2008.06.009

Hermann, J., Müntener, O., & Scambelluri, M. (2000). The importance of serpentinite mylonites for subduction and exhumation of oceanic crust. *Tectonophysics*, *327*, 225–238. doi: 10.1016/S0040-1951(00)00171-2

Hill, P. S., Schauble, E. A., Shahar, A., Tonui, E., & Young, E. D. (2009). Experimental studies of equilibrium iron isotope fractionation in ferric aquo–chloro complexes. *Geochimica et Cosmochimica Acta, 73*, 2366–2381. doi: 10.1016/j.gca.2009.01.016

Hill, P. S., Schauble, E. A., & Young, E. D. (2010). Effects of changing solution chemistry on Fe^{3+}/Fe^{2+} isotope fractionation in aqueous Fe–Cl solutions. *Geochimica et Cosmochimica Acta, 74*, 6669–6689. doi: 10.1016/j.gca.2010.08.038

Hill, P. S., & Schauble, E. A. (2008). Modeling the effects of bond environment on equilibrium iron isotope fractionation in ferric aquo-chloro complexes. *Geochimica et Cosmochimica Acta, 72*, 1939–1958.

Hirschmann, M. M., Ghiorso, M. S., & Stolper, M. (1999). Calculation of peridotite partial melting from thermodynamic models of minerals and melts. II. Isobaric variations in melts near the solidus and owing to variable source composition. *Journal of Petrology, 40*, 297–313. doi: 10.1093/petroj/40.2.297

Huang, F., Zhang, Z., Lundstrom, C. C., & Zhi, X. (2011). Iron and magnesium isotopic compositions of peridotite xenoliths from Eastern China. *Geochimica et Cosmochimica Acta, 75*, 3318–3334. doi: 10.1016/j.gca.2011.03.036

Huang, J., Guo, S., Jin, Q.-Z., & Huang, F. (2019). Iron and magneisum isotopic compositions of subduction-zone fluids and implications for arc volcanism. *Geochimica et Cosmochimica Acta, 278*, 376–391. 10.1016/j.gca.2019.06.020

Inglis, E. C., Debret, B., Burton, K. W., Millet, M.-A., Pons, M.-L., Dale, C. W., et al (2017). The behavior of iron and zinc stable isotopes accompanying the subduction of mafic oceanic crust: A case study from Western Alpine ophiolites. *Geochemistry, Geophysics, Geosystems, 18*, 2562–2579. doi: 10.1002/2016GC006735

Jackson, J. M., Sturhahn, W., Shen, G., Zhao, J., Hu, M. Y., Errandonea, D., et al. (2005). A synchrotron Mössbauer spectroscopy study of (Mg,Fe)SiO_3 perovskite up to 120 GPa. *American Mineralogist, 90*, 199–205. doi: 10.2138/am.2005.1633

Jackson, W. E., Farges, F., Yeager, M., Mabrouk, P. A., Rossano, S., Waychunas, G. A., et al. (2005). Multi-spectroscopic study of Fe(II) in silicate glasses: Implications for the coordination environment of Fe(II) in silicate melts. *Geochimica et Cosmochimica Acta, 69*, 4315–4332. doi: 10.1016/j.gca.2005.01.008

Jayasuriya, K. D., O'Neill, H. S. C., Berry, A. J., & Campbell, S. J. (2004). A Mössbauer study of the oxidation state of Fe in silicate melts. *American Mineralogist, 89*:1597–1609.

Johnson, C. M., Beard, B. L., Klein, C., Beukes, N. J., & Roden, E. E. (2008). Iron isotopes constrain biologic and abiologic processes in banded iron formation genesis. *Geochimica et Cosmochimica Acta, 72*, 151–169. doi: 10.1016/J.GCA.2007.10.013

Kelley, K. A., & Cottrell, E. (2009). Water and the oxidation state of subduction zone magmas. *Science, 325*, 605–607. doi: 10.1126/science.1174156

Kempf, E. D., & Hermann, J. (2018). Hydrogen incorporation and retention in metamorphic olivine during subduction: Implications for the deep water cycle. *Geology, 46*, 571–574. doi: 10.1130/G40120.1

Kinzler, R. J., & Grove, T. L. (1992). Primary magmas of mid-ocean ridge basalts 1. Experiments and methods. *Journal of Geophysical Research, 97*, 6885–6906.

Kirby, S. H., Engdahl, R. E., & Denlinger, R. (1996). Intermediate-depth intraslab earthquakes and arc volcanism as physical expressions of crustal and uppermost mantle metamorphism in subducting slabs. In: Bebout, G. E., Scholl, D. W., Kirby, S. H., & Platt, J. P. (eds.) *Subduction: Top to Bottom.* Washington, DC: American Geophysical Union. pp. 195–214.

Klein, F., Bach, W., Humphris, S. E., Kahl, W.A., Jöns, N., Moskowitz, B., & Berquó, T. S. (2014). Magnetite in seafloor serpentinite-Some like it hot. *Geology, 42*:135–138. doi: 10.1130/G35068.1

Knipping, J. L., Behrens, H., Wilke, M., Göttlicher, J., & Stabile, P. (2015). Effect of oxygen fugacity on the coordination and oxidation state of iron in alkali bearing silicate melts. *Chemical Geology, 411*, 143–154. doi: 10.1016/j.chemgeo.2015.07.004

Konter, J. G., Pietruszka, A. J., Hanan, B. B., Finlayson, V. A., Craddock, P. R., Jackson, M. G., & Dauphas, N. (2016). Unusual $\delta^{56}Fe$ values in Samoan rejuvenated lavas generated in the mantle. *Earth and Planetary Science Letters, 450*, 221–232. doi: 10.1016/j.epsl.2016.06.029

Kress, V. C., & Carmichael, I. S. E. (1991). The compressibility of silicate liquids containing Fe_2O_3 and the effect of composition, temperature, oxygen fugacity and pressure on their redox states. *Contributions to Mineralogy and Petrology,108*, 82–92. doi: 10.1007/BF00307328

Kwak, T. A. P., Brown, W. M., Abeysinghe, P. B., & Teong Hing Tan. (1986). Fe solubilities in very saline hydrothermal fluids: their relation to zoning in some ore deposits. *Economic Geology, 81*, 447–465. doi: 10.2113/gsecongeo.81.2.447

Lange, R. A., & Carmichael, I. S. E. (1987). Densities of Na_2O-K_2O-CaO-MgO-FeO-Fe_2O_3-Al_2O_3-TiO_2-SiO_2 liquids: New measurements and derived partial molar properties. *Geochimica et Cosmochimica Acta, 51*, 2931–2946. doi: 10.1016/0016-7037(87)90368-1

Lécuyer, C., & Ricard, Y. (1999). Long-term fluxes and budget of ferric iron: Implication for the redox states of the Earth's mantle and atmosphere. *Earth and Planetary Science Letters, 165*, 197–211. doi: 10.1016/S0012-821X(98)00267-2

Lee, C.-T. A., Leeman, W. P., Canil, D., & Li, Z.-X. A. (2005). Similar V/Sc systematics in MORB and arc basalts: Implications for the oxygen fugacities of their mantle source regions. *Journal of Petrology, 46*, 2313–2336. doi: 10.1093/petrology/egi056

Li, C., & Ripley, E. M. (2009). Sulfur contents at sulfide-liquid or anhydrite saturation in silicate melts: Empirical equations and example applications. *Economic Geology, 104*, 405–412. doi: 10.2113/gsecongeo.104.3.405

Li, J.-X., Qin, K.-Z., Li, G.-M., Evans, N. J., Huang, F., & Zhao, J.-X. (2018). Iron isotope fractionation during magmatic-hydrothermal evolution: A case study from the Duolong porphyry Cu-Au deposit, Tibet. *Geochimica et Cosmochimica Acta, 238*, 1–15. doi: 10.1016/J.GCA.2018.07.008

Li, Q.-W., Nebel, O., Nebel-Jacobsen, Y., Richter, M., Wang, R., Weinberg, R., et al. (2019). Crustal reworking at convergent margins traced by Fe isotopes in I-type intrusions from the Gangdese arc, Tibetan Plateau. *Chemical Geology, 510*, 47–55. https://doi.org/10.1016/j.chemgeo.2019.02.007

Lin, J.-F., Speziale, S., Mao, Z., & Marquardt, H. (2013). Effects of the electronic spin transition of iron in lower mantle minerals: implications for deep mantle geophysics and geochemistry. *Reviews of Geophysics*, *51*, 244–275. doi: 10.1002/rog.20010

Linnen, R. L., & Keppler, H. (2002). Melt composition control of Zr/Hf fractionation in magmatic processes. *Geochimica et Cosmochimica Acta*, *66*, 3293–3301. https://doi.org/10.1016/S0016-7037(02)00924-9

Liu, J., Dauphas, N., Roskosz, M., Hu, M. Y., Yang, H., Bi, W., et al. (2017). Iron isotopic fractionation between silicate mantle and metallic core at high pressure. *Nature Communications*, *8*, 14377. doi: 10.1038/ncomms14377

Liu, P.-P., Zhou, M.-F., Luais, B., Cividini, D., & Rollion-Bard, C. (2014). Disequilibrium iron isotopic fractionation during the high-temperature magmatic differentiation of the Baima Fe–Ti oxide-bearing mafic intrusion, SW China. *Earth and Planetary Science Letters*, *399*, 21–29. doi: 10.1016/j.epsl.2014.05.002

Lundgaard, K. L., Tegner, C. (2004). Partitioning of ferric and ferrous iron between plagioclase and silicate melt. *Contributions to Mineralogy and Petrology*, *147*, 470–483. doi: 10.1007/s00410-004-0568-0

Mallmann, G., & O'Neill, H. S. C. (2009). The crystal/melt partitioning of V during mantle melting as a function of oxygen fugacity compared with some other elements (Al, P, Ca, Sc, Ti, Cr, Fe, Ga, Y, Zr and Nb). *Journal of Petrology*, *50*, 1765–1794. doi: 10.1093/petrology/egp053

Manning, C. E. (2004). The chemistry of subduction-zone fluids. *Earth and Planetary Science Letters*, *223*, 1–16. doi: 10.1016/j.epsl.2004.04.030

Mathez, E. A. (1984). Influence of degassing on oxidation states of basaltic magmas. *Nature*, *310*, 371–375.

McCammon, C. (1997). Perovskite as a possible sink for ferric iron in the lower mantle. *Nature*, *387*, 694–696. doi: 10.1038/42685

McCammon, C., Dubrovinsky, L., Narygina, O., Kantor, I., Wu, X., Glazyrin, K., et al. (2010). Low-spin Fe^{2+} in silicate perovskite and a possible layer at the base of the lower mantle. *Physics of the Earth and Planetary Interiors*, *180*, 215–221. doi: 10.1016/J.PEPI.2009.10.012

Merkulova, M. V., Muñoz, M., Brunet, F., Vidal, O., Hattori, K., Vantelon, D., et al. (2017). Experimental insight into redox transfer by iron- and sulfur-bearing serpentinite dehydration in subduction zones. *Earth and Planetary Science Letters*, *479*, 133–143. doi: 10.1016/j.epsl.2017.09.009

Mével, C. (2003). Serpentinization of abyssal peridotites at mid-ocean ridges. *Comptes Rendus Geoscience*, *335*, 825–852. doi: 10.1016/j.crte.2003.08.006

Milesi, V., Guyot, F., Brunet, F., Richard, L., Recham, N., Benedetti, M., et al. (2015). Formation of CO_2, H_2 and condensed carbon from siderite dissolution in the 200-300 °C range and at 50 MPa. *Geochimica et Cosmochimica Acta*, *154*, 201–211. doi: 10.1016/j.gca.2015.01.015

Nebel, O., Arculus, R. J., Sossi, P. A., Jenner, F. E., & Whan, T. H. E. (2013). Iron isotopic evidence for convective resurfacing of recycled arc-front mantle beneath back-arc basins. *Geophysical Research Letters*, *40*, 5849–5853. doi: 10.1002/2013GL057976

Nebel, O., Sossi, P. A., Bénard, A., Wille, M., Vroon, P. Z., & Arculus, R. J. (2015). Redox-variability and controls in subduction zones from an iron-isotope perspective. *Earth and Planetary Science Letters*, *432*, 142–151. doi: 10.1016/j.epsl.2015.09.036

Nebel, O., Sossi, P. A., Foden, J., Bénard, A., Brandl, P. A., Stammeier, J. A., et al. (2018). Iron isotope variability in ocean floor lavas and mantle sources in the Lau back-arc basin. *Geochimica et Cosmochimica Acta*, *241*, 150–163. doi: 10.1016/j.gca.2018.08.046

Nell, J., Wood, B. J., & Mason, T. O. (1989). High-temperature cation distributions in Fe_3O_4-$MgAl_2O_4$-$MgFe_2O_4$-$FeAl_2O_4$ spinels from thermopower and conductivity measurement. *American Mineralogist*, *74*, 339–351.

O'Neill, H. S. C., Berry, A. J., McCammon, C. C., Jayasuriya, K. D., Campbell, S. J., & Foran, G. (2006). An experimental determination of the effect of pressure on the $Fe^{3+}/\sum Fe$ ratio of an anhydrous silicate melt to 3.0 GPa. *American Mineralogist*, *91*, 404–412. doi: 10.2138/am.2005.1929

O'Neill, H. S. C., & Mavrogenes, J. A. (2002). The sulfide capacity and the sulfur content at sulfide Saturation of Silicate Melts at 1400 °C and 1 bar. *Journal of Petrology*, *43*, 1049–1087.

O'Neill, H. S. C., & Navrotsky, A. (1984). Cation distributions and thermodynamic properties of binary spinel solid solutions. *American Mineralogist*, *69*, 733–753.

Padron-Navarta, J. A., Lopez Sanchez-Vizcaino, V., Garrido, C. J., & Gomez-Pugnaire, M. T. (2011). Metamorphic record of high-pressure dehydration of antigorite serpentinite to chlorite harzburgite in a subduction setting (Cerro del Almirez, Nevado-Filabride Complex, Southern Spain). *Journal of Petrology*, *52*, 2047–2078. doi: 10.1093/petrology/egr039

Paulick, H., Bach, W., Godard, M., De Hoog, J. C. M., Suhr, G., & Harvey, J. (2006). Geochemistry of abyssal peridotites (Mid-Atlantic Ridge, 15°20′N, ODP Leg 209): Implications for fluid/rock interaction in slow spreading environments. *Chemical Geology*, *234*, 179–210. doi: 10.1016/j.chemgeo.2006.04.011

Pauling, L. (1929). The principles determining the structure of complex ionic crystals. *Journal of the American Chemical Society*, *51*, 1010–1026.

Peters, B. J., Shahar, A., Carlson, R. W., Day, J. M. D., & Mock, T. D. (2019). A sulfide perspective on iron isotope fractionation during ocean island basalt petrogenesis. *Geochimica et Cosmochimica Acta*, *245*, 59–78. doi: 10.1016/J.GCA.2018.10.015

Petric, A., Jacob, K. T., & Alcock, C. B. (1981). Thermodynamic properties of Fe_3O_4-$FeAl_2O_4$ spinel solid solutions. *Journal of the American Ceramic Society*, *64*, 632–639. doi: 10.1111/j.1151-2916.1981.tb15860.x

Poitrasson, F., Delpech, G., & Grégoire, M. (2013). On the iron isotope heterogeneity of lithospheric mantle xenoliths: implications for mantle metasomatism, the origin of basalts and the iron isotope composition of the Earth. *Contributions to Mineralogy and Petrology*, *165*, 1243–1258. doi: 10.1007/s00410-013-0856-7

Poitrasson, F., & Freydier, R. (2005). Heavy iron isotope composition of granites determined by high resolution

MC-ICP-MS. *Chemical Geology*, *222*, 132–147. doi: 10.1016/j.chemgeo.2005.07.005

Poitrasson, F., Halliday, A. N., Lee, D.-C., Levasseur, S., & Teutsch. N. (2004). Iron isotope differences between Earth, Moon, Mars and Vesta as possible records of contrasted accretion mechanisms. *Earth and Planetary Science Letters*, *223*, 253–266. https://doi.org/10.1016/j.epsl.2004.04.032

Poli, S., & Schmidt, M. W. (2002). Petrology of subducted slabs. *Annual Review of Earth and Planetary Sciences*, *30*, 207–235. doi: 10.1146/annurev.earth.30.091201.140550

Polyakov, V. B. (2009). Equilibrium iron isotope fractionation at core-mantle boundary conditions. *Science*, *323*, 912–4. doi: 10.1126/science.1166329

Polyakov, V. B., Clayton, R. N., Horita, J., & Mineev, S. D. (2007). Equilibrium iron isotope fractionation factors of minerals – Reevaluation from the data of nuclear inelastic resonant X-ray scattering and Mössbauer spectroscopy. *Geochimica et Cosmochimica Acta*, *71*, 3833–3846. https://doi.org/10.1016/j.gca.2007.05.019

Polyakov, V. B., & Kharlashina, N. N. (1994). Effect of pressure on equilibrium isotopic fractionation. *Geochimica et Cosmochimica Acta*, *58*, 4739–4750. https://doi.org/10.1016/0016-7037(94)90204-6

Polyakov, V. B., & Mineev, S. D. (2000). The use of Mössbauer spectroscopy in stable isotope geochemistry. *Geochimica et Cosmochimica Acta*, *64*, 849–865. https://doi.org/10.1016/S0016-7037(99)00329-4

Poustovetov, A. A., & Roeder, P. L. (2000). The distribution of Cr between basaltic melt and chromian spinel as an oxygen geobarometer. *The Canadian Mineralogist*, *39*, 309–317.

Richards, J. P. (2011). Magmatic to hydrothermal metal fluxes in convergent and collided margins. *Ore Geology Reviews*, *40*, 1–26. doi: 10.1016/j.oregeorev.2011.05.006

Roskosz, M., Sio, C. K. I., Dauphas, N., & Bi, W. (2015). Spinel – olivine – pyroxene equilibrium iron isotopic fractionation and applications to natural peridotites. *Geochimica et Cosmochimica Acta*, *169*:184–199. doi: 10.1016/j.gca.2015.07.035

Rouxel, O., Dobbek, N., Ludden, J., & Fouquet, Y. (2003). Iron isotope fractionation during oceanic crust alteration. *Chemical Geology*, *202*, 155–182. doi: 10.1016/j.chemgeo.2003.08.011

Rustad, J. R., & Yin, Q.-Z. (2009). Iron isotope fractionation in the Earth's lower mantle. *Nature Geoscience*, *2*, 514–518. doi: 10.1038/ngeo546

Sanchez-Valle, C., Hin, R. C., Testemale, D., Borca, C., & Grolimund, D. (2017). Stability of oxidized iron species and the redox budget of slab-derived fluids. In: American Geophysical Union, Fall Meeting. p V11D–01

Sanloup, C., Drewitt, J. W. E., Konôpková, Z., Dalladay-Simpson, P., Morton, D. M., Rai, N., et al. (2013). Structural change in molten basalt at deep mantle conditions. *Nature*, *503*, 104–107. doi: 10.1038/nature12668

Saunier, G., Pokrovski, G. S., & Poitrasson, F. (2011). First experimental determination of iron isotope fractionation between hematite and aqueous solution at hydrothermal conditions. *Geochimica et Cosmochimica Acta*, *75*, 6629–6654. doi: 10.1016/j.gca.2011.08.028

Scambelluri, M., Pettke, T., & Cannaò, E. (2015). Fluid-related inclusions in Alpine high-pressure peridotite reveal trace element recycling during subduction-zone dehydration of serpentinized mantle (Cima di Gagnone, Swiss Alps). *Earth and Planetary Science Letters*, *429*, 45–59. doi: 10.1016/j.epsl.2015.07.060

Scambelluri, M., Pettke, T., Rampone, E., Godard, M., & Reusser, E. (2014). Petrology andtrace element budgets of high-pressure peridotites indicate subduction dehydration of serpentinized mantle (Cima di Gagnone, Central Alps, Switzerland). *Journal of Petrology*, *55*, 459–498. doi: 10.1093/petrology/egt068

Schauble E (2004) Applying stable isotope fractionation theory to new systems. *Rev Mineral Geochemistry* *55*:65–111. doi: 10.2138/gsrmg.55.1.65

Schauble, E. A., Rossman, G. R., & Taylor, H. P. (2001). Theoretical estimates of equilibrium Fe-isotope fractionations from vibrational spectroscopy. *Geochimica et Cosmochimica Acta*, *65*, 2487–2497. https://doi.org/10.1016/S0016-7037(01)00600-7

Schmidt, M. W., & Poli, S. (1998). Experimentally based water budgets for dehydrating slabs and consequences for arc magma generation. *Earth and Planetary Science Letters*, *163*, 361–379. doi: 10.1016/S0012-821X(98)00142-3

Schoenberg, R., Marks, M. A., Schuessler, J. A., von Blanckenburg, F., & Markl, G. (2009). Fe isotope systematics of coexisting amphibole and pyroxene in the alkaline igneous rock suite of the Ilímaussaq Complex, South Greenland. *Chemical Geology*, *258*, 65–77. https://doi.org/10.1016/j.chemgeo.2008.06.023

Schoenberg, R., & von Blanckenburg, F. (2006). Modes of planetary-scale Fe isotope fractionation. *Earth and Planetary Science Letters*, *252*, 342–359. doi: 10.1016/j.epsl.2006.09.045

Schuessler, J. A., Schoenberg, R., & Sigmarsson, O. (2009). Iron and lithium isotope systematics of the Hekla volcano, Iceland — Evidence for Fe isotope fractionation during magma differentiation. *Chemical Geology*, *258*, 78–91. https://doi.org/10.1016/j.chemgeo.2008.06.021

Severmann, S., Johnson, C. M., Beard, B. L., German, C. R., Edmonds, H. N., Chiba, H., & Green, D. R. H. (2004). The effect of plume processes on the Fe isotope composition of hydrothermally derived Fe in the deep ocean as inferred from the Rainbow vent site, Mid-Atlantic Ridge, 36°14′N. *Earth and Planetary Science Letters*, *225*, 63–76. doi: 10.1016/j.epsl.2004.06.001

Shahar, A., Elardo, S. M., & Macris, C. A. (2017). Equilibrium fractionation of non-traditional stable isotopes: an experimental perspective. *Reviews in Mineralogy and Geochemistry*, *82*, 65–83.doi: 10.2138/rmg.2017.82.3

Shahar, A., Schauble, E. A., Caracas, R., Gleason, A. E., Reagan, M. M., Xiao, Y., et al. (2016). Pressure-dependent isotopic composition of iron alloys. *Science*, *353*, 580–582. doi: 10.1126/science.aad9945

Shahar, A., Young, E. D., & Manning, C. E. (2008). Equilibrium high-temperature Fe isotope fractionation between fayalite and magnetite: An experimental calibration. *Earth and Planetary Science Letters*, *268*, 330–338. doi: 10.1016/j.epsl.2008.01.026

Shanks, W. C., Bischoff, J. L., & Rosenbauer, R. J. (1981). Seawater sulfate reduction and sulfur isotope fractionation in basaltic systems: Interaction of seawater with fayalite and magnetite at 200-350°C. *Geochimica et Cosmochimica Acta*, *45*, 1977–1995. doi: 10.1016/0016-7037(81)90054-5

Simon, A. C., Pettke, T., Candela, P. A., Piccoli, P. M., & Heinrich, C. A. (2004). Magnetite solubility and iron transport in magmatic-hydrothermal environments. *Geochimica et Cosmochimica Acta*, *68*, 4905–4914. doi: 10.1016/j.gca.2004.05.033

Smythe, D. J., Wood, B. J., & Kiseeva, E. S. (2017). The S content of silicate melts at sulfide saturation: New experiments and a model incorporating the effects of sulfide composition. *American Mineralogist*, *102*, 795–803. doi: 10.2138/am-2017-5800CCBY

Sossi, P. A., Foden, J. D., & Halverson, G. P. (2012). Redox-controlled iron isotope fractionation during magmatic differentiation: an example from the Red Hill intrusion, *S. Tasmania. Contributions to Mineralogy and Petrology*, *164*, 757–772. doi: 10.1007/s00410-012-0769-x

Sossi, P. A., Halverson, G. P., Nebel, O., & Eggins, S. M. (2015). Combined separation of Cu, Fe and Zn from rock matrices and improved analytical protocols for stable isotope determination. *Geostandards and Geoanalytical Research*, *39*, 129–149. doi: 10.1111/j.1751-908X.2014.00298.x

Sossi, P. A., Nebel, O., & Foden, J. (2016). Iron isotope systematics in planetary reservoirs. *Earth and Planetary Science Letters*, *452*, 295–308. doi: 10.1016/j.epsl.2016.07.032

Sossi, P. A., & O'Neill, H. S. C. (2017). The effect of bonding environment on iron isotope fractionation between minerals at high temperature. *Geochimica et Cosmochimica Acta*, *196*, 121–143. doi: 10.1016/j.gca.2016.09.017

Su, B. X., Teng, F.-Z., Hu, Y., Shi, R.-D., Zhou, M.-F., Zhu, B., et al. (2015). Iron and magnesium isotope fractionation in oceanic lithosphere and sub-arc mantle: Perspectives from ophiolites. *Earth and Planetary Science Letters*, *430*, 523–532. doi: 10.1016/j.epsl.2015.08.020

Syverson, D. D., Borrok, D., & Seyfried, W. E. (2013). Experimental determination of equilibrium Fe isotopic fractionation between pyrite and dissolved Fe under hydrothermal conditions. *Geochimica et Cosmochimica Acta*, *122*, 170–183. https://doi.org/10.1016/j.gca.2013.08.027

Syverson, D. D., Luhmann, A. J., Tan, C., Borrok, D. M., Ding, K., Seyfried Jr., W. E. (2017). Fe isotope fractionation between chalcopyrite and dissolved Fe during hydrothermal recrystallization: An experimental study at 350 °C and 500 bars. *Geochimica et Cosmochimica Acta*, *200*, 87–109. doi: 10.1016/J.GCA.2016.12.002

Teng, F.-Z., Dauphas, N., & Helz, R. T. (2008). Iron isotope fractionation during magmatic differentiation in Kilauea Iki lava lake. *Science*, *320*, 1620–2. doi: 10.1126/science.1157166

Teng, F.-Z., Dauphas, N., Huang, S., & Marty, B. (2013). Iron isotopic systematics of oceanic basalts. *Geochimica et Cosmochimica Acta*, *107*, 12–26. https://doi.org/10.1016/j.gca.2012.12.027

Testemale, D., Brugger, J., Liu, W., Etschmann, B., & Hazemann, J.-L. (2009). In-situ X-ray absorption study of Iron(II) speciation in brines up to supercritical conditions. *Chemical Geology*, *264*, 295–310. doi: 10.1016/j.chemgeo.2009.03.014

Toplis, M. J. (2005). The thermodynamics of iron and magnesium partitioning between olivine and liquid: criteria for assessing and predicting equilibrium in natural and experimental systems. *Contributions to Mineralogy and Petrology*, *149*, 22–39. doi: 10.1007/s00410-004-0629-4

Toplis, M. J., & Carroll, M. R. (1995). An experimental study of the influence of oxygen fugacity on Fe-Ti oxide stability, phase relations, and mineral-melt equilibria in ferro-basaltic systems. *Journal of Petrology*, *36*, 1137–1170.

Toplis, M. J., & Carroll, M. R. (1996). Differentiation of ferro-basaltic magmas under conditions open and closed to oxygen: Implications for the Skaergaard intrusion and other natural systems. *Journal of Petrology*, *37*, 837–858.

Trommsdorff, V., López Sánchez-Vizcaíno, V., Gómez-Pugnaire, M. T., & Müntener, O. (1998). High pressure breakdown of antigorite to spinifex-textured olivine and orthopyroxene, SE Spain. *Contributions to Mineralogy and Petrology*, *132*, 139–148. doi: 10.1007/s004100050412

Turner, S. P., Foden, J. D., & Morrison, R. S. (1992). Derivation of some A-type magmas by fractionation of basaltic magma: An example from the Padthaway Ridge, South Australia. *Lithos*, *28*, 151–179.

Turner, S. P., Williams, H. M., Piazolo, S., et al. (2018). Sub-arc xenolith Fe-Li-Pb isotopes and textures tell tales of their journey through the mantle wedge and crust. *Geology*, *46*, 947–950. doi: 10.1130/G45359.1

Urey, H. C. (1947). The thermodynamic properties of isotopic substances. *Journal of the Chemical Society*, 0, 562–581.

Urosevic, M., Nebel, O., Padron-Navarta, J. A., & Rubatto, D. (2018). Assessment of O and Fe isotope heterogeneity in garnet from Kakanui (New Zealand) and Erongo (Namibia). *European Journal of Mineralogy*, *30*. doi: 10.1029/EO060i020p00457-02

Vanpeteghem, C. B., Angel, R. J., Ross, N. L., Jacobsen, S. D., Dobson, D. P., Litasov, K. D., & Ohtani, E. (2006). Al, Fe substitution in the $MgSiO_3$ perovskite structure: A single-crystal X-ray diffraction study. *Physics of the Earth and Planetary Interiors*, *155*, 96–103. doi: 10.1016/J.PEPI.2005.10.003

Wallace, P. J., & Edmonds, M. (2011). The sulfur budget in magmas: Evidence from melt inclusions, submarine glasses, and volcanic gas emissions. *Reviews in Mineralogy and Geochemistry*, *73*, 215–246. doi: 10.2138/rmg.2011.73.8

Wawryk, C. M., & Foden, J. D. (2015). Fe-isotope fractionation in magmatic-hydrothermal mineral deposits: A case study from the Renison Sn-W deposit, Tasmania. *Geochimica et Cosmochimica Acta*, *150*, 285–298. doi: 10.1016/j.gca.2014.09.044

Wawryk, C. M., Foden, J. D. (2017). Iron-isotope systematics from the Batu Hijau Cu-Au deposit, Sumbawa, Indonesia. *Chemical Geology*, *466*, 159–172. doi: 10.1016/J.CHEMGEO.2017.06.004

Wechsler, B. A., Lindsley, D. H., & Prewitt, C. T. (1984). Crystal structure and cation distribution in titanomagnetites ($Fe_{3-x}Ti_xO_4$). *American Mineralogist*, *69*, 754–770.

Wechsler, B. A., & Prewitt, C. T. (1984). Crystal structure of ilmenite ($FeTiO_3$) at high temperature and at high pressure. *American Mineralogist*, *69*, 176–185.

Wentzcovitch, R. M., Justo, J. F., Wu, Z., da Silca, C. R. S., Yuen, D. A., & Kohlstedt, D. (2009). Anomalous compressibility of ferropericlase throughout the iron spin cross-over. *Proceedings of the National Academy of Sciences of the U S A 106*, 8447–8452. doi: 10.1073/pnas.0812150106

Weyer, S., Anbar, A., Brey, G., Münker, C., Mezger, K., & Woodland, A. B. (2005). Iron isotope fractionation during planetary differentiation. *Earth and Planetary Science Letters, 240*, 251–264. doi: 10.1016/j.epsl.2005.09.023

Weyer, S., & Ionov, D. A. (2007). Partial melting and melt percolation in the mantle: The message from Fe isotopes. *Earth and Planetary Science Letters, 259*, 119–133. doi: 10.1016/j.epsl.2007.04.033

Wilke, M. (2005). Fe in magma – An overview. *Annals of Geophysics, 48*, 609–617.

Wilke, M., Farges, F., Partzsch, G. M., Schmidt, C., & Behrens, H. (2007). Speciation of Fe in silicate glasses and melts by in-situ XANES spectroscopy. *American Mineralogist, 92*, 44–56. doi: 10.2138/am.2007.1976

Williams, H. M., & Bizimis, M. (2014). Iron isotope tracing of mantle heterogeneity within the source regions of oceanic basalts. *Earth and Planetary Science Letters, 404*, 396–407. doi: 10.1016/j.epsl.2014.07.033

Williams, H. M., McCammon, C. A., Peslier, A. H., Halliday, A. N., Teutsch, N., Levasseur, S., & Burg, J.-P. (2004). Iron isotope fractionation and the oxygen fugacity of the mantle. *Science, 304*, 1656–1659. doi: 10.1126/science.1095679

Williams, H. M., Nielsen, S. G., Renac, C., Griffin, W. L., O'Reilly, S. Y., McCammon, C. A., et al. (2009). Fractionation of oxygen and iron isotopes by partial melting processes: Implications for the interpretation of stable isotope signatures in mafic rocks. *Earth and Planetary Science Letters, 283*, 156–166. doi: 10.1016/j.epsl.2009.04.011

Williams, H. M., Peslier, A. H., McCammon, C., Halliday, A. N., Levasseur, S., Teutsch, N., & Burg, J.-P. (2005). Systematic iron isotope variations in mantle rocks and minerals: The effects of partial melting and oxygen fugacity. *Earth and Planetary Science Letters, 235*, 435–452.

Williams, H. M., Prytulak, J., Woodhead, J. D., Kelley, K. A., Brounce, M., & Plank, T. (2018). Interplay of crystal fractionation, sulfide saturation and oxygen fugacity on the iron isotope composition of arc lavas: An example from the Marianas. *Geochimica et Cosmochimica Acta, 226*, 224–243. doi: 10.1016/j.gca.2018.02.008

Williams, H. M., Wood, B. J., Wade, J., Frost, D. J., & Tuff, J. (2012). Isotopic evidence for internal oxidation of the Earth's mantle during accretion. *Earth and Planetary Science Letters, 321–322*, 54–63. doi: 10.1016/j.epsl.2011.12.030

Woodhead, J., Eggins, S., & Gamble, J. (1993). High field strength and transition element systematics in island arc and back-arc basin basalts: evidence for multi-phase melt extraction and a depleted mantle wedge. *Earth and Planetary Science Letters, 114*, 491–504. https://doi.org/10.1016/0012-821X(93)90078-N

Woodland, A. B., Kornprobst, J., & Wood, B. J. (1992). Oxygen thermobarometry of orogenic lherzolite massifs. *Journal of Petrology, 33*, 203–230. doi: 10.1093/petrology/33.1.203

Wu, C. C., & Mason, T. O. (1981). Thermopower measurement of cation distribution in magnetite. *Journal of the American Ceramic Society, 64*, 520–522. doi: 10.1111/j.1151-2916.1981.tb10317.x

Wu, H., He, Y., Bao, L., Zhu, C., & Li, S. (2017). Mineral composition control on inter-mineral iron isotopic fractionation in granitoids. *Geochimica et Cosmochimica Acta, 198*, 208–217. doi: 10.1016/j.gca.2016.11.008

Xia, Y., Li, S., & Huang, F. (2017). Iron and Zinc isotope fractionation during magmatism in the continental crust: Evidence from bimodal volcanic rocks from Hailar basin, NE China. *Geochimica et Cosmochimica Acta. 213*, 35–46. doi: 10.1016/j.gca.2017.06.018

Yang, H., Lin, J.-F., Hu, M. Y., Roskosz, M., Bi, W., Zhao, J., et al. (2019). Iron isotopic fractionation in mineral phases from Earth's lower mantle: Did terrestrial magma ocean crystallization fractionate iron isotopes? *Earth and Planetary Science Letters, 506*, 113–122. doi: 10.1016/J.EPSL.2018.10.034

Young, E. D., Manning, C. E., Schauble, E. A., Shahar, A., Macris, C. A., Lazar, C., & Jordan, M. (2015). High-temperature equilibrium isotope fractionation of non-traditional stable isotopes: Experiments, theory, and applications. *Chemical Geology, 395*, 176–195. doi: 10.1016/j.chemgeo.2014.12.013

Young, E. D., Tonui, E., Manning, C. E., Schauble, E., & Macris, C. A. (2009). Spinel–olivine magnesium isotope thermometry in the mantle and implications for the Mg isotopic composition of Earth. *Earth and Planetary Science Letters, 288*, 524–533.

Zhang, H. L., Hirschmann, M. M., Cottrell, E., & Withers, A. C. (2017). Effect of pressure on $Fe^{3+}/\Sigma Fe$ ratio in a mafic magma and consequences for magma ocean redox gradients. *Geochimica et Cosmochimica Acta, 204*, 83–103. doi: 10.1016/J.GCA.2017.01.023

Zhang, L., Sun, W., Zhang, Z., An, Y., & Liu, F. (2019). Iron isotopic composition of supra-subduction zone ophiolitic peridotites from northern Tibet. *Geochimica et Cosmochimica Acta, 258*, 274–289. doi: 10.1016/J.GCA.2019.04.031

Zhao, X., Zhang, H., Zhu, X., Tang, S., & Tang, Y. (2010). Iron isotope variations in spinel peridotite xenoliths from North China Craton: Implications for mantle metasomatism. *Contributions to Mineralogy and Petrology, 160*, 1–14. doi: 10.1007/s00410-009-0461-y

Zhao, X., Zhang, H., Zhu, X., Tang, S., & Yan, B. (2012). Iron isotope evidence for multistage melt–peridotite interactions in the lithospheric mantle of eastern China. *Chemical Geology, 292–293*, 127–139. doi: 10.1016/j.chemgeo.2011.11.016

Zhao, X., Zhang, Z., Huang, S., Liu, Y., Li, X., & Zhang, H. (2017). Coupled extremely light Ca and Fe isotopes in peridotites. *Geochimica et Cosmochimica Acta, 208*, 368–380. doi: 10.1016/j.gca.2017.03.024

Zhu, X. K., Guo, Y., O'Nions, R. K., Young, E. D., & Ash, R. D. (2001). Isotopic homogeneity of iron in the early solar nebula. *Nature, 412*, 311–313. doi: 10.1038/35085525

16

Zinc and Copper Isotopes as Tracers of Redox Processes

Edward C. Inglis and Frédéric Moynier

ABSTRACT

The study of natural isotopic variations of Zn and Cu has been ongoing for the past 30 years. The coupling of isotopic data with theory has improved our understanding of these two systems with respect to magmatic and redox processes. In this chapter we provide a review of this relative new field, including analytical work and theoretical isotopic fractionation calculations. As a multivalent element in nature, stable isotope theory predicts that Cu isotopes should respond to changes in oxidation state. We show that while the application of Cu isotopes as a source tracer and exploration tool has been limited, Cu isotopes can serve as good tracers of redox processes during the aqueous secondary alteration of Cu mineral deposits. Zinc is not multivalent in nature and thus changes in redox have no effect on its isotopic composition in a geological context. However, because of isotopic fractionation during C and S complexation, Zn isotopes have been used as tracers of the mobility of aqueous S and C in a number of different settings. In particular, they have yielded powerful constraints on the cycling of oxidizing fluids in subduction zones.

16.1. INTRODUCTION

The geochemistry of non-traditional stable isotopes has emerged in recent years as a powerful tool to trace different geochemical processes (e.g., Johnson et al., 2004; Teng et al., 2017). To this end, both low- and high-temperature natural systems have been explored, and the understanding of processes occurring in these domains has been greatly advanced owing to this novel toolset. Indeed, the field of stable isotope geochemistry has seen significant advancement over the last twenty years, building on the pioneering work of Alfred Nier in the mid 1900s and the determination of light or traditional stable isotopes of O, C, N, and S (Nier, 1950, 1947, 1940; Nier & Gulbransen, 1939). More recently the isotopes of heavier mass elements, termed non-traditional stable isotopes, have received much attention from geochemists seeking to understand both high- and low-temperature natural processes (Figure 16.1). This rapid expansion of the field of stable isotope geochemistry has seen the routine analysis of different major and trace element isotopes, many of which display uniquely different geochemical properties. Two such elements are those of copper (Cu) and zinc (Zn).

Copper, a group 11 d block element, is multivalent in nature, and exhibits three oxidation states, with the electron configurations Cu^0 ([Ar]$3d^{10}4s^1$); Cu^+ ([Ar]$3d^{10}$); and Cu^{2+} ([Ar]$3d^9$). It has two naturally occurring stable isotopes: ^{63}Cu (69.17%) and ^{65}Cu (30.83%), displays both siderophile and strongly chalcophile behaviour (Siebert et al., 2011) and is moderately volatile (T_C=1037 K [Lodders, 2003]; [T_C = 50% condensation temperature]). Zinc neighbors Cu as a period 4 element and is

Université de Paris, Institut de Physique du Globe de Paris, Paris, France

Magma Redox Geochemistry, Geophysical Monograph 266, First Edition.
Edited by Roberto Moretti and Daniel R. Neuville.

DOI: 10.1002/9781119473206.ch16

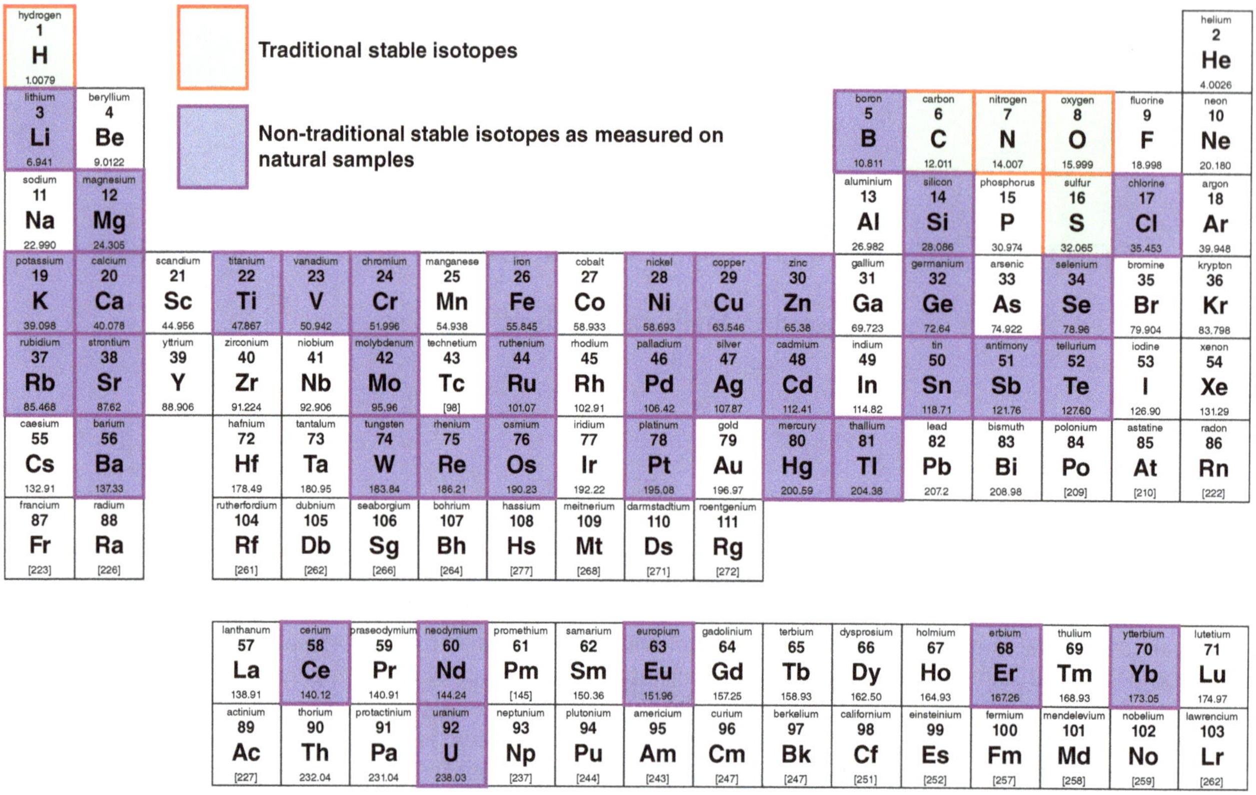

Figure 16.1 A periodic stable showing traditional and selected non-traditional stable isotope systems that have been applied to different geochemical studies. Copper and Zn occur as block d, transition metals, with two and five stable isotopes, respectively.

located in group 11 of the d block. It possesses five stable isotopes: ^{64}Zn (49.17%), ^{66}Zn (27.43%), ^{67}Zn (4.04%), ^{68}Zn (18.45%), and ^{70}Zn (0.61%), and is found naturally either as Zn(0) or Zn(II), with the respective electron configurations of Zn^0 ([Ar]$3d^{10}4s^2$) and Zn^{2+} ([Ar]$3d^{10}$). Zinc displays broadly similar geochemical behaviour to Cu, being both chalcophile and slightly siderophile in nature (Mahan et al., 2017; Siebert et al., 2011). One of the most significant characteristics of Zn is its moderately volatile behaviour. Zinc displays a T_C of 726 K (Lodders, 2003) and represents one of the most abundant moderately volatile elements in silicate reservoirs (Moynier et al., 2017).

The stable isotope composition of a given substance is controlled by two main factors: (i) the isotopic composition of the parent substance, i.e., its inherited composition; and (ii) chemical processes that can bias or fractionate the isotopes of the given element. To best illustrate this, we can consider an example of mantle-derived basalt. Partial melting of a mantle peridotite forms the basaltic melt, and it is the isotopic composition of the peridotite source that exerts a first-order control on the isotopic composition of the resultant melt. It is then the role of different processes that can control the isotopic composition of the final basaltic rock, such as the fractionation of isotopes between basaltic melt and peridotitic residue during partial melting and extraction and the given isotopic fractionation factor between the melt and different constituent minerals of the basaltic rock. The redox state of the system has a fundamental control of how these processes can proceed. As such, it is apparent that the stable isotope composition of different materials provides a powerful insight into both source and process that has influenced their evolution.

When considering different processes that may fractionate the stable isotopes of a given material, a simple set of governing qualitative principals can be considered. These are best described by Schauble (2004) and are summarized here:

1. Smaller equilibrium isotopic fractionations are expected at higher temperatures. The magnitude of such fractionations is roughly proportional to $1/T^2$.

2. Typically, fractionations are greatest for the lighter elements and isotopes of very different mass, when all else is equal.

3. At equilibrium, the heavy isotopes of a given element will be concentrated in substances where that given element forms the stiffest bonds. The magnitude of the resulting isotopic fractionation will be roughly proportional to the

differences in bond stiffness between the equilibrated substances. The shortest and strongest (thus stiffest) bonds are commonly associated with features such as high oxidation states, bonding with elements toward the top of the periodic table, and low coordination numbers. It should be noted that when the element of interest is directly bonded to a hydrogen atom, or is part of a low mass molecule, then the expected enrichment in heavy isotopes might not be apparent.

4. An exception to these rules is the nuclear field shift effect. This effect sees an isotopic shift in atomic energy levels due to the difference in nuclear size and shape and electron densities at the nucleus (Bigeleisen, 1996). This effect is dominant for heavier elements and is responsible for the isotopic variations of elements such as Tl or U (Fujii et al., 2009; Moynier et al., 2009; Schauble, 2007) and could be the source of small non-mass dependent isotopic fractionations (Fujii et al., 2006; Moynier et al., 2009).

The application of such principles, combined with the rich array of differing geochemical properties exhibited by elements with two or more stable isotopes, make for a powerful tool in the earth and planetary sciences. Of particular note for the purpose of this review are the stable isotopes systems of Cu and Zn. Both systems have received growing attention over the past twenty years as tracers of a vast array of different geological processes (e.g., Moynier et al., 2017). This review will concentrate on the systematics of these two isotope systems with respect redox processes occurring within high temperature settings.

16.2. THE DETERMINATION OF CU AND ZN ISOTOPE RATIOS

Much of the rapid expansion of non-traditional stable isotope geochemistry can largely be attributed to advances in mass spectrometric techniques. Early mass spectrometers relied upon the introduction of samples in a gas source chamber (GS) or solid source thermal ionization (TIMS). Given the high energies required to efficiently ionize many elements of the periodic table, the application of such mass spectrometers is limited. It was not until the advent of plasma source, multiple collector mass spectrometers (MC-ICPMS) in the 1990s that the field of non-traditional stable isotope geochemistry saw rapid expansion.

Because high temperature isotopic fraction of heavy elements is predicted to be relatively small, on the order of a few parts per thousand (e.g., Schauble, 2004), the determination of stable isotope ratios to high precision is necessary. One of the first comprehensive studies to present a method for determining Cu and Zn isotopes ratios at precisions appropriate to resolve variation in natural, high-temperature samples (< 40ppm at the 95% confidence interval) was that of Maréchal et al. (1999). Since this time numerous scientific papers have been published presenting Cu and Zn stable isotope data for a range of different high-temperature samples, many of which routinely measured with reproducibility better than 30 ppm (2 *sd*) by MC-ICPMS.

The determination of Cu and Zn isotope ratios from natural samples first requires the isolation of the target element or elements (i.e., Cu and/or Zn) from the sample matrix. The purpose for this is twofold – firstly it isolates any spectral interferences that may occupy the same nominal mass space as the analyte isotopes. Secondly, it reduces the effect of matrix-related mass discrimination that may be induced when the sample solution is introduced to the mass spectrometer. This is of particular importance when using sample-standard bracketing to correct for instrumental mass bias. To achieve the isolation of Cu and Zn from sample matrices it is preferable to use a method of anion exchange chromatography. Several different chromatographic exchange protocols have been suggested by previous workers. In the pioneering studies of Marechal and Albarède (2002) and Maréchal et al. (1999) the use of strongly basic, macro-porous anion exchange resin (AG MP-1) combined with HCl was suggested to effectively isolate Cu, Fe, and Zn from different sample matrices. More recently the use of bead-form anion exchange resin (AG1-X8) and HCl/HNO_3 reagents have been suggested for the combined separation of Cu, Fe, and Zn (Sossi et al., 2015). Although these methods allow the purification of Cu, Fe, and Zn from a single column the sole isolation of Zn from matrix elements is also possible using AG1-X8 and dilute HBr/HNO_3 without a risk of saturation of the column (Moynier & Le Borgne, 2015).

As previously stated, the vast majority of Zn and Cu isotope data present in the literature have been obtained by MC-ICPMS instruments. One of the major analytical issues encountered during the determination of stable isotope ratios by plasma source machines is that of mass discrimination. Mason et al. (2004) suggest that mass discrimination can be considered with respect to two different aspects: (i) instrumental mass bias; and (ii) matrix effects owing to differences in the sample composition. The mass bias associated with MC-ICPMS instruments is best described by the exponential law (Belshaw et al., 1998), and several workers have specifically suggested that exponential mass fractionation law best describes the MC-ICPMS related mass discrimination of Zn and Cu isotopes (Archer & Vance, 2004; Maréchal et al., 1999; Mason et al., 2004). For Zn isotopes this can be written as:

$$\left(\frac{^{66}Zn}{^{64}Zn}\right)_{corrected} = \left(\frac{^{66}Zn}{^{64}Zn}\right)_{measured} \times \left(\frac{65.926\ AMU}{63.929\ AMU}\right)^{\beta} \tag{16.1}$$

Where β equals the fractionation factor and 65.926 AMU and 63.929 AMU represent the absolute mass of ^{66}Zn and ^{64}Zn in atomic mass units. The same applies for Cu isotopes but the ratio $^{66}Zn/^{64}Zn$ is replaced by $^{65}Cu/^{63}Cu$ and absolute masses of ^{65}Cu and ^{63}Cu replace those of ^{66}Zn and ^{64}Zn. The fractionation factor, β, for the ratio $^{66}Zn/^{64}Zn$ can thus be derived by rearranging Equation 16.1.

Because the instrumental fractionation of stable isotopes presents a major problem when trying to investigate natural mass dependent variations, it is thus necessary to correct for this problem of mass discrimination. Commonly there are three different methods employed to correct for isotopic fractionation occurring during collection of stable isotope data by plasma source instruments: (i) standard-sample bracketing; (ii) empirical external normalization; and (iii) double-spiking.

Standard-sample bracketing (SSB) has been adopted as a method for the correction of instrumental mass bias during the analysis of stable isotopes for a number of years (e.g., Zhu et al., 2000). This type of correction involves measuring a known bracketing standard before and after analyses of individual unknowns. Assuming the instrumental mass bias is linear and constant between individual measurements, it can simply be extrapolated between points and used to correct the sample. The main drawbacks with this technique are that matrix effects remain uncorrected and the time required per analysis can be long, as for each measurement of an unknown two measurements of the bracketing standard are required.

Another method commonly utilized for the correction of instrumental mass discrimination that of empirical external normalization. Most analysts combine both SSB and external normalization methods to provide a robust means for correcting instrumental mass discrimination effects when determining Zn or Cu isotope ratios (e.g., Archer & Vance, 2004; Maréchal et al., 1999; Mason et al., 2004). Performing an external normalization to correct for mass bias effects is most commonly achieved by doping the sample unknown with a standard spike of similar elemental mass. For Zn this has been achieved using Cu, while for Cu either Zn or Ni have been suggested (Larner & Rehkämper, 2012; Maréchal et al., 1999; Mason et al., 2004). Originally it was assumed that, because the dopant elements used for normalization are of a similar mass to the analyte element, the fractionation factors (β) would be the same ($\beta^{Cu} = \beta^{Zn}$). If this were to have been true then a straightforward internal normalization could have been applied (e.g., Woodhead, 2002). Unfortunately, Maréchal et al. (1999) demonstrated differing ionization behaviour between Cu and Zn; as such it can be taken that $\beta^{Cu} \neq \beta^{Zn}$. They were able to suggest that as such mass bias follows an exponential law then the below equation best describes, and allows one to correct for, such instrumental mass discrimination:

$$\frac{\ln\left(\frac{^{66}Zn}{^{64}Zn}\right) Measured - \ln\left(\frac{^{66}Zn}{^{64}Zn}\right) True}{\ln\left(\frac{^{65}Cu}{^{63}Cu}\right) Measured - \ln\left(\frac{^{65}Cu}{^{63}Cu}\right) True} = \frac{\beta Zn}{\beta Cu}\,\frac{\left(\frac{M_{66}}{M_{64}}\right)}{\left(\frac{65.926\ AMU}{63.929\ AMU}\right)} \tag{16.2}$$

In turn, by plotting the natural logarithm of $^{66}Zn/^{64}Zn_{measured}$ against the natural logarithm of $^{65}Cu/^{63}Cu_{measured}$ the respective β values for Cu and Zn can be determined and used to calculate a $^{66}Zn/^{64}Zn_{corrected}$ or $^{65}Cu/^{63}Cu_{corrected}$. When a ratio corrected for instrumental mass discrimination effects is successfully determined it can be used to calculate a final delta value as per Equation 16.3 for Zn:

$$\delta^{66}Zn = \left(\frac{\left(\frac{^{66}Zn}{^{64}Zn}\right)_{sample}}{\left(\frac{^{66}Zn}{^{64}Zn}\right)_{JMC-Lyon}} - 1\right) \times 1000 \tag{16.3}$$

or Equation 16.4 for Cu:

$$\delta^{65}Cu = \left(\frac{\left(\frac{^{65}Cu}{^{63}Cu}\right)_{sample}}{\left(\frac{^{65}Cu}{^{63}Cu}\right)_{NIST\ SRM\ 976}} - 1\right) \times 1000 \tag{16.4}$$

It should be noted that owing to an exhausted supply of both NIST SRM 976 and JMC-Lyon standard solutions, it is not possible for most analysts to measure their unknowns relative to these standards during the course of an analytical session. As such, it has become commonplace to select an alternative standard solution that has previously been calibrated relative to these standards. This allows a simple offset correction to be applied to the calculated delta value to present it as being relative to JMC-Lyon or NIST SRM 976 (Moynier et al., 2017).

Whilst it is most common to correct for instrumental mass discrimination effects using the combined SSB and external normalization method as described here, some studies have presented data obtained using a Zn double spike (e.g., Arnold et al., 2010; Bermin et al., 2006; Conway et al., 2013). The background to the double spike technique for correcting mass discrimination effects is best explained by Rudge et al. (2009). The double spike technique has been successfully applied to numerous different stable isotope systems over the past two decades, and had proved a reliable and often ultra-precise method for correcting mass discrimination (e.g., Creech et al., 2017; Inglis et al., 2018; McCoy-West et al., 2017; Millet et al., 2012; Millet & Dauphas, 2014; Nanne et al., 2017; Ripperger & Rehkämper, 2007; Siebert et al., 2001). One of the key advantages to using a double spike to correct for instrumental mass bias is that it is capable of handling complex sample matrices and a low abundance of analyte elements (Arnold et al., 2010; Conway et al., 2013). Furthermore, providing that samples are spiked and correctly equilibrated before processing, it is also possible to correct for any isotopic fractionation that might occur during sample preparation. Despite this, only a few labs routinely use a double spike for Zn isotope measurements. This is largely because for many samples analyzed for Zn isotopes the correction of mass discrimination by SSB and external normalization is easier, more cost effective, and a perfectly adequate means for generating precise data. For example, the long-term reproducibility of the IRMM-3702 standard measured using a Thermo Scientific Neptune Plus MC-ICPMS at Durham University using a combined external normalization by Cu doping and SSB correction yielded an average $\delta^{66}Zn_{JMC\text{-}Lyon}$ of 0.28±0.043 ‰ (2SD; n = 102) (Inglis, 2018). For comparison the ETH-Zürich group report a long-term reproducibility of IRMM-3702 when measured using a Neptune instrument of $\delta^{66}Zn_{JMC\text{-}Lyon}$ of 0.300 ± 0.058 ‰ (2SD; n = 163) (Moynier et al., 2017). Consequently, it is apparent that the same magnitude of precision can be achieved by using SSB + external normalization as it can by double spike. It should be noted that because Cu only has two isotopes, mass bias correction by double spike it not possible.

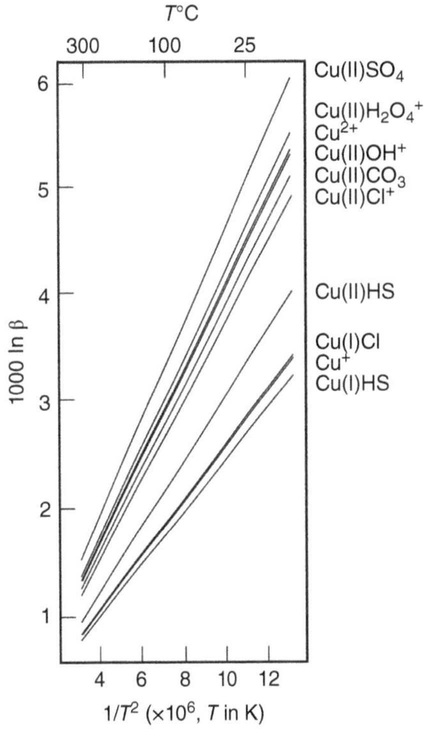

Figure 16.2 The temperature dependent of ln β or isotopic enrichment factor, between different Cu species in aqueous solutions. Figure is taken from Moynier & Fujii (2017), with data from Sherman (2013) and Fujii et al. (2013).

16.3. THEORETICAL AND EXPERIMENTAL CONSTRAINTS ON CU AND ZN ISOTOPE BEHAVIOR IN RELATION TO REDOX PROCESSES

One of the key theoretical principles in the field of stable isotope geochemistry is that changes in oxidation state are suggested to trigger isotopic fractionation of redox-sensitive elements (Schauble, 2004). This principle is well documented for other multivalent stable isotope systems, such as Fe (e.g., Dauphas et al., 2009, 2014; Hibbert et al., 2012; Polyakov & Mineev, 2000; Rouxel et al., 2005; Williams et al., 2004, 2005) and Mo (e.g., Arnold et al., 2004; Barling et al., 2001; Siebert et al., 2003), but has received relatively little attention for Cu.

Copper most commonly exists in nature in either Cu^+ or Cu^{2+} forms and, as such, there is the potential for Cu stable isotopes to serve as a tracer of redox processes. A significant contribution to the understanding of Cu isotope behaviour in nature has come from ab initio methods (e.g., Fujii et al., 2013, 2014; Moynier et al., 2017; Télouk et al., 2015; Sherman, 2013). Such studies have served to calculate a range of different isotope enrichment factors between different compounds under differing physiochemical conditions. This is best summarized in Figure 16.2, which shows the predicted isotopic enrichment factors (ln β) for different Cu species and compounds in aqueous solutions. While it is clear that distinct isotopic enrichments between Cu^{2+} and Cu^+ bearing aqueous species exists, it is apparent that the scale of such enrichments decreases drastically over the range of temperatures plotted (0–300 °C). It is important to highlight the relationship between equilibrium fractionation and $1/T^2$. Across the range of temperatures related to hydrothermal systems this predicted redox fractionation would likely be small yet observable, but at liquidus temperatures relevant to magmatic systems it is predicted that this effect would be much smaller.

In support of such theoretical constraints on Cu isotope behaviour in response to changes in redox are contributions from the experimental community. The early study of Maréchal and Albarède (2002) examined fractionation of Zn and Cu isotopes during the process of anion-exchange chromatography at ambient temperatures. Their results demonstrated that both the fractionation of Zn and Cu isotopes were the result of equilibrium effect and that their data did not suggest that the isotopic fraction observed could be ascribed to changes in oxidation states. The first study to comprehensively examine Cu isotope fractionation as a function of redox was that of

Zhu et al. (2002). As part of their work, they conducted a simple Cu redox precipitation reaction between aqueous Cu^{2+} and reduced Cu^+ precipitate. Their findings were significant as they showed, for the first time, that precipitation of a reduced Cu^+ species from Cu^{2+} solution was accompanied by isotopic fractionation. From their experimental data they calculated a fractionation factor between Cu^{2+} and Cu^+ of $\alpha_{Cu(II)\text{-}Cu(I)} = 1.00403 \pm 0.00004$. The key results from their study are summarized in Figure 16.3, which shows a Rayleigh fractionation model, based on their experimental data, for the evolution of $\delta^{65}Cu$ during the precipitation of Cu^+ species from a Cu^{2+} solution. Further experiments by Ehrlich et al. (2004) built on these early observations of Zhu et al. (2002) and demonstrated that the precipitation of Cu^{2+}-bearing phases from an aqueous Cu^+ solution was manifest with relatively small fractionations of Cu isotopes, ultimately confirming earlier suggestions that it was changes in redox that triggered Cu isotope fractionation, and not a simple mineralogical control.

Unlike Cu, Zn typically only occurs in one oxidation state (Zn^{2+}) in nature, and although large isotopic variations have been documented during electroplating experiments and the reduction of Zn^{2+} to Zn metal (Kavner et al., 2008), such redox changes are not thermodynamically favorable in the natural realm, and Zn should be considered as a monovalent element in a geological context. As such Zn isotopes will not undergo redox driven

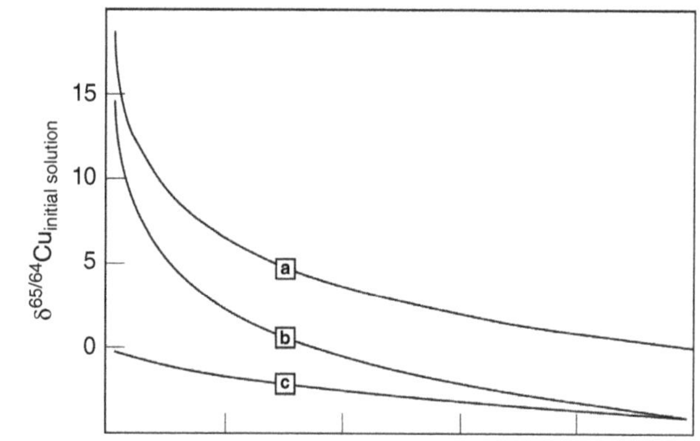

Figure 16.3 A Rayleigh fractionation model summarizing the key experimental findings of Zhu et al., 2002. Curve a shows the evolution of Cu^{2+} in the solution, while curve b shows the instantaneous variation of $\delta^{65}Cu$ within the Cu(I) solid precipitated from the Cu^{2+} solution. The curve c shows the evolution of $\delta^{65}Cu$ within the precipitate owing to accumulation. The initial starting $\delta^{65}Cu$ value of the solution was 0. The $\delta^{65}Cu$ is expressed here as the $^{65}Cu/^{64}Cu$ of the samples relative to the same ratio within the initial starting solution.

fractionations. Despite this, contributions from several ab initio and other calculative studies have predicted that Zn isotopes could in fact serve as useful tracers of the mobility of different redox-sensitive elements in nature. Such studies have suggested that strong isotopic variations are imparted when Zn is complexed with aqueous sulfate and carbonate ligands, and consequently that Zn isotopes could serve as a powerful tracer of aqueous Zn-S and Zn-C complexes (Black et al., 2011; Fujii et al., 2010, 2011; Fujii & Albarède, 2012). In Figure 16.4 theoretical isotopic enrichment factors (ln β) between different Zn-S and Zn-C complexes as a function of temperature are presented. These calculations clearly demonstrate the effect of redox-sensitive Zn-S and Zn-C complexes in aqueous solution.

Experimental studies aimed at the testing of such theoretical predictions are sparse. Consequently, no controlled ground-truthing of the sensitivity of Zn isotopes to complexation by S and C ligands is available. Nevertheless, several natural studies have examined the effect of S and C ligands on Zn isotopes and are discussed in Section 16.4.

16.4. APPLICATION OF CU AND ZN TO TRACE REDOX PROCESSES IN NATURAL SYSTEMS

16.4.1. Hydrothermal Systems

Hydrothermal deposits were amongst some of the first natural samples to be analyzed for isotope ratios of Cu and Zn (e.g., Graham et al., 2004; Larson et al., 2003; Markl et al., 2006; Mason et al., 2005; Rouxel et al., 2004a; Zhu et al., 2000) and represent some of the largest variations observed in terrestrial geological materials, e.g., -16.5 to +9.98 ‰ (Albarède, 2004; Moynier et al., 2017). Because of the obvious association, hydrothermal deposits also serve as a natural laboratory to examine the effects of redox on these isotope systems.

Early Studies. The first documented study of Cu ore minerals was presented by Zhu et al. (2000). As part of their work, they measured the Cu isotope composition of different Cu-sulfides, Cu-carbonates, and native Cu metal from different terrestrial and marine mineralizing systems. They demonstrated that magmatic chalcopyrite displays limited variation of Cu isotopes (-0.62 to 0.40), while seafloor hydrothermal chalcopyrite displays a significant range in $\delta^{65}Cu_{NIST976}$ values of between -0.49 to 1.15. Based on their findings they concluded that the variation observed in seafloor hydrothermal sulfides could be tentatively explain by the leaching of isotopically heavy Cu by hydrothermal processes and a subsequent isotopic exchange between early formed Cu-phases and

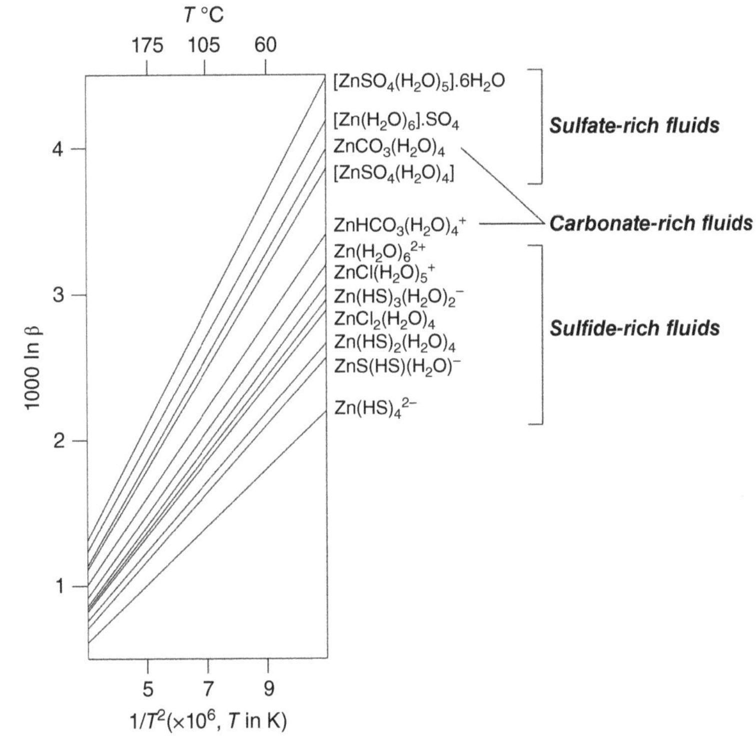

Figure 16.4 Temperature dependence of ln β between different Zn-S and Zn-C species in aqueous solution. Figure is taken from Pons et al. (2011) with the data for the Zn sulfate rich fluids species taken from Black et al. (2011) and the data for the Zn sulphide- and Zn carbonate-rich fluids taken from Fujii et al. (2011).

isotopically light late-stage hydrothermal fluids. Building on this work Jiang et al. (2002) analyzed Cu ore minerals formed at the Jinman low-temperature (150–300 °C) sediment hosted hydrothermal Cu deposit, again demonstrating a wide range of $\delta^{65}Cu$ values for Cu-sulfides of between $\delta^{65}Cu_{NIST976}$ = -3.70 – +0.30 ‰. Their data suggested that, unlike seafloor hydrothermal and terrestrial magmatic Cu-sulfides, the Cu minerals formed in the Jinman system are offset toward lighter values. They suggested a number of explanations for these values: (i) Based on previous fluid inclusion work the temperature of the different ore-forming fluids was of a relatively low temperature (< 289 °C) compared to many seafloor hydrothermal systems, meaning that deposition temperature could have exerted a control on Cu isotope partitioning. (ii) A possible source control owing to the fact that ore metal was sourced from sediments at Jinman rather than igneous rocks as concluded by Zhu et al. (2000). (iii) Different ore-forming stages, similar to the model proposed by Zhu et al. (2000), whereby ore forming fluids are progressively isotopically distilled leaving the first precipitates fractionated from the later stage precipitates. (iv) The relationship between ore genesis and the involvement of organic matter – because the role of biological fractionation has been observed for other stable isotopes (e.g., Fe), and it was known that organic matter was involved within the proposed mineralization model for the Jinman deposit. Further work by Graham et al. (2004) focused on Cu isotope ratios in the Grasberg Cu-Au porphyry deposit. The range in Cu isotope values that they observed was complex throughout the deposit. They documented a spatial and temporal evolution of Cu isotopes between successive intrusions, with the youngest most enriched in heavy Cu relative to the eldest. They sought to explain this observed variation either by successive isotopic distillation of Cu from an underlying source and the establishment of hydrothermal cells in each intrusion, or the successive distillation of Cu isotopes from the ore fluid as it migrated across the deposit. In external skarn deposits they suggested the Cu isotope behavior could be controlled by variations in metal sources and/or different mineralization events.

Although they targeted Cu isotope systematics relating to hydrothermal systems, these aforementioned studies did not specifically address the relationship between Cu isotope ratios and redox state. The first studies to examine these effects were Larson et al. (2003) and Rouxel et al. (2004b). Both of these studies systematically dissected different magmatic and hydrothermal mineralizing systems associated with Cu ore minerals to examine the behavior of Cu isotopes. Both studies documented similar observations; firstly, they showed that providing that the Cu is sourced from a single reservoir, the range of Cu isotope values for primary Cu sulfides in both magmatic and hydrothermal systems was limited, typically ranging between -0.2 to 0.15‰. Secondly, they showed that secondary Cu ore minerals formed by later-stage oxidative weathering, alteration, and redistribution of primary Cu sulfides displayed highly variable δ^{65}Cu values with a range of up to 9‰. Larson et al. (2003) were the first to note that in the supergene environment there was a systematic difference between mineral pairs for phases that incorporate Cu in different oxidation states, i.e., native copper and cuprite – a finding that was later confirmed by Mason et al. (2005) who demonstrated a large fractionation associated with the reduction of Cu(II) to Cu(I) during supergene mineralization within the Alexandrinka volcanic hosted massive sulfide deposit in Russia. These observations were fundamental in identifying the significance of redox processes in different natural ore-forming systems, especially relating to low-temperature supergene weathering and enrichment processes. Building on these significant contributions Markl et al. (2006) presented a comprehensive study of the effect of redox processes in hydrothermal mineralizing systems. Their data supported the observations of Larson et al. (2003), Mason et al. (2005), and Rouxel et al. (2004b), in that even despite various Cu sources and several mineralizing events, the $\delta^{65}Cu_{NIST976}$ values for primary Cu(I) ore minerals from the Schwarzwald mining district (Germany) were extremely restricted; in most cases being 0±0.5 ‰, while secondary Cu(II)-bearing phases displayed a significant range in $\delta^{65}Cu_{NIST976}$ values of between -1.55 to 2.41‰, but typically all having values 0.5‰. From their data they were able to conclude that generally high temperature hydrothermal processes and primary mineralization of Cu ore do not exert significant control on the Cu isotope systematics of such deposits. Instead, they concluded that much of the observed Cu isotope variation resulted from the later-stage, low-temperature redox-related processes, which enrich the secondary alteration assemblage in ^{65}Cu relative to the primary Cu phase. Based on these findings they ultimately suggested that it is redox-controlled dissolution/precipitation processes that control the fractionation of Cu isotopes in this setting.

These pioneering early papers paved the way for multiple studies of Cu isotopes within hydrothermal systems. Several of these later studies have tested the effect of redox, specifically addressing the fractionation of Cu isotopes between reduced and oxidized copper minerals. For example Asael et al. (2007) studied the Timna Valley stratiform sedimentary hosted copper deposit in Israel. They demonstrated that reduced Cu sulfides were isotopically lighter ($\delta^{65}Cu_{NIST976}$ = -3.4 to -1.2‰) than coexisiting Cu(II) carbonates and hydroxides ($\delta^{65}Cu_{NIST976}$ = -1.2 to 0.5‰), thus confirming earlier suggestions of an enrichment in the light isotope during the reduction of $Cu(II)_{aq}$ and the precipitation of Cu sulfides. They further concluded that the later in-situ oxidation of the Cu sulfides is associated with a heavy isotope enrichement during the growth of Cu(II) phases, with this isotopic variation between Cu(I) and Cu(II) minerals being controlled by reservoir effect within an open system.

Porphyry Systems. The use of Cu isotopes was tested as an exploration tool within Cu porphyries by Mathur et al. (2009) and suggested that distinct isotopic signatures exist between different portions of Cu porphyries (e.g., high-temperature hypogene, enrichment, and leach cap phases). Furthermore, they expanded on the findings of Markl et al. (2006) to factor mass balance consideration to the redox fractionation between Cu(I) and Cu(II) mineral pairs given Cu transport within the porphyry. Later work by Li et al. (2010) presented a detailed dissection of the large Northparkes orthomagmatic Cu-Au porphyry system in SE Australia. Specifically, they analyzed the Cu isotope composition of sulfide minerals from different alteration and primary zones within four drill cores in two separate ore bodies from the deposit. They found a similar result for all the drill cores, which is typified by a down-hole decrease in $\delta^{65}Cu_{NIST976}$ values from and average 0.29±0.56‰ (1σ; n=30) at the low grade phyllic-propylitic peripheral alteration zones to -0.25±0.36‰ (1σ; n=30) at the margins of the most mineralized zones. Within the primary mineralized high grade, sulfides have an average $\delta^{65}Cu_{NIST976}$ of 0.19±0.14‰ (1σ; n=40). While their main conclusion was that the Cu isotopic variation observed within the drill cores was the result of isotopic fractionation between vapor, solution, and sulfides, they also considered the effect of change of oxidation state. Much like the previous studies they suggested that because Cu is typically present as Cu(I) in hydrothermal fluids and primary sulfides then limited redox-driven fractionation would exist during primary high-temperature mineralization. However, changes in oxidation state could account for the variation seen within the different alteration zones of the deposit. Because the valency of Cu is a function of temperature, pH, and fO_2 then variations in these parameters, consistent with differing alteration stages, could trigger a shift between Cu(I) and Cu(II) and thus drive an isotopic fractionation. To investigate this, they used the reaction:

$$2CuCl^{+} + 2Cl^{-} + H_2O = 2CuCl_2^{-} + 2H^{+} + 0.5O_2 \quad (16.5)$$

Where CuCl+ and $CuCl_2$ are considered to be representative Cu(II) and Cu(I) species, respectively, in a low

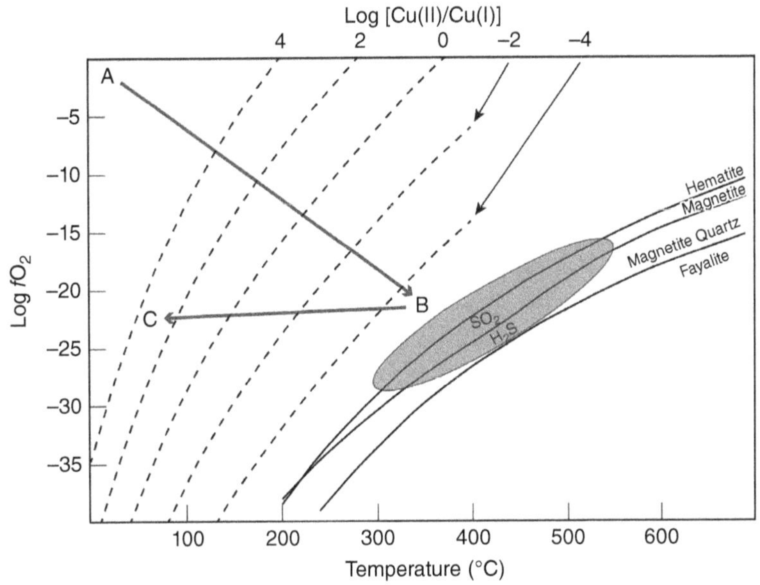

Figure 16.5 The effect of and temperature log fO_2 on the speciation of Cu within copper porphyry deposits. This model is used to explain the Cu isotope variation observed within the Northparkes orthomagmatic Cu-Au porphyry system by Li et al. (2010).

salinity hydrothermal environment. Using this equation they were able to compute a range of different Cu(II)/Cu(I) ratio for solutions of varying temperatures and fO_2 values. Their findings, summarized in Figure 16.5, demonstrated that under the physicochemical conditions typical for Cu precipitation in porphyry deposits, Cu(I) was dominant, while under conditions seen within the supergene environment, Cu(II) dominated. This study was significant for determining a quantative model for ΔCu(I)-Cu(II) to explain the redox-driven Cu isotope variation within porphyry systems.

This model for redox controls on Cu isotope varability in Cu porphyries was later called into question by data obtainted for the Bingham Canyon porphyry. In a similar approach to Li et al. (2010), others dissected the core and different alteration zones of the Bingham intrustion, but unlike the results presented for Northparkes, they showed Cu isotope compositions that were lightest within the potassic core of the deposit and became progressively heavier towards the argillic alteration zones. They tentatively explained their results to reflect changes in temperature and pH over the different alteration zones of the deposit. Given the contrasting nature of the data presented by Li et al. (2010) and Dendas (2012) there was a requirement to re-evaluate the Cu isotope systematics within porphyry environments. Mathur et al. (2013) studied the Cu isotope systematics of the Pebble porphyry Cu-Au-Mo deposit in Alaska. They found a significant range in $\delta^{65}Cu_{NIST976}$ values between -2.09 to 1.11‰ within hypogene Cu sulfides, which are correlated with $\delta^{18}O$. They suggested that the fractionation of Cu isotopes between the hydrothermal fluid and primary precipitating chalcopyrite would be positive as the fluid evolves from magmatic to magmatic-meteoric in composition. As this fluid evolved to trigger advanced argillic alteration, a negative fluid-chalcopyrite fractionation could occur. They explained this variation to be a result of changes in Cu speciation during fluid evolution, with Cu likely being transported as $CuHS^0$ during sodic-potassic, potassic, and illite alteration events, while the fluids resulting in advanced argillic alteration would most likely transport Cu as $CuCl^-_2$. This change in speciation is a function of changes temperature, fO_2, and pH as the primary magmatic fluid mixes with metoric water. Ultimately, they concluded that the dominant control on Cu isotope fractionation within the porphyry environment was changes in aqueous Cu speciation as a function of temperature, pH, and fO_2.

Magmatic Systems. Although Cu and Zn isotopes have been relatively well studied in magmatic systems (e.g., Chen et al., 2013; Savage et al., 2015), their application as tracer of redox processes has received little attention. As with hydrothermal systems, Zn is effectively monovalent within magmatic systems so is not discussed here. Conversely, Cu isotopes have received limited attention with respect to oxidation state in this setting.

Li et al. (2009) studied Cu isotopes in granites from the Lachlan fold belt in SE Australia. They found a significant variation in $\delta^{65}Cu_{NIST976}$ values for the 32 granite samples they analyzed of between −0.46 to 1.51‰. They demonstrated that significant variation of $\delta^{65}Cu_{NIST976}$ values was present within S-type granites (−0.463 to 0.205‰). They hypothesized that because this variation could result from heterogenous Cu isotopic composition of the source sediments that could be impaired owing to redox processes during weathering and diagenesis of the protolith. Using this information, it could thus be possible to comment on the nature of the sedimentary source material.

The role of mantle partial melting was examined by Huang et al. (2017), who studied peridotites from the Western Alps. They found a resovable variation of $\delta^{65}Cu_{NIST976}$ values from −0.113 to 0.379‰ and suggested that fractionation was largely controlled by precipitation of sulfide minerals during S saturation of the parent melt, thus demonstrating that true redox reactions b are unlikely to fractionate Cu isotopes during mantle melting processes.

More recently the role of redox during magmatic Ni-Cu mineralization was investigated by Zhao et al. (2017). In their study they analyzed whole rocks and single sulfides from the Tulaergen magmatic Ni-Cu deposit in NW China, with the objective to test the effect of redox-driven Cu isotope fractionation during sulfide segregation from silicate melt and later internal fractionations within the segregated sulfide melt. They found that early segregated chalcopyrite was enriched in the light Cu isotope owing to charge balance effects, while late segregated chalcopyrite was isotopically heavier. The segregated component was later fractionated into Cu-rich liquid and Fe-rich cumulate phases, which would alter chalcopyrite ionic states from $Cu^{+}Fe^{3+}S^{2-}$ to $Cu^{2+}Fe^{2+}S^{2}$, resulting in an increase in $\delta^{65}Cu_{NIST976}$ values. These results demonstrate the Cu isotopes can be fractionated by oxidation changes during high-temperature mineralizing processes.

16.4.2. Tracers of Fluid Rock Reactions and Subduction Zone Mass Transfer

Recently Zn isotopes have been applied to study the mobility of oxidizing species during fluid-rock interaction and transfer of fluids in subduction zones. Because Zn isotopes are predicted as sensitive tracers of the mobility of S- and C-bearing fluids in nature (Black et al., 2011), there is the potential for Zn isotopes to indirectly elucidate redox-mediating geological processes.

Although not strictly commenting on the redox processes, one of the first studies to look at whether Zn isotopes could be utilized to remark on the chemistry and speciation of geological fluids was that of Pons et al. (2011). In their study they examined a suite of Archean serpentinites from the Isua Supercrustal Belt, demonstrating a striking similarity between these ancient serpentinites and modern counterparts from the Mariana forearc, while both being notably offset towards lighter values than those observed in modern ridge and ophiolitic serpentinites. They ultimately concluded that, based on ab-initio calculations (Fujii et al., 2011), the low $\delta^{66}Zn$ values observed in both the Isua Archean and modern Mariana serpentinites were likely the result of infiltration and reaction of the serpentinites with a carbonate-rich, high-pH fluid.

One recent application of Zn isotopes has been to trace redox processes occurring in subduction zones. The metamorphism of hydrated oceanic lithosphere during subduction leads to the destabilization of hydrous mineral phases and the transfer of fluids between the downgoing slab and the overlying mantle wedge (e.g., Bouilhol et al., 2015). Such fluids are commonly suggested to play a pivotal role in the transfer of redox-sensitive components in subduction zones (Evans & Tompkins, 2011; Kelley & Cottrell, 2009), but the exact redox mechanism is poorly understood.

Recently Zn isotopes have been drawn upon to comment on the redox chemistry of slab-derived fluids.

Pons et al. (2016) present Zn isotope data for a suite of metamorphosed and partially dehydrated slab serpentinites. In their study they demonstrate a progressive shift in the whole-rock $\delta^{66}Zn$ values of subducted serpentinites, suggesting the release of isotopically light serpentinite-derived fluids. In strong support of this they show that high-pressure antigorite and secondary olivine-bearing serpentinites, which are interpreted to reflect the high permeability fluid pathway where serpentinite derived fluids are localized, are enriched in isotopically heavy Zn relative to the partially dehydrated antigorite serpentinites. Furthermore, they show that sub-arc mantle, fluid-precipitated olivines are also strongly enriched in isotopically heavy Zn. Linking these observations to a series of distillation models they suggested that the likely mechanism for this isotopic fractionation is the release of $Zn\text{-}SO_X$ bearing serpentinite-derived fluids. Consequently, they demonstrate the sensitivity of Zn isotopes to trace the mobility of highly oxidizing fluids during fluid cycling at subduction zones. Debret et al. (2018) built on this finding and showed how Zn and Fe stable isotopes were useful tracers of oxidizing fluids during slab devolatilization processes and this could be used to comment on the transfer of carbonate in this setting.

In a complementary study Inglis et al. (2017) tested the behavior of Zn stable isotopes in a suite of prograde metabasalts and metagabbros which were interpreted to be representative of the metamorphosed mafic oceanic crust equilibrated under different metamorphic conditions in a subduction zone. By analyzing these rocks, they demonstrated that even after high-pressure metamorphism at eclogite facies, the resulting anhydrous eclogite assemblage preserved a $\delta^{66}Zn$ indistinguishable within uncertainty from estimates of modern MORB (Wang et al.,

2017), suggesting that subduction zone dehydration does not trigger a shift in the Zn isotope composition of downgoing mafic lithologies. This result combined with dehydration modeling allowed them to conclude that the fluids released during the dehydration of the mafic oceanic crust are likely poor carriers of dissolved Zn-SO_X and/or Zn-CO_X species.

16.5. SUMMARY AND CONCLUSIONS

• Advances over the last twenty years have enabled the routine acquisition of both precise and accurate Cu and Zn stable isotope data on a range of natural and synthetic sample materials.

• As a multivalent element in nature, stable isotope theory predicts that Cu isotopes should respond to changes in redox. Isotope enrichment factors calculated by ab initio methods predict that distinct isotopic differences should occur between Cu(I) and Cu(II) bearing species at low temperatures, but this difference scales as a function of $1/T^2$, so that it would be reduced at high temperatures. Precipitation experiments confirm distinct Cu isotope fractionation between $Cu(II)_{aqueous}$ and $Cu(I)_{precipitate}$. In a geological setting the redox-related behavior of Cu isotopes has mainly been studied in ore-forming environments. Such studies have demonstrated that primary mineralization has little effect on Cu isotope variation. Despite this, several workers conclude that secondary oxidative alteration of the reduced primary assemblage can trigger a redox-related isotopic fractionation.

• Zinc is not multivalent in nature and thus changes in redox will not fractionate Zn isotopes in a geological context. Despite this, ab initio calculation suggests that Zn isotopes are highly sensitive to complexation but aqueous S and C bearing species. Building on this prediction several studies have used Zn isotopes as tracers of the mobility of aqueous S and C in a number of different settings. Recently, Zn isotopes have yielded powerful constraints on the cycling of oxidizing fluids in subduction zones.

ACKNOWLEDGMENTS

We thank two anomymous reviewers for the their useful suggestions and constructive comments and the editors Robert Moretti and Daniel Neuville for the invitation to write this chapter. FM acknowledges financial support of the UnivEarthS Labex program. Parts of this work were supported by IPGP multidisciplinary program PARI, by Region le-de-France SESAME Grants no. 12015908 and EX047016, the IdEx Université de Paris grant, ANR-18-IDEX-0001.

REFERENCES

Albarède, F. (2004). The stable isotope geochemistry of copper and zinc. *Reviews in Mineralogy and Geochemistry*, *55*, 1, 409–427. https://doi.org/10.2138/gsrmg.55.1.409

Archer, C., & Vance, D. (2004). Mass discrimination correction in multiple-collector plasma source mass spectrometry: an example using Cu and Zn isotopes. *Journal of Analytical Atomic Spectrometry*, *19*, 5, 656–665. https://doi.org/10.1039/B315853E

Arnold, T., Schönbächler, M., Rehkämper, M., Dong, S., Zhao, F. J., Kirk, G. J., et al. (2010). Measurement of zinc stable isotope ratios in biogeochemical matrices by double-spike MC-ICPMS and determination of the isotope ratio pool available for plants from soil. *Analytical and Bioanalytical Chemistry*, *398*, 7–8, 3115–3125. doi: 10.1007/s00216-010-4231-5

Barling, J., Arnold, G. L., & Anbar, A. D. (2001). Natural mass-dependent variations in the isotopic composition of molybdenum. *Earth and Planetary Science Letters*, *193*, 3–4, 447–457. https://doi.org/10.1016/S0012-821X(01)00514-3

Belshaw, N. S., Freedman, P. A., O'Nions, R. K., Frank, M., & Guo, Y. (1998) A new variable dispersion double-focusing plasma mass spectrometer with performance illustrated for Pb isotopes. *International Journal of Mass Spectrometry*, *181*, 1–3, 51–58. https://doi.org/10.1016/S1387-3806(98)14150-7

Bermin, J., Vance, D., Archer, C., & Statham, P. J. (2006). The determination of the isotopic composition of Cu and Zn in seawater. *Chemical Geology*, *226*, 3–4, 280–297. https://doi.org/10.1016/j.chemgeo.2005.09.025

Bigeleisen, J. (1996). Nuclear size and shape effects in chemical reactions. Isotope chemistry of the heavy elements. *Journal of the American Chemical Society*, *118*, 15, 3676–3680. https://doi.org/10.1021/ja954076k

Black, J. R., Kavner, A., & Schauble, E. A. (2011). Calculation of equilibrium stable isotope partition function ratios for aqueous zinc complexes and metallic zinc. *Geochimica et Cosmochimica Acta*, *75*, 3, 769–783. https://doi.org/10.1016/j.gca.2010.11.019

Blum, J. D., & Bergquist, B. A. (2007). Reporting of variations in the natural isotopic composition of mercury. *Analytical and BioanalyticalCchemistry*, *388*, 2, 353–359. https://doi.org/10.1007/s00216-007-1236-9

Bouilhol, P., Magni, V., van Hunen, J., & Kaislaniemi, L. (2015). A numerical approach to melting in warm subduction zones. *Earth and Planetary Science Letters*, *411*, 37–44. https://doi.org/10.1016/j.epsl.2014.11.043

Charbonnier, Q., Moynier, F., & Bouchez, J. (2018). Barium isotope cosmochemistry and geochemistry. *Science Bulletin*, *63*, 6, 385–394. https://doi.org/10.1016/j.scib.2018.01.018

Chen, H., Savage, P. S., Teng, F. Z., Helz, R. T., & Moynier, F. (2013). Zinc isotope fractionation during magmatic differentiation and the isotopic composition of the bulk Earth. *Earth and Planetary Science Letters*, *369*, 34–42. https://doi.org/10.1016/j.epsl.2013.02.037

Conway, T. M., Rosenberg, A. D., Adkins, J. F., & John, S. G. (2013). A new method for precise determination of iron, zinc and cadmium stable isotope ratios in seawater by double-spike mass spectrometry. *Analytica Chimica Acta*, *793*, 44–52. https://doi.org/10.1016/j.aca.2013.07.025

Creech, J. B., Moynier, F., & Badullovich, N. (2017). Tin stable isotope analysis of geological materials by double-spike MC-ICPMS. *Chemical Geology*, *457*, 61–67. https://doi.org/10.1016/j.chemgeo.2017.03.013

Dauphas, N., Roskosz, M., Alp, E. E., Neuville, D. R., Hu, M. Y., Sio, C. K., et al. (2014). Magma redox and structural controls on iron isotope variations in Earth's mantle and crust. *Earth and Planetary Science Letters*, *398*, 127–140. https://doi.org/10.1016/j.epsl.2014.04.033

Dauphas, N., Craddock, P. R., Asimow, P. D., Bennett, V. C., Nutman, A. P., & Ohnenstetter, D. (2009). Iron isotopes may reveal the redox conditions of mantle melting from Archean to Present. *Earth and Planetary Science Letters*, *288*, 1–2, 255–267. https://doi.org/10.1016/j.epsl.2009.09.029

Ehrlich, S., Butler, I., Halicz, L., Rickar, D., Oldroyd, A., & Matthews, A. (2004). Experimental study of the copper isotope fractionation between aqueous Cu (II) and covellite, *CuS. Chemical Geology*, *209*, 3–4, 259–269. https://doi.org/10.1016/j.chemgeo.2004.06.010

Evans, K. A., & Tomkins, A. G. (2011). The relationship between subduction zone redox budget and arc magma fertility. *Earth and Planetary Science Letters*, *308*, 3–4, 401–409. https://doi.org/10.1016/j.epsl.2011.06.009

Fujii, T., & Albarède, F. (2012). Ab initio calculation of the Zn isotope effect in phosphates, citrates, and malates and applications to plants and soil. *PLoS One*, *7*, 2, 30726. https://doi.org/10.1371/journal.pone.0030726

Fujii, T., Moynier, F., & Albarède, F. (2006). Nuclear field vs. nucleosynthetic effects as cause of isotopic anomalies in the early Solar System. *Earth and Planetary Science Letters*, *247*, 1–2, 1–9. https://doi.org/10.1016/j.epsl.2006.04.034

Fujii, T., Moynier, F., & Albarède, F. (2009). The nuclear field shift effect in chemical exchange reactions. *Chemical Geology*, *267*, 3–4, 139–156. https://doi.org/10.1016/j.chemgeo.2009.06.015

Fujii, T., Moynier, F., Blichert-Toft, J., & Albarède, F. (2014). Density functional theory estimation of isotope fractionation of Fe, Ni, Cu, and Zn among species relevant to geochemical and biological environments. *Geochimica et Cosmochimica Acta*, *140*, 553–576. https://doi.org/10.1016/j.gca.2014.05.051

Fujii, T., Moynier, F., Abe, M., Nemoto, K., & Albarède, F. (2013). Copper isotope fractionation between aqueous compounds relevant to low temperature geochemistry and biology. *Geochimica et Cosmochimica Acta*, *110*, 29–44. https://doi.org/10.1016/j.gca.2013.02.007

Fujii, T., Moynier, F., Pons, M. L., & Albarède, F. (2011). The origin of Zn isotope fractionation in sulfides. *Geochimica et Cosmochimica Acta*, *75*, 23, 7632–7643. https://doi.org/10.1016/j.gca.2011.09.036

Fujii, T., Moynier, F., Telouk, P., & Abe, M. (2010). Experimental and theoretical investigation of isotope fractionation of zinc between aqua, chloro, and macrocyclic complexes. *The Journal of Physical Chemistry A*, *114*, 7, 2543–2552. https://doi.org/10.1021/jp908642f

Graham, S., Pearson, N., Jackson, S., Griffin, W., & O'Reilly, S. Y. (2004). Tracing Cu and Fe from source to porphyry: in situ determination of Cu and Fe isotope ratios in sulfides from the Grasberg Cu–Au deposit. *Chemical Geology*, *207*, 3–4, 147–169. https://doi.org/10.1016/j.chemgeo.2004.02.009

Hibbert, K. E. J., Williams, H. M., Kerr, A. C., & Puchtel, I. S. (2012). Iron isotopes in ancient and modern komatiites: evidence in support of an oxidised mantle from Archean to present. *Earth and Planetary Science Letters*, *321*, 198–207. https://doi.org/10.1016/j.epsl.2012.01.011

Inglis, E. C. (2018). Elemental and isotopic insights into fluid mobility accompanying the subduction of oceanic lithosphere. Doctoral thesis, Durham University.

Inglis, E. C., Debret, B., Burton, K. W., Millet, M. A., Pons, M. L., Dale, C. W., et al. (2017). The behavior of iron and zinc stable isotopes accompanying the subduction of mafic oceanic crust: A case study from Western Alpine ophiolites. *Geochemistry, Geophysics, Geosystems*, *18*, 7, 2562–2579. https://doi.org/10.1002/2016GC006735

Jiang, S., Woodhead, J., Yu, J., Pan, J., Liao, Q., & Wu, N. (2002). A reconnaissance of Cu isotopic compositions of hydrothermal vein-type copper deposit Jinman, Yunnan, China. *Chinese Science Bulletin*, *47*, 3, 249–251. https://doi.org/10.1360/02tb9059

Johnson, C. M., Beard, B. L., & Albarède, F. (2004). *Geochemistry of Non-Traditional Stable Isotopes*. Reviews in Mineralogy and Geochemistry, 55, Mineralogical Society of America.

Kavner, A., John, S. G., Sassand, S., & Boyle, E. A. (2008). Redox-driven stable isotope fractionation in transition metals: Application to Zn electroplating. *Geochimica et Cosmochimica Acta*, *72*, 7, 1731–1741. https://doi.org/10.1016/j.gca.2008.01.023

Kelley, K. A., & Cottrell, E. (2009). Water and the oxidation state of subduction zone magmas. *Science*, *325*, 5940, 605–607. doi: 10.1126/science.1174156

Larner, F., & Rehkämper, M. (2012). Evaluation of Stable Isotope Tracing for ZnO Nanomaterials - New Constraints from High Precision Isotope Analyses and Modeling. *Environmental Science & Technology*, *46*, 7, 4149–4158. https://doi.org/10.1021/es204440d

Larson, P. B., Maher, K., Ramos, F. C., Chang, Z., Gaspar, M., & Meinert, L. D. (2003). Copper isotope ratios in magmatic and hydrothermal ore-forming environments. *Chemical Geology*, *201*, 3–4, 337–350. https://doi.org/10.1016/j.chemgeo.2003.08.006

Li, W., Jackson, S. E., Pearson, N. J., & Graham, S. (2010). Copper isotopic zonation in the Northparkes porphyry Cu–Au deposit, SE Australia. *Geochimica et Cosmochimica Acta*, *74*, 14, 4078–4096. https://doi.org/10.1016/j.gca.2010.04.003

Li, W., Jackson, S. E., Pearson, N. J., Alard, O., & Chappell, B. W. (2009). The Cu isotopic signature of granites from the Lachlan Fold Belt, SE Australia. *Chemical Geology*, *258*, 1–2, 38–49. https://doi.org/10.1016/j.chemgeo.2008.06.047

Lodders, K. (2003). Solar system abundances and condensation temperatures of the elements. *The Astrophysical Journal*, *591*, 2, 1220.

Mahan, B., Siebert, J., Pringle, E. A., & Moynier, F. (2017). Elemental partitioning and isotopic fractionation of Zn between metal and silicate and geochemical estimation of the S content of the Earth's core. *Geochimica et Cosmochimica Acta*, *196*, 252–270. https://doi.org/10.1016/j.gca.2016.09.013

Maréchal, C., & Albarède, F. (2002). Ion-exchange fractionation of copper and zinc isotopes. *Geochimica et Cosmochimica*

Acta, *66*, 9, 1499–1509. https://doi.org/10.1016/S0016-7037(01)00815-8

Maréchal, C. N., Télouk, P., & Albarède, F. (1999). Precise analysis of copper and zinc isotopic compositions by plasma-source mass spectrometry. *Chemical Geology*, *156*, 1–4, 251–273. https://doi.org/10.1016/S0009-2541(98)00191-0

Markl, G., Lahaye, Y., & Schwinn, G. (2006). Copper isotopes as monitors of redox processes in hydrothermal mineralization. *Geochimica et Cosmochimica Acta*, *70*, 16, 4215–4228. https://doi.org/10.1016/j.gca.2006.06.1369

Mason, T. F., Weiss, D. J., Chapman, J. B., Wilkinson, J. J., Tessalina, S. G., Spiro, B., et al. (2005). Zn and Cu isotopic variability in the Alexandrinka volcanic-hosted massive sulphide (VHMS) ore deposit, Urals, Russia. *Chemical Geology*, *221*, 3–4, 170–187. https://doi.org/10.1016/j.chemgeo.2005.04.011

Mason, T. F., Weiss, D. J., Horstwood, M., Parrish, R. R., Russell, S. S., Mullane, E., & Coles, B. J. (2004). High-precision Cu and Zn isotope analysis by plasma source mass spectrometry Part 2. Correcting for mass discrimination effects. *Journal of Analytical Atomic Spectrometry*, *19*, 2, 218–226. https://doi.org/10.1039/B306953B

McCoy-West, A.J., Millet, M. A., & Burton, K. W. (2017). The neodymium stable isotope composition of the silicate Earth and chondrites. *Earth and Planetary Science Letters*, *480*, 121–132. https://doi.org/10.1016/j.epsl.2017.10.004

Millet, M. A., & Dauphas, N. (2014). Ultra-precise titanium stable isotope measurements by double-spike high resolution MC-ICP-MS. *Journal of Analytical Atomic Spectrometry*, *29*, 8, 1444–1458. https://doi.org/10.1039/C4JA00096J

Millet, M. A., Baker, J. A., & Payne, C. E. (2012). Ultra-precise stable Fe isotope measurements by high resolution multiple-collector inductively coupled plasma mass spectrometry with a ^{57}Fe–^{58}Fe double spike. *Chemical Geology*, *304*, 18–25. https://doi.org/10.1016/j.chemgeo.2012.01.021

Moynier, F., & Le Borgne, M. (2015). High precision zinc isotopic measurements applied to mouse organs. *Journal of Visualized Experiments*, *99*, 52479–52479. doi: 10.3791/52479

Moynier, F., & Fujii, T. (2017). Calcium isotope fractionation between aqueous compounds relevant to low-temperature geochemistry, biology and medicine. *Scientific Reports*, *7*, 44255. doi: 10.1038/srep44255

Nanne, J. A., Millet, M. A., Burton, K.W., Dale, C. W., Nowell, G. M., & Williams, H. M. (2017). High precision osmium stable isotope measurements by double spike MC-ICP-MS and N-TIMS. *Journal of Analytical Atomic Spectrometry*, *32*, 4, 749–765. https://doi.org/10.1039/C6JA00406G

Nier, A. O. (1940). A mass spectrometer for routine isotope abundance measurements. *Review of Scientific Instruments*, *11*, 7, 212–216. https://doi.org/10.1063/1.1751688

Nier, A. O. (1947). A mass spectrometer for isotope and gas analysis. *Review of Scientific Instruments*, *18*, 6, 398–411. https://doi.org/10.1063/1.1740961

Nier, A. O. (1950) A redetermination of the relative abundances of the isotopes of carbon, nitrogen, oxygen, argon, and potassium. *Physical Review*, *77*, 6, 789. https://doi.org/10.1103/PhysRev.77.789

Nier, A. O., & Gulbransen, E. A. (1939). Variations in the relative abundance of the carbon isotopes. *Journal of the American Chemical Society*, *61*, 3, 697–698. https://doi.org/10.1021/ja01872a047

Polyakov, V. B., & Mineev, S. D. (2000). The use of Mössbauer spectroscopy in stable isotope geochemistry. *Geochimica et Cosmochimica Acta*, *64*, 5, 849–865. https://doi.org/10.1016/S0016-7037(99)00329-4

Pons, M. L., Debret, B., Bouilhol, P., Delacour, A., & Williams, H. (2016). Zinc isotope evidence for sulfate-rich fluid transfer across subduction zones. *Nature Communications*, *7*, 13794. https://doi.org/10.1038/ncomms13794

Pons, M. L., Quitté, G., Fujii, T., Rosing, M, T., Reynard, B., Moynier, F., et al. (2011). Early Archean serpentine mud volcanoes at Isua, Greenland, as a niche for early life. *Proceedings of the National Academy of Sciences*, *108*, 43, 17639–17643. https://doi.org/10.1073/pnas.1108061108

Ripperger, S., & Rehkämper, M. (2007). Precise determination of cadmium isotope fractionation in seawater by double spike MC-ICPMS. *Geochimica et Cosmochimica Acta*, *71*, 3, 631–642. doi: 10.1016/j.gca.2006.10.005

Rouxel, O. J., Bekker, A., & Edwards, K. J. (2005). Iron isotope constraints on the Archean and Paleoproterozoic ocean redox state. *Science*, *307*, 5712, 1088–1091. doi: 10.1126/science.1105692

Rouxel, O., Fouquet, Y., & Ludden, J. N. (2004a). Copper isotope systematics of the Lucky Strike, Rainbow, and Logatchev sea-floor hydrothermal fields on the Mid-Atlantic Ridge. *Economic Geology*, *99*, 3, 585–600. https://doi.org/10.2113/gsecongeo.99.3.585

Rouxel, O., Fouquet, Y., & Ludden, J. N. (2004b). Subsurface processes at the lucky strike hydrothermal field, Mid-Atlantic ridge: evidence from sulfur, selenium, and iron isotopes. *Geochimica et Cosmochimica Acta*, *68*, 10, 2295–2311. https://doi.org/10.1016/j.gca.2003.11.029

Rudge, J. F., Reynolds, B. C., & Bourdon, B. (2009). The double spike toolbox. *Chemical Geology*, *265*, 3–4, 420–431. https://doi.org/10.1016/j.chemgeo.2009.05.010

Savage, P. S., Moynier, F., Chen, H., Shofner, G., Siebert, J., Badro, J., & Puchtel, I. S. (2015). Copper isotope evidence for large-scale sulphide fractionation during Earth's differentiation. *Geochemical Perspectives Letters*, *1*, 1, 53–64. doi: 10.7185/geochemlet.1506

Schauble, E. A. (2004). Applying stable isotope fractionation theory to new systems. *Reviews in Mineralogy and Geochemistry*, *55*, 1, 65–111. https://doi.org/10.2138/gsrmg.55.1.65

Schauble, E. A. (2007). Role of nuclear volume in driving equilibrium stable isotope fractionation of mercury, thallium, and other very heavy elements. *Geochimica et Cosmochimica Acta*, *71*, 9, 2170–2189. https://doi.org/10.1016/j.gca.2007.02.004

Sherman, D. M. (2013). Equilibrium isotopic fractionation of copper during oxidation/reduction, aqueous complexation and ore-forming processes: Predictions from hybrid density functional theory. *Geochimica et Cosmochimica Acta*, *118*, 85–97. https://doi.org/10.1016/j.gca.2013.04.030

Siebert, C., Nägler, T. F., & Kramers, J. D. (2001). Determination of molybdenum isotope fractionation by double-spike multicollector inductively coupled plasma mass spectrometry. *Geochemistry, Geophysics, Geosystems*, *2*, 7. https://doi.org/10.1029/2000GC000124

Siebert, J., Corgne, A., & Ryerson, F. J. (2011). Systematics of metal–silicate partitioning for many siderophile elements applied to Earth's core formation. *Geochimica et Cosmochimica Acta*, *75*, 6, 1451–1489. https://doi.org/10.1016/j.gca.2010.12.013

Sossi, P. A., Halverson, G. P., Nebel, O., & Eggins, S. M. (2015). Combined separation of Cu, Fe and Zn from rock matrices and improved analytical protocols for stable isotope determination. *Geostandards and Geoanalytical Research*, *39*, 2, 129–149. https://doi.org/10.1111/j.1751-908X.2014.00298.x

Teng, F. Z., Watkins, J. M., & Dauphas, N. (2017). *Non-Traditional Stable Isotopes*. Reviews in Mineralogy and Geochemistry, Volume *82*, Mineralogical Society of America.

Wang, Z. Z., Liu, S. A., Liu. J., Huang, J., Xiao, Y., Chu, Z. Y., et al. (2017). Zinc isotope fractionation during mantle melting and constraints on the Zn isotope composition of Earth's upper mantle. *Geochimica et Cosmochimica Acta*, *198*, 151–167. https://doi.org/10.1016/j.gca.2016.11.014

Williams, H. M., Peslier, A. H., McCammon, C., Halliday, A. N., Levasseur, S., Teutsch, N., & Burg, J. P. (2005). Systematic iron isotope variations in mantle rocks and minerals: the effects of partial melting and oxygen fugacity. *Earth and Planetary Science Letters*, *235*, 1–2, 435–452. https://doi.org/10.1016/j.epsl.2005.04.020

Williams, H. M., McCammon, C. A., Peslier, A. H., Halliday, A. N., Teutsch, N., Levasseur, S., & Burg, J. P. (2004). Iron isotope fractionation and the oxygen fugacity of the mantle. *Science*, *304*, 5677, 1656–1659. doi: 10.1126/science.1095679

Woodhead, J. (2002). A simple method for obtaining highly accurate Pb isotope data by MC-ICP-MS. *Journal of Analytical Atomic Spectrometry*, *17*, 10, 1381–1385. https://doi.org/10.1039/B205045E

Zhu, X. K., O'Nions, R. K., Guo, Y., Belshaw, N. S., & Rickard, D. (2000). Determination of natural Cu-isotope variation by plasma-source mass spectrometry: implications for use as geochemical tracers. *Chemical Geology*, *163*, 1–4, 139–149. https://doi.org/10.1016/S0009-2541(99)00076-5

Zhu, X. K., Guo, Y., Williams, R. J. P., O'Nions, R. K., Matthews, A., Belshaw, N. S., et al. (2002). Mass fractionation processes of transition metal isotopes. *Earth and Planetary Science Letters*, *200*, 1–2, 47–62. https://doi.org/10.1016/S0012-821X(02)00615-5

17

Mineral-Melt Partitioning of Redox-Sensitive Elements

Guilherme Mallmann[1], Antony D. Burnham[1], and Raul O.C. Fonseca[2]

ABSTRACT

Elements with variable valence state (i.e., redox-sensitive) often show contrasting mineral/melt partition coefficients as a function of oxygen fugacity (fO_2) in magmatic systems. This is because trace-element incorporation into crystal lattices depends on the charge, size, and crystal-field stabilization energy of atoms, all of which differ greatly between oxidized and reduced species of the same element. This has two critical implications: (i) petrologic/geochemical modeling of partitioning behavior of redox-sensitive trace-elements in magmatic systems requires some knowledge of their oxidation state; and (ii) the oxidation state of magmatic systems may be inferred from partitioning relations of redox-sensitive trace elements preserved in mineral and melt phases of rapidly cooled magmas. The advantage of this oxybarometric approach is that mineral/melt partitioning relations are not sensitive to late-stage degassing, charge-transfer on quenching, or surficial alteration. In this chapter we discuss the theoretical treatment of experimental mineral/melt partitioning data of redox-sensitive trace elements, and review aspects concerning the partitioning behavior of well-known redox-sensitive elements, including transition metals (Ti, V, Cr, Fe), rare earth elements (Ce, Eu), U, and siderophile elements (Mo, W, Re, and platinum group elements) under planetary magmatic fO_2 conditions.

17.1. INTRODUCTION

Although it is possible to determine valence-state ratios of redox-sensitive trace elements in natural and synthetic silicate glasses using techniques such X-ray absorption spectroscopy (e.g., Sutton & Newville, 2014), it may be challenging to relate these ratios to the oxygen fugacity (fO_2) of a silicate melt. This is because many elements exchange electrons with Fe, the most abundant redox-sensitive element in most magmas, upon quenching. For example, a mid-ocean ridge basaltic (MORB) melt at 1400 °C equilibrated at fO_2 equivalent to FMQ-1.7 (i.e., 1.7 log units below the fO_2 defined by the fayalite-magnetite-quartz buffer at the same temperature) has $Cr^{2+}/\Sigma Cr$ ~0.5 (where $\Sigma Cr = Cr^{2+} + Cr^{3+}$), but upon quenching to a glass all the Cr^{2+} is converted to Cr^{3+} by reaction with Fe^{3+} (Berry et al., 2003):

$$\underset{\text{melt}}{Cr^{2+}} + \underset{\text{melt}}{Fe^{3+}} = \underset{\text{glass}}{Fe^{2+}} + \underset{\text{glass}}{Cr^{3+}} \tag{17.1}$$

Similar reactions with Fe have been documented for Eu and Ce (Burnham & Berry, 2014; Burnham et al., 2015), but because they require only the transfer of electrons, they are probably unavoidable. Consequently, valence-state ratios of trace elements in natural Fe-bearing glasses may not be representative of their high-temperature melt chemistry.

[1]*Research School of Earth Sciences, Australian National University, Canberra, ACT, Australia*

[2]*Institute of Geology, Mineralogy and Geophysics, Ruhr-University Bochum, Bochum, Germany*

Magma Redox Geochemistry, Geophysical Monograph 266, First Edition.
Edited by Roberto Moretti and Daniel R. Neuville.

DOI: 10.1002/9781119473206.ch17

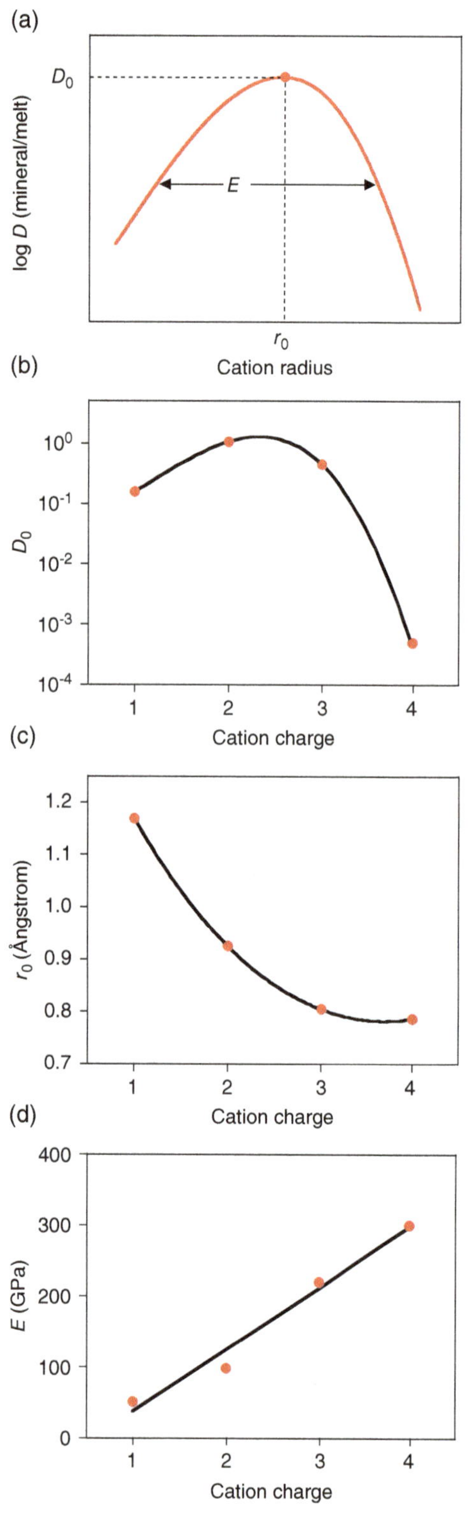

Figure 17.1 **(a)** Schematic illustration of lattice strain parameters D_0, r_0, and E for the mineral/melt partition coefficient (D) of an element i as modelled by the Brice equation (Blundy & Wood, 1994); $D_i = D_0 \exp([-4\pi EN_A(r_0 - r_i)^2 - 1/3(r_0 - r_i)^3]/RT)$, and **(b, c, d)** measured changes in these parameters with cation valence state (i.e., charge) as determined experimentally for $CaSiO_3$/melt partitioning (Law et al., 2000).

Provided that these elements are only present at low abundances relative to Fe, the exchange minimally affects the $Fe^{3+}/\Sigma Fe$ (where $\Sigma Fe = Fe^{2+} + Fe^{3+}$) of the magma. However, this ratio can still be affected by degassing of volatile species (e.g., Burgisser & Scaillet, 2007; Mathez, 1984; Sato, 1978) as well as surficial oxidation (e.g., Rhodes & Vollinger, 2005). Sulfur degassing appears to be particularly important in the Earth because of the strong capacity of S to either reduce or oxidize Fe depending on the initial oxidation state of the magma (e.g., de Moor et al., 2013; Moussallam et al., 2016). For instance, the $Fe^{3+}/\Sigma Fe$ of reduced magmas, where S^{2-} (sulfide) is the dominant melt species, may be further reduced upon degassing of SO_2:

$$\underset{\text{melt}}{S^{2-}} + 3\,\underset{\text{melt}}{Fe_2^{3+}O_3} = \underset{\text{gas}}{S^{4+}O_2} + 6\,\underset{\text{melt}}{Fe^{2+}O} + \underset{\text{melt}}{O^{2-}} \tag{17.2}$$

whereas the $Fe^{3+}/\Sigma Fe$ of oxidized magmas, where S^{6+} (sulfate) is the dominant melt species, may be further oxidized upon degassing of SO_2:

$$\underset{\text{melt}}{S^{6+}O_4^{2-}} + 2\,\underset{\text{melt}}{Fe^{2+}O} = \underset{\text{gas}}{S^{4+}O_2} + \underset{\text{melt}}{Fe_2^{3+}O_3} + \underset{\text{melt}}{O^{2-}} \tag{17.3}$$

While potentially significant in their own right, such effects obscure the primary magmatic redox characteristics. In contrast, high-temperature distributions of trace elements between minerals and melts (i.e., partitioning relations) are readily preserved in rocks that have cooled at moderate or rapid rates, and hence allow the petrologist to "see" melt chemistry that more accurately reflects the properties of the source of a magma. Furthermore, analysis of trace elements in minerals and glasses is readily achieved using benchtop equipment, unlike some spectroscopic techniques such as XANES that require access to the handful of suitably equipped synchrotron beamlines worldwide. Hence, mineral/melt partitioning allows more convenient insights into the redox states of magmas.

To derive viable oxybarometers from mineral/melt partitioning, experimental data and a thermodynamic framework are required. Experimental equipment, such as gas-mixing tube furnaces, allows access to ~25 log units of fO_2 at 1400 °C; from 10 log units below the iron-wüstite buffer (IW-10) to six log units above the hematite-magnetite buffer (HM+6). For several redox-sensitive elements this is sufficient to encompass the entire transition from one valence state to another (where the redox reaction involves the exchange of 1 electron, such as Cr^{2+} to Cr^{3+}, the transition occurs over 16 log units, whereas for elements whose redox transition involves the exchange of 8 electrons, such as S^{2-} to S^{6+}, the transition occurs over 2 log units). This allows the partitioning behavior of the end-members to be determined. Nevertheless, in several cases it is not feasible to do this. For instance, where one valence state is highly volatile (e.g., Cr^{6+} or Mo^{6+}), or where an intermediate valence state always exists alongside a higher or lower valence state (e.g., V^{3+}, V^{4+} and U^{5+}), or when extreme redox conditions necessary to constrain the partitioning behavior of a valence state alone cannot be achieved in the laboratory. In such cases, it may be necessary to model the expected behavior of a valence state.

Lattice strain theory has become one of the fundamental concepts for understanding trace element partitioning into minerals (Blundy & Wood, 1994; Wood & Blundy, 2003). The essence of the theory is that substitution of a trace element i onto a cation site in a crystal lattice is energetically neutral if the substituent ion has the same ionic radius ($r_i = r_0$) and charge as the ion it replaces, resulting in a partition coefficient D_0. Where its ionic charge and/or radius differs, electrostatic work is done in repelling or attracting the adjacent anions (usually oxygen), and this work results in the trace element being less compatible than the one with optimal charge and radius, i.e., $D_i < D_0$. The dependence of log D_i on r_i is approximately parabolic; D_0 likewise has a quadratic dependence on atomic number, Z (Figure 17.1). These two controlling variables are the reason that the various valence states of an element can have striking differences in their partitioning behavior. Consider the removal of an electron from Eu^{2+}:

$$Eu^{2+} = Eu^{3+} + e- \tag{17.4}$$

which not only increases the charge of the cation, but also (because of its higher effective nuclear charge) causes a contraction of the electronic orbitals, resulting in a decrease in ionic radius. This same principle applies to other major and trace elements capable of more than one valence state. As charge and ionic radius are principal controls on partitioning, it is almost inevitable that different valence states of an element will behave like different elements. Routine analytical techniques do not measure the speciation of redox-sensitive elements, which are usually reported as total concentrations without regard for the important differences between their valence states.

There are, however, challenges in the application of lattice strain theory to predict partition coefficients. These come mainly from crystal field effects and from the multitude of substitution mechanisms that are possible for a single cation in mineral sites.

Furthermore, there are important crystal-chemical implications for the fact that different valence states, such as Cr^{2+} and Cr^{3+}, behave like different elements. Crystals must have an overall neutral charge, hence when Cr^{2+} substitutes for Mg^{2+} in olivine, this criterion is observed, with the simple substitution $Mg^{2+} \leftrightarrow Cr^{2+}$ with the end-member Cr^{2+}-olivine forming according to the reaction:

$$\underset{\text{melt}}{2Cr^{2+}O} + \underset{\text{melt}}{SiO_2} = \underset{\text{olivine}}{Cr_2SiO_4} \tag{17.5a}$$

which, because forsterite (Mg_2SiO_4) saturation requires $2MgO + SiO_2$ and hence a fixed relationship between the activities of SiO_2 and MgO, can equivalently be written:

$$2\underset{\text{melt}}{Cr^{2+}O} + \underset{\text{olivine}}{Mg_2SiO_4} = 2\underset{\text{melt}}{MgO} + \underset{\text{olivine}}{Cr_2SiO_4} \tag{17.5b}$$

However, in order to substitute Cr^{3+} into the olivine structure, accommodation must be made for the difference in charge. One possibility involves the creation of vacancies [vac] on the octahedral site: $2\ Mg^{2+} \leftrightarrow 4/3\ Cr^{3+} + 2/3[vac]$. The reaction for formation of the corresponding end-member component is:

$$\underset{\text{melt}}{CrO_{1.5}} + 0.75\underset{\text{melt}}{SiO_2} = 0.75\underset{\text{olivine}}{Cr_{1.33}[vac]_{0.67}SiO_4} \tag{17.6}$$

Another possibility involves coupled substitution with Al and Si on the tetrahedral site: $Mg^{2+} + Si^{4+} \leftrightarrow Cr^{3+} + Al^{3+}$. The reaction for formation of the corresponding end-member component is:

$$\underset{\text{melt}}{CrO_{1.5}} + \underset{\text{melt}}{AlO_{1.5}} + 0.5\underset{\text{olivine}}{Mg_2SiO_4} = \underset{\text{olivine}}{MgCrAlO_4} + \underset{\text{melt}}{SiO_2} \tag{17.7}$$

Yet a third is tied to the incorporation of water:

$$\underset{\text{melt}}{CrO_{1.5}} + 0.5\underset{\text{melt}}{H_2O} + \underset{\text{melt}}{SiO_2} = \underset{\text{olivine}}{CrHSiO_4} \tag{17.8}$$

It can be seen from Equations 17.5–17.8 that these different substitutions each have their own dependence on the activity of SiO_2, and, additionally, some depend on the activities of Al_2O_3 and H_2O. Furthermore, other components in the melt (e.g., CaO) affect the activity coefficients of CrO and $CrO_{1.5}$, and accordingly the transition between Cr^{2+} and Cr^{3+} and their partition coefficients are also sensitive to melt composition (Berry et al., 2006). Consequently, the examples of partition coefficients quoted in this review should be taken as illustrative, rather than definitive.

In this chapter we summarize current knowledge on the effect of oxygen fugacity on the partitioning relations of redox-sensitive elements with relevant application to magmatic processes. We start by outlining some of the theoretical background behind the treatment of experimental partitioning data of redox-sensitive trace elements. This is followed by detailed sections on mineral/melt partitioning of transition metals (Ti, V, Cr, Fe), rare earth elements (Ce, Eu), uranium (U), and siderophile elements (Mo, W, Re, and platinum group elements).

17.2. THEORETICAL BACKGROUND

The bulk partition coefficient for a redox-variable element M (D_M) is given by the sum of the contributions from each valence state (a+, b+, c+), which is equivalent to the average of the partition coefficients for each valence state ($D_{M^{a+}}$, $D_{M^{b+}}$, $D_{M^{c+}}$) weighted by their relative abundances:

$$D_M = D_{M^{a+}}\left(\frac{M^{a+}}{\sum M}\right) + D_{M^{b+}}\left(\frac{M^{b+}}{\sum M}\right) + D_{M^{c+}}\left(\frac{M^{c+}}{\sum M}\right) + \ldots \tag{17.9}$$

where $\sum M = M^{a+} + M^{b+} + M^{c+} + \ldots$. Any thermodynamic treatment of mineral/melt partitioning of redox-sensitive elements, however, requires that both the homogeneous redox equilibria between the components in the melt phase and the heterogeneous equilibria of the mineral-melt partitioning reaction are considered.

17.2.1. Homogeneous Equilibria

The general reaction describing the oxidation of a redox-variable element M in a melt is:

$$\underset{\text{melt}}{M^{x+}O_{x/2}} + n/4\, O_2 = \underset{\text{melt}}{M^{(x+n)+}O_{(x+n)/2}} \tag{17.10}$$

where x is the effective valence state of the metal cation M, and n is the number of electrons involved in the redox process. The equilibrium constant (K) for this homogeneous reaction is given by:

$$K_{10} = \frac{X_{M^{(x+n)+}O_{(x+n)/2}}\,\gamma_{M^{(x+n)+}O_{(x+n)/2}}}{X_{M^{x+}O_{x/2}}\,\gamma_{M^{x+}O_{x/2}}\left(f_{O_2}\right)^{n/4}} \tag{17.11}$$

where X is mole fraction and γ is the activity coefficient of each component relative to the standard states implicit in the definition of the Gibb's free energy of the reaction at the temperature of interest. Assuming that, for the low concentrations of trace elements, the activity coefficients are constant for a given composition (Henry's Law), the expression may be simplified to:

$$K'_{10} = \left(\frac{[M^{(x+n)+}O_{(x+n)/2}]}{[M^{x+}O_{x/2}]}\right)\left(f_{O_2}\right)^{-n/4} \tag{17.12}$$

where $K'_{10} = K_{10}/(\gamma_{M^{(x+n)+}O(x+n)/2}/\gamma_{M^{x+}Ox/2})$ and the values in square brackets represent concentrations converted from mole fractions (note that the conversion factor from mole fractions to concentration cancels out between the numerator and denominator in the quotient). Taking the logarithm of this equation and rearranging gives:

$$\log\left(\frac{[M^{(x+n)+}O_{(x+n)/2}]}{[M^{x+}O_{x/2}]}\right) = \frac{n}{4}\log f_{O_2} + \log K'_{10} \tag{17.13}$$

In the case where only two valence states are possible in the melt, by defining $\sum M = M^{x+} + M^{(x+n)+}$, one obtains:

$$\frac{M^{x+}}{\sum M} = \frac{1}{1 + 10^{(n/4\log f_{O_2} + \log K'_{10})}} \tag{17.14a}$$

and

$$\frac{M^{(x+n)+}}{\sum M} = \frac{1}{1 + 10^{-(n/4\log f_{O_2} + \log K'_{10})}} \tag{17.14b}$$

From this equation, it can be seen that $M^{x+}/\sum M$ takes a sigmoidal form, constrained to vary from 0 to 1, with log fO_2. The slope is defined by $n/4$ (where n is the number of electrons in the redox reaction), and the position in log fO_2 space over which the transition from one valence state to the other occurs is determined by log K'_{10} (i.e., the simplified equilibrium constant for the homogeneous redox reaction). For elements such as U and V, which have three and four possible valence states in the melt, $\sum M \neq M^{x+} + M^{(x+n)+}$; instead, more complicated formulations are required. In the case of Cr, which can occur as Cr^{2+}, Cr^{3+}, and Cr^{6+}, there is essentially no coexistence of Cr^{2+} and Cr^{6+}, and hence Equations 17.14a,b are valid approximations.

The above treatment groups all M^{x+} together. However, from a melt structural viewpoint, it has been argued that Fe^{3+} and Eu^{3+} occur both as dissociated cations and in complexes such as FeO_2^- and EuO_2^- (Fraser, 1975; Ottonello et al., 2001). Although it would be possible to write separate reactions for oxidation of Fe^{2+} to Fe^{3+} and FeO_2^-, the resulting equations would have an identical mathematical form to the simple model we present here. Our activity coefficients, and the use of a modified equilibrium constant log K'_{10}, accommodate the terms that would be necessary to account for the speciation of the participating ions. The use of a more refined structural model may, however, prove advantageous in describing the dependence of log K'_{10} on melt composition. This is addressed in greater detail in Moretti (Chapter 6).

17.2.2. Heterogeneous Equilibria

In the mineral (min) phase, the concentration of M will be given by the sum of all possible valence states, each with its own partitioning reaction:

$$\underset{\text{mineral}}{M^{x+}} = \left[\underset{\text{melt}}{M^{x+}}O_{x/2}\right] + \sum\left[\underset{\text{melt}}{N^{x+}}O_{x/2}\right] \tag{17.15}$$

where $N^{x+}O_{x/2}$ are the "stoichiometric" control needed to form the phase component in which $M^{x+}O_{x/2}$ occurs (O'Neill & Eggins, 2002). The equilibrium constant for this heterogeneous reaction is given by:

$$K_{15} = \frac{X^{\text{min}}_{M^{x+}}\,\gamma^{\text{min}}_{M^{x+}O}}{X^{\text{melt}}_{M^{x+}O_{x/2}}\,\gamma^{\text{melt}}_{M^{x+}O_{x/2}}\prod a^{\text{melt}}_{N^{x+}O_{x/2}}} \tag{17.16}$$

where, $\prod a_{N^{x+}O_{x/2}}$ is the product of the activities of the components $N^{x+}O_{x/2}$ that make up the "stoichiometric control" in the melt. The mineral/melt partition coefficient for each valence state of M is given by:

$$D_{M^{x+}} = \frac{\gamma^{\text{min}}_{M^{x+}O}}{K_{15}\left(\gamma^{\text{melt}}_{M^{x+}O_{x/2}} \prod a^{\text{melt}}_{N^{x+}O_{x/2}} c\right)} \tag{17.17}$$

where c is a constant to convert mole fractions to concentrations by weight. From this equation, it becomes clear that the mineral/melt partitioning of M will depend on the activities of the major element oxides in the system, even if they mix ideally.

17.2.3. Mineral/Melt Partitioning as a Function of fO_2

The mineral/melt partition coefficient of a redox-sensitive element M at constant pressure, temperature, and major element composition can be defined (alternatively to Equation 17.9) as the sum of the concentrations of all valence states in the mineral phase, divided by the sum of the concentration of all valence states in the melt phase:

$$D_M = \frac{\sum\left[M^{x+}_{\text{min}}\right]}{\sum\left[M^{x+}_{\text{melt}}\right]} \tag{17.18}$$

Assuming that M occurs in the valence states 1+, 2+, 3+ ..., this equation can be rearranged to:

$$D_M = \frac{\frac{\left[M^{1+}_{\text{min}}\right]}{\left[M^{1+}_{\text{melt}}\right]} + \frac{\left[M^{2+}_{\text{min}}\right]}{\left[M^{1+}_{\text{melt}}\right]} + \frac{\left[M^{3+}_{\text{min}}\right]}{\left[M^{1+}_{\text{melt}}\right]} \ldots}{1 + \frac{\left[M^{2+}_{\text{melt}}\right]}{\left[M^{1+}_{\text{melt}}\right]} + \frac{\left[M^{3+}_{\text{melt}}\right]}{\left[M^{1+}_{\text{melt}}\right]} \ldots} \tag{17.19}$$

Homogeneous redox reactions can be written relating the different valence states to either the more reduced or oxidized end-members:

$$M^{1+}O_{0.5} + 1/2\,O_2 = M^{3+}O_{1.5} \tag{17.20a}$$

$$M^{1+}O_{0.5} + 1/4\,O_2 = M^{2+}O \tag{17.20b}$$

and the ratios of the concentrations of these related to K' and fO_2 in the following way:

$$\left[M^{3+}_{\text{melt}}\right]/\left[M^{1+}_{\text{melt}}\right] = (fO_2)^{0.5}\,K'_{20a} \tag{17.21a}$$

$$\left[M^{2+}_{\text{melt}}\right]/\left[M^{1+}_{\text{melt}}\right] = (fO_2)^{0.25}\,K'_{20b} \tag{17.21b}$$

Substituting Equation 17.21 into Equation 17.19 and rearranging gives:

$$D_M = \frac{D^{1+} + \left[D^{2+}(fO_2)^{0.25}K'_{20b}\right] + \left[D^{3+}(fO_2)^{0.5}K'_{20a}\right] + \ldots}{1 + \left[(fO_2)^{0.25}K'_{20b}\right] + \left[(fO_2)^{0.5}K'_{20a}\right] + \ldots} \tag{17.22}$$

For convenience, this expression can also be written in the following form:

$$D_M = \frac{D^{1+} + \left(D^{2+}\,10^{0.25\log fO_2 + \log K'_{20b}}\right) + \left(D^{3+}\,10^{0.5\log fO_2 + \log K'_{20a}}\right) + \ldots}{1 + \left(10^{0.25\log fO_2 + \log K'_{20b}}\right) + \left(10^{0.5\log fO_2 + \log K'_{20a}}\right) + \ldots} \tag{17.23}$$

which then allows for the log fO_2 term to be substituted more readily by oxygen fugacity terms relative to buffers in the customary way (e.g., ΔFMQ, ΔIW, etc.) as a form of minimizing the effect of temperature and pressure, if non-isobaric/isothermal data are considered.

17.3. TRANSITION METALS (Fe, Cr, Ti, V)

A large number of transition metals (those elements whose atoms have incompletely filled $3d$ orbitals) occur in more than one valence state under magmatic fO_2

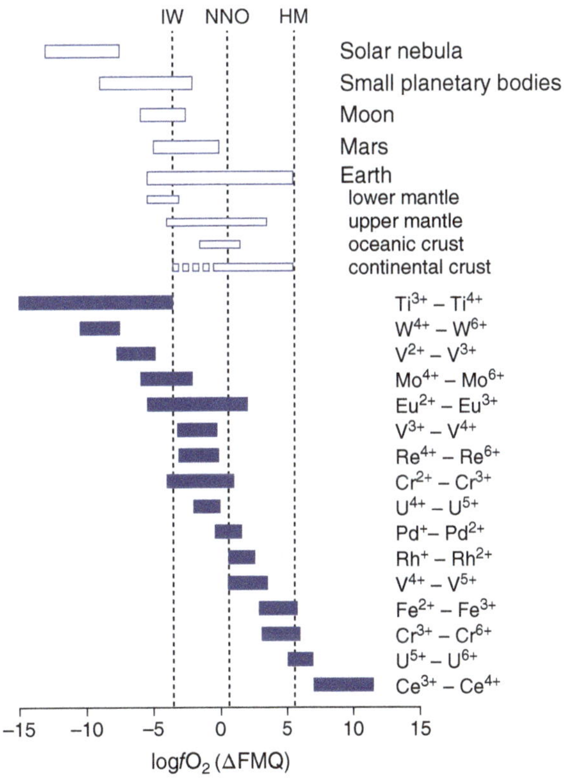

Figure 17.2 Relative oxygen fugacity conditions (in log units relative to FMQ, the fayalite-magnetite-quartz buffer; values from O'Neill, 1987) for several planetary geochemical reservoirs (Frost & McCammon, 2008; Grossman et al., 2008; Wadhwa, 2008), and valence state transition ranges for the elements discussed in this chapter. See text for references. Also plotted for reference the IW (Fe-FeO), NNO (Ni-NiO), and HM (hematite-magnetite) buffers.

conditions (Figure 17.2). Amongst the redox-sensitive first-row transition metals, Fe is a major constituent of the solar system while others like Ti, V, and Cr occur in minor-to-trace concentrations (Palme & O'Neill, 2014). Mineral/melt partition relations of transition metals are harder to constrain in comparison with other trace elements as their behavior frequently does not obey Henry's Law. Furthermore, their electronic configuration results in strong crystal-field effects (Burns, 1993) and hence partitioning behavior is often anomalous in the context of lattice strain theory.

17.3.1. Iron (Fe)

As the only redox-sensitive cation that is a major element component in the solar system, iron has a fundamental role in magmatic redox processes. Arguably, fO_2 in the solid Earth is entirely controlled by the partitioning behavior of iron, which occurs as Fe^{3+} (ferric iron) and Fe^{2+} (ferrous iron), as well as Fe^0 (metallic iron). The crystal-chemical controls on Fe partitioning are essential to understanding fO_2 throughout the Earth's mantle. For example, in the dominant upper mantle phase olivine (Mg_2SiO_4), iron occurs almost exclusively in the divalent form substituting for Mg in octahedral sites (i.e., $Fe^{3+}/\Sigma Fe \sim 0$; O'Neill et al., 1993). This means that, in a system closed to oxygen, a reduced basaltic melt that crystallizes olivine will become oxidized (i.e., its $Fe^{3+}/\Sigma Fe$ will increase). Conversely, an oxidized melt that crystallizes magnetite ($Fe^{2+}Fe_2^{3+}O_4$, with $Fe^{3+}/\Sigma Fe$ ~0.66) will become reduced (i.e., its $Fe^{3+}/\Sigma Fe$ will decrease). Other ferromagnesian minerals in the upper mantle (e.g., pyroxenes, amphiboles, garnets) will also contribute to this mass balance.

Another fundamental example that illustrates the significance of crystal-chemical controls on the partitioning behavior of Fe comes from the most abundant mineral phase in the Earth's mantle, bridgmanite (Fe,Mg)(Si,Al)O_3. Bridgmanite can contain up to 20 atomic % Fe, a substantial proportion of which is trivalent ($Fe^{3+}/\Sigma Fe$ up to 0.6; McCammon, 1997; Frost et al., 2004). This preference of bridgmanite for ferric iron occurs because the coupled substitution of Fe^{3+} on to the VIII-fold coordinated Mg site, charge balanced by Al substitution on to the VI-fold coordinated Si site, is energetically favorable (Kesson et al., 1995). Therefore, Fe^{3+} occurs in experimentally grown Al-bearing bridgmanite even at very reducing conditions, with fO_2 maintained low via a disproportionation reaction that form metallic iron (e.g., Shim et al., 2017):

$$4\underbrace{{}^{\mathrm{VIII}}\left(Fe^{2+}_{0.75}Al^{3+}_{0.25}\right){}^{\mathrm{VI}}\left(Al^{3+}_{0.25}Si^{4+}_{0.75}\right)O_3}_{\text{bridgmanite}}$$

$$= 2\underbrace{{}^{\mathrm{VIII}}\left(Fe^{3+}\right){}^{\mathrm{VI}}\left(Al^{3+}\right)O_3}_{\text{bridgmanite}} + \underbrace{3SiO_2}_{\text{stishovite}} + \underbrace{Fe^0}_{\text{metal}} \tag{17.24}$$

These types of disproportionation reactions are powerful in explaining the paradoxical evidence suggesting high Fe^{3+} contents in the very reduced lower mantle (McCammon, 2005).

In other abundant ferromagnesian minerals such as pyroxenes, the incorporation of Fe^{2+} and Fe^{3+} is also strongly dictated by stoichiometric controls. The experimental data of McCanta et al. (2004), for instance, highlight significant steric constraints on the substitution of Fe^{3+} for Fe^{2+} in pigeonite (low Ca clinopyroxene) compared to augite (high Ca clinopyroxene). The limited incorporation of Fe^{3+} into pigeonite is such that even under highly oxidizing conditions only a few percent of the total Fe will be trivalent. In contrast, alkali pyroxenes like aegirine ($NaFe^{3+}Si_2O_6$) have a strong affinity for Fe^{3+} but will not precipitate, even under oxidized conditions, unless there is sufficient Na available in the melt. As there is complete solid solution between augite and aegirine, Fe^{3+}/Fe^{2+} of a pyroxene has a complex dependence on melt composition when Fe is a major element, and lies beyond the scope of this discussion. A related issue arises with the partitioning of trace elements into magnetite, which is strongly fO_2-dependent. However, this is mostly due to the stoichiometric control by the Fe^{3+}/Fe^{2+} ratio of the melt rather than any variation in oxidation state of the trace elements themselves. For example, Ti becomes less compatible from FMQ to FMQ+4, well outside the range at which any Ti^{3+} is present, and this behavior is also exhibited by redox-invariant trace elements such as Zr and Hf (Sievwright et al., 2017). Hence, care should be taken when evaluating redox-related trends in Fe-rich systems to understand the factors contributing to variations in partition coefficients. The relationships described in this chapter are only applicable to trace components of melts and minerals.

Based on experiments with simplified synthetic compositions doped with small concentrations of Fe as a function of fO_2, Mallmann and O'Neill (2009) showed that $D_{Fe^{2+}} >> D_{Fe^{3+}}$ in olivine and orthopyroxene, but $D_{Fe^{2+}} < D_{Fe^{3+}}$ in clinopyroxene and spinel (Figure 17.3a). Amphibole/melt data from King et al. (2000) indicates that the partitioning of both Fe^{3+} and Fe^{2+} are both close to unity, with $D_{Fe^{3+}}$ about a factor of two higher than $D_{Fe^{2+}}$ (though the range of fO_2 in this study, from FMQ-6 to FMQ+3, is not sufficient to constrain either Fe^{2+} or Fe^{3+}, so the differences in Ds between the two species are likely larger). The data of Martel et al. (1999) and Krawczynski et al. (2012), which examined fO_2 over an even more limited range alongside the effects of H_2O, T and P, show higher D_{Fe}/D_{Mg} at more oxidizing

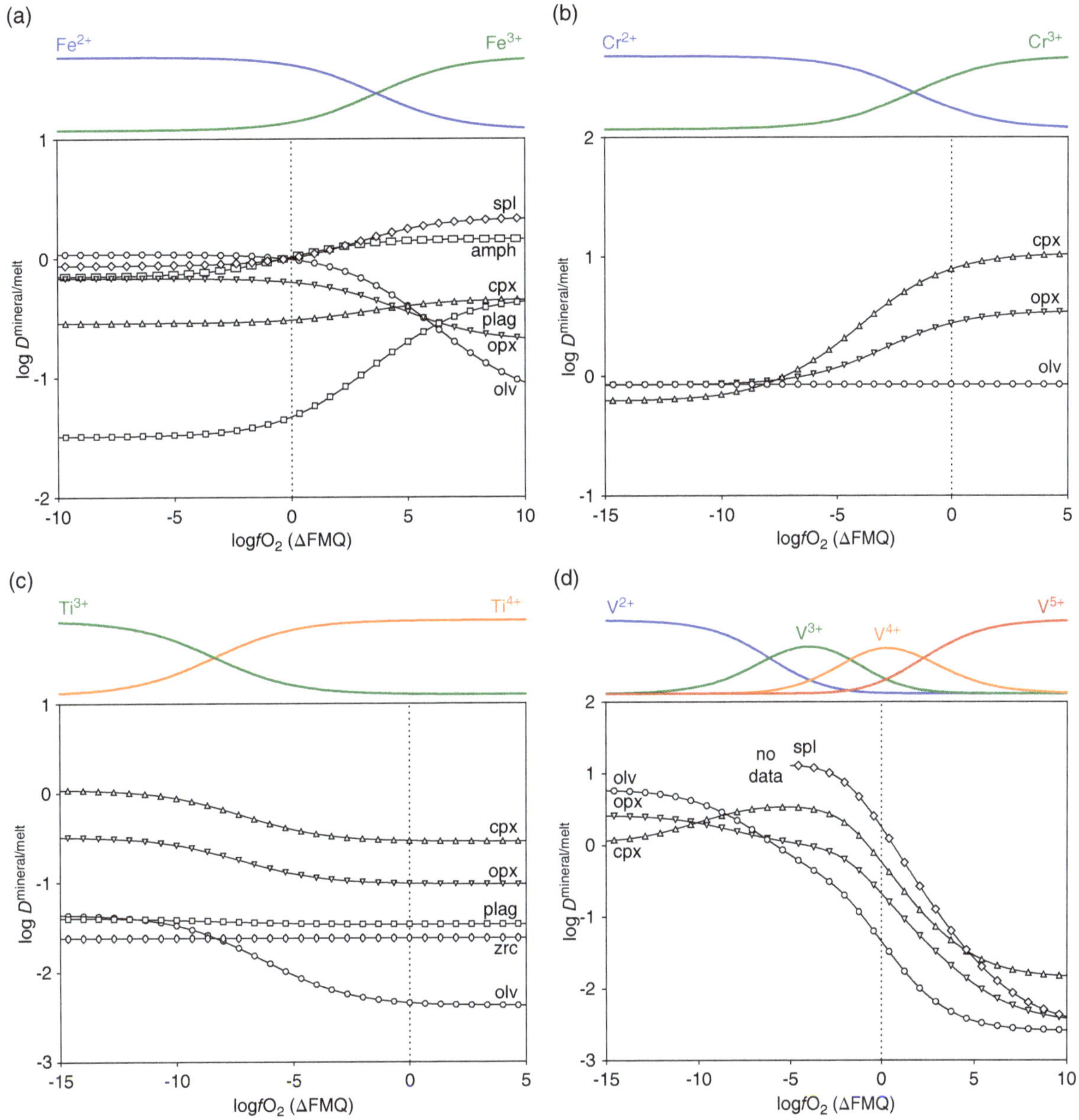

Figure 17.3 Mineral/melt partitioning of transition metals as a function of oxygen fugacity (in log units relative to FMQ, the fayalite-magnetite-quartz buffer; values from O'Neill, 1987) for selected minerals. Abbreviations: amph – amphibole, cpx – clinopyroxene, opx – orthopyroxene, olv – olivine, plag – plagioclase feldspar, spl – spinel, zrc – zircon. All curves shown in the main diagram are best fits of the experimental data to an equation with the same form as Equation 22. The fraction (from 0 to 1) of the different valence states of each element are plotted on top of the diagram. These were calculated using Equation 17.13 using available data. **(a)** Fe mineral/melt partitioning data from King et al. (2000) (amph), Mallmann and O'Neill (2009) (spl, olv, opx, cpx) and Phinney (1992) (plag). **(b)** Cr mineral/melt partitioning data from Mallmann and O'Neill (2009) (olv, opx, cpx). **(c)** Ti mineral/melt partitioning data from Mallmann and O'Neill (2009) (olv, opx, cpx), Peters et al. (1995) (plag), and Burnham and Berry (2012) (zrc). **(d)** V mineral/melt partitioning data from Mallmann and O'Neill (2009) (olv, opx, cpx, spl).

conditions, supporting a higher compatibility for Fe^{3+} in igneous amphiboles. It should be noted in passing that amphibole $Fe^{3+}/\Sigma Fe$ is readily altered by a redox reaction $Fe^{2+} + OH^{-} = Fe^{3+} + O^{2-} + \frac{1}{2} H_2$, and hence may not preserve magmatic values (Popp et al., 1995).

In non-ferromagnesian minerals such as feldspars, iron behaves as a typical trace element. In plagioclase feldspar for instance (Figure 17.3a), Fe^{3+} (r^{IV} = 0.49 Å)[1] is significantly more compatible than Fe^{2+} (Phinney, 1992; Wilke & Behrens, 1999) as it directly substitutes for

[1] Ionic radii reported in this study were taken from Shannon (1976).

Al^{3+} (r^{IV} = 0.39 Å) in tetrahedral sites, as opposed to Fe^{2+} which is significantly smaller (r^{VIII} = 0.78 Å) than the larger VIII-fold coordinated A-site (r_0 ~1.2 Å) (Longhi et al., 1976). In some oxidized peralkaline lavas, alkali feldspars with up to 2.8 wt% Fe_2O_3 have been reported (Mann et al., 2006). Lundgaard and Tegner (2004) found a strong dependence of $D_{Fe^{2+}}$ and $D_{Fe^{3+}}$ on melt composition, particularly SiO_2, but not on plagioclase composition.

17.3.2. Chromium (Cr)

Chromium can exist as Cr^{2+}, Cr^{3+}, and Cr^{6+} in magmas (e.g., Berry & O'Neill, 2004; Schreiber & Haskin, 1976), but direct detection of Cr^{2+} in quenched basaltic melts is hindered by an electron exchange with Fe on quenching (Reaction 17.1; Berry et al., 2003). Mineral-melt partitioning relations, however, have long indicated the presence of Cr^{2+} in silicate melts over typical magmatic conditions (e.g., Barnes, 1986; Hanson & Jones, 1998; Postovetov & Roeder, 2000). The divalent and trivalent species of Cr overwhelmingly dominate over typical magmatic fO_2s (Figure 17.2), with hexavalent Cr only becoming stable at very oxidizing conditions (usually > FMQ+4; Berry & O'Neill, 2004). Because such highly oxidizing conditions are uncommon in magmatic systems, and Cr^{6+} appears to be highly incompatible in high-temperature mineral phases (Mallmann & O'Neill, 2009), the presence of small amounts of Cr^{6+} in magmas should have no significant effect on the bulk partitioning of Cr (although it should be noted that the transition from Cr^{3+} to Cr^{6+} is expected to occur over a relatively narrow range of fO_2 given that the redox reaction involves three electrons). High-temperature experiments under the highly oxidizing conditions necessary to constrain the behavior of Cr^{6+} are limited (hexavalent Cr appears to be volatile at high temperature; Wijbrans et al., 2015), so the focus of our discussion will be on Cr^{2+} and Cr^{3+}.

Chromium is a major constituent of many rock-forming minerals (Burns & Burns, 1975), normally making end-member components where Cr^{3+} substitutes for Al^{3+} (and Fe^{3+}). Some of these substitutions are complete solid solutions, most notably between spinel ($MgAl_2^{3+}O_4$) and magnesiochromite ($MgCr_2^{3+}O_4$), and between grossular ($Ca_3Al_2^{3+}Si_3O_{12}$) and uvarovite ($Ca_3Cr_2^{3+}Si_3O_{12}$). In most cases, Cr^{3+} only occurs in octahedral coordination, as a result of its large crystal-field stabilization energy (CFSE; Burns, 1993). The ionic radius of Cr^{3+} is also significantly larger than Al^{3+} (and hence Si^{4+}), making it difficult for tetrahedral substitution. Incorporation of Cr^{2+} is unfavored due to its electronic configuration, which in octahedral coordination induces a destabilizing crystal-field energy and Jahn-teller distortions (Burns, 1975). Hence, Cr^{2+} can substitute for Mg^{2+} but only to a limited extent, and end-member Cr^{2+} silicates do not occur in nature (Li et al., 1995).

The importance of Cr to mantle phase equilibria (e.g., Liu & O'Neill, 2004), the link between chromite and precious metals (e.g., platinum group elements, PGE) in layered intrusions (e.g., Barnes, 1986), and its utility as an oxybarometer in magmatic systems (e.g., Postovetov & Roeder, 2000) has prompted numerous studies aimed at determining Cr solubility in silicate melts (e.g., Roeder & Reynolds, 1991) and its partitioning between associated phases. Here we focus on minerals where Cr substitutes as a trace component, as this allows us to constrain the effect of fO_2 on the partitioning behavior of both Cr^{2+} and Cr^{3+}.

Amongst minerals where Cr substitutes as trace element, olivine has the largest available experimental partitioning dataset as a function of fO_2. In olivine, Cr^{2+} (r^{VI} = 0.80 Å) is incorporated by direct replacement of Mg^{2+} (r^{VI} = 0.72 Å) in octahedral sites without significant lattice distortion (Reaction 17.5; Jollands et al., 2018). Substitution of Cr^{3+} (r^{VI} = 0.615 Å) into olivine octahedral sites, on the other hand, creates charge imbalance that can be accounted for either by octahedral vacancies (Reaction 17.6; Hanson & Jones, 1998; Jollands et al., 2018; Papike et al., 2005), or by coupled substitution with Al^{3+} and Si^{4+} in tetrahedral sites (Reaction 17.7; Jollands et al., 2018).

Olivine/melt partitioning studies from Schreiber and Haskin (1976) and Hanson and Jones (1998) in the $CaO\text{-}MgO\text{-}Al_2O_3\text{-}SiO_2$ (CMAS) show a systematic dependence of $D_{Cr^{3+}}$ on melt composition (Figure 17.4),

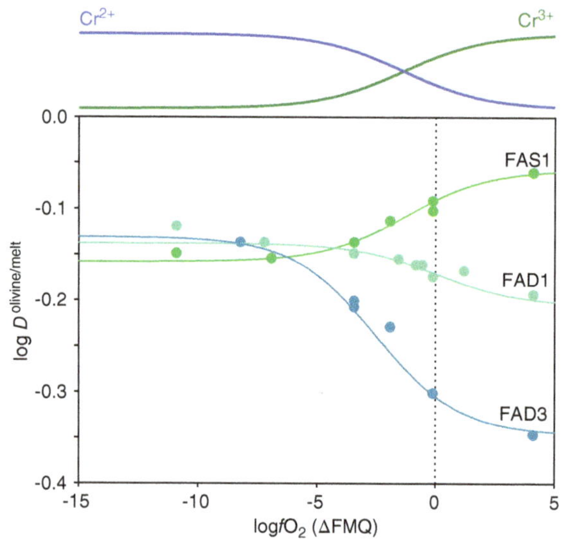

Figure 17.4 Comparison of Cr olivine(Fo_{100})/melt partitioning coefficients obtained at 1 bar and 1320 °C as a function of oxygen fugacity (in log units relative to FMQ, the fayalite-magnetite-quartz buffer; values from O'Neill, 1987) for three Fe-free compositions in the $CaO\text{-}MgO\text{-}Al_2O_3\text{-}SiO_2$ system: FAS1, FAD1 and FAD3 (Hanson & Jones, 1998). The results illustrate a mild effect of melt composition on the partitioning of Cr^{3+} yet no significant effect on the partitioning of Cr^{2+}. The fraction (from 0 to 1) of the different valence states are plotted on top of the diagram.

though the dependence is not straightforward, suggesting that both substitution mechanisms described above likely operate. Hanson and Jones (1998), however, were able to parameterize $D_{Cr^{3+}}$ between olivine and melt using the universal melt descriptor NBO/T ($D_{Cr^{3+}} = -0.39$ NBO/$T + 1.29$). For Cr^{2+}, however, partitioning between olivine and melt appears to be independent of melt composition, but varies slightly with temperature (Hanson & Jones, 1998). More often than not, $D_{Cr^{2+}}$ is nearly equal to $D_{Cr^{3+}}$ between olivine and melt (including for basaltic compositions), so that effectively no change in bulk Cr partitioning is detected as a function of fO_2 (Mallmann & O'Neill, 2009; Hanson & Jones, 1998; Figure 17.3b). Direct Cr valence measurements in olivine by XANES have confirmed these findings (Bell et al., 2014).

Mallmann and O'Neill (2009) found that Cr^{2+} is similarly partitioned amongst olivine, ortho-, and clinopyroxene, but Cr^{3+} is significantly more compatible in pyroxenes, particularly clinopyroxene (Figure 17.3b). This is not surprising given that Cr^{3+} ($r^{VI} = 0.615$ Å) has an ionic radius that is similar to the ideal strain-free octahedral M1 site of clinopyroxenes for trivalent cations ($r_0^{3+,M1}$ ~0.668 Å; Hill et al., 2000). Cr^{2+} ($r^{VI} = 0.80$ Å), on the other hand, is significantly larger than the ideal M1 site of clinopyroxenes for divalent cations ($r_0^{2+,M1}$ ~0.69 Å; Hill et al., 2000). Mallmann and O'Neill (2009) noted a significant difference between the partitioning of Cr^{3+} between clinopyroxene and melt obtained for their compositions V1 and V7 (which are identical except for the Cr-doping levels), suggesting that Cr^{3+} does not behave according to Henry's Law in the concentration range of these experiments (> 1.5 wt%. Cr_2O_3), probably because of the strong coupling of Cr^{3+} with Al^{3+} in pyroxenes (Klemme & O'Neill, 2000). Chromium is noticeably more compatible in augite compared to pigeonite and enstatite, given the availability of cations (Al^{3+} and Na^{+}) for charge balancing the substitution of Cr^{3+} for Mg in octahedral sites (Karner et al., 2007; Mallmann & O'Neill, 2009). Interestingly, Cr^{2+} appears to be dominant species entering the structure of lower mantle phases such as ferropericlase and bridgmanite (Eeckhout et al., 2007). Neither Cr^{3+} nor Cr^{2+} appears to be easily incorporated into plagioclase (Aigner-Torres et al., 2007) as Cr^{3+} is too large for the tetrahedral site, and Cr^{2+} too small for the large distorted VIII-fold coordinated site.

Finally, we note that Cr is a well-known outlier in lattice-strain parabolas due to large crystal-field effects, so the prediction of Cr^{2+} and Cr^{3+} partitioning using lattice strain models should be done with great caution.

17.3.3. Titanium (Ti)

Titanium primarily occurs as Ti^{4+} over most of the range of fO_2s relevant to planetary scientists, but spectroscopic studies indicate that under very reducing conditions (i.e., below FMQ-4) a substantial fraction of Ti occurs as Ti^{3+} (Schreiber, 1977). The transition from Ti^{3+} to Ti^{4+} appears to be strongly dependent on melt composition (Borisov, 2012; Figure 17.2). Hence, although the partitioning of Ti is a poor oxybarometer for terrestrial magmatism, it is markedly more relevant to understanding the geochemical evolution of the early solar nebula and reduced bodies such as the Moon (Figure 17.2).

Hibonite ($CaAl_{12}O_{19}$) occurs in the calcium aluminum-rich inclusions (CAIs) of chondritic meteorites. It is one of the earliest phases to condense from the solar nebula, and can contain up to 8 wt% TiO_2 (Allen et al., 1978). The blue color of many hibonite grains has been shown to be due to the presence of Ti^{3+} (Burns & Burns, 1984; Sutton et al., 2017), incorporated as a $CaTi_2^{3+}Al_{10}O_{19}$ component, but Ti^{4+} can also be accommodated in hibonite via a coupled substitution with Mg, as a $CaMg_2Ti_2^{4+}Al_8O_{19}$ component (Doyle et al., 2014). The coupled nature of the substitution of Ti^{4+} means that the availability of Mg exerts a strong control on $Ti^{3+}/\Sigma Ti$ (where $\Sigma Ti = Ti^{3+} + Ti^{4+}$) in hibonite; since it is often unclear which phases are in equilibrium in CAIs, this oxybarometer may have limited utility (Berry et al., 2017).

Ti^{3+} is more compatible than Ti^{4+} in the majority of phases, including olivine, diopside, orthopyroxene and anorthite (Mallmann & O'Neill, 2009; Peters et al., 1995; Figure 17.3c). Ti K-edge XANES measurements of both synthetic and natural lunar basalt samples have revealed that Ti^{3+} is enriched in ilmenite, armalcolite and clinopyroxene at fO_2 of about FMQ-4.5 (Leitzke et al., 2018; Simon & Sutton, 2018). However, as it is most compatible in clinopyroxene, essentially all research has focused on this phase. Indeed, a Ti^{3+}-pyroxene (grossmanite, $CaTi^{3+}AlSiO_6$) has been reported from the Allende meteorite, and $NaTi^{3+}Si_2O_6$ has been synthesized experimentally (Ohashi et al., 1982; Ma & Rossman, 2009). Although quantitative oxybarometry is hampered by multiple sources of uncertainty, including the equilibrium mineral assemblage (as with hibonite), the mixing relations of Ti-bearing pyroxenes and the temperature of formation, decreasing $Ti^{3+}/\Sigma Ti$ from core to rim of some CAIs suggests increasingly oxidizing conditions in the solar nebula over a period of up to 300 kyr, perhaps by as much as 6–7 log fO_2 units (Young et al., 2012). However, in some CAIs an oxidation trend, followed by a more reduced rim, has been observed. This is attributed to incompatible Ti^{4+} concentrating in the melt phase, which was isolated from the solar nebula by a thick melilite mantle, followed by late-stage ingress of nebular gases to reduce the residual melt prior to crystallization of the rims (Papike et al., 2016). Ti^{3+} has also been observed in lunar pyroxenes (Burns et al., 1972), but is thought to be a minor component even on such a reduced planetary body

(Papike et al., 2005). Terrestrial magmas generally form at far more oxidizing conditions and Ti is expected to occur exclusively as Ti^{4+} (Figure 17.2). Pyroxenes in the native iron-bearing lavas of Disko Island, Greenland, have been suggested to contain some Ti^{3+} (Pedersen, 1981), but this is speculative and may not be compatible with the rather high Fe contents.

17.3.4. Vanadium (V)

Vanadium is the redox-sensitive trace element that occurs in the largest number of valence states at typical magmatic fO_2s (Figure 17.2), namely V^{2+}, V^{3+}, V^{4+}, and V^{5+} (Borisov et al., 1987; Schreiber et al., 1987; Sutton et al., 2005). Given the plethora of valence states with theoretically contrasting geochemical behaviors, significant effort has been made to characterize the effect of fO_2 and other variables such as pressure, temperature, and mineral and melt composition on the partitioning of V in silicate/oxide phases relevant to mantle melting and magmatic differentiation (e.g., Arató and Audétat, 2017; Canil, 1997, 1999, 2002; Canil & Fedortchouk, 2000, 2001; Karner et al., 2008; Laubier et al., 2014; Mallmann & O'Neill, 2009, 2013; Papike et al., 2015; Righter et al. 2006a,b, 2011; Shearer et al., 2006; Sievwright et al., 2017; Toplis and Corgne 2002; Wijbrans et al., 2015). The end goal of many of these studies is to calibrate the sensitivity of V partitioning to fO_2 as an oxybarometer.

Under terrestrial magmatic fO_2 conditions (Figure 17.2), V^{3+} and V^{4+} are the dominant valence states, and spinel the major mineral group host. In pure magnetite ($Fe^{2+}\,Fe_2^{3+}\,O_4$), where the stoichiometry fixes Fe^{3+}/Fe^{2+} at 2, incorporation of vanadium occurs via an fO_2-independent electron-exchange reaction: $Fe^{3+} + V^{3+} = Fe^{2+} + V^{4+}$ (O'Neill & Navrotsky, 1984). However, experimental studies of spinel group minerals have invariably shown strong dependence of vanadium partitioning with fO_2 both in Fe-free and Fe-bearing systems (e.g., Arató & Audétat, 2017; Canil et al., 1999; Mallmann & O'Neill, 2009; Papike et al., 2015; Righter et al., 2006b; Sievwright et al., 2017; Toplis & Corgne, 2002). Importantly, because Fe is an essential component of magnetite, the partitioning of non-redox variable elements depends on fO_2 (Sievwright et al., 2017); this is not the case for Fe-poor spinels. For example, the ratio $V^{3+}/\,V^{4+}$ of pure spinel ($MgAl_2O_4$) is controlled by the reaction:

$$1/2\,\underset{\text{spinel}}{MgV_2^{3+}O_4} + 3/2\,\underset{\text{melt}}{MgO} + 1/4\,\underset{\text{gas}}{O_2} = \underset{\text{spinel}}{Mg_2V^{4+}O_4} \tag{17.25}$$

It is therefore unsurprising that the bulk partitioning of V between spinel group minerals and silicate melt is a strong function of fO_2 as well as composition (mineral and melt). Experimental data shows that $D_{V^{3+}} >> D_{V^{4+}} > D_{V^{5+}}$ between spinel and silicate melt (Figure 17.3d), and that the bulk partitioning of vanadium for Cr-spinels is higher than that for Al-spinels (Canil, 2002; Mallmann & O'Neill, 2009). The negative correlation between Al and V^{3+} in experimentally grown spinels provides strong evidence that the preferred mode of vanadium substitution into spinel is by direct replacement of Al in octahedral sites (Balan et al., 2006; Canil, 2002; Righter et al., 2006b). However, the details of the various possible substitution mechanisms are complex, given the structure of spinels and the multivalent nature of their cations (e.g., Papike et al., 2015). Melt composition also appears to play a significant role, as demonstrated by V partitioning experiments between magnetite and silicic melts (Arató & Audétat, 2017).

Owing to its high charge and small ionic radius, V^{5+} is highly incompatible in all rock-forming minerals. When sufficiently oxidized conditions are achieved, the partitioning of V^{5+} can be obtained, though this value is typically << 0.01 (e.g., Mallmann & O'Neill, 2009). V^{4+} has partition coefficients that fall between V^{3+} and V^{5+} (Figure 17.3d) and therefore is much harder to constrain. The ionic radius of V^{4+} is similar to that of Ti^{4+} in both octahedral and tetrahedral coordination (r^{VI} = 0.58 and 0.605 Å; r^{IV} = 0.46 and 0.42 Å; for V^{4+} and Ti^{4+}, respectively), and both have similar electronic configurations and thus no significant difference in CFSE, so an approximation can be obtained using lattice strain theory. In connection with that, we note that D_V^{4+} for pyroxene and olivine obtained by Mallmann and O'Neill (2009) are about an order of magnitude higher than what is inferred from lattice strain parabolas of other tetravalent cations. For some minerals, constraining the partitioning of V^{3+} may also be difficult when it shows compatibility between V^{2+} and V^{4+} (e.g., olivine or orthopyroxene). For clinopyroxene though, V^{3+} is clearly more compatible than both V^{2+} and V^{4+}, and its partitioning readily determined (Figure 17.3d). Partition coefficients for V^{2+} can be obtained if sufficiently reduced experimental data is available, such as those reported by Mallmann and O'Neill (2009), but extrapolation from lattice strain parabolas should be avoided since V^{2+} has the same electronic configuration as Cr^{3+} and therefore large CFSE in octahedral coordination (Burns, 1993).

For mineral phases that only crystallize at high pressure from natural magmas, partitioning relations are harder to constrain as a function of fO_2 since this intensive parameter needs to be buffered externally using double-capsule techniques or internally using low-solubility solid metal-oxide. Righter et al. (2011) extended the fO_2 conditions obtained by Mallmann and O'Neill (2007, 2009) to constrain the $D_{V^{3+}}$ in garnet at ~10. The compatibility order of vanadium valences in garnet is $D_{V^{3+}} >> D_{V^{4+}} > D_{V^{5+}}$, with values comparable to clinopyroxene (Mallmann and O'Neill, 2009). XANES studies by Righter et al.

(2011), however, suggest that small amounts of V^{2+} are incorporated into garnet, even under relatively oxidizing conditions.

Because of the complexities in developing a thermodynamic partitioning model that involves four different oxidation states, all of which likely varying differently with compositional parameters, Mallmann and O'Neill (2013) developed an empirical calibration of V partitioning between olivine and melt over a restricted range of fO_2s (from FMQ-4 to FMQ+4), where log D_V vary more or less linearly with log fO_2, accounting for the effects of olivine and melt composition. Similarly, Arató and Audétat (2017) developed an empirical calibration for magnetite. It would be useful to develop similar models for other common magmatic minerals, but this requires more extensive experimental databases since substitution of different V valence states in minerals like pyroxene and garnet are more complex than in olivine and magnetite.

17.4. RARE EARTHS (Ce, Eu)

The rare earth elements (REE), including Y, have favorable characteristics for examining redox processes. Apart from Ce and Eu, the REE exist exclusively in the trivalent state and their ionic radii vary smoothly from La to Lu (with Y somewhere between Ho and Er). This provides excellent coverage for fitting a lattice strain parabola to the REE^{3+}; anomalies relative to this parabola are readily quantifiable, and often visually striking. Zircon is notable because its REE patterns commonly feature both Ce and Eu anomalies (Figure 17.5). They are typically quantified as Ce/Ce∗ and Eu/Eu∗, where Ce∗ and Eu∗ are the values that would be predicted by interpolation or extrapolation. The simplest is the geometric mean of La and Pr (i.e., a linear interpolation on a logarithmic axis):

$$Ce* = (La_N \times Pr_N)^{0.5} \tag{17.26}$$

where N denotes normalization to CI chondrite, whole rock, or matrix glass. La is often below the limit of detection for analyses of zircon, and so extrapolation from Pr-Nd-Sm may be used instead. Alternatively, polynomial fits to the REE pattern may be used (e.g., O'Neill, 2016; Burnham & Berry, 2017). There are systematic deviations between the methods, so values taken from different sources may need recalculation before being compared.

Cerium exists as both Ce^{3+} and Ce^{4+} in magmas. However, at typical magmatic fO_2 the proportion of Ce^{4+} is small (Figure 17.2), with $Ce^{4+}/\Sigma Ce$ ~0.001 (Burnham & Berry, 2012, 2014). At such low levels, it can be seen from Figure 17.6a that where Ce^{4+} is less compatible than Ce^{3+} there is no perceptible effect on the bulk partitioning of Ce: even in the most extreme case, i.e., $D_{Ce^{4+}} = 0$, D_{Ce} would drop by 0.1%, an amount that is far less than the achievable analytical accuracy. By a similar line of logic, it can be calculated that in order for a mineral in a terrestrial magma to develop a noticeable Ce anomaly, e.g., Ce/Ce∗ = 1.1, $D_{Ce^{4+}}$ must be greater than $D_{Ce^{3+}}$ by a factor of 100. For minerals with a steep or strongly curved REE patterns, or larger analytical uncertainties, even a 10% excess of Ce might not be a convincing anomaly.

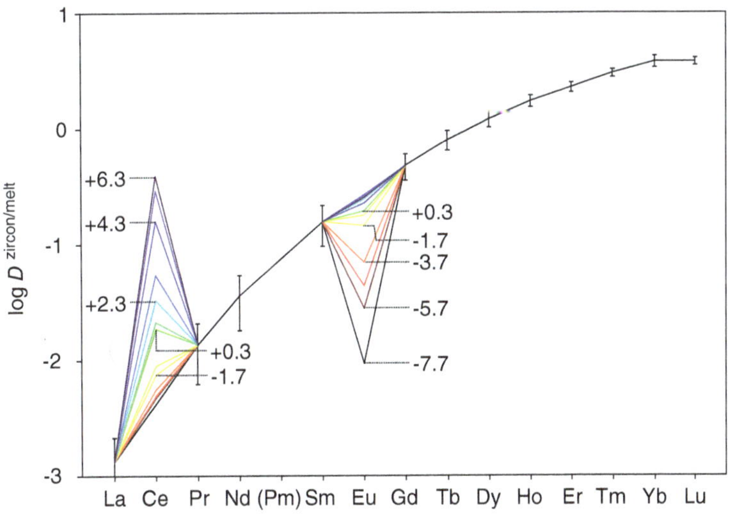

Figure 17.5 Zircon/melt partition coefficients for REE elements determined experimentally at 1 bar as a function of oxygen fugacity by Burnham and Berry (2012). The colored segments highlight systematic changes in the partitioning of Ce and Eu with oxygen fugacity (labeled relative to the FMQ buffer).

Few minerals have a strong affinity for Ce^{4+}. The most notable, zircon, will be addressed in detail here. This is a favorable substitution on account of the identical 4+ charge of Zr, and a similarly large ionic radius (r^{VIII} = 0.84 and 0.97 Å for Zr and Ce^{4+}, respectively). Positive Ce anomalies have been reported from other Zr minerals such as baddeleyite (Schärer et al., 2011), wadeite (Jaques, 2016), elpidite, and armstrongite (Gerdes et al., 2017). Thomson et al. (2016) reported both positive and negative Ce anomalies for inclusions in diamonds, but the precise interpretation of these grains is complicated by their polymineralic nature and a lack of experimental partitioning data.

Zircon is the commonest of these, and has attracted the most experimental work (Burnham & Berry, 2012; Trail et al., 2011, 2012). The most recent calibration of the Ce-in-zircon oxybarometer uses lattice strain theory to infer $D_{Ce^{3+}}$ and $D_{Ce^{4+}}$, which allow $Ce^{4+}/\Sigma Ce$ in the melt to be calculated and the corresponding fO_2 deduced (Smythe & Brenan, 2016). This approach avoids the challenge of growing analyzable well-equilibrated zircon crystals, but there is presently a discrepancy between the two available models for Ce speciation in silicate melts (Burnham & Berry, 2014; Smythe & Brenan, 2015) and hence further work is required to develop a robust version of this oxybarometer.

For other minerals, these has been no systematic study of the effect of fO_2, but this can be predicted by a combination of lattice strain modeling (and for minerals with more than one cation site, whose 4+ parabolas are underconstrained, by making the approximation $D_{Ce^{4+}} = D_{U^{4+}}$) and $Ce^{4+}/\Sigma Ce$ estimated from the optical basicity of the melt composition (Burnham & Berry, 2014). It can be seen that Ce^{4+} is less compatible than Ce^{3+} in Zr-free minerals and that, as explained above, D_{Ce} does not change appreciably in the fO_2 range of natural magmas.

Although the foregoing discussion considered mineral/melt partitioning, it is not always possible to consider mineral compositions in terms of mineral/melt partition coefficients. In particular, plutonic rocks do not retain any melt fraction, but even volcanic rocks often contain a significant cargo of crystals that are substantially older and may not be in equilibrium with their present host (e.g., Murphy et al., 1998). In the case of Ce, it is possible to use the bulk rock or even CI chondrite as a normalization factor, because magma compositions with Ce anomalies are uncommon. It is often stated that subduction zone magmas can have $Ce/Ce^* < 1$, indicating the incorporation of Ce-depleted sediments. As highlighted by O'Neill (2016), however, many reported examples of Ce anomalies in lavas are poorly constrained. Relatively small negative Ce anomalies, Ce/Ce^* ~0.9, are known in a variety of subduction settings (e.g., Hastie et al., 2009; Woodhead, 1989). As zircon Ce anomalies are typically one or two orders of magnitude larger, failure to account for a pre-existing Ce anomaly in the magma results in an underestimate of log fO_2 of < 0.2. The same is not true for Eu.

Europium can exist as Eu^{2+} and Eu^{3+} in magmas. The transition occurs most rapidly below the FMQ buffer, a range over which $Fe^{3+}/\Sigma Fe$ changes more slowly, and hence it is potentially a useful oxybarometer for reduced magmas (Figure 17.2). The crystal chemistry of Eu^{2+} (r^{VI} = 1.17 Å) can be thought of as similar to Sr^{2+} (r^{VI} = 1.18 Å), and indeed there is even a Eu^{2+} feldspar (Kimata, 1988). In contrast, Eu^{3+} behaves as a typical middle REE. Accordingly, $D_{Eu^{2+}} > D_{Eu^{3+}}$ for plagioclase and alkali feldspars and åkermanitic melilite (Kuehner et al., 1989), but for most other minerals $D_{Eu^{3+}} > D_{Eu^{2+}}$ (Figure 17.6b). Plagioclase feldspar and clinopyroxene are by far the most studied in terms of Eu partitioning as a function of fO_2 and the following discussion will be restricted to these, although phases such as titanite, with $D_{Eu^{3+}}/D_{Eu^{2+}} \approx 2500$ may be important to consider for some igneous systems (Loader et al., 2017).

There are various models that allow plagioclase/melt $D_{Eu^{2+}}$ and $D_{Eu^{3+}}$ to be predicted using lattice strain theory (Aigner-Torres et al., 2007; Dohmen & Blundy, 2014). Corresponding models for clinopyroxene and other minerals have not been developed, and even in the case of plagioclase, Eu oxybarometry will be more precise where measured values of partition coefficients for proxy elements are used for $D_{Eu^{2+}}$ and $D_{Eu^{3+}}$. Rearranging Equation 17.9, we can approximate $Eu^{3+}/\Sigma Eu$ by:

$$Eu^{3+}/\Sigma Eu = (D_{Eu} - D_{Sr})/(D_{MREE} - D_{Sr}) \qquad (17.27)$$

where D_{Eu} is the observed Eu partition coefficient and D_{MREE} is an estimate of $D_{Eu_{3+}}$, i.e., D_{Sm}, D_{Gd}, or preferably $(D_{Sm} \times D_{Gd})^{0.5}$. For values of $Eu^{3+}/\Sigma Eu$ in the range 0.1 to 0.9, the fO_2 can be calculated by rearranging the general equation from Burnham et al. (2015):

$$\log f_{O_2} = 40.4 - 25640/T - 56.8\Lambda - 4 \log \left(\Sigma Eu/Eu^{3+} - 1\right) \qquad (17.28)$$

where T is temperature in Kelvin, and Λ is optical basicity, which can be calculated as:

$$\Lambda = \frac{\frac{SiO_2}{62.6} + \frac{TiO_2}{36.3} + \frac{Al_2O_3}{56.6} + \frac{FeO}{55.8} + \frac{MgO}{51.7} + \frac{CaO}{56.1} + \frac{Na_2O}{53.2} + \frac{K_2O}{67.3}}{\frac{SiO_2}{30.04} + \frac{TiO_2}{39.9} + \frac{Al_2O_3}{34} + \frac{FeO}{55.8} + \frac{MgO}{40.3} + \frac{CaO}{56.1} + \frac{Na_2O}{62} + \frac{K_2O}{94.2}} \qquad (17.29)$$

with all oxide components in wt%, and no distinction drawn between Fe^{2+} and Fe^{3+}. The effect of H_2O was not included in the optical basicity model of Duffy (1993) and hence is not considered here. At natural concentrations, MnO and P_2O_5 have negligible influence (e.g., 0.5 wt% P_2O_5 would change log K' for Eu^{2+}–Eu^{3+} by ~0.02).

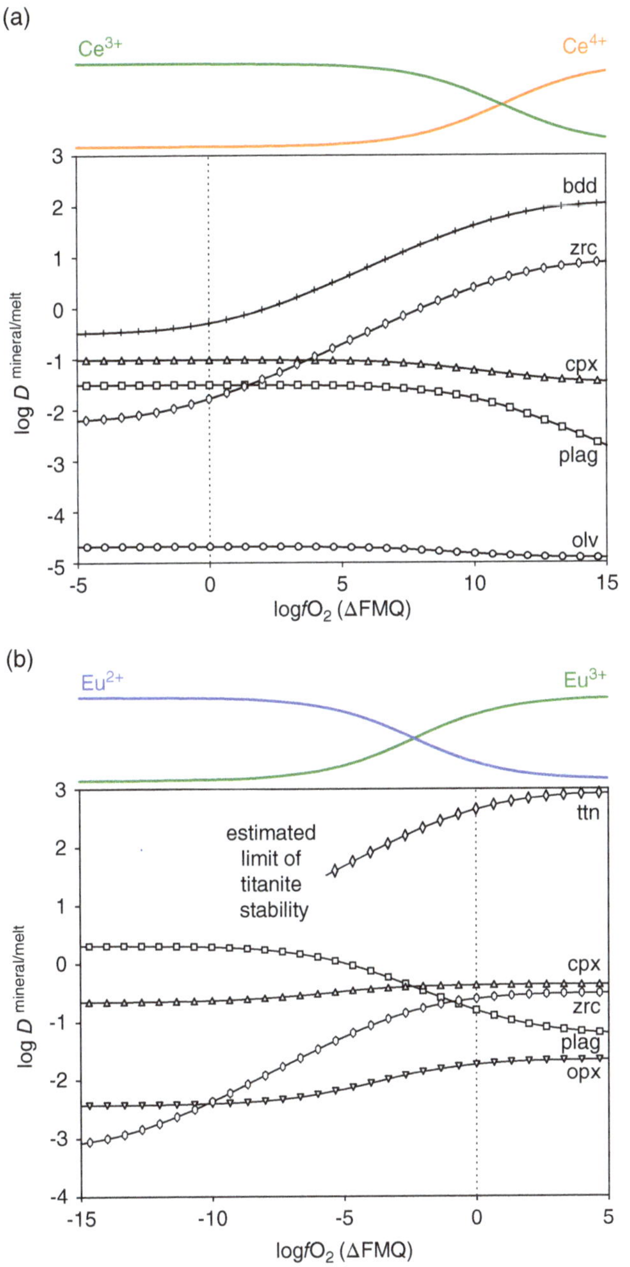

Figure 17.6 Mineral/melt partitioning of redox-sensitive rare earth elements (REE), Ce **(a)** and Eu **(b)** as a function of oxygen fugacity (in log units relative to FMQ, the fayalite-magnetite-quartz buffer; values from O'Neill, 1987) for selected minerals. Abbreviations: bdd – baddeleyite, cpx – clinopyroxene (aluminous diopside), olv – olivine, opx – orthopyroxene, plag – plagioclase feldspar, ttn – titanite, zrc – zircon. Where data were not obtained as a function of fO_2, partition coefficients for Ce^{3+}, Ce^{4+}, Eu^{2+}, and Eu^{3+} (or estimates thereof) were used in conjunction with the $Ce^{4+}/\Sigma Ce$ and $Eu^{3+}/\Sigma Eu$ models of Burnham and Berry (2014) and Burnham et al. (2015). Data from Klemme and Meyer (2003) (bdd), Leitzke et al. (2017) (cpx-Ce, plag-Ce, olv), Sun et al. (1974) (cpx-Eu, plag-Eu, opx-Eu), van Kan Parker et al. (2011) (opx-Ce), Bachmann et al. (2005) (ttn), and Burnham and Berry (2012) (zrc). See Figure 17.3 for more details.

With values of $Eu^{3+}/\Sigma Eu$ outside the range 0.1 to 0.9, the reliability of this type of oxybarometer is compromised by the shallow slope of $M^{x+}/\Sigma M$ against fO_2. An error of ±0.03 propagates to more than a log unit of fO_2. A key challenge in using Eu partitioning to determine fO_2 is the strong dependence of log K' on melt composition (Aigner-Torres et al., 2007; Burnham & Berry, 2015). The relationship given in Equation 17.28 was derived for alkali-free compositions, though experimental data for Na- and K-bearing compositions in the literature are generally consistent with this relationship. Moreover, although it lies outside the scope of this review, we note that in hydrothermal fluids the Eu^{2+}-Eu^{3+} equilibrium is 12 times more sensitive to pH than to log fO_2, and hence this variable must also be considered in low-temperature systems (Brugger et al., 2008).

17.5. URANIUM (U)

Although it has been known for many years that U can exist as U^{4+}, U^{5+}, and U^{6+} in silicate melts (e.g., Calas, 1979; Schreiber, 1983), this fact has often been rather underappreciated in the geochemical literature. The radioactivity of U generates heat as well as a series of decay products culminating in Pb, and hence understanding the geochemistry of U is particularly important. Interpretation of U-series disequilibrium depends on accurate partition coefficients for U and Th (as well as other elements in the series). Because of potential differences in redox between mid-oceanic ridges and subduction zone settings, for example, it is essential to consider the role of fO_2 in controlling the partitioning of U relative to Th.

Partitioning of U as a function of fO_2 has only been studied for a handful of minerals (Figure 17.7). Despite the importance of garnet in influencing U-series disequilibrium (Beattie, 1993), little is known about the how its partitioning of U might be affected by fO_2. For all minerals studied to date, it appears that $D_{U^{4+}} > D_{U^{5+}} > D_{U^{6+}}$. As with the intermediate oxidation states of V, the partitioning of U^{5+} is difficult to constrain without independent spectroscopic measurements of $U^{5+}/\Sigma U$ in the same melt composition. An additional complication arises because U^{6+} is volatile and is gradually lost from 1 atm experiments, leading to uncertainty on $D_{U^{6+}}$ because diffusion is unlikely to occur fast enough to keep crystals in equilibrium with the changing melt composition (Burnham & Berry, 2012).

Although there are insufficient data to develop a general expression for $U^{n+}/\Sigma U$ as a function of melt composition, temperature, pressure, and fO_2, it is known that higher oxidation states are favored by lower temperatures and melts with higher optical basicities (Schreiber, 1983; Halse, 2014). Notably, U^{5+} appears to have a reduced stability field with increasing pressure (Halse, 2014): at

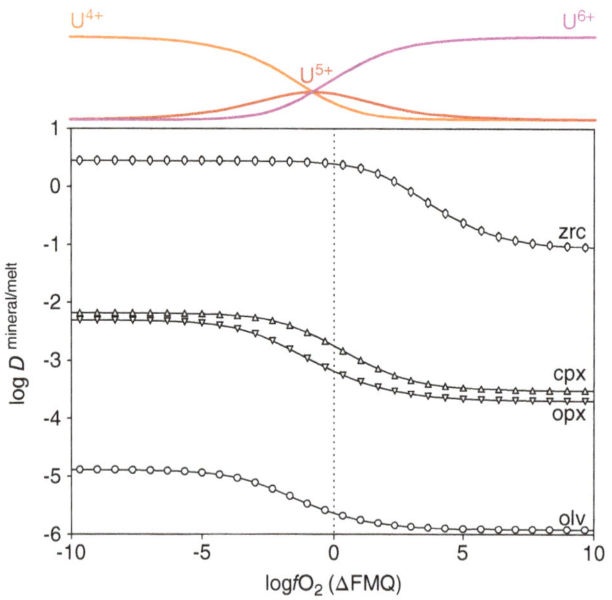

Figure 17.7 Mineral/melt partitioning of U as a function of oxygen fugacity (in log units relative to FMQ, the fayalite-magnetite-quartz buffer; values from O'Neill, 1987) for selected minerals. Abbreviations: cpx - clinopyroxene (diopside), olv - olivine, opx - orthopyroxene, zrc - zircon. Data from Fonseca et al. (2014) (opx, cpx, olv), and Burnham and Berry (2012) (zrc). See Figure 17.3 for more details.

higher pressure it appears that U^{4+} oxidizes directly to U^{6+}, leading to a sharper change in partition coefficients as a function of fO_2 (Mallmann et al., 2016).

U-Pb dating of zircon is widely regarded as one of the most reliable techniques for determining rock ages. For zircons younger than a few million years old, it becomes important to consider the contribution of ^{230}Th to the ingrowth of ^{206}Pb, because minerals for which U is less compatible than Th will develop a ^{206}Pb excess, and vice versa. Correcting for these consequences of U-series disequilibrium necessitates estimates of D_{Th}/D_U. U is more compatible than Th under reducing conditions, but much less compatible at high fO_2: the variation in terrestrial magmatic redox conditions equates to a ~100 kyr range of possible age corrections (McLean et al., 2011).

17.6. SIDEROPHILE ELEMENTS (Mo, W, Re, Pt GROUP ELEMENTS)

Siderophile elements (i.e., those with strong tendency to alloy with Fe) are important tracers of planetary differentiation owing to their propensity of fractionating into planetary cores. Because of this, a great deal of effort has been made to investigate the behavior of these elements during core formation and to determine their valence state in silicate melts.

17.6.1. Molybdenum (Mo), Tungsten (W), and Rhenium (Re)

Mo, W, and Re are known to occur in 4+ and 6+ valence states at fO_2s relevant to planetary magmatism (Cottrell et al., 2009; Ertel et al., 2001; Holzheid et al., 1994; O'Neill & Eggins, 2002; O'Neill et al., 2008; Righter et al., 2016; Wade et al., 2012, 2013). While the hexavalent state of Mo, W, and Re is dominant over most planetary conditions (Figure 17.2), the mineral/melt partitioning behavior of these elements is strongly controlled by their tetravalent state (Figure 17.8).

A common feature of the partitioning behavior of Mo, W, and Re is that their 4+ valence state is significantly more compatible in silicates and oxides (e.g., spinel) than their 6+ valence state, the latter of which being almost perfectly incompatible in crystalline phases. The relative compatibility in silicates and oxides tends to follow the order $D_{cpx}>D_{opx}>D_{olv}>D_{spl}\sim D_{plg}$, independently of valence state (Fonseca et al., 2014; Leitzke et al., 2016, 2017; Mallmann & O'Neill, 2007; Wijbrans et al., 2015). For all three elements, the difference between the mineral/melt partition coefficient of their reduced and oxidized species spans three to five orders of magnitude (Figure 17.8). The difference in partition coefficients between the valence states is most pronounced for clinopyroxene, with D_{6+} ~10^{-5} and D_{4+} ~3-6. The compatibility of tetravalent Re, Mo, and W in clinopyroxene results from the nearly ideal size of these cations (r^{VI} = 0.66 Å for W^{4+}, 0.65 Å for Mo^{4+}, and 0.63 Å for Re^{4+}) relative to the octahedral M1 site ($r_0^{4+,M1}$ ~0.66 Å; Hill et al., 2000), which makes the substitution into clinopyroxene lattice energetically favorable. While the mineral/melt partitioning of Re^{4+} and Mo^{4+} were determined directly by experimental data at reducing conditions (Mallmann & O'Neill, 2007; Leitzke et al., 2017), data for W^{4+} was constrained by Fonseca et al. (2014) using a lattice strain fit to Ti, Hf, Zr, and Re. This is because the extremely reducing conditions necessary to reduce all W^{6+} to W^{4+} cannot be achieved in the laboratory. The important conclusion of these studies is that the partitioning behavior of these elements in magmatic systems will be completely different depending on fO_2, with significant implications for comparing lunar vs. terrestrial, or oceanic vs. arc magmatism.

The effect of composition on the mineral/melt partitioning of Mo, W, and Re in pyroxenes has been addressed in some studies. Hill et al. (2000) and Righter and Shearer (2003) found a negative correlation between $D_{Mo,W}$ and the Ca-Tschermak (CaTs) component in clinopyroxene. However, the opposite trend was found by Leitzke et al. (2017), who observed that the clinopyroxenes with the highest CaTs reported by Hill et al. (2000) were formed from melt compositions with the highest CaO. Leitzke et al. (2017) speculated that, because the activity

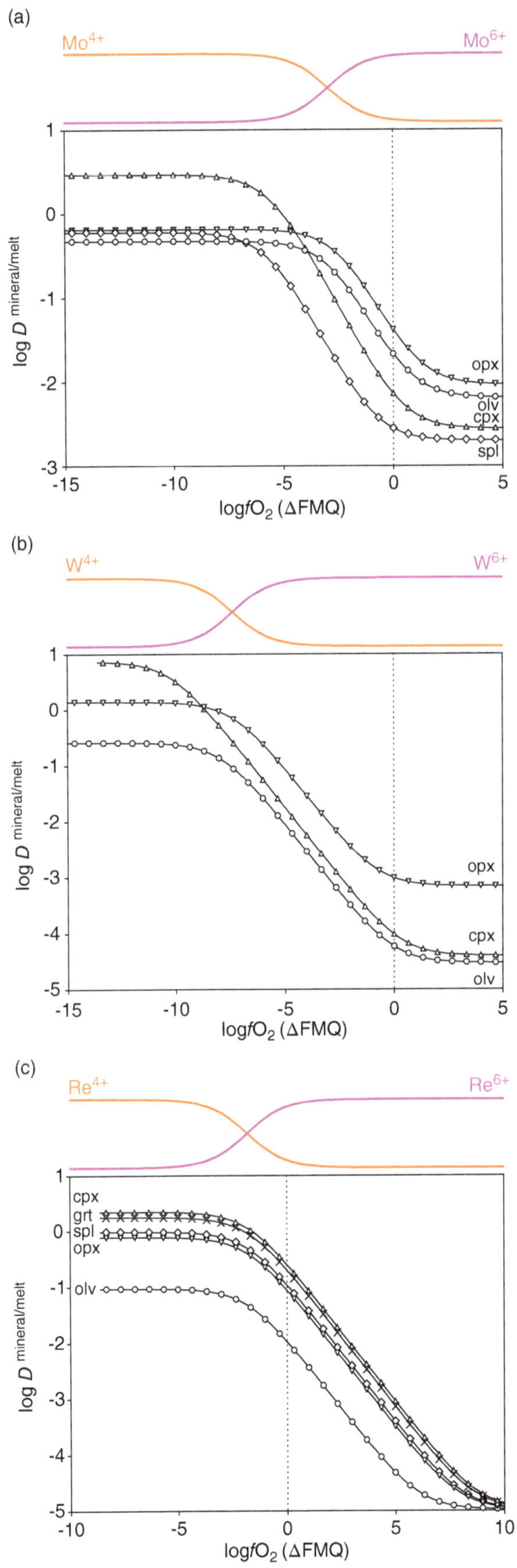

Figure 17.8 Mineral/melt partitioning of siderophile elements (Mo, W, and Re) as a function of oxygen fugacity (in log units relative to FMQ, the fayalite-magnetite-quartz buffer; values from O'Neill, 1987) for selected minerals. Abbreviations: cpx - clinopyroxene, opx - orthopyroxene, olv - olivine, spl - spinel, grt - garnet. Data from Leitzke et al. (2017) (Mo: cpx, opx, olv), Wijbrans et al. (2015) (Mo: spl), Fonseca et al. (2014) (W: opx, cpx, olv), and Mallmann and O'Neill (2007) (Re: cpx, opx, spl, grt, ol). See Figure 17.3 for more details.

coefficients of MoO_3 and MoO_2 in silicate melts decrease with increasing CaO content (O'Neill & Eggins, 2002), the negative correlation found by Hill et al. (2000) could be due to the high CaO content of the silicate melt rather than pyroxene composition. Leitzke et al. (2016) investigated the effect of TiO_2 in the silicate melt, observing a positive correlation between the TiO_2 content of the melt on the partitioning of Mo^{4+} and W^{4+}, but little to no effect on the partitioning of Mo^{6+} and W^{6+} (perhaps because these cations are very incompatible). Conversely, partitioning coefficients for Hf and Zr decrease with TiO_2 content of the melt. (D_{Ti} remained constant, as expected for Henry's Law behavior). Leitzke et al. (2016) argued that the introduction of $CaTi^{4+}Al_2O_6$ into clinopyroxene with increasing melt TiO_2 raises elastic strain in the M1 site of clinopyroxene. The increased elastic strain (which results in a tighter parabola) leads to an increase in the partitioning values of cations whose size are close to ideal for that site (e.g., W^{4+}), and a decrease in the partitioning values of those cations whose size are either smaller or larger than the ideal for that site (e.g., Zr^{4+} and Hf^{4+}).

17.6.2. Platinum Group Elements (Ru, Rh, Pd, Os, Ir, Pt)

There have been few studies investigating the redox-dependent partitioning behavior of PGEs during magmatism, with most focusing on the role of base metal sulfides and, to a lesser extent, oxides. The scarcity of partitioning data arises from the low solubility of PGE in silicate melts (typically < a few μg/g), especially under reducing conditions (O'Neill et al., 1995), and from the almost unavoidable presence of sub-micrometer metallic nuggets that plague the analyses of PGEs in synthetic silicate glasses and often lead to erroneous solubility values (Ertel et al., 2008). The solubility of a PGE in a silicate melt in equilibrium with metal can give an indication of its valence state(s), as it will dissolve according to Equation 17.10 (with $x = 0$ for the metal). However, despite significant effort in the last 20–30 years, PGE solubility results are still controversial. This leads to ambiguous interpretations regarding the PGE speciation as a function of fO_2. For example, solubility data for Ru are used to indicate that it either dissolves as Ru^{3+} (Borisov &

Nachtweyh, 1998), or as both Ru^{3+} and Ru^{4+} (Laurenz et al., 2013) in silicate melts over planetary magmatic fO_2s. Similarly, Os appears to dissolve as Os^{3+} under most conditions, though the possibility of Os^{4+} being stable at more oxidizing conditions cannot be ruled out (Borisov & Walker, 2000; Fortenfant et al., 2006). Data for Ir are also ambiguous, indicating speciation as either Ir^{3+}, or combinations of Ir^{1+}-Ir^{3+} or Ir^{2+}-Ir^{3+} (Brenan et al., 2005; Borisov & Palme, 1995; Fonseca et al., 2011; O'Neill, 2015). Pt appears to occur exclusively as Pt^{2+}, except perhaps at very oxidizing conditions (FMQ > 5) where Pt^{4+} may also be stable (Ertel et al., 1999). Of all PGEs, unambiguous evidence of redox-sensitivity over magmatic fO_2s only exists for palladium and rhodium, both of which occur as monovalent and divalent cations under typical magmatic fO_2s (e.g., Laurenz et al., 2010; Sommer, 2014; Figure 17.2). Reports of other valence states for Pd and Rh in the literature arise from interpretation of micro(nano) nugget contaminated data (e.g., Borisov et al., 1994; Ertel et al., 1999).

Of the common magmatic minerals, spinel is the only mineral capable of accommodating significant amounts of PGEs, particularly Ru, Ir, and Rh (e.g., Park et al., 2017). Wijbrans et al., (2015) reported a marked increase in D_{Rh} around FMQ in both Fe-free and Fe-rich spinels, indicating higher compatibility of Rh^{3+} relative to Rh^{2+}. Brenan et al. (2012) observed significant variations in the partitioning of several PGEs with fO_2 for Fe- and Cr-bearing spinels. However, as noted above, the partitioning of trace elements into magnetite is controlled by the Fe^{3+} and Fe^{2+} content of the melt, and it is often difficult to deconvolve this effect, and that of drastically changing site occupancies in the spinel from that of valence state changes in the trace element of interest. Nevertheless, the data in the Fe-free system are consistent with a change from Rh^{+} to Rh^{2+}, and also with olivine/melt partitioning data that shows decreasing D_{Rh} with increasing fO_2 (Brenan et al., 2003). Of the very few spinel/melt partitioning data available, Pd appears to be incompatible in spinel (Brenan et al., 2012).

17.7. CONCLUDING REMARKS

In this chapter we have shown that the partitioning behavior of redox-sensitive trace elements will almost inevitably change with fO_2 in magmatic systems. This is because the incorporation of cations into crystal lattices (at constant pressure, temperature, and composition) depends on their charge, size, and crystal-field stabilization energies, all of which likely vary substantially between the oxidized and reduced species of the same element. This dependence on fO_2 means that petrologic/geochemical modelling involving elements such as Ti, Cr, V, Ce, Eu, U, Mo, Re, W, Pd, and Rh (and probably others) must account for the oxidation state of the magma. As the only major-component redox-sensitive elements in many magmatic systems, the partitioning behavior of Fe is more likely to control rather than to reflect fO_2. Given the large range (10–15 order of magnitude; Figure 17.2) of magmatic fO_2 in planetary systems, significant differences in the partitioning behavior of individual elements ought to be expected. For instance, several studies have shown that certain trace elements behave in completely different ways during lunar compared to terrestrial magmatism, or during magmatism in mid-ocean ridge compared to subduction zone settings.

Nevertheless, if the partitioning of redox-sensitive trace elements is determined experimentally as a function of fO_2 (and the effects of other parameters such as pressure, temperature, and composition accounted for), then partitioning relations preserved in natural samples (e.g., glass and phenocryst in a volcanic rock) may provide useful ways of estimating the oxidation state of magmatic systems. This approach has the greatest advantage of being insensitive to late-stage degassing, charge-transfer (electron-exchange) reactions, or surficial alteration, all of which may potentially alter other magmatic redox proxies (e.g., $Fe^{3+}/\sum Fe$).

The preceding discussion demonstrates the importance of a good understanding of redox in modeling or interpreting trace element concentrations in melts and minerals. However, it should also be apparent that further work is required to bring our ability to model the partitioning of redox-variable elements. First, there are several elements for which very little information is available: most notably the PGEs, which have been investigated in surprisingly few minerals and whose redox transitions are poorly constrained. Copper was excluded from this review because of a paucity of data, but the data of Liu et al. (2014, 2015) indicate that its partitioning changes significantly from FMQ+1 to FMQ+5, a range that is highly relevant to arc environments where essentially all the world's economic Cu deposits are located. Likewise, experimental partitioning data by Mallmann and O'Neill (2009) indicate a potential change of phosphorus valence states from 5+ to 3+ under conditions below FMQ-4, but the results could also be explained by volatility loss from the melt after crystallization. And, under ultra-reducing conditions (~FMQ-7), where Nb and Ta begin to show siderophile tendencies, there is evidence that they may take on lower valence states in silicate melts (Cartier et al., 2014).

Furthermore, the dependence of these redox transitions and partition coefficients on temperature and melt composition is complex, and only partially understood for the majority of the redox couples addressed here. As seen in Figure 17.4, melt composition affects the partitioning of different oxidation states of the same element in differing ways, an effect that is additional to the influence of stoichiometric controls on trace element partitioning. Similarly, melt compositional effects can shift the redox transition by up to 6 log units of fO_2 (Figure 17.2). Because of these complications, extreme caution should be taken when attempting to study partitioning of redox-variable elements outside their calibrated ranges.

Many factors are important in selecting a mineral/melt partitioning oxybarometer. First, the anticipated redox state of the system will give an indication of which redox couples are relevant, but a large contrast between the compatibilities of the valence states of the element can sufficiently outweigh this consideration (for example, zircon will develop a Ce anomaly at conditions that are many orders of magnitude more reducing than the midpoint of the Ce^{3+}-Ce^{4+} equilibrium). The figures presented in this chapter may guide the geoscientist in determining whether the partition coefficient is likely to be sensitive to fO_2 for their samples. As emphasized above, the availability of appropriate experimental constraints is important, as are more general petrological considerations, for instance resistance to alteration processes such as diffusive re-equilibration. In the ideal case, multiple elements and minerals would be combined to check for internal consistency.

ACKNOWLEDGMENTS

We thank Aaron Bell and Roberto Moretti for their constructive comments. G.M. acknowledges funding of a Discovery Early Career Research Award from the Australian Research Council – ARC (grant DE160100169), A.B. acknowledges support from the ARC (grant FL130100066) and R.O.C.F. acknowledges funding of a Heisenberg Fellowship from the German Research Council – DFG (grants FO 698/5-1 and FO 698/11-1).

REFERENCES

Aigner-Torres, M., Blundy, J., Ulmer, P., & Pettke, T. (2007). Laser ablation ICPMS study of trace element partitioning between plagioclase and basaltic melts: An experimental approach. *Contributions to Mineralogy and Petrology*, *153*, 647–667. https://doi.org/10.1007/s00410-006-0168-2

Allen, J. M., Grossman, L., Davis, A. M., & Hutcheon, I. D. (1978). Mineralogy, textures and mode of formation of a hibonite-bearing Allende inclusion. 9th Lunar and Planetary Science Conference, 1209–1233.

Arató, R., & Audétat, A. (2017). Experimental calibration of a new oxybarometer for silicic magmas based on the vanadium partitioning between magnetite and silicate melt. *Geochimica et Cosmochimica Acta*, *209*, 284–295. https://doi.org/10.1016/j.gca.2017.04.020

Bachmann, O., Dungan, M. A., & Bussy, F. (2005). Insights into shallow magmatic processes in large silicic magma bodies: The trace element record in the Fish Canyon magma body, Colorado. *Contributions to Mineralogy and Petrology*, *149*, 338–349. https://doi.org/10.1007/s00410-005-0653-z

Barnes, S. J. (1986). The distribution of chromium among orthopyroxene, spinel and silicate liquid at atmospheric pressure. *Geochimica et Cosmochimica Acta*, *50*, 1889–1909. https://doi.org/10.1016/0016-7037(86)90246-2

Beattie, P. (1993). Uranium–thorium disequilibria and partitioning on melting of garnet peridotite. *Nature*, *363*, 63–65. https://doi.org/10.1038/363063a0

Bell, A. S., Burger, P. V., Le, L., Shearer, C. K., Papike, J. J., Sutton, S. R., et al. (2014). XANES measurements of Cr valence in olivine and their application to planetary basalts. *American Mineralogist*, *99*, 1404–1412.

Berry, A. J., & O'Neill, H. St.C. (2004). A XANES study of the oxidation state of chromium in silicate glasses. *American Mineralogist*, *89*, 790–798. doi: 10.2138/am-2004-5-613

Berry, A. J., O'Neill, H. St.C., Scott, D. R., Foran, G. J., & Shelley, J. M. G. (2006). The effect of composition on Cr^{2+}/Cr^{3+} in silicate melts. *American Mineralogist*, *91*, 1901–1908. https://doi.org/10.2138/am.2006.2097

Berry, A. J., Schofield, P. F., Kravtsova, A. N., Miller, L. A., Stephen, N. R., Walker, A. M., et al. (2017). The limitations of hibonite as a single-mineral oxybarometer for early solar system processes. *Chemical Geology*, *466*, 32–40. https://doi.org/10.1016/j.chemgeo.2017.03.031

Berry, A. J., Shelley, J. M. G., Foran, G. J., O'Neill, H. St.C., & Scott, D. R. (2003). A furnace design for XANES spectroscopy of silicate melts under controlled oxygen fugacities and temperatures to 1773 K. *Journal of Synchrotron Radiation*, *10*, 332–336. doi: 10.1107/s0909049503007556

Blundy J., & Wood, B. (1994). Prediction of crystal-melt partition coefficients from elastic moduli. *Nature*, *372*, 452–454. https://doi.org/10.1038/372452a0

Borisov, A. A. (2012). The Ti^{4+}/Ti^{3+} ratio of magmatic melts: application to the problem of the reduction of lunar basalts. *Petrology*, *20*, 391–398. https://doi.org/10.1134/S0869591112040030

Borisov, A. A., Kadik, A. A., Zharkova, Y. V., & Kalinichenko, N. V. (1987). Effects of oxygen fugacity on the ratio between valency forms of vanadium in magmas. *Geokhimiya*, *7*, 915–920.

Borisov, A., & Nachtweyh, K. (1998). Ru solubility in silicate melts: experimental results in oxidizing region. 29th Lunar and Planetary Science Conference, 1320.

Borisov, A., & Palme, H. (1995). The solubility of iridium in silicate melts: new data from experiments with $Ir_{10}Pt_{90}$ alloys. *Geochimica et Cosmochimica Acta*, *59*, 481–485. https://doi.org/10.1016/0016-7037(94)00358-S

Borisov, A., & Walker, R. J. (2000). Os solubility in silicate melts: New efforts and results. *American Mineralogist*, *85*, 912–917. https://doi.org/10.2138/am-2000-0704

Borisov, A., Palme, H., & Spettel, B. (1994). Solubility of palladium in silicate melts: Implications for core formation in the Earth. *Geochimica et Cosmochimica Acta*, *58*, 705–716. https://doi.org/10.1016/0016-7037(94)90500-2

Brenan, J. M., Finnigan, C. F., McDonough, W. F., & Homolova, V. (2012). Experimental constraints on the partitioning of Ru, Rh, Ir, Pt and Pd between chromite and silicate melt: The importance of ferric iron. *Chemical Geology*, *302*, 16–32. https://doi.org/10.1016/j.chemgeo.2011.05.015

Brenan, J. M., McDonough, W. F., & Ash, R. (2005). An experimental study of the solubility and partitioning of iridium, osmium and gold between olivine and silicate melt. *Earth and Planetary Science Letters*, *237*, 855–872. https://doi.org/10.1016/j.epsl.2005.06.051

Brenan, J. M., McDonough, W. F., & Dalpe, C. (2003). Experimental constraints on the partitioning of rhenium and some platinum-group elements between olivine and silicate melt. *Earth and Planetary Science Letters*, *212*, 135–150. https://doi.org/10.1016/S0012-821X(03)00234-6

Brugger, J., Etschmann, B., Pownceby, M., Liu, W., Grundler, P., & Brewe, D. (2008). Oxidation state of europium in scheelite: Tracking fluid–rock interaction in gold deposits. *Chemical Geology*, *257*, 26–33. https://doi.org/10.1016/j.chemgeo.2008.08.003

Burgisser, A., & Scaillet, B. (2007). Redox evolution of a degassing magma rising to the surface. *Nature*, *445*, 194–197. https://doi.org/10.1038/nature05509

Burnham, A. D, & Berry, A. J. (2017). Formation of Hadean granites by melting of igneous crust. *Nature Geoscience*, *10*, 457–461. https://doi.org/10.1038/ngeo2942

Burnham, A. D., & Berry, A. J. (2012). An experimental study of trace element partitioning between zircon and melt as a function of oxygen fugacity. *Geochimica et Cosmochimica Acta*, *95*, 196–212. https://doi.org/10.1016/j.gca.2012.07.034

Burnham, A. D., & Berry, A. J. (2014). The effect of oxygen fugacity, melt composition, temperature and pressure on the oxidation state of cerium in silicate melts. *Chemical Geology*, *366*, 52–60. https://doi.org/10.1016/j.chemgeo.2013.12.015

Burnham, A. D., Berry, A. J., Halse, H. R., Schofield, P. F., Cibin, G., & Mosselmans, J. F. W. (2015). The oxidation state of europium in silicate melts as a function of oxygen fugacity, composition and temperature. *Chemical Geology*, *411*, 248–259. https://doi.org/10.1016/j.chemgeo.2015.07.002

Burns, R. G. (1975). On the occurrence and stability of divalent chromium in olivine included in diamonds. *Contributions to Mineralogy and Petrology*, *51*, 213–221. https://doi.org/10.1007/BF00372081

Burns, R. G. (1993). *Mineralogical applications of crystal field theory*. 2nd Edition, Cambridge University Press, 551 pp.

Burns, R. G., & Burns, V. M. (1984). Crystal chemistry of meteoritic hibonites. *Journal of Geophysical Research*, *89*, 313–321. https://doi.org/10.1029/JB089iS01p0C313

Burns, R. G., Huggins, F. E. & Abu-Eid, R. M. (1972). Polarized absorption spectra of single crystals of lunar pyroxenes and olivines. *The Moon*, *4*, 93–102. https://doi.org/10.1007/BF00562917

Burns, V. M., & Burns, R. G. (1975). Mineralogy of chromium. *Geochimica et Cosmochimica Acta*, *39*, 903–910. https://doi.org/10.1016/B978-0-08-019954-2.50015-9

Calas, G. (1979). Etude expérimentale du comportement de l'uranium dans les magmas: États d'oxydation et coordinance. *Geochimica et Cosmochimica Acta*, *43*, 1521–1531. https://doi.org/10.1016/0016-7037(79)90145-5

Canil, D. (1997). Vanadium partitioning and the oxidation state of Archaean komatiite magmas. *Nature*, *389*, 842–845. https://doi.org/10.1038/39860

Canil, D. (1999). Vanadium partitioning between orthopyroxene, spinel and silicate melt and the redox states of mantle source regions for primary magmas. *Geochimica et Cosmochimica Acta*, *63*, 557–572. https://doi.org/10.1016/S0016-7037(98)00287-7

Canil, D. (2002). Vanadium in peridotites, mantle redox and tectonic environments: Archean to present. *Earth and Planetary Science Letters*, *195*, 75–90. https://doi.org/10.1016/S0012-821X(01)00582-9

Canil, D., & Fedortchouk, Y. (2000). Clinopyroxene-liquid partitioning for vanadium and the oxygen fugacity during the formation of cratonic and oceanic mantle lithosphere. *Journal of Geophysical Research*, *105*, 26003–26016. https://doi.org/10.1029/2000JB900221

Canil, D., & Fedortchouk, Y. (2001). Olivine-liquid partitioning of vanadium and other trace elements, with applications to modern and ancient picrites. *Canadian Mineralogist*, *39*, 319–330. https://doi.org/10.2113/gscanmin.39.2.319

Canil, D. (1999). Vanadium partitioning between orthopyroxene, spinel and silicate melt and the redox states of mantle source regions for primary magmas. *Geochimica et Cosmochimica Acta*, *63*, 557–572. https://doi.org/10.1016/S0016-7037(98)00287-7

Cartier, C., Hammouda, T., Boyet, M., Mathon, O., Testemale, D., & Moine, B. N. (2015). Evidence for Nb^{2+} and Ta^{3+} in silicate melts under highly reducing conditions: A XANES study. *American Mineralogist*, *100*, 2152–2158. https://doi.org/10.2138/am-2015-5330

Cottrell, E., Walter, M. J., & Walker, D., (2009). Metal–silicate partitioning of tungsten at high pressure and temperature: Implications for equilibrium core formation in Earth. *Earth Planetary Science Letters*, *281*, 275–287. https://doi.org/10.1016/j.epsl.2009.02.024

de Moor, J. M., Fischer, T. P., Sharp, Z. D., King, P. L., Wilke, M., Botcharnikov, R. E., et al. (2013). Sulfur degassing at Erta Ale (Ethiopia) and Masaya (Nicaragua) volcanoes: Implications for degassing processes and oxygen fugacities of basaltic systems. *Geochemistry, Geophysics, Geosystems*, *14*, doi: 10.1002/ggge.20255

Dohmen, R., & Blundy, J. (2014). A predictive thermodynamic model for element partitioning between plagioclase and melt as a function of pressure, temperature and composition. *American Journal of Science*, *314*, 1319–1372. https://doi.org/10.2475/09.2014.04

Doyle, P. M., Schofield, P. F., Berry, A. J., Walker, A. M., & Knight, K. S. (2014). Substitution of Ti^{3+} and Ti^{4+} in hibonite ($CaAl_{12}O_{19}$). *American Mineralogist*, *99*, 1369–1382. https://doi.org/10.2138/am.2014.4532

Duffy, J. A. (1993). A review of optical basicity and its applications to oxidic systems. *Geochimica et Cosmochimica Acta*, *57*, 3961–3970. https://doi.org/10.1016/0016-7037(93)90346-X

Eeckhout, S. G, Bolfan-Casanov, N., McCammon C., Klemme, S., & Amiguet, E. (2007). XANES study of the oxidation state of Cr in lower mantle phases: Periclase and magnesium silicate perovskite. *American Mineralogist*, *92*, 966–972. doi: 10.2138/am.2007.2318

Ertel W., O'Neill H. St.C., Sylvester P. J., Dingwell D. B., & Spettel B. (2001). The solubility of rhenium in silicate melts: Implications for the geochemical properties of rhenium at high temperatures. *Geochimica et Cosmochimica Acta*, *65*, 2161–2170. https://doi.org/10.1016/S0016-7037(01)00582-8

Ertel, W., Dingwell, D. B., & Sylvester, P. J. (2008). Siderophile elements in silicate melts — A review of the mechanically assisted equilibration technique and the nanonugget issue. *Chemical Geology*, *248*, 119–139. https://doi.org/10.1016/j.chemgeo.2007.12.013

Ertel, W., O'Neill, H. St.C., Sylvester, P. J., & Dingwell, D. B. (1999). Solubilities of Pt and Rh in a haplobasaltic silicate melt at 1300 °C. *Geochimica et Cosmochimica Acta*, *63*, 2439–2449.

Fonseca, R. O. C., Mallmann, G., O'Neill, H. St.C., Campbell, I. H., & Laurenz, V. (2011). Solubility of Os and Ir in sulfide melt: Implications for Re/Os fractionation during mantle melting. *Earth and Planetary Science Letters*, *311*, 339–350. https://doi.org/10.1016/j.epsl.2011.09.035

Fonseca, R. O. C., Mallmann, G., Sprung, P., Sommer, J. E., Heuser, A., Speelmanns, I. M., & Blanchard, H. (2014). Redox controls on tungsten and uranium crystal/silicate melt partitioning and implications for the U/W and Th/W ratio of the lunar mantle. *Earth and Planetary Science Letters*, *404*, 1–13. https://doi.org/10.1016/j.epsl.2014.07.015

Fortenfant, S. S., Dingwell, D. B., Ertel-Ingrisch, W., Capmas, F., Birck, J. L., & Dalpe, C. (2006). Oxygen fugacity dependence of Os solubility in haplobasaltic melt. *Geochimica et Cosmochimica Acta*, *70*, 742–756. https://doi.org/10.1016/j.gca.2005.10.004

Fraser, D. G. (1975). Activities of trace elements in silicate melts. *Geochimica et Cosmochimica Acta*, *39*, 1525–1530. https://doi.org/10.1016/0016-7037(75)90154-4

Frost, D. J., Liebske, C., Langenhorst, F., McCammon, C. A., Tronnes, R. G., & Rubie, D. C. (2004). Experimental evidence for the existence of iron-rich metal in the Earth's lower mantle. *Nature*, *428*, 409–412. doi: 10.1038/nature02413

Gerdes, A., Kogarko, L. N., & Vladykin, N. V. (2017). New data on the age and nature of the Khan-Bogd alkaline granites, Mongolia. *Doklady Earth Sciences*, *477*, 1320–1324. doi: 10.1134/S1028334X17110137

Grossman, L., Beckett, J. R., Fedkin, A. V., Simon, S. B., & Cieslan, F. J. (2008). Redox conditions in the solar nebula: Observational, experimental, and theoretical calculations. *Reviews in Mineralogy and Geochemistry*, *68*, 93–140. doi: 10.2138/rmg.2008.68.7

Halse, H. R. (2014). *Using synchrotron radiation to determine the oxidation state of uranium in magmas.* PhD thesis, Imperial College London (http://hdl.handle.net/10044/1/19423).

Hanson, B., & Jones, J. H. (1998). The systematics of Cr^{3+} and Cr^{2+} partitioning between olivine and liquid in the presence of spinel. *American Mineralogist*, *83*, 669–684.

Hastie, A. R., Kerr, A. C., Mitchell, S. F., & Millar, I. L. (2009). Geochemistry and tectonomagmatic significance of lower Cretaceous island arc lavas from the Devils Racecourse Formation, eastern Jamaica. In: James, K., Lorente, M. A., & Pindell, J. L. (eds.) *The Origin and Evolution of the Caribbean Plate*. Geological Society London, Special Publications, *328*, 339–360. doi: 10.1144/SP328.14

Hill, E., Wood, B. J., & Blundy, J. D. (2000). The effect of Ca-Tschermaks component on trace element partitioning between clinopyroxene and silicate melt. *Lithos*, *53*, 203–215. https://doi.org/10.1016/S0024-4937(00)00025-6

Holzheid, A., Borisov, A., & Palme, H. (1994). The effect of oxygen fugacity and temperature on solubilities of nickel, cobalt and molybdenum in silicate melts. *Geochimimica et Cosmochimica Acta*, *58*, 1975–1981. https://doi.org/10.1016/0016-7037(94)90429-4

Jaques, A. L. (2016). Major and trace element variations in oxide and titanate minerals in the West Kimberley lamproites, Western Australia. *Mineralogy and Petrology*, *110*, 159–197. https://doi.org/10.1007/s00710-015-0420-4

Jollands, M. C., O'Neill, H. St.C., Van Orman, J., Berry, A. J., Hermann, J., Newville, M., & Lanzirotti, A. (2018). Substitution and diffusion of Cr^{2+} and Cr^{3+} in synthetic forsterite and natural olivine at 1200-1500 °C and 1 bar. *Geochimica et Cosmochimica Acta*, *220*, 407–428. https://doi.org/10.1016/j.gca.2017.09.030

Karner, J. M, Papike, J. J., Shearer, C. K., McKay, G., Le, L., & Burger, P. (2007). Valence state partitioning of Cr and V between pyroxene-melt: Estimates of oxygen fugacity for martian basalt QUE 94201. *American Mineralogist*, *92*, 1238–1241. https://doi.org/10.2138/am.2007.2607

Karner, J. M., Papike, J. J., Sutton, S. R., Shearer, C. K., Burger, P., McKay, G., & Le, L. (2008). Valence state partitioning of V between pyroxene-melt: Effects of pyroxene and melt composition, and direct determination of V valence states by XANES. Application to Martian QUE 94201 composition. *Meteoritics & Planetary Science*, *43*, 1275–1285. https://doi.org/10.1111/j.1945-5100.2008.tb00697.x

Kesson, S. E., Fitzgeral, J. D., Shelley, J. M. G., & Withers, R. L. (1995). Phase relations, structure and crystal chemistry of some aluminous silicate perovskites. *Earth and Planetary Science Letters*, *134*, 187–201. https://doi.org/10.1016/0012-821X(95)00112-P

Kimata, M. (1988). The crystal structure of non-stoichiometric Eu-anorthite: An explanation of the Eu-positive anomaly. *Mineralogical Magazine*, *52*, 257–265.

King, P. L., Hervig, R. L., Holloway, J. R., Delaney, J. S., & Dyar, M. D. (2000). Partitioning of Fe^{3+}/Fe_{total} between amphibole and basanitic melt as a function of oxygen fugacity. *Earth and Planetary Science Letters*, *178*, 97–112. https://doi.org/10.1016/S0012-821X(00)00071-6

Klemme, S., & O'Neill, H. St.C. (2000). The effect of Cr on the solubility of Al in orthopyroxene: Experiments and thermodynamic modelling. *Contributions to Mineralogy and Petrology*, *140*, 84–98. https://doi.org/10.1007/s004100000140

Krawczynski, M. J., Grove, T. L. & Behrens, H. (2012). Amphibole stability in primitive arc magmas: Effects of temperature, H_2O content, and oxygen fugacity. *Contributions to Mineralogy and Petrology*, *164*, 317–339. https://doi.org/10.1007/s00410-012-0740-x

Kuehner, S. M., Laughlin, J. R., Grossman, L., Johnson, M. L. & Burnett, D. S. (1989). Determination of trace element mineral/liquid partition coefficients in melilite and diopside by ion and electron microprobe techniques. *Geochimica et Cosmochimica Acta*, *53*, 3115–3130. https://doi.org/10.1016/0016-7037(89)90093-8

Laubier, M., Grove, T. L., & Langmuir, C. H. (2014). Trace element mineral/melt partitioning for basaltic and basaltic andesitic melts: An experimental and laser ablation ICP-MS study with application to the oxidation state of mantle source regions. *Earth and Planetary Science Letters*, *392*, 265–278. doi: 10.1016/j.epsl.2014.01.053

Laurenz, V., Fonseca, R. O. C., Ballhaus, C., & Sylvester, P. J. (2010). Solubility of palladium in picritic melts: 1. *The effect of iron. Geochimica et Cosmochimica Acta*, *74*, 2989–2998. https://doi.org/10.1016/j.gca.2010.02.015

Laurenz, V., Fonseca, R. O. C., Ballhaus, C., Jochum, K. P., Heuser, A., & Sylvester, P. J. (2013). The solubility of palladium and ruthenium in picritic melts: 2. *The effect of sulfur. Geochimica et Cosmochimica Acta*, *108*, 172–183. https://doi.org/10.1016/j.gca.2013.01.013

Law, K. M., Blundy, J. D., Wood, B. J., & Ragnarsdottir, K. V. (2000). Trace element partitioning between wollastonite and silicate-carbonate melt. *Mineralogical Magazine*, *64*, 651–661. https://doi.org/10.1180/002646100549670

Leitzke, F. P., Fonseca, R. O. C., Michely, L. T., Sprung, P., Münker, C., Heuser, A., & Blanchard, H. (2016). The effect of titanium on the partitioning behavior of high-field strength elements between silicates, oxides and lunar basaltic melts with applications to the origin of mare basalts. *Chemical Geology*, *440*, 219–238. https://doi.org/10.1016/j.chemgeo.2016.07.011

Leitzke, F. P., Fonseca, R. O. C., Sprung, P., Mallmann, G., Lagos, M., Michely, L. T. & Münker, C. (2017). Redox dependent behaviour of molybdenum during magmatic processes in the terrestrial and lunar mantle: Implications for the Mo/W of the bulk silicate Moon. *Earth and Planetary Science Letters*, *474*, 503–515. https://doi.org/10.1016/j.epsl.2017.07.009

Leitzke, F. P., Fonseca, R. O. C., Göttlicher, J., Steininger, R., Jahn, S., Prescher, C., & Lagos, M. (2018). Ti K-edge XANES study on the coordination number and oxidation state of titanium in pyroxene, olivine, armalcolite, ilmenite, and silicate glass during mare basalt petrogenesis. *Contributions to Mineralogy and Petrology*, *173*, 103. https://doi.org/10.1007/s00410-018-1533-7

Li, J. P., O'Neill, H. St.C. & Seifert, F. (1995). Subsolidus phase relations in the system MgO-SiO_2-Cr-O in equilibrium with metallic Cr, and their significance for the petrochemistry of chromium. *Journal of Petrology*, *36*, 107–132. https://doi.org/10.1093/petrology/36.1.107

Liu, X., & O'Neill, H. St.C. (2004). The effect of Cr_2O_3 on the partial melting of spinel-lherzolite in the system CaO-MgO-Al_2O_3-SiO_2-Cr_2O_3 at 1.1 GPa. *Journal of Petrology*, *45*, 2261–2286. doi: 10.1093/petrology/egh055

Liu, X., Xiong, X., Audétat, A., Li, Y., Song, M., Li, L., et al. (2014). Partitioning of copper between olivine, orthopyroxene, clinopyroxene, spinel, garnet and silicate melts at upper mantle conditions. *Geochimica et Cosmochimica Acta*, *125*, 1–22. https://doi.org/10.1016/j.gca.2013.09.039

Liu, X., Xiong, X., Audétat, A., & Li, Y. (2015). Partitioning of Cu between mafic minerals, Fe–Ti oxides and intermediate to felsic melts. *Geochimica et Cosmochimica Acta*, *151*, 86–102. https://doi.org/10.1016/j.gca.2014.12.010

Longhi, J., Walker, D., & Hays, J. F. (1976). Fe and Mg in plagioclase. 7th Lunar and Planetary Science Conference, 1281–1300.

Lundgaard, K. L., & Tegner, C. (2004). Partitioning of ferric and ferrous iron between plagioclase and silicate melt. *Contributions to Mineralogy and Petrology*, *147*, 470–483. https://doi.org/10.1007/s00410-004-0568-0

Ma, C., & Rossman, G. R. (2009). Grossmanite, $CaTi^{3+}AlSiO_6$, a new pyroxene from the Allende meteorite. *American Mineralogist*, *94*, 1491–1494. https://doi.org/10.2138/am.2009.3310

Mallmann, G., & O'Neill, H. St.C. (2007). The effect of oxygen fugacity on the partitioning of Re between crystals and silicate melt during mantle melting. *Geochimica et Cosmochimica Acta*, *71*, 2837–2857. https://doi.org/10.1016/j.gca.2007.03.028

Mallmann, G., & O'Neill, H. St.C. (2009). The crystal/melt partitioning of V during mantle melting as a function of oxygen fugacity compared with some other elements (Al, P, Ca, Sc, Ti, Cr, Fe, Ga, Y, Zr and Nb). *Journal of Petrology*, *50*, 1765–1794. https://doi.org/10.1093/petrology/egp053

Mallmann, G., & O'Neill, H. St.C. (2013). Calibration of an empirical thermometer and oxybarometers based on the partitioning of Sc, Y and V between olivine and silicate melt. *Journal of Petrology*, *54*, 933–949. https://doi.org/10.1093/petrology/egt001

Mallmann, G., Wykes, J., Berry, A. J., O'Neill, H. St.C., Cline II, C. J., Turner, S., & Rushmer, T.A. (2016). In situ XANES of U and Th in silicate liquids at high pressure and temperature. American Geophysical Union, Fall Meeting 2015, abstract MR13A-2418.

Mann, U., Marks, M., & Markl, G. (2006). Influence of oxygen fugacities on mineral compositions in peralkaline melts: The Katzenbuckel volcano, Southwest Germany. *Lithos*, *91*, 262–285. https://doi.org/10.1016/j.lithos.2005.09.004

Martel, C., Pichavant, M., Holtz, F., Scaillet, B., Bourdier, J. L., & Traineau, H. (1999). Effects of f_{O_2} and H_2O on andesite phase relations between 2 and 4 kbar. *Journal of Geophysical Research: Solid Earth*, *104*, 29453–29470. https://doi.org/10.1029/1999JB900191

Mathez, E. A. (1984). Influence of degassing on oxidation states of basaltic magmas. *Nature*, *310*, 371–375. https://doi.org/10.1038/310371a0

McCammon, C. (1997). Perovskite as a possible sink for ferric iron in the lower mantle. *Nature*, *387*, 694–696. https://doi.org/10.1038/42685

McCammon, C. (2005). The paradox of mantle redox. *Science*, *308*, 807–808. doi: 10.1126/science.1110532

McCanta, M. C., Dyar, M. D., Rutherford, M. J., & Delaney, J. S. (2004). Iron partitioning between basaltic melts and clinopyroxene as a function of oxygen fugacity. *American Mineralogist*, *89*, 1685–1693. https://doi.org/10.2138/am-2004-11-1214

McLean, N. M., Bowring, J. F., & Bowring, S. A. (2011). An algorithm for U-Pb isotope dilution data reduction and uncertainty propagation. *Geochemistry, Geophysics, Geosystems, 12*, Q0AA18. https://doi.org/10.1029/2010GC003478

Moussallam, Y., Edmonds, M., Scaillet, B., Peters, N., Gennaro, E., Sides, I., & Oppenheimer, C. (2016). The impact of degassing on the oxidation state of basaltic magmas: A case study of Kilauea volcano. *Earth and Planetary Science Letters, 450*, 317–325. https://doi.org/10.1016/j.epsl.2016.06.031

Murphy, M. D., Sparks, R. S. J., Barclay, J., Carroll, M. R., Lejeune, A. M., Brewer, T. S., et al. (1998). The role of magma mixing in triggering the current eruption at the Soufriere Hills Volcano, Montserrat, West Indies. *Geophysical Research Letters, 25*, 3433–3436. https://doi.org/10.1029/98GL00713

O'Neill, H. St.C. (1987). Quartz-fayalite-iron and quartz-fayalite-magnetite equilibria and the free energy of formation of fayalite (Fe_2SiO_4) and magnetite (Fe_3O_4). *American Mineralogist, 72*, 67–75.

O'Neill, H. StC. (2015). The effect of sulfide dissolved in silicate melts on enhancing the solubility of the Highly Siderophile Elements. American Geophysical Union, Fall Meeting 2015, abstract V52A-04.

O'Neill, H. St.C. (2016). The smoothness and shapes of chondrite-normalized rare earth element patters in basalts. *Journal of Petrology, 57*, 1463–1508. https://doi.org/10.1093/petrology/egw047

O'Neill, H. St.C., & Eggins, S. M. (2002). The effect of melt composition on trace element partitioning: An experimental investigation of the activity coefficients of FeO, NiO, CoO, MoO_2 and MoO_3 in silicate melts. *Chemical Geology, 186*, 151–181. https://doi.org/10.1016/S0009-2541(01)00414-4

O'Neill, H. St.C., & Navrotsky, A. (1984). Cations distributions and thermodynamic properties of binary spinel solid solutions. *American Mineralogist, 69*, 733–753.

O'Neill, H. St.C., Berry, A. J., & Eggins, S. M. (2008). The solubility and oxidation state of tungsten in silicate melts: Implications for the comparative chemistry of W and Mo in planetary differentiation processes. *Chemical Geology, 255*, 346–359. https://doi.org/10.1016/j.chemgeo.2008.07.005

O'Neill, H. St.C., Dingwell, D. B., Borisov, A., Spettel, B., & Palme, H. (1995). Experimental petrochemistry of some highly siderophile elements at high temperatures, and some implications for core formation and the mantle's early history. *Chemical Geology, 120*, 255–273. https://doi.org/10.1016/0009-2541(94)00141-T

O'Neill, H. St.C., McCammon, C., Canil, D., Rubie, D. C., Ross, C. R., & Seifert, F. (1993). Mössbauer spectroscopy of mantle transition zone phases and determination of minimum Fe^{3+} content. *American Mineralogist, 78*, 456–460.

Ohashi, H., Fujita, T. & Osawa, T. (1982). The crystal structure of $NaTiSi_2O_6$ pyroxene. *The Journal of the Japanese Association of Mineralogists, Petrologists and Economic Geologists, 77*, 305–309.

Ottonello, G., Moretti, R., Marini, L., & Zuccolini, M.V. (2001). Oxidation state of iron in silicate glasses and melts: a thermochemical model. *Chemical Geology, 174*, 157–179. https://doi.org/10.1016/S0009-2541(00)00314-4

Palme, H., & O'Neill, H. St.C. (2014), Cosmochemical estimates of mantle composition. In: *Treatise on Geochemistry*, 2nd Edition, Chapter 3.1, 1–39.

Papike, J. J., Karner, J. M., & Schearer, C. K. (2005). Comparative planetary mineralogy: Valence state partitioning of Cr, Fe, Ti, and V among crystallographic sites of olivine, pyroxene, and spinel from planetary basalts. *American Mineralogist, 90*, 277–290. https://doi.org/10.2138/am.2005.1779

Papike, J. J., Burger, P. V., Bell, A. S., Shearer, C. K., Le, L., & Jones, J. (2015). Normal to inverse transition in martian spinel: Understanding the interplay between chromium, vanadium, and iron valence state partitioning through a crystal-chemical lens. *American Mineralogist, 100*, 2018–2025. https://doi.org/10.2138/am-2015-5208

Papike, J. J., Simon, S. B., Burger, P. V., Bell, A. S., Shearer, C. K., & Karner, J. M. (2016). Chromium, vanadium, and titanium valence systematics in Solar System pyroxene as a recorder of oxygen fugacity, planetary provenance, and processes. *American Mineralogist, 101*, 907–918. https://doi.org/10.2138/am-2016-5507

Park, J. W., Kamenetsky, V., Campbell, I., Park, G., Hanski, E., & Pushkarev, E. (2017). Empirical constraints on partitioning of platinum group elements between Cr-spinel and primitive terrestrial magmas. *Geochimica et Cosmochimica Acta, 216*, 393–416. https://doi.org/10.1016/j.gca.2017.05.039

Pedersen, A. K. (1981). Armalcolite-bearing Fe-Ti oxide assemblages in graphite-equilibrated sialic volcanic rocks with native iron from Disko, central West Greenland. *Contributions to Mineralogy and Petrology, 77*, 307–324. doi: 10.1007/BF00371560

Peters, M. T., Shaffer, E. E., Burnett, D. S., & Kim, S. S. (1995). Magnesium and titanium partitioning between anorthite and Type B CAI liquid: Dependence on oxygen fugacity and liquid composition. *Geochimica et Cosmochimica Acta, 59*, 2785–2796. https://doi.org/10.1016/0016-7037(95)00173-W

Phinney, W. C. (1992). Partition coefficients for iron between plagioclase and basalt as a function of oxygen fugacity: Implications for Archean and lunar anorthosites. *Geochimica et Cosmochimica Acta, 56*, 1885–1895. https://doi.org/10.1016/0016-7037(92)90318-D

Popp, R. K., Virgo, D., Yoder, H. S., Hoering, T. C. & Phillips, M. W. (1995). An experimental study of phase equilibria and Fe oxy-component in kaersutitic amphibole: Implications for the fH and aHO in the upper mantle. *American Mineralogist, 80*, 534–548. https://doi.org/10.2138/am-1995-5-614

Poustovetov, A. A., & Roeder, P. L. (2000). The distribution of Cr between basaltic melt and chromium spinel as an oxygen geobarometer. *Canadian Mineralogist, 39*, 309–317. doi: 10.2113/gscanmin.39.2.309

Rhodes, J. M., & Vollinger, M. J. (2005). Ferric/ferrous ratios in 1984 Mauna Loa lavas: A contribution to understanding the oxidation state of Hawaiian magmas. *Contributions to Mineralogy and Petrology, 149*, 666–674. https://doi.org/10.1007/s00410-005-0662-y

Righter, K., & Shearer C. K. (2003). Magmatic fractionation of Hf and W: Constraints on the timing of core formation and differentiation in the Moon and Mars. *Geochimica et*

Cosmochimica Acta, *67*, 2497–2507. https://doi.org/10.1016/S0016-7037(02)01349-2

Righter, K., Sutton, S. R., Newville, M., Le, L., Schwandt, C. S., Uchida, H., et al. (2006a). An experimental study of the oxidation state of vanadium in spinel and basaltic melt with implications for the origin of planetary basalt. *American Mineralogist*, *91*, 1643–1656.

Righter, K., Leeman, W. P, & Hervig, R. L. (2006b). Partitioning of Ni, Co and V between spinel-structured oxides and silicate melts: Importance of spinel composition. *Chemical Geology*, *227*, 1–25. https://doi.org/10.1016/j.chemgeo.2005.05.011

Righter, K., Sutton, S., Danielson, L., Pando, K., Schmidt, G., Yang, H., et al. (2011). The effect of f_{O_2} on the partitioning and valence of V and Cr in garnet/melt pairs and the relation to terrestrial mantle V and Cr content. *American Mineralogist*, *96*, 1278–1290. 10.2138/am.2011.3690

Righter, K., Danielson, L. R., Pando, K. M., Shofner, G. A., Sutton, S. R., Newville, M., & Lee, C. T. (2016). Valence and metal/silicate partitioning of Mo: implications for conditions of Earth accretion and core formation. *Earth and Planetary Science Letters*, *437*, 89–100. https://doi.org/10.1016/j.epsl.2015.12.025

Roeder, P. L. & Reynolds, I. (1991). Crystallization of chromite and chromium solubility in basaltic melts. *Journal of Petrology*, *32*, 909–934. https://doi.org/10.1093/petrology/32.5.909

Sato, M. (1978). Oxygen fugacity of basaltic magmas and the role of gas-forming elements. *Geophysical Research Letters*, *5*, 447–449. https://doi.org/10.1029/GL005i006p00447

Schärer, U., Berndt, J., & Deutsch, A. (2011). The genesis of deep-mantle xenocrystic zircon and baddeleyite megacrysts (Mbuji-Mayi kimberlite): Trace-element patterns. *European Journal of Mineralogy*, *23*, 241–255. doi: 10.1127/0935-1221/2011/0023-2088

Schreiber, H. D. (1977). Redox states of Ti, Zr, Hf, Cr, and EU in basaltic magmas-an experimental study. *Lunar and Planetary Science Conference Proceedings 8*, 1785–1807.

Schreiber, H. D. (1983). The chemistry of uranium in glass-forming aluminosilicate melts. *Journal of the Less Common Metals*, *91*, 129–147. https://doi.org/10.1016/0022-5088(83)90102-9

Schreiber, H. D., & Haskin, L. A. (1976). Chromium in basalts: Experimental determination of redox states and partitioning among synthetic silicate phases. 7th Lunar and Planetary Science Conference, 1221–1259.

Schreiber, H. D., Merkel, R. C., Schreiber, L. V., & Balazs, G. B. (1987). Mutual interactions of redox couples via electron exchange in silicate melts: Models for geochemical systems. *Journal of Geophysical Research*, *92*, 9233–9245. https://doi.org/10.1029/JB092iB09p09233

Shannon, R. D. (1976). Revised effective ionic radii and systematic studies of inter-atomic distance in halides and chalcogenides. *Acta Cyrstallographica*, *32*, 751–767. https://doi.org/10.1107/S0567739476001551

Shearer, C. K., McKay, G., Papike, J. J., & Karner, J. M. (2006). Valence state partitioning of vanadium between olivine-liquid: Estimates of the oxygen fugacity of Y980459 and application to other olivine-phyric martian basalts. *American Mineralogist*, *91*, 1657–1663. https://doi.org/10.2138/am.2006.2155

Shim, S-H., Grocholski, B., Ye, Y., Alp, E. E., Xu, S., Morgan, D., et al. (2017). Stability of ferrous-iron-rich bridgmanite under reducing midmantle conditions. *Proceedings of the National Academy of Sciences of the USA*, *114*, 6468–6473.

Sievwright, R. H., Wilkinson, J. J., O'Neill, H. St.C., & Berry, A. J. (2017). Thermodynamic controls on element partitioning between titanomagnetite and andesitic-dacitic melts. *Contributions to Mineralogy and Petrology*, *172*, 62. https://doi.org/10.1007/s00410-017-1385-6

Simon, S. B., & Sutton, S. R. (2018). Valences of Ti, Cr, and V in Apollo 17 high-Ti and very low-Ti basalts and implications for their formation. *Meteoritics & Planetary Science*, *53*, 2138–2154. https://doi.org/10.1111/maps.13123

Smythe, D. J., & Brenan, J. M. (2015). Cerium oxidation state in silicate melts: Combined f_{O_2}, temperature and compositional effects. *Geochimica et Cosmochimica Acta*, *170*, 173–187. https://doi.org/10.1016/j.gca.2015.07.016

Smythe, D. J., & Brenan, J. M. (2016). Magmatic oxygen fugacity estimated using zircon-melt partitioning of cerium. *Earth and Planetary Science Letters*, *453*, 260–266. https://doi.org/10.1016/j.epsl.2016.08.013

Sommer, J. (2014). Experimental study: Solubility of Pt and Rh in silicate melt as a function of silicate melt compositions, temperature and oxygen fugacity. MSc Thesis, University of Bonn, Germany.

Sun, C.-O., Williams, R. J., & Sun, S.-S. (1974). Distribution coefficients of Eu and Sr for plagioclase-liquid and clinopyroxene-liquid equilibria in oceanic ridge basalt: An experimental study. *Geochimica et Cosmochimica Acta*, *38*, 1415–1433. https://doi.org/10.1016/0016-7037(74)90096-9

Sutton, S. R., & Newville, M. (2014). Synchrotron x-ray spectroscopic analysis. In: H. Holland and K. Turekian (eds.), *Treatise on Geochemistry*, 2nd Edition, v. *15*, pp. 213–230.

Sutton, S. R., Goodrich, C. A., & Wirick, S. (2017). Titanium, vanadium and chromium valences in silicates of ungrouped achondrite NWA 7325 and ureilite Y-791538 record highly-reduced origins. *Geochimica et Cosmochimica Acta*, *204*, 313–330. https://doi.org/10.1016/j.gca.2017.01.036

Sutton, S. R., Karner, J., Papike, J., Delaney, J. S., Shearer, C., Newville, M., et al. (2005). Vanadium K edge XANES of synthetic and lunar basaltic glasses and application to microscale oxygen barometry. *Geochimica et Cosmochimica Acta*, *69*, 2333–2348. https://doi.org/10.1016/j.gca.2004.10.013

Thomson, A. R., Kohn, S. C., Bulanova, G. P., Smith, C. B., Araujo, D., & Walter, M. J. (2016). Trace element composition of silicate inclusions in sub-lithospheric diamonds from the Juina-5 kimberlite: Evidence for diamond growth from slab melts. *Lithos*, *265*, 108–124. https://doi.org/10.1016/j.lithos.2016.08.035

Toplis, M. J., & Corgne, A. (2002). An experimental study of element partitioning between magnetite, clinopyroxene and iron-bearing silicate liquids with particular emphasis on vanadium. *Contributions to Mineralogy and Petrology*, *144*, 22–37. https://doi.org/10.1007/s00410-002-0382-5

Trail, D., Watson, E. B., & Tailby, N. D. (2011). The oxidation state of Hadean magmas and implications for early earth's atmosphere. *Nature*, *480*, 79–82. https://doi.org/10.1038/nature10655

Trail, D., Watson, E. B., & Tailby, N. D. (2012). Ce and Eu anomalies in zircon as proxies for the oxidation state of magmas. *Geochimica et Cosmochimica Acta*, *97*, 70–87. 10.1016/j.gca.2012.08.032

van Kan Parker, M., Mason, P. R. D. & van Westrenen, W. (2011). Experimental study of trace element partitioning between lunar orthopyroxene and anhydrous silicate melt: Effects of lithium and iron. *Chemical Geology*, *285*, 1–14. https://doi.org/10.1016/j.chemgeo.2011.02.007

Wade, J., Wood, B. J., & Norris, A. (2013). The oxidation state of tungsten in silicate melt at high pressures and temperatures. *Chemical Geology*, *335*, 189–193. https://doi.org/10.1016/j.chemgeo.2012.10.011

Wade, J., Wood, B. J., & Tuff, J. (2012). Metal–silicate partitioning of Mo and W at high pressures and temperatures: evidence for late accretion of sulphur to the Earth. *Geochimica et Cosmochimica Acta*, *85*, 58–74. https://doi.org/10.1016/j.gca.2012.01.010

Wadhwa, M. (2008). Redox conditions on small bodies, the Moon and Mars. *Reviews in Mineralogy and Geochemistry*, *68*, 493–510. https://doi.org/10.2138/rmg.2008.68.17

Wijbrans, C. H., Klemme, S., Berndt, J., & Vollmer, C. (2015). Experimental determination of trace element partition coefficients between spinel and silicate melt: The influence of chemical compositions and oxygen fugacity. *Contributions to Mineralogy and Petrology*, *169*, 45. https://doi.org/10.1007/s00410-015-1128-5

Wilke, M., & Behrens, H. (1999). The dependence of the partitioning of iron and europium between plagioclase and hydrous tonalitic melt on oxygen fugacity. *Contributions to Mineralogy and Petrology*, *137*, 102–114. https://doi.org/10.1007/s004100050585

Wood, B. J., & Blundy, J. D. (2003). Trace element partitioning under crustal and upper mantle conditions: The influences of ionic radius, cation charge, pressure, and temperature. In: R. W Carlson (Ed.), *Treatise on Geochemistry*, vol. 2, chapter 2.09, pp. 395–424.

Woodhead, J. D. (1989). Geochemistry of the Mariana arc (western Pacific): Source composition and processes. *Chemical Geology*, *76*, 1–24. https://doi.org/10.1016/0009-2541(89)90124-1

Young, E. D., Dyl, K. A., & Simon, J. I. (2012). Response to the Comment by SB Simon, L. Grossman, and SR Sutton on "Valence state of titanium in the Wark-Lovering rim of a Leoville CAI as a record of progressive oxidation in the early Solar Nebula". *Geochimica et Cosmochimica Acta*, *85*, 377–382. https://doi.org/10.1016/j.gca.2012.01.029

18

Titanomagnetite – Silicate Melt Oxybarometry

Róbert Arató and Andreas Audétat

ABSTRACT

The oxidation state of magmas is a parameter of prime importance in magmatic processes. Although various techniques exist to reconstruct the oxidation state of magmas, this task remains challenging, particularly in the case of intrusive rocks. Titanomagnetite is a ubiquitous phase in most magmatic rocks and due to its ability to incorporate redox-sensitive elements it can be used to estimate the fO_2 of the melt. Traditional titanomagnetite-related oxybarometry is based on the measurement of titanomagnetite-ilmenite equilibrium pairs, which is applicable to most fresh volcanic rocks containing both titanomagnetite and ilmenite. However, it cannot be used for reconstructing fO_2 in ilmenite-free magmas, or in which Fe-Ti oxides were destroyed/reset at subsolidus conditions. Therefore, two methods were developed that can be used also in altered rocks. These two methods are based on (i) the partitioning of vanadium between titanomagnetite and silicate melt, and (ii) the exchange of iron and titanium between titanomagnetite and silicate melt. Both of these phases are commonly preserved as inclusions in phenocrysts, and their original compositions can be obtained by analyzing entire inclusions by LA-ICP-MS. In this way, titanomagnetite–silicate melt partition coefficients and corresponding fO_2 can be determined even for slowly cooled igneous rocks such as granites.

18.1. INTRODUCTION

The oxidation state of magmas is important for several reasons, as detailed in Chapter 1 of this book. Its strong effect on mineral stability has been demonstrated by many authors, starting with the pioneering studies of Muan (1955) and Eugster (1957) who provided the first experimental evidence that some mineral phases are only stable at reduced conditions (e.g., wüstite), whereas others are stable only at oxidized conditions (e.g., hematite). Some years later, Buddington and Lindsley (1964) investigated the effect of oxidation state, commonly expressed as oxygen fugacity (fO_2), on the stability of the minerals in the FeO-Fe_2O_3-TiO_2 system and found that the composition of the magnetite-ulvöspinel and hematite-ilmenite solid solution series is a strong function of fO_2. Fincham and Richardson (1954) showed that fO_2 also plays a crucial role in controlling the nature of magmatic sulfur-bearing phases, and related to that, Carroll and Rutherford (1985) could demonstrate that anhydrite occurs only at high fO_2 whereas pyrrhotite is present at more reducing conditions. However, the influence of oxygen fugacity in magmatic systems is not restricted to mineral stability. Physical properties of magmas, such as diffusion rate (e.g., Jollands et al., 2018), rheology (Kohlstedt & Zimmermann, 1996), and viscosity (Liebske et al., 2003; Di Genova et al., 2017) are also strongly influenced by fO_2. It also affects the solubility of volatiles, their corresponding fluid–melt partition coefficients, as well as mineral-melt and fluid-melt partition coefficients of various metals. For example, sulfur is dissolved in silicate melts either

Bayerisches Geoinstitut, University of Bayreuth, Bayreuth, Germany

Magma Redox Geochemistry, Geophysical Monograph 266, First Edition.
Edited by Roberto Moretti and Daniel R. Neuville.

DOI: 10.1002/9781119473206.ch18

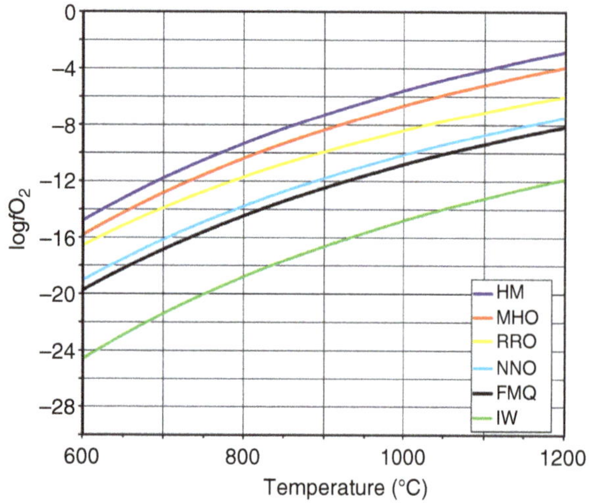

Fig. S1 Summary of commonly applied oxygen fugacity buffers and their position in the oxygen fugacity–temperature space. HM= hematite-magnetite (Frost, 1991), MHO= MnO-Mn_3O_4 (Chou, 1978), RRO= Re-ReO_2 (Pownceby, 1994), NNO= Ni-NiO (Huebner & Sato, 1970), FMQ= fayalite-magnetite-quartz (Frost, 1991), IW= iron-wüstite (Frost, 1991).

as S^{2-} or S^{6+} or a combination of both, depending on magmatic oxygen fugacity (Carroll & Rutherford, 1988). The S^{2-} to S^{6+} (sulfide to sulfate) transition takes place in a relatively narrow fO_2 range, which causes an abrupt increase in sulfur solubility in melts (Jugo et al., 2010) and a decrease in the fluid-melt partition coefficient of S (Keppler, 2010) one log unit above the fayalite-magnetite-quartz buffer (FMQ+1; Fig. S1).

The above processes either directly or indirectly control the mineralizing potential of intrusions; the knowledge of magmatic oxidation state is therefore crucial for understanding ore formation. In his landmark contribution, Ishihara (1977) categorized granites as oxidized or reduced based on their Fe-Ti oxide content. Since then, multiple studies aimed at relating the oxidation state of intrusions to their ore-forming potential and to associate specific types of ore deposits with oxidized or reduced sources. It was found that porphyry copper deposits and porphyry Mo deposits are usually associated with magnetite-series rocks (e.g., Ishihara, 1977; Candela & Bouton, 1990; Candela, 1992; Blevin & Chappell, 1992; Hedenquist & Lowenstern, 1994; Audétat et al., 2011), whereas tin and tungsten deposits are usually associated with ilmenite-series rocks (e.g., Ishihara, 1977; Lehmann, 1990).

The Fe-Ti oxide (or more simply magnetite–ilmenite) oxybarometer (e.g., Carmichael, 1967; Andersen & Lindsley, 1988; Ghiorso & Evans, 2008) is probably the best established magmatic oxybarometer to date. However, it is commonly difficult or even impossible to reconstruct magmatic fO_2 based on this method, especially in the case of intrusive and/or mineralized rocks. Most oxybarometers such as the magnetite-ilmenite oxybarometer are prone to resetting during slow cooling, whereas empirical redox indicators such as the whole rock Fe(III)/Fe(II) ratio or the presence of magmatic anhydrite rarely survive processes of hydrothermal alteration and surficial weathering (Ballard et al., 2002), meaning that their final state does not reflect magmatic conditions anymore.

Based on the studies of Irving (1978) and Canil (2002), which demontrated that the partitioning of multivalent elements such as Fe or V between titanomagnetite (and, more broadly, spinel) and melt is sensitive to fO_2, new empirical oxybarometers were calibrated experimentally. The main advantage of these calibrations over traditional approaches is that fO_2 can be calculated based on the measurement of melt inclusions and titanomagnetite inclusions – two frequently occurring inclusion types in silicic rocks – that were preserved within phenocrysts (e.g., quartz) and thus were protected from subsolidus and hydrothermal alteration. In this way the oxidation state of slowly cooled and/or altered rocks, which are commonly the only type of igneous rocks found in ore deposits, becomes also accessible.

In this chapter, titanomagnetite-based oxybarometers are briefly reviewed with a focus on two recently developed methods, starting with theoretical considerations, followed by experimental calibrations, and ending with applications to natural samples.

18.2. OXYBAROMETERS RELATED TO TITANOMAGNETITE

Titanomagnetite is a spinel group mineral, a solid solution between pure magnetite and ulvöspinel (detailed description in O'Neill & Navrotsky, 1984). Spinels have the general formula of AB_2O_4, where A and B denote cations of either 2+ and 3+ valence ($A^{2+}B^{3+}{}_2O_4$, so-called 2-3 spinels) or of 4+ and 2+ valence ($A^{4+}B^{2+}{}_2O_4$, so-called 4-2 spinels). Spinel structure (space group $Fd\bar{3}m$) is generally described as a slightly distorted cubic close packed (CCP) array of anions. The cations fill the interstices within the oxygen framework, on 16 octahedral and 8 tetrahedral sites. This cation occupancy results in two different cation distribution schemes. In normal spinel, cation A occupies the tetrahedral site and the two B cations occupy the octahedral sites (general formula: AB_2O_4), whereas in inverse spinel one of the B cations occupies the tetrahedral site and the remaining A and B cations fill the octahedral sites, resulting in the general formula $B(AB)O_4$. Both ideal magnetite (Fe_3O_4) and ulvöspinel (Fe_2TiO_4) have inverse spinel structure at room temperature, with the structural formulae $(Fe^{3+})(Fe^{2+}Fe^{3+})O_4$

and $(Fe^{2+})(Fe^{2+}Ti^{4+})O_4$, respectively, whereas at magmatic temperatures the structure gets disordered towards random distribution (Wu & Mason, 1981). A complete solid solution between magnetite and ulvöspinel exists at temperatures above 600 °C (Price, 1981). The intermediate compositions, known as titanomagnetite ($Fe^{2+}{}_{1+x}Fe^{3+}{}_{2-2x}Ti_xO_4$), are formed by the replacement of two Fe^{3+} cations by Fe^{2+} and Ti^{4+}. For the sake of simplicity, in this study we refer to the magnetite-ulvöspinel solid solution as titanomagnetite.

The Fe and Ti content of magnetite coexisting with ilmenite is extremely sensitive to oxygen fugacity (and also temperature). Several oxybarometers involving titanomagnetite are based on this principle. The most commonly used ones are the magnetite-ilmenite oxybarometer (where magnetite refers in fact to titanomagnetite) and the so-called QUILF (quartz-ulvöspinel-ilmenite-feldspar; Andersen et al., 1993) method, whereas other mineral stability/equilibria-based oxybarometers focus on biotite, amphibole, K-feldspar, and magnetite (e.g., Buddington & Lindsley, 1964). On the other hand, an important character of magnetite is that its structure can easily distort to accommodate a large range of cations, including those of transition metals such as Sc, Ti, V, Cr, Mn, Fe, Co, Ni, Cu, Zn, Ru, Rh, and Ir (Righter et al., 2006a). This opens various opportunities to estimate fO_2 via the partitioning of redox-sensitive elements between magnetite and melt (e.g., Toplis & Corgne, 2002; Righter et al., 2006a, b; Mallmann & O'Neill, 2009).

18.3. OXYBAROMETERS BASED ON MINERAL EQUILIBRIA INVOLVING TITANOMAGNETITE

By far the most commonly used oxybarometer in crustal rocks is based on the composition of coexisting Fe-Ti oxides. The compositional sensitivity of the magnetite-ulvöspinel solid solution coexisting with the hematite-ilmenite solid solution (Fig. 18.1) to temperature and oxygen fugacity was first demonstrated in the 1960s. Buddington and Lindsley (1964) experimentally determined the compositions of coexisting Fe-Ti oxides for a wide range of temperatures and oxygen fugacities and applied the experimental calibration to various natural magmatic and metamorphic rocks. They demonstrated that stable Fe-Ti oxide pairs occur according to the following evolution sequence with increasing oxygen fugacity: ulvöspinel-rich $magnetite_{ss}$ + $ilmenite_{ss}$, ulvöspinel-poor $magnetite_{ss}$ + $ilmenite_{ss}$, ulvöspinel-poor $magnetite_{ss}$ + $hematite_{ss}$, $hematite_{ss}$ + rutile, where $_{ss}$ refers to solid solution. This is in line with later studies (e.g., Ghiorso & Sack, 1991) demonstrating that the ulvöspinel content of titanomagnetite decreases whereas the hematite content of the hematite-ilmenite series increases with increasing fO_2

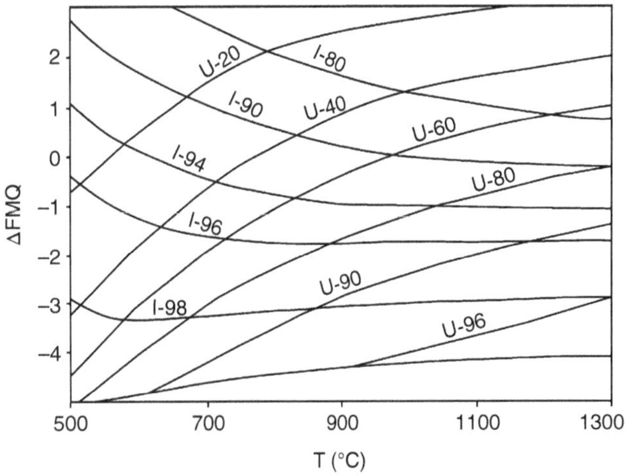

Fig. 18.1 Graphical representation of the magnetite–ilmenite oxybarometer/thermometer (simplified after Ghiorso & Sack, 1991). Oxygen fugacity is defined as relative to the fayalite-magnetite-quartz buffer (FMQ). The composition of the titanomagnetite is given in mole percent ulvöspinel (U) whereas that of the ilmenite is given as mole percent ilmenite (I).

(Fig. 18.1). On the other hand, Buddington and Lindsley (1964) already pointed out that Fe-Ti oxides in natural samples commonly show subsolidus re-equilibration features, precluding the estimation of original magmatic fO_2 in slowly cooled rocks. Subsequently, Carmichael (1967) successfully applied the titanomagnetite-ilmenite oxybarometer (and thermometer) to a series of SiO_2-rich volcanic rocks and found consistent relationships between fO_2 and the stability of ferromagnesian silicate minerals such as olivine and orthopyroxene. However, he observed that in some rocks, multiple generations of Fe-Ti oxides appear, some of which probably did not form in equilibrium with the others. The equilibrium test of Bacon and Hirschmann (1988), which is based on the Mg/Mn ratio of both oxide phases, allows distinguishing between equilibrium and non-equilibrium pairs. A corresponding thermodynamic model for the Fe-Ti oxide thermometer/ oxybarometer was developed by Andersen and Lindsley (1988) as well as by Ghiorso and Sack (1991) and Ghiorso and Evans (2008). The most commonly used calibrations are those of Andersen and Lindsley (1988) and Ghiorso and Evans (2008). It should be noted that these two calibrations commonly yield significantly different results, especially with regard to temperature (commonly > 100 °C difference). Several factors are responsible for this discrepancy. First, the model of Andersen and Lindsley (1988) is based on a less extensive dataset that the one of Ghiorso and Evans (2008), with the latter covering a broad range of experimental conditions between 800 °C and 1300 °C, and an fO_2 range from FMQ-2.3 to FMQ +3.7. Second, the model of Ghiorso and Evans (2008) accounts for the configurational entropy changes related to both the short-range cation order and the $R\bar{3}$-$R\bar{3}c$ order-disorder transition of the rhombohedral phase, whereas the model of Andersen and Lindsley (1988) does not. These modifications result in differences in fO_2 and T estimation because the latter model assumes an ordered $R\bar{3}$ structure for the hematite-ilmenite solid solution over the entire T-fO_2 range. This assumption, however, is not valid at oxygen fugacities above FMQ+1.7 and at temperatures of 700–900 °C, where ilmenite has a cation-disordered structure. Therefore, the model of Ghiorso and Evans (2008) is better established at the above-mentioned conditions, which are typical for many natural magmas.

The study of Ghiorso and Evans (2008) also shows that temperature estimates are associated with relatively large uncertainties (easily >50 °C at oxidized conditions), whereas oxygen fugacity can be accurately constrained throughout the whole temperature range. This renders Fe-Ti oxide pairs an excellent indicator of the oxidation state, provided that they are unaltered and in equilibrium with each other.

As demonstrated by Carmichael (1967), also the stability and composition of ferromagnesian silicates depends on fO_2 and T. Following up on that, Frost et al. (1988) showed that in Fe-rich metamorphic rocks and in highly evolved igneous rocks in which fayalite coexists with quartz and Fe-Ti oxides fO_2 is controlled by the following equilibrium reaction (QUILF):

$$SiO_2(qtz) + 2Fe_2TiO_4(usp) = 2FeTiO_3(ilm) + Fe_2SiO_4(fay) \quad (18.1)$$

Rocks that contain all these four minerals are scarce. However, subsequent studies (e.g., Frost & Lindsley, 1992; Lindsley & Frost, 1992) extended the QUILF calibration by additional equilibria involving augite, pigeonite, and orthopyroxene in the system FeO-CaO-MgO-SiO_2-TiO_2 (referred to as "Ca-QUILF"). This extension made QUILF applicable to a broader range of rock compositions, including ones that do not contain olivine or ilmenite (software application: Andersen et al., 1993). An advantage of the method is that it enables estimation of fO_2 also in rocks in which Fe-Ti oxides suffered subsolidus alteration, provided that a number of parameters (e.g., pressure) can be estimated independently, which is often not straightforward in practice. The QUILF model was further extended to titanite-bearing mineral assemblages by Xirouchakis et al. (2001).

A different approach to estimate fO_2 was presented by Wones and Eugster (1965) based on biotite-alkali feldspar-titanomagnetite equilibria, which was further advanced by the studies of Wones (1981, 1989), Carmichael and Ghiorso (1990), Frost (1991), Ghiorso and Sack (1991), Andersen et al. (1993), and Cygan et al. (1996). This oxybarometer is based on the following equilibrium (simplified after Wones & Eugster, 1965):

$$KFe_3AlSi_3O_{10}(OH)_2 + 1/2\,O_2 = KAlSi_3O_8 + FeFe_2O_4 + H_2O \quad (18.2)$$

Because biotite and K-feldspar are usually chemically more complex, this leads to a continuous reaction series in which biotite becomes increasingly Mg-rich with increasing fO_2, at the expense of components released to K-feldspar and titanomagnetite. From Equation 18.2 it is clear that estimation of fO_2 via this reaction requires knowledge of the magmatic, unaltered compositions of biotite, alkali feldspar, and titanomagnetite, as well as of the H_2O fugacity, which can again be challenging in altered/intrusive rocks.

For further information regarding titanomagnetite-based oxybarometry, the reader is referred to the studies of Frost and Lindsley (1991) and other chapters in the same RiMG issue, as well as to a more recent review by Herd et al (2008).

18.4. OXYBAROMETERS BASED ON ELEMENT PARTITIONING BETWEEN TITANOMAGNETITE AND SILICATE MELT

The partitioning of several elements was found to be fO_2-dependent in the titanomagnetite-melt system, although only a few of them were applied later as an oxybarometer. Iron and vanadium partitioning are of upmost importance in this context and are described in detail later. Interestingly, $D_{metal}^{mgt/melt}$ was also found to depend on fO_2 in the case of other metals (Sc, Ti, Mn, Co, Ni, Cu, Zn, Nb), which is not a direct result of changing fO_2 due to their mainly homovalent nature. Horn et al. (1994) observed that the partition coefficient of Co and Sc between Cr-bearing titanomagnetite and basaltic melt slightly increases with increasing fO_2 and explained the latter dependence with the simultaneous increase of Fe^{3+} in the oxide phase. Increasing partition coefficients of Mn, Zn, Ni, Cu, Ti, and Nb with increasing fO_2 were reported by Sievwright et al. (2017). These observations were explained by the decreasing FeO content of the melt, which in turn was assumed to change melt composition and affect element partitioning. In contrast, Toplis and Corgne (2002) found no evidence for the fO_2 dependence of divalent cation (Co, Mn, and Ni) partitioning between titanomagnetite and ferrobasaltic melt, despite the varying titanomagnetite and slightly varying melt composition in their experiments. Although Sievwright et al. (2017) argue that the partitioning of homovalent tetravalent and pentavalent cations in response to fO_2 can be predicted easier, and therefore could offer an improved proxy for fO_2, most studies focused on V, since $D_V^{mgt/melt}$ can vary several orders of magnitude under terrestrial conditions.

18.4.1. Vanadium Partitioning Oxybarometry

Vanadium has a strong affinity to titanomagnetite because of the similar ionic radii of V^{3+} (0.64 Å) and Fe^{3+} (0.645 Å), as well as of V^{4+} (0.58 Å) and Ti^{4+} (0.605 Å) in octahedral coordination. Consequently, V concentrations in natural titanomagnetites are usually easily measurable by microanalytical methods such as LA-ICP-MS and EPMA. Vanadium concentrations in silicate melts are distinctly lower than in titanomagnetite and generally decrease with increasing SiO_2 content given that the melt is titanomagnetite-saturated (Toplis & Corgne, 2002; Prytulak et al, 2016), with some highly evolved rhyolite melts containing less than 1 ppm V.

Multiple studies have shown that vanadium partitioning between spinel group minerals and silicate melt ($D_V^{sp/melt}$) can be used as a proxy for magmatic oxidation state, and several calibrations have been developed (Irving, 1978; Horn et al., 1994; Canil, 1999, 2002; Righter et al., 2006a, b; Mallmann & O'Neill, 2009). The reason why V partitioning is sensitive to fO_2 is that the average oxidation state of V in spinel group minerals is always about 1 unit lower than in the coexisting silicate melt (e.g., Righter et al., 2006b) where V is present in numerous oxidation states, hence its partitioning between the two phases involves oxygen (see Equations 18.3 and 18.4). Vanadium can principally attain valence states ranging from 0 to +5. However, as suggested by several authors (e.g., Toplis & Corgne, 2002; Sutton et al., 2006) it occurs dominantly as V^{3+}, V^{4+}, or V^{5+} in silicate melts at the conditions relevant for the Earth's crust. According to the calculations of Toplis and Corgne (2002) the dominant valence state of V at FMQ-0.3 is 4+. With increasing fO_2 the relative proportion of V^{4+} in the silicate melt decreases and the proportion of V^{5+} increases until it becomes dominant above ~FMQ+2.2 (Fig. 18.2).

As shown by Righter et al. (2006b), the valence of V in spinel group minerals is always lower than in the silicate melt. At upper crustal conditions it is predominantly V^{3+} and occurs only in the octahedral site in pure magnetite, whereas minor amounts of octahedrally coordinated V^{4+} are additionally present in titanomagnetite (Toplis & Corgne, 2002; Righter et al., 2006a). The dominant reactions describing the partitioning of V between titanomagnetite and silicate melt are the following (Arató & Audétat, 2017a):

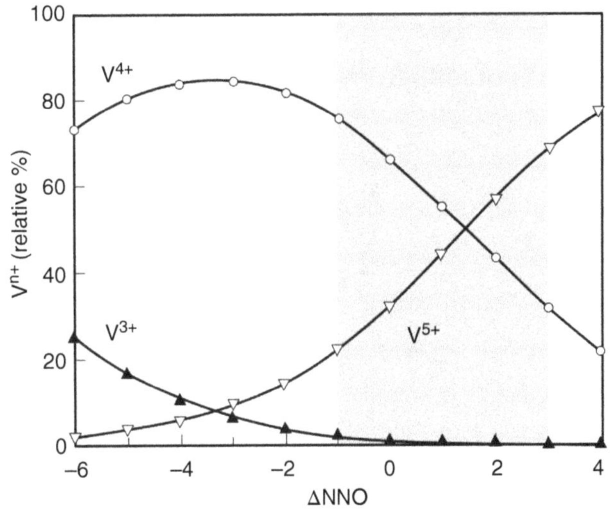

Fig. 18.2 Estimated proportions of V^{3+}, V^{4+}, and V^{5+} in a ferrobasaltic melt at 1070 °C as a function of fO_2 (slightly modified after Toplis & Corgne, 2002). The green field marks the fO_2 conditions relevant for the Earth's crust.

$$VO_{5/2}^{melt} = VO_{3/2}^{mgt} + 0.5O_2 \tag{18.3}$$

$$VO_2^{melt} = VO_{3/2}^{mgt} + 0.25O_2 \tag{18.4}$$

From these two equations it is clear that the partitioning of vanadium depends on fO_2 over the entire range of fO_2 conditions relevant for the Earth's crust. This is in agreement with previous studies (e.g., Irving, 1978; Horn et al., 1994; Papike et al., 2005; Mallmann & O'Neill, 2009, Sossi et al., 2018), which demonstrated that $D_V^{sp/melt}$ changes as a function of oxidation state over that range.

While all the above-mentioned V-partitioning studies focused on mafic systems, a recent study by Arató and Audétat (2017a) focused on rhyolitic melts. Their experiments showed that at fixed melt composition and temperature the partition coefficient of vanadium between titanomagnetite and melt ($D_V^{mgt/melt}$) decreases by 1.5–1.7 log units as fO_2 increases from FMQ+0.7 to values corresponding to the MnO-Mn_3O_4 (MHO) buffer (Fig. 18.3; Fig. S1). This strong oxygen fugacity dependence means that $D_V^{mgt/melt}$ is a sensitive proxy of fO_2 also in SiO_2-rich melts. However, their study highlighted some differences regarding the temperature and titanomagnetite composition dependence of vanadium partitioning as compared to the studies performed in mafic systems. Righter et al. (2006a) observed that $D_V^{spinel-melt}$ decreased with increasing temperature at FMQ+0.7, whereas no temperature effect was observed at FMQ+3.7. In contrast, Arató and Audétat (2017a) found that temperature significantly influences vanadium partitioning both at relatively reduced (FMQ+0.7) and at relatively oxidized

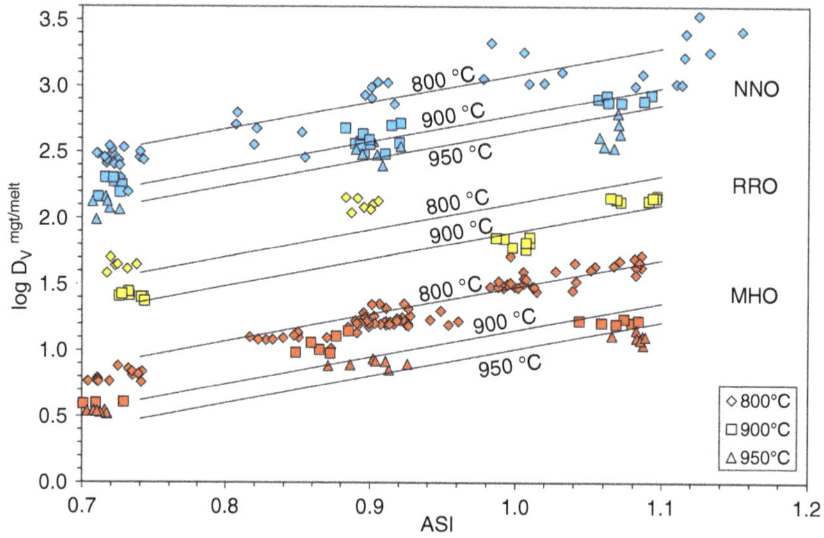

Fig. 18.3 Results of titanomagnetite-melt vanadium partitioning experiments in rhyolitic systems as a function of oxygen fugacity, temperature, and melt composition (Arató & Audétat, 2017a). ASI= molar $Al_2O_3/(CaO+Na_2O+K_2O)$. Colors indicate different fO_2 buffers (shown on the right), MHO=MnO–Mn_3O_4 buffer, NNO=Ni–NiO buffer, RRO=Re–ReO buffer; symbol shapes mark different temperatures (diamonds: 800 °C; squares: 900 °C; triangles: 950 °C). The solid lines are calculated from the regression equation given in Equation 18.5. Slightly modified after Arató and Audétat (2017a).

(MHO) conditions, even after subtraction of the temperature effect on absolute fO_2 along a given oxygen fugacity buffer. The contrasting observations may originate from either the different spinel compositions or the different temperature ranges (1150–1325 °C and 800–1000 °C) investigated in the two studies. Similarly, Arató and Audétat (2017a) found a relatively large effect of the melt composition (Fig. 18.3), whereas none of the studies conducted in mafic systems found any clear evidence for a melt compositional effect. The melt compositional effect on V partitioning was also investigated by Sievwright et al. (2017), however, the observed compositional dependence in that study was attributed to the change in FeO^{2+} content of the melt. In contrast to the melt compositional dependence, Arató and Audétat (2017a) did not observe any effect of titanomagnetite composition (0.2–14 wt% TiO_2) whereas previous studies did (Nielsen et al., 1994; Canil, 1999, 2002; Righter et al., 2006b; Mallmann & O'Neill, 2009), which can partly be explained by the much smaller compositional range of spinel in felsic versus mafic magmas.

The results of Mallmann and O'Neill (2009) suggest that over the fO_2 range covered in the current study (FMQ+0.7–FMQ+4.0) $\log D_V^{mgt/melt}$ should vary approximately linearly with oxygen fugacity. This interpretation is in good agreement with the results of Arató and Audétat (2017a). Considering this evidence and the observed temperature and melt composition dependence, vanadium partitioning between titanomagnetite and silicate melt in silicic magmas can be described by the following equation:

$$\begin{aligned} logD_V^{mgt/melt} &= \frac{0.3726*10,000}{T(K)} + 2.0465*ASI \\ &\quad - 0.4773*\Delta FMQ - 2.1214 \end{aligned}, \quad (18.5)$$

or if the regression is performed for ΔFMQ:

$$\begin{aligned} \Delta FMQ = &-2.0511*logD_V^{mgt/melt} + \frac{0.7809*10,000}{T(K)} \\ &+ 4.2367*ASI - 4.4767 \end{aligned} \quad (18.6)$$

In these equations, ASI stands for molar $Al_2O_3/(CaO + Na_2O + K_2O)$ commonly used for distinguishing between peralkaline, metaluminous, and peraluminous granitoids. ΔFMQ refers to the deviation of fO_2 (in log units) from the fayalite-magnetite-quartz buffer, T is given in Kelvin, and $D_V^{mgt/melt}$ values are calculated based on V concentrations reported as µg/g V. The observed strong melt composition dependence in silicic glasses as well as the significantly lower temperatures (mostly 800–950 °C) in the experiments of Arató and Audétat (2017a) compared to previous investigations (1100–1300 °C) are likely the reasons why the $D_V^{mgt/melt}$ values in silicic systems are 1–3 orders of magnitude higher than those obtained in more mafic systems (e.g., Canil, 1999; Righter et al., 2006a, b). Therefore, a direct comparison of these calibrations with that of Arató and

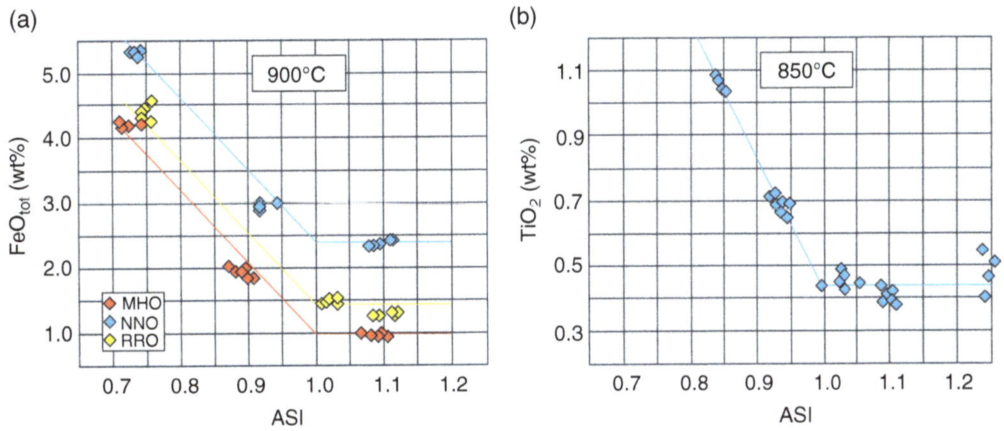

Fig. 18.4 (a) Iron solubility in rhyolite melts as a function of alumina saturation index and oxygen fugacity (data from Arató & Audétat, 2017a, b). (b) Titanium solubility as a function of alumina saturation index (modified after Kularatne & Audétat, 2014). Solid lines are guides for the eyes based on the experimental data. MHO=MnO–Mn_3O_4 buffer, NNO=Ni–NiO buffer, RRO=Re–ReO buffer.

Audétat (2017a) is not possible and the appropriate calibration should be selected based on melt composition. On the other hand, the studies by Sievwright et al. (2017) and Sossi et al. (2018) provide a better opportunity for comparing recent calibrations. Sievwright et al. (2017) report lower Dv values than Arató and Audétat (2017a) at the corresponding experimental conditions. The most plausible explanation for this discrepancy is the higher temperature and a less silicic melt composition applied by Sievwright et al. (2017). On the other hand, Sossi et al. (2018) report a good agreement with the results of Arató and Audétat (2017a) at high fO_2. At lower oxygen fugacity the discrepancy might be explained by the large difference in titanomagnetite V content in the two studies (~8 wt% vs. ~1200 ppm at FMQ+0.7) that can affect partitioning.

18.4.2. FeTiMM

FeTiMM is a recently developed oxybarometer (Arató & Audétat, 2017b) that is based on the partitioning of Fe and Ti between titanomagnetite and silicate melt. The idea of the method stemmed from iron and titanium solubility experiments. The solubility of iron in metaluminous rhyolitic glasses strongly increases with increasing oxygen fugacity (Gaillard et al., 2001). Furthermore, that study showed that melt composition also affects the Fe^{3+}/Fe^{2+} ratio and – although not stated specifically in the paper – the solubility of Fe (FeO_{tot}). Arató and Audétat (2017b) investigated the effect of melt composition on titanomagnetite solubility (and thus Fe partitioning between titanomagnetite and silicate melt) in more detail by a series of titanomagnetite-melt partitioning experiments. They found that the effect of melt composition is very strong in peralkaline melts, but diminishes/changes slope somewhere around ASI=1 (Fig. 18.4a). The increasing solubility of iron with increasing ASI in peralkaline melts is in agreement with the study of Borisov et al. (2017) who demonstrated that alkali ions can stabilize Fe in the melt. The solubility of rutile (i.e., TiO_2) in silicic melt was found to depend in a very similar manner on melt composition (Kularatne & Audétat, 2014; Fig. 18.4b). However, in contrast to iron the solubility of TiO_2 does not directly depend on fO_2. Although Sievwright et al. (2017) demonstrated that $D_{Ti}^{mgt/melt}$ changes with fO_2 through the coupled effect related to melt FeO content, by dividing the titanomagnetite–silicate melt partition coefficient of Fe ($D_{Fe}^{mgt/melt}$) by that of Ti ($D_{Ti}^{mgt/melt}$) the sensitivity to fO_2 can be maintained, whereas the effect of melt composition can be significantly reduced. The resulting Fe-Ti exchange coefficient ($D_{Fe\text{-}Ti}^{mgt/melt}$) thus depends mainly on oxygen fugacity and only to a small extent on melt composition (which is again fO_2 dependent), whereas the effects of temperature, pressure, and titanomagnetite composition were found to be negligible.

FeTiMM was calibrated on a large experimental dataset that includes 109 ilmenite-bearing and ilmenite-free experiments performed at 750–1100 °C, 0.1–700 MPa, oxygen fugacities of −1.3 to +5.5 log units relative to the FMQ (fayalite-magnetite-quartz) buffer (i.e., FMQ-1.3 to FMQ+5.5), melt compositions of 48–79 wt% SiO_2, and ASI=0.3–1.3, and titanomagnetite compositions of 0.01–28 wt% TiO_2 (Arató & Audétat, 2017b). The whole dataset could be described by a single equation, as ilmenite-saturated and ilmenite-undersaturated experiments showed the same dependence on oxygen fugacity and melt composition:

$$\Delta FMQ = \left(\log\left(D_{FeOtot}{}^{mgt/melt}/D_{TiO2}{}^{mgt/melt}\right) + 0.137*AMCNK + 0.102\right)/(0.288*AMCNK + 0.054) \quad (18.7)$$

In this equation, AMCNK stands for molar Al_2O_3/(CaO+Na_2O+K_2O+MgO), and for the calculation of $D_{FeOtot}{}^{mgt/melt}$ and $D_{TiO2}{}^{mgt/melt}$ FeO_{tot} and TiO_2 are measured in weight percent and the melt composition is reported dry. The overall uncertainty of the FeTiMM model increases from ±0.2–0.3 log units fO_2 at ≤ΔFMQ+1.5, to ±0.3–0.5 log units fO_2 at ΔFMQ+4.5. Fifty-nine out of the 109 experiments that were used to calibrate the model were ilmenite-saturated, so that fO_2 could be independently constrained via magnetite-ilmenite oxybarometry. The performance of FeTiMM on these 59 ilmenite-saturated is shown in Fig. 18.5. The uncertainty of the magnetite-ilmenite method (Ghiorso & Evans, 2008), based on their Figure 27, can be estimated at ca. ±0.3 log units at reducing conditions (ΔFMQ-1 to ΔFMQ+1) and ca. ±0.5 log units at strongly oxidizing conditions (ΔFMQ+4.5; not considering a group of outliers), i.e., similar to the uncertainty associated with FeTiMM. Including these errors, almost all fO_2 values show an agreement between the two methods within ≤0.5 log units. No correlations are evident between the degree of correspondence and fO_2, temperature, melt SiO_2 content, ASI, or titanomagnetite composition, suggesting that FeTiMM works equally well over the entire range of the investigated P-T-X conditions. This renders FeTiMM a simple and universally applicable oxybarometer. Any discrepancy in the comparison can result from various sources such as analytical errors, experimental problems, or weaknesses in the calibration of either oxybarometer.

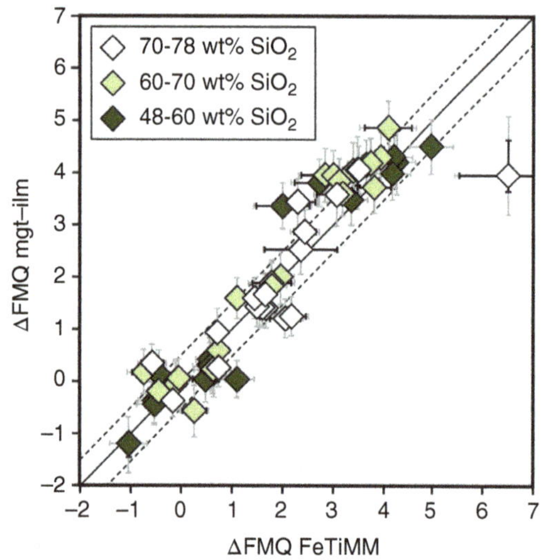

Fig. 18.5 Performance of FeTiMM on experimental samples. Oxygen fugacities (in log units relative to the FMQ buffer) obtained via FeTiMM are compared with ones obtained via magnetite-ilmenite oxybarometry using the model of Ghiorso and Evans (2008). The data are divided into three groups according to melt SiO_2 content. Black error bars (in most cases smaller than the symbol size) denote the analytical error, whereas the grey error bars show the overall error that includes both the analytical scatter and the error inherent to the model.

18.5. APPLICATION OF TITANOMAGNETITE-BASED OXYBAROMETERS TO NATURAL SILICIC ROCKS

Magnetite-ilmenite oxybarometry has been successfully applied in the last decades to various types of volcanic rocks, hence it can be used as a reference to check the accuracy of any new oxybarometer. The first application of the vanadium partitioning oxybarometer and FeTiMM to natural rhyolites and dacites (>70 wt % SiO_2) returned very promising results (Arató & Audétat 2017b, c). For the present publication FeTiMM was additionally applied to some samples of Özdemir et. al (2011) with intermediate compositions (60–70 wt% SiO_2). In both datasets, samples include ilmenite-bearing vitrophyres and holocrystalline rocks. The reader is referred to Table S1 for the sample list and for thermobarometry data and to Arató and Audétat (2017c) for a detailed sample description. In some samples the minerals and melts were preserved only in form of inclusions within phenocrysts. Oxygen fugacity was always independently constrained by magnetite-ilmenite oxybarometry. The ultimate goal in applying magnetite-ilmenite oxybarometry simultaneously was to verify that the vanadium partitioning oxybarometer and FeTiMM provide a reliable estimate of fO_2 and as such can be applied to rocks where magnetite-ilmenite oxybarometry cannot be applied. The future application of the two oxybarometers should therefore enable to constrain fO_2 in intrusive and/or altered rocks (based mainly on the measurement of titanomagnetite inclusion-melt inclusion pairs), on which magnetite-ilmenite oxybarometry is usually not applicable.

Vanadium partitioning oxybarometry additionally required a temperature constraint that was obtained either via zircon saturation thermometry, two-feldspar thermometry, or Fe-Ti-oxide thermometry. Obviously, the latter option was not available if the method was used for ilmenite-free rocks. In most samples, the fO_2 values determined via vanadium titanomagnetite-melt oxybarometry and FeTiMM agreed within 0.5 log units with the oxygen fugacity calculated from Fe-Ti-oxide pairs (Fig. 18.6), except for two data points in Fig. 18.6b which show a larger discrepancy. Since the equilibrium test described below was not yet available at the time of the Arató and Audétat (2017b) study, these two data points

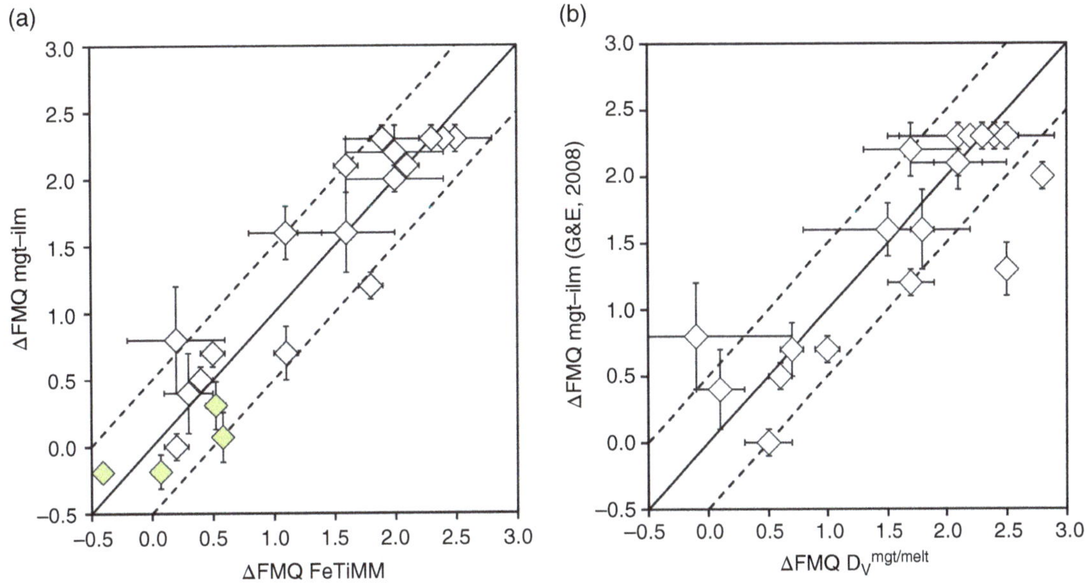

Fig. 18.6 (a) Comparison of fO_2 values (reported in log units relative to the fayalite-magnetite-quartz buffer) obtained from silica-rich (> 70 wt% SiO_2) rocks (Arató & Audétat, 2017b; white diamonds) and intermediate (60–70 wt% SiO_2) rocks (green diamonds; Özdemir et. al., 2011) via FeTiMM versus Fe-Ti oxide oxybarometry using the model of Ghiorso and Evans (2008). (b) Comparison of fO_2 values obtained from silica-rich (> 70 wt % SiO_2) rocks (Arató & Audétat, 2017c) via vanadium titanomagnetite-melt partitioning oxybarometry versus Fe-Ti oxide oxybarometry using the model of Ghiorso and Evans (2008). Error bars indicate 1 sigma standard deviations of the calculated fO_2 averages.

may be due to disequilibrium between the analyzed titanomagnetite–melt pairs.

During the application of the new oxybarometers to natural samples it was noticed that many of them contained multiple generations of Fe-Ti-oxides or showed other signs of magma mixing. To be able to check for equilibrium between a given titanomagnetite-silicate melt pair a new empirical equlibrium test was developed. This test is based on the same two elements (Mn, Mg) as the equilibrium test between titanomagnetite and ilmenite (Bacon & Hirschmann, 1988), although the rationale behind is quite different. Titanomagnetite-melt partition coefficients of Mn and Mg depend similarly on melt composition, but their absolute concentrations in the silicate melt vary strongly during magma differentiation (Fig. 18.7) so the identification of disequilibrium titanomagnetite-melt pairs should be obvious. We used the Mn-Mg exchange coefficient $D_{Mn-Mg}{}^{mgt/melt} = (D_{MnO}{}^{mgt/melt})/(D_{MgO}{}^{mgt/melt})$, with MnO and MgO given in weight percent for the test. Based on over 300 experimental data (see compilation and references in Arató & Audétat, 2017b), equilibrium seems to be attained if $D_{Mn-Mg}{}^{mgt/melt}$ ranges between 1 and 4. Disequilibrium between the two phases results in significantly (often by orders of magnitude) different exchange coefficient values, thus, melts that do not coexist with a given titanomagnetite composition are easily identifiable.

18.6. CONCLUSIONS

In this chapter we attempted to briefly summarize the current knowledge on Fe-Ti-oxide based oxybarometry. The most recent calibration of the magnetite-ilmenite oxybarometer (Ghiorso & Evans, 2008) provides reliable estimates of magmatic fO_2 for a broad range of melt compositions, given that the sample contains equilibrium pairs of titanomagnetite and ilmenite. However, it is not applicable to intrusive rocks (e.g., granites), hydrothermally altered rocks, and rocks that do not contain ilmenite. Two titanomagnetite-melt based oxybarometers, namely vanadium partitioning oxybarometry including its recent calibration for silicic rocks and the recently developed FeTiMM, bridge this gap. Both methods are based on the composition of equilibrium melt-titanomagnetite pairs. The phases can be measured either as matrix glass-titanomagnetite phenocryst pairs or as inclusions preserved in phenocrysts. The former approach is applicable only to vitreous volcanic rocks, whereas the latter approach can be used for both vitreous volcanic rocks as well for intrusive and altered rocks. Both methods provide fO_2 values that agree well with ones obtained via magnetite-ilmenite oxybarometry.

The main advantages of the two new titanomagnetite-melt based oxybarometers over classical magnetite-ilmenite oxybarometry are: (i) that they can be applied

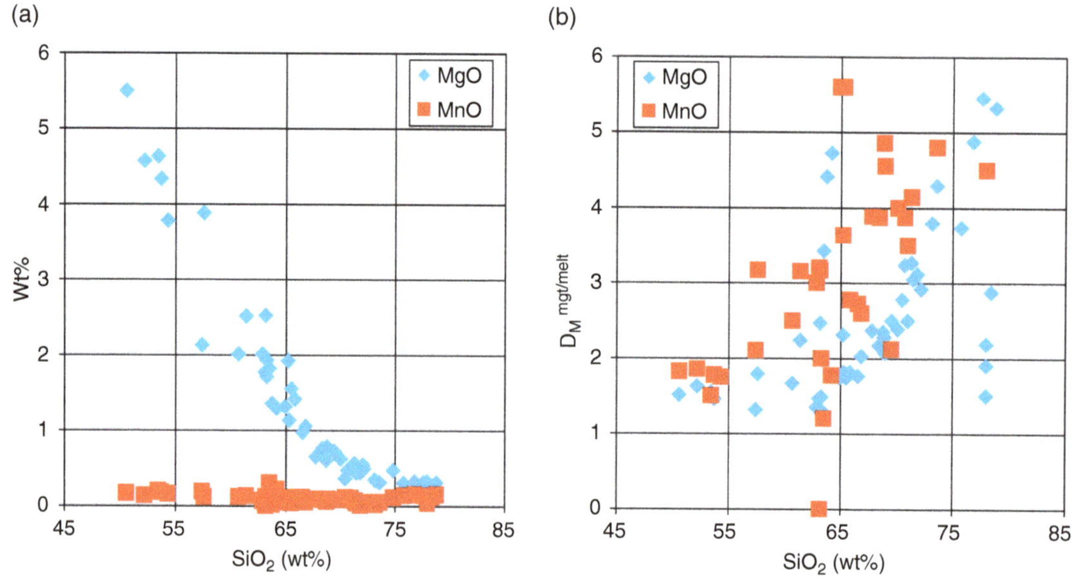

Fig. 18.7 The principle of the equilibrium test between magnetite and melt pairs. (a) Silicate melt MgO and MnO content as a function of melt SiO_2 content. Note the largely different trends. (b) Metal/melt partition coefficients as a function of melt SiO_2 content. Note that the trends are essentially identical. Data from: Arce et al. (2005), Berndt (2004), Blatter et al. (2013), Parat et al. (2008), Sisson et al. (2004), and Tomiya et al. (2010)

to both ilmenite-free and ilmenite-bearing samples; and (ii) that they are easier to apply to slowly cooled rocks such as granites by measuring titanomagnetite-melt pairs in form of inclusions. FeTiMM additionally has the advantage that (iii) it is temperature-independent; and (iv) that it has been calibrated on – and thus is applicable to – a broad range of melt compositions spanning the entire range from basalts to rhyolites using a single equation.

ACKNOWLEDGMENTS

This research was supported by the German Science Foundation grant AU 314/5-1. Paolo Sossi and an anonymous reviewer are gratefully thanked for their insightful reviews, which greatly improved the quality of the manuscript.

REFERENCES

Andersen, D. J., & Lindsley, D. H. (1988). Internally consistent solution models for Fe-Mg-Mn-Ti oxides: Fe-Ti oxides. *American Mineralogist*, *73*, 714–726.

Andersen, D. J., Lindsley, D. H., & Davidson, P. M. (1993). QUILF: A pascal program to assess equilibria among Fe–Mg–Mn–Ti oxides, pyroxenes, olivine, and quartz, *Computers & Geosciences*, *19*, 9, 1333–1350. https://doi.org/10.1016/0098-3004(93)90033-2

Arató, R., & Audétat, A. (2017a). Experimental calibration of a new oxybarometer for silicic magmas based on vanadium partitioning between magnetite and silicate melt. *Geochimica et Cosmochimica Acta*, *209*, 284–295. https://doi.org/10.1016/j.gca.2017.04.020

Arató, R., & Audétat, A. (2017b). FeTiMM–A new oxybarometer for mafic to felsic magmas. *Geochemical Perspectives Letters*, *5*, 19–23. doi: 10.7185/geochemlet.1740

Arató, R., & Audétat, A. (2017c). Vanadium magnetite–melt oxybarometry of natural, silicic magmas: a comparison of various oxybarometers and thermometers. *Contributions to Mineralogy and Petrology*, *172*, 7. 10.1007/s00410-017-1369-6

Arce, J., Macias, J., Gardner, J., & Layer, P. (2005). A 2.5 ka history of dacitic magmatism at Nevado de Toluca, Mexico: Petrological, 40Ar/39Ar dating, and experimental constraints on petrogenesis. *Journal of Petrology*, *47*, 3, 457–479. https://doi.org/10.1093/petrology/egi082

Audétat, A., Dolejs, D., & Lowenstern, J. B. (2011). Molybdenite saturation in silicic magmas: Occurrence and petrological implications. *Journal of Petrology*, *52*, 5, 891–904. https://doi.org/10.1093/petrology/egr008

Bacon, C. R., & Hirschmann, M. M. (1988). Mg/Mn partitioning as a test for equilibrium between coexisting Fe-Ti oxides. *American Mineralogist*, *73*, 1–2, 57–61.

Ballard, J. R., Palin, M. J., & Campbell, I. H. (2002). Relative oxidation states of magmas inferred from Ce(IV)/Ce(III) in zircon: application to porphyry copper deposits of northern Chile. *Contributions to Mineralogy and Petrology*, *144*, 3, 347–364. https://doi.org/10.1007/s00410-002-0402-5

Berndt, J. (2004). An experimental investigation of the influence of water and oxygen fugacity on differentiation of MORB at 200 MPa. *Journal of Petrology*, *46*, 1, 135–167. https://doi.org/10.1093/petrology/egh066

Blatter, D. L., Sisson, T. W., & Hankins, W. B. (2013). Crystallization of oxidized, moderately hydrous arc basalt at mid- to lower-crustal pressures: implications for andesite genesis.

Contributions to Mineralogy and Petrology, *166*, 3, 861–886. https://doi.org/10.1007/s00410-013-0920-3

Blevin, P. L., & Chappell, B. W. (1992). The role of magma sources, oxidation states and fractionation in determining the granite metallogeny of eastern Australia. *Transactions of the Royal Society of Edinburgh: Earth Sciences*, *83*, 1–2, 305–316. https://doi.org/10.1130/SPE272-p305

Borisov, A., Behrens, H., & Holtz, F. (2017). Effects of strong network modifiers on Fe3+/Fe2+ in silicate melts: an experimental study. *Contributions to Mineralogy and Petrology*, *172*, 5. https://doi.org/10.1007/s00410-017-1337-1

Buddington, A., & Lindsley, D. (1964). Iron-titanium oxide minerals and synthetic equivalents. *Journal of Petrology*, *5*, 2, 310–357. https://doi.org/10.1093/petrology/5.2.310

Candela, P. A. (1992). Controls on ore metal ratios in granite-related ore systems: an experimental and computational approach. *Earth and Environmental Science Transactions of The Royal Society of Edinburgh*, *83*,1–2, 317–326. https://doi.org/10.1130/SPE272-p317

Candela, P. A., & Bouton, S. L. (1990). The influence of oxygen fugacity on tungsten and molybdenum partitioning between silicate melts and ilmenite. *Economic Geology*, *85*, 3, 633–640. https://doi.org/10.2113/gsecongeo.85.3.633

Canil, D. (1999). Vanadium partitioning between orthopyroxene, spinel and silicate melt and the redox states of mantle source regions for primary magmas. *Geochimica et Cosmochimica Acta*, *63*, 3, 557–572. https://doi.org/10.1016/S0016-7037(98)00287-7

Canil, D. (2002). Vanadium in peridotites, mantle redox and tectonic environments: Archean to present. *Earth and Planetary Science Letters*, *195*, 1, 75–90. https://doi.org/10.1016/S0012-821X(01)00582-9

Carmichael, I. S. E. (1967). The iron-titanium oxides of salic volcanic rocks and their associated ferromagnesian silicates. *Contributions to Mineralogy and Petrology*, *14*, 36–64. https://doi.org/10.1007/BF00370985

Carmichael, I., & Ghiorso, M. (1990). Controls on oxidation-reduction relations in magmas, *Reviews in Mineralogy*, Washington, DC: Mineralogical Society of America, *24*, 191–212.

Carroll, M., & Rutherford, M. (1985). Sulfide and sulfate saturation in hydrous silicate melts. *Journal of Geophysical Research: Solid Earth*, *90*, S02. https://doi.org/10.1029/JB090iS02p0C601

Carroll, M., & Rutherford, M. J. (1988). Sulfur speciation in hydrous experimental glasses of varying oxidation state–results from measured wavelength shifts of sulfur X-rays. *American Mineralogist*, *73*, 7, 845–849.

Di Genova, D., Vasseur, J., Hess, K.-U., Neuville, D. R., & Dingwell, D. B. (2017). Effect of oxygen fugacity on the glass transition, viscosity and structure of silica-and iron-rich magmatic melts. *Journal of Non-Crystalline Solids*, *470*, 78–85. doi: 10.1016/j.jnoncrysol.2017.05.013

Eugster, H. P. (1957). Heterogeneous reactions involving oxidation and reduction at high pressures and temperatures. *The Journal of Chemical Physics*, *26*, 6, 1760–1761. https://doi.org/10.1063/1.1743626

Fincham, C. & Richardson, F. D. (1954). The behaviour of sulfur in silicate and aluminate melts. *Proceedings of the Royal Society of London A*, *223*, 40–62.

Frost, B. R. (1991). Introduction to oxygen fugacity and its petrologic importance. *Reviews in Mineralogy and Geochemistry*, *25*, 1, 1–9. https://doi.org/10.1515/9781501508684-004

Frost, B. R., & Lindsley, D. H. (1991). Occurrence of iron-titanium oxides in igneous rocks. *Reviews in Mineralogy and Geochemistry*, *25*, 1, 433–468.

Frost, B. R., & Lindsley, D. H. (1992) Equilibria among Fe-Ti oxides, pyroxenes, olivine, and quartz: Part II. Application. *American Mineralogist*, *77*, 1004–1004.

Frost, B. R., Lindsley, D. H., & Andersen, D. J. (1988). Fe-Ti oxide-silicate equilibria; assemblages with fayalitic olivine. *American Mineralogist*, *73*, 7–8, 727–740.

Gaillard, F., Scaillet, B., Pichavant, M., & Bény, J.-M. (2001). The effect of water and f_{O_2} on the ferric–ferrous ratio of silicic melts. *Chemical Geology*, *174*, 1, 255–273. https://doi.org/10.1016/S0009-2541(00)00319-3

Ghiorso, M. S., & Sack, O. (1991). Fe-Ti oxide geothermometry: thermodynamic formulation and the estimation of intensive variables in silicic magmas. *Contributions to Mineralogy and Petrology*, *108*, 485–510. https://doi.org/10.1007/BF00303452

Ghiorso, M. S., & Evans, B. W. (2008). Thermodynamics of rhombohedral oxide solid solutions and a revision of the Fe-Ti two-oxide geothermometer and oxygen-barometer. *American Journal of Science*, *308*, 9, 957–1039. https://doi.org/10.2475/09.2008.01

Hedenquist, J. W., & Lowenstern, J. B. (1994). The role of magmas in the formation of hydrothermal ore deposits. *Nature*, *370*, 6490, 519–527. doi: 10.1038/370519a0

Herd, C. D. (2008). Basalts as probes of planetary interior redox state. *Reviews in Mineralogy and Geochemistry*, *68*, 1, 527–553. https://doi.org/10.2138/rmg.2008.68.19

Horn, I., Foley, S. F., Jackson, S. E., & Jenner, G. A. (1994). Experimentally determined partitioning of high field strength-and selected transition elements between spinel and basaltic melt, *Chemical Geology*, *117*, 1, 193–218. https://doi.org/10.1016/0009-2541(94)90128-7

Irving, A. J. (1978). A review of experimental studies of crystal/liquid trace element partitioning. *Geochimica et Cosmochimica Acta*, *42*, 6, 743–770. https://doi.org/10.1016/0016-7037(78)90091-1

Ishihara, S. (1977). The magnetite-series and ilmenite-series granitic rocks. *Mining Geology*, *27*, 145, 293–305. https://doi.org/10.11456/shigenchishitsu1951.27.293

Jollands, M. C., Hanger, B. J., Yaxley, G. M., Hermann, J., & Kilburn, M. R. (2018). Timescales between mantle metasomatism and kimberlite ascent indicated by diffusion profiles in garnet crystals from peridotite xenoliths. *Earth and Planetary Science Letters* *481*, 143–153. https://doi.org/10.1016/j.epsl.2017.10.021

Jugo, P. J., Wilke, M., & Botcharnikov, R. E. (2010), Sulfur K-edge XANES analysis of natural and synthetic basaltic glasses: Implications for S speciation and S content as function of oxygen fugacity. *Geochimica et Cosmochimica Acta*, *74*, 20, 5926–5938. https://doi.org/10.1016/j.gca.2010.07.022

Keppler, H. (2010). The distribution of sulfur between haplogranitic melts and aqueous fluids. *Geochimica et Cosmochimica Acta*, *74*, 2, 645–660. https://doi.org/10.1016/j.gca.2009.10.010

Kohlstedt, D. L., & Zimmerman, M. E. (1996). Rheology of partially molten mantle rocks. *Annual Review of Earth and Planetary Sciences*, *24*, 41–62. https://doi.org/10.1146/annurev.earth.24.1.41

Kularatne, K., & Audétat, A. (2014). Rutile solubility in hydrous rhyolite melts at 750–900°C and 2kbar, with application to titanium-in-quartz (TitaniQ) thermobarometry. *Geochimica et Cosmochimica Acta*, *125*, 196–209. https://doi.org/10.1016/j.gca.2013.10.020

Lehmann, B. (1990). Metallogeny of tin (Lecture Notes in Earth Sciences 32), 211 pp.

Liebske, C., Behrens, H., Holtz, F., & Lange, R. A. (2003). The influence of pressure and composition on the viscosity of andesitic melts. *Geochimica et Cosmochimica Acta*, *67*, 3, 473–485. https://doi.org/10.1016/S0016-7037(02)01139-0

Lindsley, D. H., & Frost, B. R. (1992). Equilibria among Fe-Ti oxides, pyroxenes, olivine, and quartz; Part I, Theory. *American Mineralogist*, *77*, 9–10, 987–1003.

Mallmann, G., & O'Neill, H. S. C. (2009). The crystal/melt partitioning of V during mantle melting as a function of oxygen fugacity compared with some other elements (Al, P, Ca, Sc, Ti, Cr, Fe, Ga, Y, Zr and Nb). *Journal of Petrology*, *50*, 9, 1765–1794. https://doi.org/10.1093/petrology/egp053

Muan, A. (1955). Phase equilibria in the system FeO-Fe_2O_3-SiO_2. *Journal of the Minerals, Metals, and Materials Society*, *7*, 965–976. https://doi.org/10.1007/BF03377595

O'Neill, H. S. C., & Navrotsky, A. (1984). Cation distributions and thermodynamic properties of binary spinel solid solutions. *American Mineralogist*, *69*, 7–8, 733–753.

Özdemir, Y., Blundy, J., & Güleç, N. (2011). The importance of fractional crystallization and magma mixing in controlling chemical differentiation at Süphan stratovolcano, eastern Anatolia, Turkey. *Contributions to Mineralogy and Petrology*, *162*, 3, 573–597. https://doi.org/10.1007/s00410-011-0613-8

Papike, J. J., Karner, J. M., & Shearer, C. K. (2005). Comparative planetary mineralogy: Valence state partitioning of Cr, Fe, Ti, and V among crystallographic sites in olivine, pyroxene, and spinel from planetary basalts. *American Mineralogist*, *90*, 277–290. https://doi.org/10.2138/am.2005.1779

Parat, F., Holtz, F., & Feig, S. (2008). Pre-eruptive conditions of the Huerto Andesite (Fish Canyon System, San Juan Volcanic Field, Colorado): Influence of volatiles (C-O-H-S) on phase equilibria and mineral composition. *Journal of Petrology*, *49*, 5, 911–935. https://doi.org/10.1093/petrology/egn011

Price, G. (1981). Diffusion in the titanomagnetite solid solution series. *Mineralogical Magazine*, *44*, 334, 195–200.

Prytulak, J., Sossi, P., Halliday, A., Plank, T., Savage, P., & Woodhead, J. (2016). Stable vanadium isotopes as a redox proxy in magmatic systems? *Geochemical Perspectives Letters*, *3*, 1, 75–84. doi: 10.7185/geochemlet.1708

Righter, K., Leeman, W. P., & Hervig, R. L. (2006a). Partitioning of Ni, Co and V between spinel-structured oxides and silicate melts: Importance of spinel composition. *Chemical Geology*, *227*, 1–2, 1–25. https://doi.org/10.1016/j.chemgeo.2005.05.011

Righter, K., Sutton, S. R., Newville, M., Le, L., Schwandt, C. S., Uchida, H., et al. (2006b). An experimental study of the oxidation state of vanadium in spinel and basaltic melt with implications for the origin of planetary basalt. *American Mineralogist*, *91*, 10, 1643–1656.

Sievwright, R. H., Wilkinson, J. J., O'Neill, H. S. C., & Berry, A. J. (2017). Thermodynamic controls on element partitioning between titanomagnetite and andesitic–dacitic silicate melts. *Contributions to Mineralogy and Petrology*, *172*, 8. https://doi.org/10.1007/s00410-017-1385-6

Sisson, T. W., Ratajeski, K., Hankins, W. B., & Glazner, A. F. (2004). Voluminous granitic magmas from common basaltic sources. *Contributions to Mineralogy and Petrology*, *148*, 6, 635–661. https://doi.org/10.1007/s00410-004-0632-9

Sossi, P. A., Prytulak, J., & O'Neill, H. S. C. (2018). Experimental calibration of vanadium partitioning and stable isotope fractionation between hydrous granitic melt and magnetite at 800° C and 0.5 GPa. *Contributions to Mineralogy and Petrology*, *173*, 4, 27. https://doi.org/10.1007/s00410-018-1451-8

Sutton, S. R., Karner, J., Papike, J., Delaney, J. S., Shearer, C., Newville, M., et al. (2005). Vanadium K edge XANES of synthetic and natural basaltic glasses and application to microscale oxygen barometry. *Geochimica et Cosmochimica Acta*, *69*, 2333–2348. https://doi.org/10.1016/j.gca.2004.10.013

Tomiya, A., Takahashi, E., Furukawa, N., & Suzuki, T. (2010). Depth and evolution of a silicic magma chamber: Melting experiments on a low-K rhyolite from Usu Volcano, Japan. *Journal of Petrology*, *51*, 6, 1333–1354. https://doi.org/10.1093/petrology/egq021

Toplis, M. J., & Corgne, A. (2002). An experimental study of element partitioning between magnetite, clinopyroxene and iron-bearing silicate liquids with particular emphasis on vanadium. *Contributions to Mineralogy and Petrology*, *144*, 1, 22–37. https://doi.org/10.1007/s00410-002-0382-5

Wones, D. (1981). Mafic silicates as indicators of intensive variables in granitic magmas. *Mining Geology*, *31*, 4, 191–212. https://doi.org/10.11456/shigenchishitsu1951.31.191

Wones, D. R. (1989). Significance of the assemblage titanite+ magnetite+ quartz in granitic rocks. *American Mineralogist*, *74*, 7–8, 744–749.

Wones, D., & Eugster, H. (1965). Stability of biotite–experiment theory and application. *American Mineralogist*, *50*, 9, 1228–1272.

Wu, C., & Mason, T. (1981). Thermopower measurement of cation distribution in magnetite. *Journal of the American Ceramic Society*, *64*, 9, 520–522. https://doi.org/10.1111/j.1151-2916.1981.tb10317.x

Xirouchakis, D., Lindsley, D. H., & Frost, B. R. (2001). Assemblages with titanite ($CaTiOSiO_4$), Ca-Mg-Fe olivine and pyroxenes, Fe-Mg-Ti oxides, and quartz: Part II. Application. *American Mineralogist*, *86*, 2–3, 254–264. doi: 10.2138/am-2001-2-307

SUPPLEMENTARY REFERENCES

Chou, I.-M. (1978). Calibration of oxygen buffers at elevated P and T using the hydrogen fugacity sensor. *American Mineralogist*, *63*, 690–703.

Frost, B. R. (1991). Introduction to oxygen fugacity and its petrologic importance. *Reviews in Mineralogy and Geochemistry*, *25*, 1–9. https://doi.org/10.1515/9781501508684-004

Huebner, J. S., & Sato, M. (1970). The oxygen fugacity-temperature relationships of manganese oxide and nickel oxide buffers. *American Mineralogist*, *55*, 934–952.

19

The Redox Behavior of Rare Earth Elements

Maria Rita Cicconi[1,2], Charles Le Losq[2,3], Grant S. Henderson[4], and Daniel R. Neuville[2]

ABSTRACT

Elements in the lanthanides series are usually known under the name of rare earth elements (REE). In past decades, the use of these elements in industrial processes has increased exponentially with the development of novel technologies and applications. In addition to such growing industrial importance, and despite their relatively low concentrations in the Earth system, REE are sensitive markers of crystal-melt equilibria in geological systems. As a result, the study of lanthanides in materials, either as whole group or as single species, has become an important point of focus in both Geo- and Materials Sciences. Among REE, Ce and Eu are stable in many geological environments, with two different oxidation states (Ce^{3+}/Ce^{4+} and Eu^{2+}/Eu^{3+}), hence, these two redox couples have been proposed as proxies for oxygen fugacity (fO_2) in different environments (oxybarometer). However, it is essential to understand the factors influencing redox mechanisms in order to use Ce/Eu distribution as fO_2 sensors. In this chapter, we present an overview of REE properties in minerals, melts, and glasses, and illustrate their importance through examples of applications in the fields of geochemistry, petrology, and mineralogy.

19.1. INTRODUCTION

"The rare earths perplex us in our researches, baffle us in our speculations, and haunt us in our very dreams. They stretch like an unknown sea before us, mocking, mystifying and murmuring strange revelations and possibilities" (statement attributed to Sir William Crookes, 1902).

The study of trace elements in silicate glasses is a key to understanding the influence of the chemical composition of glasses on the chemical, physical, and rheological properties of a magma. Since the paper of McIntire (1963), trace elements have been widely used as monitors of magmatic processes. But, despite the huge experimental effort devoted to the comprehension of the distribution and behavior of many trace elements in natural systems, our limited knowledge on this topic prevents prediction of magmatic differentiation in most cases.

REE are particularly important in Geosciences and are the most widely used trace elements in geochemical studies of magma genesis. Their distribution in silicate melts/crystals may be used as an indicator of the genetic history of the magma. Indeed, the geochemistry of these trace elements in rocks, glasses, and aquatic systems has been a major research topic for several years, and the REE distribution in these geological settings (on Earth, but not

[1]*Department of Materials Science and Engineering, Friedrich-Alexander-Universität Erlangen-Nürnberg, Erlangen, Germany*

[2]*Université de Paris, Institut de Physique du Globe de Paris, Paris, France*

[3]*Research School of Earth Sciences, Australian National University, Canberra, ACT, Australia*

[4]*Department of Earth Sciences, University of Toronto, Toronto, Canada*

Magma Redox Geochemistry, Geophysical Monograph 266, First Edition.
Edited by Roberto Moretti and Daniel R. Neuville.

DOI: 10.1002/9781119473206.ch19

exclusively) has been of great utility in enhancing our understanding of a variety of geological, geochemical, and cosmochemical processes.

19.1.1. Trace Elements

A trace element may be defined as an element that is present in a rock at concentration less than 0.1 wt%. Trace elements can form mineral species on their own, but most commonly substitute for major cations in rock-forming minerals. These elements constitute only a small fraction of the system of interest; nevertheless, they provide important geochemical and geological information (Shaw, 2006; White, 2013). This is because:

1. relative variations in the concentrations of many trace elements are much larger than variations in the concentrations of major components;
2. in any system there are far more trace elements than major elements. Each element has chemical properties that are to some degree unique, hence there is unique geochemical information contained in the concentration variation for each element;
3. the range in behavior of trace elements is large, making them sensitive to processes to which major elements are insensitive;
4. trace elements obey Henry's Law, thus their behavior is almost always simpler than that of major elements (White, 2013).

The partition coefficient (D) is the ratio of the concentration of a trace element in a solid phase with respect to that in a liquid at equilibrium, and it is one of the most useful tools to describe magmatic processes (e.g., Bédard, 2006; O'Hara, 1995). The partition coefficient allows distinguishing between compatible and incompatible elements. Incompatible elements are those with $D << 1$ and prefer the melt phase, whereas compatible elements are those with $D \geq 1$ and prefer the mineral phase. Treuil and coworkers (e.g., Treuil & Varet, 1973; Treuil & Joron, 1975; Treuil et al., 1979) proposed another term to further differentiate incompatible elements, and coined the term hydromagmatophile to describe those elements having the highest affinity for the melt phase (Th, U, Zr, Hf, Nb, Ta, REE). The partition coefficient for a given element will vary considerably between different crystalline phases and can be less than one for one phase and greater than one for another. Hence, the terms compatible and incompatible have meaning only when the mineral phases are known. REE are not readily accepted by most silicate minerals, thus they belong to the incompatible elements group.

Incompatible and compatible elements can be divided on the basis of their charge (Z) and size (Figure 19.1a), or in terms of ionic potential (the ratio between charge and ionic ratio). Another parameter was proposed by Dietzel (1948): the Cation Field Strength CFS = $Z/(r_c+r_O)^2$, where Z is the charge of the cations and the denominator represents the cation- oxygen distance

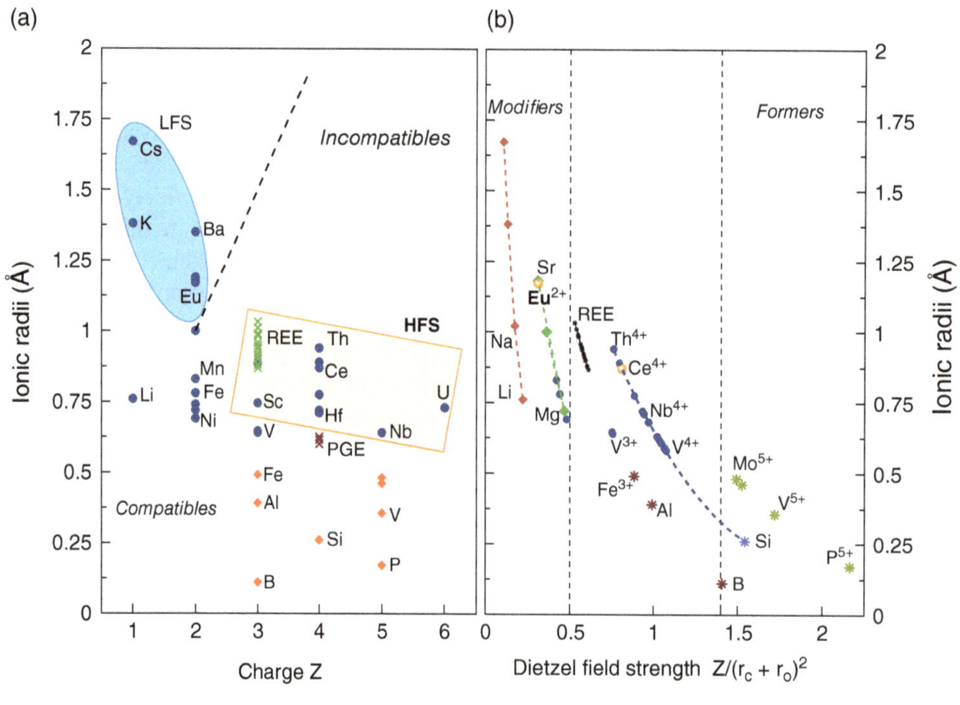

Figure 19.1 (a) Plot of ionic radius (data from Shannon, 1976) vs. ionic charge for trace elements of geological interest. Red diamonds refer to four-fold coordinated cations (modified after Rollinson, 1993). (b) Dietzel's field strength parameter for several cations (stars refer to four-fold coordinated cations).

(here considered oxygen ionic radius =1.35 Å) (Figure 19.1b).

Figure 19.1a shows a plot of ionic radii against charge for several trace elements, here considered six-fold coordinated. Small highly charged cations are known as high field strength (HFS) cations (e.g., Hf^{4+}, Th^{4+}, Ce^{4+}, Ta^{5+}, U^{6+}) and large cations of small charge are referred as low field strength (LFS) cations (Ba^{2+}, Eu^{2+}, Sr^{2+}, Pb^{2+}, K^{+}). LFS cations are also known as large ion lithophile elements (LILE). Elements with small ionic radius and a relatively lower charge tend to be compatible, and the most common four-fold coordinated network-forming cations are reported as red diamonds (Figure 19.1a). Figure 19.1b shows the relationship between the Dietzel's field strength and the ionic radii (six-fold coordinated cations) for several elements. It is similar to the charge vs. ionic radii plot, but it further shows the grouping of cations based on their structural role. Indeed, according to Dietzel (1948), the field strength is a critical parameter for glass formation, with network-forming cations having high field strength values (> 1.3–1.4), whereas network modifiers have low field strengths (< 0.5).

The two plots in Figure 19.1 highlight the main groupings of trace elements and emphasises the similarity in ionic size and charge between some elements/groups. Elements with the same ionic charge and size are expected to show very similar geochemical behaviour (Rollinson, 1993).

19.1.2. Lanthanides

The lanthanide series includes 15 elements of the periodic table (from Lanthanum to Lutetium), and they are also called rare earth elements (REE). However, often "rare earth" has been applied in a more unrestricted way, thus including Sc and Y.

Lanthanides have a $5d^1$ $6s^2$ outer electron orbital arrangement. With increasing atomic number, the inner *4f* shell is filled before the second *5d* electron shell get filled (thus lanthanides belong to the 3B group of the periodic table) and the similarity in their chemical properties is the result of their atomic structure. Lanthanides might be considered as a separated group from the rest of the periodic table, if the main interest is related to the properties that depend on the occupancy of the *4f* electron shell, from 0 (La) to 14 (Lu) (Figure 19.2).

Figure 19.2 shows the dependence of the six-fold coordinated $^{[6]}REE^{3+}$ ionic radii in going from La to Lu (data from Shannon, 1976). Moving from left to right across the periodic table (increasing atomic number), the radius of each lanthanide 3+ ion gradually decreases, and this is referred to as "the lanthanide contraction" and it is caused by the poor screening of the *4f* electrons. Beside the values for trivalent lanthanides, the ionic radii for Ce and Eu in different oxidation states are also reported in Figure 19.2.

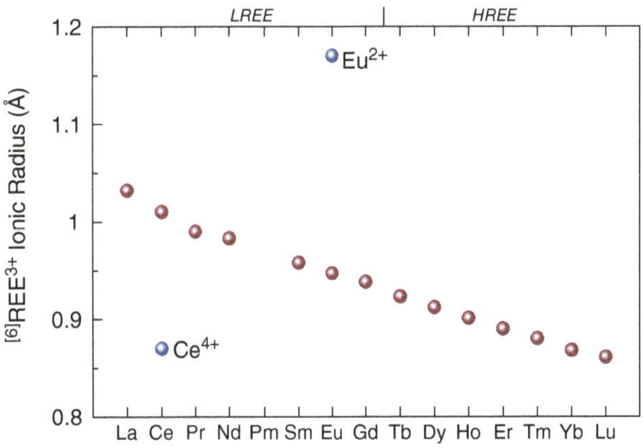

Figure 19.2 Ionic radii for six-fold coordinated REE^{3+}. (Radius values from Shannon, 1976).

Indeed, those elements are multivalent, and their electronic state in melts and minerals is determined by temperature as well as the redox state of the system[1]. Multivalent REE are stable in the trivalent state over a wide range of oxygen partial pressures. Nevertheless, at the oxygen fugacity (fO_2) of the Earth's surface Ce can be stabilized partly or entirely in the tetravalent state and Eu can be partly stabilized in the divalent state at lower oxygen fugacities (e.g., Earth's interior) (White, 2013).

REE are divided into light and heavy groups based on their atomic weight. The light rare earth elements (LREE) are, usually, from La to Gd, whereas the heavy rare earth elements (HREE) are from Tb to Lu. Because of their relative high charge and large ionic radius, the lightest REE are considered highly incompatible (Figure 19.1), whereas the heavy REE have appropriately small radii and might be accommodated in many common minerals (e.g., garnets in Figure 19.3).

The REE are typically not readily accepted by most petrogenetically important silicate minerals but can be compatible in a few major minerals. According to McKay (1989) the importance of REE for petrogenetic modeling mainly arises from four factors:

1. Their ionic radii vary systematically with atomic number. Hence, because of the strong influence of size on partitioning, REE partition coefficients also vary smoothly with atomic number. On the other hand, the shapes of the partition coefficient patterns (plots of distribution coefficient versus atomic number) are very different from one mineral to another.

2. The systematic variation of the REE partition coefficients with atomic number sometimes presents spikes related to the so-called Eu-(Ce) anomalies. Indeed, beside the trivalent state typical of REE, Ce can be partly or

[1] In this chapter, we use oxygen fugacity, fO_2, as equivalent of the partial pressure of oxygen.

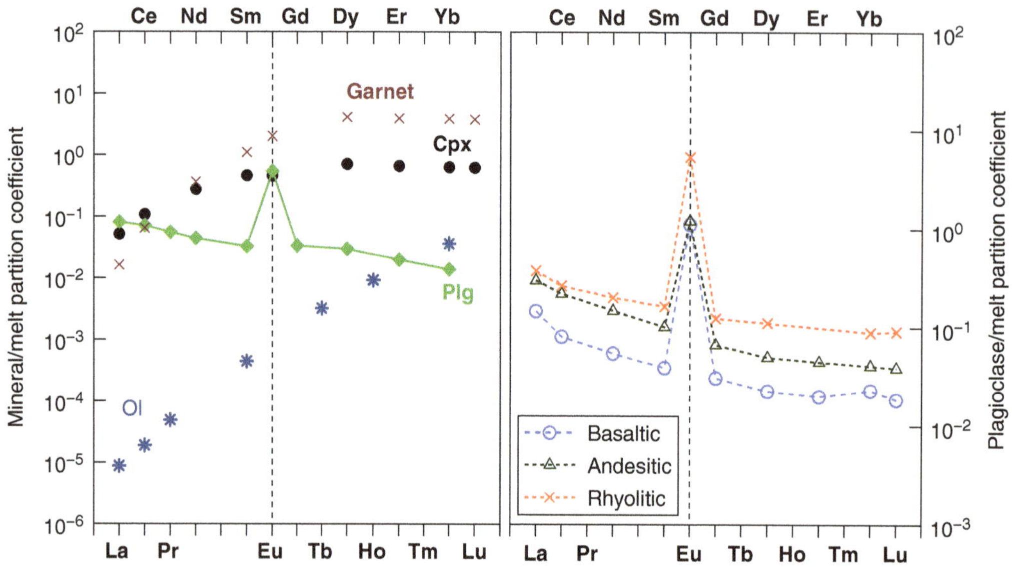

Figure 19.3 (a) Mineral/melt REE partition coefficients for plagioclase (Plg), olivine (Ol), clinopyroxene (Cpx), and garnet in basaltic melts. (b) REE plagioclase/melt partition coefficients in basalt, andesite, and rhyolite. Data compilation from Rollinson (1989) and references therein.

entirely stabilized in the oxidized form (Ce^{4+}) and Eu can be partly stabilized in the reduced form (Eu^{2+}) (Figure 19.2). The latter especially has been considered strongly diagnostic of the involvement of certain minerals in petrogenetic processes.

3. Since REE are highly refractory elements, their relative abundances, in most undifferentiated materials, are generally assumed to be proportional to their relative abundances in the sun and in chondritic meteorites. Thus, they minimize one of the free parameters in most petrogenetic models: the relative elemental abundances in the starting material (cf. McKay, 1989).

4. There are three important isotopic systems that provided a major contribution in the understanding of planetary evolution processes: Sm-Nd, Lu-Hf, and La-Ce.

19.2. GEOCHEMISTRY OF RARE EARTH ELEMENTS

The determination of the partition coefficient between coexisting phases (i.e., mineral/melt) is essential for using REE partition coefficient patterns in the understanding of magmatic differentiation. REE abundance, in natural minerals and magmas follows the Oddo-Harkins rule: a lanthanide with even atomic number is more abundant than its neighboring lanthanides with odd atomic numbers. Normalizing the REE abundances to those in chondritic meteorites is usually performed as it eliminates effects related to nuclear stability and nucleosynthesis, and further produces a smooth pattern (White, 2013). CI chondrites are a class of meteorites that are taken to be the best representative of the average concentrations of non-volatile elements in the solar system.

The mineral/melt REE partition coefficient patterns in the major minerals involved in basaltic magma petrogenesis are reported in Figure 19.3a. The partition coefficient patterns are similar in basaltic, andesitic, and rhyolitic liquids. However, differences of one to several orders of magnitude, in the absolute values, can be observed. Even when comparing the mineral/melt REE partition coefficient data for a synthetic lunar mare basalt (e.g., Lipin & McKay, 1989) it is possible to observe comparable trends. Hence, there is a strong chemical control on the REE partitioning. Olivine (Ol in Figure 19.3a) produces little REE fractionation in partial melts, and the pattern increases from LREE to HREE. For garnet/melt partition coefficient the pattern slopes upward as well, as HREEs enter into Al sites in garnets.

Comparing partition coefficient values for pyroxenes reported in the literature is quite challenging because they vary between published sources, indicating that pyroxene patterns are strongly dependent on composition. For example, McKay (1989) reported that a low-Ca (pigeonite) and a high-Ca pyroxene (augite) present quite different trends. Furthermore, Gaetani and Grove (1995) observed that the proportion of Al entering a pyroxene influences the Ce and Yb partition coefficients between clinopyroxene and melt.

The plagioclase/melt REE partition coefficients form a pattern quite different compared to that of the other minerals (Figure 19.3a). The pattern decreases from

LREE to HREE and the value of REE partition coefficients are generally low, with one exception: Eu. Indeed, the partition coefficient for Eu in plagioclase (D_{Eu}) is anomalously higher compared to the other REE. This exception is referred to as the "Eu-anomaly," and greater values of D_{Eu} are called positive anomalies. The positive Eu-anomaly in the plagioclase/melt partition coefficients means that a significant proportion of europium is entering in the crystal. Indeed, the reduced species Eu^{2+} enters the large alkali/alkali-earth site in the plagioclase structure. Figure 19.3b shows the plagioclase/melt partition coefficients for the REE vs. atomic number, respectively for basaltic, andesitic, and rhyolitic compositions. In addition to the mineral phase, the composition of the melt influences REE partition coefficient patterns. By increasing the silica content (and polymerization) of the melt, there is a clear increase in the REE partition coefficients, with almost one order of magnitude difference between basalt and rhyolite melts (Figure 19.3b).

The ionic radius of reduced europium ($^{[6]}Eu^{2+}$ = 1.17 Å) is much larger than that for trivalent europium ($^{[6]}Eu^{3+}$ = 0.947 Å; ionic radii from Shannon, 1976) (Figure 19.2). Combined with the difference in charge, this produces large variations between Eu divalent and trivalent partition coefficients in many minerals.

Based on Figure 19.1, a strong correlation between coordination and ionic radius is clear, and larger cations likely will occupy larger sites in minerals, and vice versa. Onuma et al. (1968) noted that for any particular mineral, partition coefficient values, when plotted with respect to the cation ionic radii, tend to lie along convex-upwards curves, with a different curve for each valence. Furthermore, each curve tends to parallel those for ions of different valences, with the maximum occurring at the same ionic radius for each valence. Based on these observations, Onuma et al. (1968) concluded that (i) trace elements preferentially occupy lattice sites rather than occurring as defects or as inclusions; (ii) for trace elements of the same charge, the difference between the ionic radius of the trace ion and that of the major element being substituted for, exercises the dominant control on the partition coefficient; and (iii) the effect of ionic charge on partition coefficients appears to be independent of the effect of ionic radius.

One can note that REE^{3+} partitioning could be well predicted by using a valence versus ionic radii plot, as shown in Figure 19.1a.

19.3. MULTIVALENT RARE EARTH ELEMENTS

19.3.1. Europium

Considering that the charge of trace cations influences their ionic radius, changes in their oxidation state can produce marked variations in partitioning behavior. Among the REE, Eu is the only element for which a significant proportion of the ions, in igneous systems, are likely to be present in valences other than trivalent. This leads to "anomalous" concentrations of Eu in many igneous systems. As reported, plagioclase is often anomalously rich in Eu compared to the other rare earths and this leads to pronounced positive Eu-anomalies in plagioclase/melt partition coefficient patterns (Figure 19.3b).

Since the proportion of Eu^{2+} and Eu^{3+} are influenced by the prevailing oxygen fugacity (fO_2) conditions, the presence of positive (or negative) Eu anomalies has been considered strongly diagnostic for the understanding of the evolution of petrogenetic processes (McKay, 1989). Thus, understanding the factors influencing europium redox mechanisms is very important, especially when considering using Eu geochemical behavior as a proxy for the oxygen fugacity of the environment (e.g., Papike et al., 2005; Wadhwa, 2008). Consequently, several experimental studies have been carried out on the effect of Eu partitioning as a function of fO_2 (e.g., Drake & Weill, 1975; Philpotts, 1970; Schnetzler & Philpotts, 1970; Weill & Drake, 1973). However, it has been demonstrated that chemistry also plays a key role in influencing the $Eu^{2+}/(Eu^{2+} + Eu^{3+})$ redox fraction in geological materials (e.g., Cicconi et al., 2012; Lauer & Morris, 1977; Müller & Muecke, 1984; Morris & Haskin, 1974).

In the following, we will review the parameters mostly affecting Eu redox state: bulk chemistry, temperature, and fO_2.

As Eu^{2+} and Sr^{2+} cations have identical charge and almost identical ionic radii, Philpotts (1970) suggested that the partitioning of Sr would be equivalent to that of divalent Eu ($D_{Eu^{2+}}$), whereas to obtain the Eu^{3+} partition coefficient ($D_{Eu^{3+}}$) the value could be interpolated from the values of neighbour Sm and Gd. Based on these assumptions, Philpotts (1970) estimated the Eu^{2+}/Eu^{3+} redox ratio in several phases, and proposed Eu partitioning to be used as an oxygen fugacity barometer. Following the approach of Philpotts (1970), the mineral/melt partitioning of Eu under different redox conditions has been studied, for instance, for plagioclases (e.g., Drake & Weill, 1975; Weill & Drake, 1973; Wilke & Behrens, 1999), diopside (Grutzeck et al.1974), and pyroxenes (Ching-oh et al., 1974; Shearer et al., 2006), either in terrestrial, meteorites, or lunar materials. For example, Drake (1975) experimentally studied the partitioning of Eu between plagioclase and magmatic liquid of basaltic and andesitic compositions as a function of temperature and oxygen fugacity conditions. By using the approach of Philpotts (1970), the ratios of divalent and trivalent europium were estimated both in plagioclase minerals and coexisting magmatic liquids. The author showed that the Eu anomaly (Figure 19.4) increases as fO_2 decreases, suggesting that

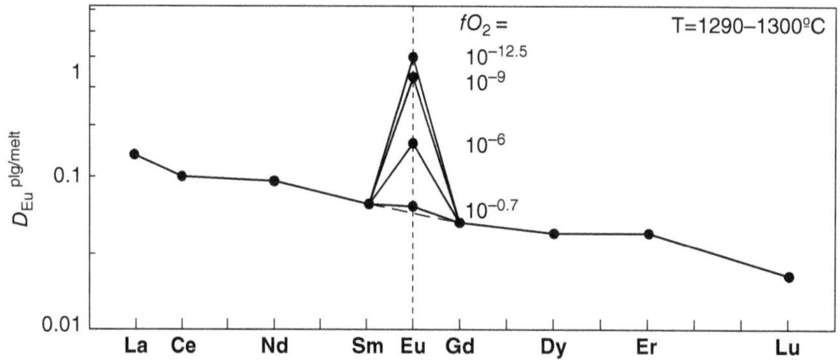

Figure 19.4 Experimental data from Drake (1975) show that the partitioning of Eu between plagioclase feldspar and magmatic liquid ($D_{Eu}^{plg/melt}$) is a strong function of oxygen fugacity conditions. Redrawn after Drake and Weill (1975).

the Eu-anomalies, and in turn, the Eu^{2+}/Eu^{3+} redox ratios in rock-forming minerals, might be used to estimate oxygen fugacity conditions.

The europium redox equilibrium can be expressed as

$$4Eu_{melt}^{3+} + 2O_{melt}^{2-} \leftrightarrow 4Eu_{melt}^{2+} + O_2, \qquad (19.1)$$

where the oxidized Eu^{3+} and reduced Eu^{2+} species and the physically dissolved oxygen of the melt, are in equilibrium. The equilibrium constant K is given by:

$$K = \frac{a\left(Eu_{melt}^{2+}\right)^4 * a(O_2)}{a\left(Eu_{melt}^{3+}\right)^4 * a\left(O_{melt}^{2-}\right)^2}, \qquad (19.2)$$

where K is controlled by oxygen fugacity and the activities (a) of oxygen and europium ions in the melt. Assuming that ionic activities are equal to 1, Equation 19.2 can then be rewritten as

$$-\log(fO_2) = 4\log\left[\frac{Eu^{2+}}{Eu^{3+}}\right] + b, \qquad (19.3)$$

where the square brackets [] denote activity and f denotes fugacity (log is to base 10). b is a term that incorporates all quantities that are function of temperature and bulk chemistry (*cf.* Schreiber, 1987). Equation 19.1 implies the release of 4 electrons during the oxidation of Eu^{2+}, explaining the coefficient 4 visible in Equation 19.3. As a result, a plot of $-\log(fO_2)$ against $\log[Eu^{2+}/Eu^{3+}]$ should provide a slope of 4/1 (for one-electron couples), when samples have the same composition and are produced at the same temperature. However, deviations from the ideal slope have been observed for Eu redox ratios (Morris et al., 1974; Cicconi et al., 2012), as well as for other multivalent elements (e.g., Paul & Douglas, 1965a, b; Schreiber et al., 1999; Cicconi et al., 2015). A possible explanation for deviations from the ideal values can be ascribed to non-equilibrium conditions between liquid and equilibrating gases (Johnston, 1964; Johnston, 1965). On the other hand, the chemical equilibrium reported above is a simplified form, because ions, logically, occur in melt as solvated species coordinated to oxygens within the silicate glass network (Schreiber, 1987). Ergo, the europium ions activities in the melt may not be directly proportional to the concentration redox ratio. In order to obtain the dependence of the redox ratio to the bulk chemistry, Schreiber proposed an alternate form of the chemical equation for redox equilibria by taking into account the number of oxygen associated with the oxidized redox ion in the melt (coordination spheres). Thus, the cation coordination, or more generically speaking, the cation speciation might be a crucial point when studying redox equilibria.

We can summarize that the main parameters that must be taken into account for determining the redox equilibria in melts are:

- the oxygen partial pressure in the melt (i.e., oxygen fugacity at ambient pressure);
- the thermodynamic equilibrium constant, which depends on temperature and standard free energy of the reaction;
- the activities (a) of ions in the melt.

Schreiber et al. (1994) described the dependence of the redox equilibria for several multivalent elements (including Eu and Ce) in simplified sodium silicate compositions, reporting a good correlation with the basicity of the melt. The optical basicity (Λ) concept was firstly introduced by Duffy and Ingram (Duffy & Ingram, 1971, 1976; Duffy, 1993) and developed in the following years by, e.g., Lebouteiller and Courtine (1998), Moretti (2005), and Ottonello et al. (2001). The optical basicity concept has been developed for quantitative determination of the acid–base properties of mixtures containing a large number of oxides, as for instance glasses, alloys, slags, and molten salts. The optical basicity value is obtained from measurements involving the electron donor power of probe oxides (such as Pb^{2+}). The red shift in the optical absorption band related to the electronic transition 6s → 6p is correlated to the increases of covalency, therefore providing the average electron density. For multicomponent systems, such as glasses and melts, the theoretical optical basicity can be calculated by using partial Λ values of oxide components and their molar proportion (Duffy & Ingram, 1976; Duffy, 1993; Lebouteiller & Courtine, 1998; see Le Losq et al., Chapter 12, this volume).

Many multivalent element redox equilibria follow the general rule of redox controlled by melt basicity, with the oxidation-reduction equilibrium shifting toward more oxidized species with increasing basicity of the melt (i.e., by increasing the amount of network-modifying oxides). Since a greater network-modifying ion activity results in increasing melt depolymerization, this parameter can be broadly related to the ratio of non-bridging oxygens (NBO) per tetrahedral cations (T) in glasses and melts (NBO/T parameter).

The dependence of the Eu^{2+}/Eu^{3+} ratio on glass/magma composition, temperature, and redox conditions has been experimentally investigated by using several techniques: Electron Paramagnetic Resonance (EPR), Voltammetry, Optical Absorption and Luminescence Spectroscopy, Mössbauer, and more recently X-Ray Absorption Spectroscopy (XAS).

EPR spectroscopy data (e.g., Lauer & Morris, 1977; Morris & Haskin, 1974; Morris et al., 1974; Schreiber, 1987) confirmed that oxygen fugacity, temperature, and composition are all important parameters that affect the Eu^{2+}/Eu^{3+} redox ratio in melts. This technique was extensively used to evaluate the Eu redox ratios in glasses, though its use is inadequate for natural samples since the presence of iron overshadows/alters the signal related to europium. In addition, the presence of titanium is also problematic for EPR measurements, because the Ti^{3+} signal partially overlaps the Eu^{2+} signal (Lauer & Morris, 1977).

Morris & Haskin (1974) investigated by EPR the dependence of the concentration ratio of Eu^{2+} to Eu^{3+} on bulk chemistry in Ca-Mg-Al silicate melts. The authors determined that Eu^{2+}/Eu^{3+} redox ratio increases with the [(Al+Si)/O] molar ratio, i.e., with a decrease in NBO. In a following study, Morris et al. (1974) studied the Eu^{2+}/Eu^{3+} redox ratio in Ca-Mg silicate and Ca-Al silicate glasses depending on temperature and oxygen fugacity conditions. The authors confirmed the endothermic nature of the chemical equation for europium redox equilibrium. For diopsidic and anorthitic compositions they further reported deviations from the ideal slope of 4 (from Equation 19.3) in the $-\log(fO_2)$ vs. $\log [Eu^{2+}/Eu^{3+}]$ plot. On the contrary, Virgo et al. (1981) reported a slope very close to the ideal one for diopside glass composition. Diopsidic composition was studied also via Linear Sweep Voltammetry (LSV) by Colson et al. (1990). This technique was used to measure Eu redox potentials, providing an enthalpy of reduction for Eu^{3+} of 31±5 kcal/mol. Moreover, the authors determined the Eu diffusion rate in diopsidic melt (1.5E^{-6} cm^2/s at 1500 °C).

Additional studies of Eu redox ratios have been performed using Mössbauer spectroscopy (e.g., Concas et al., 1998; Virgo et al., 1981). This technique is useful for the determination of Eu oxidation states since the isomer shift differs significantly between Eu^{3+} and Eu^{2+}. However, it does not allow a precise determination of the occupancy/mixing on the europium sites (Harmening et al., 2010). Using Mössbauer, Virgo et al. (1981) reported that the Eu^{2+}/Eu^{3+} redox ratio decreases by a factor of 4 upon replacement of alkali by alkaline-earth elements in a diopside-like melt composition. Concas et al. (1998) carried out Mössbauer investigations of rare earth sites in europium-bearing glasses of different compositions (phosphate, borate, and silicate). They observed that there is a deviation of the Eu^{3+} sites from cubic symmetry. Thus, the different glass networks influence the Eu^{3+} local environment.

Photoluminescence spectroscopy has been extensively used to study Eu-bearing materials, and in particular the effect of bulk composition on luminescence properties (e.g., Cicconi et al., 2017; Wang et al., 2007). However, most of these studies concern synthetic glasses, glass-ceramics, and crystals for technological applications (e.g., phosphors and LEDs). For geological materials, the photoluminescence techniques have mainly been used for detecting divalent and trivalent Eu in minerals (Gaft et al, 2015). Even if this technique has a high sensitivity (ideal to study trace elements), and is extremely sensitive to variations in the surrounding ligands, it does not allow for quantification of the $Eu^{2+}/(Eu^{2+} + Eu^{3+})$ ratios.

XAS has been used to quantify the Eu^{2+}/Eu^{3+} redox ratio in many minerals because of the high sensitivity and spatial resolution of this technique (Brugger et al., 2006; Rakovan et al., 2001; Shearer et al., 2011; Sutton et al., 2005; Takahashi et al., 2005; Tanaka et al., 1995). This technique has also been recently used on natural and synthetic glasses. Karner et al. (2010) studied the partitioning of Eu between augite and melt, in analogues of martian basalts, as a function of fO_2 (from FMQ to IW-1) [2]. The REE partition coefficients patterns between melt and pyroxenes represent well the different imposed fugacities used for the experiments, with increasing negative Eu anomalies (Karner et al., 2010). The quantification of the Eu^{2+}/Eu^{3+} redox ratio either in residual glass or mineral was obtained by analysis of the X-ray

[2] respectively, Fayalite-Magnetite-Quartz and Iron-Wüstite buffer.

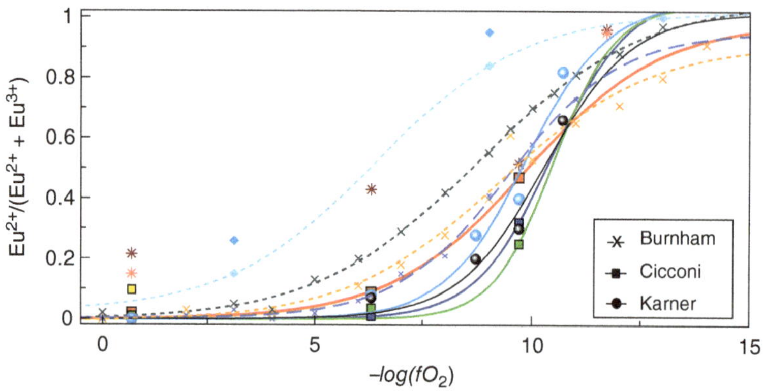

Figure 19.5 $Eu^{2+}/(Eu^{2+} + Eu^{3+})$ redox fraction as determined by XANES vs. oxygen fugacity values, for different bulk compositions. Solid lines represent the evolution of Eu redox ratio with fO_2, extrapolated from data reported by Karner et al. (2010) and Cicconi et al. (2012). Other compositions are haplogranitic glasses (stars; Cicconi et al. 2012) or CAS, MAS compositions (diamonds; Burnham et al., 2015). Crosses (and dotted lines) represent modified diopside-anorthite compositions from Burnham et al. (2015) (AD and AD+qz). Blue dashed line follows the trend of Eu redox in a synthetic MORB glass. See text for details.

Absorption Near Edge Structure (XANES) portion of the XAS spectra. The authors reported that, at the FMQ buffer, europium is mostly present in its oxidized form, both in the melt and in the augite crystal, whereas under more reduced conditions (IW-1) both phases are extremely enriched in reduced europium (circles in Figure 19.5).

Cicconi et al. (2012) synthetized six set of anhydrous glasses with compositions spanning from simplified (Na-silicate) to multicomponent ones resembling haplobasaltic (diopside-anorthite, DiAn) and haplogranitic (HPG8) melts. The authors observed that in glasses synthetized in air the $Eu^{2+}/(Eu^{2+} + Eu^{3+})$ redox fraction markedly increased in going from basaltic to granitic composition (respectively, squares and stars in Figure 19.5), varying from 0.0 up to 0.22 (±0.06). Moreover, Cicconi and coauthors (2012) reported that, in three basaltic compositions equilibrated at 1400 °C under oxygen fugacity conditions close to the FMQ buffer, the Eu oxidation state has only slight variations, with Eu^{3+} still being the predominant species. Only at more reducing conditions (IW buffer), was there evidence of a significant increase in divalent Eu in mafic compositions (squares in Figure 19.5). These observations were in agreement with the previous study of Karner et al. (2010). Felsic samples (haplogranitic glasses; stars in Figure 19.5) always stabilized higher amounts of divalent Eu species. Despite the strong variation of the Eu redox ratio as a function of the bulk melt chemistry, and the correlation between Eu redox state and melt polymerization, Cicconi et al. (2012) suggested that there is not a simple relationship with simplified chemical parameters, such as the non-bridging oxygen per tetrahedron (NBO/T) parameter, the alkali content or the Na/(Na+K) molar ratio. Hence, the dependence of the Eu^{2+}/Eu^{3+} ratio on glass/magma composition might be controlled by a combination of several factors.

Burnham et al. (2015) extended the dataset previously reported on Fe-free, alkali-free Ca-Mg aluminosilicate melts with experiments covering a very wide range of fO_2 (from FMQ-13.7 to FMQ+6.3). Moreover, the authors also reported the data for a synthetic Mid-Ocean Ridge Basalt (MORB). Using this extended set of compositions and oxygen fugacities, the authors corroborate all previous findings, with the $Eu^{2+}/(Eu^{2+} + Eu^{3+})$ redox fraction varying systematically with imposed fO_2 conditions and synthesis temperatures, as well as with melt composition. In Figure 19.5, points related to two diopside-anorthite compositions (AD and AD+qz, equilibrated at 1400 °C) from Burnham et al. (2015) are reported (crosses in Figure 19.5) along the MORB glasses (blue dashed line). Moreover, two compositions in the ternary systems CAS and MAS ($CaO-Al_2O_3-SiO_2$ and $MgO-Al_2O_3-SiO_2$, respectively) are shown as well (diamonds in Figure 19.5). Interestingly, the values obtained by Burnham and coauthors (2015) for the latter two compositions approach the trend reported by Cicconi et al. (2012) for haplogranitic glasses.

All studies cited above agree on the strong influence of bulk composition on the Eu redox ratio, which could be partially rationalized in terms of the influence of the degrees of covalency of the cations building the glass network.

The dependence of Eu redox ratio as described by Morris and Haskin (1974) (on the dependence of Eu redox on melt polymerization and on the increases of the Eu^{2+}/Eu^{3+} as the radius of the network-modifying cation decreases) has been interpreted by Fraser (1975) in terms of the amphoteric character of europium oxides in silicate melts, and the occurrence of different structural units. Thus, Eu_2O_3 may be seen either as basic or acidic oxide, with two possible reactions occurring in silicate melts (Fraser 1975):

$$Eu_2O_3\ _{melt} + O^{2-}_{melt} \leftrightarrow 2EuO_2^-\ _{melt} \quad (19.4)$$

$$Eu_2O_3\ _{melt} \leftrightarrow 2Eu^{3+}_{melt} + 3O^{2-}_{melt} \quad (19.5)$$

The formation of a structural unit, and the acidic or basic nature of the oxide imply that O^{2-} may occur either as product or reactant (similarly to what observed for Fe-S interaction in silicate melts; Moretti & Ottonello, 2003), thus explaining the dependence on the oxide ion activity.

The structural unit EuO_2^- should have an independent geometry site that does not implicate a different structural role in the silicate network. Indeed, taking into account the Dietzel's Cation Field Strength parameter for divalent and trivalent europium (see Figure 19.1b) it is unlikely for Eu^{3+} to behave as a network-former cation. Experimental evidence for the structural role of both trivalent and divalent europium in silicate glasses of geological interest have been provided by EXAFS (Extended X-Ray Absorption Fine Structure) spectroscopy (Cicconi et al., 2012). This study confirmed that Eu ions will behave as network-modifying cations in silicate melts and that they are coordinated with 6 to 9 oxygens (respectively for Eu^{3+} and Eu^{2+}), in agreement with the coordination suggested by Morris & Haskin, 1974, and field strength considerations.

19.3.2. Cerium

Cerium may be stabilized in glasses, melts, and rocks either as Ce^{3+} or Ce^{4+}, and Ce-anomalies (deviations to the general trend) have been detected in several geological systems (terrestrial and lunar rocks, meteorites, minerals, seawater…) and proposed as a redox proxy (e.g., Ballard et al., 2002; Elderfield & Greaves, 1981; Evensen et al., 1978; Masuda et al., 1974; Takahashi et al., 2002; Tostevin et al., 2016; Trail et al., 2011). Indeed, as previously reported for europium, and in general for many multivalent elements (e.g., Cr, V; Papike et al., 2005) the element partitioning between coexisting phases might reflect the average oxygen fugacity conditions at the time of crystallization.

Drake & Weill (1975) observed that the behavior of Ce (as Ce^{3+} + Ce^{4+}) does not differ significantly from the value of Ce^{3+} obtained by linear interpolation between the values of La^{3+} and Nd^{3+}. This implies that the partition coefficient of the trivalent species ($D_{Ce}{}^{3+}$) approximates that of the tetravalent one ($D_{Ce}{}^{4+}$), resulting in $D_{Ce^{3+}} \sim D_{Ce^{4+}}$ in plagioclase/magmatic liquid systems. More recently, Ballard et al. (2002) determined the Ce^{4+}/Ce^{3+} ratio in zircons and used it to infer the relative oxidation state of calc-alkaline intrusions (associated with porphyry copper mineralization) at the time of crystallization. The authors' method was based on the approach of Philpotts (1970) for europium (see Section 19.3.1). Many following studies were focused on the partitioning of Ce (and on Ce oxidation state) into zircon ($ZrSiO_4$) in order to establish the prevailing redox conditions. Zircons are widespread accessory minerals physically and chemically stable, hence they are considered to preserve the original chemistry (Ballard et al., 2002; Trail et al., 2011).

Ce^{4+} is highly compatible in $ZrSiO_4$, much more than Ce^{3+} (like Hf^{4+}, Th^{4+}; see Figure 19.1) such that its high partitioning into the crystals could indicate highly oxidizing conditions in the melt at the time of crystallization. The occurrence of Ce^{4+} in zircons is emphasised by the positive Ce anomalies. However, as in the case of europium in plagioclase, it is important to obtain a quantitative understanding of the influence of bulk composition, fO_2 conditions and synthesis temperatures on the cerium redox ratio. In the following we report a brief review of the experimental studies carried out in order to assess the effect of several parameters (chemistry, temperature, fO_2) on the Ce redox state.

Earlier investigations used chemical titration for evaluation of Ce redox ratio in glasses. Johnston (1965) estimated the Ce^{3+}/Ce_{tot} redox fraction to vary, in air, from 0.26 at 1000 °C to 0.40 at 1300 °C, from experiments on Ce^{3+}-Ce^{4+} oxidation- reduction equilibria in a binary $Na_2Si_2O_5$ melt. Moreover, the author studied the Ce redox ratio in glasses synthetized at 1085 °C under different fO_2 conditions, reporting the equilibrium dependence of log [Ce^{3+}/Ce^{4+}] on -log(fO_2) with values very close to the theoretical slope of 4.

Paul & Douglas (1965a) also investigated the Ce^{3+}-Ce^{4+} redox equilibria in air, in simple binary borate and alkali silicate glasses. The main goal of this study was to understand the influence of cerium content, alkali cation type, and synthesis temperature on redox. The main results obtained by Paul and Douglas for alkali silicate glasses, synthetized in air, are shown in Figure 19.6, where it is possible to appreciate how the bulk composition affects the Ce redox ratio. In going from Li to K (increasing cation ionic radii), at equal amounts of alkali content, there is an increase of the oxidized species. However, this trend is not linear since the slope of the lines increases in the order Li < Na~K. Similar changes in slope with different alkali cation type were also reported for Fe and Cr redox pairs (Paul & Douglas, 1965b). By increasing the amount of network modifiers (increasing melt basicity) there is a marked increase of the oxidized species for K- and Na-silicate glasses and smaller increases for Li-silicate glasses (Figure 19.6). Further observations by Paul and Douglas (1965a) for alkali borate glasses regard the variation of the Ce redox ratio with the total amount of cerium, and

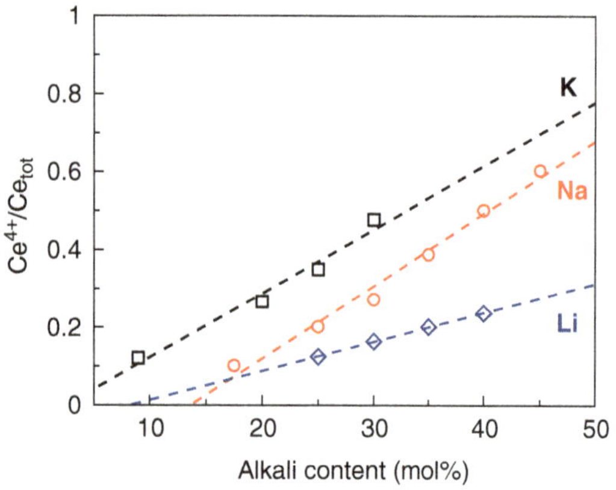

Figure 19.6 Dependence of the Ce^{4+}/Ce_{tot} redox ratio on alkali content, and alkali cation type. Lines are only guide for the eyes. Data from Paul and Douglas (1965) for synthesis in air, at 1400 °C. Errors not provided.

the increase of Ce^{4+} species with gradual increase of cerium concentration. Finally, the effect of temperature was also evaluated for Ce-bearing alkali borate glasses. When plotting the log $[Ce^{4+}/Ce^{3+}]$ vs. reciprocal temperature the authors observed the expected linear correlation, but interestingly they observed that the slope of the straight lines (considered proportional to the thermodynamic standard enthalpy of reduction[3]) was increasing as the glass basicity increased (from 0.16 for Li up to 0.4 for K-glasses).

Schreiber and coauthors (Schreiber, 1980; Schreiber et al., 1994; Schreiber et al., 1980) described the Ce^{4+}-Ce^{3+} equilibria in glasses with respect to bulk chemistry, temperature, oxygen fugacity conditions, and Ce amount (from 0.5 to 1.3wt%). The authors reported that the dependence of the Ce equilibria with respect to the -log (fO_2) to be ~4, in Ca-Mg aluminosilicate glasses. However, contrary to Paul and Douglas (1965a), Schreiber and coauthors observed no dependence of Ce redox ratio on the Ce concentration.

Baucke and Duffy (1993) considered existing redox equilibria in alkali silicate melts (at 1400 °C, in air) and report their correlation with the optical basicity Λ of the melt for several multivalent elements. The authors stated that it is possible to predict the redox equilibria from the composition. By plotting the log $[Ce^{3+}/Ce^{4+}]$ against optical basicity values, they determined a relationship from the straight line obtained, suggesting that it would be feasible to express the ratio of ionic concentrations merely in terms of the optical basicity (Baucke & Duffy, 1993). For Ce equilibrium, the authors proposed the following relationship:

$$\log\left[{}^{Ce^{3+}}/_{Ce^{4+}}\right] = 5.4 - 8.3\Lambda, \qquad (19.6)$$

More recently, Pinet et al. (2006) proposed a model relating the Ce redox ratio to oxygen fugacity, temperature, and optical basicity in silicate glasses (suitable for nuclear waste management). Moreover, the authors report that cerium solubility is strongly affected by redox state, and that Ce solubility is improved when incorporated in the reduced Ce^{3+} form. Pinet and coauthors (2006) developed a general model (similar to the one used for iron in the well-known Kress and Carmichael model; Kress & Carmichael, 1991) applicable to various silicate glass compositions (optical basicity values from 0.52 to 0.65) and temperatures in the range ~900–1240 °C.

A few recent studies have been carried out, aiming to understand the evolution of cerium redox state on temperature, pressure, and bulk composition, by using XAS (Burnham & Berry, 2014; Smythe & Brenan, 2015, 2016; Smythe et al., 2013). Interestingly, these studies report contradictory findings regarding the dependence of the $[Ce^{3+}/Ce^{4+}]$ redox ratio on melt/glass polymerization. Indeed, Burnham and Berry's (2014) results corroborate findings from previous studies (Schreiber et al., 1980; Baucke & Duffy, 1993; Pinet et al., 2006), indicating that the content of reduced species increases with increasing melt polymerization or temperature. On the contrary, Smythe et al. (2013, 2015) observed that the abundance of Ce^{4+} decreases with decreasing polymerization.

Smythe and coauthors (2013, 2015) studied anhydrous alkali-aluminosilicate glasses (synthetic basalt, andesite, dacite, and rhyolite glasses; all sets include alkali and TiO_2, and are Fe-free) that contained ~1.0 wt% CeO_2. Other sets of hydrous glasses were prepared at 0.001GPa and 1GPa to assess the effect of water (hydrous compositions favour the stabilization of reduced species). $[Ce^{3+}/Ce^{4+}]$ redox ratio assessment was carried out both by potentiometric titration and XAS ($M_{4,5}$-edges). Burnham and Berry (2014) studied the influence of bulk composition on the oxidation state of cerium, for a larger range of oxygen fugacity values (from FMQ-10 to FMQ+11) in alkali-free and Fe-free aluminosilicate compositions (the same diopside-anorthite compositions described above for Eu) doped with ~0.5 wt% CeO_2. Both authors report the possible occurrence of Ce redox modifications under the beam during XAS measurements.

From the available data, Smythe and Brenan (2015) ascribed the contrasting effect of glass composition on Ce redox ratio to the compositional differences. In the studies from Smythe and coauthors (2013, 2015) the four compositions span from basaltic to granitic, covering wide ranges of SiO_2 contents (from ~52 to ~78 mol% SiO_2) and (Si+Al)/alkali molar ratios (Smythe et al., 2013; Smythe & Brenan, 2015). Nevertheless, only a

[3] ΔH^0, generally speaking, represents the energy required for the reduction reaction.

limited dataset is reported. On the other hand, Burnham and Berry (2014) studied a much larger dataset (11 different compositions), and a wider range of oxygen fugacity conditions tested. However, all glasses in the diopside-anorthite system have SiO_2 < 65 mol%, extremely high amounts of (CaO+MgO), and no alkalies. Another point that could be source of the differences between the studies lies in the experimental technique employed for the estimation of the Ce^{3+}/Ce^{4+} redox ratio. Both authors used XAS spectroscopy even if at different edges (hence, by looking at different electronic transitions). Smythe et al. (2013) reported beam-damage and overestimation of the Ce^{4+} compared to wet chemistry analyses, thus proposing an empirical relation for correction of the Ce^{3+}/Ce^{4+} redox ratios, based on the potentiometric data. Burnham and Berry (2014) asserted that beam-induced changes were observed only for samples prepared at high pressures. The combination of XAS techniques with other analytical methods could have provided more insights on the effect of the X-ray beam on Ce redox ratio, however the authors did not report any further investigation. For example, optical absorption (OA) and photoluminescence (PL) measurements or XPS analysis, before and after the beam exposure, would have been useful to understand to what extent the X-ray beam influenced cerium ions.

More recently, Cicconi et al. (2017, 2020, 2021) investigated cerium speciation in alkali-silicate, aluminosilicate glasses and Ce-activated optical silica fibers. The authors also used XAS to quantify the Ce redox ratio, but also combined Raman and optical spectroscopy (PL) techniques. Cicconi and coauthors (2017), in agreement with Burnham and Berry (2014) and other previous studies observed that Ce redox behavior, in Na-silicate and aluminosilicate glasses, is linked to the optical basicity of the glasses, and the substitution of Al for Na strongly stabilizes reduced Ce species. However, for optical silica fibers Ce/Al codoped, the relationship bulk chemistry/redox was found to be more complex. In fact, in these glasses, the theoretical optical basicity parameter cannot be used anymore to predict the Ce redox changes, because the different Al structural role (coordinations) in the silica fibers network induces modifications in the Ce^{3+} structural environment that the optical basicity concept cannot predict (variations of the distance between Ce^{3+} and its ligands, a/o variations of the <Ce-O> bonds covalency). Thus, Ce redox state is critically affected by the nature of the cations present in the silica glass network (Cicconi et al., 2017, 2021). Raman spectroscopy analyses of Ce-bearing sodium-silicate and aluminosilicate glasses highlight the structural variations occurring when cerium is stabilized with different valences in the glass network. These observation reported by Cicconi et al. (2017) could explain the contradictory results of Smythe and coauthors

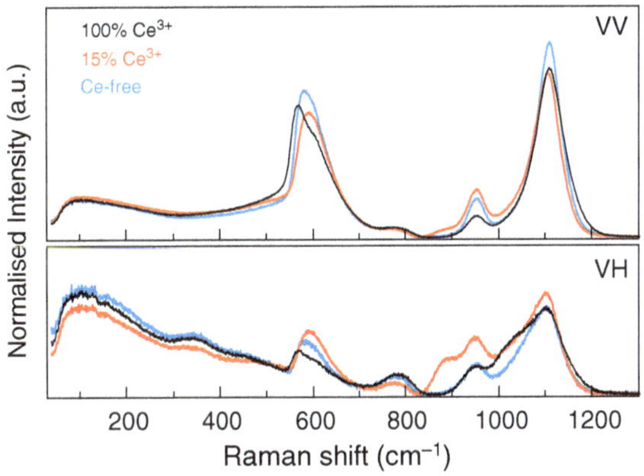

Figure 19.7 Parallel (VV) and cross-polarized (VH) Raman spectra (normalized to the total area) for a NS2 glass (blue line), and for Ce doped glasses. Black lines refer to a NS2 glass having 100%Ce^{3+} ions, and red lines to the glass with 15% Ce^{3+} ions (error on redox quantification: ±7; data from Cicconi et al., 2021).

(Smythe et al., 2013; Smythe & Brenan, 2015) and emphasize the need for understanding an element structural role, rather than limiting studies to redox changes.

Raman spectroscopy is a powerful, complementary technique that may be used in order to understand the ion structural behavior. Parallel (VV) and cross-polarized (VH) Raman spectra for Ce-free and Ce-bearing sodium-disilicate glasses (NS2) are reported in Figure 19.7 (from Cicconi et al., 2021). Silica vibrations related either to Si-O-Si ring motions or Si-O stretching motions are highly polarized bands (Hehlen & Neuville, 2015), thus their intensity drastically decreases when collected as cross-polarized (factor 1:9 in Figure 19.7). For example, the bands in the high frequency portion of the Raman spectra (~800–1300cm^{-1}), related to symmetric stretching, significantly change their intensity, though without changing the frequency position. The VV and VH spectra for the glass containing only reduced species (100±7% Ce^{3+}: black lines in Figure 19.7) indicate a higher polymerization of the network, additionally confirmed by the increase of the glass transition temperature (Cicconi et al., 2021). On the contrary, the prevalent presence of oxidized species (15±7% Ce^{3+}: red lines in Figure 19.7) induces an increase of Q^2 and Q^3 species[4], and the appearance of an additional band around 880 cm^{-1}, which is polarized (indication a non-symmetric vibration). From the available data the authors concluded that, depending on the Ce^{3+}/Ce^{4+} redox ratio, strong variations occur in the silica glass network, arising from Ce-NBO(BO) interactions.

[4] Q^n species: Q represents a Si centred tetrahedron and n the number of bridging oxygen atoms (n= 0 – 4).

Optical spectroscopy is widely used to investigate REE- and transition element-bearing glasses, because of the extremely high sensitivity, and because these techniques are able to detect small variations in the surrounding ligand (e.g., variations of the covalency). OA and PL spectroscopies have been extensively used to study Ce-doped materials (e.g., Assefa et al., 2004; Brandily-Anne et al., 2010; Fasoli et al., 2009; Herrmann et al., 2015), even if the quantification of Ce redox state is very controversial. As a matter of fact, Ce species present absorption in the UV-region with an overlap of the broad bands, which increase or decrease depending on glass chemistry, and Ce content. Moreover, only Ce^{3+} is optically active and has fluorescence emission, whereas Ce^{4+} does not show luminescence due to its closed electronic shell ([Xe] $4f^0$).

A combination of wet-chemistry, voltammetry, Raman, and XAS with optical techniques could provide a better understanding of the factors influencing the Ce redox ratio and speciation, in silicate glasses and melts.

19.3.3. Influence of Other Elements: Mutual Interactions

In the glass industry, controlling the color of the final product is a very important issue along with the fining process (the removal of gaseous inclusions). Hence, several investigations in glass and ceramic manufacturing involve understanding the effect of multivalent elements on the final color (as well as on properties). REE have been widely used in the production of special glasses, or as decolorizing agents (Ce). For example, in common window glasses (soda-lime-silica) production, the presence of divalent iron is unwanted because of a strong blue hue, and Ce_2O_3 is considered a particularly effective oxidizing agent. Glass with iron impurities up to 0.1wt% might be successfully decolorized by adding sulfates (as fining agent) and cerium oxide (Henderson, 1984).

In the past, cerium has been used in the production of special glasses for protection against UV light. Indeed, earlier studies on the ceria/ceric equilibria were done on simple binary systems in order to determine the effect of composition on cerium absorption bands and, in turn, on the UV cut-off (Paul et al., 1976). More recently, the interest for REE redox control derives from the use of glasses as nuclear waste immobilizer. For this purpose, the parameters influencing cerium redox state have been studied in many silicate and borate compositions (Pinet et al., 2014; Pinet et al., 2006). Lanthanides have properties similar to the transuranic actinides, so REE have been considered as actinide analogues for preliminary modelling of the characteristics of the waste forms. Indeed, REE are the most favored elements to be used as surrogates since they already occur as fission products; moreover, because of the geochemical similarity with actinides, they provide a way to forecast the long-term behavior of waste forms.

Another important point in glass/ceramic manufacturing concerns the understanding of the mutual interactions between multivalent elements in glasses. Both Ce and Eu redox ratios are influenced by the presence of other multivalent elements, as described by Schreiber et al. (1987). A complete understanding of such interactions between transition elements and REE is as important for glass manufacturing, as it is for geochemical and petrological interpretations of element partitioning between melt and crystals.

The use of Ce_2O_3 as decolorizing agent implies that an interaction between different multivalent elements occurs, following the reaction (Schreiber et al. 1987):

$$Fe^{2+}_{melt} + Ce^{4+}_{melt} = Fe^{3+}_{melt} + Ce^{3+}_{melt}, \quad (19.7)$$

The interaction between redox species has been argued to derive either from electron exchanges at high temperatures (melts as analogs of aqueous solution) or exclusively during the quench, implying a charge transfer process due to presence of Ce-O-Fe complexes. Unfortunately, the study of mutual interaction and of the redox behavior in melts has been limited to relatively few systems because of the experimental difficulties related with the measurements.

Fe-Eu Mutual Interactions. Since iron is the most common transition element in natural magmas, it provides a first order control on properties (see Le Losq et al., 2021, Chapter 12, this volume, for details), and since it is present in amounts orders of magnitude higher than REE (or other transition elements), it also buffers the magma redox state.

Most of the studies described in this chapter refer to Fe-free glasses, and only a few report data for Fe-bearing glasses and discuss the possible differences. Schreiber and coauthors, in their studies, attempted to understand the redox reactions not only in iron-free compositions, but also in more realistic systems in order to develop models applicable to natural magmas (e.g., Schreiber, 1980; Schreiber et al., 1982; Schreiber et al., 1980, 1987). A chemical equation for the interaction by oxidation-reduction of Eu and Fe ions in silicate melts, similar to Equation 19.7, could be expressed as:

$$Fe^{3+}_{melt} + Eu^{2+}_{melt} = Fe^{2+}_{melt} + Eu^{3+}_{melt}, \quad (19.8)$$

The occurrence of this interaction in glasses has been experimentally verified both for Eu-bearing simplified systems (CMAS, Na-silicate, DiAn) and multicomponent ones (basalts). The presence of both multivalent elements determines different redox kinetics, and the stabilization of higher amounts of reduced iron species (Schreiber et al., 1980, 1982; Cicconi, 2010; Cicconi et al., 2015).

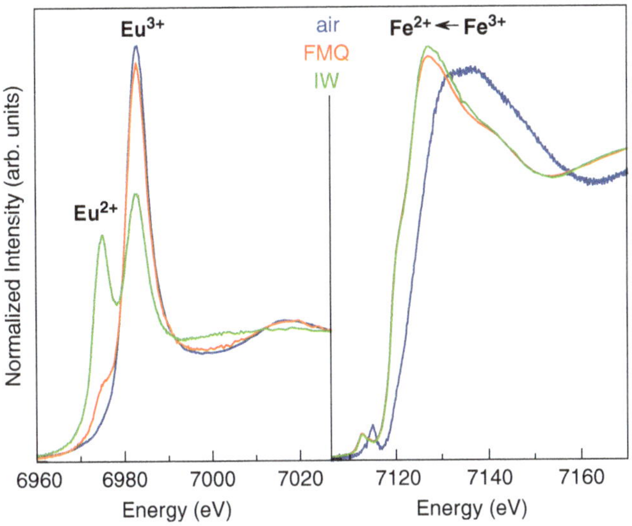

Figure 19.8 X-Ray Absorption Spectroscopy data at the Eu L_{III}-edge (left) and at the Fe K-edge (right) for glasses of DiAn composition synthetized under different conditions of fO_2 (data from Cicconi, 2010)

For example, XAS data collected for a series of DiAn glass compositions (with Eu/Fe molar ratio = 0.05) show that, for both species, the evolution of the redox ratios with fO_2 changes with respect to the glasses containing only a single multivalent element. Indeed, Fe-doped DiAn glasses prepared at ambient conditions (1400 °C, in air) should have trivalent iron as the predominant species. However, the presence of europium (Figure 19.8), leads to the stabilization of > 30% Fe^{2+} in air (±5; Cicconi, 2010). Moreover, at oxygen fugacity values close to the FMQ buffer, Fe is already almost completely reduced (> 90% Fe^{2+}±5) and the Eu^{2+}/Eu^{3+} redox ratio is lower than expected (~5% Eu^{2+}±6). On the contrary, close to the IW buffer, when all iron is completely reduced, the $Eu^{2+}/(Eu^{2+} + Eu^{3+})$ estimated for the Fe/Eu glass is 0.49 (±0.06), very close to the one estimated for the Fe-free glass (0.47±0.06). Thus, it is reasonable to consider that europium redox equilibrium goes back to the expected value when almost all trivalent iron disappears from the melt.

From the data reported in Figure 19.5 it is possible to observe that for oxygen fugacity values close to the FMQ buffer (-log(fO_2) ~6 in Figure 19.5) the Eu redox ratio in Fe-free glasses is highly variable depending on the bulk chemistry. However, the introduction of iron into the system strongly decreases this variability.

As previously reported, two mechanisms could be responsible for the lower Eu^{2+} contents: electron exchanges in the melt, or the occurrence of oxidation via charge transfer processes during glass quenching. In order to monitor the possible mutual interaction at high temperatures (and avoid possible quenching effects), a study of molten materials where both Fe and Eu redox kinetics were simultaneously monitored under imposed oxygen fugacity conditions was carried out by Cicconi et al. (2015). The studied composition was a Fe-rich basalt containing ~5 wt% Eu_2O_3 (to obtain an Eu/Fe molar ratio = 0.066).

Some of the in-situ dispersive X-ray Absorption Spectroscopy data, obtained at high temperatures (HT=1500 °C) for a basaltic glass are reported in Figure 19.9a, and the resulting redox kinetics for both Eu and Fe are shown in Figure 19.9b. At HT, both in air and under pure Ar, the stabilization of small amount of divalent europium would be expected. However, the presence of iron hampers any Eu reduction. Only under more reducing conditions (Ar/H_2), and after almost complete reduction of all iron present, do Eu ions start to be stabilized as Eu^{2+} (Figs. 19.9a,b). From the in-situ study, Cicconi et al. (2015) affirm that there is strong competition between the two redox couples in the melt, implying that electron exchange occurs also in the molten material. Moreover, they observed differences in the Eu redox ratio from the melt and the in-situ quenched glass, emphasizing that the quench effect must not be neglected in future studies of Eu/Fe-bearing glasses.

19.3.4. Other Multivalent Rare Earth Elements

As stated above, in geological systems (aqueous solution, glass, and rocks) the most stable valence for lanthanides is the trivalent one. However, REE can be stabilized with other oxidation states. In the past decades REE-bearing glass, glass-ceramic, or crystalline materials have been developed for the characteristic luminescence properties provided by the lanthanides (Lucas et al., 2014). Based on these studies, several other lanthanides have been observed to be stable as divalent or tetravalent.

• divalent REE: Eu^{2+}. Sm^{2+}, Yb^{2+}

Eu is able to lose two electrons in order to obtain the stabilization of the half-full *4f* shell: Eu^{2+} [Xe]$4f^7$, and similar considerations can be done for Sm (Sm^{2+} [Xe] $4f^6$). Yb^{2+} can be formed due to the stabilization of the full *4f* shell [Xe]$4f^{14}$.

The stabilization of divalent REE ions in different host materials has been realized either by using a highly reducing atmosphere (H_2), by electrolytic reduction, or by exposing the material to ionizing radiation (Fong, 1964; Nicoara et al., 2008; Nogami et al., 1995).

• Tetravalent REE: Ce^{4+}, Pr^{4+}, Tb^{4+}

The known forms, obtained under oxidizing conditions, are:

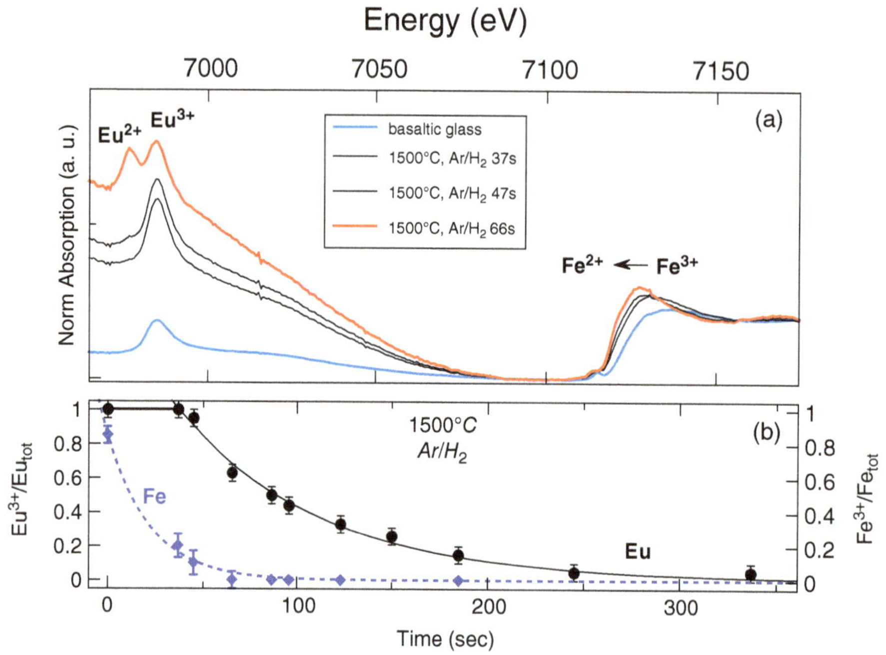

Figure 19.9 (a) XAS spectra, normalized at the Fe K-edge, collected for the starting basaltic glass and for the melt (1500°C), after 37, 47 and 66s under reducing conditions (Ar/H_2 gas). (b) Time dependence of Eu and Fe redox kinetics in the basaltic melt, under reducing conditions (data from Cicconi et al., 2015).

Ce, which lose all four outer electrons to form Ce^{4+} ions [Xe]$4f^0$. Similarly Pr^{4+} and Tb^{4+} can be formed (Morgan et al., 1987; Ueda et al., 2017).

19.4. CONCLUSIONS AND PERSPECTIVES

Trace element geochemistry gives important information in Earth and planetary sciences for investigating problems ranging from geochemistry to igneous petrology (crystal-melt equilibria, magmatic differentiation, etc.). REE distributions in minerals are strongly dependent on (i) ionic radius, and (ii) the suitable valence state and bonding forces (as in the case of monazite and xenotime) and the lack of constraints regarding element partition coefficients is a major source of error in modelling crystal/melt fractionation (Blundy & Wood, 2003, and references therein).

Many geologically important trace elements are potentially redox variable (e.g., Ti, V, Cr, Cu, Eu, Ce), however the factors influencing redox mechanisms sometimes are poorly known, thus limiting our understanding of trace element patterns, abundances, and partitioning. Determining an element oxidation state, coordination environment, and speciation provide the basic knowledge on the parameters controlling the element geochemical behavior.

The data reported here emphasize the need for understanding the redox behavior in many other systems, but also emphasize the importance of an often-underestimated parameter: the elements' structural role. There are only a few examples, in the Earth Sciences literature, where systematic studies of Ce and Eu redox are complemented with data regarding their surroundings. The combination of advanced element specific techniques (such as XAS) with other analytical methods might provide more insights on the effect of the element speciation in glasses and melts.

ACKNOWLEDGMENTS

Some of the data here shown have been acquired at LISA and at ODE beamlines, and we acknowledge ESRF (France) and SOLEIL (France) for provision of synchrotron radiation facilities.

REFERENCES

Assefa, Z., Haire, R. G., Caulder, D. L., & Shuh, D. K. (2004). Correlation of the oxidation state of cerium in sol--gel glasses as a function of thermal treatment via optical spectroscopy and XANES studies. *Spectrochimica Acta Part A: Molecular and Biomolecular Spectroscopy*, *60*, 8, 1873–1881. https://doi.org/10.1016/j.saa.2003.10.005

Ballard, J. R., Palin, M. J., & Campbell, I. H. (2002). Relative oxidation states of magmas inferred from Ce(IV)/Ce(III) in zircon: application to porphyry copper deposits of northern

Chile. *Contributions to Mineralogy and Petrology*, *144*, 3, 347–364. https://doi.org/10.1007/s00410-002-0402-5

Baucke, F., & Duffy, J. (1993). Redox reactions between cations of different polyvalent elements in glass melts: an optical basicity study. *Physics and Chemistry of Glasses*

Bédard, J. H. (2006). Trace element partitioning in plagioclase feldspar. *Geochimica et Cosmochimica Acta*, *70*, 14, 3717–3742. https://doi.org/10.1016/J.GCA.2006.05.003

Blundy, J., & Wood, B. (2003). Partitioning of trace elements between crystals and melts. *Earth and Planetary Science Letters*, *210*, 3–4, 383–397. https://doi.org/10.1016/S0012-821X(03)00129-8

Brandily-Anne, M. L., Lumeau, J., Glebova, L., & Glebov, L. B. (2010). Specific absorption spectra of cerium in multicomponent silicate glasses. In *Journal of Non-Crystalline Solids*, *356*, 2337–2343). https://doi.org/10.1016/j.jnoncrysol.2010.02.020

Brugger, J., Etschmann, B., Chu, Y. S., Harland, C., Vogt, S., Ryan, C., & Jones, H. (2006). The oxidation state of europium in hydrothermal scheelite: In situ measurement by XANES spectroscopy. *Canadian Mineralogist*, *44*, 5, 1079–1087. https://doi.org/10.2113/gscanmin.44.5.1079

Burnham, A. D., Berry, A. J., Halse, H. R., Schofield, P. F., Cibin, G., & Mosselmans, J. F. W. (2015). The oxidation state of europium in silicate melts as a function of oxygen fugacity, composition and temperature. *Chemical Geology*, *411*, 248–259. https://doi.org/10.1016/j.chemgeo.2015.07.002

Burnham, A. D., & Berry, A. J. (2014). The effect of oxygen fugacity, melt composition, temperature and pressure on the oxidation state of cerium in silicate melts. *Chemical Geology*, *366*, 52–60. https://doi.org/10.1016/j.chemgeo.2013.12.015

Ching-oh, S., Williams, R. J., & Shine-soon, S. (1974). Distribution coefficients of Eu and Sr for plagioclase-liquid and clinopyroxene-liquid equilibria in oceanic ridge basalt: an experimental study. *Geochimica et Cosmochimica Acta*, *38*, 9, 1415–1433. https://doi.org/10.1016/0016-7037(74)90096-9

Cicconi, M. R., Neuville, D. R., Blanc, W., Lupi, J.-F., Vermillac, M., & de Ligny, D. (2017). Cerium/aluminum correlation in aluminosilicate glasses and optical silica fiber preforms. *Journal of Non-Crystalline Solids*, *475*, 85–95. https://doi.org/10.1016/j.jnoncrysol.2017.08.035

Cicconi, M. R., Blanc, W., de Ligny, D., & Neuville, D. R. (2020). The influence of codoping on optical properties and glass connectivity of silica fiber preforms. *Ceramics International*, *46*, 26231–26259. https://doi.org/10.1016/j.ceramint.2020.05.233

Cicconi, M. R., Veber, A., Neuville, D. R., Baudelet, F., & de Ligny, D. (2021). Cerium speciation in silicate glasses: Structure-property relationships. Journal of Non-Crystalline Solids, 563, 120785. https://doi.org/10.1016/j.jnoncrysol.2021.120785

Cicconi, M. R. (2010). *Europium (Eu) and cerium (Ce) in silicate glasses: XAS study of their oxidation states and local environments*. PhD thesis, Camerino University, Italy.

Cicconi, M. R., Giuli, G., Paris, E., Ertel-Ingrisch, W., Ulmer, P., & Dingwell, D. B. (2012). Europium oxidation state and local structure in silicate glasses. *American Mineralogist*, *97*, 5–6, 918–929. https://doi.org/10.2138/am.2012.4041

Cicconi, M. R., Neuville, D. R., Tannou, I., Baudelet, F., Floury, P., Paris, E., & Giuli, G. (2015). Competition between two redox states in silicate melts: An in-situ experiment at the Fe K-edge and Eu L<inf>3</inf>-edge. *American Mineralogist*, *100*, 4, 1013–1016. https://doi.org/10.2138/am-2015-5172

Cicconi, M. R., Veber, A., de Ligny, D., Rocherullé, J., Lebullenger, R., & Tessier, F. (2017). Chemical tunability of europium emission in phosphate glasses. *Journal of Luminescence*, *183*. https://doi.org/10.1016/j.jlumin.2016.11.019

Cicconi, M. R., Giuli, G., Ertel-Ingrisch, W., Paris, E., & Dingwell, D. B. (2015). The effect of the [Na/(Na+K)] ratio on Fe speciation in phonolitic glasses. *American Mineralogist*, *100*, 7, 1610–1619. https://doi.org/10.2138/am-2015-5155

Colson, R. O., Haskin, L. A., & Crane, D. (1990). Electrochemistry of cations in diopsidic melt: Determining diffusion rates and redox potentials from voltammetric curves. *Geochimica et Cosmochimica Acta*, *54*, 12, 3353–3367. https://doi.org/10.1016/0016-7037(90)90290-2

Concas, G., Congiu, F., Spano, G., Speghini, A., & Gatterer, K. (1998). Mössbauer investigation of rare earth sites in europium containing glasses. *Journal of Non-Crystalline Solids*, *232–234*, 341–345. https://doi.org/10.1016/S0022-3093(98)00390-1

Dietzel, A. (1948). Glasstruktur und Glaseigenschaften. *Glastechnische Berichte*, *22*, 41–50.

Drake, M. J. (1975). The oxidation state of europium as an indicator of oxygen fugacity. *Geochimica et Cosmochimica Acta*, *39*, 1, 55–64. https://doi.org/10.1016/0016-7037(75)90184-2

Drake, M. J., & Weill, D. F. (1975). Partition of Sr, Ba, Ca, Y, Eu^{2+}, Eu^{3+}, and other REE between plagioclase feldspar and magmatic liquid: an experimental study. *Geochimica et Cosmochimica Acta*, *39*, 5, 689–712. https://doi.org/10.1016/0016-7037(75)90011-3

Duffy, J. A., & Ingram, M. D. (1971). Establishment of an optical scale for Lewis basicity in inorganic oxyacids, molten salts, and glasses. *Journal of the American Chemical Society*, *93*, 24, 6448–6454. https://doi.org/10.1021/ja00753a019

Duffy, J. A., & Ingram, M. D. (1976). An interpretation of glass chemistry in terms of the optical basicity concept. *Journal of Non-Crystalline Solids*, *21*, 3, 373–410. https://doi.org/10.1016/0022-3093(76)90027-2

Duffy, J. A. (1993). A review of optical basicity and its applications to oxidic systems. *Geochimica et Cosmochimica Acta*, *57*, 16, 3961–3970. https://doi.org/10.1016/0016-7037(93)90346-X

Elderfield, H., & Greaves, M. J. (1981). Negative cerium anomalies in the rare earth element patterns of oceanic ferromanganese nodules. *Earth and Planetary Science Letters*, *55*, 1, 163–170. https://doi.org/10.1016/0012-821X(81)90095-9

Evensen, N. M., Hamilton, P. J., & O'Nions, R. K. (1978). Rare-earth abundances in chondritic meteorites. *Geochimica et Cosmochimica Acta*, *42*, 8, 1199–1212. https://doi.org/10.1016/0016-7037(78)90114-X

Fasoli, M., Vedda, A., Lauria, A., Moretti, F., Rizzelli, E., Chiodini, N., et al. (2009). Effect of reducing sintering atmosphere on Ce-doped sol-gel silica glasses. *Journal of Non-Crystalline Solids*, *355*, 18–21, 1140–1144. https://doi.org/10.1016/j.jnoncrysol.2009.01.043

Fong, F. K. (1964). Electrolytic Reduction of Trivalent Rare-Earth Ions in Alkaline-Earth Halides. *The Journal of*

Chemical Physics, *41*, 8, 2291–2296. https://doi.org/10.1063/1.1726262

Fraser, D. G. (1975). Activities of trace elements in silicate melts. *Geochimica et Cosmochimica Acta*, *39*, 11, 1525–1530. https://doi.org/10.1016/0016-7037(75)90154-4

Gaetani, G. A., & Grove, T. L. (1995). Partitioning of rare earth elements between clinopyroxene and silicate melt Crystal-chemical controls. *Geochimica et Cosmochimica Acta*, *59*, 10, 1951–1962. https://doi.org/10.1016/0016-7037(95)00119-0

Gaft, M., Reisfeld, R., & Panczer, G. G. (2015). *Modern Luminescence Spectroscopy of Minerals and Materials*. Cham: Springer International Publishing. https://doi.org/10.1007/978-3-319-24765-6

Grutzeck, M., Kridelbaugh, S., & Weill, D. (1974). The distribution of Sr and REE between diopside and silicate liquid. *Geophysical Research Letters*, *1*, 6, 273–275. https://doi.org/10.1029/GL001i006p00273

Harmening, T., Hermes, W., Eul, M., & Pöttgen, R. (2010). Mixed valent stannide EuRuSn3 – Structure, magnetic properties, and Mössbauer spectroscopic investigation. *Solid State Sciences*, *12*, 2, 284–290. https://doi.org/10.1016/J.SOLIDSTATESCIENCES.2009.11.009

Hehlen, B., & Neuville, D. R. (2015). Raman response of network modifier cations in alumino-silicate glasses. *Journal of Physical Chemistry B*, *119*, 10, 4093–4098. https://doi.org/10.1021/jp5116299

Henderson, P. (1984). *Rare earth element geochemistry*. Elsevier.

Herrmann, A., Othman, H. A., Assadi, A. A., Tiegel, M., Kuhn, S., & Rüssel, C. (2015). Spectroscopic properties of cerium-doped aluminosilicate glasses. *Optical Materials Express*, *5*, 4, 720. https://doi.org/10.1364/OME.5.000720

Johnston, W. D. (1964). Oxidation-Reduction Equilibria in Iron-Containing Glass. *Journal of the American Ceramic Society*, *47*, 4, 198–201. https://doi.org/10.1111/j.1151-2916.1964.tb14392.x

Johnston, W. D. (1965). Oxidation-Reduction Equilibria in Molten Na2O.2SiO2 Glass. *Journal of the American Ceramic Society*, *48*(4), 184–190. https://doi.org/10.1111/j.1151-2916.1965.tb14709.x

Karner, J. M., Papike, J. J., Sutton, S. R., Burger, P. V., Shearer, C. K., Le, L., et al. (2010). Partitioning of Eu between augite and a highly spiked martian basalt composition as a function of oxygen fugacity (IW-1 to QFM): Determination of Eu2+/Eu3+ ratios by XANES. *American Mineralogist*, *95*, 2–3, 410–413. https://doi.org/10.2138/am.2010.3394

Kress, V. C., & Carmichael, I. S. E. (1991). The compressibility of silicate liquids containing Fe2O3 and the effect of composition, temperature, oxygen fugacity and pressure on their redox states. *Contributions to Mineralogy and Petrology*, *108*, 1–2, 82–92. https://doi.org/10.1007/BF00307328

Lauer, H. V., & Morris, R. V. (1977). Redox Equilibria of Multivalent Ions in Silicate Glasses. *Journal of the American Ceramic Society*, *60*, 9–10, 443–451. https://doi.org/10.1111/j.1151-2916.1977.tb15530.x

Lebouteiller, A., & Courtine, P. (1998). Improvement of a bulk optical basicity table for oxidic systems. *Journal of Solid State Chemistry*, *137*, 94–103. https://doi.org/10.1006/jssc.1997.7722

Le Losq, C., Cicconi, M. R., & Neuville, D. R. (2021). Iron in silicate glasses and melts: Implications for volcanological processes. In Moretti, R., & Neuville, D. R. (Eds.), *Redox Magma Geochemistry*. Geophysical Monograph Series 266. American Geophysical Union.

Lipin, B. R., & McKay, G. A. (1989). *Geochemistry and mineralogy of rare earth elements*. Mineralogical Society of America.

Lucas, J., Lucas, P., Le Mercier, T., Rollat, A., & Davenport, W. G. (2014). *Rare earths: science, technology, production and use*. Elsevier.

Masuda, A., Tanaka, T., Nakamura, N., & Kurasawa, H. (1974). Possible REE anomalies of Apollo 17 REE patterns. In: *Lunar Science Conference, 5th, Houston, Tex., March 18-22, 1974, Proceedings. Volume 2. (A75-39540 19-91)* New York, Pergamon Press, Inc., 1974, pp. 1247–1253.

McIntire, W. L. (1963). Trace element partition coefficients—a review of theory and applications to geology. *Geochimica et Cosmochimica Acta*, *27*, 12, 1209–1264. https://doi.org/10.1016/0016-7037(63)90049-8

McKay, G. A. (1989). Partitioning of rare earth elements between major silicate minerals and basaltic melts. In *Reviews in Mineralogy and Geochemistry*, *21*, 45–77.

Moretti, R. (2005). Polymerisation, basicity, oxidation state and their role in ionic modelling of silicate melts. *Annals of Geophysics*, *48*, 4/5, 583–608.

Moretti, R., & Ottonello, G. (2003). Polymerization and disproportionation of iron and sulfur in silicate melts: Insights from an optical basicity-based approach. In *Journal of Non-Crystalline Solids*, *323*, 111–119. https://doi.org/10.1016/S0022-3093(03)00297-7

Morgan, S. H., Magruder, R. H., & Silberman, E. (1987). Raman Spectra of Rare-Earth Phosphate Glasses. *Journal of the American Ceramic Society*, *70*, 12, C-378-C–380. https://doi.org/10.1111/j.1151-2916.1987.tb04925.x

Morris, R., & Haskin, L. (1974). EPR measurement of the effect of glass composition on the oxidation states of europium. *Geochimica et Cosmochimica Acta*, *38*, 9, 1435–1445. https://doi.org/10.1016/0016-7037(74)90097-0

Morris, R., Haskin, L., Biggar, G., & O'Hara, M. (1974). Measurement of the effects of temperature and partial pressure of oxygen on the oxidation states of europium in silicate glasses. *Geochimica et Cosmochimica Acta*, *38*, 9, 1447–1459. https://doi.org/10.1016/0016-7037(74)90098-2

Müller, P., & Muecke, G. K. (1984). Significance of Europium anomalies in silicate melts and crystal-melt equilibria: a re-evaluation. *Contributions to Mineralogy and Petrology*, *87*, 3, 242–250. https://doi.org/10.1007/BF00373057

Nicoara, I., Pecingina-Garjoaba, N., & Bunoiu, O. (2008). Concentration distribution of Yb2+ and Yb3+ ions in YbF3: CaF2 crystals. *Journal of Crystal Growth*, *310*, 7–9, 1476–1481. https://doi.org/10.1016/J.JCRYSGRO.2007.11.024

Nogami, M., Abe, Y., Hirao, K., & Cho, D. H. (1995). Room temperature persistent spectra hole burning in Sm2+-doped silicate glasses prepared by the sol-gel process. *Applied Physics Letters*, *66*, 22, 2952–2954. https://doi.org/10.1063/1.114240

O'Hara, M. J. (1995). Trace Element Geochemical Effects of Integrated Melt Extraction and "Shaped" Melting Regimes.

Journal of Petrology, *36*, 4, 1111–1132. https://doi.org/10.1093/petrology/36.4.1111

Onuma, N., Higuchi, H., Wakita, H., & Nagasawa, H. (1968). Trace element partition between two pyroxenes and the host lava. *Earth and Planetary Science Letters*, *5*, 47–51. https://doi.org/10.1016/S0012-821X(68)80010-X

Ottonello, G., Moretti, R., Marini, L., & Vetuschi Zuccolini, M. (2001). Oxidation state of iron in silicate glasses and melts: A thermochemical model. *Chemical Geology*, *174*, 1–3, 157–179. https://doi.org/10.1016/S0009-2541(00)00314-4

Papike, J. J., Karner, J. M. M., & Shearer, C. K. K. (2005). Comparative planetary mineralogy: Valence state partitioning of Cr, Fe, Ti, and V among crystallographic sites in olivine, pyroxene, and spinel from planetary basalts. *American Mineralogist*, *90*, 2–3, 277–290. https://doi.org/10.2138/am.2005.1779

Paul, A., & Douglas, R. W. (1965a). Cerous-ceric equilibrium in binary alkali borate and alkali silicate glasses. *Physics and Chemistry of Glasses*, *6*, 6, 212–215.

Paul, A., & Douglas, R. W. (1965b). Ferrous-ferric equilibrium in binary alkali silicate glasses. *Physics and Chemistry of Glasses*, *6*, 6, 207–211.

Paul, A., Mulholland, M., & Zaman, M. S. (1976). Ultraviolet absorption of cerium(III) and cerium(IV) in some simple glasses. *Journal of Materials Science*, *11*, 11, 2082–2086. https://doi.org/10.1007/PL00020336

Philpotts, J. A. (1970). Redox estimation from a calculation of Eu2+ and Eu3+ concentrations in natural phases. *Earth and Planetary Science Letters*, *9*, 3, 257–268. https://doi.org/10.1016/0012-821X(70)90036-1

Pinet, O., Hugon, I., & Mure, S. (2014). Redox Control of Nuclear Glass. *Procedia Materials Science*, *7*, 124–130. https://doi.org/10.1016/J.MSPRO.2014.10.017

Pinet, O., Phalippou, J., & Di Nardo, C. (2006). Modeling the redox equilibrium of the Ce4+/Ce3+ couple in silicate glass by voltammetry. *Journal of Non-Crystalline Solids*, *352*, 50–51, 5382–5390. https://doi.org/10.1016/j.jnoncrysol.2006.08.034

Rakovan, J., Newville, M., & Sutton, S. (2001). Evidence of heterovalent europium in zoned llallgua apatite using wavelength dispersive XANES. *American Mineralogist*, *86*, 5–6, 697–700. https://doi.org/10.2138/am-2001-5-610

Rollinson, H. R. (1993). *Using geochemical data : evaluation, presentation, interpretation*. Longman Scientific & Technical.

Schnetzler, C., & Philpotts, J. A. (1970). Partition coefficients of rare-earth elements between igneous matrix material and rock-forming mineral phenocrysts—II. *Geochimica et Cosmochimica Acta*, *34*, 3, 331–340. https://doi.org/10.1016/0016-7037(70)90110-9

Schreiber, H. D. (1980). Properties of redox ions in glasses: An interdisciplinary perspective. *Journal of Non-Crystalline Solids*, *42*, 1–3, 175–183. https://doi.org/10.1016/0022-3093(80)90019-8

Schreiber, H. D. (1987). An Electrochemical Series of Redox Couples in Silicate Melts: a review and applications to geochemistry. *Journal of Geophysical Research*, *92*, 9225–9232. https://doi.org/10.1029/JB092iB09p09225

Schreiber, H. D., Balazs, G. B., Shaffer, A. P., & Jamison, P. L. (1982). Iron metal production in silicate melts through the direct reduction of Fe(II) by Ti(III), Cr(II), and Eu(II). *Geochimica et Cosmochimica Acta*, *46*, 10, 1891–1901. https://doi.org/10.1016/0016-7037(82)90127-2

Schreiber, H. D., Kochanowski, B. K., Schreiber, C. W., Morgan, A. B., Coolbaugh, M. T., & Dunlap, T. G. (1994). Compositional dependence of redox equilibria in sodium silicate glasses. *Journal of Non-Crystalline Solids*, *177*, 340–346. https://doi.org/10.1016/0022-3093(94)90548-7

Schreiber, H. D., Lauer, H. V., & Thanyasiri, T. (1980). The redox state of cerium in basaltic magmas: an experimental study of iron-cerium interactions in silicate melts. *Geochimica et Cosmochimica Acta*, *44*, 10, 1599–1612. https://doi.org/10.1016/0016-7037(80)90120-9

Schreiber, H. D., Merkel, R. C., Schreiber, V. L., & Balazs, G. B. (1987). Mutual interactions of redox couples via electron exchange in silicate melts: Models for geochemical melt systems. *Journal of Geophysical Research*, *92*, B9, 9233. https://doi.org/10.1029/JB092iB09p09233

Schreiber, H. D., Wilk Jr, N. R., & Schreiber, C. W. (1999). A comprehensive electromotive force series of redox couples in soda–lime–silicate glass. *Journal of Non-Crystalline Solids*, *253*, 1–3, 68–75. https://doi.org/10.1016/S0022-3093(99)00344-0

Shannon, R. D. (1976). Revised effective ionic radii and systematic studies of interatomic distances in halides and chalcogenides. *Acta Crystallographica Section A*, *32*, 5, 751–767. https://doi.org/10.1107/S0567739476001551

Shaw, D. M. (2006). *Trace elements in magmas : a theoretical treatment*. Cambridge University Press.

Shearer, C. K., Papike, J. J., Burger, P. V., Sutton, S. R., McCubbin, F. M., & Newville, M. (2011). Direct determination of europium valence state by XANES in extraterrestrial merrillite: Implications for REE crystal chemistry and martian magmatism. *American Mineralogist*, *96*, 8–9, 1418–1421. https://doi.org/10.2138/am.2011.3860

Shearer, C. K., Papike, J. J., & Karner, J. M. (2006). Pyroxene europium valence oxybarometer: Effects of pyroxene composition, melt composition, and crystallization kinetics. *American Mineralogist*, *91*, 10, 1565–1573. https://doi.org/10.2138/am.2006.2098

Smythe, D. J., & Brenan, J. M. (2015). Cerium oxidation state in silicate melts: Combined fO2, temperature and compositional effects. *Geochimica et Cosmochimica Acta*, *170*, 173–187. https://doi.org/10.1016/j.gca.2015.07.016

Smythe, D. J., & Brenan, J. M. (2016). Magmatic oxygen fugacity estimated using zircon-melt partitioning of cerium. *Earth and Planetary Science Letters*, *453*. Elsevier. https://doi.org/10.1016/j.epsl.2016.08.013

Smythe, D. J., Brenan, J. M., Bennett, N. R., Regier, T., & Henderson, G. S. (2013). Quantitative determination of cerium oxidation states in alkali-aluminosilicate glasses using M4,5-edge XANES. *Journal of Non-Crystalline Solids*, *378*, 258–264. https://doi.org/10.1016/J.JNONCRYSOL.2013.07.017

Sutton, S. R. R., Karner, J., Papike, J., Delaney, J. S. S., Shearer, C., Newville, M., et al. (2005). Vanadium K edge XANES of synthetic and natural basaltic glasses and application to microscale oxygen barometry. *Geochimica et Cosmochimica Acta*, *69*, 9, 2333–2348. https://doi.org/10.1016/j.gca.2004.10.013

Takahashi, Y., Kolonin, G. R., Shironosova, G. P., Kupriyanova, I. I., Uruga, T., & Shimizu, H. (2005). Determination of the Eu(II)/Eu(III) ratios in minerals by X-ray absorption near-edge structure (XANES) and its application to hydrothermal deposits. *Mineralogical Magazine*, *69*, 2, 179–190. https://doi.org/10.1180/0026461056920245

Takahashi, Y., Sakami, H., & Nomura, M. (2002). Determination of the oxidation state of cerium in rocks by Ce LIII-edge X-ray absorption near-edge structure spectroscopy. *Analytica Chimica Acta*, *468*, 2, 345–354. https://doi.org/10.1016/S0003-2670(02)00709-2

Tanaka, T., Yoshida, T., Yoshida, S., Baba, T., & Ono, Y. (1995). XANES study of Eu species encapsulated in potassium Y-type zeolite. *Physica B: Condensed Matter*, *208–209*, 687–688. https://doi.org/10.1016/0921-4526(94)00790-3

Tostevin, R., Shields, G. A., Tarbuck, G. M., He, T., Clarkson, M. O., & Wood, R. A. (2016). Effective use of cerium anomalies as a redox proxy in carbonate-dominated marine settings. *Chemical Geology*, *438*, 146–162. https://doi.org/10.1016/J.CHEMGEO.2016.06.027

Trail, D., Watson, E. B., & Tailby, N. D. (2011). The oxidation state of Hadean magmas and implications for early Earth's atmosphere. *Nature*, *480*, 7375, 79–82. https://doi.org/10.1038/nature10655

Treuil, M, & Joron, J. L. (1975). Utilisation des elements hygromagmatophiles pour la simplification de la modélisation quantitative des processus magmatiques: exemples de l'Afar et de la dorsale médio-atlantique. *Società Italiana Mineralogia e Petrologia*, *31*, 1, 125–174.

Treuil, M., Joron, J.-L., Jaffrezic, H., Villemant, B., & Calas, G. (1979). Géochimie des éléments hygromagmatophiles, coefficients de partage minéraux /liquides et propriétés structurales de ces éléments dans les liquides magmatiques. *Bulletin de Minéralogie*, *102*, 4, 402–409. https://doi.org/10.3406/bulmi.1979.7336

Treuil, M., & Varet, J. (1973). Criteres volcanologiques, petrologiques et geochimiques de la genese et de la differenciation des magmas basaltiques; exemple de l'Afar. *Bulletin de La Societe Geologique de France*, *S7-XV*, 5–6, 506–540. https://doi.org/10.2113/gssgfbull.S7-XV.5-6.506

Ueda, K., Shimizu, Y., Nagamizu, K., Matsuo, M., & Honma, T. (2017). Luminescence and Valence of Tb Ions in Alkaline Earth Stannates and Zirconates Examined by X-ray Absorption Fine Structures. *Inorganic Chemistry*, *56*, 20, 12625–12630. https://doi.org/10.1021/acs.inorgchem.7b02165

Virgo, D., Seifert, F. A., & Mysen, B. O. (1981). The oxidation state of europium in albite and alkali-earth silicate glasses. *Carnegie Institute of Washington Yearbook*, *80*, 313–316.

Wadhwa, M. (2008). Redox Conditions on Small Bodies, the Moon and Mars. *Reviews in Mineralogy and Geochemistry*, *68*, 1, 493–510. https://doi.org/10.2138/rmg.2008.68.17

Wang, C., Peng, M., Jiang, N., Jiang, X., Zhao, C., & Qiu, J. (2007). Tuning the Eu luminescence in glass materials synthesized in air by adjusting glass compositions. *Materials Letters*, *61*, 17, 3608–3611. https://doi.org/10.1016/j.matlet.2006.11.133

Weill, D. F. F., & Drake, M. J. J. (1973). Europium anomaly in plagioclase feldspar: experimental results and semiquantitative model. *Science*, *180*, 4090, 1059–1060. https://doi.org/10.1126/science.180.4090.1059

White, W. M. (2013). *Geochemistry*. John Wiley & Sons.

Wilke, M., & Behrens, H. (1999). The dependence of the partitioning of iron and europium between plagioclase and hydrous tonalitic melt on oxygen fugacity. *Contributions to Mineralogy and Petrology*, *137*, 1–2, 102–114. https://doi.org/10.1007/s004100050585

INDEX

Magma Redox Geochemistry, Geophysical Monograph 266, First Edition.
Edited by Roberto Moretti and Daniel R. Neuville.

DOI: 10.1002/9781119473206.index

www.ingramcontent.com/pod-product-compliance
Lightning Source LLC
LaVergne TN
LVHW082110200926
848484LV00008B/1934